UNDERSTANDING EARTH

To our colleague and friend Raymond Siever: whose vision and dedication as a scientist, educator, and author have helped us all to better understand Earth's environment.

Meet the Authors

JOHN GROTZINGER is a field geologist interested in the evolution of Earth's surficial environments and biosphere. His research addresses the chemical development of the early oceans and atmosphere, the environmental context of early animal evolution, and the geologic factors that regulate sedimentary basins. He has contributed to developing the basic geologic framework of a number of sedimentary basins and orogenic belts in northwestern Canada, northern Siberia, southern Africa, and the western United States. He received a B.S. in geoscience from Hobart College in 1979, an M.S. in geology from the University of Montana in 1981, and a Ph.D. in geology from Virginia Tech in 1985. He spent three years as a research scientist at the Lamont-Doherty Geological Observatory before joining the Massachusetts Institute of Technology (MIT) faculty in 1988. From 1979 to 1990, he was engaged in regional mapping for the Geological Survey of Canada. He currently works as a geologist on the *Mars Exploration Rover* team. This mission is the first to conduct ground-based exploration of the bedrock geology of another planet, resulting in the discovery of sedimentary rocks formed in aqueous depositional environments.

In 1998, Dr. Grotzinger was named the Waldemar Lindgren Distinguished Scholar at MIT and in 2000 became the Robert R. Shrock Professor of Earth and Planetary Sciences. In 2005, he moved from MIT to the California Institute of Technology (Caltech), where he is the Fletcher Jones Professor of Geology. He received the Presidential Young Investigator Award of the National Science Foundation in 1990, the Donath Medal of the Geological Society of America in 1992, and the Henno Martin Medal of the Geological Society of Namibia in 2001. He is a member of the American Academy of Arts and Sciences and the U.S. National Academy of Sciences.

TOM JORDAN is a geophysicist interested in the composition, dynamics, and evolution of the solid Earth. He has conducted research into the nature of deep subduction, the formation of thickened keels beneath the ancient continental cratons, and the question of mantle stratification. He has developed a number of seismological techniques for investigating Earth's interior that bear on geodynamic problems. He has also worked on modeling plate motions, measuring tectonic deformations, quantifying seafloor morphology, and characterizing large earthquakes. He received his Ph.D. in geophysics and applied mathematics at Caltech in 1972 and taught at Princeton University and the Scripps Institution of Oceanography before joining the MIT faculty as the Robert R. Shrock Professor of Earth and Planetary Sciences in 1984. He served as the head of MIT's Department of Earth, Atmo-spheric and Planetary Sciences for the decade 1988–1998. He moved from MIT to the University of Southern California (USC) in 2000, where he is University Professor, W. M. Keck Professor of Earth Sciences, and Director of the Southern California Earthquake Center.

Dr. Jordan received the Macelwane Medal of the American Geophysical Union in 1983, the Woollard Award of the Geological Society of America in 1998, and the Lehmann Medal of the American Geophysical Union in 2005. He is a member of the American Academy of Arts and Sciences, the U.S. National Academy of Sciences, and the American Philosophical Society.

FRANK PRESS has made pioneering contributions to the fields of geophysics, oceanography, lunar and planetary sciences, and natural resource exploration. He was a member of the team that discovered the fundamental difference between oceanic and continental crust and built the instruments used in the research. Dr. Press was on the faculties of Columbia University, Caltech, and MIT. In addition, he served as president of the U.S. National Academy of Sciences and as a senior fellow at the Department of Terrestrial Magnetism, Carnegie Institution of Washington. He is currently with The Washington Advisory Group. In 1993, Frank Press was awarded the Japan Prize by the emperor for his work in the Earth sciences.

Dr. Press has advised four presidents on scientific issues. Jimmy Carter appointed him Science Advisor to the President. Bill Clinton awarded him the National Medal of Science. Three times, *U.S. News & World Report* surveys named him the country's most influential scientist.

RAYMOND SIEVER was an internationally known expert in sedimentary petrology, geochemistry, and the evolution of oceans and the atmosphere. Dr. Siever was a long-time member of Harvard University's Department of Earth and Planetary Sciences, and he chaired the geology department for eight years. He was one of the first sedimentologists to apply the techniques of geochemistry to the study of sedimentary rocks, especially sandstones and cherts.

In addition to cowriting the popular geology text *Earth* with Frank Press, Dr. Siever wrote *Sand* (Scientific American Library) and (with F. J. Pettijohn and Paul Potter) the classic textbook *Sand and Sandstone* (Springer-Verlag). Dr. Siever was a fellow of the Geological Society of America and the American Academy of Arts and Sciences and has been honored with distinguished awards from the Society of Sedimentary Geology, the Geochemical Society, and the American Association of Petroleum Geologists.

Brief Contents

Contents

OUR VISION

When the first edition of *Earth* was published, the concept of plate tectonics was still new. For the first time, an all-encompassing theory could be used as a framework for learning about the immense forces at work in Earth's interior. Given this new paradigm, our strategy was to make the learning of Earth science as process-based as possible. This new picture of Earth as a dynamic, coherent system was central to *Earth* and to its successor, *Understanding Earth*.

Now, with *Understanding Earth*, Fifth Edition, we are taking another step forward. We present geology as a unified, process-based science with the power to convey global meaning to geologic features wherever they are found. **Our approach to *Understanding Earth* emphasizes global-scale "system" interactions,** but we also use this approach to characterize the components of the global system, such as the dynamics of ocean crust formation ("Spreading Centers as Magmatic Geosystems," p. 92) and the dynamics of soil formation ("Soils as Geosystems," p. 381).

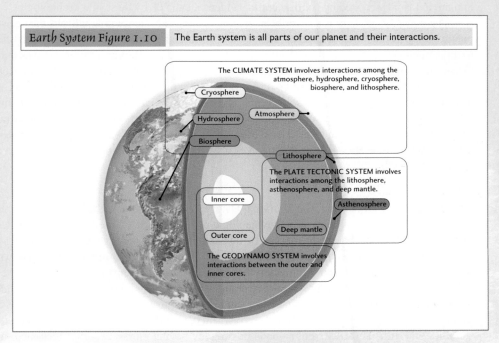

Earth System Figure 1.10   The Earth system is all parts of our planet and their interactions.

The CLIMATE SYSTEM involves interactions among the atmosphere, hydrosphere, cryosphere, biosphere, and lithosphere.

Cryosphere
Hydrosphere
Biosphere
Atmosphere
Lithosphere

The PLATE TECTONIC SYSTEM involves interactions among the lithosphere, asthenosphere, and deep mantle.

Inner core
Outer core
Asthenosphere
Deep mantle

The GEODYNAMO SYSTEM involves interactions between the outer and inner cores.

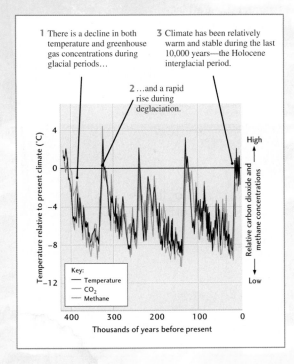

1 There is a decline in both temperature and greenhouse gas concentrations during glacial periods...

3 Climate has been relatively warm and stable during the last 10,000 years—the Holocene interglacial period.

2 ...and a rapid rise during deglaciation.

Temperature relative to present climate (°C)

Relative carbon dioxide and methane concentrations

High

Low

Key:
— Temperature
— CO_2
— Methane

400   300   200   100   0
Thousands of years before present

Geology is as much about climate as it is about tectonics, and in this edition we describe the geologic processes that unite these historically separate fields. Our discussions of such core concepts as the rock cycle, metamorphism and exhumation, and landscape evolution help illustrate important links. Further, **we consider the concept of geologic time and the historical development of Earth (and the other terrestrial planets) to be as important a process.** We illustrate this principle by grouping together material that underscores long-term evolution, whether it involves our solar system (**Chapter 9**, "Early History of the Terrestrial Planets"), Earth itself (**Chapter 10**, "Evolution of the Continents"), or life (**Chapter 11**, "Geobiology: Life Interacts with the Earth"). Our preparation of **Chapters 9 and 11**—new to this edition of *Understanding Earth*—was motivated by our desire to expose students to the fresh perspective on our solar system provided by the increasing number of geologically oriented

Figure 15.10 A graph showing three types of data recovered from the Vostok ice core in East Antarctica: temperatures inferred from oxygen isotope (black line) and the concentrations of two important greenhouse gases, carbon dioxide (blue line) and methane (red line). The gas concentrations come from measurements of air samples trapped as tiny bubbles within the Antarctic ice. [IPCC, *Climate Change 2001: The Scientific Basis*.]

missions to other planets (especially Mars) and the recognition that life (especially microbial life) has had a profound effect on Earth.

In writing this edition of *Understanding Earth*, **we have drawn on powerful new laboratory and field tools and new theoretical approaches.** New technology such as GPS and continuous satellite monitoring of Earth from space allows us to view the motion of plates, the raising and erosion of mountains, the buildup of crustal strain before an earthquake, global warming, the retreat of glaciers, and rising sea level—all in almost real time. It is remarkable that we can now use earthquake waves to image the flow of the solid mantle hundreds and thousands of kilometers deep in the Earth, revealing patterns of rising plumes and subducting plates. And we can now send robotic geologists to other planets to investigate geologic processes, such as the role of water in shaping the early surface of Mars and constraining its climate history. Finally, rapid scanning of the genetic makeup of even the tiniest microorganism allows us to determine how organisms control geologically important processes. The view of Earth as a system of interacting components subject to interference by humankind is now backed by solid scientific evidence, enabled by these new technologies.

The power of geology has never been greater. Geological science now informs the decisions of public policy leaders in government, industry, and community organizations. But there is still work to do in helping to make this flow of information more effective. Recent disasters such as the Indian Ocean tsunami of 2004 (see **Chapter 13**) and the flooding of New Orleans by Hurricane Katrina in 2005 (see **Chapter 20**) underscore the need for vigilance in planning for and limiting exposure to these rare but powerful and inevitable events. The planning process begins with understanding how Earth works and by appreciating the history of similar events in the geologic past. The devastation caused by both of these events could have been substantially reduced had public policy leaders been more knowledgeable about the geologic forces involved and the likelihood (history) of such events.

Early Coverage of Plate Tectonics

Chapter 2, "Plate Tectonics: The Unifying Theory," allows us to take full advantage of tectonic theory as a framework for discussing key geologic processes. Early coverage of the basic tenets of tectonic theory means that the theory can be invoked throughout the text, providing the big picture as well as the links connecting geologic phenomena. For instance, **Chapter 5** offers an expanded section on plate tectonics and sedimentary basins. **Chapter 6** presents metamorphism in terms of plate interactions, describing pressure-temperature-time paths and their significance for interpreting tectonic processes, including exhumation and uplift. **Chapter 9**—new to this edition of *Understanding Earth*—allows a process-based comparison between "flake tectonics" on Venus and plate tectonics on Earth. This treatment is made possible by having introduced plate tectonics early on. Finally, the section of the book dedicated to surface processes is capped with **Chapter 22**, in which the discussion of landscape evolution integrates previous chapters and makes the case for significant interactions between climate and tectonics.

Viewing Earth as a System

We begin with an improved discussion of the Earth system in **Chapter 1**. The components of the Earth system are described, and the exchanges of energy and matter through the system are illustrated. This discussion serves as the springboard for the Earth systems perspective that pervades the text. In **Chapter 3**, the rock cycle first illustrates how plate tectonic and climate processes interact as a system to generate the three basic classes of rocks. In **Chapter 4** (in the section entitled "Spreading Centers as Magmatic Geosystems"), we then show how one of the components of the

Figure 15.7 Earth's greenhouse atmosphere balances incoming and outgoing radiation. [IPCC, *Climate Change 2001: The Scientific Basis.*]

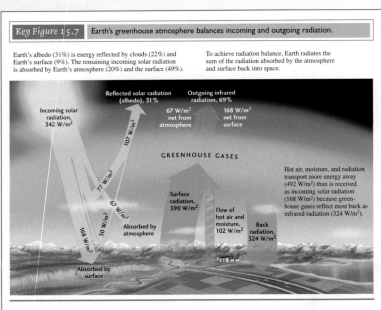

Key Figure 15.7 Earth's greenhouse atmosphere balances incoming and outgoing radiation.

Earth's albedo (31%) is energy reflected by clouds (22%) and Earth's surface (9%). The remaining incoming solar radiation is absorbed by Earth's atmosphere (20%) and the surface (49%).

To achieve radiation balance, Earth radiates the sum of the radiation absorbed by the atmosphere and surface back into space.

Reflected solar radiation (albedo), 31%

Outgoing infrared radiation, 69%

Incoming solar radiation, 342 W/m²

67 W/m² net from atmosphere

168 W/m² net from surface

107 W/m²

GREENHOUSE GASES

77 W/m²

67 W/m²

30 W/m²

168 W/m²

Surface radiation, 390 W/m²

Absorbed by atmosphere

Flow of hot air and moisture, 102 W/m²

Back radiation, 324 W/m²

Hot air, moisture, and radiation transport more energy away (492 W/m²) than is received as incoming solar radiation (168 W/m²) because greenhouse gases reflect most back as infrared radiation (324 W/m²).

Absorbed by surface

Earth system (spreading centers) can be studied as a system itself, with inputs, processes, and outputs. This pedagogical approach is reinforced later in the same chapter when we discuss "Subduction Zones as Magmatic Geosystems." Throughout the book, we use this construct to illustrate particularly well-defined geosystems, as in the case of volcanoes (**Chapter 12**) and soils (**Chapter 16**).

Traditionally isolated geologic processes such as metamorphism (**Chapter 6**) and landscape evolution (**Chapter 22**) are discussed in the context of system interactions between tectonics and climate, illustrated with examples of pressure-temperature paths and models for denudation and exhumation. In **Chapter 13**, the relationship among stress, strain, and the behavior of regional fault systems in controlling earthquakes illustrates how geosystems behave. The convective processes of Earth's deep interior are discussed in **Chapter 14** as the driving mechanisms behind the plate tectonic and geodynamo systems.

Our discussion of the climate system (**Chapter 15**) is updated to provide a more comprehensive understanding of the major processes that control Earth's climate, as well as an understanding of Earth's climate history. Subsequent chapters in this section of the book—which discusses surface processes—are rewritten to emphasize not just the particular process (such as the flow of groundwater or wind) but how these processes are controlled by the climate system and, where relevant, the plate tectonic system. Finally, the role of biology in Earth system behavior is presented in **Chapter 11** and then reinforced later with examples from many chapters.

New Chapters
Comparative Planetology

Chapter 9, "Early History of the Terrestrial Planets," explores the solar system not only in the vast reaches of interplanetary space but also backward in time to its earliest history. It discusses how Earth and the other planets formed around the Sun and how they differentiated into layered bodies. The geologic processes that have shaped Earth are compared with those of Mercury, Mars, Venus, and the Moon. The history of climate as well as tectonics is emphasized in the discussion of Venus and Mars. Mars receives special attention owing to the spectacular new results from recent geologically oriented missions, culminating with the first in situ outcrop studies on another planet. Finally, we discuss how exploration of the solar system by

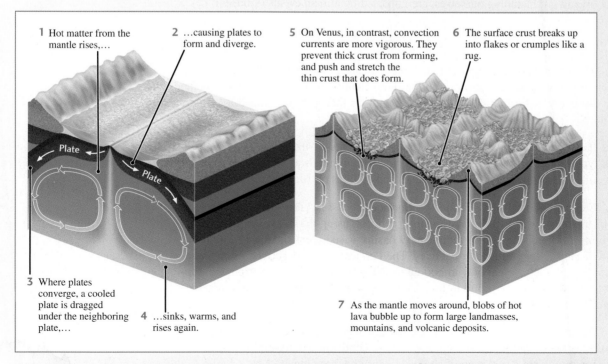

1 Hot matter from the mantle rises,…

2 …causing plates to form and diverge.

3 Where plates converge, a cooled plate is dragged under the neighboring plate,…

4 …sinks, warms, and rises again.

5 On Venus, in contrast, convection currents are more vigorous. They prevent thick crust from forming, and push and stretch the thin crust that does form.

6 The surface crust breaks up into flakes or crumples like a rug.

7 As the mantle moves around, blobs of hot lava bubble up to form large landmasses, mountains, and volcanic deposits.

Figure 9.15 Plate tectonics on Earth versus flake tectonics on Venus.

spacecraft might answer fundamental questions about the evolution of our planet and the life it harbors.

Concepts in planetary geology also appear in other chapters. Sulfate minerals thought to be of importance on Mars are described in **Chapter 3**; impact (shock)-related metamorphism is discussed in **Chapter 6**; and the search for extraterrestrial life is presented in **Chapter 11**.

Geobiology

Chapter 11, "Geobiology: Life Interacts with the Earth," explores the links between life and environment on Earth. We describe how the biosphere works as a system and what gives Earth its ability to support life through the cycling of biologically important elements. The chapter discusses the remarkable role of microorganisms in Earth processes, including mineral and rock formation (and destruction). We also discuss geobiological interactions throughout Earth's history and some of the major evolutionary events that changed our planet. Finally, we consider the key ingredients for sustaining life on Earth and searching for life on other planets and in other solar systems.

Geobiological processes are discussed in other chapters as well, including their role in the formation of sedimentary rocks (**Chapter 5**); the occurrence of extremophiles living in volcanoes and mid-ocean ridges (**Chapter 12**); how the Earth's magnetic field helps some organisms to navigate (**Chapter 14**); the role of organisms in modulating climate (**Chapter 15**) and in influencing weathering (**Chapter 16**) and groundwater potability (**Chapter 17**); and in the way that organisms, including ourselves, are impacted by human events (**Chapter 23**).

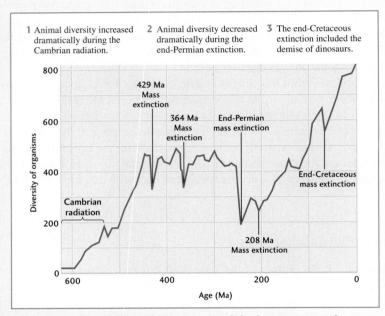

1 Animal diversity increased dramatically during the Cambrian radiation.

2 Animal diversity decreased dramatically during the end-Permian extinction.

3 The end-Cretaceous extinction included the demise of dinosaurs.

Figure 11.21 The diversity of organisms reveals both extinctions and radiations. During a radiation, such as the Cambrian explosion, the number of new groups of organisms increases. During a mass extinction, the number of groups decreases.

Major Revisions

• **Chapter 1**, "The Earth System," features a new section called "Peeling the Onion," a complete discussion of Earth's layered internal structure. The scientific method is introduced in the context of the discovery of Earth's shape and the hypotheses that were developed to explain early observations.

• **Chapter 3**, "Earth Materials: Minerals and Rocks," presents a streamlined overview of minerals, rock types, and the rock cycle. It now includes ore minerals as well. A new feature box discusses sulfides, sulfates, acid mine drainage, and sulfates on Mars.

• **Chapter 5**, "Sedimentation: Rocks Formed by Surface Processes," now develops the sedimentary part of the rock cycle as a "source-to-sink" concept, using the Mississippi River and delta as the example. The discussion of sedimentary basins as sediment sinks now follows as the logical end of the sediment transport pathway. The Persian Gulf is described as an example of a flexural basin, making clear how important such basins are to the generation and trapping of oil reserves, such as those found in Iraq. The section on "Sedimentary Environments" now emphasizes process up front.

• **Chapter 7**, "Deformation: Modification of Rocks by Folding and Fracturing," is now solidly grounded in plate tectonics and includes a current view of continental deformation. The chapter contains a more complete description of geologic maps and mapping.

- **Chapter 13**, "Earthquakes," includes new material on the 2004 Sumatra earthquake and tsunami.

- **Chapter 14**, "Exploring Earth's Interior," has a new section on Earth's magnetic field and the biosphere.

- **Chapter 15**, "The Climate System," now appears earlier in the text and with a greater focus on climate as a global geosystem that can be understood in terms of interacting components. Climate variability is also discussed, with special emphasis on ancient ice ages and Milankovitch cycles.

- **Chapter 16**, "Weathering, Erosion, and Mass Wasting: Interface Between Climate and Tectonics," now integrates material previously separated in different parts of the book. Weathering is discussed in the context of its important variables, including climate. The chapter has a completely revised and updated discussion of soils and takes a process-oriented approach in the new section entitled "Soil as a Geosystem." The chapter now also uses a widely accepted classification scheme for soils (U.S. soil taxonomy system).

A small headland near Banda Ache on the west coast of Sumatra previously covered by dense jungle to the waterline but now stripped clean to height of about 15 m by the tsunami. [José Borrero, University of Southern California/Tsunami Research Group.]

- **Chapter 18**, "Stream Transport: From Mountains to Oceans," is completely reorganized so that the larger-scale, geomorphic aspects of stream systems come first, followed by the discussion of fluid flow and sediment transport. The final section, called "Streams as Geosystems," pulls it all together. The chapter begins with the Lewis and Clark expedition, which is referred to throughout the chapter (for example, in the discussion of stream networks).

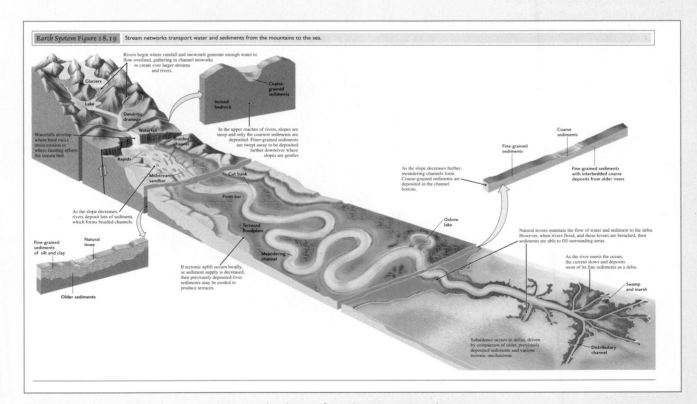

Figure 18.19 Stream networks transport water and sediments from the mountains to the sea.

• **Chapter 20**, "Coastlines and Ocean Basins," has a new section on hurricanes and storm surge, along with a new feature box on the great New Orleans flood of 2005 and a new table on the hurricane intensity scale. The chapter has been reorganized to progress from the coast to the continental margin to the deep ocean.

• **Chapter 21**, "Glaciers: The Work of Ice," includes a new section called "Glaciations and the Climate System," which discusses the Wisconsin glaciation.

• **Chapter 23**, "The Human Impact on Earth's Environment," now treats human civilization as a global geosystem that interacts strongly with the carbon cycle and other components of the Earth system, focusing on the depletion of petroleum as an energy resource and the potential for climate change that arises from fossil-fuel burning.

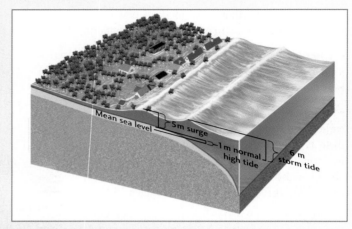

Figure 20.8 A storm tide is the combination of storm surge and the normal astronomical tide.

Telling Stories with Words and Pictures

Our enduring goal is to tell a story rather than provide an aggregated set of facts, and the illustrations of this edition of *Understanding Earth* reflect that goal. Each chapter features several **Key Figures**, illustrations of the chapter's most important concepts and geologic processes. Key Figures not only represent the key geologic processes but also help the reader to focus on the chapter's key ideas.

Most chapters also feature **Earth System Figures** that illustrate geologic processes as part of global, regional, or even local geosystems. These illustrations should reinforce the reader's understanding of Earth as a system of interacting components, with inputs, processes, and outputs.

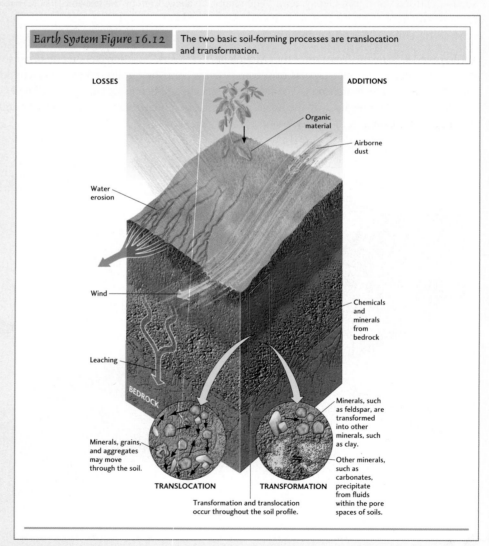

Figure 16.12 There are two important soil-forming processes: transformation and translocation.

MEDIA AND SUPPLEMENTS PACKAGE

A selection of electronic media and printed supplementary materials designed to support both instructors and students is available to users of this new edition of *Understanding Earth*. By focusing primarily on the importance of visualizing key concepts in geology, we are providing instructors with the presentation tools they need to help their students truly understand Earth's processes and students with the study tools they need to study geology effectively and apply their newly acquired knowledge.

For Instructors

The **Instructor's Resource CD-ROM** (ISBN 0-7167-4509-7) allows instructors to **search** and **export** all the resources listed below by key term or chapter:

- All text images
- Animations, videos, flashcards, and more
- *Instructor's Resource Manual*
- PowerPoint files (lecture slides)
- *Test Bank* files
- Expeditions in Geology
- Geology in the News

The **Test Bank** ([print] ISBN 0-7167-4516-X and [CD-ROM] ISBN 0-7167-3803-1) includes approximately 50 multiple-choice questions for each chapter (over 1000 total).

The **Computerized Test Bank CD-ROM** provides the *Test Bank* files in an electronic format that allows instructors to edit, resequence, and add questions.

The **Instructor's Resource Manual** (ISBN 0-7167-4507-0), written by Peter L. Kresan and Reed Mencke, formerly of the University of Arizona, includes chapter-by-chapter sample lecture outlines, ideas for cooperative learning activities and exercises that can be easily copied and used as handouts and quizzes, and guides to the Web resources. The *Instructor's Resource Manual* also includes an instructional design section that contains teaching tips from many instructors at the University of Arizona Learning Center. The *Instructor's Resource Manual* is also available on both the Instructor's Resource CD-ROM and the Companion Web Site.

The **Overhead Transparency Set** (ISBN 0-7167-3842-2) includes *all* textbook illustrations in full-color acetate transparencies.

The **Companion Web Site,** at **www.whfreeman.com/understandingearth5e,** provides access to all student materials on the Web site in addition to a password-protected Instructor's site that contains all the PowerPoint presentations and text art available on the Instructor's Resource CD-ROM, the *Instructor's Resource Manual*, and the Quiz Gradebook (which keeps track of students' Graded Online Quiz scores).

Online Course Materials (WebCT, Blackboard): As a service for adopters, we will provide content files in the appropriate online course format, including the instructor and student resources for this text.

Expeditions in Geology: A virtual field trip for your whole class! Explore geological phenomena with these brief video tutorials. Accompany Jerry Magloughlin of Colorado State University as he flies across the continent and around the world filming extraordinary examples of Earth in action. Your lecture comes to life as you provide an up-close examination of the splendor and intrigue of various geological processes and landmarks. (Available in high-definition format on the Instructor's Resource CD-ROM.)

Geology in the News (on the **Instructor's Resource CD-ROM**): This series of one-to three-minute news features (all recently aired on prime-time news) offers expert analysis of geological events that have headline-making impacts on our lives. By connecting what students see and hear every day in the mainstream media to the processes of geology, this series allows you to convey the relevance of geology and geologists to our lives.

For Students

Understanding Earth, Fifth Edition eBook! The *Understanding Earth*, Fifth Edition, eBook is a complete online version of the textbook that provides a rich learning experience by taking full advantage of the electronic medium. This online version integrates all existing media resources and adds features unique to the eBook, such as

- Easy access from any Internet-connected computer via a standard Web browser
- Quick, intuitive navigation to any section or subsection, as well as any printed book page number
- Integration of all student Companion Web Site animated tutorials and activities
- In-text self-quiz questions
- In-text links to all glossary entries
- Interactive chapter summary exercises
- Text highlighting, down to the level of individual phrases
- A bookmarking feature that allows for quick reference to any page
- A powerful Notes feature that allows students or instructors to add notes to any page
- A full glossary and index
- Full-text search, including the glossary and index
- Automatic saving of all notes, highlighting, and bookmarks

Additional features for instructors include:

- Custom Chapter Selection: Instructors can choose the chapters that correspond to their syllabus, and students will get a custom version of the eBook with the selected chapters only.
- Instructor Notes: Instructors can choose to create an annotated version of the eBook with their notes on any page. When students in their course log in, they will see the instructor's version.
- Custom Content: Instructor notes can include text, Web links, and even images, allowing instructors to place any content they choose exactly where they want it.
- Online Quizzing: The online quizzes from the student Companion Web Site are integrated into the eBook.

The eBook is available FREE with the text (use special package ISBN: 0-7167-76650), or online at http://ebooks.bfwpub.com or at the **Companion Web Site** at **www.whfreeman.com /understandingearth5e.**

The **Companion Web Site,** at **www.whfreeman.com/understandingearth5e,** includes many study tools that allow students to visualize geological processes and practice their newly acquired knowledge. The Companion Web Site contains

- Animations, including more than 40 animated figures from the textbook
- Online Review Exercises, which include interactive exercises, virtual reality field trips, drag-and-drop exercises, and matching exercises
- Flashcards
- Online Quizzing
- Concept Self-Checker
- Geology in Practice exercises: inquiry-based learning activities that ask students to apply their newly acquired knowledge and think like geologists
- Current Events in Geology: an archive of geologically relevant articles from popular news sources, updated monthly

The **Student Study Guide** (ISBN 0-7167-3981-X), written by Peter L. Kresan and Reed Mencke, formerly of the University of Arizona, includes tips on studying geology, chapter summaries, practice exams, and practice exercises that incorporate figures from the text and Web resources.

The **EarthInquiry** series, developed by the **American Geological Institute** in collaboration with experienced geology instructors, is a collection of Web-based investigative activities that provides a direct way for students to explore and work with the vast amount of geological data now accessible via the Web. Covering such diverse topics as earthquakes and plate boundaries and the recurrence interval of floods, each EarthInquiry module asks students to analyze real-time data to develop a deeper understanding of fundamental geoscience concepts. Each module consists of a password-protected Web component and an accompanying workbook.

For more information about EarthInquiry or to read about the various modules currently available, please visit **www.whfreeman.com/earthinquiry.**

ACKNOWLEDGMENTS

It is a challenge both to geology instructors and to authors of geology textbooks to compress the many important aspects of geology into a single course and to inspire interest and enthusiasm in their students. To meet this challenge, we have called on the advice of many colleagues who teach in all kinds of college and university settings. From the earliest planning stages of each edition of this book, we have relied on a consensus of views in designing an organization for the text and in choosing which topics to include. As we wrote and rewrote the chapters, we again relied on our colleagues to guide us in making the presentation pedagogically sound, accurate, accessible, and stimulating to students. To each one we are grateful.

The following instructors were involved assisted in the planning or reviewing stages of this new edition:

Kathryn A. Baldwin
 Washington State University
Charly Bank
 Colorado College
Ray Beiersdorfer
 Youngstown State University
Larry Benninger
 University of North Carolina at Chapel Hill
Elisa Bergslien
 State University of New York at Buffalo
Grenville Draper
 Florida International University
Eric Essene
 University of Michigan
Sharon Gabel
 State University of New York at Oswego

Steve Gao
 Kansas State University
David H. Griffing
 Hartwick College
Scott P. Hippensteel
 University of North Carolina at Charlotte
Linda C. Ivany
 Syracuse University
Thomas J. Kalakay
 Rocky Mountain College
Haraldur R. Karlsson
 Texas Tech University
David T. King Jr.
 Auburn University
Andrew H. Knoll
 Harvard University

Jeffrey R. Knott
 University of California at Fullerton
Kelly Liu
 Kansas State University
Sakiko N. Olsen
 Johns Hopkins University
Dr. Leslie Reid
 University of Calgary
Steven Semken
 Arizona State University
Eric Small
 University of Colorado at Boulder
John Waldron
 University of Alberta

We remain indebted to the following instructors who helped shape earlier editions of *Understanding Earth*:

Wayne M. Ahr
 Texas A & M University
Gary Allen
 University of New Orleans
Jeffrey M. Amato
 New Mexico State University
N. L. Archbold
 Western Illinois University
Allen Archer
 Kansas State University
Richard J. Arculus
 University of Michigan, Ann Arbor

Philip M. Astwood
 University of South Carolina
R. Scott Babcock
 Western Washington University
Evelyn J. Baldwin
 El Camino Community College
Suzanne L. Baldwin
 Syracuse University
Charles W. Barnes
 Northern Arizona University
Carrie E. S. Bartek
 University of North Carolina, Chapel Hill

John M. Bartley
 University of Utah
Lukas P. Baumgartner
 University of Wisconsin, Madison
Richard J.Behl
 California State University, Long Beach
Kathe Bertine
 San Diego State University
David M. Best
 Northern Arizona University
Roger Bilham
 University of Colorado

Dennis K. Bird
Stanford University

Stuart Birnbaum
University of Texas, San Antonio

David L. S. Blackwell
University of Oregon

Arthur L. Bloom
Cornell University

Phillip D. Boger
State University of New York, Geneseo

Stephen K. Boss
University of Arkansas

Michael D. Bradley
Eastern Michigan University

David S. Brembaugh
Northern Arizona University

Robert L. Brenner
University of Iowa

Edward Buchwald
Carleton College

David Bucke
University of Vermont

Robert Burger
Smith College

Timothy Byrne
University of Connecticut

J. Allan Cain
University of Rhode Island

F. W. Cambray
Michigan State University

Ernest H. Carlson
Kent State University

Max F. Carman
University of Houston

James R. Carr
University of Nevada

L. Lynn Chyi
University of Akron

Allen Cichanski
Eastern Michigan University

George R. Clark
Kansas State University

G. S. Clark
University of Manitoba

Mitchell Colgan
College of Charleston

Roger W. Cooper
Lamar University

Spencer Cotkin
University of Illinois

Peter Dahl
Kent State University

Jon Davidson
University of California, Los Angeles

Larry E. Davis
Washington State University

Craig Dietsch
University of Cincinnati

Yildirim Dilek
Miami University

Bruce J. Douglas
Indiana University

Grenville Draper
Florida International University

Carl N. Drummond
*Indiana University/Purdue University,
Fort Wayne*

William M. Dunne
University of Tennessee, Knoxville

R. Lawrence Edwards
University of Minnesota

C. Patrick Ervin
Northern Illinois University

Stanley Fagerlin
Southwest Missouri State University

Pow-foong Fan
University of Hawaii

Jack D. Farmer
University of California, Los Angeles

Mark D. Feigenson
Rutgers University

Stanley C. Finney
California State University, Long Beach

Charlie Fitts
University of Southern Maine

Tim Flood
Saint Norbert College

Richard M. Fluegeman, Jr.
Ball State University

Michael F. Follo
Colby College

Richard L. Ford
Weaver State University

Nels F. Forsman
University of North Dakota

Charles Frank
Southern Illinois University

William J. Frazier
Columbus College

Robert B. Furlong
Wayne State University

Sharon L. Gabel
State University of New York, Oswego

Alexander E. Gates
Rutgers University

Dennis Geist
University of Idaho

Katherine A. Giles
New Mexico State University

Gary H. Girty
San Diego State University

Michelle Goman
Rutgers University

William D. Gosnold
University of North Dakota

Richard H. Grant
University of New Brunswick

Julian W. Green
University of South Carolina, Spartanburg

Jeffrey K. Greenberg
Wheaton College

Bryan Gregor
Wright State University

G. C. Grender
*Virginia Polytechnic Institute and
State University*

David H. Griffing
University of North Carolina

Mickey E. Gunter
University of Idaho

David A. Gust
University of New Hampshire

Kermit M. Gustafson
Fresno City College

Bryce M. Hands
Syracuse University

Ronald A. Harris
West Virginia University

Douglas W. Haywick
University of Southern Alabama

Michael Heaney III
Texas A&M University

Richard Heimlich
Kent State University

Tom Henyey
University of Southern California

Eric Hetherington
University of Minnesota

J. Hatten Howard III
University of Georgia

Herbert J. Hudgens
Tarrant County Junior College

Warren D. Huff
University of Cincinnati

Ian Hutcheon
University of Calgary

Alisa Hylton
Central Piedmont Community College

Mohammad Z. Igbal
University of Northern Iowa

Neil Johnson
Appalachian State University

Ruth Kalamarides
Northern Illinois University

Frank R. Karner
University of North Dakota

Alan Jay Kaufman
University of Maryland

Phillip Kehler
University of Arkansas, Little Rock

James Kellogg
*University of South Carolina
at Columbia*

David T. King, Jr.
Auburn University

Cornelius Klein
Harvard University

Jeff Knott
California State University at Fullerton

Peter L. Kresan
University of Arizona

Albert M. Kudo
University of New Mexico

Richard Law
Virginia Tech

Robert Lawrence
Oregon State University

Don Layton
Cerritos College

Peter Leavens
University of Delaware

Patricia D. Lee
University of Hawaii, Manoa

Barbara Leitner
University of Montevallo

Laurie A. Leshin
Arizona State University

Kelly Liu
Kansas State University

John D. Longshore
Humboldt State University

Stephen J. Mackwell
Pennsylvania State University

J. Brian Mahoney
University of Wisconsin, Eau Claire

Erwin Mantei
Southwest Missouri State University

Bart S. Martin
Ohio Wesleyan University

Gale Martin
Community College of Southern Nevada

Peter Martini
University of Guelph

G. David Mattison
Butte College
Florentin Maurrassee
Florida International University
George Maxey
University of North Texas
Joe Meert
Indiana State University
Lawrence D. Meinert
Washington State University, Pullman
Robert D. Merrill
California State University, Fresno
Jonathan S. Miller
University of North Carolina
James Mills
DePauw University
Kula C. Misra
University of Tennessee, Knoxville
Roger D. Morton
University of Alberta
Peter D. Muller
State University of New York, Oneonta
Henry Mullins
Syracuse University
John E. Mylroie
Mississippi State University
J. Nadeau
Rider University
Stephen A. Nelson
Tulane University
Andrew Nyblade
Pennsylvania State University
Peggy A. O'Day
Arizona State University
Kieran O'Hara
University of Kentucky
William C. Parker
Florida State University
Simon M. Peacock
Arizona State University
E. Kirsten Peters
Washington State University, Pullman
Philip Piccoli
University of Maryland
Donald R. Prothero
Occidental College

Terrence M. Quinn
University of South Florida
C. Nicholas Raphael
Eastern Michigan University
Loren A. Raymond
Appalachian State University
J. H. Reynolds
West Carolina University
Mary Jo Richardson
Texas A&M University
Robert W. Ridkey
University of Maryland
James Roche
Louisiana State University
Gary D. Rosenberg
Indiana University/Purdue University, Indianapolis
William F. Ruddiman
University of Virginia
Malcolm Rutherford
University of Maryland
William E. Sanford
Colorado State University
Charles K. Schamberger
Millersville University
James Schmitt
Montana State University
Fred Schwab
Washington and Lee University
Donald P. Schwert
North Dakota State University
Jane Selverstone
University of New Mexico
Steven C. Semken
Navajo Community College
D. W. Shakel
Pima Community College
Thomas Sharp
Arizona State University
Charles R. Singler
Youngstown State University
David B. Slavsky
Loyola University of Chicago
Douglas L. Smith
University of Florida
Richard Smosma
West Virginia University

Donald K. Sprowl
University of Kansas
Steven M. Stanley
Johns Hopkins University
Don Steeples
University of Kansas
Randolph P. Steinen
University of Connecticut
Dorothy L. Stout
Cypress College
Sam Swanson
University of Georgia, Athens
Bryan Tapp
University of Tulsa
John F. Taylor
Indiana University of Pennsylvania
Kenneth J. Terrell
Georgia State University
Thomas M. Tharps
Purdue University
Nicholas H. Tibbs
Southeast Missouri State University
Jody Tinsley
Clemson University
Herbert Tischler
University of New Hampshire
Jan Tullis
Brown University
James A. Tyburczy
Arizona State University
Kenneth J. Van Dellen
Macomb Community College
Michael A. Velbel
Michigan State University
J. M. Wampler
Georgia Tech
Donna Whitney
University of Minnesota
Elisabeth Widom
Miami University, Oxford
Rick Williams
University of Tennessee
Lorraine W. Wolf
Auburn University

Others have worked with us more directly in writing and preparing manuscript for publication. At our side always were the editors at W. H. Freeman and Company: Randi Rossignol and Valerie Raymond. Mary Louise Byrd supervised the process from final manuscript to printed text. Diana Siemens and Eleanor Wedge were our copyeditor and proofreader. Victoria Anderson coordinated the media supplements. Blake Logan designed the text, and Ted Szczepanski and Christina Micek obtained and edited the many beautiful photographs. We thank Sheridan Sellers, our compositor and layout artist; Susan Wein, our production manager; and Bill Page, our illustration coordinator.

UNDERSTANDING EARTH

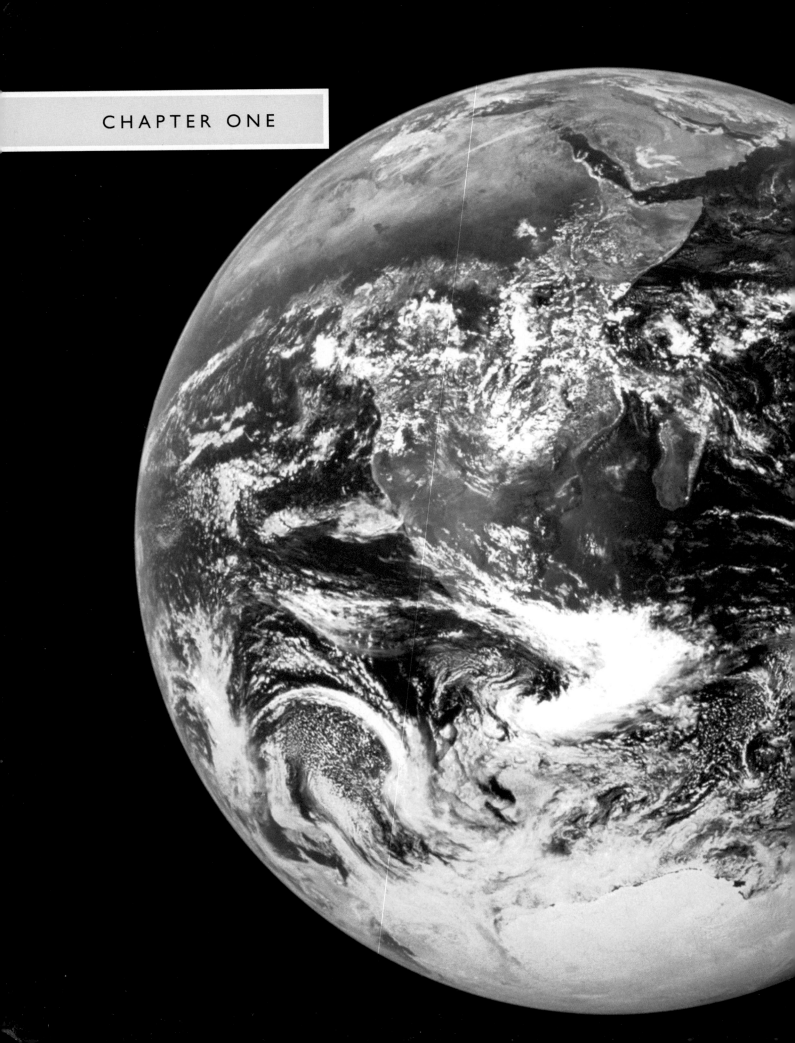

THE EARTH SYSTEM

Earth is a unique place, home to millions of organisms, including ourselves. No other planet we've yet discovered has the same delicate balance of conditions necessary to sustain life. *Geology* is the science that studies Earth: how it was born, how it evolved, how it works, and how we can help preserve its habitats for life. Geologists try to answer many questions about Earth's surface and interior. Why do the continents expose dry land? Why are the oceans so deep? How did the Himalaya, Alps, and Rocky Mountains reach their great heights? What process generated island chains such as Hawaii in the middle of the Pacific Ocean and the deep trenches near the ocean's margins? More generally, how does the face of our planet change over time, and what forces drive these changes? We think you will find the answers to these questions quite fascinating—they will allow you to look at the world around you with new eyes. Welcome to the science of geology!

We have organized the discussion of geology in this book around three basic concepts that will appear in almost every chapter: (1) Earth as a system of interacting components, (2) plate tectonics as a unifying theory of geology, and (3) changes in the Earth system through geologic time.

This chapter gives a broad picture of how geologists think. It starts with the scientific method, the observational approach to the physical universe on which all scientific inquiry is based. Throughout the book, you will see the scientific method in action as you discover how Earth scientists gather and interpret information about our planet. In this first chapter, we will illustrate how the scientific method was applied to discover some of Earth's basic features—its shape and internal layering.

We will also introduce you to a geologist's view of time. You may start to think about time differently as you begin to comprehend the immense span of geologic history. Earth and the other planets in our solar system formed about 4.5 billion years ago. More than 3 billion years ago, living cells developed on Earth's surface, and life has been evolving ever since. Yet our human origins date back only a few million years—a mere few hundredths of a percent of Earth's existence. The scales that measure individual lives in decades and

First image of the whole Earth showing the Antarctic and African continents, taken by the *Apollo 17* astronauts on December 7, 1972. [NASA.]

mark off periods of human history in hundreds or thousands of years are inadequate to study Earth.

To explain features that are millions or even billions of years old, we look at what is happening on Earth today. We study our complex natural world as an *Earth system* involving many interacting components, some beneath its solid surface, others in its atmosphere and oceans. Many of these components—for example, the Los Angeles air basin, the Great Lakes, Hawaii's Mauna Loa volcano, and the continent of North America—are themselves complex subsystems or *geosystems*. To understand the various parts of Earth, geologists often study its geosystems separately, as if each existed alone. To get a complete perspective on how Earth works, however, scientists must learn how its geosystems interact with one another—how gases from volcanic systems can trigger changes in the climate system, for exam-

ple, or how living organisms can modify the climate system and, in turn, be affected by climate changes.

THE SCIENTIFIC METHOD

The goal of all science is to explain how the universe works. The **scientific method,** on which all scientists rely, is a general plan based on methodical observations and experiments (**Figure 1.1**). Scientists believe that physical events have physical explanations, even if they may be beyond our present capacity to understand them.

When scientists propose a *hypothesis*—a tentative explanation based on data collected through observations and experiments—they present it to the community of scientists for criticism and repeated testing. A hypothesis that is confirmed by other scientists gains credibility, particularly if it explains new data or predicts the outcome of new experiments.

A set of hypotheses that has survived repeated challenges and accumulated a substantial body of experimental support can be elevated to the status of a *theory*. Although a theory can explain and predict observations, it can never be considered finally proved. The essence of science is that no explanation, no matter how believable or appealing, is closed to question. If convincing new evidence indicates that a theory is wrong, scientists may modify it or discard it. The longer a theory holds up to all scientific challenges, however, the more confidently it is held.

Knowledge based on many hypotheses and theories can be used to create a *scientific model*—a precise representation of how a natural system is built or should behave. Models combine a set of related ideas to make predictions, allowing scientists to test the consistency of their knowledge. Like a good hypothesis or theory, a good model makes predictions that agree with observations. These days, scientific models are often formulated as computer programs that simulate the behavior of natural systems through numerical calculations. In the virtual reality of a computer, numerical simulations can reproduce phenomena that are just too difficult to replicate in a real laboratory, including the behavior of natural systems that operate over long periods of time or large expanses of space.

To encourage discussion of their ideas, scientists share them and the data on which they are based. They present their findings at professional meetings, publish them in professional journals, and explain them in informal conversations with colleagues. Scientists learn from one another's work as well as from the discoveries of the past. Most of the great concepts of science, whether they emerge as a flash of insight or in the

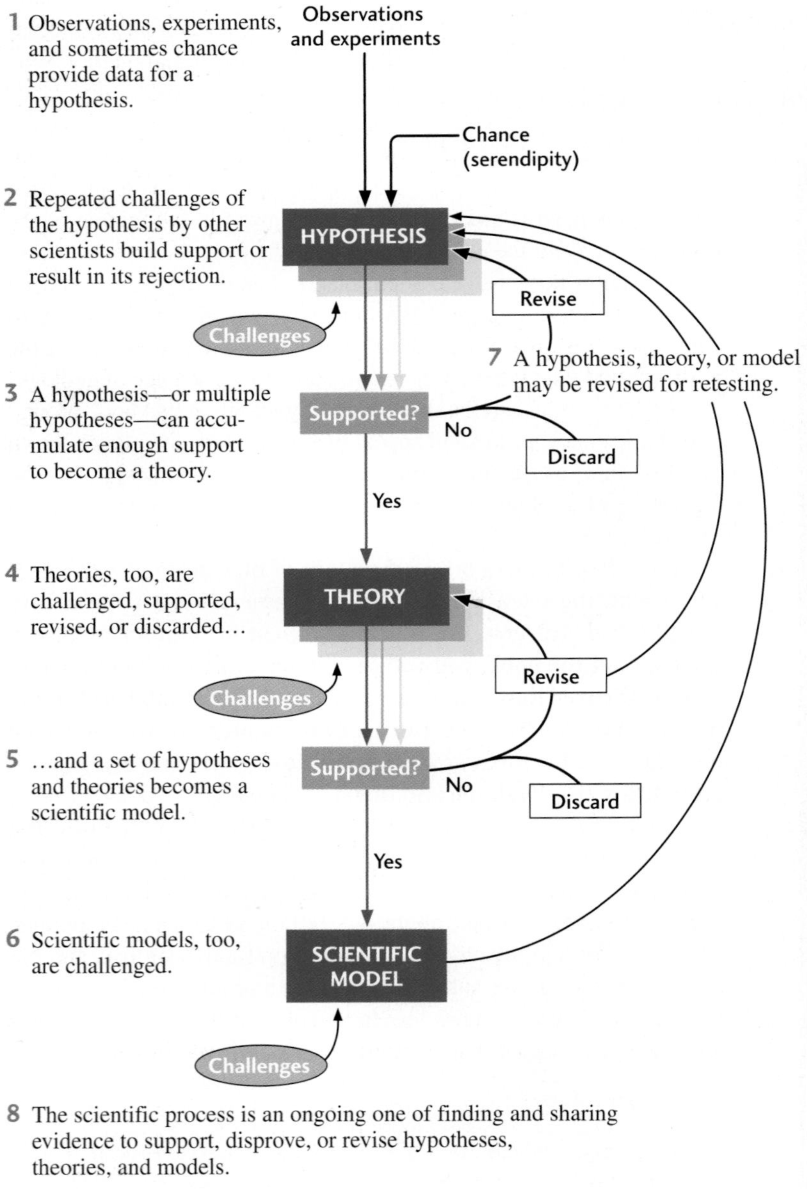

1 Observations, experiments, and sometimes chance provide data for a hypothesis.

2 Repeated challenges of the hypothesis by other scientists build support or result in its rejection.

3 A hypothesis—or multiple hypotheses—can accumulate enough support to become a theory.

4 Theories, too, are challenged, supported, revised, or discarded...

5 ...and a set of hypotheses and theories becomes a scientific model.

6 Scientific models, too, are challenged.

7 A hypothesis, theory, or model may be revised for retesting.

8 The scientific process is an ongoing one of finding and sharing evidence to support, disprove, or revise hypotheses, theories, and models.

Figure 1.1 An outline of the scientific method.

course of painstaking analysis, result from untold numbers of such interactions. Albert Einstein put it this way: "In science . . . the work of the individual is so bound up with that of his scientific predecessors and contemporaries that it appears almost as an impersonal product of his generation."

Because such free intellectual exchange can be subject to abuses, a code of ethics has evolved among scientists. Scientists must acknowledge the contributions of all others on whose work they have drawn. They must not falsify data, use the work of others without recognizing them, or be otherwise deceitful in their work. They must also accept responsibility for training the next generation of researchers and teachers. These principles are supported by the basic values of scientific cooperation, which a president of the National Academy of Sciences, Bruce Alberts, has aptly described as "honesty, generosity, a respect for evidence, openness to all ideas and opinions."

EARTH'S SHAPE AND SURFACE

The scientific method has its roots in *geodesy,* a very old branch of Earth science that studies Earth's shape and surface. In 1492, Columbus set a westward course for India because he believed in a theory of geodesy favored by Greek philosophers: *we live on a sphere.* His math was poor, how-

ever, so he badly underestimated Earth's circumference. Instead of a shortcut, he took the long way around, finding a New World instead of the Spice Islands! Had Columbus properly understood the ancient Greeks, he might not have made this fortuitous mistake, because they had accurately measured Earth's size more than 17 centuries earlier.

The credit for determining Earth's size goes to Eratosthenes, a Greek librarian who lived in Alexandria, Egypt. Sometime around 250 B.C., a traveler told him about a very interesting observation. At noon on the first day of summer (June 21), a deep well in the city of Syene, about 800 km south of Alexandria, was completely lit up by sunlight because the Sun was directly overhead. Acting on a hunch, Eratosthenes did an experiment. He set up a vertical pole in his own city, and at high noon on the summer solstice, the pole cast a shadow. By assuming the Sun was very far away so that the light rays falling on the two cities were parallel, Eratosthenes could demonstrate from simple geometry that the ground surface must be curved. The most perfect curved surface was a sphere, so he hypothesized that Earth had a spherical shape (the Greeks admired geometrical perfection). By measuring the length of the pole's shadow in Alexandria, he calculated that if vertical lines through the two cities could be extended to Earth's center, they would intersect at an angle of about 7°, which is about 1/50 of 360°, a full circle (**Figure 1.2**). Multiplying 50 times the distance between

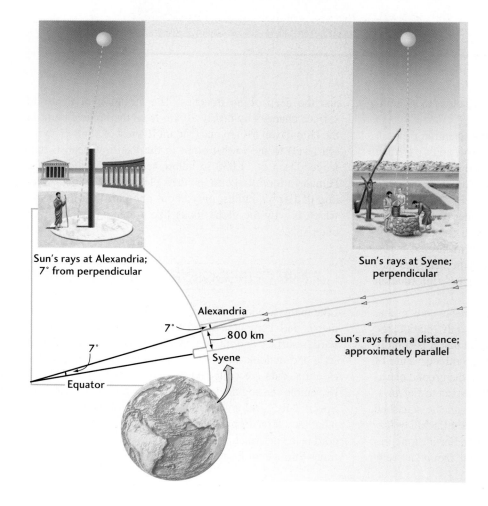

Sun's rays at Alexandria; 7° from perpendicular

Sun's rays at Syene; perpendicular

Alexandria

7°

800 km

Sun's rays from a distance; approximately parallel

7°

Syene

Equator

Figure 1.2 How Eratosthenes measured Earth's circumference. At noon on the summer solstice (June 21), a vertical well in the southern Egyptian town of Syene (near modern Aswan) was filled with light, indicating that the Sun was directly overhead, whereas a vertical pole in his hometown of Alexandria cast a shadow. From the length of the shadow, he found that Syene and Alexandria were separated by about 1/50 of Earth's circumference, about 7° out of 360°. Since the measured distance between the two towns was approximately 800 km, he calculated a circumference close to its modern value of 40,000 km.

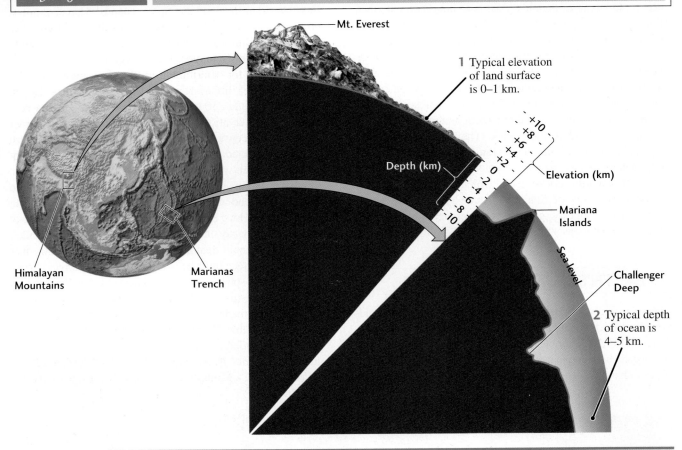

Key Figure 1.3 — Earth's topography is measured with respect to sea level.

the two cities, he deduced a circumference close to its modern value of 40,000 km.

In this powerful demonstration of the scientific method, Eratosthenes made observations (the shadow angle), formed a hypothesis (spherical shape), and applied some mathematical theory (spherical geometry) to propose a remarkably accurate model of Earth's physical form. His model was a good one because it correctly predicted other types of measurements, such as the distance at which a ship's tall mast disappears over the horizon. Moreover, it makes clear why well-designed experiments and good measurements are central to the scientific method: they give us new information about the natural world.

Much more precise measurements have shown that Earth is not a perfect sphere. Owing to its daily rotation, the planet bulges out slightly at its equator, so that it is slightly squashed at the poles. In addition, the smooth curvature of Earth's surface is disturbed by changes in the ground elevation. This **topography** is measured with respect to *sea level,* a smooth surface that conforms closely with the squashed spherical shape expected for the rotating Earth. Many features of geological significance stand out in Earth's topography (**Figure 1.3**), such as the continental mountain belts

and the deep ocean trenches. The elevation of the solid surface changes by nearly 20 km from the highest point in the Himalayan Mountains (Mount Everest at 8848 m above sea level) to the lowest point in the Pacific Ocean (Challenger Deep at 11,030 m below sea level). Although the Himalaya loom large to us, their elevation is a small fraction of Earth's radius, only about one part in a thousand, which is why the globe looks like a smooth sphere from outer space.

THE GEOLOGIC RECORD

Like many sciences, geology depends on laboratory experiments and computer simulations to describe and study Earth's surface and interior. Geology has its own particular style and outlook, however. It is an outdoor science in that essential data are collected by geologists in the field and by remote sensing devices, such as Earth-orbiting satellites. Specifically, geologists compare direct observations with what they infer from the *geologic record.* The geologic record is the information preserved in rocks formed at various times throughout Earth's long history.

In the eighteenth century, the Scottish physician and geologist James Hutton advanced a historic principle of geology that can be summarized as "the present is the key to the past." Hutton's concept became known as the **principle of uniformitarianism,** and it holds that the geologic processes we see in action today have worked in much the same way throughout geologic time.

The principle of uniformitarianism does not mean that all geologic phenomena are slow. Some of the most important processes happen as sudden events. A large meteorite that impacts Earth can gouge out a vast crater in a matter of seconds. A volcano can blow its top and a fault can rupture the ground in an earthquake almost as quickly. Other processes do occur much more slowly. Millions of years are required for continents to drift apart, for mountains to be raised and eroded, and for river systems to deposit thick layers of sediments. Geologic processes take place over a tremendous range of scales in both space and time (**Figure 1.4**).

Nor does the principle of uniformitarianism mean that we have to observe geologic phenomena directly to know that they are important in the current Earth system. In recorded history, humans have never witnessed a large meteorite impact, but we know they have occurred many times in the geologic past and will certainly happen again. The same can be said for the vast volcanic outpourings that have covered areas bigger than Texas with lava and poisoned the global atmosphere with volcanic gases. The long-term evolution of Earth is punctuated by many extreme, though infrequent, events involving rapid changes in the Earth system. Geology is the study of extreme events as well as progressive change.

From Hutton's day onward, geologists have observed nature at work and used the principle of uniformitarianism to interpret features found in old rock formations. This approach has been very successful. However, Hutton's principle is too confining for geologic science as it is now practiced. Modern

2 Over millions of years, layers of sediments built up over that rock. The most recent layer—the top—is about 250 million years old.

About 50,000 years ago, the explosive impact of a meteorite (perhaps weighing 300,000 tons) created this 1.2-km-wide crater in just a few seconds.

1 The rocks at the bottom of the Grand Canyon are 1.7–2.0 billion years old.

Figure 1.4 Geologic phenomena can stretch over thousands of centuries or can occur with dazzling speed. (*left*) The Grand Canyon, Arizona. [John Wang/PhotoDisc/Getty Images.] (*right*) Meteor Crater, Arizona. [John Sanford/Photo Researchers.]

geology must deal with the entire range of Earth's history, which began more than 4.5 billion years ago. As we will see, the violent processes that shaped Earth's early history were distinctly different from those that operate today. To understand that history, we will need some information about Earth's deep interior, which is layered like an onion.

PEELING THE ONION: Discovery of a Layered Earth

Ancient thinkers divided the universe into two parts, the Heavens above and Hades below. The sky was transparent and full of light, and they could directly observe its stars and track its wandering planets. In places, the ground quaked and erupted hot lava. Surely something terrible was going on down there! But Earth's interior was dark and closed to human view.

So it remained until about a century ago, when geologists began to look downward into Earth's interior, not with waves of light but with waves produced by earthquakes. An earthquake occurs when geologic forces cause brittle rocks to fracture, sending out vibrations like those sent out by the cracking of ice in a river. These **seismic waves** (from the Greek word for earthquake, *seismos*) illuminate the interior and can be recorded on *seismometers,* sensitive instruments that allow geologists to make pictures of Earth's inner workings, much as doctors use ultrasound and CAT scans to image the inside of your body. When the first networks of seismometers were installed around the world at the end of the nineteenth century, geologists began to discover that Earth's interior was divided into concentric layers of different compositions, separated by sharp, nearly spherical boundaries (**Figure 1.5**).

Earth's Density

Evidence for Earth's layering was first proposed at the end of the nineteenth century by the German physicist Emil Wiechert, before much seismic data had become available. He wanted to understand why our planet is so heavy, or more precisely, so *dense.* The density of a substance is easy to calculate: just measure its mass on a scale and divide by its volume. A typical rock, such as the granite used for tombstones, has a density of about 2.7 g/cm³. Estimating the density of the entire planet is a little harder, but not much. Eratosthenes had shown how to measure Earth's volume in 250 B.C., and sometime around 1680, the great English scientist Isaac Newton figured out how to calculate its mass from the force of gravity that pulls objects to its surface. The details, which involved careful laboratory experiments to calibrate Newton's law of gravity, were worked out by another Englishman, Henry Cavendish. In 1798, he calculated Earth's average density to be about 5.5 g/cm³, twice as dense as tombstone granite.

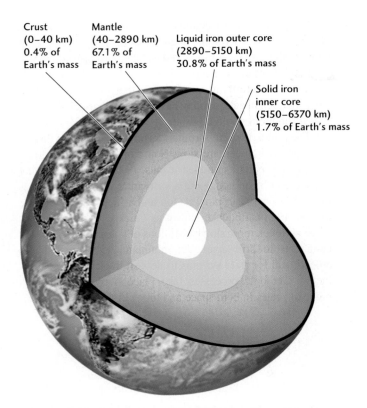

Crust (0–40 km) 0.4% of Earth's mass

Mantle (40–2890 km) 67.1% of Earth's mass

Liquid iron outer core (2890–5150 km) 30.8% of Earth's mass

Solid iron inner core (5150–6370 km) 1.7% of Earth's mass

Figure 1.5 Earth's major layers, showing their volume and mass expressed as a percentage of Earth's total volume and mass.

Wiechert was puzzled. He knew that a planet made entirely of common rocks, which are *silicates* (contain SiO_2), could not have such a high density. Some iron-rich rocks brought to the surface by volcanoes have densities as high as 3.5 g/cm³, but no ordinary rock approached Cavendish's value. He also knew that, going downward into Earth's interior, the pressure on rock increases from the weight of the overlying mass. The pressure squeezes the rock into a smaller volume, making its density higher. But Wiechert found that the pressure effect was too small to account for the density Cavendish had calculated.

The Mantle and Core

In thinking about what lay beneath him, Wiechert turned outward to the solar system and, in particular, to meteorites, which are pieces of the solar system that have fallen to Earth. He knew that some meteorites are made of a mixture of two heavy metals, iron and nickel, and thus had densities as high as 8 g/cm³ (**Figure 1.6**). He also knew that these elements are relatively abundant throughout our solar system. So, in 1896, he stated a grand hypothesis. Sometime in Earth's past, most of the iron and nickel in its interior had dropped inward to its center under the force of gravity. This created a dense **core,** which was surrounded by a shell of silicate-rich rocks that he called the **mantle** (using the German word for "coat"). With this hypothesis, he could

(a)

(b)

Figure 1.6 Two common types of meteorites. (a) A stony meteorite similar in composition to Earth's silicate mantle; (b) an iron-nickel meteorite similar in composition to Earth's core. The former has a density of about 3 g/cm^3, the latter a density of about 8 g/cm^3. [John Grotzinger/Ramón Rivera-Moret/Harvard Mineralogical Museum.]

come up with a two-layer Earth model that agreed with Cavendish's value for the average density. Moreover, he could also explain the existence of iron-nickel meteorites: they were chunks of the core from an Earthlike planet (or planets) that had broken apart, most likely by collisions with other planets.

Wiechert got busy testing his hypothesis using waves recorded by seismometers located around the globe (he designed one himself). The first results showed a shadowy inner mass that he took to be the core, but he had problems identifying some of the seismic waves. These waves come in two basic types: *compressional waves,* which expand and compress as they travel through solid, liquid, or gas; and *shear waves,* which involve side-to-side motion (shearing). Shear waves can propagate only through solids, which resist

shearing, and not through fluids such as air and water, which have no resistance to this type of motion.

In 1906, a British seismologist, Robert Oldham, was able to sort out the paths traveled by the various types of seismic waves and show that shear waves did not propagate through the core. The core, at least in its outer part, is liquid! This turns out to be not too surprising. Iron melts at a lower temperature than silicates, which is why metallurgists can use containers made of ceramic (a type of silicate) to hold molten iron. Earth's deep interior is hot enough to melt the iron-nickel alloy but not silicate rock. Beno Gutenberg, one of Wiechert's students, confirmed Oldham's observations that the outer part of the core is liquid and, in 1914, determined that the depth to the *core-mantle boundary* is just shy of 2900 km (see Figure 1.5).

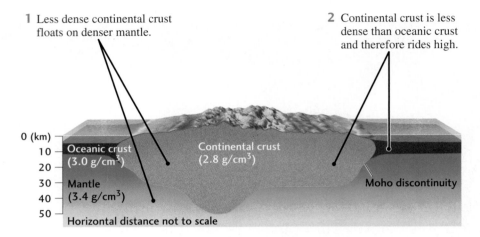

1 Less dense continental crust floats on denser mantle.

2 Continental crust is less dense than oceanic crust and therefore rides high.

Oceanic crust (3.0 g/cm^3)

Continental crust (2.8 g/cm^3)

Mantle (3.4 g/cm^3)

Moho discontinuity

Horizontal distance not to scale

Figure 1.7 Continents float high because they are made of rocks with lower densities than rocks of the mantle or oceanic crust.

The Crust

Five years earlier, a Croatian scientist had detected another boundary at the relatively shallow depth of 40 km beneath the European continent. This boundary, named the *Mohorovičić discontinuity* ("Moho" for short) after its discoverer, separates a **crust** composed of low-density silicates, which are rich in aluminum and potassium, from mantle silicates of higher density, which contain more magnesium and iron.

Like the core-mantle boundary, the Moho boundary is global. However, it was found to be substantially shallower beneath the oceans than beneath the continents. On a global basis, the average thickness of oceanic crust is only about 7 km, compared to almost 40 km for the continents. Moreover, rocks in the oceanic crust contain more iron and are therefore denser than continental rocks. Because the continental crust is thicker but less dense than oceanic crust, the continents ride high by floating like buoyant rafts on the denser mantle (**Figure 1.7**), much as icebergs float on the ocean. Continental buoyancy explains the most striking feature of Earth's surface: why the elevations shown in Figure 1.3 fall in two main groups, 0–1 km above sea level for much of the land surface and 4–5 km below sea level for much of the deep oceans.

Shear waves travel well through the mantle and crust, so we know that both are solid rock. How can continents float on solid rock? Rocks can be solid and strong over the short term (seconds to years) but weak over the long term (thousands to millions of years). Over very long intervals, the mantle below a depth of about 100 km has little strength and flows when it must adjust to support the weight of continents and mountains.

The Inner Core

Because the mantle is solid and the outer part of the core is liquid, the core-mantle boundary reflects seismic waves just as a mirror reflects light waves. In 1936, Danish seismolo-gist Inge Lehmann discovered another sharp spherical surface at the much greater depth of 5150 km, indicating a central mass with a higher density than the liquid core. Later studies showed that this inner core can transmit both shear waves and compressional waves. The **inner core** is therefore a solid metallic sphere with a radius of 1220 km—about two-thirds the size of the Moon—suspended within the liquid **outer core.**

Geologists were puzzled by the existence of a "frozen" inner core. From other considerations, they knew that temperatures inside Earth should increase with depth. According to the best current estimates, the temperature rises from about 3500°C at the core-mantle boundary to almost 5000°C at its center. If the inner core is hotter, how could it be frozen while the outer core is molten? The mystery was eventually solved by laboratory experiments on iron-nickel alloys, which showed that the "freezing" was due to higher pressures rather than lower temperatures at Earth's center.

Chemical Composition of Earth's Major Layers

By the mid-twentieth century, geologists had discovered all of Earth's major layers—crust, mantle, outer core, and inner core—plus a number of more subtle features in its interior. They found, for example, that the mantle itself is layered into an *upper mantle* and a *lower mantle,* separated by a *transition zone* where the rock density increases in a series of steps. These density steps are not caused by changes in the rock's chemical composition but rather by changes in its compactness due to the increasing pressure with depth. The two largest density jumps in the transition zone are located at depths of about 400 km and 650 km, but they are smaller than the density increases across the Moho discontinuity and core-mantle boundary, which are due to changes in composition (see Figure 1.5).

Geologists were also able to show that Earth's outer core could not be made of a pure iron-nickel alloy, because the

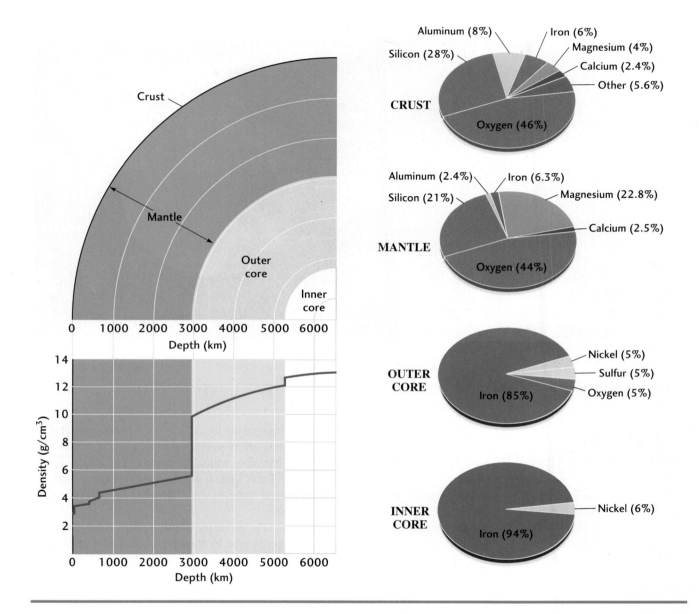

Key Figure 1.8 | Jumps in density between Earth's major layers are caused by changes in their chemical composition.

densities of these metals are higher than the observed density of the outer core. About 10 percent of the outer core's mass must be made of lighter elements, such as oxygen and sulfur. On the other hand, the density of the solid inner core is slightly higher than that of the outer core and is consistent with a nearly pure iron-nickel alloy.

By bringing together many lines of evidence, geologists have put together a model of the composition of Earth and its various layers. The data include the composition of crustal and mantle rocks as well as the compositions of meteorites, thought to be samples of the cosmic material from which planets like Earth were originally made.

Only 8 elements, out of more than 100, make up 99 percent of Earth's mass (**Figure 1.8**). In fact, about 90 percent of the Earth consists of only four elements: iron, oxygen, silicon, and magnesium. The first two are the most abundant elements, each accounting for nearly a third of the planet's overall mass, but they are distributed very differently. Iron, the densest common element, is concentrated in the core, whereas oxygen, the lightest common element, is concentrated in the crust and mantle. These relationships show that the different compositions of Earth's layers are primarily the work of gravity. As you can see in Figure 1.8, the crustal rocks on which we stand are almost 50 percent oxygen.

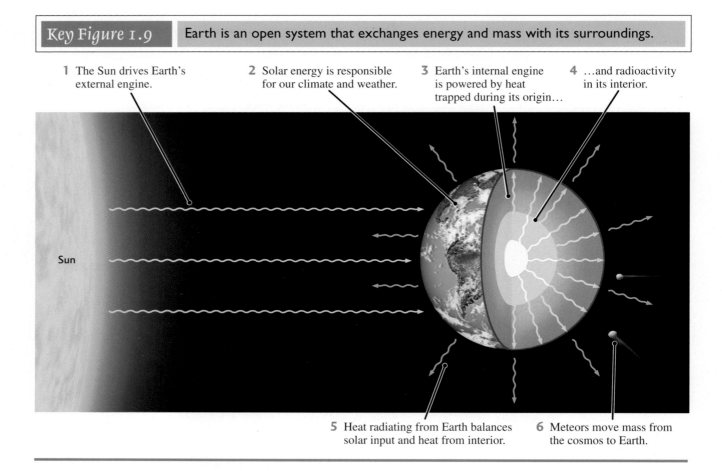

Key Figure 1.9 Earth is an open system that exchanges energy and mass with its surroundings.

1 The Sun drives Earth's external engine.

2 Solar energy is responsible for our climate and weather.

3 Earth's internal engine is powered by heat trapped during its origin…

4 …and radioactivity in its interior.

Sun

5 Heat radiating from Earth balances solar input and heat from interior.

6 Meteors move mass from the cosmos to Earth.

EARTH AS A SYSTEM OF INTERACTING COMPONENTS

Earth is a restless planet, continually changing through geologic activity such as earthquakes, volcanoes, and glaciation. This activity is powered by two heat engines: one internal, the other external (**Figure 1.9**). A heat engine—for example, the gasoline engine of an automobile—transforms heat into mechanical motion or work. Earth's internal engine is powered by the heat energy trapped during the planet's violent origin and generated by radioactivity in its deep interior. The internal heat drives motions in the mantle and core, supplying the energy to melt rock, move continents, and lift up mountains. Earth's external engine is driven by solar energy—heat supplied to Earth's surface by the Sun. Heat from the Sun energizes the atmosphere and oceans and is responsible for our climate and weather. Rain, wind, and ice erode mountains and shape the landscape, and the shape of the landscape, in turn, changes the climate.

All the parts of our planet and all their interactions, taken together, constitute the **Earth system.** Although Earth scientists have long thought in terms of natural systems, it was not until the late twentieth century that they had the tools to investigate how the Earth system actually works. Networks of instruments and Earth-orbiting satellites now collect information about the Earth system on a global scale, and computers are powerful enough to calculate the mass and energy transfers within the system. The major components of the Earth system are depicted in **Figure 1.10**. We have discussed some of them already; we will define the others shortly.

We will talk about the Earth system throughout this text. Let's get started by thinking about some of its basic features. Earth is an *open system* in the sense that it exchanges mass and energy with the rest of the cosmos. Radiant energy from the Sun energizes the weathering and erosion of Earth's surface, as well as the growth of plants, which feed almost all living things. Our climate is controlled by the balance between the solar energy coming into the Earth system and the energy Earth radiates back into space. These days, the exchange of material between Earth and space is relatively small—only about a million tons of meteorites, equivalent to a cube 70 m on a side, fall to Earth each year—but the mass transfer was much greater during the early life of the solar system.

Although we think of Earth as a single system, it is a challenge to study the whole thing all at once. Instead, we will focus our attention on parts of the system (subsystems) we are trying to understand. For instance, in the discussion of recent climate changes, we will primarily consider interactions among the atmosphere, hydrosphere, and biosphere that

Earth System Figure 1.10   The Earth system is all parts of our planet and their interactions.

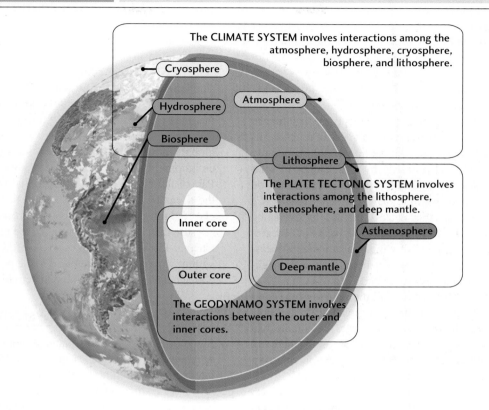

The CLIMATE SYSTEM involves interactions among the atmosphere, hydrosphere, cryosphere, biosphere, and lithosphere.

Cryosphere

Hydrosphere

Atmosphere

Biosphere

Lithosphere

The PLATE TECTONIC SYSTEM involves interactions among the lithosphere, asthenosphere, and deep mantle.

Inner core

Asthenosphere

Outer core

Deep mantle

The GEODYNAMO SYSTEM involves interactions between the outer and inner cores.

Spheres

Solar Radiation Energizes These Components

Atmosphere	Gaseous envelope extending from Earth's surface to an altitude of about 100 km
Hydrosphere	Surface waters comprising all oceans, lakes, rivers, and groundwaters
Cryosphere	Polar ice caps, glaciers, and other surface ice and snow
Biosphere	All organic matter related to life near Earth's surface

Earth's Internal Heat Energizes These Components

Lithosphere	Strong, rocky outer shell of the solid Earth that comprises the crust and uppermost mantle down to an average depth of about 100 km; forms the tectonic plates
Asthenosphere	Weak, ductile layer of mantle beneath the lithosphere that deforms to accommodate the horizontal and vertical motions of plate tectonics
Deep mantle	Mantle beneath the asthenosphere, extending from about 400 km deep to the core-mantle boundary (about 2900 km deep)
Outer core	Liquid shell composed primarily of molten iron, extending from about 2900 km to 5150 km in depth
Inner core	Inner sphere composed primarily of solid iron, extending from about 5150 km deep to Earth's center at 6370 km

are driven by solar energy. Our coverage of how the continents are deformed to make mountains will focus on interactions between the crust and the deeper mantle that are driven by Earth's internal energy. Specialized subsystems that describe specific types of terrestrial behavior, such as climate changes or mountain building, are called **geosystems.** The Earth system can be thought of as the collection of all these open, interacting (and often overlapping) geosystems.

In this chapter, we will introduce three important geosystems that operate on a global scale: the climate system, the plate tectonic system, and the geodynamo. Later in the book, we will have occasion to discuss a number of smaller geosystems. Here are three examples: volcanoes that erupt hot lava (Chapter 12), hydrologic systems that give us our drinking water (Chapter 17), and petroleum reservoirs that produce oil and gas (Chapter 23).

The Climate System

Weather is the term we use to describe the temperature, precipitation, cloud cover, and winds observed at a particular location and time on Earth's surface. We all know how variable the weather can be—hot and rainy one day, cool and dry the next—depending on the movements of storm systems, warm and cold fronts, and other atmospheric disturbances. Because the atmosphere is so complex, even the best forecasters have a hard time predicting the weather more than four or five days in advance. However, we can guess in rough terms what our weather will be much further into the future, because the weather is governed primarily by the changes in solar energy input on seasonal and daily cycles: summers are hot, winters cold; days are warmer, nights cooler. *Climate* is a description of these weather cycles obtained by averaging temperature and other variables over many years of observation. A complete description of climate also includes measures of how variable the weather has been, such as the highest and lowest temperatures ever recorded on a given day.

The **climate system** includes all the Earth system components that determine climate on a global scale and how climate changes with time. In other words, the climate system describes not only the behavior of the atmosphere but also how climate is influenced by the hydrosphere, cryosphere, biosphere, and lithosphere (see Figure 1.10).

When the Sun warms Earth's surface, some of the heat is trapped by water vapor, carbon dioxide, and other gases in the atmosphere, much as heat is trapped by frosted glass in a greenhouse, This *greenhouse effect* explains why Earth has a pleasant climate that makes life possible. If its atmosphere contained no greenhouse gases, its surface would be frozen solid! Therefore, greenhouse gases, particularly carbon dioxide, play an essential role in regulating climate. As we will learn in later chapters, the concentration of carbon dioxide in the atmosphere is a balance between the amount spewed out of Earth's interior in volcanic eruptions and the amount withdrawn during the weathering of silicate rocks. In this way, the climate system is regulated by interactions with the solid Earth.

To understand these types of interactions, scientists build numerical models—virtual climate systems—on large computers, and they compare the results of their computer simulations with observed data. They hope to improve the models by testing them against additional observations, so that they can accurately predict how climate will change in the future. A particularly urgent problem is to understand the global warming that might be caused by human-generated emissions of carbon dioxide and other greenhouse gases. Part of the public debate about global warming centers on the accuracy of computer predictions. Skeptics argue that even the most sophisticated computer models are unreliable because they lack many features of the real Earth system. In Chapter 15, we will discuss some aspects of how the climate system works and, in Chapter 23, the practical problems of climate change caused by human activities.

The Plate Tectonic System

Some of Earth's more dramatic geologic events—volcanic eruptions and earthquakes, for example—also result from interactions within the Earth system. These phenomena are driven by Earth's internal heat, which escapes through the circulation of material in Earth's solid mantle.

We have seen that Earth is zoned by chemistry: its crust, mantle, and core are chemically distinct layers. Earth is also zoned by strength, a property that measures how much an Earth material can resist being deformed. Material strength depends on chemical composition (bricks are strong, soap bars are weak) and temperature (cold wax is strong, hot wax is weak). In some ways, the outer part of the solid Earth behaves like a ball of hot wax. Cooling of the surface forms

---

Key Figure 1.11 | Convection carries heat upward by the motion of matter.

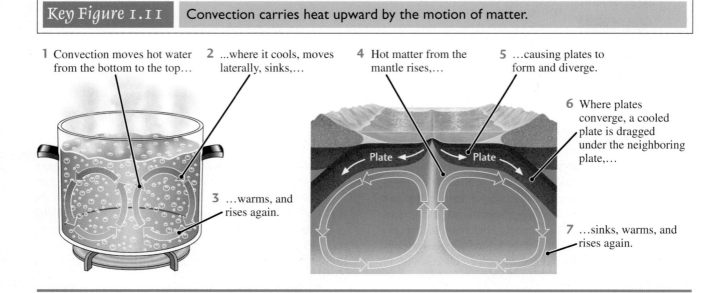

1 Convection moves hot water from the bottom to the top…

2 …where it cools, moves laterally, sinks,…

3 …warms, and rises again.

4 Hot matter from the mantle rises,…

5 …causing plates to form and diverge.

6 Where plates converge, a cooled plate is dragged under the neighboring plate,…

7 …sinks, warms, and rises again.

Plate

Plate

the strong outer shell or **lithosphere** (from the Greek *lithos,* meaning "stone") that encases a hot, weak **asthenosphere** (from the Greek *asthenes,* meaning "weak"). The lithosphere includes the crust and the top part of the mantle down to an average depth of about 100 km. When subjected to force, the lithosphere tends to behave as a nearly rigid and brittle shell, whereas the underlying asthenosphere flows as a moldable, or ductile, solid.

According to the remarkable theory of *plate tectonics,* the lithosphere is not a continuous shell; it is broken into about a dozen large plates that move over Earth's surface at rates of a few centimeters per year. Each plate is a rigid unit that rides on the asthenosphere, which also is in motion. The lithosphere that forms a plate may be just a few kilometers thick in volcanically active areas and perhaps 200 km thick or more beneath the older, colder parts of the continents. The discovery of plate tectonics in the 1960s led to the first unified theory that explained the worldwide distribution of earthquakes and volcanoes, continental drift, mountain building, and many other geologic phenomena. Chapter 2 will be devoted to a detailed description of plate tectonics.

Why do the plates move across Earth's surface instead of locking up into a completely rigid shell? The forces that push and pull the plates around the surface come from the heat engine in Earth's solid mantle. Driven by internal heat, hot mantle material rises where plates separate. The lithosphere cools and becomes more rigid as it moves away, eventually sinking into the mantle under the pull of gravity

at boundaries where plates converge. This general process, in which hotter material rises and cooler material sinks, is called *convection* (**Figure 1.11**). We note that the flow in ductile solids is usually slower than fluid flow, because even "weak" solids (say, wax or taffy) are more resistant to deformation than ordinary fluids (say, water or mercury).

The convecting mantle and its overlying mosaic of lithospheric plates constitute the **plate tectonic system.** As with the climate system (which involves a wide range of convective processes in the atmosphere and oceans), scientists use computer simulations to study plate tectonics, and they revise the models when their implications disagree with actual data.

The Geodynamo System

The third global geosystem involves interactions that produce a **magnetic field** deep inside the Earth, in its fluid outer core. This magnetic field reaches far into outer space, causing compass needles to point north and shielding the biosphere from the Sun's harmful radiation. When rocks form, they become slightly magnetized by this field, so geologists can study how the magnetic field behaved in the past and use it to help them decipher the geologic record.

Earth's internal magnetic field behaves as if a powerful bar magnet were located at Earth's center and inclined about 11° from its axis of rotation. The magnetic force points into Earth at the north magnetic pole and outward at the south magnetic pole (**Figure 1.12**). A compass needle free to

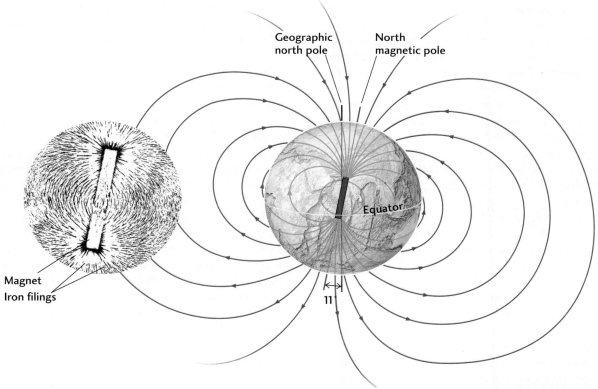

Figure 1.12 *(left)* The magnetic field of a bar magnet is revealed by the alignment of iron filings on paper. [After *PSSC Physics,* 3d ed. (Lexington, Mass.: D. C. Heath, 1971).] *(right)* Earth's magnetic field is much like the field that would be produced if a giant bar magnet were placed at the Earth's center and slightly inclined (11°) from the axis of rotation. Lines of magnetic force produced by such a bar magnet are shown. A compass needle points to the north magnetic pole because it orients in the direction of the local line of force.

swing under the influence of the magnetic field will rotate into a position parallel to the local line of force, approximately in the north-south direction.

Although a permanent magnet at Earth's center can explain the dipolar ("two-pole") nature of the observed magnetic field, this hypothesis can be easily rejected. Laboratory experiments demonstrate that the field of a permanent magnet is destroyed when the magnet is heated above about 500°C. We know that the temperatures in Earth's deep interior are much higher than that—thousands of degrees at its center—so, unless the magnetism is constantly regenerated, it cannot be maintained.

Scientists theorize that heat flowing out of Earth's core causes convection that generates and maintains the magnetic field. Why is a magnetic field created by convection in the outer core but not by convection in the mantle? First, the outer core is made primarily of iron, which is a very good electrical conductor, whereas the silicate rocks of the mantle are very poor electrical conductors. Second, the convective motions are a million times more rapid in the liquid outer core than in the solid mantle. These rapid motions stir up electric currents in the iron to create a **geodynamo** with a strong magnetic field.

A dynamo is an engine that produces electricity by rotating a coil of conducting wire through a magnetic field. The magnetic field can come from a permanent magnet or be generated by passing electricity through another coil—an electromagnet. The big dynamos in all commercial power plants use electromagnets (permanent magnets are too weak). The energy needed to keep the magnetic field going, as well as the electricity sent out to customers, comes from the mechanical work required to rotate the coil. In most power plants, this work is done by steam or falling water. The geodynamo in Earth's outer core operates on the same basic principles, except that the work comes from convection powered by the core's internal heat. Similar convective dynamos are thought to generate the strong magnetic fields observed on Jupiter and the Sun.

For some 400 years, scientists have known that a compass needle points to the north because of Earth's magnetic field. Imagine how stunned they were a few decades ago when they found geologic evidence that the magnetic field can completely reverse itself—that is, it can flip its north magnetic pole with its south magnetic pole. Over about half of geologic time, a compass needle would have pointed to the south!

These *magnetic reversals* occur at irregular intervals ranging from tens of thousands to millions of years. The processes that cause them are not well understood, but computer models of the geodynamo show sporadic reversals in the absence of any other external factors—that is, purely through internal interactions. As we will see in the next chapter, geologists have found magnetic reversals to be very useful, because they can use their imprint on the geologic record to help them figure out the motions of the tectonic plates.

AN OVERVIEW OF GEOLOGIC TIME

So far, we have discussed Earth's size and shape, its internal layering and composition, and the operation of its three major geosystems. How did Earth get its layered structure in the first place? How have the global geosystems evolved through geologic time? To begin to answer these questions, we present a brief overview of geologic time from the birth of the planet to the present. Later chapters will fill in the details.

Comprehending the immensity of geologic time is a challenge. The popular writer John McPhee has eloquently noted that geologists look into the "deep time" of Earth's early history (measured in billions of years), just as astronomers look into the "deep space" of the outer universe (measured in billions of light-years). **Figure 1.13** presents geologic time as a ribbon marked with some major events and transitions.

Origin of Earth and Its Global Geosystems

From meteorites, geologists have been able to show that Earth and the other planets formed about 4.56 billion years ago by the rapid condensation of a dust cloud that circulated around the young Sun. This violent process, which involved the aggregation and collision of progressively larger clumps of matter, will be described in more detail in Chapter 9. Within just 100 million years (a relatively short period of time, geologically speaking), the Moon had formed and Earth's core had separated from its mantle. Exactly what happened during the next several hundred million years is hard to figure out, because very little of the rock record survived the intense bombardment by the large meteorites that were constantly smashing into Earth. This early period of Earth's history can be appropriately called the geologic "dark ages."

The oldest rocks now found on Earth's surface are about 4 billion years old. Rocks as ancient as 3.8 billion years show evidence of erosion by water, indicating the existence of a hydrosphere and the operation of a climate system not too different from that of the present. Rocks only slightly younger, 3.5 billion years old, record a magnetic field about as strong as the one we see today, which puts a bound on the age of the geodynamo. By 2.5 billion years ago, enough low-density crust had collected at Earth's surface to form large continental masses. The geologic processes that then modified these continents were very similar to those we see operating today in plate tectonics.

The Evolution of Life

Life also began very early in Earth's history, as we can tell from the study of **fossils,** traces of organisms preserved in the geologic record. Fossils of primitive bacteria have been

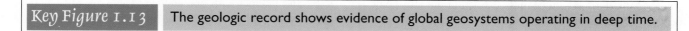

Key Figure 1.13   The geologic record shows evidence of global geosystems operating in deep time.

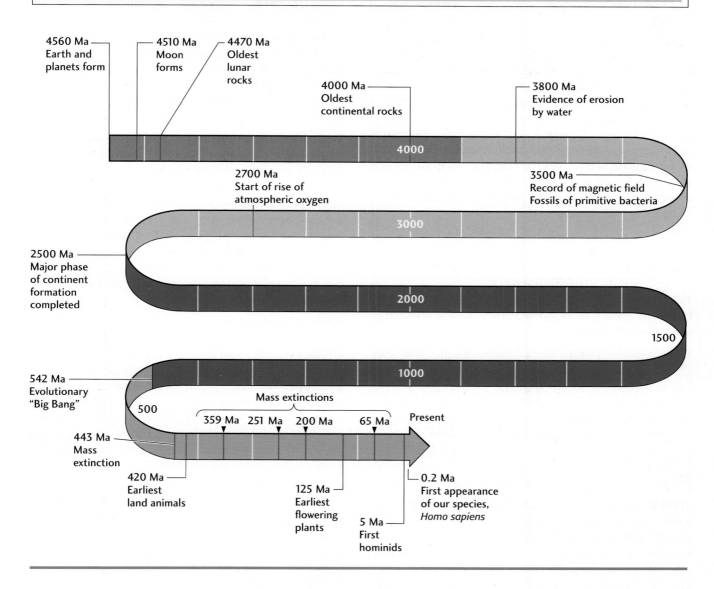

found in rocks dated at 3.5 billion years. A key event was the evolution of organisms such as plants that release oxygen into the atmosphere and oceans. The buildup of oxygen in the atmosphere was under way by 2.5 billion years ago. The increase to modern levels of atmospheric oxygen most likely occurred in a series of steps over a period perhaps as long as 2 billion years.

Life on early Earth was primitive, consisting mostly of small, single-celled organisms that floated near the surface of the oceans or lived on the seafloor. Between 1 billion and 2 billion years ago, more complex life-forms such as algae and seaweed evolved. The first animals appeared about 600 million years ago, evolving in a series of waves. In a period starting 542 million years ago and probably lasting less than 10 million years, eight entirely new branches of the animal kingdom were established, including ancestors to

nearly all animals inhabiting the Earth today. It was during this evolutionary explosion, sometimes called biology's Big Bang, that animals with shells first left their shelly fossils.

Although biological evolution is often viewed as a very slow process, it is punctuated by brief periods of rapid change. Spectacular examples are major mass extinctions, during which many types of animals and plants suddenly disappeared from the geologic record. Five of these huge turnovers are marked on the time ribbon in Figure 1.13. The last was caused by a major meteorite impact 65 million years ago. The meteorite, not much larger than about 10 km in diameter, caused the extinction of half of Earth's species, including all dinosaurs. This extreme event may have made it possible for mammals to become the dominant species and paved the way for humankind's emergence in the last 200,000 years.

The causes of the other mass extinction are still being debated. In addition to meteorite impacts, scientists have proposed other types of extreme events, such as rapid climate changes brought on by glaciations and massive eruptions of volcanic material. The evidence is often ambiguous or inconsistent. For example, the largest extinction event of all time took place about 250 million years ago, wiping out nearly 95 percent of all species. A meteorite impact has been proposed by some investigators, but the geologic record shows that the ice sheets expanded at this time and seawater chemistry changed, consistent with a major climate crisis. At the same time, an enormous volcanic eruption covered an area in Siberia almost half the size of the United States with 2 or 3 million cubic kilometers of lava. This mass extinction has been dubbed "Murder on the Orient Express," because there are so many suspects!

SUMMARY

What is geology? Geology is the science that deals with Earth—its history, its composition and internal structure, and its surface features.

How do geologists study Earth? Geologists, like other scientists, use the scientific method. They share the data that they develop and check one another's work. A hypothesis is a tentative explanation of a body of data. A set of related hypotheses confirmed by other data and experiments may be elevated to a theory. A theory may be abandoned or modified when subsequent observations shows it to be false. Confidence grows in those theories that withstand repeated tests and are able to predict the results of new experiments.

What is Earth's size and shape? Earth's overall shape is a sphere with an average radius of 6370 km that bulges slightly at the equator and is slightly squashed at the poles, owing to the planet's rotation. Its solid surface has topography that deviates from this overall shape by about 10 km. Elevations fall into two main groups: 0–1 km above sea level for much of the land surface and 4–5 km below sea level for much of the deep oceans.

What are Earth's major layers? Earth's interior is divided into concentric layers of different compositions, separated by sharp, nearly spherical boundaries. The outer layer is the crust, which varies from about 40 km thick beneath continents to about 8 km thick beneath oceans. Below the crust is the mantle, a thick shell of denser rock that extends to the core-mantle boundary at a depth of 2900 km. The central core, which is composed primarily of iron and nickel, is divided into two layers: a liquid outer core and a solid inner core, separated by a boundary at a depth of 5150 km.

How do we study Earth as a system of interacting components? When we try to understand a complex system such as Earth, we find that it is often easier to break the system down into subsystems (geosystems) to see how they work and interact with one another. There are three major global geosystems: the climate system, which mainly involves interactions among the atmosphere, hydrosphere, and biosphere; the plate tectonic system, which mainly involves interactions among Earth's solid components (lithosphere, asthenosphere, and deep mantle); and the geodynamo system, which mainly involves interactions within Earth's central core. The climate system is driven by heat from the Sun, whereas the plate tectonic and geodynamo systems are driven by Earth's internal heat.

What are the basic elements of plate tectonics? The lithosphere is not a continuous shell; it is broken into about a dozen large plates. Driven by convection in the mantle, plates move over Earth's surface at rates of a few centimeters per year. Each plate acts as a rigid unit, riding on the asthenosphere, which also is in motion. The lithosphere begins to form from rising hot mantle material where plates separate, cooling and becoming more rigid as it moves away from this divergent boundary. Eventually, it sinks into the asthenosphere, dragging material back into the mantle at boundaries where plates converge.

What are some major events in Earth's history? Earth formed as a planet 4.56 billion years ago. Rocks as old as 4 billion years have survived in Earth's crust. Liquid water existed on Earth's surface by 3.8 billion years ago, and the geodynamo was generating a magnetic field by 3.5 billion years ago. The earliest evidence of life has been found in rocks of this latter age. By 2.5 billion years ago, the oxygen content of the atmosphere was rising because of oxygen production by early plant life, and the geologic processes at Earth's surface were very similar to those operating today in plate tectonics. Animals appeared suddenly about 600 million years ago, diversifying rapidly in a great evolutionary explosion. The subsequent evolution of life was marked by a series of mass extinctions, the last caused by a large meteorite impact 65 million years ago, which killed off the dinosaurs. Our species, *Homo sapiens,* first appeared about 160,000 years ago.

KEY TERMS AND CONCEPTS

asthenosphere (p. 13)

climate system (p. 12)

core (p. 6)

crust (p. 8)

Earth system (p. 10)

fossil (p. 14)

geodynamo (p. 14)

geosystem (p. 11)

inner core (p. 8)

lithosphere (p. 13)

magnetic field (p. 13)

mantle (p. 6)

outer core (p. 8)

plate tectonic system (p. 13)

principle of uniformitarianism (p. 5)

scientific method (p. 2)

seismic wave (p. 6)

topography (p. 4)

EXERCISES

1. Illustrate the differences between a hypothesis, a theory, and a model with some examples drawn from this chapter.

2. Give an example of how the model of Earth's spherical shape developed by Eratosthenes can be experimentally tested.

3. Give two reasons why Earth's shape is not a perfect sphere.

4. If you made a model of Earth's spherical shape that was 10 cm in radius, how high would Mount Everest rise above sea level?

5. It is thought that a large meteorite impact 65 million years ago caused the extinction of half of Earth's living species, including all the dinosaurs. Does this event disprove the principle of uniformitarianism? Explain your answer.

6. How does the chemical composition of Earth's crust differ from that of its deeper interior? From that of its core?

7. Explain how the outer core can be a liquid while the deep mantle is a solid.

8. How do the terms *weather* and *climate* differ? Express the relationship between climate and weather using examples from your experience.

9. Earth's mantle is solid, but it convects as part of the plate tectonic system. Explain why these statements are not contradictory.

THOUGHT QUESTIONS

1. How does science differ from religion as a way to understand the world?

2. Imagine you are a tour guide on a journey from Earth's surface to its center. How would you describe the material that your tour group encounters on the way down? Why is the density of the material always increasing as you go deeper?

3. How does viewing Earth as a system of interacting components help us to understand our planet? Give an example of an interaction between two or more geosystems that could affect the geologic record.

4. In what general ways are the climate system, the plate tectonic system, and the geodynamo system similar? In what ways are they different?

5. Not every planet has a geodynamo. Why not? If Earth did not have a magnetic field, what might be different about our planet?

6. Based on the material presented in this chapter, what can we say about how long ago the three major global geosystems began to operate?

7. If no theory can be proved true, why do almost all geologists believe strongly in Darwin's theory of evolution?

PLATE TECTONICS
The Unifying Theory

The lithosphere—Earth's strong, rigid outer shell of rock—is broken into about a dozen plates, which slide by, converge with, or separate from each other as they move over the weaker, ductile asthenosphere. Plates are created where they separate and recycled where they converge, in a continuous process of creation and destruction. Continents, embedded in the lithosphere, drift along with the moving plates. The theory of **plate tectonics** describes the movement of plates and the forces acting between them. It also explains volcanoes; earthquakes; and the distribution of mountain chains, rock assemblages, and structures on the seafloor—all of which result from movements at plate boundaries. Plate tectonics provides a conceptual framework for a large part of this book and, indeed, for much of geology. **This chapter lays out the plate tectonics theory and examines how the forces that drive plate motions arise from the mantle convection system.**

THE DISCOVERY OF PLATE TECTONICS

In the 1960s, a great revolution in thinking shook the world of geology. For almost 200 years, geologists had developed various theories of *tectonics* (from the Greek *tekton,* meaning "builder")—the general term used to describe mountain building, volcanism, and other processes that construct geologic features on Earth's surface. It was not until the discovery of plate tectonics, however, that a single theory could satisfactorily explain the whole range of geologic processes. Physics had a comparable revolution at the beginning of the twentieth century, when the theory of relativity unified the physical laws that govern space, time, mass, and motion. Biology had a comparable revolution in the middle of the twentieth century, when the discovery of DNA allowed biologists to explain how organisms transmit the information that controls their growth, development, and functioning from generation to generation.

Mount Everest, Nepal, the highest mountain in the world, as viewed from Kala Pattar. [Michael C. Klesius/National Geographic/Getty Images.]

The basic ideas of plate tectonics were put together as a unified theory of geology about 40 years ago. The scientific synthesis that led to plate tectonics, however, really began much earlier in the twentieth century, with the recognition of evidence for continental drift.

 ## Continental Drift

Such changes in the superficial parts of the globe seemed to me unlikely to happen if the earth were solid to the center. I therefore imagined that the internal parts might be a fluid more dense, and of greater specific gravity than any of the solids we are acquainted with, which therefore might swim in or upon that fluid. Thus the surface of the earth would be a shell, capable of being broken and disordered by the violent movements of the fluid on which it rested.

(Benjamin Franklin, 1782, in a letter to French geologist
Abbé J. L. Giraud-Soulavie)

The concept of **continental drift**—large-scale movements of continents over the globe—has been around for a long time. In the late sixteenth century and in the seventeenth century, European scientists noticed the jigsaw-puzzle fit of the coasts on both sides of the Atlantic, as if the Americas, Europe, and Africa had been part of a single continent and had subsequently drifted apart. By the close of the nineteenth century, the Austrian geologist Eduard Suess had put together some of the pieces of the puzzle. He postulated that the present-day southern continents had once formed a single giant continent called *Gondwanaland* (or *Gondwana*). In 1915, Alfred Wegener, a German meteorologist who was recovering from wounds suffered in World War I, wrote a book on the breakup and drift of continents. In it, he laid out the remarkable similarity of rocks, geologic structures, and fossils on opposite sides of the Atlantic (**Figure 2.1**). In the years that followed, Wegener postulated a supercontinent, which he called **Pangaea** (Greek for "all lands"), that broke up into the continents as we know them today.

Although Wegener was correct in asserting that the continents had drifted apart, his hypotheses about how fast they were moving and what forces were pushing them across Earth's surface turned out to be wrong, which reduced his credibility among other scientists. After about a decade of spirited debate, physicists convinced geologists that Earth's outer layers were too rigid for continental drift to occur, and Wegener's ideas fell into disrepute among all except a few geologists.

The advocates of the drift hypothesis pointed not only to geographic matching but also to similarities in rock ages and trends in geologic structures on opposite sides of the Atlantic (see Figure 2.1). They also offered arguments, accepted now as good evidence of drift, based on fossil and climate data. Identical 300-million-year-old fossils of the reptile *Mesosaurus,* for example, are found only in Africa and South America, suggesting that the two continents were joined at that time (**Figure 2.2**). The animals and plants on differ-

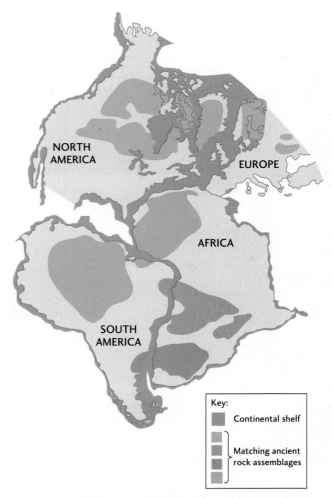

Figure 2.1 The jigsaw-puzzle fit of continents bordering the Atlantic Ocean formed the basis of Alfred Wegener's theory of continental drift. In his book *The Origin of Continents and Oceans,* Wegener cited as additional evidence the similarity of geologic features on opposite sides of the Atlantic. The matchup of ancient crystalline rocks is shown in adjacent regions of South America and Africa and of North America and Europe. [Geographic fit from data of E. C. Bullard; geological data from P. M. Hurley.]

ent continents showed similarities in evolution until the postulated breakup time. After that, they followed different evolutionary paths, presumably because of the isolation and changing environments of the separating continents. In addition, rocks deposited by glaciers that existed 300 million years ago are now distributed across South America, Africa, India, and Australia. If the southern continents had once been part of Gondwanaland near the South Pole, a single continental glacier could account for these glacial deposits.

Seafloor Spreading

The geologic evidence did not convince the skeptics, who maintained that continental drift was physically impossible. No one had yet come up with a plausible driving force that could have split Pangaea and moved the continents apart. Wegener, for example, thought the continents floated like

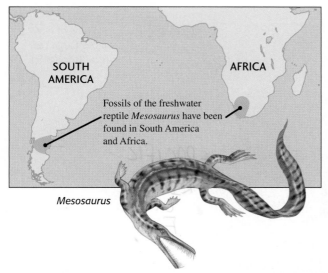

Figure 2.2 Fossils of the reptile *Mesosaurus,* 300 million years old, are found in South America and Africa and nowhere else in the world. If *Mesosaurus* could swim across the South Atlantic Ocean, it could have crossed other oceans and should have spread more widely. The observation that it did not suggests that South America and Africa must have been joined 300 million years ago. [After A. Hallam, "Continental Drift and the Fossil Record," *Scientific American* (November 1972): 57–66.]

boats across the solid oceanic crust, dragged along by the tidal forces of the Sun and Moon. His hypothesis was quickly rejected, however, because it could be shown that tidal forces are much too weak to move continents.

The breakthrough came when scientists realized that convection in Earth's mantle (discussed in Chapter 1) could push and pull the continents apart, creating new oceanic crust through the process of **seafloor spreading.** In 1928, the British geologist Arthur Holmes proposed that convection currents "dragged the two halves of the original continent apart, with consequent mountain building in the front where the currents are descending, and the ocean floor development on the site of the gap, where the currents are ascending." Given the physicists' arguments that Earth's crust and mantle are rigid and immobile, Holmes conceded that "purely speculative ideas of this kind, specially invented to match the requirements, can have no scientific value until they acquire support from independent evidence."

Convincing evidence emerged from extensive exploration of the seafloor after World War II. The mapping of the undersea Mid-Atlantic Ridge and the discovery of the deep, cracklike valley, or rift, running down its center sparked much speculation (**Figure 2.3**). Geologists found that almost

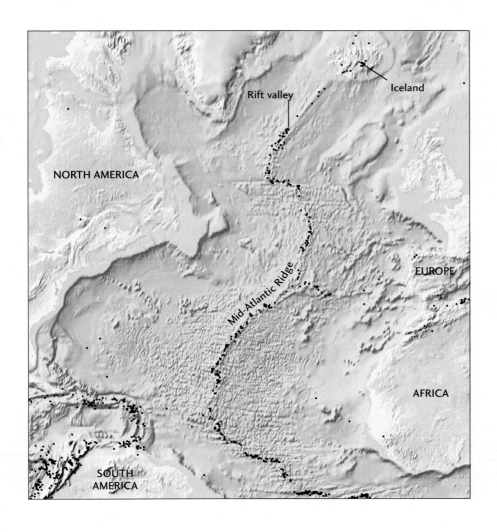

Figure 2.3 The North Atlantic Ocean floor, showing the cracklike rift valley running down the center of the Mid-Atlantic Ridge and associated earthquakes (black dots).

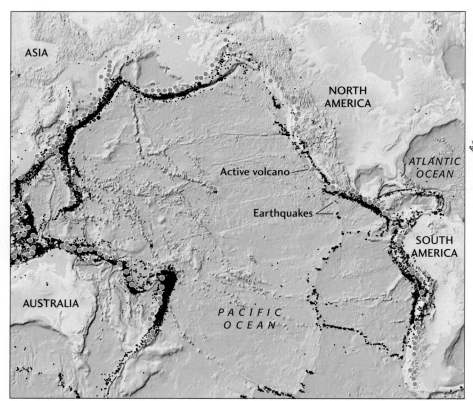

Figure 2.4 The Pacific Ring of Fire, showing active volcanoes (large red circles) and earthquakes (small black dots).

* PACIFIC RING of Fire!

all earthquakes in the Atlantic Ocean occur near this rift valley. Because tectonic faulting generates most earthquakes, these results indicated that the rift was a tectonically active feature. Other mid-ocean ridges with similar rifts and earthquake activity were found in the Pacific and Indian oceans.

In the early 1960s, Harry Hess of Princeton University and Robert Dietz of the Scripps Institution of Oceanography proposed that the crust separates along the rifts in mid-ocean ridges and that new seafloor forms by upwelling of hot new crust into these cracks. The new seafloor—actually the top of newly created lithosphere—spreads laterally away from the rift and is replaced by even newer crust in a continuing process of plate creation.

The Great Synthesis: 1963–1968

The seafloor spreading hypothesis put forward by Hess and Dietz in 1962 explained how the continents could drift apart through the creation of new lithosphere at mid-ocean rifts. Could the seafloor and its underlying lithosphere be destroyed by recycling back into Earth's interior? If not, Earth's surface area would have to increase over time. For a period in the early 1960s, some physicists and geologists actually believed in this idea of an expanding Earth. Other geologists recognized that the seafloor was indeed being recycled in regions of intense volcanic and earthquake activity around the margins of the Pacific Ocean basin, known

collectively as the Ring of Fire (**Figure 2.4**). The details of this process, however, remained unclear.

In 1965, the Canadian geologist J. Tuzo Wilson first described tectonics around the globe in terms of rigid plates moving over Earth's surface. He characterized the three basic types of boundaries where plates move apart, come together, or slide past each other. Soon after, other scientists showed that almost all current tectonic deformations—the processes by which rocks are folded, faulted, sheared, or compressed by Earth stresses—are concentrated at these boundaries. They measured the rates and directions of the tectonic motions and demonstrated that these motions are mathematically consistent with a system of rigid plates moving over the planet's spherical surface. The basic elements of the plate tectonics theory were established by the end of 1968. By 1970, the evidence for plate tectonics had become so persuasive that almost all Earth scientists embraced the theory. Textbooks were revised, and specialists began to consider the implications of the new concept for their own fields.

THE MOSAIC OF PLATES

According to the theory of plate tectonics, the rigid lithosphere is not a continuous shell but is broken into a mosaic of about a dozen large, rigid plates that move over Earth's surface. Each plate moves as a distinct unit, riding on the asthenosphere, which is also in motion. The largest is the Pacific Plate, which comprises much (though not all) of the Pacific Ocean basin. Some of the plates are named after the continents they include, but in no case is a plate identical with a continent. The North American Plate, for instance, extends from the Pacific coast of North America to the middle of the Atlantic Ocean, where it meets the Eurasian and African plates. The major plates and their present-day motions are represented in **Figure 2.5**.

In addition to the major plates, there are a number of smaller ones. An example is the tiny Juan de Fuca Plate, a piece of oceanic lithosphere trapped between the giant Pacific and North American plates just offshore of the northwestern United States. Others are continental fragments, such as the small Anatolian Plate, which includes much of Turkey. (Not all of the smaller plates are shown in Figure 2.5.)

To see geology in action, go to a plate boundary. Depending on which boundary you visit, you will find earthquakes; volcanoes; mountains; long, narrow rifts; folding; and faulting. Many geologic features develop through the interactions of plates at their boundaries. The three basic types of plate boundaries are depicted in **Figure 2.6** (pages 26–27) and discussed in the following pages.

• At *divergent boundaries,* plates move apart and new lithosphere is created (plate area increases).

• At *convergent boundaries,* plates come together and one is recycled back into the mantle (plate area decreases).

• At *transform-fault boundaries,* plates slide horizontally past each other (plate area remains constant).

Like many models of nature, the three types of plates shown in Figure 2.6 are idealized. Besides these basic types, there are "oblique" boundaries that combine divergence or convergence with some amount of transform faulting. Moreover, what actually goes on at a plate boundary depends on the type of lithosphere involved, because continental and oceanic lithosphere behave differently. The continental crust is made of rocks that are both lighter and weaker than either the oceanic crust or the mantle beneath the crust. Later chapters will examine these differences in more detail, so for now you need to keep in mind only two consequences: (1) because it is lighter, continental crust is not as easily recycled as oceanic crust, and (2) because continental crust is weaker, plate boundaries that involve continental crust tend to be more spread out and more complicated than oceanic plate boundaries.

Divergent Boundaries

Divergent boundaries within the ocean basins are narrow rifts that approximate the idealization of plate tectonics. Divergence within the continents is usually more compli-

cated and distributed over a wider area. This difference is illustrated in Figure 2.6.

Oceanic Plate Separation On the seafloor, the boundary between separating plates is marked by a mid-ocean ridge that exhibits active volcanism, earthquakes, and rifting caused by tensional (stretching) forces that are pulling the two plates apart. Figure 2.6a shows what happens in one example, the Mid-Atlantic Ridge. Here seafloor spreading is at work as the North American and Eurasian plates separate and new Atlantic seafloor is created by mantle upwelling. (A more detailed portrait of the Mid-Atlantic Ridge is shown in Figure 2.3.) The island of Iceland exposes a segment of the otherwise submerged Mid-Atlantic Ridge, allowing geologists to view the process of plate separation and seafloor spreading directly (**Figure 2.7**, page 28). The Mid-Atlantic Ridge continues in the Arctic Ocean north of Iceland and connects to a nearly globe-encircling system of mid-ocean ridges that winds through the Indian and Pacific oceans, ending along the western coast of North America. These **spreading centers** have created the millions of square kilometers of oceanic crust that now floor the world's oceans.

Continental Plate Separation Early stages of plate separation, such as the Great Rift Valley of East Africa (see Figure 2.6b), can be found on some continents. These divergent boundaries are characterized by rift valleys, volcanic activity, and earthquakes distributed over a wider zone than is found at oceanic spreading centers. The Red Sea and the Gulf of California are rifts that are further along in the spreading process (**Figure 2.8**, page 29). In these cases, the continents have separated enough for new seafloor to form along the spreading axis, and the ocean has flooded the rift valleys. Sometimes continental rifting slows or stops before the continent splits apart and a new ocean basin opens. The Rhine Valley along the border of Germany and France is a weakly active continental rift that may be this type of "failed" spreading center. Will the East African Rift continue to open, causing the Somali Subplate to split away from Africa completely and form a new ocean basin, as happened between Africa and the island of Madagascar? Or will the spreading slow and eventually stop, as appears to be happening in western Europe? Geologists don't know the answers.

Convergent Boundaries

Plates cover the globe, so if they separate in one place, they must converge somewhere else, to conserve Earth's surface area. (As far as we can tell, our planet is not expanding!) Where plates collide, they form convergent boundaries. The profusion of geologic events resulting from plate collisions makes convergent boundaries the most complex type observed.

Key Figure 2.5 Earth's lithosphere is made of moving plates.

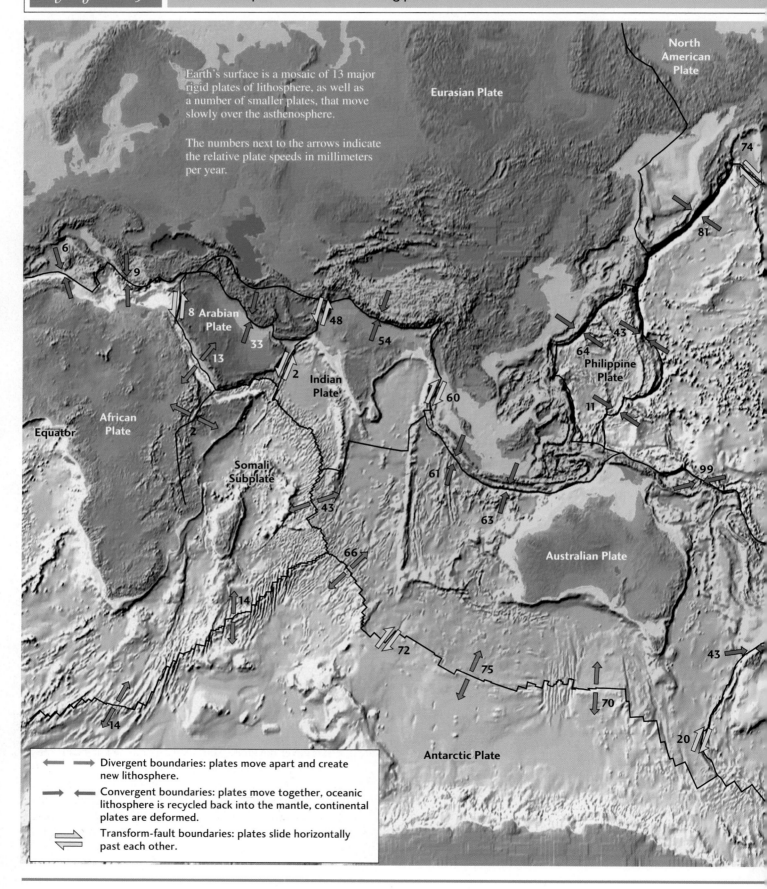

Earth's surface is a mosaic of 13 major rigid plates of lithosphere, as well as a number of smaller plates, that move slowly over the asthenosphere.

The numbers next to the arrows indicate the relative plate speeds in millimeters per year.

North American Plate

Eurasian Plate

74

81

6

9

8 Arabian Plate

48

54

33

43

13

2

64

Philippine Plate

Indian Plate

2

60

11

African Plate

Equator

2

Somali Subplate

61

99

43

63

66

Australian Plate

14

72

75

43

14

70

20

Antarctic Plate

Divergent boundaries: plates move apart and create new lithosphere.

Convergent boundaries: plates move together, oceanic lithosphere is recycled back into the mantle, continental plates are deformed.

Transform-fault boundaries: plates slide horizontally past each other.

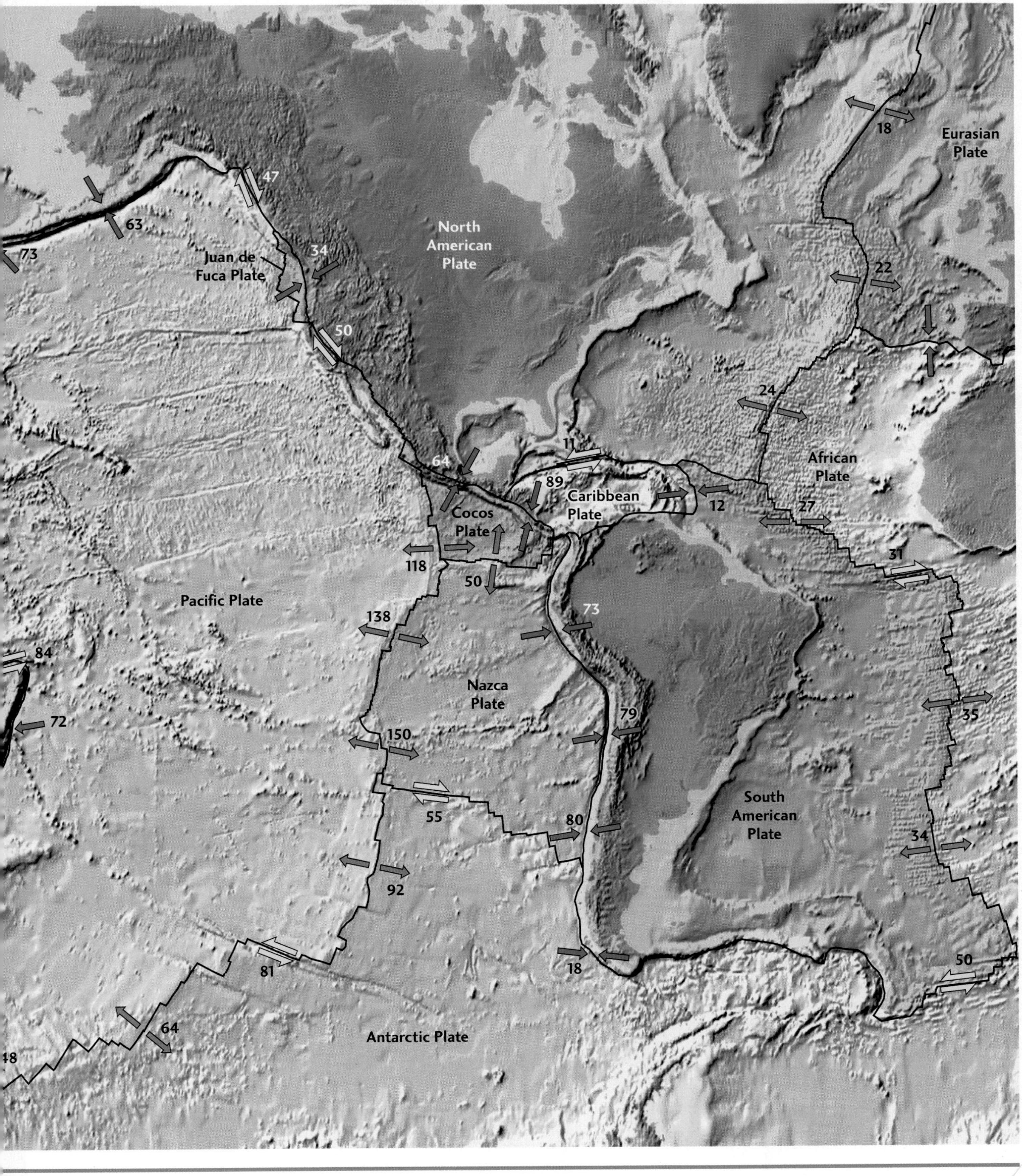

North American Plate

Eurasian Plate

Juan de Fuca Plate

African Plate

Caribbean Plate

Cocos Plate

Pacific Plate

Nazca Plate

South American Plate

Antarctic Plate

47

63

73

34

50

64

11

89

12

27

24

22

18

31

35

34

50

118

50

138

73

84

72

150

79

80

55

92

81

18

64

48

[Plate boundaries by Peter Bird, UCLA.]

Earth System Figure 2.6

Interactions at plate boundaries depend on the direction of relative plate motion and the type of crust.

DIVERGENT BOUNDARIES

(a) Oceanic Plate Separation

Rifting and spreading along a narrow zone have created the Mid-Atlantic Ridge, a mid-ocean mountain chain where volcanoes and earthquakes are concentrated.

(b) Continental Plate Separation

In East Africa, an earlier stage of rifting and spreading has created parallel valleys in a zone with volcanoes and earthquakes.

CONVERGENT BOUNDARIES

(c) Ocean–Ocean Convergence

When two oceanic plates converge, they form a deep-sea trench and a volcanic island arc.

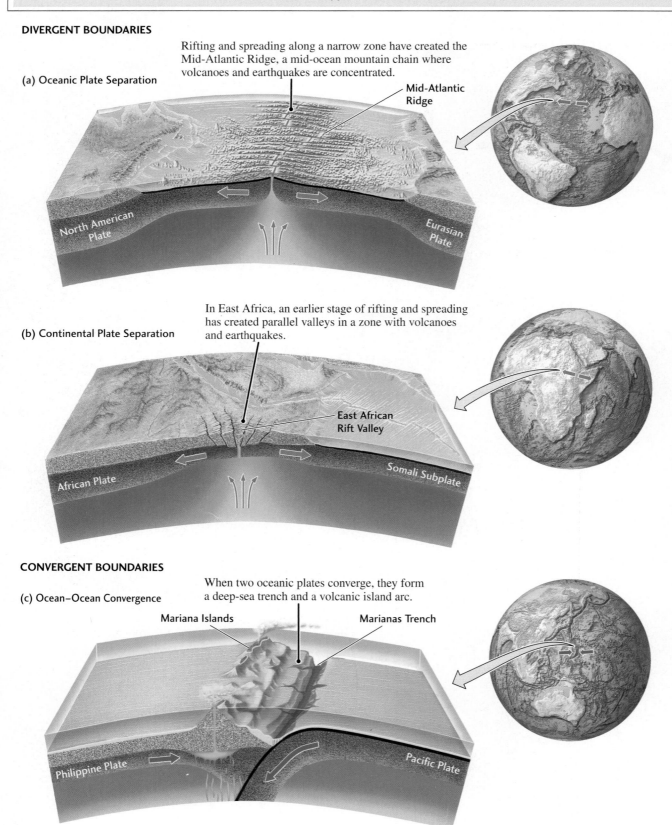

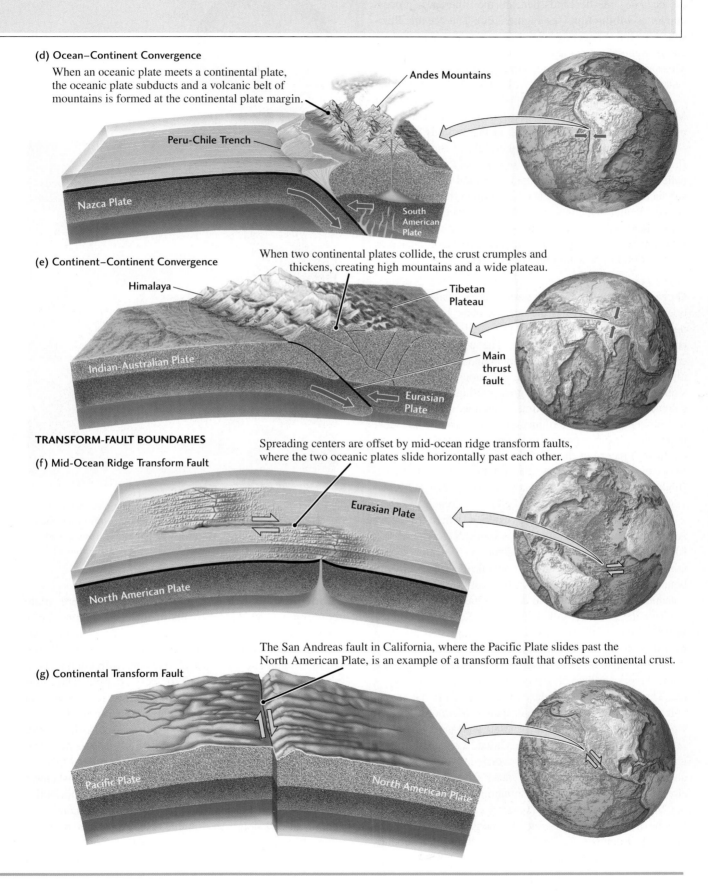

(d) Ocean–Continent Convergence

When an oceanic plate meets a continental plate, the oceanic plate subducts and a volcanic belt of mountains is formed at the continental plate margin.

Andes Mountains

Peru-Chile Trench

Nazca Plate

South American Plate

(e) Continent–Continent Convergence

When two continental plates collide, the crust crumples and thickens, creating high mountains and a wide plateau.

Himalaya

Tibetan Plateau

Indian-Australian Plate

Main thrust fault

Eurasian Plate

TRANSFORM-FAULT BOUNDARIES

(f) Mid-Ocean Ridge Transform Fault

Spreading centers are offset by mid-ocean ridge transform faults, where the two oceanic plates slide horizontally past each other.

Eurasian Plate

North American Plate

The San Andreas fault in California, where the Pacific Plate slides past the North American Plate, is an example of a transform fault that offsets continental crust.

(g) Continental Transform Fault

Pacific Plate

North American Plate

Ocean-Ocean Convergence If the two plates involved are oceanic, one descends beneath the other in a process known as **subduction** (see Figure 2.6c). The oceanic lithosphere of the subducting plate sinks into the asthenosphere and is eventually recycled by the mantle convection system. This sinking produces a long, narrow deep-sea trench. In the Marianas Trench of the western Pacific, the ocean reaches its greatest depth, about 11 km—deeper than the height of Mount Everest. As the cold lithospheric slab descends, the pressure increases. Water trapped in the rocks is squeezed out and rises into the asthenosphere above the slab. This fluid melts the mantle, producing a chain of volcanoes, called an **island arc,** on the seafloor behind the trench. The subduction of the Pacific Plate has formed the volcanically active Aleutian Islands west of Alaska as well as the abundant island arcs of the western Pacific. The cold slabs of lithosphere descending into the mantle cause earthquakes as deep as 690 km beneath these island arcs.

Ocean-Continent Convergence If one plate has a continental edge, it overrides the oceanic plate, because continental crust is lighter and much less easily subducted than oceanic crust (see Figure 2.6d). The continental margin crumples and is uplifted into a mountain chain roughly parallel to the deep-sea trench. The enormous forces of collision and subduction produce great earthquakes along the subduction interface. Over time, materials are scraped off the descending slab and incorporated into the adjacent mountains, leaving geologists with a complex (and often confusing) record of the subduction process. As in the case of ocean-ocean convergence, the water carried down by the subducting oceanic plate melts the mantle wedge and forms volcanoes in the mountain belts behind the trench.

The western coast of South America, where the South American Plate collides with the oceanic Nazca Plate, is a subduction zone of this type. A great chain of high mountains, the Andes, rises on the continental side of the collision boundary, and a deep-sea trench lies just off the coast. The volcanoes here are active and deadly. One of them, Nevado del Ruiz in Colombia, killed 25,000 people when it erupted in 1985. Some of the world's greatest earthquakes have been recorded along this boundary. Another example occurs where the small Juan de Fuca Plate subducts beneath the North American Plate off the coast of western North America. This convergent boundary gives rise to the dangerous volcanoes of the Cascade Range, such as Mount St. Helens, which had a major eruption in 1980 and a minor one in 2004. As our understanding of the Cascadia subduction zone grows, scientists are increasingly worried that a great earthquake could occur there and cause devastating damage along the coasts of Oregon, Washington, and British Columbia. Such an earthquake could possibly cause a large tsunami like the disastrous one generated by the great Sumatra earthquake of December 26, 2004, which occurred in a subduction zone in the Indian Ocean.

Figure 2.7 The Mid-Atlantic Ridge, a divergent plate boundary, surfaces above sea level in Iceland. The cracklike rift valley filled with new volcanic rocks indicates that plates are being pulled apart. [Gudmundur E. Sigvaldason, Nordic Volcanological Institute.]

Continent-Continent Convergence Where plate convergence involves two continents (see Figure 2.6e), oceanic-type subduction cannot occur. The geologic consequences of such a collision are impressive. The collision of the Indian and Eurasian plates, both with continents at their leading edges, provides the best example. The Eurasian Plate overrides the Indian Plate, but India and Asia remain afloat. The collision creates a double thickness of crust forming the highest mountain range in the world, the Himalaya, as well as the vast high plateau of Tibet. Severe earthquakes occur in the crumpling crust of this and other continent-continent collision zones. Geologists have been able to show that many episodes of mountain building throughout Earth's history were caused by continent-continent collisions. An example is the Appalachian Mountains that run along the eastern coast of North America. This chain was uplifted when North America, Eurasia, and

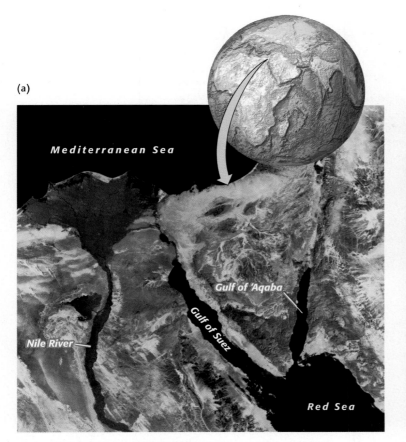

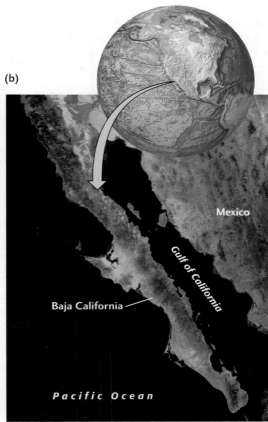

Figure 2.8 (a) The Red Sea (*lower right*) divides to form the Gulf of Suez on the left and the Gulf of 'Aqaba on the right. The Arabian Peninsula, on the right, splitting away from Africa on the left, has opened these great rifts, which are now flooded by the sea. The Nile River (*far left*) flows north into the Mediterranean Sea (*top*). [Earth Satellite Corporation.] (b) The Gulf of California, an opening ocean resulting from plate motions, marks a widening rift between Baja California and the Mexican mainland. [Jeff Schmaltz, MODIS Rapid Response Team, NASA/GSFC.]

Africa collided to form the supercontinent of Pangaea about 300 million years ago.

Transform-Fault Boundaries

At boundaries where plates slide past each other, lithosphere is neither created nor destroyed. Such boundaries are **transform faults:** fractures along which relative displacement occurs as horizontal slip between the adjacent blocks (see Figure 2.6f, g). Transform-fault boundaries are typically found along mid-ocean ridges where the continuity of a divergent boundary is broken and the boundary is offset in a steplike pattern.

The San Andreas fault in California, where the Pacific Plate slides by the North American Plate, is a prime example of a transform fault on land, as shown in **Figure 2.9**. Because the plates have been sliding past each other for millions of years, rocks facing each other on the two sides of the fault are of different types and ages. Large earthquakes, such as the one that destroyed San Francisco in 1906, can occur on transform-fault boundaries. There is much concern that within the next several decades, a sudden slip could occur along the San Andreas fault or related faults near Los Angeles and San Francisco, resulting in an extremely destructive earthquake.

Transform faults can connect divergent plate boundaries with convergent boundaries and convergent boundaries with other convergent boundaries. Can you find examples of these types of transform-fault boundaries in Figure 2.5?

Combinations of Plate Boundaries

Each plate is bordered by some combination of divergent, convergent, and transform-fault boundaries. As we can see in Figure 2.5, the Nazca Plate in the Pacific is bounded on three sides by divergence zones, where new lithosphere is generated along mid-ocean ridge segments offset in a stepwise pattern by transform faults. It is bounded on one side by the Peru-Chile subduction zone, where lithosphere is consumed at a deep-sea trench. The North American Plate is bounded on the east by the Mid-Atlantic Ridge, a divergence zone; on the west by the San Andreas fault and other transform-fault boundaries; and on the northwest by subduction zones and transform-fault boundaries that run from Oregon to the Aleutians.

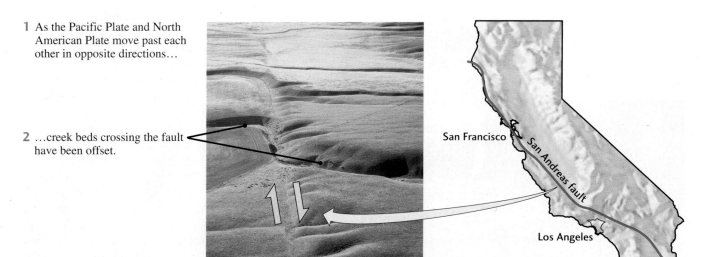

1 As the Pacific Plate and North American Plate move past each other in opposite directions…

2 …creek beds crossing the fault have been offset.

San Francisco

San Andreas fault

Los Angeles

Figure 2.9 The view northwest along the San Andreas fault in the Carrizo Plain of central California. The San Andreas is a transform fault, forming a portion of the sliding boundary between the Pacific Plate on the left and the North American Plate on the right. Notice how movement on the fault has offset the streams flowing across it. [John Shelton.]

RATES AND HISTORY OF PLATE MOTIONS

How fast do plates move? Do some plates move faster than others, and if so, why? Is the velocity of plate movements today the same as it was in the geologic past? Geologists have developed ingenious methods to answer these questions and thereby gain a better understanding of plate tectonics. In this section, we will examine three of these methods.

The Seafloor as a Magnetic Tape Recorder

In World War II, extremely sensitive instruments were developed to detect submarines by the magnetic fields emanating from their steel hulls. Geologists modified these instruments slightly and towed them behind research ships to measure the local magnetic field created by magnetized rocks beneath the sea. Steaming back and forth across the ocean, seagoing scientists discovered regular patterns in the strength of the local magnetic field that completely surprised them. In many areas, the magnetic field alternated between high and low values in long, narrow parallel bands, called **magnetic anomalies,** that were almost perfectly symmetrical with respect to the crest of the mid-ocean ridge. An example is shown in **Figure 2.10.** The detection of these patterns was one of the great discoveries that confirmed seafloor spreading and led to the plate tectonics theory. It also allowed geologists to measure plate motions far back into geologic time. To understand these advances, we need to look more closely at how rocks become magnetized.

The Rock Record of Magnetic Reversals on Land

Magnetic anomalies are evidence that Earth's magnetic field does not remain constant over time. At present, the north magnetic pole is closely aligned with the geographic north pole (see Figure 1.12), but small changes in the geodynamo can flip the orientation of the north and south magnetic poles by 180°, causing a *magnetic reversal.*

In the early 1960s, geologists discovered that a precise record of this peculiar behavior can be obtained from layered flows of volcanic lava. When iron-rich lavas cool, they become slightly magnetized in the direction of Earth's magnetic field. This phenomenon is called *thermoremanent magnetization,* because the rock "remembers" the magnetization long after the magnetizing field existing at the time it formed has changed.

In layered lava flows, each layer of rock from the top down represents a progressively earlier period of geologic time: layers deeper in the stack are older. The age of each layer can then be determined by various dating methods (described in Chapter 8). Measuring the thermoremanent magnetization of rock samples from each layer reveals the direction of Earth's magnetic field when that layer cooled. By repeating these measurements at hundreds of places around the world, geologists have worked out the detailed history of reversals going back into geologic time. The **magnetic time scale** of the past 5 million years is given in Figure 2.10.

About half of all rocks studied are found to be magnetized in a direction opposite that of Earth's present magnetic field. Apparently, the field has flipped frequently over geologic time, and normal fields (same as now) and reversed fields (opposite to now) are equally likely. Major periods when the field is normal or reversed are called *magnetic chrons;* they seem to last about half a million years, although the pattern of reversals becomes highly irregular as we move back in geologic time. Within the major chrons are short-lived reversals of the field, known as *magnetic subchrons*, which may last anywhere from several thousand to 200,000 years.

Magnetic Anomaly Patterns on the Seafloor The peculiar banded magnetic patterns found on the seafloor (see Figure 2.10) puzzled scientists until 1963, when two Englishmen, F. J. Vine and D. H. Mathews—and, independently, two Canadians, L. Morley and A. Larochelle—made a startling proposal. Based on the new evidence for magnetic reversals that land geologists had collected from lava flows, they reasoned that the high and low magnetic bands on the seafloor corresponded to bands of rock that were magnetized during ancient episodes of normal and reversed magnetism. That is, when a research ship was above rocks magnetized in the normal direction, it would record a locally stronger field, or a *positive magnetic anomaly*. When it was above rocks magnetized in the reversed direction, it would record a locally weaker field, or a *negative magnetic anomaly*.

This idea provided a powerful test of the seafloor spreading hypothesis, which states that new seafloor is created along the rift at the crest of a mid-ocean ridge as the plates move apart (see Figure 2.10). Magma flowing up from the interior solidifies in the crack and becomes magnetized in the direction of Earth's field at the time. As the seafloor splits and moves away from the ridge, approximately half of the newly magnetized material moves to one side and half to the other, forming two symmetrical magnetized bands. Newer material fills the crack, continuing the process. In this way, the seafloor acts like a tape recorder that encodes the history of the opening of the oceans by imprinting the reversals of Earth's magnetic field.

Within a few years, marine scientists were able to show that this model provides a consistent explanation for the symmetrical patterns of seafloor magnetic anomalies found on mid-ocean ridges around the world. Moreover, it gave them a precise tool for measuring the rates of seafloor spreading now and in the geologic past. This evidence contributed substantially to the discovery and confirmation of plate tectonics.

Inferring Seafloor Ages and Relative Plate Velocity By using the ages of reversals that had been worked out from magnetized lavas on land, geologists could assign ages to the bands of magnetized rocks on the seafloor. They could then calculate how fast the seafloor opened by using the formula *speed = distance/time,* where distance is measured from the ridge axis and time equals seafloor age. For instance, the magnetic anomaly pattern in Figure 2.10 shows that the boundary between the Gauss normal polarity chron and the Gilbert reverse polarity chron, which was dated from lava flows at 3.3 million years, is located about 30 km away from the Reykjanes Ridge crest. Here, seafloor spreading moved

the North American and Eurasian plates apart by about 60 km in 3.3 million years, giving a spreading rate of 18 km per million years or, equivalently, 18 mm/year.

On a divergent plate boundary, the combination of the spreading rate and the spreading direction gives the **relative plate velocity:** the velocity at which one plate moves relative to the other.

If you look at Figure 2.5, you will see that the spreading rate at the Mid-Atlantic Ridge south of Iceland is fairly low compared to the rate at many other places on the mid-ocean ridges. The speed record for spreading can be found on the East Pacific Rise just south of the equator, where the Pacific and Nazca plates are separating at a rate of about 150 mm/year—an order of magnitude faster than the rate in the North Atlantic. A rough average for mid-ocean ridges around the world is 50 mm/year. This is approximately the rate at which your fingernails grow—so, geologically speaking, the plates move very fast indeed. These spreading rates provide important data for the study of the mantle convection system, a topic we will return to later in this chapter.

We can follow the magnetic time scale through many reversals of Earth's magnetic field. The corresponding magnetic bands on the seafloor, which can be thought of as age bands, have been mapped in detail from the ridge crests across the ocean basins over a time span of almost 200 million years.

The power and convenience of using seafloor magnetization to work out the history of ocean basins cannot be overemphasized. Simply by steaming back and forth over the ocean, measuring the magnetic fields of the seafloor rocks and correlating the pattern of reversals with the time sequence worked out by the methods just described, geologists determined the ages of various regions of the seafloor without even examining rock samples. In effect, they learned how to "replay the tape."

Although seafloor magnetization is a very effective tool, it is an indirect, or remote, sensing method in that rocks are not recovered from the seafloor and their ages are not directly determined in the laboratory. Direct evidence of seafloor spreading and plate movement was still needed to convince the few remaining skeptics. Deep-sea drilling supplied it.

Deep-Sea Drilling

In 1968, a program of drilling into the seafloor was launched as a joint project of major oceanographic institutions and the National Science Foundation. Later, many nations joined the effort. This global experiment aimed to drill through, retrieve, and study seafloor rocks from many places in the world's oceans. Using hollow drills, scientists brought up cores containing sections of seafloor rocks. In some cases, the drilling penetrated thousands of meters below the seafloor surface. Geologists now had an opportunity to work out the history of the ocean basins from direct evidence.

One of the most important facts geologists sought was the age of each sample. Small particles falling through the

Key Figure 2.10 | Magnetic mapping can measure the rate of seafloor spreading.

An oceanographic survey over the Reykjanes Ridge, part of the Mid-Atlantic Ridge southwest of Iceland, showed an oscillating pattern of magnetic field strength. This figure illustrates how scientists worked out the explanation of this pattern.

1 A ship towing a sensitive magnetometer recorded magnetic anomalies,...

2 ...alternating bands of high and low magnetism.

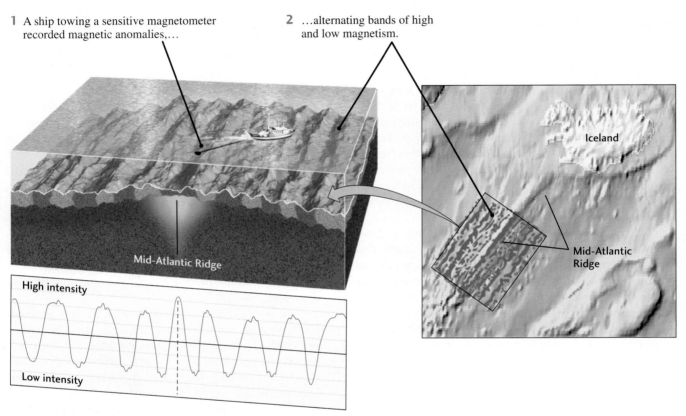

3 The bands proved to be roughly symmetrical on both sides of the Mid-Atlantic Ridge. But what was the meaning of these anomalies? Volcanoes would provide a clue.

4 Scientists studying volcanic lavas also observed magnetic anomalies. When iron-rich lava cools, it becomes magnetized in the direction of Earth's magnetic field.

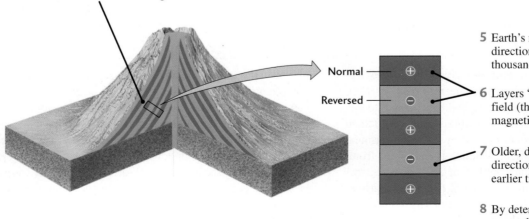

5 Earth's magnetic field reverses direction at intervals of hundreds of thousands of years.

6 Layers "remember" the magnetic field (thermoremanent magnetization).

7 Older, deeper layers preserve the direction of the magnetic field at earlier times.

8 By determining the ages of magnetic reversals at many volcanoes, scientists constructed a magnetic time scale.

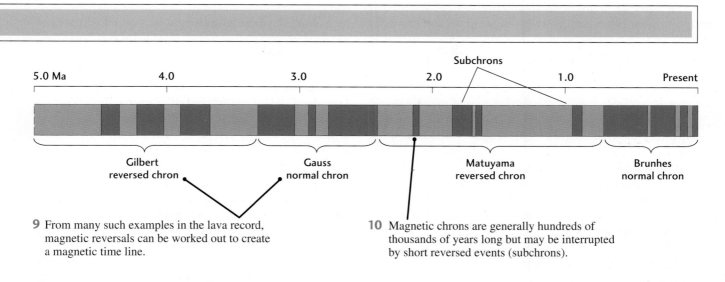

5.0 Ma   4.0   3.0   Subchrons   2.0   1.0   Present

Gilbert reversed chron   Gauss normal chron   Matuyama reversed chron   Brunhes normal chron

9 From many such examples in the lava record, magnetic reversals can be worked out to create a magnetic time line.

10 Magnetic chrons are generally hundreds of thousands of years long but may be interrupted by short reversed events (subchrons).

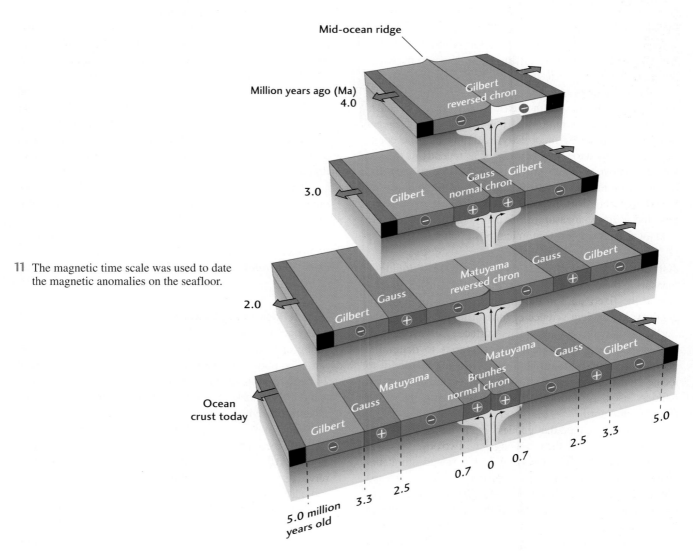

11 The magnetic time scale was used to date the magnetic anomalies on the seafloor.

Mid-ocean ridge

Million years ago (Ma)

Ocean crust today

Scientists concluded that the bands of anomalies on the seafloor were a recording of seafloor spreading. Seafloor spreading acts like a tape recorder. The speed of seafloor spreading can be calculated from *speed = distance/time*.

ocean water—dust from the atmosphere, organic material from marine plants and animals—begin to accumulate as seafloor *sediments* as soon as new oceanic crust forms. Therefore, the age of the oldest sediments in the core, those immediately on top of the crust, tells the geologist how old the ocean floor is at that spot. The age of sediments is obtained primarily from the fossil skeletons of tiny, single-celled animals that live in the ocean and sink to the bottom when they die (see Chapter 8). Geologists found that the sediments in the cores become older with increasing distance from mid-ocean ridges and that the age of the seafloor at any one place agrees almost perfectly with the age determined from magnetic reversal data. This agreement validated magnetic dating of the seafloor and clinched the concept of seafloor spreading.

Measurements of Plate Motions by Geodesy

In his publications advocating continental drift, Alfred Wegener made a big mistake: he proposed that North America and Europe were drifting apart at a rate of nearly 30 meters per year—a thousand times faster than the Atlantic seafloor is actually spreading! This unbelievably high speed was one of the reasons that many scientists roundly rejected his notions of continental drift. Wegener made his estimate by incorrectly assuming that the continents were joined together as Pangaea as recently as the last ice age (which occurred only about 20,000 years ago). His belief in a rapid rate also involved some wishful thinking: he hoped that the drift hypothesis could be confirmed by repeated accurate measurements of the distance across the Atlantic Ocean using astronomical positioning.

Astronomical Positioning Astronomical positioning—measuring the positions of points on Earth's surface in relation to the fixed stars in the night sky—is a technique of **geodesy,** the ancient science of measuring the shape of the Earth and locating points on its surface. Surveyors have used astronomical positioning for centuries to determine geographic boundaries on land, and sailors have used it to locate their ships at sea. Four thousand years ago, Egyptian builders used astronomical positioning to aim the Great Pyramid due north.

Wegener imagined that geodesy could be used to measure continental drift in the following way. Two observers, one in Europe and the other in North America, would simultaneously determine their positions relative to the fixed stars. From these positions, they would calculate the distance between their two observing posts at that instant. They would then repeat this distance measurement from the same observing posts sometime later—say, after 1 year. If the continents are drifting apart, then the distance should have increased, and the value of the increase would determine the speed of the drift.

For this technique to work, however, one must determine the relative positions of the observing posts accurately enough to measure the motion. In Wegener's day, the accuracy of astronomical positioning was poor; uncertainties in fixing intercontinental distances exceeded 100 m. Therefore, even at the high rates of motion he was proposing, it would take a number of years to observe drift. He claimed that two astronomical surveys of the distance between Europe and Greenland (where he worked as a meteorologist), taken 6 years apart, supported his high rate, but he was wrong again. We now know that the spreading of the Mid-Atlantic Ridge from one survey to the next was only about a tenth of a meter—a thousand times too small to be observed by the techniques that were then available.

Owing to the high accuracy required to observe plate motions directly, geodetic techniques did not play a significant role in the discovery of plate tectonics. Geologists had to rely on the evidence for seafloor spreading from the geologic record—the magnetic stripes and ages from fossils described earlier. Beginning in the late 1970s, however, an astronomical positioning method was developed that used signals from distant "quasi-stellar radio sources" (quasars) recorded by huge dish antennas. This method can measure intercontinental distances to an amazing accuracy of 1 mm. In 1986, a team of scientists using this method showed that the distance between antennas in Europe (Sweden) and North America (Massachusetts) had increased 19 mm/year over a period of 5 years, very close to the rate predicted by geologic models of plate tectonics. Wegener's dream of directly measuring continental drift by astronomical positioning was realized at last.

Postscript: Today, the Great Pyramid of Egypt is not aimed directly north, as stated previously, but slightly east of north. Did the ancient Egyptian astronomers make a mistake in orienting the pyramid 40 centuries ago? Archaeologists think probably not. Over this period, Africa drifted enough to rotate the pyramid out of alignment with true north.

Global Positioning System Doing geodesy with big radio telescopes is expensive and is not a practical tool for investigating plate tectonic motions in remote areas of the world. Since the mid-1980s, geologists have used a constellation of 24 Earth-orbiting satellites, called the Global Positioning System (GPS), to make the same types of measurements with the same astounding accuracy using inexpensive, portable radio receivers not much bigger than this book (**Figure 2.11**). GPS receivers record high-frequency radio waves keyed to precise atomic clocks aboard the satellites. The satellite constellation serves as an outside frame of reference, just as the fixed stars and quasars do in astronomical positioning.

The changes in distance between land-based GPS receivers placed on different plates, recorded over several years, agree in both magnitude and direction with those found from magnetic anomalies on the seafloor. These experiments indicate that plate motions are remarkably steady over periods of time ranging from a few years to millions of years. Geologists are now using GPS to measure plate motions on a yearly basis at many locations around the globe.

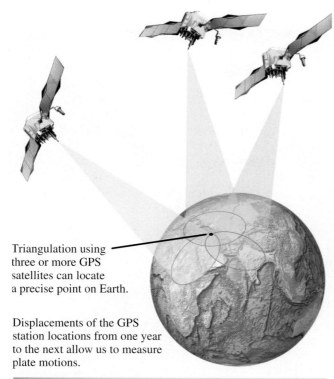

Triangulation using three or more GPS satellites can locate a precise point on Earth.

Displacements of the GPS station locations from one year to the next allow us to measure plate motions.

A GPS station

Figure 2.11 Signals from GPS satellites can be used to monitor plate motions. [Photo courtesy Southern California Earthquake Center.]

Postscript: GPS receivers are now used in automobiles, as part of a navigating system that will lead the driver to a specific street address. It is interesting that the scientists who developed the atomic clocks used in GPS did so for research in fundamental physics and had no idea they would be creating a multibillion-dollar industry. Along with the transistor, laser, and many other technologies, GPS demonstrates the serendipitous manner in which basic research repays the society that supports it.

THE GRAND RECONSTRUCTION

The supercontinent of Pangaea was the only major landmass that existed 250 million years ago. One of the great triumphs of modern geology is the reconstruction of events that led to the assembly of Pangaea and to its later fragmentation into the continents we know today. Let's use what we have learned about plate tectonics to see how this feat was accomplished.

Seafloor Isochrons

The color map in **Figure 2.12** shows the ages of the world's ocean floors as determined by magnetic reversal data and fossils from deep-sea drilling. Each colored band represents a span of time corresponding to the age of the crust within that band. The boundaries between bands, called **isochrons,** are contours that connect rocks of equal age. Isochrons tell us the time that has elapsed since the crustal rocks were injected as magma into a mid-ocean rift and, therefore, the amount of spreading that has occurred since they formed. Notice how the seafloor becomes progressively older on both sides of the mid-ocean rifts. For example, the distance from a ridge axis to a 140-million-year isochron (boundary between green and blue bands) indicates the extent of new ocean floor created over that time span. The more widely spaced isochrons (the wider colored bands) of the eastern Pacific signify faster spreading rates than those in the Atlantic.

In 1990, after a 20-year search, geologists found the oldest oceanic rocks by drilling into the seafloor of the western Pacific. These rocks turned out to be about 200 million years old, only about 4 percent of Earth's age. This date indicates how geologically young the seafloor is compared with the continents. Over a period of 100 million to 200 million years in some places and only tens of millions of years in others, the ocean lithosphere forms, spreads, cools, and subducts back into the underlying mantle. In contrast, the oldest continental rocks are about 4 billion years old.

Reconstructing the History of Plate Motions

Earth's plates behave as rigid bodies. That is, the distances between three points on the same rigid plate—say, New York, Miami, and Bermuda on the North American Plate—do not change very much, no matter how far the plate moves. But the distance between, say, New York and Lisbon increases because the two cities are on different plates that are separating along a narrow zone of spreading on the Mid-Atlantic Ridge. The direction of the movement of one plate in relation to another depends on geometric principles that govern the behavior of rigid plates on a sphere. Two primary principles are

1. *Transform-fault boundaries indicate the directions of relative plate movement.* With few exceptions, no overlap, buckling, or separation occurs along typical transform-fault boundaries in the oceans. The two plates merely slide past each other without creating or destroying plate material. Look for a transform-fault boundary if you want to deduce the direction of relative plate motion, because the orientation of the fault is the direction in which one plate slides with respect to the other, as Figure 2.6 shows.

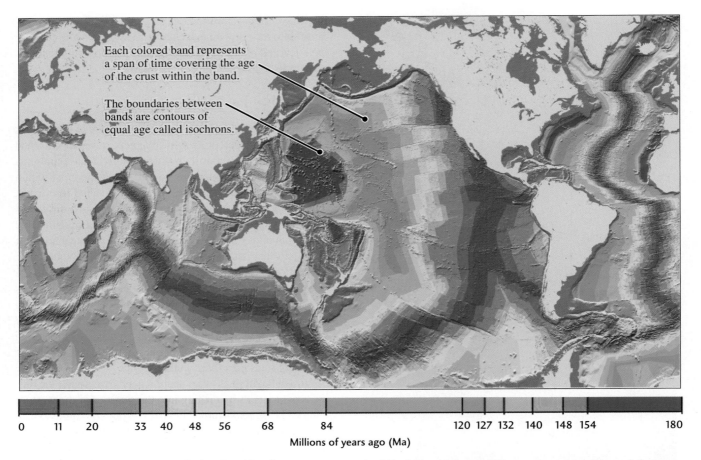

Each colored band represents a span of time covering the age of the crust within the band.

The boundaries between bands are contours of equal age called isochrons.

| 0 | 11 | 20 | 33 | 40 | 48 | 56 | 68 | 84 | 120 | 127 | 132 | 140 | 148 | 154 | 180 |

Millions of years ago (Ma)

Figure 2.12 Age of seafloor crust. Each colored band represents a span of time covering the age of the crust within the band. The boundaries between bands are contours of equal age called isochrons. Isochrons give the age of the seafloor in millions of years since its creation at mid-ocean ridges. Light gray indicates land. Dark gray indicates shallow water over continental shelves. Mid-ocean ridges, along which new seafloor is extruded, coincide with the youngest seafloor (red). [*Journal of Geophysical Research* 102 (1997): 3211–3214. Courtesy of R. Dietmar Müller.]

2. *Seafloor isochrons reveal the positions of divergent boundaries in earlier times.* Isochrons on the seafloor are roughly parallel and symmetrical with the ridge axis along which they were created (see Figure 2.12). Because each isochron was at the boundary of plate separation at an earlier time, isochrons that are of the same age but on opposite sides of an ocean ridge can be brought together to show the positions of the plates and the configuration of the continents embedded in them as they were in that earlier time.

The Breakup of Pangaea

Using these principles, geologists have reconstructed the opening of the Atlantic Ocean and the breakup of Pangaea. **Figure 2.13a** shows the supercontinent of Pangaea as it existed 240 million years ago. It began to break apart when North America rifted away from Europe about 200 million years ago (Figure 2.13b). The opening of the North Atlantic was accompanied by the separation of the northern continents (Laurasia) from the southern continents (Gondwana) and the rifting of Gondwana along what is now the eastern coast of Africa (Figure 2.13c). The breakup of Gondwana separated South America, Africa, India, and Antarctica, creating the South Atlantic and Southern oceans and narrowing the Tethys Ocean (Figure 2.13d). The separation of Australia from Antarctica and the ramming of India into Eurasia closed the Tethys Ocean, giving us the world we see today (Figure 2.13e).

The plate motions have not ceased, of course, so the configuration of the continents will continue to evolve. A plausible scenario for the distribution of continents and plate boundaries 50 million years in the future is displayed in Figure 2.13f.

The Assembly of Pangaea by Continental Drift

The isochron map in Figure 2.12 tells us that all of the seafloor on Earth's surface today has been created since the breakup of Pangaea. We know from the geologic record in older continental mountain belts, however, that plate tectonics was operating for billions of years before this breakup. Evidently, seafloor spreading took place as it does today, and there were previous episodes of continental drift and collision. Subduction back into the mantle has destroyed the sea-floor created in these earlier times, so we must rely on the older evidence preserved on continents to identify and chart the movements of ancient continents (paleocontinents).

Old mountain belts such as the Appalachians of North America and the Urals, which separate Europe from Asia, help us locate ancient collisions of the paleocontinents. In many places, the rocks reveal ancient episodes of rifting and subduction. Rock types and fossils also indicate the distribution of ancient seas, glaciers, lowlands, mountains, and climates. Knowledge of ancient climates enables geologists to locate the latitudes at which the continental rocks formed, which in turn helps them to assemble the jigsaw puzzle of paleocontinents. When volcanism or mountain building produces new continental rocks, these rocks also record the direction of Earth's magnetic field, just as oceanic rocks do when they are created by seafloor spreading. Like a compass frozen in time, the fossil magnetism of a continental fragment records its ancient orientation and position.

The left side of Figure 2.13 shows one of the latest efforts to depict the pre-Pangaean configuration of continents. It is truly impressive that modern science can recover the geography of this strange world of hundreds of millions of years ago. The evidence from rock types, fossils, climate, and paleomagnetism has allowed scientists to reconstruct an earlier supercontinent, called **Rodinia,** that formed about 1.1 billion years ago and began to break up about 750 million years ago. They have been able to chart its fragments over the subsequent 500 million years as these fragments drifted and reassembled into the supercontinent of Pangaea. Geologists continue to sort out more details of this complex jigsaw puzzle, whose individual pieces change shape over geologic time.

Implications of the Grand Reconstruction

Hardly any branch of geology remains untouched by this grand reconstruction of the continents. Economic geologists have used the fit of the continents to find mineral and oil deposits by correlating the rock formations in which they exist on one continent with their predrift continuations on another continent. Paleontologists have rethought some aspects of evolution in light of continental drift. Geologists have broadened their focus from the geology of a particular region to a world-encompassing picture. The concept of plate tectonics provides a way to interpret, in global terms, such geologic processes as rock formation, mountain building, and climate change.

Oceanographers are reconstructing currents as they might have existed in the ancestral oceans to understand the modern circulation better and to account for the variations in deep-sea sediments that are affected by such currents. Scientists are "forecasting" backward in time to describe temperatures, winds, the extent of continental glaciers, and the level of the sea as they were in ancient times. They hope to learn from the past so that they can predict the future of the

climate system better—a matter of great urgency because of the possibility of greenhouse warming triggered by human activity. What better testimony to the triumph of this once outrageous hypothesis than its ability to revitalize and shed light on so many diverse topics?

MANTLE CONVECTION: The Engine of Plate Tectonics

Everything discussed so far might be called descriptive plate tectonics. But a description is hardly an explanation. We need a more comprehensive theory that explains *why* plates move. Finding such a theory remains one of the major challenges confronting scientists who study the Earth system. In this section, we will discuss several aspects of the problem that have been central to recent research by these scientists.

As Arthur Holmes and other early advocates of continental drift realized, mantle convection is the "engine" that drives the large-scale tectonic processes operating on Earth's surface. In Chapter 1, we described the mantle as a hot solid capable of flowing like a sticky fluid (warm wax or cold syrup, for example). Heat escaping from Earth's deep interior causes this material to convect (circulate upward and downward) at speeds of a few tens of millimeters per year.

Almost all scientists now accept that the lithospheric plates somehow participate in the flow of this mantle convection system. As is often the case, however, "the devil is in the details." Many different hypotheses have been advanced on the basis of one piece of evidence or another, but no one has yet come up with a satisfactory, comprehensive theory that ties everything together. In what follows, we will pose three questions that get at the heart of the matter and give you our opinions about their answers. But you should be careful not to accept these tentative answers as facts. Our understanding of the mantle convection system remains a work in progress, which we may have to alter as new evidence becomes available. Future editions of this book may contain different answers!

Where Do the Plate-Driving Forces Originate?

Here's an experiment you can do in your kitchen: heat a pan of water until it is about to boil, then sprinkle some dry tea leaves in the center of the pan. You will notice that the leaves move across the surface of the water, dragged along by the convection currents in the hot water. Is this the way plates move about, passively dragged to and fro on the backs of convection currents rising up from the mantle?

The answer appears to be no. The main evidence comes from the rates of plate motion we discussed earlier in this chapter. From Figure 2.5, we see that the faster-moving plates (the Pacific, Nazca, Cocos, and Indian plates) are being subducted along a large fraction of their boundaries. In contrast, the slower-moving plates (the North American, South American, African, Eurasian, and Antarctic plates) do

| Key Figure 2.13 | Continental rifting, drifting, and collisions assembled and dispersed Pangaea. |

ASSEMBLY OF PANGAEA

RODINIA Late Proterozoic, 750 Ma

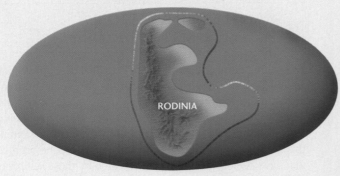

1 The supercontinent of Rodinia formed about 1.1 billion years ago and began to break up about 750 million years ago.

Late Proterozoic, 650 Ma

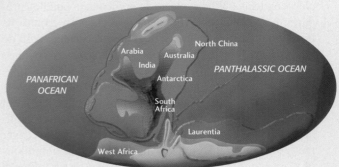

2 Geologists have used a variety of evidence, including paleomagnetism and information about ancient climates, to reconstruct the pre-Pangaean pattern of continental drift.

Middle Ordovician, 458 Ma

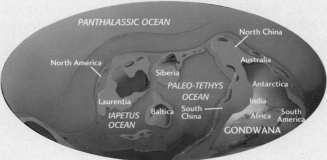

Early Devonian, 390 Ma

PANGAEA (a) Early Triassic, 237 Ma

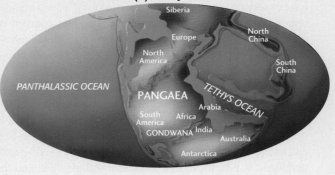

3 The supercontinent Pangaea was mostly assembled by 237 Ma, surrounded by a superocean called Panthalassa (Greek for "all seas"), the ancestral Pacific Ocean. The Tethys Ocean, between Africa and Eurasia, was the ancestor of the Mediterranean Sea.

BREAKUP OF PANGAEA

(b) Early Jurassic, 195 Ma

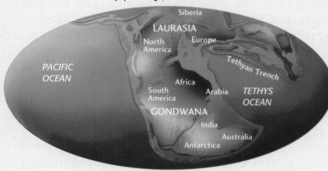

4 The breakup of Pangaea was signaled by the opening of rifts from which lava poured. Rock assemblages that are relics of this great event can be found today in 200-million-year-old volcanic rocks from Nova Scotia to North Carolina.

(c) Late Jurassic, 152 Ma

5 By about 150 million years ago, Pangaea was in the early stages of breakup. The Atlantic Ocean had partially opened, the Tethys Ocean had contracted, and the northern continents (Laurasia) had all but split away from the southern continents (Gondwana). India, Antarctica, and Australia began to split away from Africa.

(d) Late Cretaceous, Early Tertiary, 66 Ma

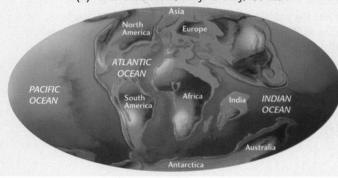

6 By 66 million years ago, the South Atlantic had opened and widened. India was well on its way northward toward Asia, and the Tethys Ocean was closing to form the Mediterranean.

THE PRESENT-DAY AND FUTURE WORLD

(e) PRESENT-DAY WORLD

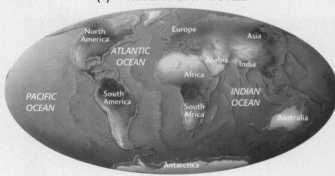

7 The modern world has been produced over the past 65 million years. India collided with Asia, ending its trip across the ocean, and is still pushing northward into Asia. Australia has separated from Antarctica.

(f) 50 million years in the future

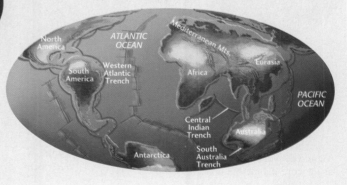

[Paleogeographic maps by Christopher R. Scotese, 2003 PALEOMAP Project (www.scotese.com).]

not have significant attachments of downgoing slabs. These observations suggest that the gravitational pull exerted by the cold (and thus heavy) slabs of old lithosphere cause rapid plate motions. In other words, the plates are not dragged along by convection currents from the deep mantle but rather "fall back" into the mantle under their own weight. According to this hypothesis, seafloor spreading is the passive up-welling of mantle material where the plates have been pulled apart by subduction forces.

But if the only important force in plate tectonics is the gravitational pull of subducting slabs, why did Pangaea break apart and the Atlantic Ocean open up? The only subducting slabs of lithosphere currently attached to the North and South American plates are found in the small island arcs that bound the Caribbean and Scotia seas, which are thought to be too small to drag the Atlantic apart. One possibility is that the overriding plates, as well as the subducting plates, are pulled toward their convergent boundaries. For example, as the Nazca Plate subducts beneath South America, it may cause the plate boundary at the Peru-Chile Trench to retreat toward the Pacific, "sucking" the South American Plate to the west.

Another possibility is that Pangaea acted as an insulating blanket, preventing heat from getting out of Earth's mantle (as it otherwise would through the process of seafloor spreading). The heat presumably built up over time, forming hot bulges in the mantle beneath the supercontinent. These bulges raised Pangaea slightly and caused it to rift apart in a kind of "landslide" off the top of the bulges. Gravitational forces continued to drive subsequent seafloor spreading as the plates "slid downhill" off the crest of the Mid-Atlantic Ridge. Earthquakes that sometimes occur in plate interiors show direct evidence of the compression of plates by these "ridge push" forces.

The driving forces of plate tectonics are manifestations of convection in the mantle, in the sense that they involve hot matter rising in one place and cold matter sinking in another. Although many questions remain, we can be reasonably sure that (1) the plates themselves play an active role in this system, and (2) the forces associated with the sinking slabs and elevated ridges are probably the most important in governing the rates of plate motion. Scientists are attempting to resolve other issues raised in this discussion by comparing observations with detailed computer models of the mantle convection system. Some results will be discussed in Chapter 14.

How Deep Does Plate Recycling Occur?

For plate tectonics to work, the lithospheric material that goes down in subduction zones must be recycled through the mantle and eventually come back up as new lithosphere created along the spreading centers of the mid-ocean ridges. How deep into the mantle does this recycling process extend? That is, where is the lower boundary of the mantle convection system?

The deepest the boundary can be is about 2900 km below Earth's outer surface, where a sharp boundary separates the mantle from the core. As we saw in Chapter 1, the iron-rich liquid below this core-mantle boundary is much denser than the solid rocks of the mantle, preventing any significant exchange of material between the two layers. We can thus imagine a system of whole-mantle convection in which the material from the plates circulates all the way through the mantle, down as far as the core-mantle boundary (**Figure 2.14a**).

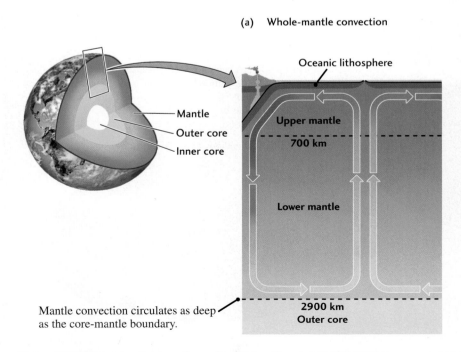

(a) **Whole-mantle convection**

Oceanic lithosphere

Mantle
Outer core
Inner core

Upper mantle
700 km

Lower mantle

2900 km
Outer core

Mantle convection circulates as deep as the core-mantle boundary.

(b) **Stratifed convection**

Plate convection is confined to the upper mantle.

Boundary near 700 km separates the two convection systems.

Lower mantle convects more sluggishly than upper mantle.

Figure 2.14 Two competing hypotheses for the mantle convection system.

In the early days of plate tectonics theory, however, many scientists were convinced that plate recycling takes place at much shallower depths in the mantle. The evidence came from deep earthquakes that mark the descent of lithospheric slabs in subduction zones. The greatest depth of these earthquakes varies among subduction zones, depending on how cold the descending slabs are, but geologists found that no earthquakes were occurring below about 700 km. Moreover, the properties of earthquakes at these great depths indicated that the slabs were encountering more rigid material that slowed and perhaps blocked their downward progress.

Based on this and other evidence, scientists hypothesized that the mantle might be divided into two layers: an upper mantle system in the outer 700 km, where the recycling of lithosphere takes place, and a lower mantle system, from 700 km deep to the core-mantle boundary, where convection is much more sluggish. According to this hypothesis, called stratified convection, the separation of the two systems is maintained because the upper system consists of lighter rocks than the lower system and thus floats on top, in the same way the mantle floats on the core (Figure 2.14b).

The way to test these two competing hypotheses is to look for "lithospheric graveyards" below the convergent zones where old plates have been subducted. Old subducted lithosphere is colder than the surrounding mantle and can therefore be "seen" using earthquake waves (much as doctors use ultrasound waves to look into your body). Moreover, there should be lots of it down there. From our knowledge of past plate motions, we can estimate that, just since the breakup of Pangaea, lithosphere equivalent to the surface area of Earth has been recycled back into the mantle. Sure enough, scientists have found regions of colder material in the deep mantle under North and South America, eastern

Asia, and other sites adjacent to plate collision boundaries. These zones occur as extensions of descending lithospheric slabs, and some appear to go down as far as the core-mantle boundary. From this evidence, most scientists have concluded that plate recycling takes place through whole-mantle convection rather than stratified convection.

What Is the Nature of Rising Convection Currents?

Mantle convection implies that what goes down must come up. Scientists have learned a lot about downgoing convection currents because they are marked by narrow zones of cold subducted lithosphere that can be detected by earthquake waves. What about the rising currents of mantle material needed to balance subduction? Are there concentrated, sheetlike upwellings directly beneath the mid-ocean ridges? Most scientists who study the problem think not. Instead, they believe that the rising currents are slower and spread out over broader regions. This view is consistent with the idea, discussed above, that seafloor spreading is a rather passive process: pull the plates apart almost anywhere, and you will generate a spreading center.

There is one big exception, however: a type of narrow, jetlike upwelling called a *mantle plume* (**Figure 2.15**). The best evidence for mantle plumes comes from regions of intense, localized volcanism (called *hot spots*), such as Hawaii, where huge volcanoes are forming in the middle of plates, far away from any spreading center. The plumes are thought to be slender cylinders of fast-rising material, less than 100 km across, that come from the deep mantle, perhaps forming in very hot regions near the core-mantle boundary. Mantle plumes are so intense that they can literally burn holes in the

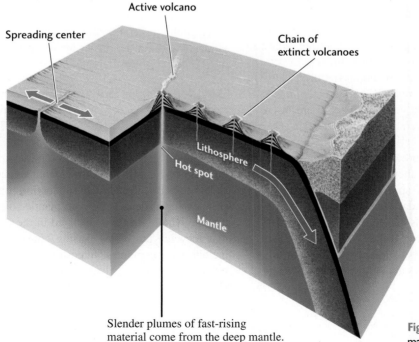

Spreading center

Active volcano

Chain of extinct volcanoes

Lithosphere

Hot spot

Mantle

Slender plumes of fast-rising material come from the deep mantle.

Figure 2.15 An illustration of the mantle plume hypothesis.

plates and erupt tremendous volumes of lava. Plumes may be responsible for the massive outpourings of lava—millions of cubic kilometers—found in such places as Siberia and the Columbia Plateau of eastern Washington and Oregon. Some of these lava floods were so large and occurred so quickly that they may have changed Earth's climate and killed off many life-forms in mass extinction events (see Chapter 1). We will describe plume volcanism in more detail in Chapter 12.

The plume hypothesis was first put forward in 1970, soon after the plate theory had been established, by one of the founders of plate tectonics, W. Jason Morgan of Princeton University. Like other aspects of the mantle convection system, however, the observations that bear on rising convection currents are indirect, and the plume hypothesis remains very controversial.

THE THEORY OF PLATE TECTONICS AND THE SCIENTIFIC METHOD

Earlier, we considered the scientific method and how it guides the work of geologists. In the context of the scientific method, plate tectonics is not a dogma but a confirmed theory whose strength lies in its simplicity, its generality, and its consistency with many types of observations. Theories can always be overturned or modified. As we have seen, competing hypotheses have been advanced about how convection generates plate tectonics. But the theory of plate tectonics—like the theories of Earth's age, the evolution of life, and genetics—explains so much so well and has survived so many efforts to prove it false that geologists treat it as fact.

The question remains, why wasn't plate tectonics discovered earlier? Why did it take the scientific establishment so long to move from skepticism about continental drift to acceptance of plate tectonics? Scientists approach their subjects differently. Scientists with particularly inquiring, uninhibited, and synthesizing minds are often the first to perceive great truths. Although their perceptions frequently turn out to be false (think of the mistakes Wegener made in proposing continental drift), these visionary people are often the first to see the great generalizations of science. Deservedly, they are the ones history remembers.

Most scientists, however, proceed more cautiously and wait out the slow process of gathering supporting evidence. Continental drift and seafloor spreading were slow to be accepted largely because the audacious ideas came far ahead of the firm evidence. Scientists had to explore the oceans, develop new instruments, and drill the seafloor before the majority could be convinced. Today, many scientists are still waiting to be convinced of ideas about how the mantle convection system really works.

SUMMARY

What is the theory of plate tectonics? According to the theory of plate tectonics, the lithosphere is broken into about a dozen rigid, moving plates. Three types of plate boundaries are defined by the relative motion between plates: divergent, convergent, and transform fault. The area of Earth's surface does not change through geologic time; therefore, the area of new plate created at divergent boundaries—the spreading centers of mid-ocean ridges—equals the plate area consumed at convergent boundaries by the process of subduction.

What are some of the geologic characteristics of plate boundaries? In addition to earthquake belts, many large-scale geologic features, such as narrow mountain belts and chains of volcanoes, are associated with plate boundaries. Convergent boundaries are marked by deep-sea trenches, earthquake belts, mountains, and volcanoes. The Andes and the trenches of the western coast of South America are modern examples. Old mountain belts, such as the Appalachians and the Urals, are the remnants of ancient continental collisions. Divergent boundaries are typically marked by volcanic activity and earthquakes at the crest of a mid-ocean ridge, such as the Mid-Atlantic Ridge. Transform-fault boundaries, along which plates slide past each other, can be recognized by their linear topography, earthquake activity, and, in the oceans, offsets in magnetic anomaly bands.

How can the age of the seafloor be determined? We can measure the age of the ocean's floor by comparing magnetic anomaly bands mapped on the seafloor with the sequence of magnetic reversals worked out on land. The procedure has been verified and extended by deep-sea drilling. Geologists can now draw isochrons for most of the world's oceans, enabling them to reconstruct the history of seafloor spreading over the past 200 million years. Using this method and other geologic data, geologists have developed a detailed model of how Pangaea broke apart and the continents drifted into their present configuration.

What is the engine that drives plate tectonics? The plate tectonic system is driven by mantle convection, and the energy comes from Earth's internal heat. The plates themselves play an active role in this system. For example, the most important forces in plate tectonics come from the cooling lithosphere as it slides away from spreading centers and sinks back into the mantle in subduction zones. Lithospheric slabs extend as deep as the core-mantle boundary, indicating that the whole mantle is involved in the convection system that recycles the plates. Rising convection currents may include mantle plumes, intense upwellings from the deep mantle that cause localized volcanism at hot spots such as Hawaii.

KEY TERMS AND CONCEPTS

continental drift (p. 20)

geodesy (p. 34)

island arc (p. 28)

isochron (p. 35)

magnetic anomaly (p. 30)

magnetic time scale (p. 30)

Pangaea (p. 20)

plate tectonics (p. 19)

relative plate velocity (p. 31)

Rodinia (p. 37)

seafloor spreading (p. 21)

spreading center (p. 23)

subduction (p. 28)

transform fault (p. 29)

EXERCISES

1. From Figure 2.5, trace the boundaries of the South American Plate on a sheet of paper and identify segments that are divergent, convergent, and transform-fault boundaries. Approximately what fraction of the plate area is occupied by the South American continent? Is the fraction of the South American Plate occupied by oceanic crust increasing or decreasing over time? Explain your answer using the principles of plate tectonics.

2. In Figure 2.5, identify an example of a transform-fault boundary that (a) connects a divergent plate boundary with a convergent plate boundary and (b) connects a convergent plate boundary with another convergent plate boundary.

3. From the isochron map in Figure 2.12, estimate how long ago the continents of Australia and Antarctica were separated by seafloor spreading. Did this happen before or after South America separated from Africa?

4. Name three mountain belts that formed by continental collisions that are occurring now or have occurred in the past.

5. Most active volcanoes are located on or near plate boundaries. Give an example of a volcano that is not on a plate boundary and describe a hypothesis consistent with plate tectonics that can explain it.

THOUGHT QUESTIONS

1. Why are there active volcanoes along the Pacific coast in Washington and Oregon but not along the eastern coast of the United States?

2. What mistakes did Wegener make in formulating his theory of continental drift? Do you think the geologists of his era were justified in rejecting his theory?

3. Would you characterize plate tectonics as a hypothesis, a theory, or a fact? Why?

4. In Figure 2.12, the isochrons are symmetrically distributed in the Atlantic Ocean but not in the Pacific. For example, the oldest seafloor (in darkest blue) is found in the western Pacific Ocean but not in the eastern Pacific. Why?

5. The theory of plate tectonics was not widely accepted until the magnetic striping of the ocean floor was discovered. In light of earlier observations—the jigsaw-puzzle fit of the continents, the occurrence of fossils of the same life-forms on both sides of the Atlantic, and paleoclimatic conditions—why is the magnetic striping such a key piece of evidence?

6. How do the differences between continental and oceanic crust affect the way plates interact?

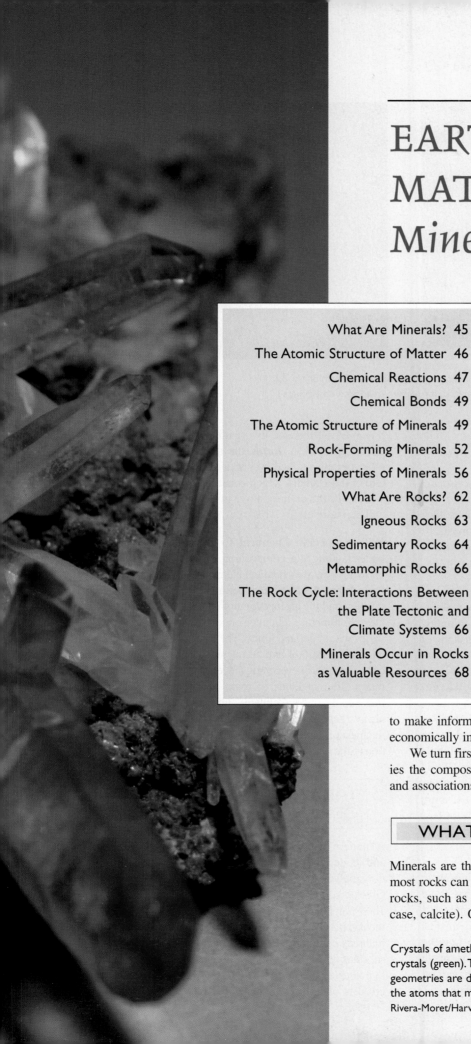

EARTH MATERIALS
Minerals and Rocks

In Chapter 2, we saw how plate tectonics describes Earth's large-scale structure and dynamics, but we touched only briefly on the wide variety of materials that appear in plate tectonic settings. **In this chapter, we focus on rocks, the records of geologic processes, and minerals, the building blocks of rocks.**

Rocks and minerals help determine the structure of the Earth system, much as concrete, steel, and plastic determine the structure, design, and architecture of large buildings. To tell Earth's story, geologists often adopt a "Sherlock Holmes" approach: they use current evidence to deduce the processes and events that occurred in the past at some particular place. The kinds of minerals found in volcanic rocks, for example, give evidence of eruptions that brought molten rock to Earth's surface. The minerals of a granite reveal that it crystallized deep in the crust under the very high temperatures and pressures that occur when two continental plates collide and form mountains such as the Himalaya. Understanding the geology of a region allows us to make informed guesses about where undiscovered deposits of economically important mineral resources might lie.

We turn first to **mineralogy**—the branch of geology that studies the composition, structure, appearance, stability, occurrence, and associations of minerals.

WHAT ARE MINERALS?

Minerals are the building blocks of rocks: with the proper tools, most rocks can be separated into their constituent minerals. A few rocks, such as limestone, contain only a single mineral (in this case, calcite). Other rocks, such as granite, are made of several

Crystals of amethyst and quartz, growing on top of epidote crystals (green). The planar surfaces are crystal faces, whose geometries are determined by the underlying arrangement of the atoms that make up the crystals. [John Grotzinger/Ramón Rivera-Moret/Harvard Mineralogical Museum.]

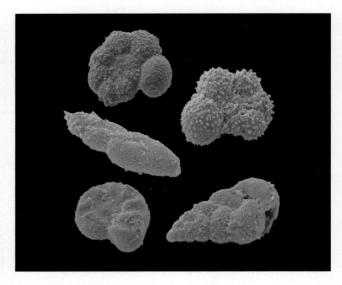

Figure 3.1 The mineral calcite is found in the shells of many organisms, such as foraminifera. [*left:* John Grotzinger/Ramón Rivera-Moret/Harvard Mineralogical Museum; *right:* Andrew Syred/Photo Researchers.]

different minerals. To identify and classify the many kinds of rocks that compose the Earth and understand how they formed, we must know about minerals.

Geologists define a **mineral** as a *naturally occurring solid crystalline substance, generally inorganic, with a specific chemical composition.* Minerals are homogeneous: they cannot be divided mechanically into smaller components.

Let's examine each part of our definition of a mineral in a little more detail.

Naturally Occurring . . . To qualify as a mineral, a substance must be found in nature. Diamonds mined in South Africa are minerals. Synthetic versions produced in industrial laboratories are not minerals. Nor are the thousands of laboratory products invented by chemists.

Solid Crystalline Substance . . . Minerals are solid substances—they are neither liquids nor gases. When we say that a mineral is *crystalline,* we mean that the tiny particles of matter, or atoms, that compose it are arranged in an orderly, repeating, three-dimensional array. Solid materials that have no such orderly arrangement are referred to as *glassy* or *amorphous* (without form) and are not conventionally called minerals. Windowpane glass is amorphous, as are some natural glasses formed during volcanic eruptions. Later in this chapter, we will explore in detail the process by which crystalline materials form.

Generally Inorganic . . . Minerals are defined as inorganic substances and so exclude the organic materials that make up plant and animal bodies. Organic matter is composed of organic carbon, the form of carbon found in all organisms, living or dead. Decaying vegetation in a swamp may be geologically transformed into coal, which also is made of organic carbon; but although it is found as a natural deposit, coal is not considered a mineral. Many minerals,

however, are secreted by organisms. One such mineral, calcite (**Figure 3.1**), forms the shells of oysters and many other organisms, and it contains inorganic carbon. The calcite of these shells, which constitute the bulk of many limestones, fits the definition of a mineral because it is inorganic and crystalline.

With a Specific Chemical Composition . . . The key to understanding the composition of Earth's materials lies in knowing how the chemical elements are organized into minerals. What makes each mineral unique is its chemical composition and the arrangement of its atoms in an internal structure. A mineral's chemical composition either is fixed or varies within defined limits. The mineral quartz, for example, has a fixed ratio of two atoms of oxygen to one atom of silicon. This ratio never varies, even though quartz is found in many different kinds of rock. The chemical elements that make up the mineral olivine—iron, magnesium, and silicon—always have a fixed ratio. Although the number of iron and magnesium atoms may vary, the sum of those atoms in relation to the number of silicon atoms always forms a fixed ratio.

THE ATOMIC STRUCTURE OF MATTER

A modern dictionary lists many meanings for the word *atom* and its derivatives. One of the first is "anything considered the smallest possible unit of any material." To the ancient Greeks, *atomos* meant "indivisible." John Dalton (1766–1844), an English chemist and the father of modern atomic theory, proposed that atoms are particles of matter so small that they cannot be seen with any microscope and so universal that they compose all substances. In 1805, Dalton hypothesized that each of the various chemical elements consists of a different kind of atom, that all atoms of any given element are identical, and that chemical compounds are formed by various combinations of atoms of different elements in definite proportions.

By the early twentieth century, physicists, chemists, and mineralogists, building on Dalton's ideas, had come to understand the structure of matter much as we do today. We now know that an **atom** is the smallest unit of an element that retains the physical and chemical properties of that element. We also know that atoms are the small units of matter that combine in chemical reactions and that atoms themselves are divisible into even smaller units.

The Structure of Atoms

Understanding the structure of atoms allows us to predict how chemical elements will react with one another and form new crystal structures. For more detailed information about the structure of atoms, see Appendix 4.

The Nucleus: Protons and Neutrons At the center of every atom is a dense **nucleus** containing virtually all the mass of the atom in two kinds of particles: protons and neutrons (**Figure 3.2**). A **proton** has a positive electrical charge of +1. A **neutron** is electrically neutral—that is, uncharged. Atoms of the same chemical element may have different numbers of neutrons, but the number of protons does not vary. For instance, all carbon atoms have six protons.

Electrons Surrounding the nucleus is a cloud of moving particles called **electrons,** each with a mass so small that it is conventionally taken to be zero. Each electron carries a negative electrical charge of –1. The number of protons in the nucleus of any atom is balanced by the same number of electrons in the cloud surrounding the nucleus, so an atom is electrically neutral. Thus the nucleus of the carbon atom is surrounded by six electrons (see Figure 3.2).

Atomic Number and Atomic Mass

The number of protons in the nucleus of an atom is called its **atomic number.** Because all atoms of the same element have the same number of protons, they also have the same atomic number. All atoms with six protons, for example, are carbon atoms (atomic number 6). In fact, the atomic number of an element can tell us so much about an element's behavior that the periodic table organizes elements according to their atomic number (see Appendix 4). Elements in the same vertical group, such as carbon and silicon, tend to react similarly.

The **atomic mass** of an element is the sum of the masses of its protons and neutrons. (Electrons, because they have so little mass, are not included in this sum.) Atoms of the same chemical element always have the same number of protons but may have different numbers of neutrons and therefore different atomic masses. Atoms with different numbers of neutrons are called **isotopes.** Isotopes

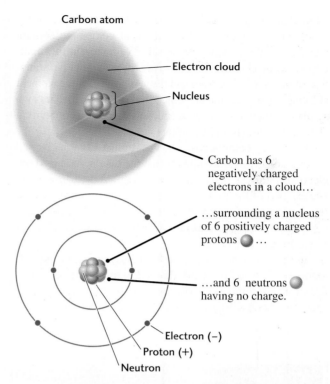

Carbon atom

Electron cloud

Nucleus

Carbon has 6 negatively charged electrons in a cloud…

…surrounding a nucleus of 6 positively charged protons …

…and 6 neutrons having no charge.

Electron (–)

Proton (+)

Neutron

Figure 3.2 Electron structure of the carbon atom (carbon-12). The electrons, each with a charge of –1, are represented as a negatively charged cloud surrounding the nucleus, which contains six protons, each with a charge of +1, and six neutrons, each with zero charge. The size of the nucleus is greatly exaggerated; it is much too small to show at a true scale.

of the element carbon, for example, all with six protons, may have six, seven, or eight neutrons, giving atomic masses of 12, 13, and 14.

In nature, the chemical elements exist as mixtures of isotopes, so their atomic masses are never whole numbers. Carbon's atomic mass, for example, is 12.011. It is close to 12 because the isotope carbon-12 is overwhelmingly abundant. The relative abundance of the various isotopes of an element on Earth is determined by processes that enhance the abundance of some isotopes over others. Carbon-12, for example, is favored by some reactions, such as photosynthesis, in which organic carbon compounds are produced from inorganic carbon compounds.

CHEMICAL REACTIONS

The structure of an atom determines its chemical reactions with other atoms. **Chemical reactions** are interactions of the atoms of two or more chemical elements in certain fixed proportions that produce chemical compounds. For example, when two hydrogen atoms combine with one oxygen atom, they form a new chemical compound, water (H_2O). The properties of a chemical compound may be entirely different from those of its constituent elements. For example, when an atom of sodium, a metal, combines with an atom of chlorine, a noxious gas, they form the chemical compound sodium chloride, better known as table salt. We represent this compound by the chemical formula *NaCl*, the symbol *Na* standing for the element sodium and the symbol *Cl* for the element chlorine. (Every chemical element has been assigned its own symbol, which we use as a kind of shorthand for writing chemical formulas and equations.)

Chemical compounds, such as minerals, are formed either by **electron sharing** between the reacting atoms or by **electron transfer** between the reacting atoms. Carbon and silicon, two of the most abundant elements in Earth's crust, tend to form compounds by electron sharing. Diamond is a compound composed entirely of carbon atoms sharing electrons (**Figure 3.3**).

In the reaction between sodium (Na) and chlorine (Cl) atoms to form sodium chloride (NaCl), electrons are transferred. The sodium atom loses one electron, which the chlorine atom gains (**Figure 3.4**). Because the chlorine atom

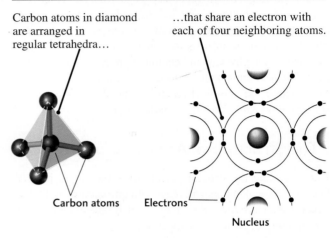

Key Figure 3.3 | Shared electrons form covalent bonds.

Carbon atoms in diamond are arranged in regular tetrahedra…

…that share an electron with each of four neighboring atoms.

Carbon atoms

Electrons

Nucleus

Key Figure 3.4 | Some atoms transfer electrons, forming ionic bonds.

(a) **1** When sodium (Na) and chlorine (Cl) react, the sodium atom loses one electron…

2 …and the chlorine atom acquires that electron. In this reaction, an orderly array of ions is formed.

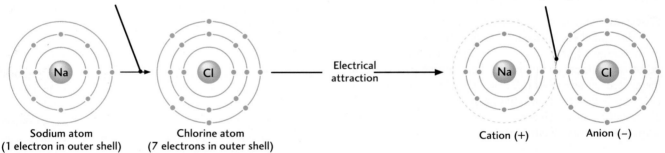

Sodium atom
(1 electron in outer shell)

Chlorine atom
(7 electrons in outer shell)

Electrical attraction

Cation (+)

Anion (−)

(b) **3** Sodium and chloride ions pack together in a cubic structure.

4 Each sodium ion (circled in red) is surrounded by six chloride ions (circled in yellow), and vice versa.

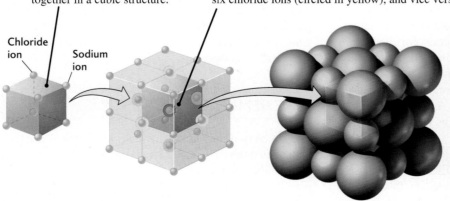

Chloride ion

Sodium ion

NaCl crystal

Table salt, halite

has gained a negatively charged electron, it is now negatively charged, Cl⁻. Likewise, the loss of an electron gives sodium a positive charge, Na⁺. The compound NaCl itself remains electrically neutral because the positive charge on Na⁺ is exactly balanced by the negative charge on Cl⁻. A positively charged ion is a **cation,** and a negatively charged ion is an **anion.**

CHEMICAL BONDS

When a chemical compound is formed by either electron sharing or electron transfer, the ions or atoms that make up the compound are held together by electrical forces of attraction between electrons and protons. These electrical attractions, or *chemical bonds,* between shared electrons or between gained or lost electrons may be strong or weak, and the bonds created by these attractions are correspondingly strong or weak. Strong bonds keep a substance from decomposing into its elements or into other compounds. They also make minerals hard and keep them from cracking or splitting. Two major types of bonds are found in most rock-forming minerals: ionic bonds and covalent bonds.

Ionic Bonds

The simplest form of chemical bond is the **ionic bond.** Bonds of this type form by electrical attraction between ions of opposite charge, such as Na⁺ and Cl⁻ in sodium chloride (see Figure 3.4). This attraction is of exactly the same nature as the static electricity that can make clothing of nylon or silk cling to the body. The strength of an ionic bond decreases greatly as the distance between ions increases. Bond strength increases as the electrical charges of the ions increase. Ionic bonds are the dominant type of chemical bonds in mineral structures; *about 90 percent of all minerals are essentially ionic compounds.*

Covalent Bonds

Elements that do not readily gain or lose electrons to form ions and instead form compounds by sharing electrons are held together by **covalent bonds.** These are generally stronger than ionic bonds. One mineral with a covalently bonded crystal structure is diamond, consisting of the single element carbon. Carbon atoms have four electrons and acquire four more by electron sharing. In diamond, every carbon atom (not an ion) is surrounded by four others arranged in a regular *tetrahedron,* a four-sided pyramidal form, each side a triangle (see Figure 3.3). In this configuration, each carbon atom shares an electron with each of its four neighbors, resulting in a very stable configuration. Figure 3.3 shows a network of carbon tetrahedra linked together.

Metallic Bonds

Atoms of metallic elements, which have strong tendencies to lose electrons, pack together as cations, and the freely mobile electrons are shared and dispersed among the ions. This free electron sharing results in a kind of covalent bond that we call a **metallic bond.** It is found in a small number of minerals, among them the metal copper and some sulfides.

The chemical bonds of some minerals are intermediate between pure ionic and pure covalent bonds because some electrons are exchanged and others are shared.

THE ATOMIC STRUCTURE OF MINERALS

Minerals can be viewed in two complementary ways: as crystals (or grains) that we can see with the naked eye and as assemblages of submicroscopic atoms organized in an ordered three-dimensional array. We will now look more closely at the orderly forms that characterize mineral structure and at the conditions under which minerals form. Later in this chapter, we will see that the crystal structures of minerals are manifested in their physical properties. First, however, we turn to the question of how minerals form.

How Do Minerals Form?

Minerals form by the process of **crystallization,** in which the atoms of a gas or liquid come together in the proper chemical proportions and crystalline arrangement. (Remember that the atoms in a mineral are arranged in an ordered three-dimensional array.) The bonding of carbon atoms in diamond, a covalently bonded mineral, is one example of crystal structure. Carbon atoms bond together in tetrahedra, each tetrahedron attaching to another and building up a regular three-dimensional structure from a great many atoms (see Figure 3.3). As a diamond crystal grows, it extends its tetrahedral structure in all directions, always adding new atoms in the proper geometric arrangement. Diamonds can be synthesized under very high pressures and temperatures that mimic conditions in Earth's mantle.

The sodium and chloride ions that make up sodium chloride, an ionically bonded mineral, also crystallize in an orderly three-dimensional array. In Figure 3.4a, we can see the geometry of their arrangement, with each ion of one kind surrounded by six ions of the other in a series of *cubic* structures extending in three directions. We can think of ions as solid spheres, packed together in close-fitting structural units. Figure 3.4b shows the relative sizes of the ions in NaCl. There are six neighboring ions in NaCl's basic structural unit. The relative sizes of the sodium and chloride ions allow them to fit together in a closely packed arrangement.

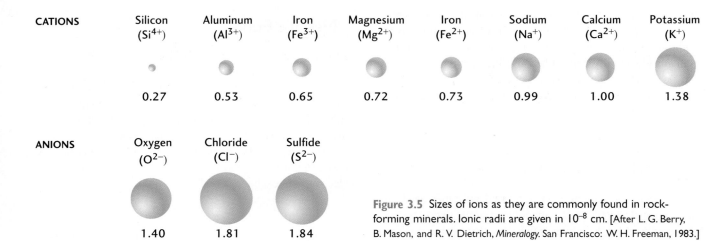

CATIONS	Silicon (Si^{4+})	Aluminum (Al^{3+})	Iron (Fe^{3+})	Magnesium (Mg^{2+})	Iron (Fe^{2+})	Sodium (Na^+)	Calcium (Ca^{2+})	Potassium (K^+)
	0.27	0.53	0.65	0.72	0.73	0.99	1.00	1.38

ANIONS	Oxygen (O^{2-})	Chloride (Cl^-)	Sulfide (S^{2-})
	1.40	1.81	1.84

Figure 3.5 Sizes of ions as they are commonly found in rock-forming minerals. Ionic radii are given in 10^{-8} cm. [After L. G. Berry, B. Mason, and R. V. Dietrich, *Mineralogy*. San Francisco: W. H. Freeman, 1983.]

Many of the cations of abundant minerals are relatively small; most anions are large (**Figure 3.5**). This is the case with the most common Earth anion, oxygen. Because anions tend to be larger than cations, most of the space of a crystal is occupied by the anions and the cations fit into the spaces between them. As a result, crystal structures are determined largely by how the anions are arranged and how the cations fit between them.

Cations of similar sizes and charges tend to substitute for one another and to form compounds having the same crystal structure but differing chemical composition. *Cation substitution* is common in minerals containing the silicate ion (SiO_4^{4-}), such as olivine, which is abundant in many volcanic rocks. Iron (Fe) and magnesium (Mg) ions are similar in size, and both have two positive charges, so they easily substitute for each other in the structure of olivine. The composition of pure magnesium olivine is Mg_2SiO_4; the pure iron olivine is Fe_2SiO_4. The composition of olivine with both iron and magnesium is given by the formula $(Mg,Fe)_2SiO_4$, which simply means that the number of iron and magnesium cations may vary, but their combined total (expressed as a subscript 2) does not vary in relation to each SiO_4^{4-} ion. The proportion of iron to magnesium is determined by the relative abundance of the two elements in the molten material from which the olivine crystallized. In many silicate minerals, aluminum (Al) substitutes for silicon (Si). Aluminum and silicon ions are so similar in size that aluminum can take the place of silicon in many crystal structures. The difference in charge between aluminum (3+) and silicon (4+) ions is balanced by an increase in the number of other cations, such as sodium (1+).

Crystallization starts with the formation of microscopic single **crystals,** ordered three-dimensional arrays of atoms in which the basic arrangement is repeated in all directions. The boundaries of crystals are natural flat (plane) surfaces called *crystal faces.* The crystal faces of a mineral are the external expression of the mineral's internal atomic structure. **Figure 3.6** pairs a drawing of a perfect crystal (which

are very rare in nature) with a photograph of the actual mineral The six-sided (hexagonal) shape of the quartz crystal corresponds to its hexagonal internal atomic structure.

During crystallization, the initially microscopic crystals grow larger, maintaining their crystal faces as long as they are free to grow. Large crystals with well-defined faces form when growth is slow and steady and space is adequate to allow growth without interference from other crystals nearby. For this reason, most large mineral crystals form in open spaces in rocks, such as fractures or cavities.

Often, however, the spaces between growing crystals fill in, or crystallization proceeds too rapidly. Crystals then grow over one another and coalesce to become a solid mass of crystalline particles, or *grains*. In this case, few or no grains show crystal faces. Large crystals that can be seen with the naked eye are relatively unusual, but many microscopic minerals in rocks display crystal faces.

Unlike crystalline minerals, glassy materials—which solidify from liquids so quickly that they lack any internal atomic order—do not form crystals with plane faces. Instead they are found as masses with curved, irregular surfaces. The most common glass is volcanic glass.

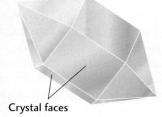

Crystal faces

Quartz, a hexagonal crystal

Figure 3.6 A perfect crystal. A perfect crystal is rare, but no matter how irregular the shapes of the faces may be, the angles are always exactly the same. [Breck P. Kent.]

When Do Minerals Form?

Lowering the temperature of a liquid below its freezing point is one way to start the process of crystallization. In water, for example, 0°C is the temperature below which crystals of ice, a mineral, start to form. Similarly, a magma —hot, molten liquid rock—crystallizes solid minerals when it cools. As a magma falls below its melting point, which may be higher than 1000°C, crystals of silicate minerals such as olivine or feldspar begin to form. (Geologists usually refer to melting points of magmas rather than freezing points, because freezing implies cold.)

Crystallization can also occur as liquids evaporate from a solution. A solution is a homogeneous mixture of one chemical substance with another, such as salt and water. As the water evaporates from a salt solution, the concentration of salt eventually gets so high that the solution can hold no more salt and is said to be saturated. If evaporation continues, the salt starts to **precipitate,** or drop out of solution as crystals. Deposits of halite or table salt form under just these conditions when seawater evaporates to the point of saturation in some hot, arid bays or arms of the ocean (**Figure 3.7**).

Diamond and graphite exemplify the dramatic effects that temperature and pressure can have on mineral formation. Diamond and graphite (the material used as the "lead" in pencils) are **polymorphs,** alternative structures for a single chemical compound (**Figure 3.8**). These two minerals, both formed from carbon, have different crystal structures and very different appearances (see Figure 3.8). From experimentation and geological observation, we know

Figure 3.7 Halite crystals precipitating within a modern hypersaline lagoon. San Salvador Island, Bahamas. Note cubic shape of salt crystals. [John Grotzinger.]

| Key Figure 3.8 | Atoms are the building blocks of minerals. |

Carbon forms two polymorphs, graphite and diamond, that are alternative structures of a single chemical compound.

Graphite is formed at lower pressures and temperatures than diamond. Its carbon forms sheets whose atoms are more loosely packed than those in diamond.

Within its sheets, carbon atoms are joined by strong covalent bonds.

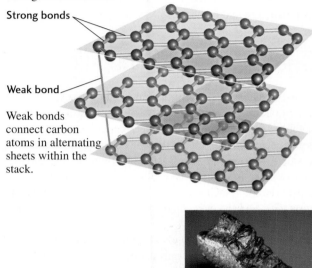

Strong bonds

Weak bond

Weak bonds connect carbon atoms in alternating sheets within the stack.

Graphite

Natural **diamond** is formed by very high pressures and temperatures in Earth's mantle. Its carbon atoms are closely packed.

All carbon atoms in diamond are covalently bonded to four other carbon atoms.

The carbon atoms are closely packed, and all the bonds are very strong.

Electrons   Nucleus

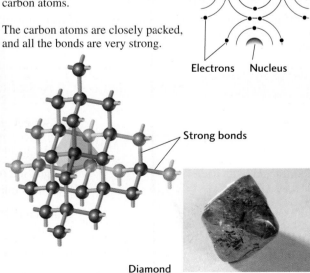

Strong bonds

Diamond

[John Grotzinger/Ramón Rivera-Moret/Harvard Mineralogical Museum.]

that diamond forms and remains stable at the very high pressures and temperatures of Earth's mantle. High pressure forces the atoms in diamond to be closely packed. Diamond therefore has a higher density (mass per unit volume), 3.5 g/cm³, than graphite, which is less closely packed and has a density of only 2.1 g/cm³. Graphite forms and is stable at moderate pressures and temperatures, such as those in Earth's crust.

Low temperatures also can produce closer packing. Quartz and cristobalite are polymorphs of silica (SiO_2). Quartz forms at low temperatures and is relatively dense (2.7 g/cm³). Cristobalite, which forms at higher temperatures, has a more open structure and is therefore less dense (2.3 g/cm³).

ROCK-FORMING MINERALS

All minerals have been grouped into eight classes according to their chemical composition; six of those classes are listed in Table 3.1. Some minerals, such as copper, occur naturally as un-ionized pure elements, and they are classified as *native elements*. Most others are classified by their anions. Olivine, for example, is classed as a silicate by its silicate anion, SiO_4^{4-}. Halite (NaCl) is classed as a halide by its chloride anion, Cl^-. So is its close relative sylvite, potassium chloride (KCl).

Although many thousands of minerals are known, geologists commonly encounter only about 30 of them. These minerals are the building blocks of most crustal rocks and are called *rock-forming minerals*. Their relatively small number corresponds to the small number of elements found in major abundance in Earth's crust. As we learned in Chapter 1, 99 percent of the crust is made up of only nine elements.

In the following pages, we consider the most common rock-forming minerals:

• *Silicates,* the most abundant minerals in Earth's crust, are composed of oxygen (O) and silicon (Si)—the two most abundant elements in the crust—mostly in combination with the cations of other elements.

• *Carbonates* are minerals made of carbon and oxygen in the form of the carbonate anion (CO_3^{2-}) in combination with calcium and magnesium. Calcite ($CaCO_3$) is one such mineral.

• *Oxides* are compounds of the oxygen anion (O^{2-}) and metallic cations; an example is the mineral hematite (Fe_2O_3).

• *Sulfides* are compounds of the sulfide anion (S^{2-}) and metallic cations, a group that includes the mineral pyrite (FeS_2).

• *Sulfates* are compounds of the sulfate anion (SO_4^{2-}) and metallic cations, a group that includes the mineral anhydrite ($CaSO_4$).

The other chemical classes of minerals, including native elements and halides, are not as common as the rock-forming minerals.

Silicates

The basic building block of all silicate mineral structures is the *silicate ion*. It is a tetrahedron—a pyramidal structure with four sides—composed of a central silicon ion (Si^{4+}) surrounded by four oxygen ions (O^{2-}), giving the formula SiO_4^{4-} (**Figure 3.9**). Because the silicate ion has a negative charge, it often bonds to cations to form electrically neutral minerals. The silicate ion typically bonds with cations such as sodium (Na^+), potassium (K^+), calcium (Ca^+), magnesium (Mg^{2+}), and iron (Fe^{2+}). Alternatively, it can share oxygen ions with other silicon-oxygen tetrahedra. Tetrahedra may be isolated (linked only to cations), or they may be linked to other silica tetrahedra in rings, single chains, double chains, sheets, or frameworks, some of which are shown in Figure 3.9.

Isolated Tetrahedra Isolated tetrahedra are linked by the bonding of each oxygen ion of the tetrahedron to a cation (see Figure 3.9a). The cations, in turn, bond to the oxygen ions of other tetrahedra. The tetrahedra are thus isolated from one another by cations on all sides. Olivine is a rock-forming mineral with this structure.

Table 3.1 Some Chemical Classes of Minerals

Class	Defining Anions	Example
Native elements	None: no charged ions	Copper metal (Cu)
Oxides and hydroxides	Oxygen ion (O^{2-}) Hydroxyl ion (OH^-)	Hematite (Fe_2O_3) Brucite ($Mg[OH]_2$)
Halides	Chloride (Cl^-), fluoride (F^-), bromide (Br^-), iodide (I^-)	Halite (NaCl)
Carbonates	Carbonate ion (CO_3^{2-})	Calcite ($CaCO_3$)
Sulfates	Sulfate ion (SO_4^{-2})	Anhydrite ($CaSO_4$)
Silicates	Silicate ion (SiO_4^{-4})	Olivine (Mg_2SiO_4)

Key Figure 3.9 Common silicate minerals are polymorphs of silicate ions and often other elements.

1 The silicate ion forms tetrahedra with a central silicon ion surrounded by four oxygen ions.

Silicate ion (SiO$_4^{4-}$)

Oxygen ions (O^{2-})

Silicon ion (Si^{4+})

Quartz structure

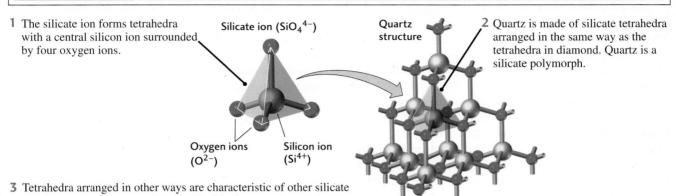

2 Quartz is made of silicate tetrahedra arranged in the same way as the tetrahedra in diamond. Quartz is a silicate polymorph.

3 Tetrahedra arranged in other ways are characteristic of other silicate minerals and determine their cleavage directions.

Mineral	Chemical formula	Cleavage planes and number of cleavage directions	Silicate structure	Specimen
Olivine (a)	(Mg, Fe)$_2$SiO$_4$	1 plane	Isolated tetrahedra	
Pyroxene (b)	(Mg, Fe)SiO$_3$	2 planes at 90°	Single chains	
Amphibole (c)	Ca$_2$(Mg, Fe)$_5$Si$_8$O$_{22}$(OH)$_2$	2 planes at 60° and 120°	Double chains	
Micas (d)	Muscovite: KAl$_2$(AlSi$_3$O$_{10}$)(OH)$_2$ Biotite: K(Mg, Fe)$_3$AlSi$_3$O$_{10}$(OH)$_2$	1 plane	Sheets	
Feldspars (e)	Orthoclase feldspar: KAlSi$_3$O$_8$ Plagioclase feldspar: (Ca, Na) AlSi$_3$O$_8$	2 planes at 90°	Three-dimensional frameworks	

Single-Chain Linkages Single chains also form by sharing oxygen ions. Two oxygen ions of each tetrahedron bond to adjacent tetrahedra in an open-ended chain (see Figure 3.9b). Single chains are linked to other chains by cations. Minerals of the pyroxene group are single-chain silicate minerals. Enstatite, a pyroxene, is composed of iron or magnesium ions, or both, and is limited to a chain of tetrahedra in which the two cations may substitute for each other, as in olivine. The formula $(Mg,Fe)SiO_3$ represents this structure.

Double-Chain Linkages Two single chains may combine to form double chains linked to each other by shared oxygen ions (see Figure 3.9c). Adjacent double chains linked by cations form the structure of the amphibole group of minerals. Hornblende, a member of this group, is an extremely common mineral in both igneous and metamorphic rocks. It has a complex composition that includes calcium (Ca^{2+}), sodium (Na^+), magnesium (Mg^{2+}), iron (Fe^{2+}), and aluminum (Al^{3+}).

Sheet Linkages In sheet structures, each tetrahedron shares three of its oxygen ions with adjacent tetrahedra to build stacked sheets of tetrahedra (see Figure 3.9d). Cations may be interlayered with tetrahedral sheets. The micas and clay minerals are the most abundant sheet silicates. Muscovite, $KAl_3Si_3O_{10}(OH)_2$, is one of the most common sheet silicates and is found in many types of rocks. It can be separated into extremely thin, transparent sheets. Kaolinite, $Al_2Si_2O_5(OH)_4$, which also has this structure, is a common clay mineral found in sediments and is the basic raw material for pottery.

Frameworks Three-dimensional frameworks form as each tetrahedron shares all its oxygen ions with other tetrahedra. Feldspars, the most abundant minerals in Earth's crust, are framework silicates (see Figure 3.9e), as is another of the most common minerals, quartz (SiO_2).

Silicate Compositions Chemically, the simplest silicate is silicon dioxide, also called silica (SiO_2), which is found most often as the mineral quartz. When the silicate tetra-

hedra of quartz are linked, sharing two oxygen ions for each silicon ion, the total formula adds up to SiO_2.

In other silicate minerals, the basic units—rings, chains, sheets, and frameworks—are bonded to such cations as sodium (Na^+), potassium (K^+), calcium (Ca^{2+}), magnesium (Mg^{2+}), and iron (Fe^{2+}). As noted in the discussion of cation substitution, aluminum (Al^{3+}) substitutes for silicon in many silicate minerals.

Carbonates

The nonsilicate mineral calcite (calcium carbonate, $CaCO_3$) is one of the abundant minerals in Earth's crust and is the chief constituent of a group of rocks called limestones (**Figure 3.10**). Its basic building block, the carbonate ion (CO_3^{2-}), consists of a carbon ion surrounded by three oxygen ions in a triangle, as shown in Figure 3.10b. The carbon atom shares electrons with the oxygen atoms. Groups of carbonate ions are arranged in sheets somewhat like the sheet silicates and are bonded by layers of cations (see Figure 3.10c). The sheets of carbonate ions in calcite are separated by layers of calcium ions. The mineral dolomite, $CaMg(CO_3)_2$, another major mineral of crustal rocks, is made up of the same carbonate sheets separated by alternating layers of calcium ions and magnesium ions.

Oxides

Oxide minerals are compounds in which oxygen is bonded to atoms or cations of other elements, usually metallic ions such as iron $(Fe^{2+}$ or $Fe^{3+})$. Most oxide minerals are ionically bonded, their structures varying with the size of the metallic cations. This group is of great economic importance because it includes the ores of most of the metals, such as chromium and titanium, used in the industrial and technological manufacture of metallic materials and devices. Hematite (Fe_2O_3), shown in **Figure 3.11**, is a chief ore of iron.

Another of the abundant minerals in this group, spinel, is an oxide of two metals, magnesium and aluminum $(MgAl_2O_4)$. Spinel (see Figure 3.11) has a closely packed cubic structure and a high density (3.6 g/cm^3), reflecting the

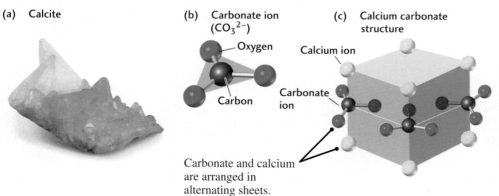

(a) Calcite

(b) Carbonate ion (CO_3^{2-})
  Oxygen
  Carbon

(c) Calcium carbonate structure
  Calcium ion
  Carbonate ion

Carbonate and calcium are arranged in alternating sheets.

Figure 3.10 Carbonate minerals, such as calcite (calcium carbonate, $CaCO_3$), have a layered structure. (a) Calcite. [John Grotzinger/Ramón Rivera-Moret/Harvard Mineralogical Museum.] (b) Top view of the carbonate building block, a carbon ion surrounded in a triangle by three oxygen ions, with a net charge of –2. (c) View of the alternating layers of calcium and carbonate ions.

Figure 3.11 Nonsilicate minerals. *left:* hematite; *right:* spinel. [John Grotzinger/Ramón Rivera-Moret/Harvard Mineralogical Museum.]

conditions of high pressure and temperature under which it forms. Transparent gem-quality spinel resembles ruby and sapphire and is found in the crown jewels of England and Russia.

Sulfides

The chief ores of many valuable minerals—such as copper, zinc, and nickel—are members of the sulfide group. This group includes compounds of the sulfide ion (S^{2-}) with metallic cations. In the sulfide ion, a sulfur atom has gained two electrons in its outer shell. Most sulfide minerals look like metals, and almost all are opaque. The most common sulfide mineral is pyrite (FeS_2), often called "fool's gold" because of its yellowish metallic appearance (**Figure 3.12**).

Sulfates

The basic building block of all sulfates is the sulfate ion (SO_4^{2-}). It is a tetrahedron made up of a central sulfur atom surrounded by four oxygen ions (O^{2-}). One of the most abundant minerals of this group is gypsum, the primary component of plaster (**Figure 3.13**). Gypsum forms when seawater evaporates. During evaporation, Ca^{2+} and SO_4^{2-}, two ions that are abundant in seawater, combine and precipitate as layers of sediment, forming calcium sulfate ($CaSO_4 \cdot 2H_2O$). (The dot in this formula signifies that two water molecules are bonded to the calcium and sulfate ions.)

Another calcium sulfate, anhydrite ($CaSO_4$), differs from gypsum in that it contains no water. Its name is derived from the word *anhydrous*, meaning "free from water." Gypsum is stable at the low temperatures and pressures found at Earth's surface, whereas anhydrite is stable at the higher temperatures and pressures where sedimentary rocks are buried.

As we discovered in 2004, sulfate minerals were precipitated from water early in the history of Mars, the Red Planet. Much as on Earth, these sulfate minerals were precipitated when lakes and shallow seas dried up and formed sedimentary layers. However, many of these sulfate minerals are quite different from the sulfate mineral precipitates

Figure 3.12 A sample of pyrite, also known as "fool's gold." [John Grotzinger/Ramón Rivera-Moret/Harvard Mineralogical Museum.]

Figure 3.13 Gypsum is a sulfate formed when seawater evaporates. [John Grotzinger/Ramón Rivera-Moret/Harvard Mineralogical Museum.]

commonly found on Earth and include strange iron-bearing sulfates that precipitated from very harsh, acidic waters.

PHYSICAL PROPERTIES OF MINERALS

Geologists use their knowledge of mineral composition and structure to understand the origins of rocks. First they must identify the minerals that make up a rock. To do so, they rely greatly on chemical and physical properties that can be observed relatively easily. In the nineteenth and early twentieth centuries, geologists carried field kits for the rough chemical analysis of minerals that would help in identification. One such test is the origin of the phrase "the acid test." It consists of dropping diluted hydrochloric acid (HCl) on a mineral to see if it fizzes (**Figure 3.14**). The fizzing indicates that carbon dioxide (CO_2) is escaping, which means that the mineral is likely to be calcite, a carbonate.

We will now review the physical properties of minerals, many of which contribute to their practical and decorative value.

Hardness

Hardness is a measure of the ease with which the surface of a mineral can be scratched. Just as a diamond, the hardest mineral known, scratches glass, so a quartz crystal, which is harder than feldspar, scratches a feldspar crystal. In 1822, Friedrich Mohs, an Austrian mineralogist, devised a scale (now known as the **Mohs scale of hardness**) based on the ability of one mineral to scratch another. At one extreme is the softest mineral (talc); at the other, the hardest (diamond) (Table 3.2). The Mohs scale is still one of the best practical

Figure 3.14 The acid test. One easy but effective way to identify certain minerals is to drop diluted hydrochloric acid (HCl) on the substance. If it fizzes, indicating the escape of carbon dioxide, the mineral is likely to be calcite. [Chip Clark.]

Table 3.2	Mohs Scale of Hardness	
Mineral	**Scale Number**	**Common Objects**
Talc	1	
Gypsum	2	——— Fingernail
Calcite	3	——— Copper coin
Fluorite	4	
Apatite	5	——— Knife blade
Orthoclase	6	——— Window glass
Quartz	7	——— Steel file
Topaz	8	
Corundum	9	
Diamond	10	

tools for identifying an unknown mineral. With a knife blade and a few of the minerals on the hardness scale, a field geologist can gauge an unknown mineral's position on the scale. If the unknown mineral is scratched by a piece of quartz but not by the knife, for example, it lies between 5 and 7 on the scale.

Recall that covalent bonds are generally stronger than ionic bonds. The hardness of any mineral depends on the strength of its chemical bonds: the stronger the bonds, the harder the mineral. Crystal structure and hardness vary in the silicate group of minerals. For example, hardness varies from 1 in talc, a sheet silicate, to 8 in topaz, a silicate with isolated tetrahedra. Most silicates fall in the 5 to 7 range on the Mohs scale. Only sheet silicates are relatively soft, with hardnesses between 1 and 3.

Within groups of minerals having similar crystal structures, increasing hardness is related to other factors that also increase bond strength:

• *Size:* The smaller the atoms or ions, the smaller the distance between them and the greater the electrical attraction—and thus the stronger the bond.

• *Charge:* The larger the charge of ions, the greater the attraction between them and thus the stronger the bond.

• *Packing of atoms or ions:* The closer the packing of atoms or ions, the smaller the distance between them and thus the stronger the bond.

Size is an especially important factor for most metallic oxides and for most sulfides of metals with high atomic numbers—such as gold, silver, copper, and lead. Minerals of these groups are soft, with hardnesses of less than 3, because their metallic cations are so large. The larger the atoms or ions, the greater the distance between them, and the weaker

EARTH ISSUES

3.1 Sulfide Minerals React to Form Acid Waters on Earth and Mars

Many economically significant mineral deposits are associated with high concentrations of sulfide minerals. During the course of mining, rainwater and groundwater may interact with these minerals to produce highly acidic surface water and groundwater. Unfortunately, acid water is lethal to most organisms. As these acid waters spread throughout the environment, extensive devastation may occur. In extreme cases, the only organisms that can survive are acidophilic (acid-loving) microbes that are specially adapted to live (and actually thrive) in such an extreme environment. These microbes live by eating sulfide!

In a few places on Earth, where sulfide minerals occur in concentrations high enough to start this process, these acid waters are produced naturally. One of these places is the Rio Tinto in Spain. Here, geologists have been able to study a system in which a naturally occurring ore deposit, almost 400 million years old, is interacting with groundwater that flows through the ore deposit by hydrothermal circulation. Sulfide minerals such as pyrite (FeS_2) in the ore deposit react with oxygen in the groundwater to produce sulfuric acid, sulfate ions (SO_4^{-2}), and iron ions (Fe^{3+}) dissolved in warm spring water that flows out as a river (*rio* in Spanish). Microorganisms help to stimulate this reaction.

The river is red (*tinto* in Spanish) because of the dissolved Fe^{3+} ions. The Fe^{3+} ions combine with oxygen to produce the iron oxide minerals goethite and hematite, which may be reddish or brownish in color. In addition, unusual iron sulfate minerals such as jarosite (yellow-brown in color) form abundantly in the Rio Tinto. When geologists encounter this mineral on Earth, we know that the water from which it precipitated must have been extremely acidic.

What is a rare—and environmentally damaging—geologic setting on Earth may once have been widespread on Mars. Recent exploration of Mars has revealed abundant sulfate minerals similar to those found in the Rio Tinto, including jarosite. Understanding how this unusual mineral forms on Earth allows geologists to make inferences about past environments on Mars. In this case, the presence of jarosite indicates that early waters on Mars may have been very acidic and were likely formed through the interaction of oxidizing groundwater with igneous rocks composed of basalt with trace amounts of sulfide.

In turn, this scenario has implications for how we think about the possibility of life—past or present—on other planets. Environments such as the Rio Tinto on Earth show that microorganisms have learned to adapt to such acidic conditions, and they help motivate the search for ancient life on Mars. Some scientists, however, think that although life may have learned to adapt to such harsh conditions, it may not have been able to originate under these conditions. In any event, future exploration for life on other planets will be strongly guided by our understanding of rocks, minerals, and extreme environments on Earth.

Microorganisms thrive in acid water. Rio Tinto, Spain.
[Courtesy of Andrew H. Knoll.]

---

the bond. Carbonates and sulfates, groups in which the structures are packed less closely, also are soft, with hardnesses of less than 5.

Cleavage

Cleavage is the tendency of a crystal to break along flat planar surfaces. The term is also used to describe the geometric pattern produced by such breakage. Cleavage varies inversely with bond strength: high bond strength produces poor cleavage; low bond strength produces good cleavage. Because of their strength, covalent bonds generally give poor or no cleavage. Ionic bonds are relatively weak, so they give excellent cleavage.

If the bonds between some of the planes of atoms or ions in a crystal are weak, the mineral can be made to split along those planes. Muscovite, a mica sheet silicate, breaks along smooth, lustrous, flat, parallel surfaces, forming

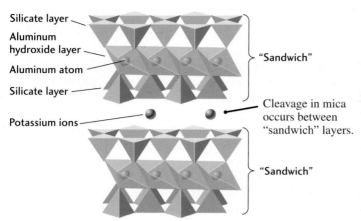

Silicate layer
Aluminum hydroxide layer
Aluminum atom
Silicate layer
Potassium ions

"Sandwich"

Cleavage in mica occurs between "sandwich" layers.

"Sandwich"

Figure 3.15 Cleavage of mica. The diagram shows the cleavage planes in the mineral structure, oriented perpendicular to the plane of the page. Horizontal lines mark the interfaces of silica-oxygen tetrahedral sheets and sheets of aluminum hydroxide bonding the two tetrahedral layers into a sandwich. Cleavage takes place between composite tetrahedral–aluminum hydroxide sandwiches. The photograph shows thin sheets separating along the cleavage planes. [Chip Clark.]

transparent sheets less than a millimeter thick. Mica's excellent cleavage results from weakness of the bonds between the sandwiched layers of cations and tetrahedral silica sheets (**Figure 3.15**).

Cleavage is classified according to two primary sets of characteristics: (1) the number of planes and pattern of cleavage, and (2) the quality of surfaces and ease of cleaving.

Number of Planes; Pattern of Cleavage The number of planes and patterns of cleavage are identifying hallmarks of many rock-forming minerals. Muscovite, for example, has only one plane of cleavage, whereas calcite and dolomite have three excellent cleavage directions that give them a rhomboidal shape (**Figure 3.16**).

A crystal's structure determines its cleavage planes and its crystal faces. Crystals have fewer cleavage planes than possible crystal faces. Faces may be formed along any of numerous planes defined by rows of atoms or ions. Cleavage occurs along any of those planes across which the bonding is weak. All crystals of a mineral exhibit its characteristic cleavage, whereas only some crystals display particular faces.

Galena (lead sulfide, PbS) and halite (sodium chloride, NaCl) cleave along three planes, forming perfect cubes. Distinctive angles of cleavage help identify two important groups of silicates, the pyroxenes and amphiboles, that otherwise often look alike (**Figure 3.17**). Pyroxenes have a single-chain linkage and are bonded so that their cleavage planes are almost at right angles (about 90°) to each other. In cross section, the cleavage pattern of pyroxene is nearly a square. In contrast, amphiboles, the double chains, bond to give two cleavage planes, at about 60° and 120° to each other. They produce a diamond-shaped cross section.

Quality of Surfaces; Ease of Cleaving A mineral's cleavage is assessed as perfect, good, or fair, according to the quality of surfaces produced and the ease of cleaving. Muscovite can be cleaved easily, producing extremely high quality, smooth surfaces; its cleavage is *perfect*. The single- and double-chain silicates (pyroxenes and amphiboles, respectively) show *good* cleavage. Although these minerals break easily along the cleavage plane, they also break across it, producing cleavage surfaces that are not as smooth as those of mica. *Fair* cleavage is shown by the ring silicate beryl. Beryl's cleavage is less regular, and the mineral breaks relatively easily along directions other than cleavage planes.

Many minerals are so strongly bonded that they lack even fair cleavage. Quartz, a framework silicate, is so strongly bonded in all directions that it breaks only along irregular

Figure 3.16 Example of rhomboidal cleavage in calcite. [Charles D. Winters/Photo Researchers.]

Pyroxene

Pyroxene is a single-chain silicate mineral. When the weak bonds between chains break, the angle is about 90°.

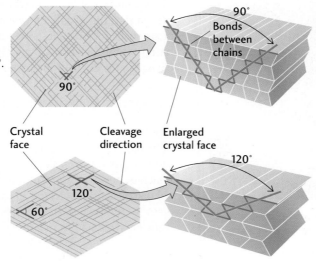

Amphibole

Amphibole is a double-chain silicate. When the bonds between the double chains break, the angle is about 120°.

Figure 3.17 Comparison of cleavage directions and typical crystal faces in pyroxene and amphibole. These two minerals often look very much alike, but their angles of cleavage differ. These angles are frequently used to identify and classify them.

surfaces. Garnet, an isolated tetrahedral silicate, also is bonded strongly in all directions and so has no cleavage. This absence of a tendency to cleave is found in most framework silicates and in silicates with isolated tetrahedra.

Fracture

Fracture is the tendency of a crystal to break along irregular surfaces other than cleavage planes. All minerals show fracture, either across cleavage planes or—in such minerals as quartz—with no cleavage in any direction. Fracture is related to how bond strengths are distributed in directions that cut across crystal planes. Breakage of these bonds results in irregular fractures. Fractures may be *conchoidal,* showing smooth, curved surfaces like those of a thick piece of broken glass. A common fracture surface with an appearance like split wood is described as *fibrous* or *splintery.* The shape and appearance of many kinds of irregular fractures depend on the particular structure and composition of the mineral.

Luster

How the surface of a mineral reflects light gives it a characteristic **luster.** Mineral lusters are described by the terms listed in Table 3.3. Luster is controlled by the kinds of atoms present and their bonding, both of which affect the way light passes through or is reflected by the mineral. Ionically bonded crystals tend to be glassy, or vitreous, but covalently bonded materials are more variable. Many tend to have an adamantine luster, like that of diamond. Metallic luster is shown by pure metals, such as gold, and by many sulfides, such as galena (lead sulfide, PbS). Pearly luster results from multiple reflections of light from planes beneath the sur-

faces of translucent minerals, such as the mother-of-pearl inner surfaces of many clam shells, which are made of the mineral aragonite. Luster, although an important criterion for field classification, depends heavily on the visual perception of reflected light. Textbook descriptions fall short of the actual experience of holding the mineral in your hand.

Color

The **color** of a mineral is imparted by light—either transmitted through or reflected by crystals or irregular masses—or a streak. **Streak** refers to the color of the fine deposit of mineral dust left on an abrasive surface, such as a tile of unglazed porcelain, when a mineral is scraped across it. Such

Table 3.3	Mineral Luster
Luster	**Characteristics**
Metallic	Strong reflections produced by opaque substances
Vitreous	Bright, as in glass
Resinous	Characteristic of resins, such as amber
Greasy	The appearance of being coated with an oily substance
Pearly	The whitish iridescence of such materials as pearl
Silky	The sheen of fibrous materials such as silk
Adamantine	The brilliant luster of diamond and similar minerals

Figure 3.18 Hematite may be black, red, or brown, but it always leaves a reddish brown streak when scratched along a ceramic plate. [Breck P. Kent.]

tiles are called *streak plates* (**Figure 3.18**). A streak plate is a good diagnostic tool because the uniform small grains of mineral that are present in the powder are revealed on the plate. A mineral formed of hematite (Fe_2O_3), for example, may be black, red, or brown, but this mineral will always leave a trail of reddish brown dust on a streak plate.

Color is a complex and not yet fully understood property of minerals. It is determined both by the kinds of ions found in the pure mineral and by trace impurities.

Ions and Mineral Color The color of pure substances depends on the presence of certain ions, such as iron or chromium, that strongly absorb portions of the light spectrum. Olivine that contains iron, for example, absorbs all colors except green, which it reflects, so we see this type of olivine as green. We see pure magnesium olivine as white (transparent and colorless).

Trace Impurities and Mineral Color All minerals contain impurities. Instruments can now measure even very small quantities of some elements—as little as a billionth of a gram in some cases. Elements that make up much less than 0.1 percent of a mineral are reported as "traces," and many of them are called trace elements.

Some trace elements can be used to interpret the origins of the minerals in which they are found. Others, such as the trace amounts of uranium in some granites, contribute to local natural radioactivity. Still others, such as small dispersed flakes of hematite that color a feldspar crystal brownish or reddish, are notable because they give a general color to an otherwise colorless mineral. Many of the gem varieties of minerals, such as emerald (green beryl) and sapphire (blue corundum), get their color from trace impurities dissolved in the solid crystal (**Figure 3.19**). Emerald derives its color from chromium; the sources of sapphire's blue color are iron and titanium.

The color of a mineral may be distinctive, but it is not the most reliable clue to its identity. Some minerals always show the same color; others may have a range of colors. Many minerals show a characteristic color only on freshly broken surfaces or only on weathered surfaces. Some—precious opals, for example—show a stunning display of colors on reflecting surfaces. Others change color slightly with a change in the angle of the light shining on their surfaces.

Specific Gravity and Density

You can easily feel the difference in weight between a piece of hematite iron ore and a piece of sulfur of the same size by lifting the two pieces. A great many common rock-forming minerals, however, are too similar in **density**—mass per unit volume (usually expressed in grams per cubic centimeter, g/cm^3)—for such a simple test. Scientists therefore need some easy method to measure this property of minerals. A standard measure of density is **specific gravity,** which is the weight of a mineral divided by the weight of an equal volume of pure water at 4°C.

Density depends on the atomic mass of a mineral's ions and how closely they are packed in its crystal structure. Consider the iron oxide magnetite, with a density of 5.2 g/cm^3. This high density results partly from the high atomic mass of iron and partly from the closely packed structure that magnetite has in common with the other members of the spinel group of minerals (see page 54). The density of the iron silicate olivine, 4.4 g/cm^3, is lower than that of magnetite for two reasons. First, the atomic mass of silicon, one of the elements that make up olivine, is lower than that of iron. Second, this olivine has a more openly packed structure than minerals of the spinel group. The density of magnesium olivine is even lower, 3.32 g/cm^3, because magnesium's atomic mass is much lower than that of iron.

Increases in density caused by increases in pressure affect the way minerals transmit light, heat, and earthquake waves. Experiments at extremely high pressures have shown that olivine converts into the denser structure of the spinel group at pressures corresponding to a depth of 400 km. At a

Figure 3.19 Minerals can be gems. Rubies and sapphires are formed of the same common mineral, corundum (aluminum oxide). Small amounts of impurities produce the intense color that we value. Ruby, for example, is red because of small amounts of chromium, the same element that gives emeralds their green color. [John Grotzinger/Ramón Rivera-Moret/Harvard Mineralogical Museum.]

greater depth, 670 km, mantle materials are further transformed into silicate minerals with the even more densely packed structure of the mineral perovskite (calcium titanate, $CaTiO_3$). Because of the huge volume of the lower mantle, perovskite is probably the most abundant mineral in the Earth as a whole. Some perovskite minerals have been synthesized to be used as high-temperature semiconductors, which conduct electricity without loss of current and may have great commercial potential. Mineralogists experienced with natural perovskites helped unravel the structure of these newly created materials. Temperature also affects density: the higher the temperature, the more open and expanded the structure and thus the lower the density.

Crystal Habit

A mineral's **crystal habit** is the shape in which its individual crystals or aggregates of crystals grow. Crystal habits are often named after common geometric shapes, such as blades, plates, and needles. Some minerals have such a distinctive crystal habit that they are easily recognizable. An example is quartz, with its six-sided column topped by a pyramid-like set of faces. These shapes indicate not only the planes of atoms or ions in the mineral's crystal structure but also the typical speed and direction of crystal growth. Thus, a needlelike crystal is one that grows very quickly in one direction and very slowly in all other directions. In contrast, a plate-shaped crystal (often referred to as *platy*) grows fast in all directions that are perpendicular to its single direction of slow growth. Fibrous crystals take shape as multiple long, narrow fibers, essentially aggregates of long needles (**Figure 3.20**).

Figure 3.20 Asbestos (chrysotile). Fibers are readily combed from the solid mineral. [Runk/Schoenberger/Grant Heilman Photography.]

Asbestos is a generic name for a group of silicates with a more or less fibrous habit that allows the crystals to become embedded in the lungs after having been inhaled. Other minerals with deleterious properties include arsenic-containing pyrites, some of which are poisonous when ingested and others of which release toxic fumes when heated. Mineral dust diseases are found in many miners, who may face large occupational exposures. An example is silicosis, a disease of the lungs caused by inhaling quartz dust.

Table 3.4 summarizes the mineral physical properties that we discussed in this section.

Table 3.4	Physical Properties of Minerals
Property	**Relation to Composition and Crystal Structure**
Hardness	Strong chemical bonds give high hardness. Covalently bonded minerals are generally harder than ionically bonded minerals.
Cleavage	Cleavage is poor if bonds in crystal structure are strong, good if bonds are weak. Covalent bonds generally give poor or no cleavage; ionic bonds are weak and so give excellent cleavage.
Fracture	Type is related to distribution of bond strengths across irregular surfaces other than cleavage planes.
Luster	Tends to be glassy for ionically bonded crystals, more variable for covalently bonded crystals.
Color	Determined by kinds of atoms or ions and trace impurities. Many ionically bonded crystals are colorless. Iron tends to color strongly.
Streak	Color of fine powder is more characteristic than that of massive mineral because of uniformly small size of grains.
Density	Depends on atomic weight of atoms or ions and their closeness of packing in crystal structure. Iron minerals and metals have high density; covalently bonded minerals have more open packing and so have lower density.
Crystal habit	Depends on planes of atoms or ions in a mineral's crystal structure and the typical speed and direction of crystal growth.

WHAT ARE ROCKS?

A geologist's primary aim is to understand rock properties and to deduce their geologic origins from these properties. Such deductions further our understanding of our planet and provide important information about fuel reserves. For example, knowing that oil forms in certain kinds of sedimentary rocks that are rich in organic matter allows us to explore for new oil reserves more intelligently. Similarly, our knowledge of the properties of rocks will help us find new reserves of other useful and economically valuable mineral and energy resources, such as gas, coal, and metal ores.

Understanding how rocks form also guides us in solving environmental problems. Will this rock be prone to earthquake-triggered landslides? How might it transmit polluted waters in the ground? The underground storage of radioactive and other wastes depends on analysis of the rock to be used as a repository.

In the rest of this chapter, we turn our attention to the major groups of rocks, the minerals that form them, and the plate tectonic environments in which rocks form.

A **rock** is a naturally occurring solid aggregate of minerals or, in some cases, nonmineral solid matter. Some rocks, such as white marble, are composed of just one mineral, in this case calcite. A few rocks are composed of nonmineral matter. These include the noncrystalline, glassy volcanic rocks obsidian and pumice and coal, which is compacted plant remains. In an aggregate, minerals are joined so that they retain their individual identity (**Figure 3.21**).

What determines the physical appearance of a rock? Rocks vary in color, in the sizes of their crystals or grains, and in the kinds of minerals that compose them. Along a road cut, for example, we might find a rough white and

---

Key Figure 3.21 | Minerals are the building blocks of rocks.

Constituent minerals

Orthoclase feldspar   Quartz   Biotite   Plagioclase feldspar

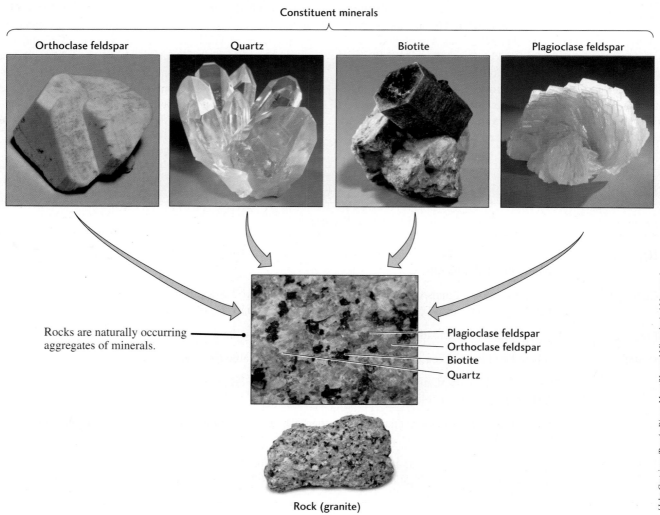

Rocks are naturally occurring aggregates of minerals.

Plagioclase feldspar
Orthoclase feldspar
Biotite
Quartz

Rock (granite)

[John Grotzinger/Ramón Rivera-Moret/Harvard Mineralogical Museum.]

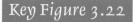

Key Figure 3.22   The three rock groups are formed in different environments by different geologic processes.

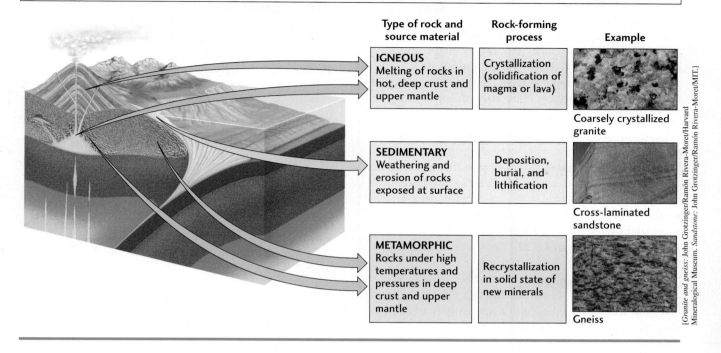

Type of rock and source material	Rock-forming process	Example
IGNEOUS Melting of rocks in hot, deep crust and upper mantle	Crystallization (solidification of magma or lava)	Coarsely crystallized granite
SEDIMENTARY Weathering and erosion of rocks exposed at surface	Deposition, burial, and lithification	Cross-laminated sandstone
METAMORPHIC Rocks under high temperatures and pressures in deep crust and upper mantle	Recrystallization in solid state of new minerals	Gneiss

[*Granite and gneiss*: John Grotzinger/Ramón Rivera-Moret/Harvard Mineralogical Museum. *Sandstone*: John Grotzinger/Ramón Rivera-Moret/MIT.]

pink-speckled rock composed of interlocking crystals large enough to be seen with the naked eye (**Figure 3.22**). Nearby, we might see a grayish rock with many large glittering crystals of mica and some grains of quartz and feldspar. Overlying both the white and pink rock and the gray one we might see the remains of a former beach: horizontal layers of striped white and mauve rock that appear to be made up of sand grains cemented together. And these rocks may all be overlain by a dark fine-grained rock, with tiny white dots in it.

The identity of a rock is determined partly by its mineralogy and partly by its texture. Mineralogy is the relative proportions of a rock's constituent minerals. **Texture** describes the sizes and shapes of its mineral crystals and the way they are put together. If the crystals (or grains), which are only a few millimeters in diameter in most rocks, are large enough to be seen with the naked eye, they are categorized as *coarse*. If they are not large enough to be seen, they are categorized as *fine*. The mineralogy and texture that determine a rock's appearance are themselves determined by the rock's geologic origin—where and how it formed (see Figure 3.22).

The dark rock that caps the sequence of rocks in our road cut, called basalt, was formed by a volcanic eruption. Its mineralogy and texture depend on the chemical composition of rocks that were melted deep in the Earth. All rocks formed by the solidification of molten rock are called **igneous rocks.**

The striped white and mauve layered rock of the road cut, a sandstone, formed as sand particles accumulated, per-

haps on a beach, and eventually were covered over, buried, and cemented together. All rocks formed as the burial products of layers of sediments (such as sand, mud, and calcium carbonate shells), whether they were laid down on the land or under the sea, are called **sedimentary rocks.**

The grayish rock of our road cut, a schist, contains crystals of mica, quartz, and feldspar. It formed deep in Earth's crust as high temperatures and pressures transformed the mineralogy and texture of a buried sedimentary rock. All rocks formed by the transformation of preexisting solid rocks under the influence of high pressure and temperature are called **metamorphic rocks.**

The three types of rocks seen in our road cut represent the three great families of rock: igneous, sedimentary, and metamorphic. We will now look more closely at each rock type and trace the rock cycle—the set of geologic processes that convert each type of rock into the other two types. Finally, we will see how these processes are all driven by plate tectonics and climate.

IGNEOUS ROCKS

Igneous rocks (from the Latin *ignis*, meaning "fire") form by crystallization from a magma, a mass of melted rock that originates deep in the crust or upper mantle. Here temperatures reach the 700°C or more needed to melt most rocks. When a magma cools slowly in the interior, microscopic crystals start to form. As the magma cools below the melting point, some of these crystals have time to grow to several millimeters in diameter or larger before the whole mass crystallizes as a coarse-grained igneous rock. But when a magma erupts from a volcano onto Earth's surface, it cools and solidifies so rapidly that individual crystals have no time to grow gradually. In that case, many tiny crystals form simultaneously, and the result is a fine-grained igneous rock. Geologists distinguish two major types of igneous rocks—intrusive and extrusive—on the basis of the sizes of their crystals.

Intrusive and Extrusive Igneous Rocks

Intrusive igneous rocks crystallize when magma intrudes into unmelted rock masses deep in Earth's crust. Large crystals grow as the magma cools, producing coarse-grained rocks. Intrusive igneous rocks can be recognized by their interlocking large crystals, which grew slowly as the magma gradually cooled (**Figure 3.23**). *Granite* is an intrusive igneous rock.

Extrusive igneous rocks form from rapidly cooled magmas that erupt at the surface through volcanoes. Extrusive igneous rocks, such as *basalt,* are easily recognized by their glassy or fine-grained texture (see Figure 3.23).

Common Minerals

Most of the minerals of igneous rocks are silicates, partly because silicon is so abundant and partly because many silicate minerals melt at the high temperatures and pressures reached in deeper parts of the crust and in the mantle. The

Table 3.5	Some Common Minerals of Igneous, Sedimentary, and Metamorphic Rocks		
Igneous Rocks	**Sedimentary Rocks**	**Metamorphic Rocks**	
*Quartz	*Quartz	*Quartz	
*Feldspar	*Clay minerals	*Feldspar	
*Mica	*Feldspar	*Mica	
*Pyroxene	Calcite	*Garnet	
*Amphibole	Dolomite	*Pyroxene	
*Olivine	Gypsum	*Staurolite	
	Halite	*Kyanite	

Note: Asterisk indicates that the mineral is a silicate.

common silicate minerals found in igneous rocks include quartz, feldspar, mica, pyroxene, amphibole, and olivine (Table 3.5).

SEDIMENTARY ROCKS

Sediments, the precursors of sedimentary rocks, are found on Earth's surface as layers of loose particles, such as sand, silt, and the shells of organisms. These particles form as

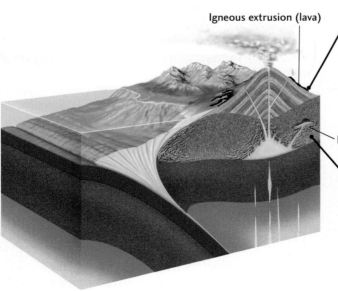

Igneous extrusion (lava)

Extrusive igneous rocks form when magma erupts at the surface, rapidly cooling to fine ash or lava and developing tiny crystals.

The resulting rock, such as the basalt sample here, is fine-grained or has a glassy texture.

Igneous intrusion

Intrusive igneous rocks crystallize when molten rock intrudes into unmelted rock masses in Earth's crust.

Large crystals grow during the slow cooling process, producing coarsely grained rocks such as the granite sample shown here.

Figure 3.23 The formation of *extrusive igneous rocks* (basalt is shown here) and *intrusive igneous rocks* (granite is shown here). [John Grotzinger/Ramón Rivera-Moret/ Harvard Mineralogical Museum.]

rocks undergo weathering and erosion. **Weathering** is all of the chemical and physical processes that break up and decay rocks into fragments of various sizes. The fragmented rock particles are then transported by **erosion,** the set of processes that loosen soil and rock and move them to the spot where they are deposited as layers of sediment (**Figure 3.24**). Weathering and erosion produce two types of sediments:

• **Siliciclastic sediments** are physically deposited particles, such as grains of quartz and feldspar derived from a weathered granite. (*Clastic* is derived from the Greek word *klastos,* meaning "broken.") These sediments are laid down by running water, wind, and ice and form layers of sand, silt, and gravel.

• **Chemical and biochemical sediments** are new chemical substances that form by precipitation when some of a rock's components dissolve during weathering and are carried in river waters to the sea. These sediments include layers of such minerals as halite (sodium chloride) and calcite (calcium carbonate, most often found in the form of reefs and shells).

From Sediment to Solid Rock

Lithification is the process that converts sediments into solid rock, and it occurs in one of two ways:

• By *compaction,* as grains are squeezed together by the weight of overlying sediment into a mass denser than the original.

• By *cementation,* as minerals precipitate around deposited particles and bind them together.

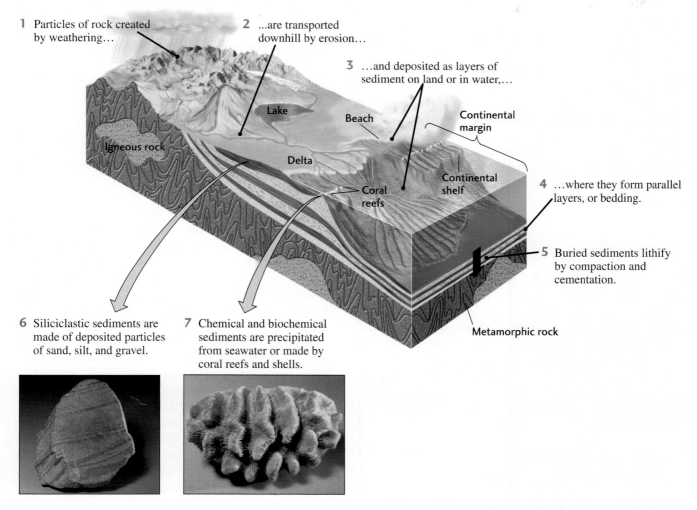

1 Particles of rock created by weathering…

2 …are transported downhill by erosion…

3 …and deposited as layers of sediment on land or in water,…

Lake

Beach

Continental margin

Igneous rock

Delta

Coral reefs

Continental shelf

4 …where they form parallel layers, or bedding.

5 Buried sediments lithify by compaction and cementation.

Metamorphic rock

6 Siliciclastic sediments are made of deposited particles of sand, silt, and gravel.

7 Chemical and biochemical sediments are precipitated from seawater or made by coral reefs and shells.

Figure 3.24 Weathering breaks down rock into smaller particles that are then carried downhill and downstream by erosion to be deposited as layers of sediment along continental margins. Other sediment is produced by biochemical precipitation, such as in the formation of coral reefs. Some corals are lithified during their growth phase and therefore may experience little change during deep burial. *left:* Cross-laminated sandstone; *right:* fossilized coral. [John Grotzinger/Ramón Rivera-Moret/MIT.]

Sediments are compacted and cemented after burial under additional layers of sediment. Thus sandstone forms by the lithification of sand particles, and limestone forms by the lithification of shells and other particles of calcium carbonate.

Layers of Sediment

Sediments and sedimentary rocks are characterized by **bedding,** the formation of parallel layers of sediment as particles settle to the bottom of the sea, a river, or a land surface. Because sedimentary rocks are formed by surface processes, they cover much of Earth's land surface and seafloor. In terms of surface area, most rocks found at Earth's surface are sedimentary, but they are difficult to preserve, and so their volume is small compared to that of the igneous and metamorphic rocks that make up the main volume of the crust.

Common Minerals

The common minerals of siliciclastic sediments are silicates, because silicate minerals predominate in rocks that weather to form sedimentary particles (see Table 3.5). The most abundant minerals in siliciclastic sedimentary rocks are quartz, feldspar, and clay minerals.

The most abundant minerals of chemically or biochemically precipitated sediments are carbonates, such as calcite, the main constituent of limestone. Dolomite, also found in limestone, is a calcium-magnesium carbonate formed by precipitation during lithification. Two other chemical sediments—gypsum and halite—form by precipitation as seawater evaporates.

METAMORPHIC ROCKS

Metamorphic rocks take their name from the Greek words for "change" (*meta*) and "form" (*morphe*). These rocks are produced when high temperatures and pressures deep in the Earth cause any kind of rock—igneous, sedimentary, or other metamorphic rock—to change its mineralogy, texture, or chemical composition while maintaining its solid form. The temperatures of metamorphism are below the melting points of the rocks (about 700°C) but high enough (above 250°C) for the rocks to change by recrystallization and chemical reactions.

Regional and Contact Metamorphism

Metamorphism may take place over a widespread area or a limited one (**Figure 3.25**). **Regional metamorphism** occurs where high pressures and temperatures extend over large regions, as happens where plates collide. Regional metamorphism accompanies plate collisions that result in moun-

tain building and the folding and breaking of sedimentary layers that were once horizontal. Where high temperatures are restricted to smaller areas, such as the rocks near and in contact with an intrusion, rocks are transformed by **contact metamorphism.**

Many regionally metamorphosed rocks, such as schists, have characteristic **foliation,** wavy or flat planes produced when the rock was structurally deformed into folds. Granular textures are more typical of most contact metamorphic rocks and of some regional metamorphic rocks formed by very high pressure and temperature.

Common Minerals

Silicates are the most abundant minerals of metamorphic rocks because the parent rocks are also rich in silicates (see Table 3.5). Typical minerals of metamorphic rocks are quartz, feldspar, mica, pyroxene, and amphibole—the same kinds of silicates characteristic of igneous rocks. Several other silicates—kyanite, staurolite, and some varieties of garnet—are characteristic of metamorphic rocks alone. These minerals form under conditions of high pressure and temperature in the crust and are not characteristic of igneous rocks. They are therefore good indicators of metamorphism. Calcite is the main mineral of marbles, which are metamorphosed limestones.

THE ROCK CYCLE: Interactions Between the Plate Tectonic and Climate Systems

Earth scientists have known for over 200 years that the three basic groups of rocks—igneous, metamorphic, and sedimentary—all can evolve from one to another and that this signifies important processes operating in and on the Earth. This knowledge has given rise to the concept of a **rock cycle,** which is now known to be the result of interactions between two of the three fundamental Earth systems: plate tectonics and climate. Driven by interactions between these two systems, material and energy are transferred among the Earth's interior, the land surface, the oceans, and the atmosphere. For example, the melting of subducting lithospheric slabs and the formation of magma result from processes operating within the plate tectonic system. When these molten rocks erupt, matter and energy are transferred to the land surface, where the materials (newly formed rocks) are subject to weathering by the climate system. The same process injects volcanic ash and carbon dioxide gas high into the atmosphere, where they may affect global climate. As global climate changes, perhaps becoming warmer or cooler, the rate of rock weathering changes, which in turn influences the rate at which material (sediment) is returned to Earth's interior.

The rock cycle begins with the creation of new oceanic lithosphere at a mid-ocean ridge spreading center as two

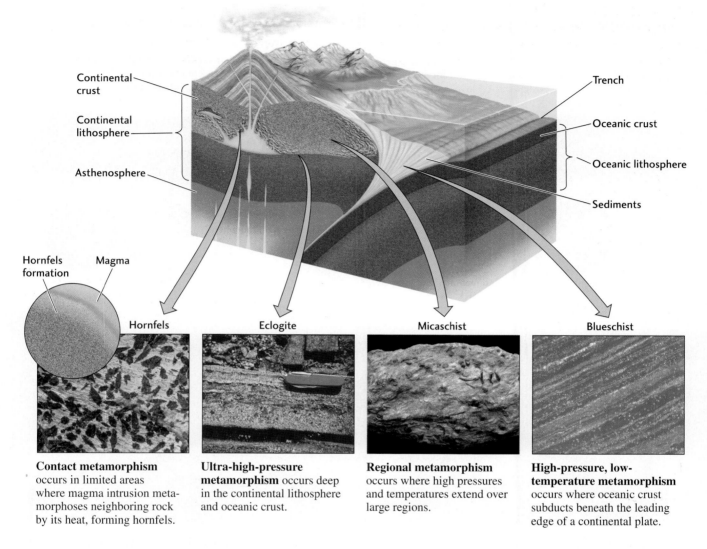

Contact metamorphism occurs in limited areas where magma intrusion metamorphoses neighboring rock by its heat, forming hornfels.

Ultra-high-pressure metamorphism occurs deep in the continental lithosphere and oceanic crust.

Regional metamorphism occurs where high pressures and temperatures extend over large regions.

High-pressure, low-temperature metamorphism occurs where oceanic crust subducts beneath the leading edge of a continental plate.

Figure 3.25 Metamorphic rocks form under four main conditions. Examples of rocks shown here are (from left to right)

hornfels [Biophoto Associates/Photo Researchers], *eclogite* [Julie Baldwin], *micaschist* [John Grotzinger], and *blueschist* [Mark Cloos].

continents drift apart (**Figure 3.26**). The ocean gets wider and wider until at some point the process reverses itself and the ocean closes. As the basin closes, igneous rocks created at the mid-ocean ridge eventually descend into a subduction zone beneath a continental plate. Sediments that were formed on the continent and transported to its edge may also be dragged down into the subduction zone. Ultimately, the two continents, which were once drifting apart, may now collide. As the igneous rocks and sediments that descend into the subduction zone go deeper and deeper into Earth's interior, they begin to melt to form a new generation of igneous rocks. The great heat associated with the intrusion of these igneous rocks, coupled with the heat and pressure that comes with being pushed to levels deep in the Earth, transforms these igneous rocks—and other surrounding rocks—into metamorphic rocks. During the collision process, these igneous and metamorphic rocks are then

uplifted into a high mountain chain as a section of Earth's crust crumples and deforms. The uplifted igneous and metamorphic rocks slowly weather.

Interaction between the plate tectonic system (uplift of rocks) and climate system (weathering of uplifted rocks) results in the transformation of igneous and metamorphic rocks into loose sediment that erosion then strips away. Water and wind transport some of these sediments across the continents and eventually to the edges of the continents, where the land meets the ocean. The sediments laid down in the sea are buried under successive layers of sediment, where they slowly lithify into sedimentary rock. These oceans, like those mentioned at the beginning of the cycle, may also have formed by spreading along mid-ocean ridges, thus completing the rock cycle.

The particular pathway illustrated here—that of a continent breaking apart, forming new seafloor, then closing back

Earth System Figure 3.26   The rock cycle is the interaction of plate tectonic and climate systems.

1 The cycle begins with rifting and development of a divergent margin within a continent. Sediments erode from the continental interior and are deposited in rift basins, where they are buried to form sedimentary rocks.

2 Rifting and spreading continue, and a new ocean basin develops. Magma rises from the asthenosphere at mid-ocean ridges and chills to form basalt, an igneous rock.

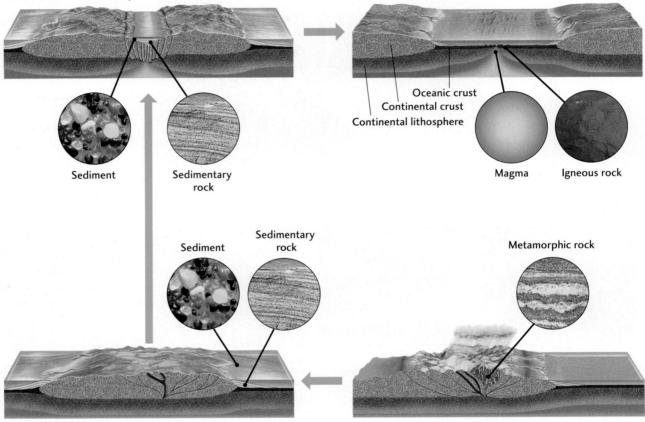

Sediment

Sedimentary rock

Oceanic crust
Continental crust
Continental lithosphere

Magma   Igneous rock

Sedimentary rock

Sediment

Metamorphic rock

6 Streams transport sediment away from collision zones to oceans, where it is deposited as layers of sand and silt. Layers of sediment are buried and lithify to form sedimentary rock.

5 Further closing of the ocean basin leads to continental collision, forming high mountain ranges. Where continents collide, rocks are buried deeper or modified by heat and pressure, forming metamorphic rocks. Uplifted mountains force moisture-laden air to rise, cool, condense, and precipitate. Weathering creates loose material—soils and sediment—that erosion strips away.

up again—is only one variation among the many that may take place in the rock cycle. Any type of rock—igneous, sedimentary, or metamorphic—can be uplifted during a mountain-building event and weathered and eroded to form new sediments. Some stages may be omitted: as a sedimentary rock is uplifted and eroded, for example, metamorphism and melting are skipped. Also, we know from deep drilling that some igneous and metamorphic rocks many kilometers deep in the crust may never have been uplifted or exposed to weathering and erosion.

The rock cycle never ends. It is always operating at different stages in various parts of the world, forming and eroding mountains in one place and laying down and burying sediments in another. The rocks that make up the solid Earth are recycled continuously, but we can see only the surface

parts of the cycle. We must deduce the recycling of the deep crust and the mantle from indirect evidence.

MINERALS OCCUR IN ROCKS AS VALUABLE RESOURCES

We have now learned the basics of what minerals and rocks are and how the plate tectonic and climate systems interact to form a rock cycle. The rock cycle also turns out to be crucial in creating economically important concentrations of valuable minerals in Earth's crust. Finding these minerals and extracting them is a vital job for Earth scientists, and so we now turn our attention to learning about where and how some of these geological prizes formed.

High concentrations of elements are found in a limited number of specific geological settings. Some examples are given later in this chapter. These settings are of economic interest, because the higher the concentration of a resource in a given deposit, the lower the cost to recover it.

Ore Minerals

Ores are rich deposits of minerals from which valuable metals can be recovered profitably. The minerals containing these metals are *ore minerals*. Ore minerals include sulfides (the main group), oxides, and silicates. Ore minerals in each of these groups are compounds of metallic elements with sulfur, oxygen, and silicon oxide, respectively. The copper ore mineral covelite, for example, is a copper sulfide (CuS). The iron ore mineral hematite (Fe_2O_3) is an iron oxide. The nickel ore mineral garnierite is a nickel silicate, $Ni_3Si_2O_5$ $(OH)_4$. In addition, some metals, such as gold, are found in their native state—that is, uncombined with other elements (**Figure 3.27**).

Recall in our discussion of the rock cycle that continental margins where subduction occurs may be associated with melting of oceanic lithosphere to form igneous rocks. Very large ore deposits can be formed in such a tectonic setting, when hot solutions—also known as **hydrothermal solutions**—are formed around bodies of molten rock. This happens when circulating groundwater comes into contact with a hot intrusion, reacts with it, and carries off significant quantities of elements and ions released by the reaction. These elements and their ions then interact with one another to deposit ore minerals, usually as the fluid cools.

3 Subsidence of the continental margin—sinking of Earth's lithosphere—leads to accumulation of sediment and formation of sedimentary rock during burial.

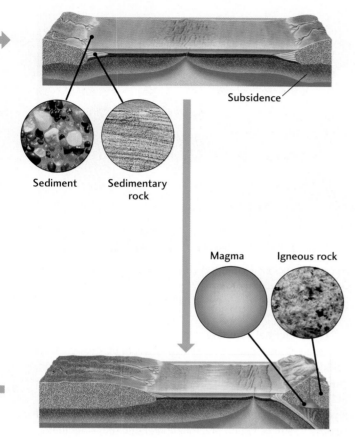

Sediment   Sedimentary rock

Magma   Igneous rock

Subsidence

4 Oceanic crust subducts beneath a continent, building a volcanic mountain chain. The subducting plate melts as it descends. Magma rises from the melting plate and mantle and cools to make granitic igneous rocks.

The chemical elements of Earth's crust are widely distributed in many kinds of minerals, and those minerals are found in a great variety of rocks. In most places, any given element will be found homogenized with other elements in amounts close to its average concentration in the crust. An ordinary granitic rock, for example, may contain a few percentage points of iron, close to the average concentration of iron in Earth's crust.

When an element is present in higher concentrations, it means the rock underwent some geologic process that concentrated larger quantities of the element than normal. In some cases, the rock cycle contributes to this concentration of important minerals. The *concentration factor* of an element in a mineral deposit is the ratio of the element's abundance in the deposit to its average abundance in the crust.

Figure 3.27 Gold occurring in the free state (native gold) on a quartz crystal. [Chip Clark.]

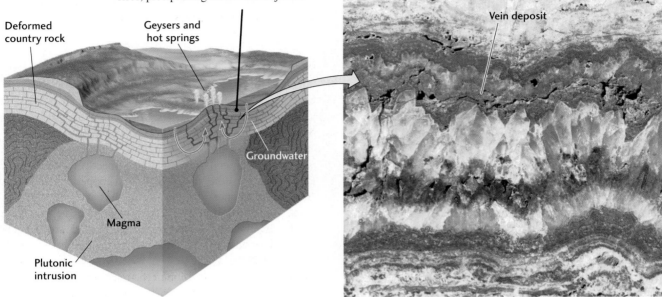

Groundwater percolating through cracks and jointed rock dissolves metal oxides and sulfides. Heated by the magma, it rises, precipitating metal ores in joints.

Deformed country rock

Geysers and hot springs

Groundwater

Magma

Plutonic intrusion

Vein deposit

Figure 3.28 Many ore deposits are found in hydrothermal veins formed by hot solutions rising from magmatic intrusions. Quartz vein deposit (about 1 cm thick) containing gold and silver ores. Oatman, Arizona. [Peter Kresan.]

Vein Deposits

Hydrothermal fluids moving through fractured rocks often deposit ore minerals. These fluids flow easily through the fractures, cooling rapidly in the process. Quick cooling causes fast precipitation of the ore minerals. The tabular (sheetlike) deposits of precipitated minerals in the fractures and joints are called **vein deposits** or simply **veins.** Some ores are found in veins; others are found in the country rock adjacent to the veins, which were altered when the hot solutions heated and infiltrated the country rock. As the solutions react with surrounding rocks, they may precipitate ore minerals together with quartz, calcite, or other common vein-filling minerals. Vein deposits are a major source of gold (**Figure 3.28**).

Hydrothermal vein deposits are among the most important sources of metal ores. Typically, metallic ores exist as sulfides, such as iron sulfide (pyrite), lead sulfide (galena), zinc sulfide (sphalerite), mercury sulfide (cinnabar)—shown in **Figure 3.29**—and copper sulfide (covelite and chalcocite). Hydrothermal solutions reach the surface as hot springs and geysers, many of which precipitate metallic ores—including ores of lead, zinc, and mercury—as they cool.

Disseminated Deposits

Mineral deposits that are scattered through volumes of rock much larger than veins are called *disseminated deposits.* In both igneous and sedimentary rocks, minerals are disseminated along abundant cracks and fractures. Among the eco-

Galena (lead sulfide) Cinnabar (mercury sulfide) Pyrite (iron sulfide) Sphalerite (zinc sulfide)

Figure 3.29 Metal sulfide ores. Sulfides are the most common types of metallic ores. [Chip Clark.]

Figure 3.30 Copper ores. Chalcopyrite and chalcocite are copper sulfide ores. Malachite is a carbonate of copper found in association with sulfides of copper. [Chip Clark.]

Chalcopyrite (a copper sulfide)　　Malachite (a copper carbonate)　　Chalcocite (a copper sulfide)

nomically important disseminated deposits are the copper deposits of Chile and the southwestern United States. These deposits develop in geologic provinces with abundant igneous rocks, usually emplaced as large intrusive bodies. In Chile, these intrusive igneous rocks are related to subduction of oceanic lithosphere beneath the Andes Mountains, very similar to what was described in our example of the rock cycle. The most common copper mineral is chalcopyrite, a copper sulfide (**Figure 3.30**). The copper was deposited when ore-forming minerals were introduced into a great number of tiny fractures in granitic intrusive rocks and in the country rocks surrounding the higher parts of igneous intrusions. Some unknown process associated with the intrusion or its aftermath broke the rocks into millions of pieces. Hydrothermal solutions penetrated and recemented the rocks by precipitating ore minerals throughout the extensive network of tiny fractures. This widespread dispersal produced a low-grade but very large resource of many millions of tons of ore, which can be mined economically by large-scale methods (**Figure 3.31**).

Extensive disseminated hydrothermal deposits may also be present in sedimentary rocks. This is the case in the lead-zinc province of the Upper Mississippi Valley, which extends from southwestern Wisconsin to Kansas and Oklahoma. The ores in this province are not associated with a known magmatic intrusion that could have been a source of hydrothermal fluids, and so their origin is very different. Some geologists speculate that the ores were deposited by groundwater that was driven out of the ancestral Appalachian Mountains when they were much higher. Continental collision between North America and Africa may have created a continental-scale squeegee that pushed fluids from deep within the collision zone all the way into the continental interior of North America. Groundwater may have penetrated hot crustal rocks at great depths and dissolved soluble ore minerals, then moved upward into the overlying sedimentary rocks, where it precipitated the minerals as fillings in cavities. In some cases, it appears that ore fluids infiltrated limestone formations and dissolved some carbonates, then replaced the carbonates with equal volumes

Figure 3.31 Kennecott Copper Mine, Utah, an open-pit mine. Open-pit mining is typical of the large-scale methods used to exploit widely disseminated ore deposits. [David R. Frazier/The Image Works.]

of new crystals of sulfide. The major minerals of the hydrothermal deposits in this province are lead sulfide (galena) and zinc sulfide (sphalerite).

Igneous Deposits

The most important *igneous ore deposits*—deposits of ore in igneous rocks—are found as segregations of ore minerals near the bottom of intrusions. The deposits form when minerals crystallize from molten magma, settle, and accumulate on the floor of a magma chamber. Most of the chromium and platinum ores in the world, such as the deposits in South Africa and Montana, are found as layered accumulations of minerals that formed in this way (**Figure 3.32**). One of the richest ore bodies ever found, at Sudbury, Ontario, is a large mafic intrusion containing great quantities of layered nickel, copper, and iron sulfides near its base. Geologists believe that these sulfide deposits formed from crystallization of a dense, sulfide-rich liquid that separated from the rest of the cooling magma and sank to the bottom of the chamber before it congealed.

Pegmatites are extremely coarse grained intrusive rocks of granitic composition. As the magma in a large granitic intrusion cools, the last melt to freeze solidifies as pegmatites, in which minerals present in only trace amounts in the parent rock are concentrated. Pegmatites may contain rare mineral deposits rich in such elements as boron, lithium, fluorine, niobium, and uranium and in such gem minerals as tourmaline.

Sedimentary Deposits

Sedimentary mineral deposits include some of the world's most valuable mineral sources. Many economically important minerals segregate as an ordinary result of sedimentary processes. Sedimentary mineral deposits are also important sources of copper, iron, and other metals. These deposits were chemically precipitated in sedimentary environments to which large quantities of metals were transported in solution. Some of the important sedimentary copper ores, such as those of the Permian Kupferschiefer (German for "copper slate") beds of Germany, may have precipitated from hot brines of hydrothermal origin, rich in metal sulfides, that interacted with sediments on the ocean bottom. The tectonic setting of these deposits may have been more like the mid-ocean ridge described in our example of the rock cycle, except that it developed on a continent. Here, rifting of the continental crust led to development of a deep trough, where sediments and ore minerals were deposited in a very still, narrow ocean.

Many rich deposits of gold, diamonds, and other heavy minerals such as magnetite and chromite are found in *placers,* ore deposits that have been concentrated by the mechanical sorting action of river currents. These ore deposits owe their origin to processes associated with weathering and sediment transport during the surface phase of the rock cycle. Uplifted rocks weather to form grains of sediment, which are then sorted by weight when currents of water flow over them. Because heavy minerals settle out of a current more quickly than lighter minerals such as quartz and feldspar, the heavy minerals tend to accumulate on river bottoms and sandbars, where the current is strong enough to keep the lighter minerals suspended and in transport but is too weak to move the heavier minerals. Similarly, ocean waves preferentially deposit heavy minerals on beaches or shallow offshore bars. The gold panner accomplishes the same thing: the shaking of a water-filled pan allows the lighter minerals to be washed away, leaving the heavier gold in the bottom of the pan.

Some placers can be traced upstream to the location of the original mineral deposit, usually of igneous origin, from which the minerals were eroded. Erosion of the Mother Lode, an extensive gold-bearing vein system lying along the western flanks of the Sierra Nevada batholith, produced the placers that were discovered in 1848 and led to the Califor-

Figure 3.32 Chromite (chrome ore, dark layers) in a layered igneous intrusive. Bushveldt Complex, South Africa. [Spence Titley.]

nia gold rush. The placers were found before their source was discovered. Placers also led to the discovery of the Kimberley diamond mines of South Africa two decades later.

SUMMARY

What is a mineral? Minerals, the building blocks of rocks, are naturally occurring, inorganic solids with specific crystal structures and chemical compositions that either are fixed or vary within a defined range. A mineral is constructed of atoms, the small units of matter that combine in chemical reactions. An atom is composed of a nucleus of protons and neutrons, surrounded by electrons. The atomic number of an element is the number of protons in its nucleus, and its atomic mass is the sum of the masses of its protons and neutrons.

How do atoms combine to form the crystal structures of minerals? Chemical substances react with one another to form compounds either by gaining or losing electrons to become ions or by sharing electrons. The ions in a chemical compound are held together by ionic bonds, which form by electrostatic attraction between positive ions (cations) and negative ions (anions). Atoms that share electrons to form a compound are held together by covalent bonds. When a mineral crystallizes, atoms or ions come together in the proper proportions to form a crystal structure—an orderly three-dimensional geometric array in which the basic arrangement is repeated in all directions.

What are the major rock-forming minerals? Silicates, the most abundant minerals in Earth's crust, are crystal structures built of silicate tetrahedra linked in various ways. Tetrahedra may be isolated (olivines) or in single chains (pyroxenes), double chains (amphiboles), sheets (micas), or frameworks (feldspars). Carbonate minerals are made of carbonate ions bonded to calcium or magnesium or both. Oxide minerals are compounds of oxygen and metallic elements. Sulfide and sulfate minerals are composed of sulfur atoms in combination with metallic elements.

What are the physical properties of minerals? A mineral's physical properties, which indicate its composition and structure, include hardness—the ease with which its surface is scratched; cleavage—its ability to split or break along flat surfaces; fracture—the way in which it breaks along irregular surfaces; luster—the nature of its reflection of light; color—imparted by transmitted or reflected light to crystals, irregular masses, or a streak (the color of a fine powder); density—the mass per unit volume; and crystal habit—the shapes of individual crystals or aggregates.

What determines the properties of the various kinds of rocks that form in and on Earth's surface? Mineralogy (the kinds and proportions of minerals that make up a rock) and texture (the sizes, shapes, and spatial arrangement of its crystals or grains) define a rock. The mineralogy and texture of a rock are determined by the geologic conditions, including chemical composition, under which it formed, either in the interior under various conditions of high temperature and pressure or at the surface, where temperatures and pressures are low.

What are the three types of rock and how do they form? Igneous rocks form by the crystallization of magmas as they cool. Intrusive igneous rocks form in Earth's interior and have large crystals. Extrusive igneous rocks, which form at the surface where lavas and ash erupt from volcanoes, have a glassy or fine-grained texture. Sedimentary rocks form by the lithification of sediments after burial. Sediments are derived from the weathering and erosion of rocks exposed at Earth's surface. Metamorphic rocks form by alteration in the solid state of igneous, sedimentary, or other metamorphic rocks as they are subjected to high temperatures and pressures in Earth's interior.

How does the rock cycle describe the formation of rocks as the products of geologic processes? The rock cycle relates geologic processes to the formation of the three types of rocks from one another. We can view the processes by starting at any point in the cycle. We began with the creation of new oceanic lithosphere at a mid-ocean ridge as two continents drift apart. The ocean gets wider until at some point the process reverses itself and the ocean closes. As the basin closes, igneous rocks created at the mid-ocean ridge eventually descend into a subduction zone beneath a continental plate. Ultimately, the two continents, which were once drifting apart, may now collide. As the igneous rocks and sediments that descend into the subduction zone go deeper and deeper into Earth's interior, they begin to melt to form a new generation of igneous rocks. The great heat associated with the intrusion of these igneous rocks, coupled with the heat and pressure that comes with being pushed to levels deep in the Earth, transforms these igneous rocks—and other surrounding rocks—into metamorphic rocks. During the collision process, these igneous and metamorphic rocks are then uplifted into a high mountain chain as a section of Earth's crust crumples and deforms. Igneous and metamorphic rocks are then uplifted to the surface in the mountain-building process. The uplifted igneous and metamorphic rocks slowly weather. Plate tectonics is the mechanism by which the cycle operates.

What is hydrothermal mineral deposition? Hydrothermal deposits, which are some of the most important ore deposits, are formed by hot water that emanates from igneous intrusions or by heated circulating groundwater or seawater. The heated water leaches soluble minerals in its path and transports them to cooler rocks, where they are precipitated in fractures, joints, and voids. These ores may be found in veins or in disseminated deposits, such as copper sulfides.

How do igneous ore deposits form? Igneous ore deposits typically form when minerals crystallize from molten magma, settle, and accumulate on the floor of a magma chamber. They are often found as layered accumulations of minerals. The rich ore body at Sudbury, Ontario, for example, is a mafic intrusion that contains great quantities of layered nickel, copper, and iron sulfides near its base.

KEY TERMS AND CONCEPTS

anion (p. 49)

atom (p. 47)

atomic mass (p. 47)

atomic number (p. 47)

bedding (p. 66)

cation (p. 49)

chemical and biochemical sediments (p. 65)

chemical reaction (p. 47)

cleavage (p. 57)

color (p. 59)

contact metamorphism (p. 66)

covalent bond (p. 49)

crystal (p. 50)

crystal habit (p. 61)

crystallization (p. 49)

density (p. 60)

electron (p. 47)

electron sharing (p. 48)

electron transfer (p. 48)

erosion (p. 65)

extrusive igneous rock (p. 64)

foliation (p. 66)

fracture (p. 59)

hardness (p. 56)

hydrothermal solution (p. 69)

igneous rock (p. 63)

intrusive igneous rock (p. 64)

ionic bond (p. 49)

isotope (p. 47)

lithification (p. 65)

luster (p. 59)

metallic bond (p. 49)

metamorphic rock (p. 63)

mineral (p. 46)

mineralogy (p. 45)

Mohs scale of hardness (p. 56)

neutron (p. 47)

nucleus (p. 47)

ore (p. 69)

polymorph (p. 51)

precipitate (p. 51)

proton (p. 47)

regional metamorphism (p. 66)

rock (p. 62)

rock cycle (p. 66)

sediment (p. 64)

sedimentary rock (p. 63)

siliciclastic sediments (p. 65)

specific gravity (p. 60)

streak (p. 59)

texture (p. 63)

vein deposit (vein) (p. 70)

weathering (p. 65)

EXERCISES

1. Define a mineral.

2. What is the difference between an atom and an ion?

3. Draw the atomic structure of sodium chloride.

4. What are two types of chemical bonds?

5. List the basic structures of silicate minerals.

6. Name three groups of minerals, other than silicates, based on their chemical composition.

7. How would a field geologist measure hardness?

8. What is the difference between the carbonate minerals calcite and dolomite?

9. What are the differences between extrusive and intrusive igneous rocks?

10. What are the differences between regional and contact metamorphism?

11. What are the differences between siliciclastic and chemical or biochemical sedimentary rocks?

12. List three common silicate minerals found in each group of rocks: igneous, sedimentary, and metamorphic.

13. Of the three groups of rocks, which form at Earth's surface and which in the interior of the crust?

14. What are the characteristics of an economical ore deposit?

15. Describe the creation of an ore body by hydrothermal activity.

THOUGHT QUESTIONS

1. Draw a simple diagram to show how silicon and oxygen in silicate minerals share electrons.

2. Diopside, a pyroxene, has the formula $(Ca,Mg)_2Si_2O_6$. What does this formula tell you about its crystal structure and cation substitution?

3. In some bodies of granite, we can find very large crystals, some as much as a meter across, yet these crystals tend to have few crystal faces. What can you deduce about the conditions under which these large crystals grew?

4. What physical properties of sheet silicates are related to their crystal structure and bond strength?

5. Choose two minerals from Appendix 5 that you think might make good abrasive or grinding stones for sharpening steel, and describe the physical property that causes you to believe they would be suitable for this purpose.

6. Aragonite, with a density of 2.9 g/cm^3, has exactly the same chemical composition as calcite, with a density of 2.7 g/cm^3. Other things being equal, which of these two minerals is more likely to have formed under high pressure?

7. There are at least eight physical properties one can use to identify an unknown mineral. Which ones are most useful in discriminating between minerals that look similar? Describe a strategy that would allow you to determine that an unknown clear calcite crystal is not the same as a known clear crystal of quartz.

8. Coal, which forms from decaying vegetation and is, therefore, a natural substance, is not considered to be a mineral. However, when coal is heated to high temperatures and buried in high-pressure areas, it transforms into the mineral graphite. Why is it, then, that coal is not considered a mineral but graphite is? Explain your reasoning.

9. What geologic processes transform a sedimentary rock into an igneous rock?

10. Which igneous intrusion would you expect to have a wider contact metamorphic zone: one intruded by a very hot magma or one intruded by a cooler magma?

11. Describe the geologic processes by which an igneous rock is transformed into a metamorphic rock and then exposed to erosion.

12. Using the rock cycle, trace the path from a magma to a granite intrusion to a metamorphic gneiss to a sandstone. Be sure to include the role of tectonics and the specific processes that create the rocks.

13. Where are igneous rocks most likely to be found? How could you be certain that the rock is igneous and not sedimentary or metamorphic?

14. Back in the late 1800s, gold miners used to pan for gold by placing sediment from rivers into a pan and filtering water through the pan while swirling the pan's content. Specifically, the miners wanted to be certain that they had found real gold and not pyrite ("fool's gold"). Why did this method work? What mineral property does the process of panning for gold use? What is another possible method for distinguishing between gold and pyrite?

SHORT-TERM PROJECTS

Why Is the Hope Diamond Blue?

In late 1955, Robert H. Wentorf, Jr., achieved something close to alchemy. He bought a jar of peanut butter at a local food co-op, took it to his lab, and then turned a glob of the spread into a few tiny (green) diamonds.* After all, peanuts are rich in proteins, which are rich in nitrogen. Synthetic diamonds are often black due to inclusions of graphite, and traces of nitrogen within the crystal structure can turn diamonds brown, yellow, or green.

Diamonds form in the upper mantle at depths below about 200 km where temperatures and pressures cause graphite, diamond's sister mineral, to collapse into a more tightly packed crystal structure (see Figure 3.8). Most diamonds weigh less than 1 carat (0.2 gram) and are colorless, pale yellow, or brown. Larger diamonds, especially colored

---

* Ivan Amato, Diamond fever. *Science News,* August 4, 1990, p. 72.

ones, are rare. The largest faceted diamond is the yellow 545.67-carat Golden Jubilee, unveiled in 1995. The Hope diamond, on display at the Smithsonian Institution, may be the best-known gemstone in the world. It is a very rare 45-carat blue diamond.

What produces the color of gems and minerals? Color is produced by the interaction of light with matter. You know how a prism or even water drops break white light into a spectrum or rainbow of colored light. When light strikes the surface of a crystal or penetrates it, it interacts with atoms; some components of the light may be absorbed and others transmitted. As white light passes through the Hope diamond, the crystal absorbs red light and transmits blue light.

Why are many diamonds colorless and others blue, red, yellow, green, or brown? Trace amounts of impurities and imperfections in the crystal structure of a mineral can change how the mineral interacts with light and cause it to be colored. Although diamonds are pure carbon, it takes only one atom of another element among the millions of atoms to turn a diamond blue. Rubies and sapphires are both crystals of aluminum oxide; yet rubies are red and sapphires are blue. Go to **http://www.whfreeman.com/understandingearth** to figure out what causes the Hope diamond to be blue, rubies to be red, and sapphires to be blue.

Identifying Building Stones

Building stones often are a clue to local geology. Local stone is both cost-effective and a source of community pride. With a partner, examine the building stones that you find on campus or in your community. Choose four to six types of stone that look different and note their locations on a local map. Draw each stone on a separate piece of paper and note such features as color, grain size, the presence or absence of layering, and whether the stone appears to contain one mineral or more than one. Also describe any evidence of chemical or physical weathering and judge how good the stone is for building. Then decide whether the stone is most likely to be igneous, metamorphic, or sedimentary, and explain why. Finally, compare a geologic map of the area with your findings in the field and explain why the stones you described do or do not record the local geology. Submit an organized folder containing your drawings, observations, and inferences.

IGNEOUS ROCKS
Solids from Melts

More than 2000 years ago, the Greek scientist and geographer Strabo traveled to Sicily to view the volcanic eruptions of Mount Etna. He observed that the hot liquid lava spilling down from the volcano onto Earth's surface cooled and hardened into solid rock within a few hours. By the eighteenth century, geologists began to understand that some sheets of rock that cut across other rock formations also formed by the cooling and solidifying of magma. In these cases, the magma cooled slowly because it had remained buried in Earth's crust. Today we know that deep in Earth's hot crust and mantle, rocks melt and rise toward the surface. Some magmas solidify before they reach the surface, and some break through and solidify on the surface. Both processes produce igneous rocks.

As we saw in Chapter 3, much of Earth's crust is composed of igneous rock, some metamorphosed and some not. Understanding the processes that melt and resolidify rocks is a key to understanding how Earth's crust forms. **In this chapter, we examine the wide range of igneous rocks, both intrusive and extrusive, and the processes by which they form.**

We also learned in Chapter 3 that plate tectonics creates a wide variety of igneous rocks. Specifically, igneous rocks form at spreading centers where plates move apart and at convergent boundaries where one plate descends beneath another. Although we still have much to learn about the exact *mechanisms* of melting and solidification, we do have good answers to some fundamental questions: How do igneous rocks differ from one another? Where do igneous rocks form? How do rocks solidify from a melt? Where do melts form?

In answering these questions, we will focus on the central role of igneous processes in the Earth system. When melted rock is transported from magma chambers in Earth's interior to volcanoes, for example, a variety of gases are also carried along. These gases, especially carbon dioxide and sulfur gases, affect the atmosphere and oceans. In this way, magmas may alter climate—an unexpected relationship drawn from analysis of the Earth system.

Columnar basalts, Devil's Post Pile National Monument, eastern Sierra Nevada. Masses of this kind of extrusive igneous rock fracture along columnar joints when they cool. [Jerry L. Ferrara/Photo Researchers.]

HOW DO IGNEOUS ROCKS DIFFER FROM ONE ANOTHER?

Today we classify rock samples in the same way that some geologists did in the late nineteenth century:

- By texture
- By mineral and chemical composition

Texture

Two hundred years ago, the first division of igneous rocks was made on the basis of texture, an aspect of rocks that largely reflects differences in mineral *crystal size*. Geologists classified rocks as either coarsely or finely crystalline (see Chapter 3). Crystal size is a simple characteristic that geologists can easily see in the field. A coarse-grained rock such as granite has separate crystals that are easily visible to the naked eye. In contrast, the crystals of fine-grained rocks such as basalt are too small to be seen, even with the aid of a magnifying lens. **Figure 4.1** shows samples of granite and basalt, accompanied by photomicrographs of very thin, transparent slices of each rock. Photomicrographs, which are simply photographs taken through a microscope, give an enlarged view of minerals and their textures. Textural differences were clear to early geologists, but more work was needed to unravel the meaning of those differences.

First Clue: Volcanic Rocks Early geologists observed volcanic rocks forming from lava during volcanic eruptions. (*Lava* is the term that we apply to magma flowing out on the surface.) Geologists noted that when lava cooled rapidly, it formed either a finely crystalline rock or a glassy one in which no crystals could be distinguished. Where lava cooled more slowly, as in the middle of a thick flow many meters high, somewhat larger crystals were present.

Second Clue: Laboratory Studies of Crystallization
Just over one-hundred years ago, experimental scientists began to understand the nature of crystallization. Anyone who has frozen an ice cube knows that water solidifies to ice in a few hours as its temperature drops below the freezing point. If you have ever attempted to retrieve your ice cubes before they were completely solid, you may have seen thin ice crystals forming at the surface and along the sides of the tray. During crystallization, the water molecules take up fixed positions in the solidifying crystal structure, and they are no longer able to move freely, as they did when the water was liquid. All other liquids, including magmas, crystallize in this way.

The first tiny crystals form a pattern. Other atoms or ions in the crystallizing liquid then attach themselves in such a way that the tiny crystals grow larger. It takes some time for the atoms or ions to "find" their correct places on a growing crystal, and large crystals form only if they have time to grow slowly. If a liquid solidifies very quickly, as a magma does when it erupts onto the cool surface of the Earth, the crystals have no time to grow. Instead, a large number of tiny crystals form simultaneously as the liquid cools and solidifies.

Third Clue: Granite—Evidence of Slow Cooling By studying volcanoes, early geologists determined that finely crystalline textures indicate quick cooling at Earth's surface. Moreover, finely crystalline igneous rocks are evidence of former volcanism. But in the absence of direct observation, how could geologists deduce that coarse-grained rocks form by *slow* cooling deep in the interior? Granite—one of the commonest rocks of the continents—turned out to be the crucial clue (**Figure 4.2**). James Hutton, one of geology's founding fathers, saw granites cutting across and disrupting layers of sedimentary rocks as he worked in the field in Scotland. He noticed that the granite had somehow fractured and invaded the sedimentary rocks, as though the granite had been forced into the fractures as a liquid.

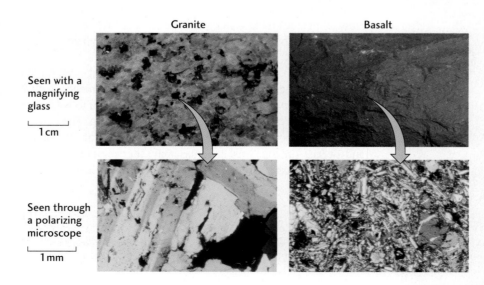

Granite   Basalt

Seen with a magnifying glass

1 cm

Seen through a polarizing microscope

1 mm

Figure 4.1 Igneous rocks were first classified by texture. Early geologists assessed texture with a small hand-held magnifying lens. Modern geologists have access to high-powered polarizing microscopes, which produce photomicrographs of thin, transparent rock slices like those shown here. [John Grotzinger/Ramón Rivera-Moret/ Harvard Mineralogical Museum. Photomicrographs by Raymond Siever.]

Granite intrusion

Metamorphosed
sedimentary rock

Figure 4.2 Granite intrusion (light colored) cutting across metamorphosed sedimentary rock. [Tom Bean/DRK PHOTO.]

As Hutton looked at more and more granites, he began to focus on the sedimentary rocks bordering them. He observed that the minerals of the sedimentary rocks in contact with the granite were different from those found in sedimentary rocks at some distance from the granite. He concluded that the changes in the sedimentary rocks must have resulted from great heat and that the heat must have come from the granite. Hutton also noted that granite was composed of interlocked crystals (see Figure 4.1). By this time, chemists had established that a slow crystallization process produces this pattern.

With these three lines of evidence, Hutton proposed that granite forms from a hot molten material that solidifies deep in the Earth. The evidence was conclusive, because no other explanation could accommodate all the facts. Other geologists, who saw the same characteristics of granites in widely separated places in the world, came to recognize that granite and many similar coarsely crystalline rocks were the products of magma that had crystallized slowly in the interior of the Earth.

Intrusive Igneous Rocks The full significance of an igneous rock's texture is now clear. As we have seen, texture is linked to the rate and therefore the place of cooling. Slow cooling of magma in Earth's interior allows adequate time for the growth of the interlocking large, coarse crystals that characterize intrusive igneous rocks (**Figure 4.3**). An **intrusive igneous rock** is one that has forced its way into surrounding rock. This surrounding rock is called **country rock.** (Later in this chapter, we will examine some special forms of intrusive igneous rocks.)

Extrusive Igneous Rocks Rapid cooling at Earth's surface produces the finely grained texture or glassy appearance of the **extrusive igneous rocks** (see Figure 4.3). These rocks, composed partly or largely of volcanic glass, form when lava or other material erupts from volcanoes. For this reason, they are also known as volcanic rocks. They fall into two major categories:

• *Lavas* Volcanic rocks formed from lavas range in appearance from smooth and ropy to sharp, spiky, and jagged, depending on the conditions under which the rocks formed.

• *Pyroclastic rocks* In more violent eruptions, **pyroclasts** form when broken pieces of lava are thrown high into the air. The finest pyroclasts are **volcanic ash,** extremely small fragments, usually of glass, that form when escaping gases force a fine spray of magma from a volcano. *Bombs* are larger particles, hurled from the volcano and streamlined by the air as they hurtle through it. All volcanic rocks lithified from these pyroclastic materials are called **tuff** (see Chapter 12 for more details).

One pyroclastic rock is **pumice,** a frothy mass of volcanic glass with a great number of *vesicles.* Vesicles are holes that remain after trapped gas has escaped from the solidifying melt. Another wholly glassy volcanic rock is **obsidian;** unlike pumice, it contains only tiny vesicles and so is solid and dense. Chipped or fragmented obsidian produces very sharp edges, and Native Americans and many other hunting groups used it for arrowheads and a variety of cutting tools.

Key Figure 4.3	Igneous rock types can be identified by texture.

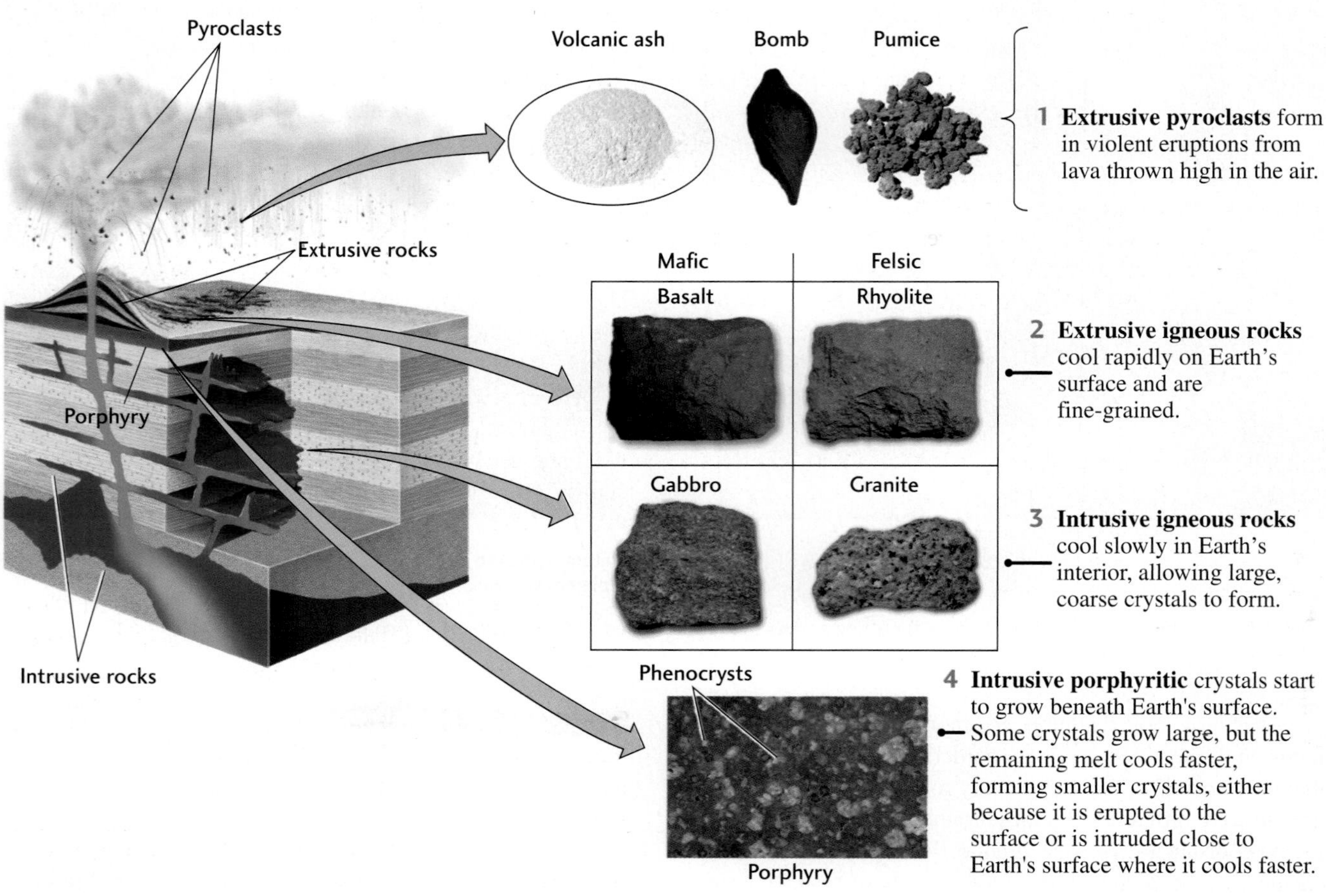

Pyroclasts

Volcanic ash   Bomb   Pumice

1 Extrusive pyroclasts form in violent eruptions from lava thrown high in the air.

Extrusive rocks

Porphyry

Intrusive rocks

Mafic	Felsic
Basalt	Rhyolite
Gabbro	Granite

2 Extrusive igneous rocks cool rapidly on Earth's surface and are fine-grained.

3 Intrusive igneous rocks cool slowly in Earth's interior, allowing large, coarse crystals to form.

Phenocrysts

Porphyry

4 Intrusive porphyritic crystals start to grow beneath Earth's surface. Some crystals grow large, but the remaining melt cools faster, forming smaller crystals, either because it is erupted to the surface or is intruded close to Earth's surface where it cools faster.

[John Grotzinger/Ramón Rivera-Moret/Harvard Mineralogical Museum.]

A **porphyry** is an igneous rock that has a mixed texture in which large crystals "float" in a predominantly fine crystalline matrix (see Figure 4.3). The large crystals, called *phenocrysts,* formed while the magma was still below Earth's surface. Then, before other crystals could grow, a volcanic eruption brought the magma to the surface, where it cooled quickly to a finely crystalline mass. In some cases, porphyries form as intrusive igneous rocks; for example, where magmas cool quickly at very shallow levels in the crust. Porphyry textures are very important to geologists because they indicate that different minerals grow at different rates, a point that will be emphasized later in this chapter. In Chapter 12, we will look more closely at how these volcanic rocks and others form during volcanism. Now, however, we turn to the second way in which the family of igneous rocks is subdivided.

Chemical and Mineral Composition

We have seen that igneous rocks can be subdivided according to their texture. They can also be classified on the basis of their chemical and mineral compositions. Volcanic glass, which is formless even under the microscope, is often classified by chemical analysis. One of the earliest classifications of igneous rocks was based on a simple chemical analysis of their silica (SiO_2) content. Silica, as noted in Chapter 3, is abundant in most igneous rocks and accounts for 40 to 70 percent of their total weight. We still refer to rocks rich in silica, such as granite, as silicic.

Modern classifications now group igneous rocks according to their relative proportions of silicate minerals (Table 4.1). These minerals are described in Appendix 4. The silicate minerals—quartz, feldspar (both orthoclase and plagioclase), muscovite and biotite micas, the amphibole and pyroxene groups, and olivine—form a systematic series. Felsic minerals are high in silica; mafic minerals are low in silica. The adjectives *felsic* (from *fel*dspar and *si*lica) and *mafic* (from *ma*gnesium and *f*erric, from the Latin *ferrum,* "iron") are applied to both the minerals and the rocks that have high contents of these minerals. Mafic minerals crystallize at higher temperatures—that is, earlier in the cooling of a magma—than felsic minerals.

As the mineral and chemical compositions of igneous rocks became known, geologists soon noticed that some extrusive and intrusive rocks were identical in composition and differed only in texture. Basalt, for example, is an extrusive rock formed from lava. Gabbro has exactly the same mineral composition as basalt but forms deep in Earth's crust (see Figure 4.3). Similarly, rhyolite and granite are identical in composition but differ in texture. Thus, extrusive and intrusive rocks form two chemically and mineralogically parallel sets of igneous rocks. Conversely, most chemical and mineral compositions can appear in either extrusive or intrusive rocks. The only exceptions are very highly mafic rocks that rarely appear as extrusive igneous rocks.

Figure 4.4 is a model that portrays these relationships. The horizontal axis plots silica content as a percentage of a given rock's weight. The percentages given—from high silica content at 70 percent to low silica content at 40 percent—cover the range found in igneous rocks. The vertical axis displays a scale measuring the mineral content of a given rock as a percentage of its volume. If you know the silica content of a rock sample, you can determine its mineral composition and, from that, the type of rock.

We can use Figure 4.4 to help with the discussion of the intrusive and extrusive igneous rocks. We begin with the felsic rocks, on the extreme left of the model.

Felsic Rocks Felsic rocks are poor in iron and magnesium and rich in minerals that are high in silica. Such minerals include quartz, orthoclase feldspar, and plagioclase feldspar. Orthoclase feldspar is more abundant than plagioclase feldspar. Plagioclase feldspars contain both calcium and sodium. As Figure 4.4 indicates, they are richer in sodium near the felsic end and richer in calcium near the mafic end. Thus, just as mafic minerals crystallize at higher temperatures than felsic minerals, calcium-rich plagioclases crystallize at higher temperatures than sodium-rich plagioclases.

Felsic minerals and rocks tend to be light in color. **Granite,** one of the most abundant intrusive igneous rocks, contains about 70 percent silica. Its composition includes abundant quartz and orthoclase feldspar and a smaller amount of plagioclase feldspar (see the left side of Figure 4.4). These light-colored felsic minerals give granite its pink or gray color. Granite also contains small amounts of muscovite and biotite micas and amphibole.

Rhyolite is the extrusive equivalent of granite. This light brown to gray rock has the same felsic composition and light coloration as granite, but it is much more finely grained. Many rhyolites are composed largely or entirely of volcanic glass.

Intermediate Igneous Rocks Midway between the felsic and mafic ends of the series are the **intermediate igneous rocks.** As their name indicates, these rocks are neither as rich in silica as the felsic rocks nor as poor in it as the mafic rocks. We find the intermediate intrusive rocks to the right of

Table 4.1	Common Minerals of Igneous Rocks		
Compositional Group	**Mineral**	**Chemical Composition**	**Silicate Structure**
FELSIC	Quartz	SiO_2	Frameworks
	Potassium feldspar	$KAlSi_3O_8$	
	Plagioclase feldspar	$NaAlSi_3O_8$; $CaAl_2Si_2O_8$	
	Muscovite (mica)	$KAl_3Si_3O_{10}(OH)_2$	Sheets
MAFIC	Biotite (mica)	K Mg Fe Al $\}$ $Si_3O_{10}(OH)_2$	
	Amphibole group	Mg Fe Ca Na $\}$ $Si_8O_{22}(OH)_2$	Double chains
	Pyroxene group	Mg Fe Ca Al $\}$ SiO_3	Single chains
	Olivine	$(Mg,Fe)_2SiO_4$	Isolated tetrahedra

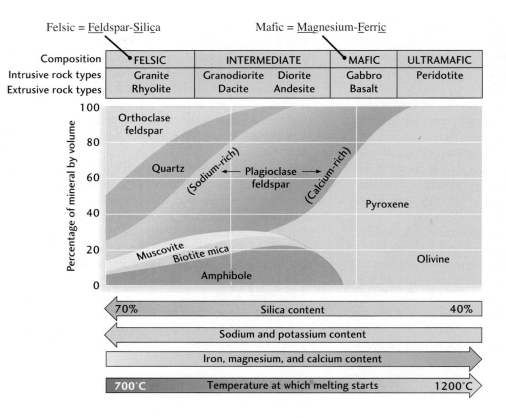

Felsic = <u>Fel</u>dspar-<u>Silic</u>a   Mafic = <u>Ma</u>gnesium-<u>Fer</u>ric

Composition	FELSIC	INTERMEDIATE		MAFIC	ULTRAMAFIC
Intrusive rock types	Granite	Granodiorite	Diorite	Gabbro	Peridotite
Extrusive rock types	Rhyolite	Dacite	Andesite	Basalt	

Figure 4.4 Classification model of igneous rocks. The vertical axis shows the mineral composition of a given rock as a percentage of its volume. The horizontal axis is a scale of silica content by weight. Thus, if you know by chemical analysis that a coarsely textured rock sample is about 70 percent silica, you could determine that its composition is about 6 percent amphibole, 3 percent biotite, 5 percent muscovite, 14 percent plagioclase feldspar, 22 percent quartz, and 50 percent orthoclase feldspar. Your rock would be granite. Although rhyolite has the same mineral composition, its fine texture would eliminate it.

granite in Figure 4.4. The first is **granodiorite,** a light-colored felsic rock that looks something like granite. It is also similar to granite in having abundant quartz, but its predominant feldspar is plagioclase, not orthoclase. To its right is **diorite,** which contains still less silica and is dominated by plagioclase feldspar, with little or no quartz. Diorites contain a moderate amount of the mafic minerals biotite, amphibole, and pyroxene. They tend to be darker than granite or granodiorite.

The volcanic equivalent of granodiorite is **dacite.** To its right in the extrusive series is **andesite,** the volcanic equivalent of diorite. Andesite derives its name from the Andes, the volcanic mountain chain in South America.

Mafic Rocks **Mafic rocks** are high in pyroxenes and olivines. These minerals are relatively poor in silica but rich in magnesium and iron, from which they get their characteristic dark colors. **Gabbro,** with even less silica than is found in the intermediate igneous rocks, is a coarsely grained, dark gray intrusive igneous rock. Gabbro has an abundance of mafic minerals, especially pyroxene. It contains no quartz and only moderate amounts of calcium-rich plagioclase feldspar.

Basalt, as we have seen, is dark gray to black and is the fine-grained extrusive equivalent of gabbro. Basalt is the most abundant igneous rock of the crust, and it underlies virtually the entire seafloor. In some places on the continents, extensive thick sheets of basalt make up large plateaus. The Columbia Plateau of Washington State and the remarkable formation known as the Giant's Causeway in Northern Ireland are two examples. The Deccan basalts of India and the Siberian basalts of northern Russia are enormous outpourings of basalt that appear to coincide

closely with two of the greatest periods of mass extinction in the fossil record. These great episodes of basalt formation, and the mechanisms responsible for them, are discussed further in Chapter 12.

Ultramafic Rocks **Ultramafic rocks** consist primarily of mafic minerals and contain less than 10 percent feldspar. Here, at the very low silica composition of only about 45 percent, we find **peridotite.** It is a coarsely grained, dark greenish gray rock made up primarily of olivine with smaller amounts of pyroxene. Peridotites are the dominant rocks in Earth's mantle and are the source of basaltic rocks formed at mid-ocean ridges. Ultramafic rocks are rarely found as extrusives. Because they form at such high temperatures, through the accumulation of crystals at the bottom of a magma chamber, they are rarely liquid and hence do not form typical lavas.

The names and exact compositions of the various rocks in the felsic-to-mafic series are less important than the trends shown in Table 4.2. There is a strong correlation between mineralogy and the temperatures of crystallization or melting. As Table 4.2 indicates, mafic minerals melt at higher temperatures than felsic minerals. At temperatures below the melting point, minerals crystallize. Therefore, mafic minerals also crystallize at higher temperatures than felsic minerals. We can also see from the table that silica content increases as we move from the mafic group to the felsic group. Increasing silica content results in increasingly complex silicate structures (see Table 4.1), which interfere with a melted rock's ability to flow. As a structure grows more complex, the ability to flow decreases. Thus **viscosity**—the measure of a

Table 4.2	Changes in Some Major Chemical Elements from Felsic to Mafic Rocks			
	Felsic	**Intermediate**		**Mafic**
Coarse-Grained (intrusive)	Granite	Granodiorite	Diorite	Gabbro
Fine-Grained (extrusive)	Rhyolite	Dacite	Andesite	Basalt

← Silica increasing

← Sodium increasing

← Potassium increasing

Calcium increasing →

Magnesium increasing →

Iron increasing →

← (Viscosity increasing)

(Melting temperature increasing) →

liquid's *resistance* to flow—increases as silica content increases. Viscosity is an important factor in the behavior of lavas, as we will see in Chapter 12.

It is clear that a rock's minerals are an indication of the conditions under which the rock's parent magma formed and crystallized. To interpret this information accurately, however, we must understand more about igneous processes. We turn to that topic next.

HOW DO MAGMAS FORM?

We know from the way Earth transmits earthquake waves that the bulk of the planet is solid for thousands of kilometers down to the core-mantle boundary (see Chapter 14). The evidence of volcanic eruptions, however, tells us that there must also be liquid regions where magmas originate. How do we resolve this apparent contradiction? The answer lies in the processes that melt rocks and create magmas.

How Do Rocks Melt?

Although we do not yet understand the exact mechanisms of melting and solidification, geologists have learned a great deal from laboratory experiments. From these experiments, we know that a rock's melting point depends on its composition and on conditions of temperature and pressure (Table 4.3).

Temperature and Melting One hundred years ago, geologists discovered that a rock does not melt completely at any

given temperature. Instead, rocks undergo **partial melting** because the minerals that compose them melt at different temperatures. As temperatures rise, some minerals melt and others remain solid. If the same conditions are maintained at any given temperature, the same mixture of solid rock and melt is maintained. The fraction of rock that has melted at a given temperature is called a *partial melt*. To visualize a partial melt, think of how a chocolate chip cookie would look if you heated it to the point at which the chocolate chips melted while the main part of the cookie stayed solid.

The ratio of liquid to solid in a partial melt depends on the composition and melting temperatures of the minerals that make up the original rock. It also depends on the temperature at the depth in the crust or mantle where melting takes place.

Table 4.3	Factors Affecting Melting Temperatures

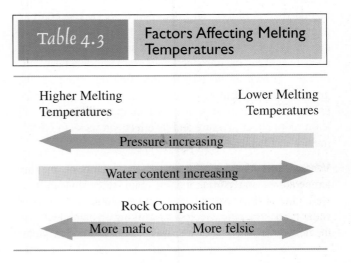

Higher Melting Temperatures   Lower Melting Temperatures

← Pressure increasing

Water content increasing →

Rock Composition

← More mafic   More felsic →

At the lower end of its melting range, a partial melt might be less than 1 percent of the volume of the original rock. Much of the hot rock would still be solid, but appreciable amounts of liquid would be present as small droplets in the tiny spaces between crystals throughout the mass. In the upper mantle, for example, some basaltic partial melts can be produced by only 1 to 2 percent melting of peridotite. However, 15 to 20 percent partial melting of mantle peridotite to form basaltic magmas is common beneath mid-ocean ridges. At the high end of the melting temperature range, much of the rock would be liquid, with lesser amounts of unmelted crystals in it. An example would be a reservoir of a basaltic magma and crystals just beneath a volcano such as the island of Hawaii.

Geologists used the new knowledge of partial melts to help them determine how different kinds of magma form at different temperatures and in different regions of Earth's interior. As you can imagine, the composition of a partial melt in which only the minerals with the lowest melting points have melted may be significantly different from the composition of a completely melted rock. Thus, basaltic magmas that form in different regions of the mantle may have somewhat different compositions. From this observation, geologists deduced that the different magmas come from different proportions of partial melt.

Pressure and Melting To get the whole story on melting, we must consider pressure, which increases with depth in the Earth as a result of the increased weight of overlying rock. Geologists found that as they melted rocks under various pressures, higher pressures led to higher melting temperatures. Thus, rocks that would melt at Earth's surface would remain solid at the same temperature in the interior. For example, a rock that melts at 1000°C at Earth's surface might have a much higher melting temperature, perhaps 1300°C, at depths in the interior. There, pressures are many thousands of times greater than the pressure at Earth's surface. It is the effect of pressure that explains why rocks in most of the crust and mantle do not melt. Rock can melt only when both its mineral composition and the temperature and pressure conditions are right. Just as an increase in pressure can keep a rock solid, a decrease in pressure can make a rock melt, given a suitably high temperature. Because of convection, Earth's mantle rises at mid-ocean ridges—at more or less constant temperature. As the mantle material rises and the pressure decreases below a critical point, solid rocks melt spontaneously, without the introduction of any additional heat. This process, known as **decompression melting,** produces the greatest volume of molten rock anywhere on Earth. It is the process by which most basalts form on the seafloor. You will learn more about pressure and its effects on rocks in Earth's interior in Chapter 14.

Water and Melting The many experiments on melting temperatures and partial melting paid other dividends as well. One of them was a better understanding of the role of water in rock melting. Geologists working on natural lavas in the field determined that water was present in some magmas.

This gave them the idea to add water to their experimental melts back in the laboratory. By adding small but differing amounts of water, they discovered that the compositions of partial and complete melts vary not only with temperature and pressure but also with the amount of water present.

Consider, for example, the effect of dissolved water on pure albite, the high-sodium plagioclase feldspar, at the low pressures of the Earth's surface. If only a small amount of water is present, pure albite will remain solid at temperatures just over 1000°C, hundreds of degrees above the boiling point of water. At these temperatures, the water in the albite is present as a vapor (gas). If large amounts of water are present, the melting temperature of the albite will decrease, dropping to as low as 800°C. This behavior follows the general rule that dissolving some of one substance (in this case, water vapor) in another (in this case, albite) lowers the melting point of the solution. If you live in a cold climate, you are probably familiar with this principle because you know that towns and municipalities sprinkle salt on icy roads to lower the melting point of the ice.

By the same principle, the melting temperature of the albite—and of all the feldspars and other silicate minerals—drops considerably in the presence of large amounts of water. In this case, the melting points of various silicates decrease in proportion to the amount of water dissolved in the molten silicate. This is an important point in our knowledge of how rocks melt. Water content is a significant factor in determining the melting temperatures of mixtures of sedimentary and other rocks. Sedimentary rocks contain an especially large volume of water in their pore spaces, more than is found in igneous or metamorphic rocks. As we will discuss later in this chapter, the water in sedimentary rocks plays an important role in melting in Earth's interior.

The Formation of Magma Chambers

Most substances are less dense in the liquid form than in the solid form. The density of a melted rock is lower than the density of a solid rock of the same composition. In other words, a given volume of melt would weigh less than the same volume of solid rock. Geologists reasoned that large bodies of magma could form in the following way. If the less dense melt were given a chance to move, it would move upward—just as oil, which is less dense than water, rises to the surface of a mixture of oil and water. Being liquid, the partial melt could move slowly upward through pores and along the boundaries between crystals of the overlying rocks. As the hot drops of melted rock moved upward, they would mix with other drops, gradually forming larger pools of molten rock within Earth's solid interior.

The ascent of magmas through the mantle and crust may be slow or rapid. Magmas rise at rates of 0.3 m/year to almost 50 m/year, over periods of tens of thousands or even hundreds of thousands of years. As they ascend, magmas may mix with other melts and may also affect the melting of lithospheric crust. We now know that the large pools of molten rock envisioned by early geologists form **magma chambers**—magma-filled cavities in the lithosphere that form as rising drops of melted rock push aside surrounding solid rock. A magma chamber may encompass a volume as large as several cubic kilometers. Geologists are still studying how magma chambers form, and we cannot yet say exactly what they look like in three dimensions. We think of them as large, liquid-filled cavities in solid rock, which expand as more of the surrounding rock melts or as liquid migrates through cracks and other small openings between crystals. Magma chambers contract as they expel magma to the surface in eruptions. We know for sure that magma chambers exist because earthquake waves can show us the depth, size, and general outlines of the chambers underlying some active volcanoes.

With this knowledge of how rocks melt to form magmas, we can now consider where in Earth's interior various kinds of magmas form.

WHERE DO MAGMAS FORM?

Our understanding of igneous processes stems from geological inferences as well as laboratory experimentation. Our inferences are based mainly on data from two sources. The first is volcanoes on land and under the sea—everywhere that molten rock erupts. Volcanoes give us information about where magmas are located. The second source of data is the record of temperatures measured in deep drill holes and mine shafts. This record shows that the temperature of Earth's interior increases with depth. Using these measurements, scientists have been able to estimate the rate at which temperature rises as depth increases.

The temperatures recorded at a given depth in some locations are much higher than the temperatures recorded at the same depth in other locations. These results indicate that some parts of Earth's mantle and crust are hotter than others. For example, the Great Basin of the western United States is an area where the North American continent is being stretched and thinned, with the result that temperature increases at an exceptionally rapid rate, reaching 1000°C at a depth of 40 km, not far below the base of the crust. This temperature is almost high enough to melt basalt. By contrast, in tectonically stable regions, such as the central parts of continents, temperature increases much more slowly, reaching only 500°C at the same depth.

We now know that various kinds of rock can solidify from magmas formed by partial melting. And we know that increasing temperatures in Earth's interior could create magmas. Let's turn now to the question of why there are so many different types of igneous rocks.

MAGMATIC DIFFERENTIATION

The processes we've discussed so far demonstrate how rocks melt to form magmas. But what accounts for the variety of igneous rocks? Are magmas of different chemical compositions made by the melting of different kinds of rocks? Or do some processes produce a variety of rocks from an originally uniform parent material?

Again, the answers to these questions came from laboratory experiments. Geologists mixed chemical elements in proportions that simulated the compositions of natural igneous rocks, then melted these mixtures in high-temperature furnaces. As the melts cooled and solidified, the experimenters carefully observed and recorded the temperatures at which crystals formed as well as the chemical compositions of those crystals. This research gave rise to the theory of **magmatic differentiation,** a process by which rocks of varying composition can arise from a uniform parent magma. Magmatic differentiation occurs because different minerals crystallize at different temperatures. During crystallization, the composition of the magma changes as it is depleted of the chemical elements that form the crystallized minerals.

In a kind of mirror image of partial melting, the first minerals to crystallize from a cooling magma are the ones that were the last to melt as determined in experiments of partial melting. This initial crystallization withdraws chemical elements from the melt, changing the magma's composition. Continued cooling crystallizes the minerals that had melted at the next lower temperature range during the partial melting experiments. Again, the magma's chemical composition changes as various elements are withdrawn. Finally, as the magma solidifies completely, the last minerals to crystallize are the ones that melted first. This is how the same parent magma, because of its changing chemical composition throughout the crystallization process, can give rise to different igneous rocks.

Fractional Crystallization: Laboratory and Field Observations

Fractional crystallization is the process by which the crystals formed in a cooling magma are segregated from the remaining liquid. This segregation happens in several ways (**Figure 4.5**). In the simplest scenario, crystals formed in a magma chamber settle to the chamber's floor and are thus removed from further reaction with the remaining liquid. The magma then migrates to new locations, forming new chambers. Crystals that had formed early segregate from the remaining magma, which continues to crystallize as it cools.

A good test case for the theory of fractional crystallization is provided by the Palisades, a line of imposing cliffs that faces the city of New York on the west bank of the Hudson River. This igneous formation is about 80 km long and, in places, more than 300 m high. It formed as a melt of basaltic composition intruded into almost horizontal sedimentary rocks. It contains abundant olivine near the bottom, pyroxene and plagioclase feldspar in the middle, and mostly plagioclase feldspar near the top. These variations in mineral composition from top to bottom made the Palisades a perfect site for testing the theory of fractional crystallization.

Key Figure 4.5 Fractional crystallization explains the composition of a basaltic intrusion.

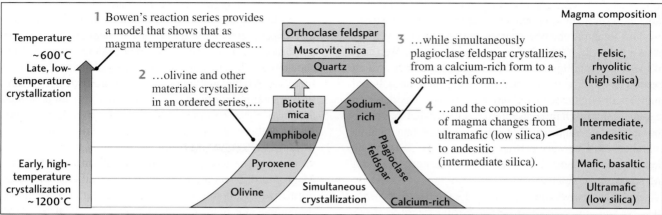

1 Bowen's reaction series provides a model that shows that as magma temperature decreases…

2 …olivine and other materials crystallize in an ordered series,…

3 …while simultaneously plagioclase feldspar crystallizes, from a calcium-rich form to a sodium-rich form…

4 …and the composition of magma changes from ultramafic (low silica) to andesitic (intermediate silica).

Temperature
~600°C
Late, low-temperature crystallization

Early, high-temperature crystallization
~1200°C

Orthoclase feldspar
Muscovite mica
Quartz
Biotite mica
Amphibole
Pyroxene
Olivine
Simultaneous crystallization

Sodium-rich
Plagioclase feldspar
Calcium-rich

Magma composition
Felsic, rhyolitic (high silica)
Intermediate, andesitic
Mafic, basaltic
Ultramafic (low silica)

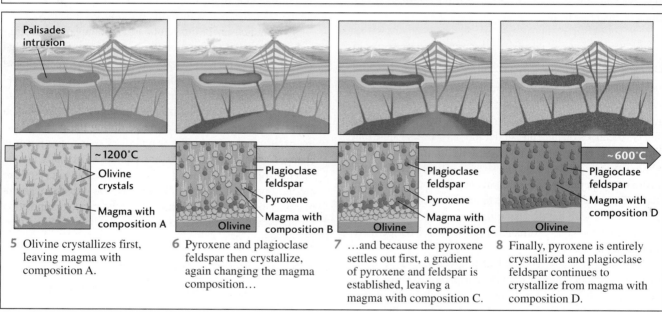

Palisades intrusion

~1200°C — Olivine crystals — Magma with composition A

Plagioclase feldspar — Pyroxene — Olivine — Magma with composition B

Plagioclase feldspar — Pyroxene — Olivine — Magma with composition C

Plagioclase feldspar — Magma with composition D — Olivine

~600°C

5 Olivine crystallizes first, leaving magma with composition A.

6 Pyroxene and plagioclase feldspar then crystallize, again changing the magma composition…

7 …and because the pyroxene settles out first, a gradient of pyroxene and feldspar is established, leaving a magma with composition C.

8 Finally, pyroxene is entirely crystallized and plagioclase feldspar continues to crystallize from magma with composition D.

9 Laboratory experiments have established an order for crystallization of minerals in magma: e.g., olivine, then pyroxene, then plagioclase feldspar.

10 As magma cools, minerals crystallize at different temperatures and settle out of the magma, leaving the remaining magma with a different composition.

11 Minerals in the Palisades are ordered with olivine at the bottom, a gradient of pyroxene and plagioclase feldspar in the center, and plagioclase feldspar at the top.

12 These findings explain the composition of rocks in the Palisades of New Jersey, a basaltic intrusion.

13 Layers of finely grained basalt which cooled quickly at the edges of the intrusion surround the slowly cooled interior of the intrusion.

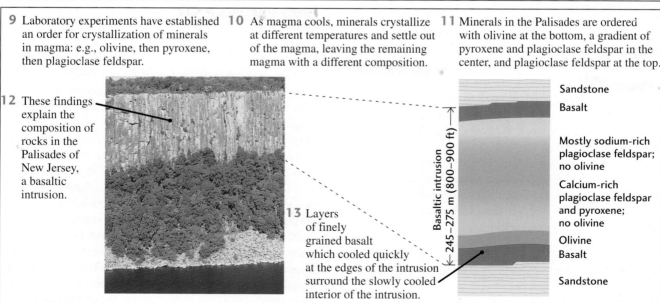

Basaltic intrusion 245–275 m (800–900 ft)

Sandstone
Basalt
Mostly sodium-rich plagioclase feldspar; no olivine
Calcium-rich plagioclase feldspar and pyroxene; no olivine
Olivine
Basalt
Sandstone

[Breck P. Kent.]

Such testing showed how laboratory experiments could help explain field observations.

Geologists melted rocks with about the same mineral compositions as found in the Palisades intrusion and determined that the temperature of the melt had to have been about 1200°C. The parts of the magma within a few meters of the relatively cold upper and lower contacts with the surrounding sedimentary rocks cooled quickly. The quick cooling formed a fine-grained basalt and preserved the chemical composition of the original melt. The hot interior cooled more slowly, as evidenced by the slightly larger crystals found in the intrusion's interior.

The ideas of fractional crystallization lead us to think that the first mineral to crystallize from the slowly cooling interior would have been olivine. This heavy mineral would sink through the melt to the bottom of the intrusion. It can be found today in the Palisades intrusion as a coarse-grained, olivine-rich layer just above the chilled, fine-grained basaltic layer along the bottom contact. Continued cooling would have produced pyroxene crystals, followed almost immediately by calcium-rich plagioclase feldspar. These minerals, too, would have settled out through the magma and accumulated in the lower third of the Palisades intrusion. The abundance of plagioclase feldspar in the upper parts of the intrusion is evidence that the melt continued to change composition until successive layers of settled crystals were topped off by a layer of mostly sodium-rich plagioclase feldspar crystals.

Being able to explain the layering of the Palisades intrusion as the result of fractional crystallization was an early success of the first version of the theory of magmatic differentiation. It firmly tied field observations to laboratory experiments and was solidly based on chemical knowledge. More than two-thirds of a century of geologic research has passed since the Palisades was first seen as a test case, and we now know that this intrusion actually has a more complex history. This history includes several injections of magma and a more complicated process of olivine settling. Nevertheless, the Palisades intrusion remains a valid example of fractional crystallization.

Granite and Basalt: Magmatic Differentiation

Studies of the lavas of volcanoes showed that basaltic magmas are common—far more common than the rhyolitic magmas that correspond in composition to granites. How could the granites that are so abundant in the crust have been derived from basaltic magmas?

The original idea of magmatic differentiation was that a basaltic magma would gradually cool and differentiate into a cooler, more silicic melt by fractional crystallization. The early stages of this differentiation would produce andesitic magma, which might erupt to form andesitic lavas or solidify by slow crystallization to form diorite intrusives. Intermediate stages would make magmas of granodiorite composition. If this process were carried far enough, its late stages would form rhyolitic lavas and granite intrusions (see Figure 4.5).

Field and laboratory work in the latter part of the twentieth century revealed that magmatic differentiation is a more complex process than had been originally described. One line of research showed that so much time would be needed for small crystals of olivine to settle through a dense, viscous magma that they might never reach the bottom of a magma chamber. Other researchers demonstrated that many layered intrusions—similar to but much larger than the Palisades—do not show the simple progression of layers predicted by the original magmatic theory.

The first sticking point, however, was the source of granite. The great volume of granite found on Earth could not have formed from basaltic magmas by magmatic differentiation, because large quantities of liquid volume are lost by crystallization during successive stages of differentiation. To produce the existing amount of granite, an initial volume of basaltic magma 10 times the size of a granitic intrusion would be required. That abundance would imply the crystallization of huge quantities of basalt underlying granite intrusions. But geologists could not find anything like that amount of basalt. Even where great volumes of basalt are found—at mid-ocean ridges—there is no wholesale conversion into granite through magmatic differentiation.

Most in question is the original idea that all granitic rocks evolve from the differentiation of a single type of magma, a basaltic melt. Instead, geologists discovered that the melting of varied source rocks of the upper mantle and crust is responsible for much variation in magma composition:

1. Rocks in the upper mantle might partially melt to produce basaltic magma.

2. A mixture of sedimentary rocks and basaltic oceanic rocks such as those found in subduction zones might melt to form andesitic magma.

3. A melt of sedimentary, igneous, and metamorphic continental crustal rocks might produce granitic magma.

Magmatic differentiation does operate, but its mechanisms are much more complex than first recognized:

• Magmatic differentiation can be achieved by the partial melting of mantle and crustal rocks over a range of temperatures and water contents.

• Magmas do not cool uniformly; they may exist transiently at a range of temperatures within a magma chamber. Such differences in temperature in and among magma chambers may cause the chemical composition of the magma to vary from one region to another.

• A few magmas are immiscible—they do not mix with one another, just as oil and water do not mix. When such magmas coexist in one magma chamber, each forms its own crystallization products. Magmas that are miscible—that *do* mix—may give rise to a crystallization path different from that followed by any one magma alone.

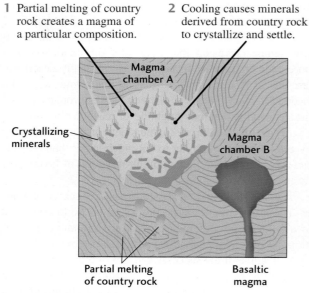

1 Partial melting of country rock creates a magma of a particular composition.

2 Cooling causes minerals derived from country rock to crystallize and settle.

Magma chamber A

Crystallizing minerals

Magma chamber B

Partial melting of country rock

Basaltic magma

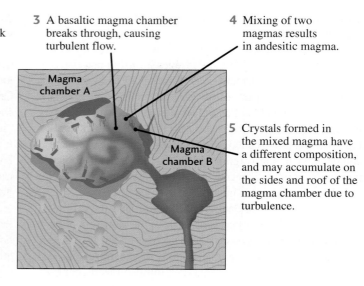

3 A basaltic magma chamber breaks through, causing turbulent flow.

4 Mixing of two magmas results in andesitic magma.

Magma chamber A

Magma chamber B

5 Crystals formed in the mixed magma have a different composition, and may accumulate on the sides and roof of the magma chamber due to turbulence.

Figure 4.6 Modern ideas of magmatic differentiation, thought to be a more complex process than first recognized. Melting is usually partial. Some magmas derived from rocks of varying compositions may mix, whereas other magmas are immiscible. Crystals may be transported to various parts of the magma chamber by turbulent currents in the liquid.

We now know more about the physical processes that interact with crystallization within magma chambers (**Figure 4.6**). Magma at various temperatures in different parts of a magma chamber may flow turbulently, crystallizing as it circulates. Crystals may settle, then be caught up in currents again, and eventually be deposited on the chamber's walls. The margins of such a magma chamber may be a "mushy" zone of crystals and melt lying between the solid rock border of the chamber and the completely liquid magma within the main part of the chamber. And, at some mid-ocean ridges, such as the East Pacific Rise, a mushroom-shaped magma chamber may be surrounded by hot basaltic rock with only small amounts (1 to 3 percent) of partial melt.

FORMS OF MAGMATIC INTRUSIONS

As noted earlier, geologists cannot directly observe the shapes of intrusive igneous rocks formed when magmas intrude the crust. We can deduce their shapes and distributions only by observing them today where intrusive rocks have been uplifted and exposed by erosion, millions of years after the magma intruded and cooled.

We do have indirect evidence of current magmatic activity. Earthquake waves, for example, show us the general outlines of magma chambers that underlie some active volcanoes. But they cannot reveal the detailed shapes or sizes of intrusions arising from those magma chambers. In some nonvolcanic but tectonically active regions, such as an area near the Salton Sea in southern California, measurements of temperatures in deep drill holes reveal a crust much hotter than normal, which may be evidence of an intrusion at depth.

But in the end, most of what we know about intrusive igneous rock is based on the work of field geologists who have examined and compared a wide variety of outcrops and have reconstructed their history. In the following pages, we consider some of these bodies: plutons, sills and dikes, and veins. **Figure 4.7** illustrates a variety of extrusive and intrusive structures.

Plutons

Plutons are large igneous bodies formed at depth in Earth's crust. They range in size from a cubic kilometer to hundreds of cubic kilometers. We can study these large bodies when uplift and erosion uncover them or when mines or drill holes cut into them. Plutons are highly variable, not only in size but also in shape and in their relationship to the surrounding country rock.

This wide variability is due in part to the different ways in which magma makes space for itself as it rises through the crust. Most magmas intrude at depths greater than 8 to 10 km. At these depths, few holes or openings exist, because the great pressure of the overlying rock would close them. But the upwelling magma overcomes even that high pressure.

Magma rising through the crust makes space for itself in three ways (**Figure 4.8**) that may be referred to collectively as *magmatic stoping:*

1. *Wedging open the overlying rock.* As the magma lifts that great weight, it fractures the rock, penetrates the cracks, wedges them open, and so flows into the rock. Overlying rocks may bow up during this process.

2. *Breaking off large blocks of rock.* Magma can push its way upward by breaking off blocks of the invaded crust.

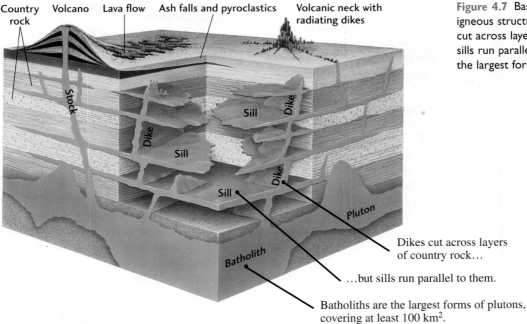

Figure 4.7 Basic extrusive and intrusive igneous structures. Notice that dikes cut across layers of country rock, but sills run parallel to them. Batholiths are the largest forms of plutons.

Dikes cut across layers of country rock…

…but sills run parallel to them.

Batholiths are the largest forms of plutons, covering at least 100 km².

These blocks, known as *xenoliths,* sink into the magma, melt, and blend into the liquid, in some places changing the composition of the magma.

3. *Melting surrounding rock.* Magma also makes its way by melting walls of country rock.

Most plutons show sharp contacts with country rock and other evidence of the intrusion of a liquid magma into solid rock. Other plutons grade into country rock and have structures vaguely resembling those of sedimentary rocks. The features of these plutons suggest that they formed by partial or complete melting of preexisting sedimentary rocks.

Batholiths, the largest plutons, are great irregular masses of coarse-grained igneous rock that by definition cover at least 100 km² (see Figure 4.7). Batholiths are found in the cores of tectonically deformed mountain belts. Accumulating field evidence shows that batholiths are thick, horizontal, sheetlike or lobe-shaped bodies extending from a funnel-shaped central region. Their bottoms may extend 10 to 15 km deep, and a few are estimated to go even deeper. The coarse grains of batholiths result from slow cooling at great depths. The rest of the plutons, similar but smaller, are called **stocks.** Both batholiths and stocks are **discordant intrusions;** that is, they cut across the layers of the country rock that they intrude.

Rising magma wedges open and fractures overlying country rock.

The overlying rock may bow up.

The magma melts surrounding rock,…

…which mixes and changes the magma's composition.

The magma also breaks off blocks of overlying rock— xenoliths— that sink into the magma.

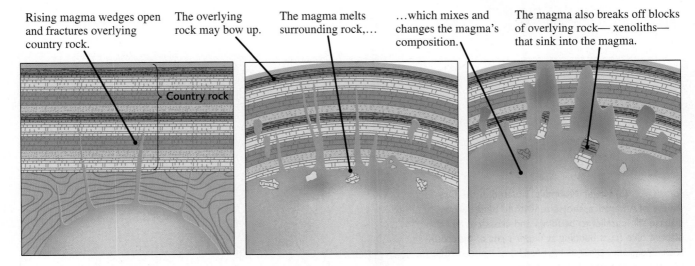

Country rock

Figure 4.8 Magmas make their way into country rock in three basic ways: by invading cracks and wedging open overlying rock, by breaking off rock, and by melting surrounding rock. Pieces of broken-off country rock, called xenoliths, can become completely dissolved in the magma. If many xenoliths are dissolved and the country rock differs in composition from the magma, the composition of the magma will change.

Sills and Dikes

Sills and dikes are similar to plutons in many ways, but they are smaller and have a different relationship to the layering of the surrounding intruded rock. A **sill** is a sheetlike body formed by the injection of magma between parallel layers of preexisting bedded rock (**Figure 4.9**). Sills are **concordant intrusions;** that is, their boundaries lie parallel to these layers, whether or not the layers are horizontal. Sills range in thickness from a single centimeter to hundreds of meters, and they can extend over considerable areas. Figure 4.9a shows a large sill at Finger Mountain, Antarctica. The 300-m-thick Palisades intrusion (see Figure 4.5) is another large sill.

Sills may superficially resemble layers of lava flows and pyroclastic material, but they differ from these layers in four ways:

1. They lack the ropy, blocky, and vesicle-filled structures that characterize many volcanic rocks (see Chapter 12).

2. They are more coarsely grained than volcanics because the sills have cooled more slowly.

3. Rocks above and below sills show the effects of heating: their color may have been changed or they may have been mineralogically altered by contact metamorphism.

4. Many lava flows overlie weathered older flows or soils formed between successive flows; sills do not.

Dikes are the major route of magma transport in the crust. They are like sills in being sheetlike igneous bodies, but dikes cut across layers of bedding in country rock and so are discordant (see Figure 4.9b). Dikes sometimes form by forcing open existing fractures, but they more often create channels through new cracks opened by the pressure of magmatic injection. Some individual dikes can be followed in the field for tens of kilometers. Their widths vary from many meters to a few centimeters. In some dikes, xenoliths provide good evidence of disruption of the surrounding rock during

(a)

1 A sill runs parallel to country rock layers.

Sill

(b)

2 A dike cuts across layers.

Dike

Figure 4.9 (a) Finger Mountain, situated in the Dry Valleys of Antarctica, shows sandstone beds split by diabase sill (parallel to bedding) and cross-cut by diabase dike (cuts bedding at upper right of outcrop). [Colin Monteath/AUSCAPE.] (b) A dike of igneous rock (dark) intruded into shaley sedimentary rock (reddish brown) in Grand Canyon National Park, Arizona. [Tom Bean/DRK PHOTO.]

the intrusion process. Dikes rarely exist alone; more typically, swarms of hundreds or thousands of dikes are found in a region that has been deformed by a large igneous intrusion.

The texture of dikes and sills varies. Many are coarsely crystalline, with an appearance typical of intrusive rocks. Many others are finely grained and look much more like volcanic rocks. Because texture corresponds to the rate of cooling, we know that the fine-grained dikes and sills invaded country rock nearer the Earth's surface, where rocks are cold in comparison with intrusions. Their fine texture is the result of fast cooling. The coarse-grained ones formed at depths of many kilometers and invaded warmer rocks whose temperatures were much closer to their own.

Veins

Veins are deposits of minerals found within a rock fracture that are foreign to the host rock. Irregular pencil- or sheet-shaped veins branch off the tops and sides of many intrusive bodies. They may be a few millimeters to several meters across, and they tend to be tens of meters to kilometers long or wide. The well-known Mother Lode of the Gold Rush of 1849 in California is a vein of quartz-bearing crystals of gold. Veins of extremely coarse-grained granite cutting across a much finer grained country rock are called **pegmatites** (**Figure 4.10**). They crystallized from a water-rich magma in the late stages of solidification. Pegmatites provide ores of many rare elements, such as lithium and beryllium.

Some veins are filled with minerals that contain large amounts of chemically bound water and are known to crystallize from hot-water solutions. From laboratory experiments, we know that these minerals crystallize at high temperatures—typically 250° to 350°C—but not nearly as high as the temperatures of magmas. The solubility and composition of the minerals in these **hydrothermal veins** indicate that abundant water was present as the veins formed. Some of the water may have come from the magma itself, but some may be from underground water in the cracks and pore spaces of the intruded rocks. Groundwaters originate as rainwater seeps into the soil and surface rocks. Hydrothermal veins are abundant along mid-ocean ridges. In these areas, seawater infiltrates cracks in basalt and circulates down into hotter regions of the basalt ridge, emerging at hot vents on the seafloor in the rift valley between the spreading plates. Hydrothermal processes at mid-ocean ridges are examined in more detail in Chapter 12.

IGNEOUS ACTIVITY AND PLATE TECTONICS

Since the theory of plate tectonics arose in the 1960s, geologists have been trying to fit the facts and theories of igneous rock formation into its framework. We noted that batholiths, for example, are found in the cores of many mountain ranges. These ranges were formed by the convergence of two plates. This observation implies a connection between plutonism and the mountain-building process and between both of them and plate tectonics—the forces responsible for plate movements.

Laboratory experiments have established the temperatures and pressures at which different kinds of rock melt. This information gives us some idea of where melting takes place. For example, mixtures of sedimentary rocks, because of their composition and water content, melt at temperatures several hundred degrees below the melting point of basalt. This leads us to expect that basalt starts to melt near the base of the crust in tectonically active regions of the upper mantle and that sedimentary rocks melt at shallower depths. The geometry of plate motions is the link we need to tie tectonic activity and rock composition to melting (**Figure 4.11**). Magma forms at two types of plate boundaries: mid-ocean ridges, where two plates diverge and the seafloor spreads, and subduction zones, where one plate dives beneath another. Mantle plumes, though not associated with plate boundaries, are also the result of partial melting and form near the core-mantle boundary deep within Earth's interior (see Chapter 12).

Most igneous rocks form at the globally extensive mid-ocean ridge network. Throughout its great length, decompression melting of the mantle creates magmas that well up along rising convection currents. Magma is extruded as

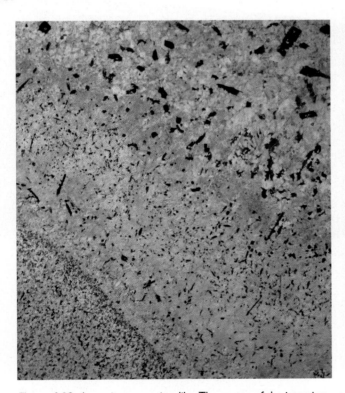

Figure 4.10 A granite pegmatite dike. The center of the intrusion (upper right) cools more slowly and developed coarser crystals. The margin of the intrusion (lower left) has finer crystals due to more rapid cooling. [John Grotzinger/Ramón Rivera-Moret/Harvard Mineralogical Museum.]

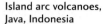

Earth System Figure 4.11 Magmatic activity is related to plate tectonic settings.

Island arc volcanoes,
Java, Indonesia

Plate divergence boundary,
Mid-Atlantic Ridge, Iceland

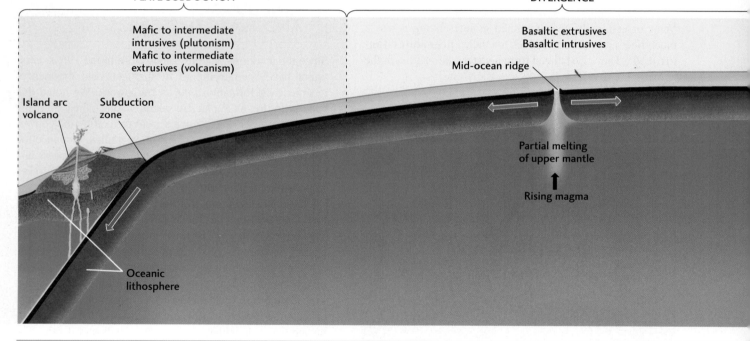

lavas, fed by magma chambers below the ridge axis. At the same time, intrusions of gabbro (the intrusive equivalent of basalt) are emplaced at depth.

Spreading Centers as Magmatic Geosystems

Before the advent of plate tectonics, geologists were puzzled by unusual assemblages of rocks that were characteristic of the seafloor but were found on land. Known as

ophiolite suites, these assemblages consist of deep-sea sediments, submarine basaltic lavas, and mafic igneous intrusions (**Figure 4.12**). Using data gathered from deep-diving submarines, dredging, deep-sea drilling, and seismic exploration, geologists now explain these rocks as fragments of oceanic crust that were transported by seafloor spreading and then raised above sea level and thrust onto a continent in a later episode of plate collision. On some of the more complete ophiolite sequences preserved on land, we can literally walk across rocks that used to lie along the boundary

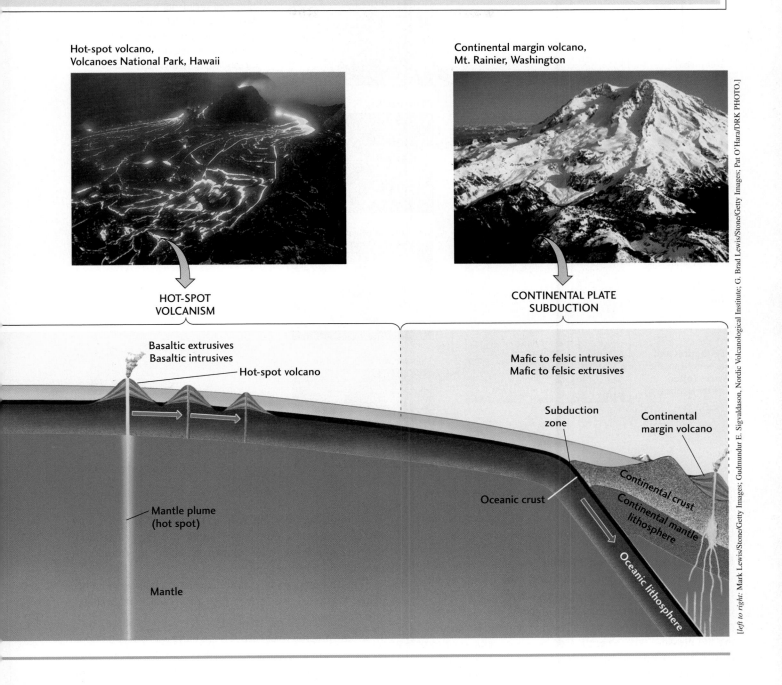

Hot-spot volcano,
Volcanoes National Park, Hawaii

Continental margin volcano,
Mt. Rainier, Washington

HOT-SPOT
VOLCANISM

CONTINENTAL PLATE
SUBDUCTION

Basaltic extrusives
Basaltic intrusives

Hot-spot volcano

Mafic to felsic intrusives
Mafic to felsic extrusives

Subduction
zone

Continental
margin volcano

Oceanic crust

Continental crust

Continental mantle
lithosphere

Mantle plume
(hot spot)

Oceanic lithosphere

Mantle

[left to right: Mark Lewis/Stone/Getty Images; Gudmundur E. Sigvaldason, Nordic Volcanological Institute; G. Brad Lewis/Stone/Getty Images; Pat O'Hara/DRK PHOTO.]

between Earth's oceanic crust and mantle. Ocean drilling has penetrated to the gabbro layer of the seafloor but not to the crust-mantle boundary below. Sound-wave profiles have found several small magma chambers similar to the one shown in **Figure 4.13**.

How does seafloor spreading work as a magmatic geosystem? We can think of this system as a huge machine that processes mantle material to produce oceanic crust. Figure 4.13 is a highly schematic and simplified representation of what may be happening, based in part on studies of ophio-

lites found on land and information gleaned from ocean drilling and sound-wave profiling.

Input Material: Peridotite in the Mantle The raw material fed into the machinery of seafloor spreading comes from the asthenosphere of the convecting mantle. The dominant rock type in the asthenosphere is peridotite. The mineral composition of the average peridotite in the mantle is chiefly olivine, with smaller amounts of pyroxene and garnet. Temperatures in the asthenosphere are hot enough to melt a

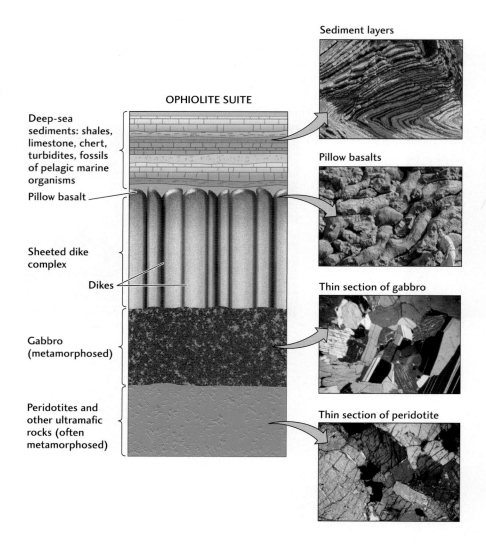

Sediment layers

Pillow basalts

Thin section of gabbro

Thin section of peridotite

OPHIOLITE SUITE

Deep-sea sediments: shales, limestone, chert, turbidites, fossils of pelagic marine organisms

Pillow basalt

Sheeted dike complex

Dikes

Gabbro (metamorphosed)

Peridotites and other ultramafic rocks (often metamorphosed)

Figure 4.12 Idealized section of an ophiolite suite. The combination of deep-sea sediments, submarine pillow lavas, sheeted basaltic dikes, and mafic igneous intrusions indicates a deep-sea origin. Ophiolites are fragments of ocean lithosphere emplaced on a continent as a result of plate collisions. Peridotite—a dominant rock in the mantle—undergoes decompression melting to form gabbro, which is then erupted to form volcanic pillow basalt (see Figure 4.13). [Photos courtesy of John Grotzinger. Thin sections courtesy of T. L. Grove.]

small fraction of this peridotite (less than 1 percent), but not hot enough to generate substantial volumes of magma.

Process: Decompression Melting Decompression melting generates magma from peridotite in the mantle at seafloor spreading centers. Recall from earlier in this chapter that the melting temperature of a mineral depends on the pressure at which it melts as well as its composition: decreasing the pressure will generally decrease a mineral's melting temperature. Consequently, if a mineral is near its melting point and the pressure on it is lowered while its temperature is kept constant, the mineral will melt.

As the plates pull apart, the partially molten peridotites are sucked inward and upward toward the spreading centers. Because the peridotites rise too fast to cool, the decrease in pressure melts a large fraction of the rock (up to 15 percent). The buoyancy of the melt causes it to rise faster than the denser surrounding rocks, separating the liquid from the remaining crystal mush to produce large volumes of magma.

Output Material: Oceanic Crust Plus Mantle Lithosphere The peridotites subjected to this process do not melt evenly; the garnet and pyroxene minerals melt more

than the olivine. For this reason, the magma generated by decompression melting is not peridotitic in composition; rather, it is enriched in silica and iron and has the same composition as basalt (see Figure 4.13).

This basaltic melt accumulates in a magma chamber below the mid-ocean ridge crest, from which it separates into three layers:

1. Some magma rises through the narrow cracks that open where the plates separate and erupts into the ocean, forming the basaltic pillow lavas that cover the seafloor (see Figure 4.13).

2. Some magma freezes in the cracks as vertical, sheeted dikes of gabbro.

3. The remaining magma freezes as massive gabbros as the underlying magma chamber is pulled apart by seafloor spreading.

These igneous units—pillow lavas, sheeted dikes, and massive gabbros—are the basic layers of the crust that geologists have found throughout the world's oceans.

Seafloor spreading results in another layer beneath this oceanic crust: the residual peridotite from which the basaltic

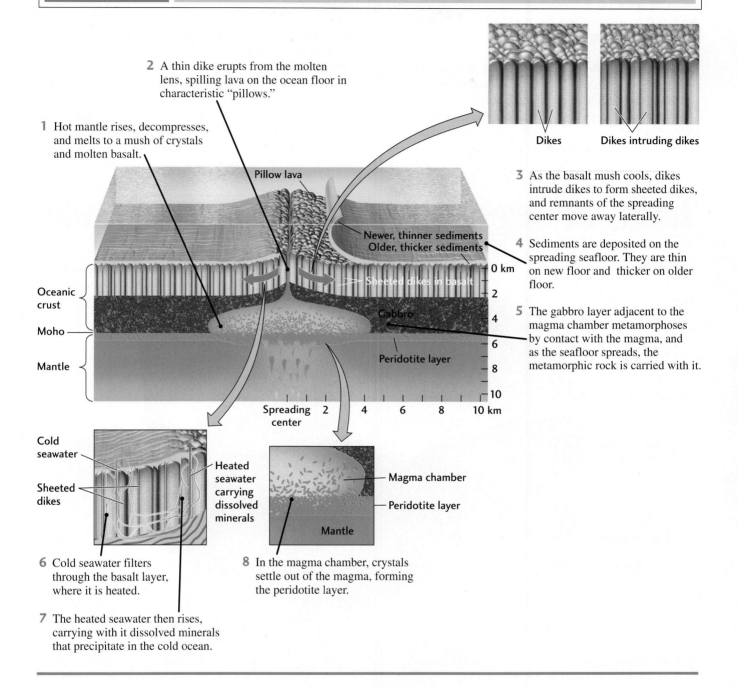

Key Figure 4.13 Decompression melting creates a magmatic geosystem at seafloor spreading centers.

2 A thin dike erupts from the molten lens, spilling lava on the ocean floor in characteristic "pillows."

1 Hot mantle rises, decompresses, and melts to a mush of crystals and molten basalt.

Dikes   Dikes intruding dikes

Pillow lava

3 As the basalt mush cools, dikes intrude dikes to form sheeted dikes, and remnants of the spreading center move away laterally.

Newer, thinner sediments
Older, thicker sediments

Sheeted dikes in basalt

4 Sediments are deposited on the spreading seafloor. They are thin on new floor and thicker on older floor.

Oceanic crust

Moho

Mantle

Gabbro

Peridotite layer

5 The gabbro layer adjacent to the magma chamber metamorphoses by contact with the magma, and as the seafloor spreads, the metamorphic rock is carried with it.

Spreading center   2   4   6   8   10 km

Cold seawater

Sheeted dikes

Heated seawater carrying dissolved minerals

Magma chamber

Peridotite layer

Mantle

6 Cold seawater filters through the basalt layer, where it is heated.

8 In the magma chamber, crystals settle out of the magma, forming the peridotite layer.

7 The heated seawater then rises, carrying with it dissolved minerals that precipitate in the cold ocean.

magma was originally derived. Geologists consider this layer to be part of the mantle, but its composition is different from that of the convecting asthenosphere. In particular, the extraction of basaltic melt makes the residual peridotite richer in olivine and stronger than ordinary mantle material. Geologists now believe it is this olivine-rich layer at the top of the mantle that gives the oceanic plates their great rigidity.

A thin blanket of deep-sea sediments begins to cover the newly formed ocean crust. As the seafloor spreads, the layers of sediments, lavas, dikes, and gabbros are transported away from the mid-ocean ridge where this characteristic sequence of rocks that make up the oceanic crust is assembled—almost like a production line.

Subduction Zones as Magmatic Geosystems

Other kinds of magmas underlie regions in which volcanoes are highly concentrated, such as the Andes Mountains of South America and the Aleutian Islands of Alaska. Both of these regions resulted from the sinking of one plate under another. Subduction zones are also major magmatic geosystems (**Figure 4.14**) that generate magmas of varying composition depending on how much and what kinds of materials from the seafloor are subducted.

Input Material: A Mixed Bag Input materials include mixtures of seafloor sediments, mixtures of oceanic basaltic crust and felsic continental crust, mantle peridotite, and water (see Figure 4.14). These chemical variations are clues that the volcanic geosystems at convergent plate boundaries operate differently from those at divergent boundaries. When an ocean plate collides with and overrides another ocean plate, several complex processes are set in motion.

Process: Fluid-Induced Melting The basic mechanism is **fluid-induced melting,** rather than decompression melting. The fluid is primarily water. We learned earlier in this chapter that water lowers the melting temperature of rock. Before the oceanic lithosphere is subducted at a convergent boundary, a lot of water has been incorporated into its outer layers. We have already discussed one process responsible for this—hydrothermal activity during the formation of the lithosphere.

Key Figure 4.14 Fluid-induced melting creates a magmatic geosystem in subduction zones.

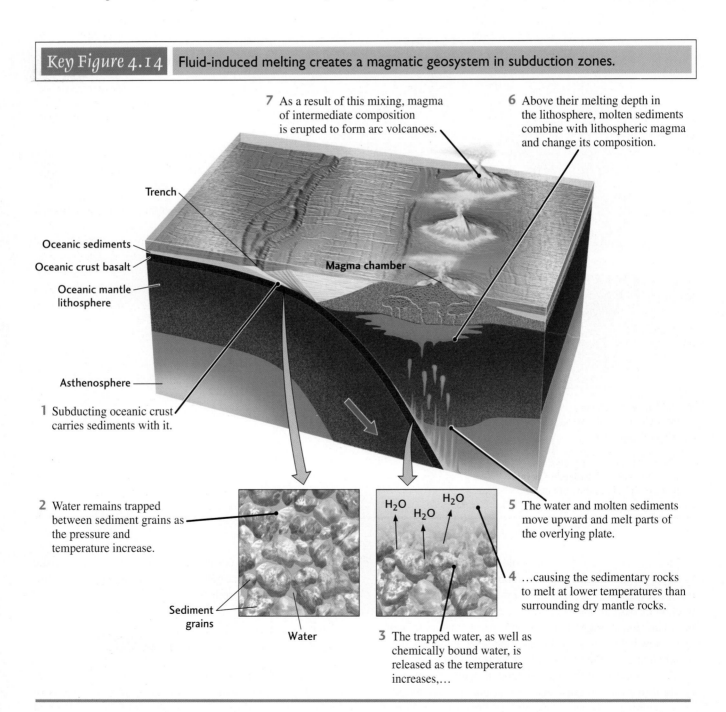

7 As a result of this mixing, magma of intermediate composition is erupted to form arc volcanoes.

6 Above their melting depth in the lithosphere, molten sediments combine with lithospheric magma and change its composition.

Trench

Oceanic sediments

Oceanic crust basalt

Oceanic mantle lithosphere

Magma chamber

Asthenosphere

1 Subducting oceanic crust carries sediments with it.

2 Water remains trapped between sediment grains as the pressure and temperature increase.

Sediment grains

Water

H_2O H_2O

H_2O

5 The water and molten sediments move upward and melt parts of the overlying plate.

4 ...causing the sedimentary rocks to melt at lower temperatures than surrounding dry mantle rocks.

3 The trapped water, as well as chemically bound water, is released as the temperature increases,...

Some of the seawater circulating through the crust near a spreading center reacts with basalt to form new minerals with water bound into their structures. In addition, as the lithosphere ages and moves across the ocean basin, sediments containing water are deposited on its surface. These sediments include shales, which are very high in clay minerals containing much water chemically bound in their crystal structures. Some of the sediments get scraped off at the deep-ocean trench where the plate subducts, but much of this water-laden material is carried downward into the subduction zone.

As the pressure increases, the water is squeezed out of the minerals in the outer layers of the descending slab and rises buoyantly into the mantle wedge above the slab. At moderate depths of about 5 km, the temperature increases to about 150°C and some of this water is released by metamorphic chemical reactions as basalt is converted to *amphibolite,* which is composed of amphibole and plagioclase feldspar (see Chapter 6). As other chemical reactions take place, additional water is released at depths ranging from 10 to 20 km. Finally, at depths greater than 100 km, the temperature increases to 1200° to 1500°C and the subducted slab undergoes an additional metamorphic transition induced by the increased pressure. Amphibolite is converted to *eclogite,* which is composed of pyroxene and garnet (see Chapter 6). The increase in both pressure and temperature in the subducting slab releases all the remaining water in addition to other materials.

During subduction, the released water induces melting of the descending basalt-rich oceanic crust and overlying peridotite-rich mantle wedge. Most of the mafic magma accumulates at the base of the crust of the overriding plate, and some of it intrudes into the crust to form magma chambers within volcanic arcs developed on oceanic crust (Figure 4.14) as well as continental crust (for example, the Andes Mountains).

Output: Volcanic Arc Magmas of Varying Composition The magmas produced by this type of fluid-induced melting are essentially basaltic in composition, although their chemistry is more variable than that of mid-ocean ridge basalts. The composition of the magmas is further altered during their residence in the crust. Within the magma chambers, the process of fractional crystallization increases the magma's silica content, producing eruptions of andesitic lavas. Where the overlying plate is continental, the heat from the magmas can melt the felsic rocks in the crust, forming magmas with even higher silica contents, such as dacitic and rhyolitic compositions (see Table 4.2). The contribution of slab fluids to the magma is inferred because trace elements known to be present in ocean crust and sediments are found in the magma.

Mantle Plumes

Basalts similar to those produced at mid-ocean ridges are found in thick accumulations over some parts of continents distant from plate boundaries. In the states of Washington, Oregon, and Idaho, the Columbia and Snake rivers flow over a great area covered by this kind of basalt, which solidified from lavas that flowed out millions of years ago. Large quantities of basalt are also erupted in isolated volcanic islands far from plate boundaries, such as the Hawaiian Islands. In such places, slender, pencil-like plumes of hot mantle rise from deep in the Earth, perhaps as deep as the core-mantle boundary. Mantle plumes that reach the surface, most of them far from plate boundaries, form the "hot spots" of the Earth and are responsible for the outpouring of huge quantities of basalt. The basalt is produced during decompression melting of the mantle. Mantle plumes and hot spots are discussed in more detail in Chapter 12.

In summary, basaltic magmas form in the upper mantle beneath mid-ocean ridges and from plumes of deep origin that give rise to interplate and intraplate hot spots. Magmas of varying composition form over subduction zones, depending on how much felsic material and water are incorporated into the mantle wedge rocks overlying the subduction zone.

SUMMARY

How are igneous rocks classified? All igneous rocks can be divided into two broad textural classes: (1) the coarsely crystalline rocks, which are intrusive and therefore cooled slowly; and (2) the finely crystalline rocks, which are extrusive and cooled rapidly. Within each of these broad categories, the rocks are classified chemically as felsic, mafic, or intermediate on the basis of their silica content, or mineralogically, based on their proportions of lighter-colored, felsic minerals and darker, mafic minerals.

How and where do magmas form? Magmas form at places in the lower crust and mantle where temperatures and pressures are high enough for at least partial melting of water-containing rock. Basalt can partially melt in the upper mantle, where convection currents bring hot rock upward at mid-ocean ridges. Mixtures of basalt and other igneous rocks with sedimentary rocks, which contain significant quantities of water, have lower melting points than dry igneous rocks. Thus, different source rocks may melt at different temperatures and thereby affect magma compositions.

How does magmatic differentiation account for the variety of igneous rocks? If a melt underwent fractional crystallization because the crystals were separated and therefore did not react with the melt, the final rocks may be more silicic than the earlier, more mafic crystals. Fractional crystallization can produce mafic igneous rocks from earlier stages of crystallization and differentiation and felsic rocks from later stages, but it does not adequately explain the abundance of granite. Magmatic differentiation of basalt does not explain the composition and abundance of igneous rocks. Different kinds of igneous rocks may be produced by variations in the compositions of magmas caused by the melting of different mixtures of sedimentary and other rocks and by mixing of magmas.

What are the forms of intrusive igneous rocks? Large igneous bodies are plutons. The largest plutons are batholiths, which are thick tabular masses with a central funnel. Stocks are smaller plutons. Less massive than plutons are sills, which are concordant with the intruded rock, lying parallel to its layering, and dikes, which are discordant with the layering, cutting across it. Hydrothermal veins form where water is abundant, either in the magma or in surrounding country rock.

How are igneous rocks related to plate tectonics? The two major magmatic geosystems are the mid-ocean ridges, where basalt wells up from the upper mantle and melts during decompression to form oceanic crust, and subduction zones, where subducting oceanic lithosphere partially melts by addition of fluid to generate differentiated magmas that rise through the crust and form island or continental volcanic arcs.

KEY TERMS AND CONCEPTS

andesite (p. 82)

basalt (p. 82)

batholith (p. 89)

concordant intrusion (p. 90)

country rock (p. 79)

dacite (p. 82)

decompression melting (p. 84)

dike (p. 90)

diorite (p. 82)

discordant intrusion (p. 89)

extrusive igneous rock (p. 79)

felsic rock (p. 81)

fluid-induced melting (p. 96)

fractional crystallization (p. 85)

gabbro (p. 82)

granite (p. 81)

granodiorite (p. 82)

hydrothermal vein (p. 91)

intermediate igneous rock (p. 81)

intrusive igneous rock (p. 79)

mafic rock (p. 82)

magma chamber (p. 85)

magmatic differentiation (p. 85)

obsidian (p. 79)

ophiolite suite (p. 92)

partial melting (p. 83)

pegmatite (p. 91)

peridotite (p. 82)

pluton (p. 88)

porphyry (p. 80)

pumice (p. 79)

pyroclast (p. 79)

rhyolite (p. 81)

sill (p. 90)

stock (p. 89)

tuff (p. 79)

ultramafic rock (p. 82)

vein (p. 91)

viscosity (p. 82)

volcanic ash (p. 79)

EXERCISES

1. Why are intrusive igneous rocks coarsely crystalline and extrusive rocks finely crystalline?

2. What kinds of minerals would you find in a mafic igneous rock?

3. What kinds of igneous rock contain quartz?

4. Name two intrusive igneous rocks with a higher silica content than that of gabbro.

5. What is the difference between a magma formed by fractional crystallization and one formed by ordinary cooling?

6. How does fractional crystallization lead to magmatic differentiation?

7. Where in the crust, mantle, or core would you find a partial melt of basaltic composition?

8. In which plate tectonic settings would you expect magmas to form?

9. Why do melts migrate upward?

10. Where on the ocean floor would you find basaltic magmas being extruded?

11. Much of Earth's crustal area, and nearly all of its mantle, are composed of basaltic or ultramafic rocks. Why are granitic and andesitic rocks as plentiful as they are on Earth? Where do the materials that constitute these rocks come from?

THOUGHT QUESTIONS

1. How would you classify a coarse-grained igneous rock that contains about 50 percent pyroxene and 50 percent olivine?

2. What kind of rock would contain some plagioclase feldspar crystals about 5 mm long "floating" in a dark gray matrix of crystals of less than 1 mm?

3. What differences in crystal size might you expect to find between two sills, one intruded at a depth of about 12 km, where the country rock was very hot, and the other at a depth of 0.5 km, where the country rock was moderately warm?

4. If you were to drill a hole through the crust of a mid-ocean ridge, what intrusive or extrusive igneous rocks might you expect to encounter at or near the surface? What intrusive or extrusive igneous rocks might you expect at the base of the crust?

5. Assume that a magma with a certain ratio of calcium to sodium starts to crystallize. If fractional crystallization occurs during the solidification process, will the plagioclase feldspars formed after complete crystallization have the same ratio of calcium to sodium that characterized the magma?

6. What observations might you make to show that a pluton solidified during fractional crystallization?

7. Why are plutons more likely than dikes to show the effects of fractional crystallization?

8. What might be the origin of a rock composed almost entirely of olivine?

9. What processes create the unequal sizes of crystals in porphyries?

10. Water is abundant in the sedimentary rocks and oceanic crust of subduction zones. How would the water affect melting in these zones?

11. Much of Earth's crustal area, and nearly all of its mantle, are composed of basaltic or ultramafic rocks. Why are granitic and andesitic rocks as plentiful as they are on Earth? Where do the materials that constitute these rocks come from?

SHORT-TERM PROJECT

**Reading the Geologic Story of Igneous Rocks:
Solids from Melts**

All rocks tell a story. The story is deciphered from various clues: texture, mineral and chemical composition, association with other rocks, and geologic setting. With careful analysis and interpretation of the rock record, the geologic history of a region can be deciphered—we can read the rock record like words on a page.

Igneous rocks are formed from a melt (a *magma*), so crystals of minerals can grow in the melt just as ice crystals grow in freezing water. As the melt cools, it becomes a crys-

tal slush and finally transforms into a solid with interlocking crystals. The size of the crystals depends largely on the cooling rate. Typically, volcanic rocks cool relatively quickly on Earth's surface and therefore contain smaller crystals. Plutonic rocks solidify slowly within the crust and contain larger crystals. Igneous rocks may also contain gas bubbles, inclusions of other rock fragments, glass, or a fine-grained matrix of ash and pumice.

Magma generation on Earth is largely restricted to active tectonic plate boundaries and hot spots. Processes and physical conditions associated with plate boundaries melt rocks in the mantle and crust. The composition of the melt depends on the rocks from which it was generated and what happens in the magma chamber before it totally solidifies. Therefore, magma composition is strongly linked to where within the Earth the melt formed and, in turn, to the type of active plate boundary. For example, decompression of rising hot bodies of soft, plastic ultramafic rock within the mantle is thought to produce partial melts of basaltic composition beneath hot spots and divergent plate boundaries. No two igneous rock bodies have exactly the same texture and composition. In fact, the characteristics of igneous rock often vary within one rock body because of all the variables that can affect rock composition and texture.

Information about various types of igneous rock, as well as images, are available on the text's Web site: **www.whfreeman.com/understandingearth**. The information provided is essentially the same information that a professional geologist would gather from field and laboratory work. Given these observations, decipher the story each rock has to tell.

CHAPTER FIVE

SEDIMENTATION
Rocks Formed by Surface Processes

Much of Earth's surface, including its seafloor, is covered with sediments. These layers of loose particles have diverse origins. Most sediments are created by weathering of the continents. Some are the remains of organisms that secreted mineral shells. Yet others consist of inorganic crystals that precipitated when dissolved chemicals in oceans and lakes combined to form new minerals.

Sedimentary rocks were once sediments, and so they are records of the conditions at Earth's surface when and where the sediments were deposited. Geologists can work backward to infer the sources of the sediments from which the rocks were formed and the kinds of places in which the sediments were originally deposited. For example, the top of Mount Everest is composed of fossiliferous (fossil-containing) limestones. Because such limestones are formed by carbonate minerals in seawater, Mount Everest must once have been part of an ocean floor!

The analysis used to determine the history of rock formations at the top of Mount Everest applies just as well to ancient shorelines, mountains, plains, deserts, and swamps. In one area, for example, sandstone may record an earlier time when beach sands accumulated along a shoreline that no longer exists. In a bordering area, carbonate reefs may have been laid down along the perimeter of a tropical island. Beyond, there may have been a nearshore area in which the sediments were shallow marine carbonate muds that later became thin-bedded limestone. By reconstructing such environments, we can map the continents and oceans of long ago.

Sedimentary rocks exposed at El Capitan, Guadalupe Mountains, New Mexico, were formed in an ancient ocean, about 260 million years ago. The lower slopes of the mountains contain siliciclastic sedimentary rocks, formed in deep sea environments. The overlying cliffs of El Capitan are limestone and dolostone, which formed in a shallow sea when calcifying animals and plants died, leaving their shells as sediment. [John Grotzinger.]

Sedimentary rocks also reveal former plate tectonic events and processes by their presence within or adjacent to volcanic arcs, rift valleys, or collisional mountains. In some cases, where the components of sediments and sedimentary rocks are derived from the weathering of preexisting rocks, we can form hypotheses about ancient climates and environments. We can also use sedimentary rocks formed by precipitation from seawater to read the history of changes in Earth's climate and seawater chemistry.

The study of sediments and sedimentary rocks has great practical value as well. Oil and gas, our most valuable sources of energy, are found in these rocks. These precious resources are becoming increasingly difficult to find, and so it is more important than ever to understand how sedimentary rocks form. As oil and gas decline in abundance, coal—which is a distinct type of sedimentary rock—will be used increasingly to generate energy. Another important energy source that may accumulate in sedimentary rocks is uranium, which is used for nuclear power. Phosphate rock used for fertilizer is sedimentary, as is much of the world's iron ore. Knowing how these kinds of sediments form helps us to find and use these limited resources.

Finally, because virtually all sedimentary processes take place at or near Earth's surface where we humans live, they provide a background for our understanding of environmental problems. We once studied sedimentary rocks primarily to understand how to exploit the natural resources just mentioned. Increasingly, however, we study these rocks to improve our understanding of Earth's environment.

In this chapter, we will see how geologic surface processes such as weathering, transportation, sedimentation, and diagenesis produce sediments and sedimentary rocks. We will describe the compositions, textures, and structures of sediments and sedimentary rocks and examine how they correlate with the kinds of environments in which the sediments and rocks are laid down. Throughout the chapter, we will apply our understanding of sediment origins to the study of human environmental problems and to the exploration for energy and mineral resources.

SEDIMENTARY ROCKS ARE PRODUCED BY SURFACE PROCESSES IN THE ROCK CYCLE

Sediments, and the sedimentary rocks formed from them, are produced by the surface processes of the rock cycle (discussed in Chapter 3). They form after rocks have been moved from Earth's interior to its surface by mountain building and before they are returned to Earth's interior by subduction.

These processes involve a source area, where the sediment particles are created, and a sink area where they are deposited in layers. The path that the sediment particles follow from source to sink may be a very long journey—one that involves several important processes resulting from

interactions between the plate tectonic and climate geosystems that govern Earth's surface and shallow crust.

The Mississippi River illustrates a typical process. Tectonic plate movement lifts up rocks in the Rocky Mountains. If rainfall increases in the Rocky Mountains, weathering of the rocks there—one of the Mississippi River's source areas—will increase. Faster weathering will produce more sediment to be released into the river and transported downhill and downstream. At the same time, if the flow in the river also increases because of the higher rainfall, transportation of the sediment down the length of the river will increase. This will increase the volume of sediment delivered to sites of deposition, known as **sedimentary basins,** in the Mississippi delta and Gulf of Mexico. And in these sedimentary basins, the sediments will pile up on top of one another—layer after layer—and be buried to depths where they may become filled with valuable oil and gas.

The rock cycle processes that are important in the formation of sedimentary rocks are reviewed in **Figure 5.1** and summarized here.

- **Weathering** is the general process by which rocks are broken down at Earth's surface to produce sediment particles. There are two types of weathering. **Physical weathering** takes place when solid rock is fragmented by mechanical processes that do not change its chemical composition. The rubble of broken stone at the tops of mountains and hills is primarily the result of physical weathering. Physical weathering also caused the cracks and breaks in the ancient tombs and monuments of Egypt. **Chemical weathering** occurs when the minerals in a rock are chemically altered or dissolved. The blurring or disappearance of lettering on old gravestones and monuments is caused mainly by chemical weathering.

- *Erosion* mobilizes the particles produced by weathering, most commonly by rainwater running downhill.

- *Transportation* occurs when currents of wind and water and the moving ice of glaciers transport particles to new locations—sediment sinks—downhill or downstream.

- *Deposition* (also called *sedimentation*) occurs when sedimentary particles settle out as winds die down, water currents slow, or glacier edges melt. These particles form layers of sediment on land or under the sea in sedimentary basins. In the ocean or in land aquatic environments, chemical precipitates form and are deposited, and the shells of dead organisms are broken up and deposited.

- *Burial* occurs as layers of sediment accumulate in sedimentary basins and older, previously deposited sediments are compacted and progressively buried deep within the basin. These sediments will remain at depth, as part of Earth's crust, until tectonic processes lift them and they return to Earth's surface.

- *Diagenesis* refers to the physical and chemical changes—including pressure, heat, and chemical reactions—by which sediments buried within sedimentary basins are lithified, or converted into sedimentary rocks.

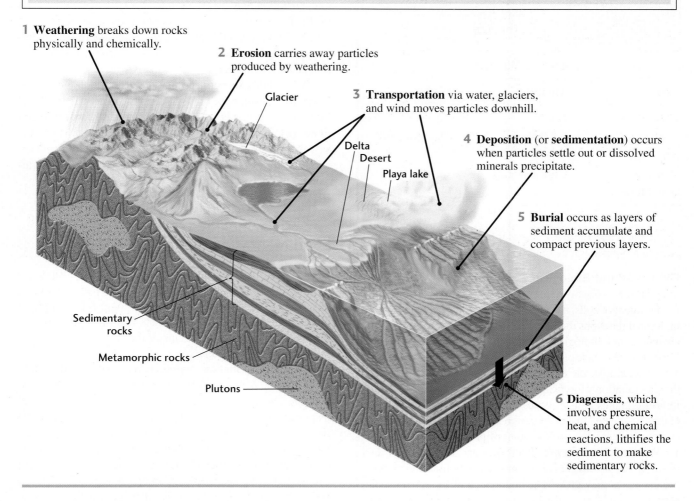

Earth System Figure 5.1 The sedimentary stages of the rock cycle comprise several overlapping processes.

1 **Weathering** breaks down rocks physically and chemically.

2 **Erosion** carries away particles produced by weathering.

Glacier

3 **Transportation** via water, glaciers, and wind moves particles downhill.

Delta
Desert
Playa lake

4 **Deposition** (or **sedimentation**) occurs when particles settle out or dissolved minerals precipitate.

5 **Burial** occurs as layers of sediment accumulate and compact previous layers.

Sedimentary rocks

Metamorphic rocks

Plutons

6 **Diagenesis**, which involves pressure, heat, and chemical reactions, lithifies the sediment to make sedimentary rocks.

The Source of Sediment: Weathering and Erosion Produce Particles and Dissolved Substances

Chemical and physical weathering reinforce each other. Chemical decay weakens rocks and makes them more susceptible to breakage and the formation of fragments. The smaller the fragments produced by physical weathering, the greater the surface area available for chemical weathering. Chemical weathering and mechanical fragmentation of rock at the surface make both solid particles and dissolved products, and erosion carries away these materials. The end products are grouped either as clastic sediments or as chemical and biological sediments. We will explain more about weathering in Chapter 16.

Siliciclastic Sediments The physical and chemical weathering of preexisting rocks forms **clastic particles** that are transported and deposited as **clastic sediments.** Clastic particles range in size from boulders and pebbles to particles of

sand, silt, and clay. They also vary widely in shape. Natural breakage along joints, bedding planes, and other fractures in the parent rock determines the shapes of boulders, cobbles, and pebbles. Sand grains tend to inherit their shapes from the individual crystals formerly interlocked in the parent rock.

The great majority of clastic sediments are produced by the weathering of rocks composed largely of silicate minerals and are called **siliciclastic sediments.** The mixture of minerals in siliciclastic sediments varies. Minerals such as quartz resist weathering and thus are found unaltered in siliciclastic sediments. There may be partly altered fragments of minerals, such as feldspar, that are less resistant to weathering and so less stable. Still other minerals in siliciclastic sediments, such as clay minerals, may be newly formed. Varying intensities of weathering can produce different sets of minerals in sediments derived from the same parent rock. Where weathering is intense, the sediment will contain only clastic particles made of chemically stable minerals, mixed with clay minerals. Where weathering is slight, many minerals that are unstable under surface conditions will survive as

Table 5.1	Minerals Present in a Granite Outcrop Under Varying Intensities of Weathering

INTENSITY OF WEATHERING

Low	Medium	High
Quartz	Quartz	Quartz
Feldspar	Feldspar	Clay minerals
Mica	Mica	
Pyroxene	Clay minerals	
Amphibole		

clastic particles. Table 5.1 shows three possible sets of minerals in a typical granite outcrop.

Chemical and Biological Sediments Chemical weathering produces dissolved ions and molecules that accumulate in the waters of soils, rivers, lakes, and oceans. Chemical and biological reactions then precipitate these substances to form chemical and biological sediments. **Chemical sediments** form at or near their place of deposition, usually from seawater. For example, the evaporation of seawater often leads to the precipitation of gypsum or halite. These sediments form in arid climates and in places where an arm of the sea becomes so isolated that evaporation concentrates the dissolved chemicals in seawater to the point of precipitation.

Biological sediments also form near their place of deposition but are the result of mineral precipitation within organisms as they grow. The abundance of biological sediments depends strongly on climate. Most are restricted to the subtropics and tropics, where carbonate-secreting organisms grow well. After the organisms die, their mineral remains, such as shells, accumulate as sediment. In the case of shells, or corals, the organism *directly* controls mineral precipitation. However, in a second but equally important process, the organism may control mineral precipitation only *indirectly*. Instead of taking minerals from the water to form a shell, the organism changes its surrounding environment so that mineral precipitation occurs on the outside of the organism, or even away from the organism. In sedimentary rocks, the mineral pyrite is often precipitated by this process (see Chapter 11).

In shallow marine environments, biological sediments directly precipitated by organisms consist of layers of particles, such as whole or fragmented shells from marine organisms. Many different types of organisms, ranging from corals to clams to algae, can contribute sediment. Sometimes shells can be transported, further broken up, and deposited as **bioclastic sediments.** These shallow-water sediments consist predominantly of two calcium carbonate minerals—calcite and aragonite—in variable proportions. Other minerals such as phosphates and sulfates are only locally abundant.

In the deep ocean, biological sediments are made of the shells of only a few kinds of organisms. They are composed predominantly of the calcium carbonate mineral calcite, but silica may be precipitated broadly over some parts of the deep ocean. Because these biological particles accumulate in very deep water where agitation by sediment-transporting currents is uncommon, the shells rarely form bioclastic sediments.

We distinguish between chemical and biological sediments for convenience only; in practice, many chemical and biological sediments overlap. In most of the world, much more rock is fragmented by physical weathering than is dissolved by chemical weathering. Thus, clastic sediments are about 10 times more abundant in Earth's crust than chemical and biological sediments.

Transportation and Deposition: The Downhill Journey to Sedimentation Sites

After clastic particles and dissolved ions have formed by weathering and erosion, they start a journey to a sedimentary basin. This journey may be very long; for example, it might span thousands of kilometers from the tributaries of the Mississippi in the highlands of the Rocky Mountains to the swamps of Louisiana.

Most transportation agents carry material downhill. Rocks falling from a cliff, sand carried by a river flowing to the sea, and glacial ice slowly creeping downhill are all responses to gravity. Although winds may blow material from a low elevation to a higher one, in the long run the effects of gravity prevail. When a windblown particle drops to the ocean and settles through the water, it is trapped. It can be picked up again only by an ocean current, which can transport it and deposit it in another site on the seafloor. Eventually, all the sediment transport paths, as complicated as they may be, lead downhill into a sedimentary basin. Marine currents such as tidal currents (see Chapter 20) transport sediments over a shorter distance than do big rivers on land. The short transport distance for chemical or biological sediments contrasts with the much greater distances over which siliciclastic sediments are transported.

Currents as Transport Agents for Clastic Particles Most sediments are transported by air and water currents. The enormous quantities of all kinds of sediment found in the oceans result primarily from the transporting capabilities of rivers, which annually carry a solid and dissolved sediment load of about 25 billion tons (25×10^{15} g).

Air currents move material, too, but in far smaller quantities than rivers or ocean currents. As particles are lifted into the fluid air or water, the current carries them downwind or downriver. The stronger the current—that is, the faster it flows—the larger the particles it can transport.

Current Strength, Particle Size, and Sorting Sedimentation starts where transportation stops. For clastic particles, gravity is the driving force of sedimentation. Particles tend to settle under the pull of gravity. This tendency works against a current's ability to carry a particle. The settling velocity is proportional to the density of the particle and to its size. Because all particles have roughly the same density, we use particle size as the best indicator of how quickly a particle will settle. In water, large grains settle faster than small ones. This is also true in air, but the difference is much smaller.

As wind and water currents begin to slow, they can no longer keep the largest particles suspended, and these settle. As the current slows even more, smaller particles settle. When the current stops completely, even the smallest particles settle. Currents segregate particles in the following ways:

- *Strong currents* (faster than 50 cm/s) carry gravels, along with an abundant supply of coarse and fine detritus. Such currents are common in swiftly flowing streams in mountainous terrains, where erosion is rapid. Beach gravels are deposited where ocean waves erode rocky shores.

- *Moderately strong currents* (20–50 cm/s) lay down sand beds. Currents of moderate strength are common in most rivers, which carry and deposit sand in their channels. Rapidly flowing floodwaters may spread sands over the width of a river valley. Winds also blow and deposit sand, especially in deserts, and waves and currents deposit sand on beaches and in the ocean. However, because air is much less dense than water, much higher current velocities are required to move sediment of the same size and density in air.

- *Weak currents* (slower than 20 cm/s) carry muds composed of the finest clastic particles. Weak currents are found on the floor of a river valley when floodwaters recede slowly or stop flowing entirely. Generally, muds are deposited in the ocean some distance from shore, where currents are too slow to keep even fine particles in suspension. Much of the floor of the open ocean is covered with mud particles originally transported by surface waves and currents or by the wind. These particles slowly settle to depths where currents

and waves are stilled and, ultimately, all the way to the bottom of the ocean.

As you can see, currents may begin by carrying particles of widely varying size, which then become separated as the strength of the current varies. A strong, fast current may lay down a bed of gravel, while keeping sands and muds in suspension. If the current weakens and slows, it will lay down a bed of sand on top of the gravel. If the current then stops altogether, it will deposit a layer of mud on top of the sand bed. This tendency for variations in current velocity to segregate sediments according to size is called **sorting.** A well-sorted sediment consists mostly of particles of a uniform size. A poorly sorted sediment contains particles of many sizes (**Figure 5.2**).

Particles are generally transported intermittently rather than steadily. A river may transport large quantities of sand and gravel when it floods but will drop them as the flood recedes, only to pick them up again and carry them even farther in the next flood. Likewise, strong winds may carry large amounts of dust for a few days and then die down and deposit the dust as a layer of sediment. Strong tidal currents along some ocean margins may transport broken shell fragments in calcium carbonate sediments to places farther offshore and drop them there.

While water and wind currents are transporting particles, the particles become abraded. *Abrasion* affects particles in two ways: it reduces particle size, and it rounds off the rough edges (**Figure 5.3**). As pebbles and large grains are transported, they tumble and strike one another or rub against bedrock. These effects apply mostly to the larger particles; there is little abrasion of sand and silt by impact.

The total time that clastic debris is transported may be many hundreds or thousands of years, depending on the distance to the final depositional area and the number of stopoffs along the way. Clastic particles eroded by the headwaters of the Missouri River in the mountains of western Montana, for example, take hundreds of years to travel the

strong currents carry gravel → weak currents mud.

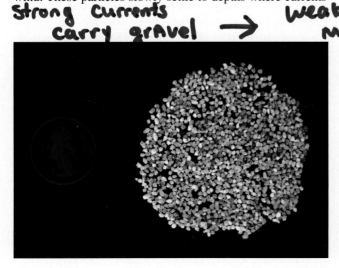

Well-sorted sand

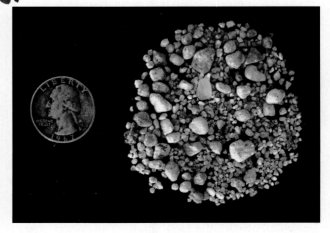

Poorly sorted sand

Figure 5.2 As currents decrease in velocity, sediment is segregated according to particle size. The relatively homogeneous group of sand grains on the left is well sorted; the group on the right is poorly sorted. [Bill Lyons.]

Distance of transport

Short   Moderate   Long

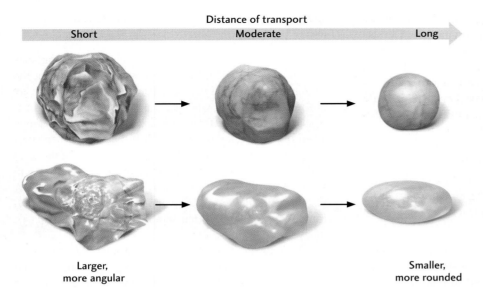

Larger,
more angular

Smaller,
more rounded

Figure 5.3 Transportation reduces the size and angularity of clastic particles. Grains become rounded and slightly smaller as they are transported, although the general shape of the grain may not change significantly.

3200 km down the Missouri and Mississippi rivers to the Gulf of Mexico.

Oceans and Lakes: Chemical Mixing Vats

The driving force of chemical and biological sedimentation is mineral precipitation rather than gravity. Chemical substances dissolved in water during weathering are carried along with the water. Materials such as dissolved calcium ions are part of the water solution itself, so gravity cannot cause them to settle out. As dissolved materials flow down rivers, they ultimately enter lake waters or the ocean.

Oceans may be thought of as huge chemical mixing tanks. Rivers, rain, wind, and glaciers constantly bring in dissolved materials. Smaller quantities of dissolved materials enter the ocean by hydrothermal chemical reactions between seawater and hot basalt at mid-ocean ridges. The ocean loses water continuously by evaporation at the surface. The inflow and outflow of water to and from the oceans are so exactly balanced that the amount of water in the oceans remains constant over such geologically short times as years, decades, or even centuries. Over a time scale of thousands to millions of years, however, the balance may shift. During the Pleistocene Ice Age, for example, significant quantities of seawater were converted into glacial ice, and sea level was drawn down by more than 100 m.

The entry and exit of dissolved materials, too, are balanced. Each of the many dissolved components of seawater participates in some chemical or biological reaction that eventually precipitates it out of the water and onto the seafloor. As a result, the ocean's **salinity**—the total amount of dissolved substances in a given volume of water—remains constant. Totaled over all the oceans of the world, precipitation balances the total inflow of dissolved material from continental weathering and from hydrothermal activity at mid-ocean ridges—yet another way in which the Earth system maintains balance.

We can better understand this chemical balance by considering the element calcium. Calcium is a component of the most abundant biological precipitate formed in the oceans, calcium carbonate ($CaCO_3$). Calcium dissolves when limestone and silicates containing calcium, such as some feldspars and pyroxenes, are weathered on land and brought to the oceans as calcium ions (Ca^{2+}). There, a wide variety of marine organisms combine the calcium ions with carbonate ions (CO_3^{2}), also present in seawater, to form their calcium carbonate shells. The calcium that entered the ocean as dissolved ions leaves it as solid sediment when the organisms die and their shells settle and accumulate as calcium carbonate sediment on the seafloor. Ultimately, the calcium carbonate sediment will be buried and transformed into limestone. The chemical balance that keeps the levels of calcium dissolved in the ocean constant is thus controlled in part by the activities of organisms.

Nonbiological mechanisms also maintain chemical balance in the oceans. For example, sodium ions (Na^+) brought into the oceans react chemically with chloride ions (Cl^-) to form the precipitate sodium chloride ($NaCl$). This happens when evaporation raises the amounts of sodium and chloride ions past the point of saturation. As we saw in Chapter 3, solutions crystallize minerals when they become so saturated with dissolved materials that they can hold no more. The intense evaporation required to crystallize salt takes place in warm, shallow arms of the sea.

SEDIMENTARY BASINS: THE SINKS FOR SEDIMENTS

The currents that move sediment across Earth's surface generally flow downhill. Therefore, sediments tend to accumulate in depressions in the Earth's crust. Depressions are formed by **subsidence,** in which a broad area of the crust sinks (subsides) relative to the surrounding crust. Subsidence is induced partly by the additional weight of sediments on the crust but is caused mostly by tectonic forces.

Sedimentary Basins

Sedimentary basins are regions of at least 10,000 km² where the combination of deposition and subsidence has formed thick accumulations of sediment and sedimentary rock. Sedimentary basins are the Earth's primary sources of oil and gas. Commercial exploration for these resources has helped

us better understand the deep structure of basins and the continental lithosphere.

Rift Basins and Thermal Subsidence Basins

When plate separation begins within a continent, basin subsidence involves stretching, thinning, and heating of the underlying lithosphere by the forces of plate separation (**Figure 5.4**). A long, narrow rift develops, bounded by great

---

Key Figure 5.4   Sedimentary basins develop on rifted continental margins.

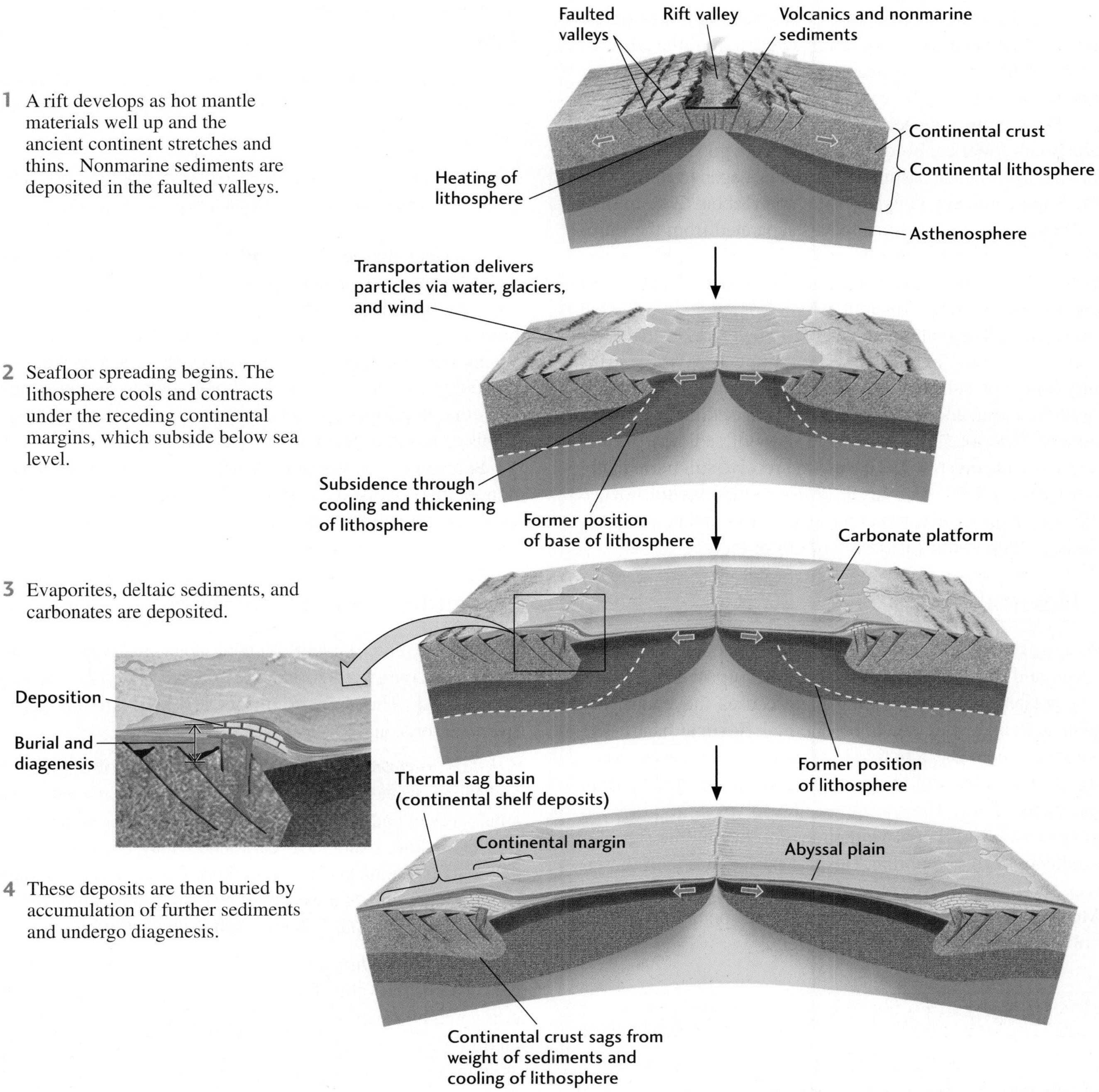

1 A rift develops as hot mantle materials well up and the ancient continent stretches and thins. Nonmarine sediments are deposited in the faulted valleys.

2 Seafloor spreading begins. The lithosphere cools and contracts under the receding continental margins, which subside below sea level.

3 Evaporites, deltaic sediments, and carbonates are deposited.

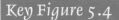

4 These deposits are then buried by accumulation of further sediments and undergo diagenesis.

Faulted valleys · Rift valley · Volcanics and nonmarine sediments

Heating of lithosphere

Continental crust · Continental lithosphere

Asthenosphere

Transportation delivers particles via water, glaciers, and wind

Subsidence through cooling and thickening of lithosphere

Former position of base of lithosphere

Carbonate platform

Deposition

Burial and diagenesis

Thermal sag basin (continental shelf deposits)

Continental margin

Former position of lithosphere

Abyssal plain

Continental crust sags from weight of sediments and cooling of lithosphere

downdropped crustal blocks. Hot ductile mantle rises and fills the space created by the thinned lithosphere and crust, initiating the volcanic eruption of basaltic rocks in the rift zone. **Rift basins** are deep, narrow, and long, with thick successions of sedimentary rocks and extrusive and intrusive igneous rocks. The rift valleys of East Africa, the Rio Grande, and the Jordan Valley in the Middle East are examples of rift basins.

At later stages, when rifting has led to seafloor spreading and the newly formed continental plates are drifting away from each other, basin subsidence continues through the cooling of the lithosphere that was thinned and heated during the earlier rifting stage (see Figure 5.4). Cooling leads to an increase in the density of lithosphere, which in turn leads to its subsidence below sea level, where sediments can accumulate. Because cooling of the lithosphere is the main process creating these basins, they are called **thermal subsidence basins.** Sediments are supplied from erosion of the adjacent land and fill the basin to sea level along the edge of the continent, thus creating the **continental shelf.**

The continental shelves off the Atlantic coasts of North and South America, Europe, and Africa are good examples of thermal subsidence basins. These basins began to form when the supercontinent Pangaea split apart about 200 million years ago and the American plates separated from the European and African plates. Figure 5.4 shows the wedge-shaped deposit of sediments underlying the Atlantic continental shelf and margin of North America, which formed during thermal subsidence. The continental shelf continues to receive sediments for a long time because the trailing edge of the drifting continent subsides slowly and because the continents provide a tremendous area from which sediments can be derived. The load of the growing mass of sediment further depresses the crust, so the basins can receive still more material from the land. As a result of the continuous subsidence and sediment supply, the deposits can accumulate in an orderly fashion to thicknesses of 10 km or more.

Flexural Basins

A third type of basin develops where tectonic plates converge and one lithospheric plate pushes up over the other. The weight of the overriding plate causes the underlying plate to bend or flex down, producing a **flexural basin.** The Mesopotamian Basin in Iraq is a flexural basin formed when the Arabian Plate collided with and was subducted beneath the Iranian Plate. The enormous oil reserves in Iraq (second only to Saudi Arabia) owe their size to having the right ingredients in this important flexural basin. In effect, the oil was squeezed out from the rocks now beneath the Zagros Mountains in Iran, forming several great pools of oil with volumes larger than 10 billion barrels.

SEDIMENTARY ENVIRONMENTS

Between the source area where sediments are formed and a sedimentary basin where they are buried and converted to sedimentary rocks, sediments travel a path through many sedimentary environments. A **sedimentary environment** is a geographic location characterized by a particular combination of climate conditions and physical, chemical, and biological processes (**Figure 5.5**). Important characteristics of sedimentary environments include

- The type and amount of water (ocean, lake, river, arid land)

- The type and strength of transport agent (water, air, ice)

- The topography (lowland, mountain, coastal plain, shallow ocean, deep ocean)

- Biological activity (precipitation of shells, growth of coral reefs, churning of sediments by worms and other burrowing organisms)

- The tectonic setting of sediment source areas (volcanic arc, collision zone) and sedimentary basins (rift, thermal subsidence, flexural)

- The climate (cold climates may form glaciers; arid climates form deserts and precipitate evaporite minerals)

Consider the beaches of Hawaii, famous for their unusual green sands, which are a result of their distinct sedimentary environment. Hawaii is a volcanic island made of olivine-bearing basalt, which is released during weathering. Rivers transport the olivine to the beach, where waves approaching and breaking on the shore, and the resulting currents that develop, concentrate the olivine and remove fragments of basalt to form olivine-rich sand deposits.

Sedimentary environments are often grouped by location: on the continents, near shorelines, or in the ocean. This very general subdivision highlights the processes that give sedimentary environments their distinct identities.

Continental Environments

Sedimentary environments on continents are diverse, owing to the wide range of temperature and rainfall on the surface of the land. These environments are built around lakes, rivers, deserts, and glaciers (see Figure 5.5).

- A *lake environment* includes inland bodies of both fresh and saline water in which the transport agents are relatively small waves and moderate currents. Chemical sedimentation of organic matter and carbonates may occur in freshwater lakes. Saline lakes such as those found in deserts evaporate and precipitate a variety of evaporite minerals, such as halite. The Great Salt Lake in Utah is an example.

- An *alluvial environment* includes a river channel, its borders, and the flat valley floor on either side of the channel that is covered by water when the river floods. Rivers are present on all the continents but Antarctica, and so alluvial deposits are widespread. Organisms are abundant in the muddy flood deposits and are responsible for organic sediments that

Key Figure 5.5 Multiple factors interact to create sedimentary environments.

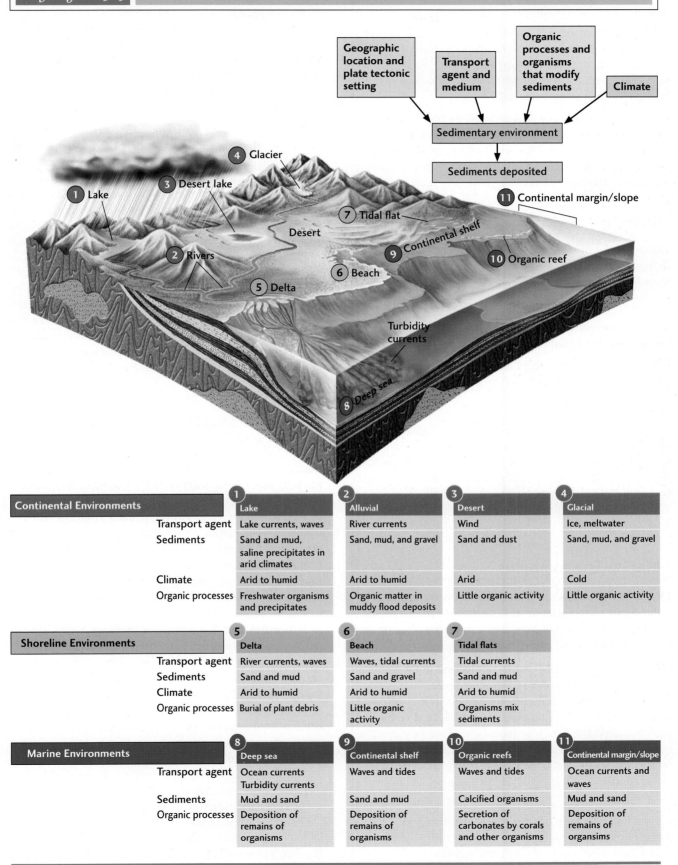

Continental Environments		**1** Lake	**2** Alluvial	**3** Desert	**4** Glacial
	Transport agent	Lake currents, waves	River currents	Wind	Ice, meltwater
	Sediments	Sand and mud, saline precipitates in arid climates	Sand, mud, and gravel	Sand and dust	Sand, mud, and gravel
	Climate	Arid to humid	Arid to humid	Arid	Cold
	Organic processes	Freshwater organisms and precipitates	Organic matter in muddy flood deposits	Little organic activity	Little organic activity

Shoreline Environments		**5** Delta	**6** Beach	**7** Tidal flats
	Transport agent	River currents, waves	Waves, tidal currents	Tidal currents
	Sediments	Sand and mud	Sand and gravel	Sand and mud
	Climate	Arid to humid	Arid to humid	Arid to humid
	Organic processes	Burial of plant debris	Little organic activity	Organisms mix sediments

Marine Environments		**8** Deep sea	**9** Continental shelf	**10** Organic reefs	**11** Continental margin/slope
	Transport agent	Ocean currents Turbidity currents	Waves and tides	Waves and tides	Ocean currents and waves
	Sediments	Mud and sand	Sand and mud	Calcified organisms	Mud and sand
	Organic processes	Deposition of remains of organisms	Deposition of remains of organisms	Secretion of carbonates by corals and other organisms	Deposition of remains of organisms

accumulate in swamplands adjacent to river channels. Climates vary from arid to humid. An example is the Mississippi River and its floodplains.

• A *desert environment* is arid. Wind and the rivers that flow intermittently through deserts transport sand and dust. The dry climate inhibits abundant organic growth, so organisms have little effect on the sediment. Desert sand dunes are an example of such an environment.

• A *glacial environment* is dominated by the dynamics of moving masses of ice and is characterized by a cold climate. Vegetation is present but has small effects on sediment. At the melting border of a glacier, meltwater streams form a transitional alluvial environment.

Shoreline Environments

The dynamics of waves, tides, and river currents on sandy shores dominate shoreline environments (see Figure 5.5). Shoreline environments include

• *Deltaic environments,* where rivers enter lakes or the sea

• *Tidal flat environments,* where extensive areas exposed at low tide are dominated by tidal currents

• *Beach environments,* where the strong waves approaching and breaking on the shore distribute sediments on the beach, depositing strips of sand or gravel

In most cases, the sediments that accumulate are of siliciclastic composition. Organisms affect these sediments mostly by burrowing into them. However, in some tropical and subtropical settings, sediment particles, particularly carbonate sediments, may be of biological origin. These biological carbonate sediments are also subject to waves and tidal currents.

Marine Environments

Marine environments are usually classified on the basis of water depth, which determines the kinds of currents present (see Figure 5.5). Alternatively, they can be classified on the basis of distance from land.

• *Continental shelf environments* are located in the shallow waters off continental shores, where sedimentation is controlled by relatively gentle currents. Sediments may be composed of either siliciclastic particles or biological carbonate particles, depending on how much siliciclastic sediment is supplied by rivers and the abundance of carbonate-producing organisms. Sedimentation may also be chemical if the climate is arid and an arm of the sea becomes isolated from the rest of the sea.

• *Organic reefs* are composed of carbonate structures formed by carbonate-secreting organisms built up on continental shelves or on oceanic volcanic islands.

• *Continental margin and slope environments* are found in the deeper waters at and off the edges of the continents, where sediment is deposited by turbidity currents. A *turbidity current* is a turbulent submarine avalanche of sediment and water that moves downslope. Sediments deposited by turbidity currents are almost always siliciclastic, except for sites where organisms produce a lot of carbonate sediment. In this case, continental margin and slope sediments may be rich in carbonates.

• *Deep-sea environments* include all the floors of the deep ocean, far from the continents, where the waters are much deeper than the reach of wave-generated currents and other shallow-water currents, such as tides. These environments include the continental slope, which is built up by turbidity currents traveling far from continental margins; the abyssal plains, which accumulate carbonate sediments provided mostly by the skeletons of plankton; and the mid-ocean ridges.

We have seen that sedimentary environments can be defined by location. They can also be categorized according to the kinds of sediments found in them or according to the dominant type of sedimentation. Grouping in this manner produces two broad classes: siliciclastic sedimentary environments and chemical and biological sedimentary environments.

Siliciclastic versus Chemical and Biological Sedimentary Environments

Siliciclastic sedimentary environments are those dominated by siliciclastic sediments. They include the continental alluvial (stream), desert, lake, and glacial environments, as well as the shoreline environments transitional between continental and marine: deltas, beaches, and tidal flats. They also include oceanic environments of the continental shelf, continental margin, and deep-ocean floor where siliciclastic sands and muds are deposited. The sediments of these siliciclastic environments are often called **terrigenous sediments,** to indicate their origin on land.

Chemical and biological sedimentary environments are characterized principally by chemical and biological precipitation (Table 5.2). By far the most abundant are *carbonate environments*—marine settings where calcium carbonate, mostly secreted by organisms, is the main sediment. Hundreds of species of mollusks and other invertebrate organisms, as well as calcareous (calcium-containing) algae, secrete carbonate shell materials. Various populations of these organisms live at different depths of water, both in quiet areas and in places where waves and currents are strong. As they die, their shells accumulate to form sediment.

Except for those of the deep sea, carbonate environments are found mostly in the warmer tropical or subtropical regions of the oceans, where carbonate-secreting organisms flourish. These regions include organic reefs, carbonate sand

Table 5.2	Major Chemical and Biological Sedimentary Environments	
Environment	**Agent of Precipitation**	**Sediments**
SHORELINE AND MARINE		
Carbonate (includes reef, bank, deep sea, etc.)	Shelled organisms, some algae; inorganic precipitation from seawater	Carbonate sands and muds, reefs
Evaporite	Evaporation of seawater	Gypsum, halite, other salts
Siliceous: deep sea	Shelled organisms	Silica
CONTINENTAL		
Evaporite	Evaporation of lake water	Halite, borates, nitrates, carbonates, other salts
Swamp	Vegetation	Peat

beaches, tidal flats, and shallow carbonate banks. In a few places, carbonate sediments may form in cooler waters that are supersaturated with carbonate—waters that are generally below 20°C, such as some regions of the Antarctic Ocean south of Australia. Carbonate sediments in cool waters are formed by a very limited group of organisms that mainly secrete calcite shell materials.

Siliceous environments are special deep-sea environments named for the remains of silica shells deposited in them. The organisms that secrete silica grow in surface waters where nutrients are abundant. Their shells settle to the ocean floor and accumulate as layers of siliceous sediment.

An *evaporite environment* is created when the warm seawater of an arid inlet or arm of the sea evaporates more rapidly than it can mix with the connected open marine seawater. The degree of evaporation and the length of time it has proceeded control the salinity of the evaporating seawater and thus the kinds of chemical sediment formed. Evaporite environments also form in lakes with no outlet rivers. Such lakes may produce sediments of halite, borate, nitrates, and other salts.

SEDIMENTARY STRUCTURES

Sedimentary structures include all kinds of features formed at the time of deposition. Sediments and sedimentary rocks are characterized by **bedding,** or *stratification,* which occurs when layers of different grain sizes or compositions are deposited on top of one another. Bedding ranges from only millimeters or centimeters thick to meters or even many

meters thick. Most bedding is horizontal, or nearly so, at the time of deposition. Some types of bedding, however, form at a high angle relative to horizontal.

Cross-Bedding

Cross-bedding consists of sets of bedded material deposited by wind or water and inclined at angles as large as 35° from the horizontal (**Figure 5.6**). Cross-beds form when

Figure 5.6 Cross-bedding in a desert environment. The varying directions of cross-bedding in this sandstone are due to changes in wind direction at the time the sand dunes were deposited. Navajo sandstone, Zion National Park, southwestern Utah. [Peter Kresan.]

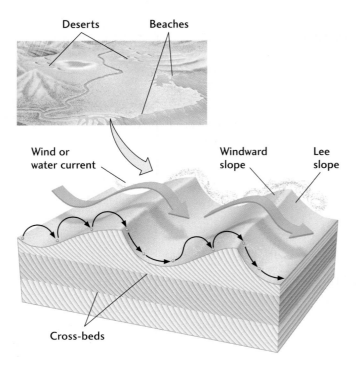

Figure 5.7 Cross-beds form when grains are deposited on the steeper, downcurrent (lee) slope of a dune or ripple.

grains are deposited on the steeper, downcurrent (lee) slopes of sand dunes on land or of sandbars in rivers and under the sea (**Figure 5.7**). Cross-bedding of wind-deposited sand dunes may be complex, a result of rapidly changing wind directions. Cross-bedding is common in sandstones and is also found in gravels and some carbonate sediments. Cross-

bedding is easier to see in sandstones than in sands, which must be excavated to see a cross section.

Graded Bedding

Graded bedding is most abundant in continental slope and deep-sea sediments deposited by dense, muddy currents called turbidity currents, which hug the bottom topography of the ocean as they move downhill. Each layer in a graded bed progresses from coarse grains at the base to fine grains at the top. As the current progressively slows, it drops progressively finer particles. The grading indicates a weakening of the current that deposited the grains. A graded bed comprises one set of coarse-to-fine beds, normally ranging from a few centimeters to several meters thick, that formed horizontal or nearly horizontal layers at the time of deposition. Accumulations of many individual graded beds can reach a total thickness of hundreds of meters. A bed formed as a result of deposition from a turbidity current is called a *turbidite.*

Ripples

Ripples are very small dunes of sand or silt whose long dimension is at right angles to the current. They form low, narrow ridges, most only a centimeter or two high, separated by wider troughs. These sedimentary structures are common in both modern sands and ancient sandstones (**Figure 5.8**). Ripples can be seen on the surfaces of windswept dunes, on underwater sandbars in shallow streams, and under the waves at beaches. Geologists can distinguish the symmetrical ripples made by waves moving back and forth on a beach from

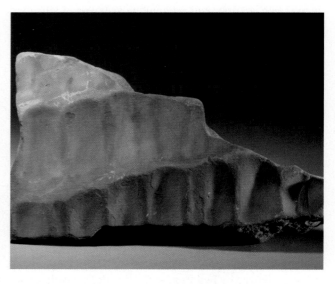

Figure 5.8 *left:* Ripples in modern sand on a beach. [John Grotzinger.] *right:* Ancient ripple-marked sandstone. [John Grotzinger/Ramón Rivera-Moret/MIT.]

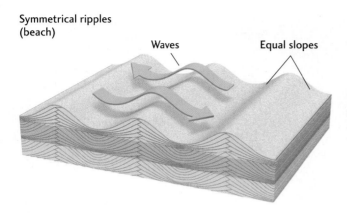

Symmetrical ripples
(beach)

Waves

Equal slopes

Asymmetrical ripples
(dune)

Gentler
slope

Steeper
slope

Wind or water

Figure 5.9 The shapes of ripples on beach sand, produced by the back-and-forth movements of waves, are symmetrical. Ripples on dunes and river sandbars, produced by the movement of a current in one direction, are asymmetrical.

the asymmetrical ripples formed by currents moving in a single direction over river sandbars or windswept dunes (**Figure 5.9**).

Bioturbation Structures

Bedding in many sedimentary rocks is broken or disrupted by roughly cylindrical tubes a few centimeters in diameter that extend vertically through several beds. These sedimentary structures are remnants of burrows and tunnels excavated by clams, worms, and many other marine organisms that live on the bottom of the sea. These organisms burrow through muds and sands—a process called **bioturbation.** They ingest sediment for the bits of organic matter it contains and leave behind the reworked sediment, which fills the burrow (**Figure 5.10**). From bioturbation structures, geologists can determine the behavior of organisms that

burrowed the sediment. Since the behavior of burrowing organisms is controlled partly by environmental processes, such as the strength of currents or the availability of nutrients, bioturbation structures help us reconstruct past sedimentary environments.

Bedding Sequences

Bedding sequences are built of interbedded and vertically stacked layers of sandstone, shale, and other sedimentary rock types. A bedding sequence may consist of cross-bedded sandstone, overlain by bioturbated siltstone, overlain in turn by rippled sandstone—in any combination of thicknesses for each rock type in the sequence.

Bedding sequences help geologists reconstruct how all the sediments were deposited and so give insight into the history of events that occurred at Earth's surface long ago.

Figure 5.10 Bioturbation structures. This rock is crisscrossed with fossilized tunnels originally made as the organisms burrowed through the mud. [John Grotzinger/Ramón Rivera-Moret/Harvard Mineralogical Museum.]

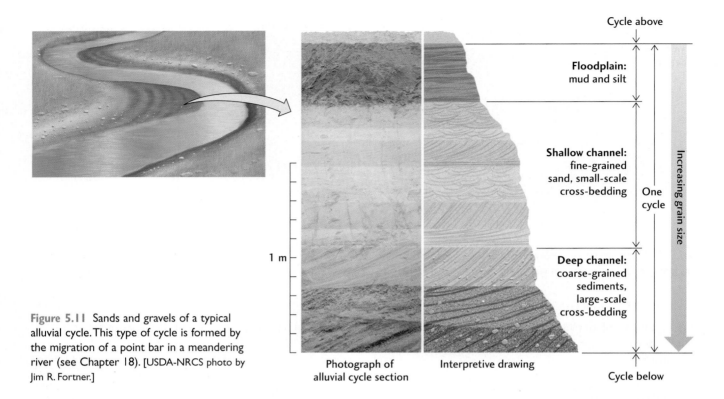

Cycle above

Floodplain:
mud and silt

Shallow channel:
fine-grained
sand, small-scale
cross-bedding

One
cycle

Increasing grain size

Deep channel:
coarse-grained
sediments,
large-scale
cross-bedding

Cycle below

1 m

Photograph of
alluvial cycle section

Interpretive drawing

Figure 5.11 Sands and gravels of a typical alluvial cycle. This type of cycle is formed by the migration of a point bar in a meandering river (see Chapter 18). [USDA-NRCS photo by Jim R. Fortner.]

Figure 5.11 shows a bedding sequence typically formed by rivers. A river lays down sequences that form as its channel migrates back and forth across the valley floor. The lower part of each sequence contains the sediments deposited in the deepest part of the channel, where the current was strongest. The upper part contains the sediments deposited in the shallow parts of the channel, where the current was weakest. Typically, a bedding sequence formed in this manner will consist of sediments that grade upward from coarse to fine.

Most bedding sequences consist of a number of small-scale subdivisions. In the example shown in Figure 5.11, the basal layers of the bedding contain cross-bedding. These layers are overlain by more cross-bedded layers, but the cross-beds are smaller scale. Horizontal bedding occurs at the top of the bedding sequence. Today, computer models are used to analyze how bedding sequences of sands were deposited in alluvial environments. Other types of bedding sequences— which consist of different arrangements of sedimentary structures—tell us about different sedimentary environments. (Bedding sequences are discussed further in Chapter 20.)

BURIAL AND DIAGENESIS: FROM SEDIMENT TO ROCK

Most of the siliciclastic particles produced by weathering and erosion of the land end up deposited in various sedimentary basins in the world's oceans, brought there by rivers, wind, and glaciers. A smaller amount of siliciclastic sediment is deposited in sedimentary basins on land. Most

chemical and biological sediments are also deposited in ocean basins, although some are deposited in continental basins containing lakes and swamps.

Burial

Once sediments reach the ocean floor, they are trapped there. The deep ocean is the ultimate sedimentary basin and, for most sediments, their final resting place. Therefore, a larger fraction of sediment deposited on the ocean floor is buried and preserved, compared to the fraction of sediment deposited on land.

Diagenesis: Heat, Pressure, and Chemistry Transform Sediment into Rock

After sediments are deposited and buried, they are subject to **diagenesis**—the many physical and chemical changes that continue until the sediment or sedimentary rock is either exposed to weathering or metamorphosed by heat and pressure (**Figure 5.12**). Burial promotes diagenesis because buried sediments are subjected to increasingly high temperatures and pressures in Earth's interior.

Temperature increases with depth in the Earth's crust at an average rate of 30°C for each kilometer of depth. At a depth of 4 km, buried sediments may reach 120°C or more, the temperature at which certain types of organic matter may be converted to oil and gas. Pressure also increases with depth—on average, about 1 atmosphere for each 4.4 meters

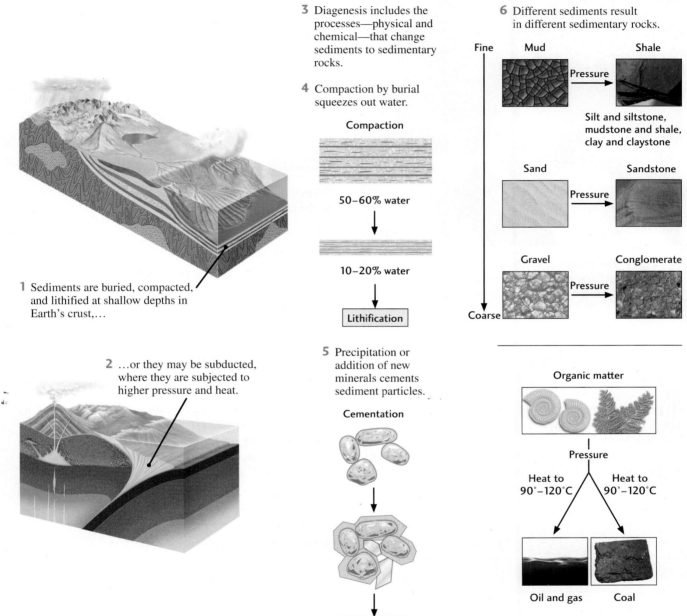

Key Figure 5.12 Diagenesis is the physical and chemical change that converts sediments to sedimentary rocks.

1 Sediments are buried, compacted, and lithified at shallow depths in Earth's crust,…

2 …or they may be subducted, where they are subjected to higher pressure and heat.

3 Diagenesis includes the processes—physical and chemical—that change sediments to sedimentary rocks.

4 Compaction by burial squeezes out water.

Compaction

50–60% water

10–20% water

Lithification

5 Precipitation or addition of new minerals cements sediment particles.

Cementation

Lithification

6 Different sediments result in different sedimentary rocks.

Fine

Mud → Pressure → Shale

Silt and siltstone, mudstone and shale, clay and claystone

Sand → Pressure → Sandstone

Gravel → Pressure → Conglomerate

Coarse

Organic matter

Pressure

Heat to 90°–120°C   Heat to 90°–120°C

Oil and gas   Coal

[*Shale:* John Grotzinger/Ramón Rivera-Moret/Harvard Mineralogical Museum; *Sandstone, conglomerate, coal:* John Grotzinger/Ramón Rivera-Moret/MIT, *coal and gas:* John Woolsey.]

of depth. This increased pressure is responsible for the compaction of buried sediments.

Buried sediments are also continuously bathed in groundwater full of dissolved minerals, which can precipitate in the pores between the sediment particles and bind them together, a chemical change called **cementation.** Cementation decreases **porosity,** the percentage of a rock's volume consisting of open pores between grains. In some sands, for

example, calcium carbonate is precipitated as calcite, which acts as a cement that binds the grains and hardens the resulting mass into sandstone (**Figure 5.13**). Other minerals, such as quartz, may cement sands, muds, and gravels into sandstone, mudstone, and conglomerate.

The major physical diagenetic change is **compaction,** a decrease in the volume and porosity of a sediment. Compaction occurs when the grains are squeezed closer

Quartz sand grains   Calcite cement

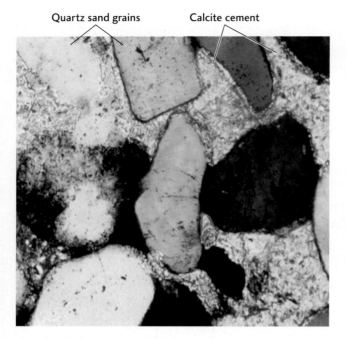

Figure 5.13 This photomicrograph of sandstone shows quartz grains (white and gray) cemented by calcite (brightly colored and variegated) introduced after deposition. [Peter Kresan.]

together by the weight of overlying sediment. Sands are fairly well packed during deposition, so they do not compact much. However, newly deposited muds, including carbonate muds, are highly porous. Often, more than 60 percent of the sediment is water in pore spaces. As a result, muds compact greatly after burial, losing more than half their water.

Both cementation and compaction result in **lithification,** the hardening of soft sediment into rock.

CLASSIFICATION OF SILICICLASTIC SEDIMENTS AND SEDIMENTARY ROCKS

We can now use our knowledge of sedimentation to classify sediments and their lithified counterparts, sedimentary rocks. The major divisions are the siliciclastic sediments and sedimentary rocks and the chemical and biological sediments and sedimentary rocks. Siliciclastic sediments and sedimentary rocks constitute more than three-quarters of the total mass of all types of sediments and sedimentary rocks in the Earth's crust. We therefore begin with them.

Classification by Particle Size

Siliciclastic sediments and rocks are categorized primarily by the size of their grains (Table 5.3):

- *Coarse:* gravel and conglomerate
- *Medium:* sand and sandstone
- *Fine:* silt and siltstone; mud, mudstone, and shale; clay and claystone

We classify siliciclastic sediments and rocks on the basis of their particle size because it distinguishes them by one of the most important conditions of sedimentation: current strength. As we have seen, the larger the particle, the stronger the current needed to move and deposit it. This relationship between current strength and particle size is the reason like-sized particles tend to accumulate in sorted beds. That is, most sand beds do not contain pebbles or mud, and most muds consist only of particles finer than sand.

Of the various types of siliciclastic sediments and sedimentary rocks, the fine-grained clastics are by far the most

Table 5.3	Major Classes of Clastic Sediments and Sedimentary Rocks	
Particle Size	**Sediment**	**Rock**
COARSE	GRAVEL	
Larger than 256 mm	Boulder	
256–64 mm	Cobble	Conglomerate
64–2 mm	Pebble	
MEDIUM		
2–0.062 mm	SAND	Sandstone
FINE	MUD	
0.062–0.0039 mm	Silt	Siltstone
Finer than 0.0039 mm	Clay	Mudstone (blocky fracture) Shale (breaks along bedding) Claystone

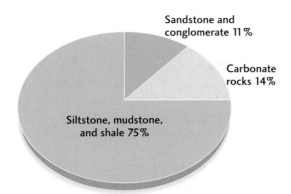

Figure 5.14 The relative abundance of the major sedimentary rock types. In comparison with these three types, all other sedimentary rock types—including evaporites, cherts, and other chemical sedimentary rocks—exist in only minor amounts.

abundant—about three times more common than the coarser clastics (**Figure 5.14**). The abundance of the fine-grained siliciclastics, which contain large amounts of clay minerals, is due to the chemical weathering of the large quantities of feldspar and other silicate minerals in Earth's crust into clay minerals. We turn now to a consideration of each of the three groups of siliciclastic sediments and sedimentary rocks in more detail.

Coarse-Grained Siliciclastics: Gravel and Conglomerate

Gravel is the coarsest siliciclastic sediment, consisting of particles larger than 2 mm in diameter and including pebbles, cobbles, and boulders (see Table 5.3). **Conglomerates** are the lithified equivalents of gravel (**Figure 5.15**). Pebbles, cobbles, and boulders are easy to study and identify because of their large size. Their size can tell us the speed of the currents that transported them. In addition, their composition can tell us about the nature of the distant terrain where they were produced.

There are relatively few environments—mountain streams, rocky beaches with high waves, and glacier melt-waters—in which currents are strong enough to transport pebbles. Strong currents also carry sand, and we almost always find sand between the pebbles. Some of it was deposited with the gravel, and some infiltrated the spaces between fragments after the gravel was deposited. Pebbles and cobbles become rounded very quickly by abrasion in the course of transport on land or in water.

Medium-Grained Siliciclastics: Sand and Sandstone

Sand consists of medium-sized particles, ranging from 0.062 to 2 mm in diameter (see Table 5.3). These sediments are moved even by moderate currents, such as those of rivers, waves at shorelines, and the winds that blow sand into dunes. Sand particles are large enough to be seen with the naked eye, and many of their features are easily discerned with a low-power magnifying glass. The lithified equivalent of sand is **sandstone** (see Figure 5.15).

Sizes and Shapes of Sand Grains The medium-sized siliciclastics, sand particles, are subdivided into fine, medium, and coarse. The average size of the grains in any one sandstone can be an important clue to both the strength of the current that carried them and the sizes of the crystals eroded from the parent rock. The range and relative abundance of the various sizes are also significant. If all the grains are close to the average size, the sand is well sorted. If many grains are much larger or smaller than the average, the sand is poorly sorted. The degree of sorting can help distinguish, for example, between sands of beaches (well sorted) and muddy sands deposited by glaciers (poorly sorted). The shapes of sand grains can also be important clues to their origin. Sand grains, like pebbles, are rounded during transport. Angular grains imply short transport distances; rounded ones indicate long journeys down a large river system.

(a) Conglomerate (b) Sandstone (c) Shale

Figure 5.15 Clastic sedimentary rocks. [*Conglomerate and sandstone:* John Grotzinger/Ramón Rivera-Moret/MIT. *Shale:* John Grotzinger/Ramón Rivera-Moret/Harvard Mineralogical Museum.]

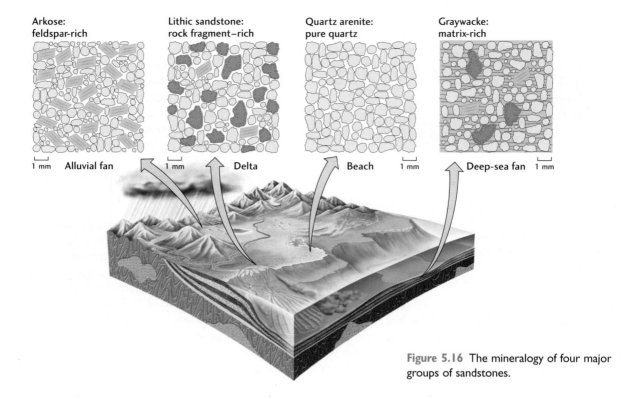

Figure 5.16 The mineralogy of four major groups of sandstones.

Mineralogy of Sands and Sandstones Siliciclastics can be further subdivided by mineralogy, which can help identify the parent rocks. Thus, there are quartz-rich and feldspar-rich sandstones. Some sands are bioclastic; they formed when material such as carbonate originally precipitated as a shell but then was broken and transported by currents. Thus, the mineral composition of sands and sandstones indicates the source areas that were eroded to produce the sand grains. Sodium- and potassium-rich feldspars with abundant quartz, for example, might indicate that the sediments were eroded from a granitic terrain. Other minerals, as we will see in Chapter 6, would indicate metamorphic parent rocks.

The mineral content of parent rocks depends on plate tectonic settings. Sandstones containing abundant fragments of mafic volcanic rocks, for example, are derived from the volcanic arcs of subduction zones.

Major Kinds of Sandstone Sandstones fall into several major groups on the basis of their mineralogy and texture (**Figure 5.16**):

• **Quartz arenites** are made up almost entirely of quartz grains, usually well sorted and rounded. These pure quartz sands result from extensive weathering that occurred before and during transport and removed everything but quartz, the most stable mineral.

• **Arkoses** are more than 25 percent feldspar. The grains tend to be poorly rounded and less well sorted than those of pure quartz sandstones. These feldspar-rich sandstones come from rapidly, eroding granitic and metamorphic terrains where chemical weathering is subordinate to physical weathering.

• **Lithic sandstones** contain many fragments derived from fine-grained rocks, mostly shales, volcanic rocks, and fine-grained metamorphic rocks.

• **Graywacke** is a heterogeneous mixture of rock fragments and angular grains of quartz and feldspar, the sand grains being surrounded by a fine-grained clay matrix. Much of this matrix is formed by relatively soft rock fragments, such as shale and some volcanic rocks, that are chemically altered and physically compacted after deep burial of the sandstone formation.

Both groundwater geologists and petroleum geologists have a special interest in sandstones. Groundwater geologists examine the origins of sandstones to predict possible supplies of water in areas of porous sandstone, such as those found in the western plains of North America. Petroleum geologists must know about the porosity and cementation of sandstones because much of the oil and gas discovered in the past 150 years has been found in buried sandstones. In addition, much of the uranium used for nuclear power plants and weapons has come from diagenetic uranium in sandstones.

Fine-Grained Siliciclastics: Silt and Siltstone; Mud, Mudstone, and Shale; Clay and Claystone

The finest-grained siliciclastic sediments and sedimentary rocks are the silts and siltstones; the muds, mudstones, and shales; and the clays and claystones. These sediments consist of particles that are less than 0.062 mm in diameter, but the sediments vary widely in their range of grain sizes and mineral compositions. Fine-grained sediments are deposited by the gentlest currents, which allow the finest particles to settle slowly to the bottom in quiet waves.

Silt and Siltstone Siltstone is the lithified equivalent of **silt,** a siliciclastic sediment in which most of the grains are between 0.0039 and 0.062 mm in diameter. Siltstone looks similar to mudstone or very fine grained sandstone.

Mud, Mudstone, and Shale Mud is a siliciclastic sediment, mixed with water, in which most of the particles are less than 0.062 mm in diameter (see Table 5.3). Thus, mud can be made of silt- or clay-sized sediments, or varying quantities of both. This general term is very useful in field-work because it is often difficult to distinguish between silt- and clay-sized sediment without a microscope. Muds are deposited by rivers and tides. After a river has flooded its lowlands and the flood recedes, the current slows and mud settles, some of it containing abundant organic matter. This mud contributes to the fertility of river bottomlands. Muds are left behind by ebbing tides along many tidal flats where wave action is mild. Much of the deep-ocean floor, where currents are weak or absent, is blanketed by mud.

The fine-grained rock equivalents of muds are mudstones and shales. **Mudstones** are blocky and show poor or no bedding. Bedding may have been well marked when the sediments were first deposited but then was lost by bioturbation. **Shales** (see Figure 5.15) are composed of silt plus a significant component of clay, which causes them to break readily along bedding planes. Many muds, mudstones, and shales are more than 10 percent carbonate, forming deposits of calcareous shales. Black, or organic, shales contain abundant organic matter. Some, called oil shales, contain large quantities of oily organic material, which makes them a potentially important source of oil. (We consider the oil shales in more detail in Chapter 23.)

Clay and Claystone Clay is the most abundant component of fine-grained sediments and sedimentary rocks and consists largely of clay minerals. Clay-sized particles are less than 0.0039 mm in diameter (see Table 5.3). Rocks made up exclusively of clay-sized particles are called **claystones.**

CLASSIFICATION OF CHEMICAL AND BIOLOGICAL SEDIMENTS AND SEDIMENTARY ROCKS

We divide nonsiliciclastic sediments into chemical and biological sediments to emphasize the importance of organisms as the chief mediators of this kind of sedimentation (Table 5.4). Chemical and biological sediments tell us about chemical conditions in the ocean, the predominant environment of sedimentation. **Carbonate environments,** by far the most abundant biological sedimentation environments, occur in marine settings where calcium carbonate is the main sediment. The shells of organisms account for much of this carbonate sediment. Chemical sedimentation

Table 5.4 — Classification of Biological and Chemical Sediments and Sedimentary Rocks

Sediment	Rock	Chemical Composition	Minerals
BIOLOGICAL			
Sand and mud (primarily bioclastic)	Limestone	Calcium carbonate ($CaCO_3$)	Calcite (aragonite)
Siliceous sediment	Chert	Silica (SiO_2)	Opal, chalcedony, quartz
Peat, organic matter	Organics	Carbon compounds; Carbon compounded with oxygen and hydrogen	(coal), (oil), (gas)
No primary sediment (formed by diagenesis)	Phosphorite	Calcium phosphate ($Ca_3[PO_4]_2$)	Apatite
CHEMICAL			
No primary sediment (formed by diagenesis)	Dolostone	Calcium-magnesium carbonate ($CaMg[CO_3]_2$)	Dolomite
Iron oxide sediment	Iron formation	Iron silicate; oxide (Fe_2O_3); limonite, carbonate	Hematite, siderite
Evaporite sediment	Evaporite	Sodium chloride ($NaCl$); calcium sulfate ($CaSO_4$)	Gypsum, anhydrite, halite, other salts

occurs in marine settings when evaporation of seawater exceeds replenishment. Chemical sedimentation also takes place in some lakes, particularly those of arid regions where evaporation is intense, such as the Great Salt Lake of Utah. Such sediments account for only a very small fraction relative to the amounts deposited along the ocean's shorelines, on continental shelves, and in the deep ocean. Chemical sediments are less abundant than biological sediments.

Biological Sediments: Carbonate Sediments and Rocks

Carbonate sediments and **carbonate rocks** form mostly from the accumulation of carbonate minerals that are directly secreted by organisms. However, in some cases the organisms do not secrete carbonate minerals but indirectly help stimulate precipitation of carbonate mineral in the organism's external environment. This process occurs because organisms can change the chemistry of their surrounding environment. In a third process—during burial and diagenesis—deposited carbonate sediments react with water to form a new suite of carbonate minerals. In all these processes, the minerals precipitated are either calcium carbonates (calcite or aragonite) or calcium-magnesium carbonate (dolomite).

The dominant biological sedimentary rock lithified from carbonate sediments is **limestone,** which is composed mainly of calcium carbonate ($CaCO_3$) in the form of the mineral calcite (**Figure 5.17a**; see Table 5.4). Limestone is formed from carbonate sand and mud and, in some cases, ancient reefs.

Another abundant carbonate rock is **dolostone,** made up of the mineral dolomite, which is composed of calcium-magnesium carbonate, $CaMg(CO_3)_2$ (see Table 5.4). Dolostones are diagenetically altered carbonate sediments and limestones. The mineral dolomite does not form as a primary precipitate from ordinary seawater, and no organisms secrete shells of dolomite. Instead, the original calcite or aragonite of a carbonate sediment is converted into dolomite after deposition. Some calcium ions in the calcite or aragonite are exchanged for magnesium ions from seawater (or magnesium-rich groundwater) slowly passing through the pores of the sediment. This exchange converts the calcium carbonate mineral, $CaCO_3$, into dolomite, $CaMg(CO_3)_2$.

Direct Biological Precipitation of Carbonate Sediments Carbonate rocks are abundant because of the large amounts of calcium and carbonate present in seawater, which organisms can directly convert into shells. Calcium is supplied by weathering of feldspars and other minerals in igneous and metamorphic rocks. Carbonate is derived from the carbon dioxide in the atmosphere. Calcium and carbonate also come from the easily weathered limestone on the continents.

(a) Limestone

(b) Gypsum

(c) Halite

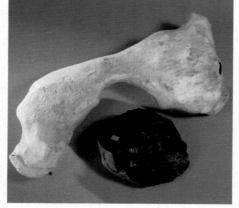

(d) Chert

Figure 5.17 Chemical and biological sedimentary rocks. (a) Limestone, lithified from carbonate sediments; (b) gypsum and (c) halite, marine evaporites that crystallize out of shallow seawater basins; and (d) chert, made up of silica sediment. [John Grotzinger/ Ramón Rivera-Moret/Harvard Mineralogical Museum.]

Most carbonate sediments of shallow marine environments are bioclastic. They were originally secreted biologically as shells by organisms living near the surface or on the bottom of the oceans. After they die, the organisms break apart, producing shells or fragments of shells that constitute individual pieces or clasts of carbonate sediment. These sediments are found from the coral reefs of the Pacific and Caribbean to the shallow banks of the Bahama Islands. Carbonate is more accessible for study in these spectacular vacation spots, but the oceanic abyssal plain is where most carbonate is deposited today.

Most of the carbonate sediments deposited on the ocean's abyssal plains are derived from the calcite shells of **foraminifera,** tiny single-celled organisms that live in surface waters, and from other organisms that secrete calcium carbonate. When the organisms die, their shells and skeletons settle to the seafloor and accumulate there as sediment (**Figure 5.18**). In addition to calcite, most carbonate

---

Key Figure 5.18 Organisms create carbonate platform systems.

1 Carbonate platform building involves interactions of the hydrosphere, biosphere, and lithosphere.

2 The Bahamas are a carbonate platform system in the Atlantic Ocean east of Florida.

3 Carbonate platforms are built in warm shallow seas by reef-building organisms such as coral and tiny foraminifera that precipitate calcium carbonate as calcite aragonite.

Needle

Foraminiferan

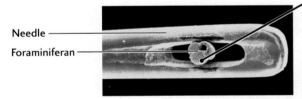
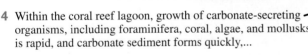
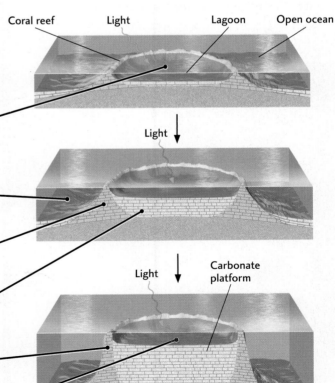

Coral reef   Light   Lagoon   Open ocean

4 Within the coral reef lagoon, growth of carbonate-secreting organisms, including foraminifera, coral, algae, and mollusks, is rapid, and carbonate sediment forms quickly,...

Light

5 ...whereas in the open ocean outside the reef, sedimentation is much slower.

6 If the sea level rises, the reef continues to grow toward the light at sea level,...

Light   Carbonate platform

7 ...and lagoon sedimentation outpaces sedimentation in the open ocean.

8 Eventually, a carbonate platform grows, with steep sides falling away to the open ocean.

9 Inorganic carbonate also precipitates out of the supersaturated lagoon water and adds to platform sedimentation.

sediments contain aragonite, a less stable form of calcium carbonate. As noted earlier in this chapter, some organisms precipitate calcite, some precipitate aragonite, and some precipitate both.

Reefs are moundlike or ridgelike organic structures composed of the carbonate skeletons of millions of organisms. In the warm seas of the present, most reefs are built by corals and hundreds of other organisms such as algae and the familiar clams and snails of our shorelines. In contrast with the soft, loose sediment produced in other environments, the calcium carbonate of the corals and other organisms forms a rigid, wave-resistant structure of solid limestone that is built up to and slightly above sea level (see Figure 5.18 and Feature 5.1). The solid limestone of the reef is produced directly by the action of organisms; there is no soft sediment stage.

Indirect Biological Precipitation of Carbonate Sediments

A significant fraction of the carbonate mud in lagoons and on shallow banks such as those of the Bahama Islands is precipitated indirectly from seawater. Microorganisms may be involved in this process, but their role is still uncertain. Their potential role would be to help shift the balance of calcium (Ca^{2+}) and carbonate (CO_3^{-}) ions in the seawater surrounding the organism so that calcium carbonate ($CaCO_3^{2-}$) is formed. Microbes can precipitate carbonate only if the external environment already contains abundant calcium and carbonate ions. In this case, the chemicals that the microbe emits into the seawater cause the minerals

to precipitate. In contrast, shelled organisms will always secrete carbonates as a normal part of their life cycle.

Carbonate platforms, both in past geological ages and at present, are a major carbonate environment. Like the Bahama Banks, these platforms are extensive flat, shallow areas where both biological and nonbiological carbonates are deposited. Below the level of the platform are carbonate ramps, gentle slopes to deeper waters that also accumulate carbonate sediment, much of it fine-grained. In other times and places, platforms may be rimmed carbonate shelves in which there is a clear demarcation of the shelf margin by reefs of various organisms and by buildups of shoals of bioclastic and other materials. Below the rims are steep slopes covered with detritus derived from the rim materials. The role of now-extinct organisms in building ancient reefs will be discussed next.

Reefs and Evolutionary Processes Today, reefs are constructed mainly by corals; but at earlier times in Earth's history, they were constructed by other organisms—such as a now-extinct variety of mollusk (**Figure 5.19**). The diversification and extinction of reef-building organisms over geologic time show how ecology and environmental change help regulate the process of evolution. Today, natural and human-generated effects threaten the growth of coral reefs, which are very sensitive to environmental change. In 1998, an El Niño event raised sea surface temperatures to the point where many reefs in the western Indian Ocean were killed. The Florida Keys reefs are dying off for a completely different reason: they're getting too much of a good thing. It turns

Figure 5.19 Reefal limestone made of extinct clams (rudists) in the Cretaceous Shuiba formation, Sultanate of Oman. [John Grotzinger.]

out that groundwaters originating in the farmlands of the Florida Peninsula are seeping out near the reefs and exposing them to lethal concentrations of nutrients.

The marine environments where carbonate sedimentation produces rigid limestone structures—including reefs, carbonate banks, and deep-water deposits in the open ocean—are considered further in Chapter 20.

Chemical Sediments: Evaporation of Water to Produce Halite, Gypsum, and Other Salts

Evaporite sediments and **evaporite rocks** are chemically precipitated from evaporating seawater and from water in arid-region lakes that have no river outlets.

Marine Evaporites Marine evaporites are the chemical sediments and sedimentary rocks formed by the evaporation of seawater. This evaporite environment is created when the warm seawater of an arid inlet or arm of the sea evaporates more rapidly than it can mix with the connected open marine seawater. The degree of evaporation controls the salinity of the evaporating seawater and thus the kinds of sedi-

ments formed. The sediments and rocks produced in these environments contain minerals formed by the crystallization of sodium chloride (halite), calcium sulfate (gypsum and anhydrite), and other combinations of the ions commonly found in seawater. As evaporation proceeds, seawater becomes more concentrated and minerals crystallize in a set sequence. As dissolved ions precipitate to form each mineral, the evaporating seawater changes composition.

Seawater has the same composition in all the oceans, which explains why marine evaporites are so similar the world over. No matter where seawater evaporates, the same sequence of minerals always forms. The history of evaporite minerals shows that the composition of the world's oceans has stayed more or less constant over the past 1.8 billion years. Before that time, however, the precipitation sequence may have been different, indicating that seawater composition changed.

The great volume of many marine evaporites, some hundreds of meters thick, shows that they could not have formed from the small amount of water that could be held in a shallow bay or pond. A huge amount of seawater must have evaporated. The way in which such large quantities of seawater evaporate is very clear in bays or arms of the sea that meet the following conditions (**Figure 5.20**):

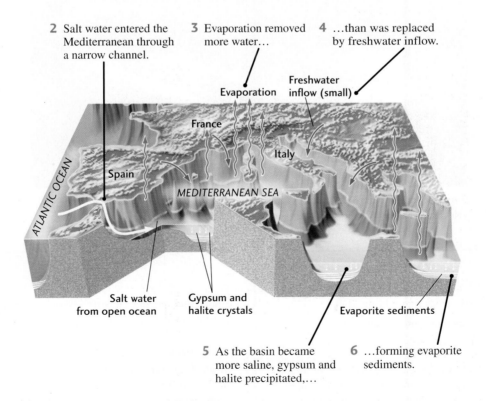

1 During the Miocene epoch, the Mediterranean Sea became a shallow evaporite basin. (We have greatly exaggerated the basin depth for this drawing.)

2 Salt water entered the Mediterranean through a narrow channel.

3 Evaporation removed more water…

4 …than was replaced by freshwater inflow.

Evaporation

Freshwater inflow (small)

France

Italy

Spain

ATLANTIC OCEAN

MEDITERRANEAN SEA

Salt water from open ocean

Gypsum and halite crystals

Evaporite sediments

5 As the basin became more saline, gypsum and halite precipitated,…

6 …forming evaporite sediments.

Figure 5.20 A marine evaporite environment. When seawater evaporated in a shallow basin, such as in the Mediterranean Sea, with a restricted connection to the open ocean, gypsum formed as an evaporite sediment. A further increase in salinity led to the crystallization of halite.

EARTH ISSUES

5.1 Darwin's Coral Reefs and Atolls

For more than 200 years, coral reefs have attracted explorers and travel writers. Ever since Charles Darwin sailed the oceans on the *Beagle* from 1831 to 1836, these reefs have been a matter of scientific discussion, too. Darwin was one of the first to analyze the geology of coral reefs, and his theory of their origin is still accepted today.

The coral reefs that Darwin studied were atolls, islands in the open ocean with circular lagoons. The outermost part of a reef is a slightly submerged, wave-resistant reef front: a steep slope facing the ocean. The reef front is composed of the interlaced skeletons of actively growing coral and calcareous algae, forming a tough, hard limestone. Behind the reef front is a flat platform extending into a shallow lagoon. An island may lie at the center of the lagoon. Parts of the reef, as well as a central island, are above water and may become forested. A great many plant and animal species inhabit the reef and the lagoon.

Coral reefs are generally limited to waters less than about 20 m deep because, below that depth, seawater does not transmit enough light to enable reef-building corals to grow. (Exceptions are some kinds of individual—noncolonial—corals that grow in much deeper waters.) Darwin explained how coral reefs could be built up from the bottom of the dark, deep ocean. The process starts with a volcano building up to the surface from the seafloor. As the volcano becomes dormant, temporarily or permanently, coral and algae colonize the shore and build fringing reefs—coral reefs similar to atolls that grow around the edges of a central volcanic island. Erosion may then lower the volcanic island almost to sea level.

Darwin reasoned that if such a volcanic island were to subside slowly beneath the waves, actively growing

Bora Bora atoll, South Pacific Ocean. Reefal organisms have built a barrier around the volcanic island, forming a protected lagoon. [Jean-Marc Truchet/Stone/Getty Images.]

- The freshwater supply from rivers is small.
- Connections to the open sea are constricted.
- The climate is arid.

In such locations, water evaporates steadily, but the openings allow seawater to flow in to replenish the evaporating waters of the bay. As a result, those waters stay at constant volume but become more saline than the open ocean. The evaporating bay waters remain more or less constantly supersaturated and steadily deposit evaporite minerals on the floor of the evaporite basin.

As seawater evaporates, the first precipitates to form are the carbonates. Continued evaporation leads to the precipitation of gypsum, calcium sulfate ($CaSO_4 \times 2H_2O$) (see Figure 5.17b). By the time gypsum precipitates, almost no carbonate ions are left in the water. Gypsum is the principal component of plaster of Paris and is used in the manufacture of wallboard, which lines the walls of most new houses.

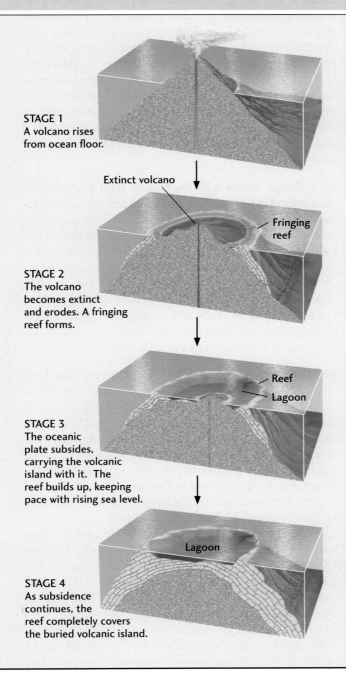

STAGE 1
A volcano rises
from ocean floor.

Extinct volcano

Fringing
reef

STAGE 2
The volcano
becomes extinct
and erodes. A fringing
reef forms.

Reef

Lagoon

STAGE 3
The oceanic
plate subsides,
carrying the volcanic
island with it. The
reef builds up, keeping
pace with rising sea level.

Lagoon

STAGE 4
As subsidence
continues, the
reef completely covers
the buried volcanic island.

coral and algae might keep pace with the subsidence, continuously building up the reef so that the island remained. In this way, the volcanic island would disappear and we would be left with an atoll. More than 100 years after Darwin proposed his theory, deep drilling on several atolls penetrated volcanic rock below the coralline limestone and confirmed the theory. And, some decades later, the theory of plate tectonics explained both volcanism and the subsidence that resulted from plate cooling and contraction.

Evolution of a coral reef from a subsiding volcanic island, first proposed by Charles Darwin in the nineteenth century.

After still further evaporation, the mineral halite (NaCl)—one of the most common chemical sediments precipitated from evaporating seawater—starts to form (see Figure 5.17c). Halite, you may remember from Chapter 3, is table salt. Deep under the city of Detroit, Michigan, beds of salt laid down by an evaporating arm of an ancient ocean are commercially mined.

In the final stages of evaporation, after the sodium chloride is gone, magnesium and potassium chlorides and sul-

fates precipitate. The salt mines near Carlsbad, New Mexico, contain commercial quantities of potassium chloride. Potassium chloride is often used as a substitute for table salt (sodium chloride) by people with certain dietary restrictions.

This sequence of precipitation has been studied in the laboratory and is matched by the bedding sequences found in certain natural salt formations. Most of the world's evaporites consist of thick sequences of dolomite, gypsum, and halite and do not contain the final-stage precipitates.

Many do not go even as far as halite. The absence of the final stages indicates that the water did not evaporate completely but was replenished by normal seawater as evaporation continued.

Nonmarine Evaporites Evaporite sediments also form in arid-region lakes that typically have few or no river outlets. In such lakes, evaporation controls the lake level, and incoming salts derived from chemical weathering accumulate. The Great Salt Lake is one of the best known of these lakes. River waters enter the lake, bringing salts dissolved in the course of weathering. In the dry climate of Utah, evaporation has more than balanced the inflow of fresh water from rivers and rain. As a result, concentrated dissolved ions in the lake make it one of the saltiest bodies of water in the world—eight times saltier than seawater.

In arid regions, small lakes may collect unusual salts, such as borates (compounds of the element boron), and some become alkaline. The water in this kind of lake is poisonous. Economically valuable resources of borates and nitrates (minerals containing the element nitrogen) are found in the sediments beneath some of these lakes.

Other Biological and Chemical Sediments

Carbonate minerals secreted by organisms are the principal source of biological sediments, and the minerals precipitated from evaporating water are the principal source of chemical sediments. However, there are several less abundant biological and chemical sediments that are locally abundant. These include chert, phosphorite, iron formations, coal, and the organic-rich sediments that produce oil and gas. The role of biological versus chemical processes in forming these sediments is variable.

Silica Sediment: Source of Chert One of the first sedimentary rocks to be used for practical purposes by our prehistoric ancestors was **chert.** Chert is made up of silica (SiO_2) (see Figure 5.17d). Early hunters used it for arrowheads and other tools because it could be chipped and shaped to form hard, sharp implements. A common name for chert is *flint,* and the terms are virtually interchangeable. The silica in most cherts is in the form of extremely fine crystalline quartz. Some geologically young cherts consist of opal, a less well crystallized form of silica.

Like calcium carbonate, much silica sediment is precipitated biologically, secreted by ocean-dwelling organisms. These organisms grow in surface waters where nutrients are abundant. When they die, they sink to the deep-ocean floor, where their shells accumulate as layers of silica sediment. After these silica sediments are buried by later sediments, they are diagenetically cemented into chert. Chert may also form as diagenetic nodules and irregular masses replacing carbonate in limestones and dolostones.

Phosphorite Sediment Among the many other kinds of chemical and biological sediments deposited in the sea are phosphorites. Sometimes called phosphate rock, **phosphorite** is composed of calcium phosphate precipitated from phosphate-rich seawater in places where currents of deep, cold water containing phosphate and other nutrients rise along continental margins. The phosphorite forms diagenetically by the interaction between muddy or carbonate sediments and the phosphate-rich water. Organisms play an important role in creating phosphate-rich water, and bacteria that live on sulfur may play a key role in precipitating phosphate minerals.

Iron Oxide Sediment: Source of Iron Formations **Iron formations** are sedimentary rocks that usually contain more than 15 percent iron in the form of iron oxides and some iron silicates and iron carbonates. Most of these rocks formed early in Earth's history, when there was less oxygen in the atmosphere and, as a result, iron dissolved more easily. Iron was transported to the sea in soluble form and precipitated where microorganisms were producing oxygen. Once thought to be of chemical origin, there is now some evidence that iron formations may have been precipitated indirectly by microorganisms (see Chapter 11).

Organic Particles: Source of Coal, Oil, and Gas Coal is a biologically produced sedimentary rock composed almost entirely of organic carbon formed by the diagenesis of swamp vegetation. Vegetation may be preserved from decay and accumulate as a rich organic material, **peat,** which contains more than 50 percent carbon. Peat is ultimately buried and transformed by diagenesis into coal. Coal is classified as an **organic sedimentary rock,** a group that consists entirely or partly of organic carbon-rich deposits formed by the decay of once-living material that has been buried.

In both lake and ocean waters, the remains of algae, bacteria, and other microscopic organisms may accumulate in sediments as organic matter that can be transformed into oil and gas. However, **oil** and **gas** are fluids that are not normally classed with sedimentary rocks. They can be considered organic sediments, however, because they form by the diagenesis of organic material in the pores of sedimentary rocks. Deep burial changes organic matter originally deposited along with inorganic sediments into a fluid that then escapes to porous formations and becomes trapped there. As noted earlier in this chapter, oil and gas are found mainly in sandstones and limestones (see Chapter 23).

SUMMARY

What are the major processes that form sedimentary rock? Weathering and erosion produce the particles that compose siliciclastic sediments and the dissolved ions that

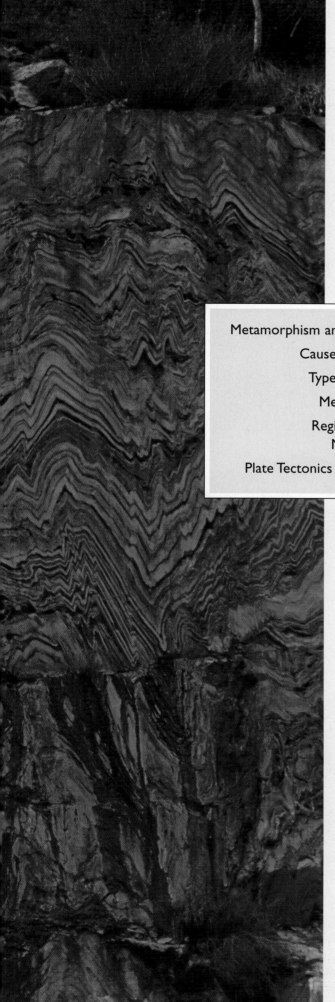

METAMORPHISM
Modification of Rocks
by Temperature
and Pressure

We are all familiar with some ways in which heat and pressure can transform materials. Frying raw ground meat changes it into a hamburger composed of chemical compounds very different from those in the raw meat. Cooking batter in a waffle iron not only heats up the batter but also puts pressure on it, transforming it into a rigid solid. In similar ways, rocks change as they encounter high temperatures and pressures. Deep in Earth's crust, tens of kilometers below the surface, temperatures and pressures are high enough to transform rock without being high enough to melt it. Increases in heat and pressure and changes in the chemical environment can alter the mineral compositions and crystalline textures of sedimentary and igneous rocks, *even though they remain solid all the while.* The result is the third large class of rocks: the **metamorphic,** or "changed form," **rocks,** which have undergone changes in mineralogy, texture, chemical composition, or all three.

Metamorphic changes occur when a rock is subjected to new temperatures and pressures. Given enough time—short by geologic standards but usually a million years or more—the rock changes mineralogically and texturally until it is in equilibrium with the new temperatures and pressures. A limestone filled with fossils, for example, might be transformed into a white marble in which no trace of fossils remains. The mineral and chemical composition of the rock may be unaltered, but its texture may have changed drastically from small calcite crystals to large, intergrown calcite crystals that erase such former features as fossils. Shale, a well-bedded rock so finely grained that no individual mineral crystal can be seen with the naked eye, might become a schist in which the original bedding is obscured and the texture is dominated by large crystals of mica. In this metamorphic transformation, both mineral composition and texture have changed, but the overall chemical composition of the rock has remained the same.

Clay minerals are silicates but differ from micas in that they contain lots of water molecules trapped between silicate sheets in the crystal structure. During metamorphism, most of this water is lost as the clay minerals are transformed to mica. Mineralogy, texture, and chemical composition

These rocks show both the layering and the deformation into folds characteristic of sedimentary rocks metamorphosed into marble, schist, and gneiss. Sequoia National Forest, California. [Gregory G. Dimijian/Photo Researchers.]

change in rocks altered by heat or by fluids derived from igneous activity. Some silicate minerals are found mostly in metamorphic rocks. These minerals include kyanite, andalusite, and sillimanite; staurolite; garnet; and epidote. However, these minerals and others such as quartz, muscovite, amphibole, and feldspar can also be found in igneous rocks. Therefore, geologists must use distinctive textures as well as mineral composition to help guide their studies of metamorphic rocks.

Geologists study metamorphic rocks for many reasons, but all relate to one common objective: *to understand how Earth's crust has evolved over geologic time.*

This chapter examines the causes of metamorphism, the types of metamorphism that take place under certain sets of conditions, and the origins of the various textures that characterize metamorphic rocks.

METAMORPHISM AND THE EARTH SYSTEM

Metamorphism, like all other geologic processes, is part of the Earth system. Earth's internal heat drives metamorphism that is caused by high temperature. Thus, Earth's interior heat

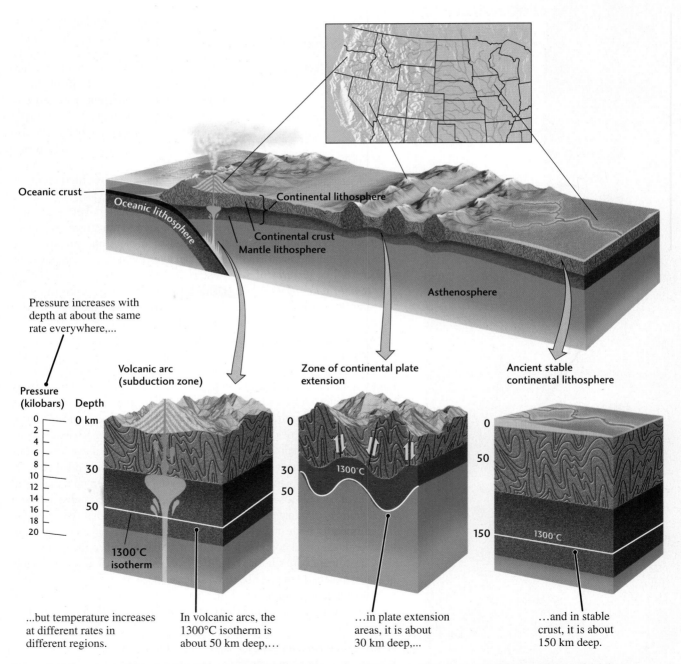

Figure 6.1 Pressure and temperature increase with depth in all regions, as shown in this cross section of a volcanic region, a continental region, and a region of ancient stable continental lithosphere. (Pressure is measured in kilobars; 1 kilobar is approximately equal to 1000 times the atmospheric pressure at Earth's surface.)

powers the parts of the Earth system that govern metamorphic —and igneous—processes. Plate tectonic processes push rocks formed at Earth's surface down to great depths, thereby subjecting them to high pressures as well as high temperatures. As we will see later in this chapter, metamorphism results in the release of water vapor, carbon dioxide, and other gases. These gases leak to the surface and contribute to the atmosphere, affecting processes that depend on atmospheric composition, such as weathering.

CAUSES OF METAMORPHISM

Sediments and sedimentary rocks belong to Earth's surface environments, whereas igneous rocks belong to the melts of the lower crust and mantle. Metamorphic rocks exposed at the surface are mainly the products of processes acting on rocks at depths ranging from the upper to the lower crust. Most have formed at depths of 10 to 30 km, the middle to lower half of the crust. Although most metamorphism takes place at depth, it can also occur at Earth's surface. We can see metamorphic changes in the baked surfaces of soils and sediments just beneath volcanic lava flows.

The internal heat of the Earth, its pressure, and its fluid composition are the three principal factors that drive metamorphism. The contribution of pressure is the result of vertically oriented forces exerted by the weight of overlying rocks and horizontally oriented forces developed as the rocks are deformed.

Temperature increases with depth at different rates in different regions of the Earth, ranging from 20° to 60°C per kilometer of depth (**Figure 6.1**). In much of Earth's crust, temperature increases at a rate of 30°C per kilometer of depth. Thus, at a depth of 15 km, the temperature will be about 450°C —much higher than the average temperature of the surface, which ranges from 10° to 20°C in most regions. The pressure at a depth of 15 km comes from the weight of all the overlying rock and amounts to about 4000 times the pressure at the surface.

High as these temperatures and pressures may seem, they are only in the middle range of metamorphism, as **Figure 6.2** shows. We refer to the metamorphic rocks formed under the lower temperatures and pressures of shallower crustal regions as *low-grade rocks* and the ones formed at the higher temperatures and pressures of deeper zones as *high-grade rocks*. As the grade of metamorphism changes, the mineral assemblages within metamorphic rocks also change.

The Role of Temperature

Heat greatly affects a rock's chemical composition, mineralogy, and texture. In Chapter 4, we learned how important the influence of heat can be in breaking chemical bonds and altering the existing crystal structures of igneous rocks. Heat has an equally important role in the formation of metamorphic rocks. For example, plate tectonic processes may move sediments and rocks from Earth's surface to its interior,

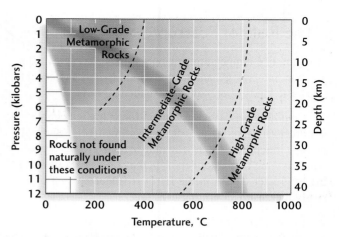

Figure 6.2 Temperatures, pressures, and depths at which low- and high-grade metamorphic rocks form. The dark band shows common rates at which temperature and pressure increase with depth over much of the continents.

where temperatures are higher. As the rock adjusts to the new temperature, its atoms and ions recrystallize, linking up in new arrangements and creating new mineral assemblages. Many new crystals will grow larger than the crystals in the original rock.

The increase of temperature with increasing depth is called a *geothermal gradient* (see Chapter 21 for further discussion of geotherms). The geothermal gradient varies depending on plate tectonic setting, but on average it is about 30°C per kilometer of depth. In areas where plate extension has thinned the continental lithosphere, such as in Nevada's Great Basin, the geothermal gradient is *steep* (for example, 50°C per kilometer of depth). In areas where the continental lithosphere is old and thick, such as beneath central North America, the geothermal gradient is *shallow* (for example, 20°C per kilometer of depth) (see Figure 6.1). As sedimentary rocks containing clay minerals are buried deeper and deeper, the clay minerals begin to recrystallize and form new minerals such as mica. With additional burial to greater depths—and temperatures—the micas become unstable and begin to recrystallize into new minerals such as garnet. Because different minerals crystallize and remain stable at different temperatures, the metamorphic geologist, like the igneous geologist, can use a rock's composition as a kind of *geothermometer* to gauge the temperature at which the rock formed. Given a specific assemblage of minerals in a metamorphic rock, the geologist can infer the temperature at which the rock formed.

Plate tectonic processes such as subduction and continental collision, which transport rocks and sediments into the hot depths of the crust, are the primary mechanisms that form most metamorphic rocks. In addition, limited metamorphism may occur where rocks are subjected to elevated temperatures near recently intruded plutons. The heat is locally intense but does not penetrate deeply. Heat pulses produced by intruding plutons can metamorphose the surrounding country rock, but the effect is local in extent.

The Role of Pressure

Pressure, like temperature, changes a rock's chemical composition, mineralogy, and texture. Solid rock is subjected to two basic kinds of pressure, also called **stress:**

1. *Confining pressure* is a general force applied equally in all directions, like the pressure a swimmer feels when submerged in a pool. Just as a swimmer feels greater pressure when moving to greater depths in the pool, a rock descending to greater depths in the Earth is subjected to progressively increasing confining pressure.

2. *Directed pressure* is force exerted in a particular direction, such as when a ball of clay is squeezed between thumb and forefinger. Directed pressure, or *differential stress,* is usually concentrated within zones or along discrete planes. The compressive force that occurs where plates converge is a form of directed pressure, and it results in deformation of the rocks near the plate boundary. Heat reduces the strength of a rock, so directed pressure is likely to cause severe folding and deformation of metamorphic rocks in mountain belts where temperatures are high. Rocks subjected to differential stress may be severely distorted, becoming flattened in the direction the force is applied and elongated in the direction perpendicular to the force.

Metamorphic minerals may be compressed, elongated, or rotated to line up in a particular direction, depending on the kind of stress applied to the rocks. Thus, directed pressure guides the shape and orientation of the new metamorphic crystals formed as the minerals recrystallize under the influence of both heat and pressure. During the recrystallization of micas, for example, the crystals grow with the planes of their sheet-silicate structures aligned perpendicular to the directed stress. During deformation, the rock may become banded as minerals of different compositions are segregated into separate planes. (See the chapter opening photograph.)

Pressure, like temperature, increases with depth in the Earth. Pressure is usually recorded in *kilobars* (1000 bars, abbreviated as kbar) and increases at a rate of 0.3 to 0.4 kbar per kilometer of depth (see Figure 6.1). One bar is approximately equivalent to the pressure of air at the surface of the Earth. A diver who is touring the deeper part of a coral reef at a depth of 10 m would experience another 1 bar of pressure. The pressure to which a rock is subjected deep in the Earth is related to both the *thickness* of the overlying rocks and the *density* of those rocks.

Minerals that are stable at the lower pressure near Earth's surface become unstable and recrystallize to new minerals under the increased pressure at depth in the crust. Using laboratory data on the pressures required for these changes, we can examine the mineralogy and texture of metamorphic rock samples and infer what the pressures were in the area where they formed. Thus, metamorphic mineral assemblages can be used as pressure gauges, or *geobarometers.* Given a specific assemblage of minerals in a metamorphic rock, the geologist can determine the range of pressures, and therefore depths, at which the rock formed.

The Role of Fluids

Metamorphism can significantly alter a rock's mineralogy by introducing or removing chemical components that dissolve in water. Hydrothermal fluids produced during metamorphism carry dissolved carbon dioxide as well as chemical substances—such as sodium, potassium, silica, copper, and zinc—that are soluble in hot water under pressure. As hydrothermal solutions percolate up to the shallower parts of the crust, they react with the rocks they penetrate, changing their chemical and mineral compositions and sometimes completely replacing one mineral with another without changing the rock's texture. This kind of change in a rock's bulk composition by fluid transport of chemical substances into or out of the rock is called **metasomatism.** Many valuable deposits of copper, zinc, lead, and other metal ores are formed by this kind of chemical substitution.

Hydrothermal fluids accelerate metamorphic chemical reactions. Atoms and ions dissolved in the fluid can migrate through a rock and react with the solids to form new minerals. As metamorphism proceeds, the water itself reacts with the rock when chemical bonds between minerals and water molecules form or break.

Where do these chemically reactive fluids originate? Although most rocks appear to be completely dry and of extremely low porosity, they characteristically contain fluid in minute pores (the spaces between grains). This water comes from chemically bound water in clay, not from sedimentary pore waters, which are largely expelled during diagenesis. In other hydrous minerals, such as mica and amphibole, water forms part of their crystalline structures. The carbon dioxide dissolved in hydrothermal fluids is derived largely from sedimentary carbonates—limestones and dolostones.

TYPES OF METAMORPHISM

Geologists can duplicate metamorphic conditions in the laboratory and determine the precise combinations of pressure, temperature, and chemical composition under which transformations might take place. But to understand how any particular combination relates to the geology of metamorphism—when, where, and how these conditions came about in the Earth—geologists categorize metamorphic rocks on the basis of the geological circumstances of their origins. We describe these categories next; **Figure 6.3** locates them in relation to major plate tectonic settings.

Regional Metamorphism

Regional metamorphism, the most widespread type, takes place where both high temperature and high pressure are

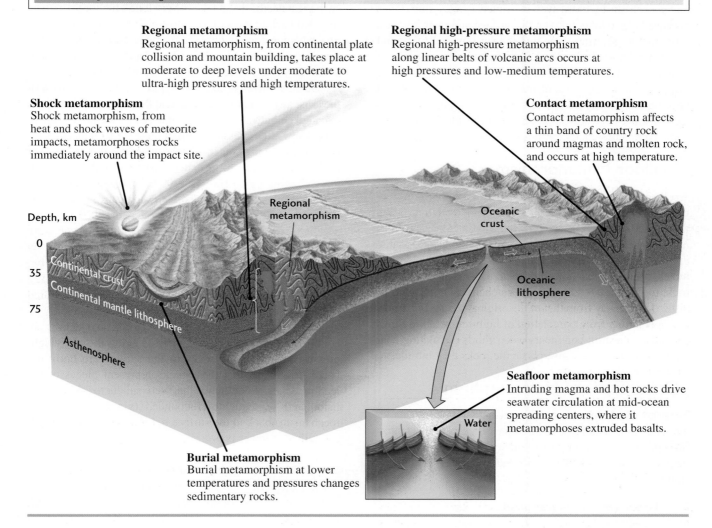

Earth System Figure 6.3   The lithosphere and asthenosphere interact to metamorphose rock.

Regional metamorphism
Regional metamorphism, from continental plate collision and mountain building, takes place at moderate to deep levels under moderate to ultra-high pressures and high temperatures.

Regional high-pressure metamorphism
Regional high-pressure metamorphism along linear belts of volcanic arcs occurs at high pressures and low-medium temperatures.

Shock metamorphism
Shock metamorphism, from heat and shock waves of meteorite impacts, metamorphoses rocks immediately around the impact site.

Contact metamorphism
Contact metamorphism affects a thin band of country rock around magmas and molten rock, and occurs at high temperature.

Regional metamorphism

Oceanic crust

Depth, km

0

35

75

Continental crust

Continental mantle lithosphere

Asthenosphere

Oceanic lithosphere

Burial metamorphism
Burial metamorphism at lower temperatures and pressures changes sedimentary rocks.

Seafloor metamorphism
Intruding magma and hot rocks drive seawater circulation at mid-ocean spreading centers, where it metamorphoses extruded basalts.

Water

imposed over large parts of the crust. We use this term to distinguish this type of metamorphism from more localized changes near igneous intrusions or faults. Regional metamorphism is a characteristic feature of convergent plate tectonic settings. It occurs in volcanic island arcs, such as the Andes of South America, and in the cores of mountain chains produced during the collision of continents, such as the Himalaya of central Asia. These mountain belts are often linear features, so ancient (and modern) zones of regional metamorphism are often linear in their distribution. In fact, geologists usually interpret belts of regionally extensive metamorphic rocks as representing sites of former mountain chains that were eroded over millions of years, exposing the rocks at Earth's surface.

Some regional metamorphic belts are created by high temperatures and moderate to high pressures near the volcanic arcs formed where subducted plates sink deep into the mantle. Regional metamorphism under very high pressures and temperatures takes place at deeper levels of the crust along boundaries where colliding continents deform rock

and raise high mountain belts. During regional metamorphism, rocks are typically transported to significant depths in the Earth's crust, then uplifted and eroded at Earth's surface. However, a full understanding of the patterns of regional metamorphism, including how rocks respond to systematic changes in temperature and pressure over time, depends on the specific tectonic setting. We will discuss this topic later in the chapter.

Contact Metamorphism

The heat from igneous intrusions metamorphoses the immediately surrounding rock. This type of localized transformation, called **contact metamorphism,** normally affects only a thin region of country rock along the contact. In many contact metamorphic rocks, especially at the margins of shallow intrusions, the mineral and chemical transformations are largely related to the high temperature of the magma. Pressure effects are important only where the magma was intruded at great depths. Here, pressure results not from the intrusion forcing its way into the country rock but from the presence of regional confining pressure. Contact metamorphism by extrusives is limited to very thin zones because lavas cool quickly at the surface and their heat has little time to penetrate deep into the surrounding rocks and cause metamorphic changes. Contact metamorphism may also affect large blocks of rock up to several meters wide that are torn off the sides of magma chambers. The blocks become completely surrounded by hot magma, heat projects inward from all directions, and the blocks may become completely metamorphosed.

Seafloor Metamorphism

Another type of metamorphism, called **seafloor metamorphism** or metasomatism, is often associated with mid-ocean ridges (see Chapter 4). Hot, fractured basalts heat infiltrating seawater, which starts to circulate through the basaltic upper crust by convection. The increase in temperature promotes chemical reactions between the seawater and the rock, forming altered basalts whose chemical compositions differ distinctively from that of the original basalt. Metamorphism resulting from percolation of high-temperature fluids also takes place on continents when fluids circulating near igneous intrusions metamorphose the rocks they intrude.

Other Types of Metamorphism

There are other types of metamorphism that produce smaller amounts of metamorphic rock. Some of these, such as ultrahigh-pressure metamorphism, are extremely important in helping geologists understand conditions deep within the Earth.

Low-Grade (Burial) Metamorphism Recall from Chapter 5 that when sedimentary rocks are gradually buried during subsidence of the crust, they slowly heat up as they come into equilibrium with the temperature of the crust around them. In this process, diagenesis alters their mineralogy and texture. Diagenesis grades into **low-grade,** or **burial, metamorphism,** which is caused by the progressive increase in pressure exerted by the growing pile of overlying sediments and sedimentary rocks and by the increase in heat associated with increased depth of burial in the Earth.

Depending on the local geothermal gradient, low-grade metamorphism typically begins at depths of 6 to 10 km, where temperatures range between 100° and 200°C and pressures are less than 3 kbar. This fact is of great importance to the oil and gas industry, which defines "economic basement" as the depth where low-grade metamorphism begins. Oil and gas wells are rarely drilled below this depth because temperatures above 130°C convert organic matter trapped in sedimentary rocks into carbon dioxide rather than crude oil and natural gas.

High-Pressure and Ultra-High-Pressure Metamorphism Metamorphic rocks formed by **high-pressure** (8 to 12 kbar) and **ultra-high-pressure** (greater than 28 kbar) **metamorphism** are rarely exposed at the surface for geologists to study. These rocks are unusual because they form at such great depths that it takes a very long time for them to be recycled back to the surface. Most high-pressure rocks form in subduction zones where sediments scraped from subducting oceanic plates are plunged to depths of over 30 km, where they experience pressures of up to 12 kbar.

Unusual metamorphic rocks once located at the base of the crust can sometimes be found at the surface. These rocks —called **eclogites**—may contain minerals such as *coesite* (a very dense, high-pressure form of quartz) that indicate pressures of greater than 28 kbar, suggesting depths of over 80 km. Such rocks form under moderate to high temperatures, ranging up to 800° to 1000°C. In a few cases, these rocks contain *microscopic diamonds,* indicative of pressures greater than 40 kbar and depths greater than 120 km! Surprisingly, outcrop exposures of these ultra-high-pressure metamorphic rocks cover areas greater than 400 km by 200 km. The only other two rocks known to come from these depths are diatremes and kimberlites (see Chapter 12), igneous rocks that form narrow pipes just a few hundred meters wide. Geologists agree that these rocks form by "volcanic" eruption, albeit from very unusual depths. In contrast, the mechanisms required to bring the ultra-high-pressure metamorphic rocks to the surface are hotly debated. It appears that these rocks represent pieces of the leading edges of continents that were subducted during collision and subsequently rebounded (via some unknown mechanism) back to the surface before they had time to recrystallize at lower pressures.

Shock Metamorphism **Shock metamorphism** occurs when a meteorite collides with Earth. Meteorites are fragments of comets or asteroids that have been brought to Earth by its gravitational field. Upon impact, the energy represented by the meteorite's mass and velocity is transformed to heat and shock waves that pass through the impacted country rock. The country rock can be shattered and even partially melted to produce *tektites.* The smallest tektites look like droplets of glass. In some cases, quartz is transformed into coesite and *stishovite,* two of its high-pressure forms.

Most large impacts on Earth have left no trace of a meteorite because these bodies are usually destroyed in the collision with Earth. The occurrence of coesite and craters with distinctive fringing fracture textures, however, is evidence of these collisions. Earth's dense atmosphere causes most meteorites to burn up before they strike its surface, so shock metamorphism is rare on Earth. On the surface of the Moon, however, shock metamorphism is pervasive. It is characterized by extremely high pressures of many tens to hundreds of kilobars.

METAMORPHIC TEXTURES

Metamorphism imprints new textures on the rocks that it alters (**Figure 6.4**). The texture of a metamorphic rock is

Key Figure 6.4 — Regional metamorphism changes rock texture.

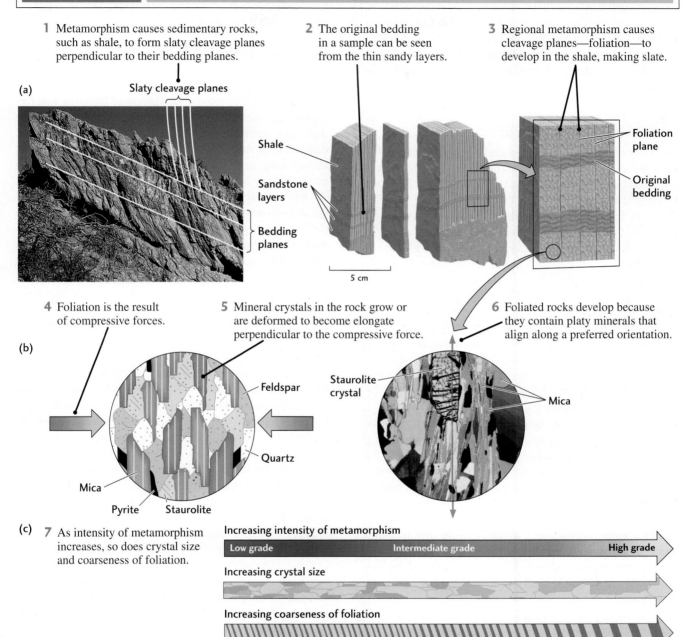

1 Metamorphism causes sedimentary rocks, such as shale, to form slaty cleavage planes perpendicular to their bedding planes.

(a)

Slaty cleavage planes

2 The original bedding in a sample can be seen from the thin sandy layers.

Shale

Sandstone layers

Bedding planes

5 cm

3 Regional metamorphism causes cleavage planes—foliation—to develop in the shale, making slate.

Foliation plane

Original bedding

4 Foliation is the result of compressive forces.

(b)

5 Mineral crystals in the rock grow or are deformed to become elongate perpendicular to the compressive force.

Feldspar

Quartz

Mica

Pyrite   Staurolite

Staurolite crystal

Mica

6 Foliated rocks develop because they contain platy minerals that align along a preferred orientation.

(c) **7** As intensity of metamorphism increases, so does crystal size and coarseness of foliation.

Increasing intensity of metamorphism

| Low grade | Intermediate grade | High grade |

Increasing crystal size

Increasing coarseness of foliation

(d) **8** Foliated rocks are classified by the degree of cleavage, schistosity, and banding, which corresponds to the intensity of metamorphism.

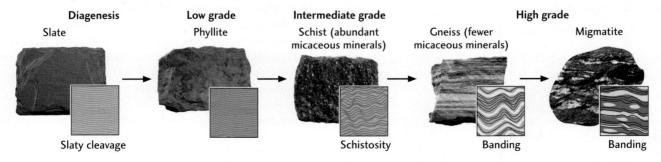

Diagenesis	**Low grade**	**Intermediate grade**	**High grade**	
Slate	Phyllite	Schist (abundant micaceous minerals)	Gneiss (fewer micaceous minerals)	Migmatite
Slaty cleavage		Schistosity	Banding	Banding

determined by the sizes, shapes, and arrangement of its constituent crystals. Some metamorphic textures depend on the particular kinds of minerals formed, such as the micas, which are platy. Variation in grain size is also important. In general, geologists find that the grain size of crystals increases as metamorphic grade increases. Each textural variety tells us something about the metamorphic process that created it.

Foliation and Cleavage

The most prominent textural feature of regionally metamorphosed rocks is **foliation,** a set of flat or wavy parallel planes produced by deformation. These foliation planes may cut the bedding at any angle or be parallel to the bedding (see Figure 6.4a). In general, as the grade of regional metamorphism increases, the foliation will become more pronounced.

A major cause of foliation is the presence of platy minerals, chiefly the micas and chlorite. Platy minerals tend to crystallize as thin platelike crystals. The planes of all the platy crystals are aligned parallel to the foliation, an alignment called the *preferred orientation* of the minerals (see Figure 6.4b). As platy minerals crystallize, the preferred orientation is usually perpendicular to the main direction of the deformation forces squeezing the rock during metamorphism. Preexisting minerals may acquire a preferred orientation and thus produce foliation by rotating until they lie parallel to the developing plane.

Minerals whose crystals have an elongate, pencil-like shape also tend to assume a preferred orientation during metamorphism: the crystals normally line up parallel to the foliation plane. Rocks that contain abundant amphiboles, typically metamorphosed mafic volcanics, have this kind of texture.

The most familiar form of foliation is seen in slate, a common metamorphic rock, which is easily split into thin sheets along smooth, parallel surfaces. This *slaty cleavage* (not to be confused with the cleavage of a mineral such as muscovite) develops along moderately thin, regular intervals in the rock.

Classification of Foliated Rocks

The **foliated rocks** are classified according to four main criteria (Figure 6.4c):

1. The size of their crystals

2. The nature of their foliation

3. The degree to which their minerals are segregated into lighter and darker bands

4. Their metamorphic grade

Figure 6.4d shows examples of the major types of foliated rocks. In general, foliation progresses from one texture

to another, reflecting the increase in temperature and pressure. In this progression, a shale may metamorphose first to a slate, then to a phyllite, then to a schist, then to a gneiss, and finally to a migmatite.

Slate **Slates** are the lowest grade of foliated rocks. These rocks are so fine-grained that their individual minerals cannot be seen easily without a microscope. They are commonly produced by the metamorphism of shales or, less frequently, of volcanic ash deposits. Slates usually range from dark gray to black, colored by small amounts of organic material originally present in the parent shale. Slate splitters learned long ago to recognize this foliation and use it to make thick or thin slates for roofing tiles and blackboards. We still use flat slabs of slate for flagstone walks in parts of the country where slate is abundant.

Phyllite The **phyllites** are of a slightly higher grade than the slates but are similar in character and origin. They tend to have a more or less glossy sheen resulting from crystals of mica and chlorite that have grown a little larger than those of slates. Phyllites, like slates, tend to split into thin sheets, but less perfectly than slates.

Schist At low grades of metamorphism, crystals of platy minerals are generally too small to be seen, foliation is closely spaced, and layers are very thin. As rocks are more intensely metamorphosed into higher grades, the platy crystals grow large enough to be visible to the naked eye, and the minerals may tend to segregate in lighter and darker bands. This parallel arrangement of sheet minerals produces the pervasive coarse, wavy foliation known as *schistosity,* which characterizes **schists.** Schists, which are intermediate-grade rocks, are among the most abundant metamorphic rock types. They contain more than 50 percent platy minerals, mainly the micas muscovite and biotite. Schists may contain thin layers of quartz, feldspar, or both, depending on the quartz content of the parent shale.

Gneiss Even coarser foliation is shown by high-grade **gneisses,** light-colored rocks with coarse bands of segregated light and dark minerals throughout the rock. The banding of gneisses into light and dark layers results from the segregation of lighter-colored quartz and feldspar and darker amphiboles and other mafic minerals. Gneisses are coarse grained, and the ratio of granular to platy minerals is higher than it is in slate or schist. The result is poor foliation and thus little tendency to split. Under conditions of high pressure and temperature, mineral assemblages of the lower-grade rocks containing micas and chlorite change into new assemblages dominated by quartz and feldspars, with lesser amounts of micas and amphiboles.

Migmatite Temperatures higher than those necessary to produce gneiss may begin to melt the country rock. In this case, as with igneous rocks (see Chapter 4), the first minerals to melt will be those with the lowest melting temperatures. Therefore only part of the country rock melts, and the melt may migrate only a short distance before freezing again. Rocks produced in this way are badly deformed and contorted, and they are penetrated by many veins, small pods, and lenses of melted rock. The result is a mixture of igneous and metamorphic rock called **migmatite.** Some migmatites are mainly metamorphic, with only a small proportion of igneous material. Others have been so affected by melting that they are considered almost entirely igneous.

Granoblastic Rocks

Granoblastic rocks are composed mainly of crystals that grow in equant (equidimensional) shapes, such as cubes and spheres, rather than in platy or elongate shapes. These rocks result from metamorphism in which deformation is absent, such as contact metamorphism. Granoblastic (nonfoliated) rocks include hornfels, quartzite, marble, greenstone, amphibolite, and granulite (**Figure 6.5**). All granoblastic rocks, except hornfels, are defined by their mineral composition rather than their texture because all of them are massive in appearance.

Hornfels is a high-temperature contact metamorphic rock of uniform grain size that has undergone little or no deformation. Its platy or elongate crystals are oriented randomly, and foliated texture is absent. Hornfels has a granular texture overall, even though it commonly contains pyroxene, which makes elongate crystals, and some micas.

Quartzites are very hard, nonfoliated white rocks derived from quartz-rich sandstones. Some quartzites are massive, unbroken by preserved bedding or foliation (see Figure 6.5a). Others contain thin bands of slate or schist, relics of former interbedded layers of clay or shale.

Marbles are the metamorphic products of heat and pressure acting on limestones and dolomites. Some white, pure marbles, such as the famous Italian Carrara marbles prized by sculptors, show a smooth, even texture of intergrown calcite crystals of uniform size. Other marbles show irregular banding or mottling from silicate and other mineral impurities in the original limestone (see Figure 6.5b).

Greenstones are metamorphosed mafic volcanic rocks. Many of these low-grade rocks form when mafic lavas and ash deposits react with percolating seawater or other solutions. Large areas of the seafloor are covered with basalts slightly or extensively altered in this way at mid-ocean ridges. An abundance of chlorite gives these rocks their greenish cast.

Amphibolite is generally a nonfoliated rock made up of amphibole and plagioclase feldspar. It is typically the product of medium- to high-grade metamorphism of mafic volcanics. Foliated amphibolites are produced when deformation occurs.

(a) Quartzite

(b) Marble

Figure 6.5 Granoblastic (nonfoliated) metamorphic rocks. (a) Quartzite [Breck P. Kent]; (b) marble [Diego Lezama Orezzoli/ Corbis].

The high-grade metamorphic rock **granulite** has a granoblastic texture; these rocks are often referred to as *granofels.* Granofels are medium- to coarse-grained rocks in which the crystals are equant and show only faint foliation at most. They are formed by the metamorphism of shale, impure sandstone, and many kinds of igneous rock.

Figure 6.6 Garnet porphyroblasts in a schist matrix. Matrix minerals are continuously recrystallized as pressure and temperature change and therefore grow only to small size. In contrast, porphyroblasts grow to large size because they are stable over a broad range of pressures and temperatures. [Chip Clark.]

Large-Crystal Textures

New metamorphic minerals may grow into large crystals surrounded by a much finer grained matrix of other minerals. These large crystals are **porphyroblasts** and are found in both contact and regionally metamorphosed rocks (**Figure 6.6**). They grow as the chemical components of the matrix are reorganized and thus replace parts of the matrix. Porphyroblasts form when there is a strong contrast between the chemical and crystallographic properties of the matrix and those of the porphyroblast minerals. This contrast causes the porphyroblast crystals to grow faster than the slow-growing minerals of the matrix, at the expense of the matrix. Porphyroblasts vary in size, ranging from a few millimeters to several centimeters in diameter. Their composition also varies. Garnet and staurolite are two common minerals that form porphyroblasts, but many others are also found. The precise composition and distribution of porphyroblasts of these two minerals can be used to infer the paths of pressure and temperature that occurred during metamorphism.

Table 6.1 summarizes the textural classes of metamorphic rocks and their main characteristics.

Table 6.1	Classification of Metamorphic Rocks by Texture		
Classification	**Characteristics**	**Rock Name**	**Typical Parent Rock**
Foliated	Distinguished by slaty cleavage, schistosity, or gneissic foliation; mineral grains show preferred orientation	Slate Phyllite Schist Gneiss	Shale, sandstone
Granoblastic (nonfoliated)	Granular, characterized by coarse or fine interlocking grains; little or no preferred orientation	Hornfels Quartzite Marble Argillite Greenstone Amphibolite[a] Granulite[b]	Shale, volcanics Quartz-rich sandstone Limestone, dolomite Shale Basalt Shale, basalt Shale, basalt
Porphyroblastic	Large crystals set in fine matrix	Slate to gneiss	Shale

[a]Typically contains much amphibole, which may show alignment of long, narrow crystals.
[b]High-temperature, high-pressure rock.

REGIONAL METAMORPHISM AND METAMORPHIC GRADE

Metamorphic rocks form under a wide range of conditions, and their minerals and textures are clues to the pressures and temperatures in the crust where and when they formed. Geologists who study the formation of metamorphic rocks constantly seek to determine the intensity and character of metamorphism more precisely than is indicated by a designation of "low grade" or "high grade." To make these finer distinctions, geologists read minerals as though they were pressure gauges and thermometers. The techniques are best illustrated by their application to regional metamorphism.

Mineral Isograds: Mapping Zones of Change

When geologists study broad belts of regionally metamorphosed rocks, they can see many outcrops, some showing one set of minerals, some showing others. Different zones within these belts may be distinguished by their index minerals, the predominant minerals that define the zones. The index minerals all formed under a limited range of pressures and temperatures (**Figure 6.7**). For example, one may cross from a region of unmetamorphosed shales to a zone of weakly metamorphosed slates and then to a zone of high-grade schists (Figure 6.7a). As the margin of the slate zone is encountered, a new mineral—chlorite—appears. Moving in the direction of increasing metamorphism, the geologist may successively encounter other metamorphic mineral zones, and the schists will become progressively more foliated (Figure 6.7b).

We can make a map of these zones where one metamorphic grade changes to another. To do so, geologists define the zones by drawing lines called *isograds* that plot the transition from one zone to the next. Isograds are used in Figure 6.7a to show a series of rocks produced by the regional metamorphism of a shale. A pattern of isograds tends to follow the trend of deformation features of a region, as outlined by folds and faults. An isograd based on a single index mineral, such as the biotite isograd, is a good approximate measure of metamorphic pressure and temperature.

To determine pressure and temperature more precisely, geologists examine a group of two or three minerals that crystallized together. For example, a sillimanite isograd would contain potassium feldspar (K-feldspar) and sillimanite, which formed by the reaction of muscovite and quartz at temperatures of about 600°C and pressures of about 5 kbar, liberating water (as water vapor) in the process:

muscovite $\quad+\quad$ quartz $\quad\rightarrow$

$$KAl_3Si_3O_{10}(OH) \qquad SiO_2$$

K-feldspar $\quad+\quad$ sillimanite $\quad+\quad$ water

$$KAlSi_3O_8 \qquad Al_2SiO_5 \qquad H_2O$$

Isograds reveal the pressures and temperatures at which minerals form, so the isograd sequence in one metamorphic belt may differ from that in another. The reason for this difference is that pressure and temperature do not increase at the same rate in all geologic settings. As we discussed earlier in this chapter, pressure increases more rapidly than temperature in some places and more slowly in others (see Figures 6.1 and 6.2).

Metamorphic Grade and Parent-Rock Composition

The kind of metamorphic rock that results from a given grade of metamorphism depends partly on the mineral composition of the parent rock. The metamorphism of slate shown in Figure 6.7b reveals the effects of pressure and temperature on rocks rich in clay minerals, quartz, and perhaps some carbonate minerals. The metamorphism of mafic volcanic rocks, composed predominantly of feldspars and pyroxene, follows a different course (**Figure 6.8**).

In the regional metamorphism of a basalt, for example, the lowest-grade rocks characteristically contain various **zeolite** minerals. The silicate minerals in this class contain water in cavities within the crystal structure. Zeolite minerals form at very low temperatures and pressures. Rocks that include this group of minerals are thus identified as zeolite grade.

Overlapping with the zeolite grade is a higher grade of metamorphosed mafic volcanic rocks, the **greenschists,** whose abundant minerals include chlorite. Next are the *amphibolites,* which contain large amounts of amphiboles. The granulites, coarse-grained rocks containing pyroxene and calcium plagioclase, are the highest grade of metamorphosed mafic volcanics. Greenschist, amphibolite, and granulite grade rocks are also formed during metamorphism of sedimentary rocks, as shown in Figure 6.7c.

These pyroxene-bearing granulites are the products of high-grade metamorphism in which the temperature is high and the pressure is moderate. The opposite situation, in which the pressure is high and the temperature moderate, produces rocks of **blueschist** grade with various starting compositions, from mafic volcanic rocks to shaley sedimentary rocks. The name comes from the abundance of glaucophane, a blue amphibole, in these rocks. Still another metamorphic rock, formed at extremely high pressures and moderate to high temperatures, is *eclogite,* which is rich in garnet and pyroxene.

Metamorphic Facies

We can put all this information about metamorphic grades —derived from parent rocks of many different chemical compositions—on a graph of temperature and pressure (see Figure 6.7d). **Metamorphic facies** are groupings of rocks of various mineral compositions formed under different grades of metamorphism from different parent rocks. By

Key Figure 6.7   Index minerals, grade, and facies describe metamorphism.

1 Index minerals define metamorphic zones. Laboratory studies have determined the temperature and pressure at which various rocks and minerals have formed.

2 Isograds—lines that plot the transition from one mineral to another—can be used to plot the degree of metamorphism (temperature and pressure) over an area such as New England.

3 As rocks such as slate metamorphose, they progress from low-grade rocks to high-grade rocks.

(a)

Canada

ME

NY

Isograds

VT   NH

MA

CT

RI

200 km

Key:

Not metamorphosed

Low grade { Chlorite zone
Biotite zone

Medium grade { Garnet zone
Staurolite zone

High grade   Sillimanite zone

(b)

Pressure (kilobars)

Low Grade

Slate

Increasing metamorphic grade

Phyllite

Intermediate Grade

Schist

High Grade

Blueschist

Gneiss

Migmatite

Temperature (°C)

Depth (km)

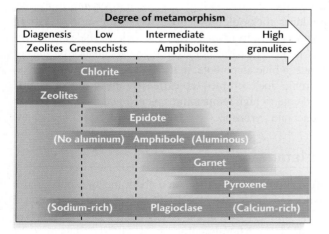

Degree of metamorphism			
Diagenesis	Low	Intermediate	High
Zeolites	Greenschists	Amphibolites	granulites

Chlorite

Zeolites

Epidote

(No aluminum) Amphibole (Aluminous)

Garnet

Pyroxene

(Sodium-rich)   Plagioclase   (Calcium-rich)

Figure 6.8 Changes in the mineral composition of mafic rocks, metamorphosed under conditions ranging from low grade to high grade.

designating particular metamorphic facies, we can be more specific about the degree of metamorphism preserved in rocks. Two essential points characterize the concept of metamorphic facies:

1. Different kinds of metamorphic rocks form from parent rocks of different composition at the same grade of metamorphism.

2. Different kinds of metamorphic rocks form at different grades of metamorphism from parent rocks of the same composition.

Table 6.2 lists the major minerals of the metamorphic facies produced from shale and basalt. Because parent rocks vary so greatly in composition, there are no sharp boundaries between metamorphic facies (see Figure 6.7d). Perhaps the most important reason for analyzing metamorphic

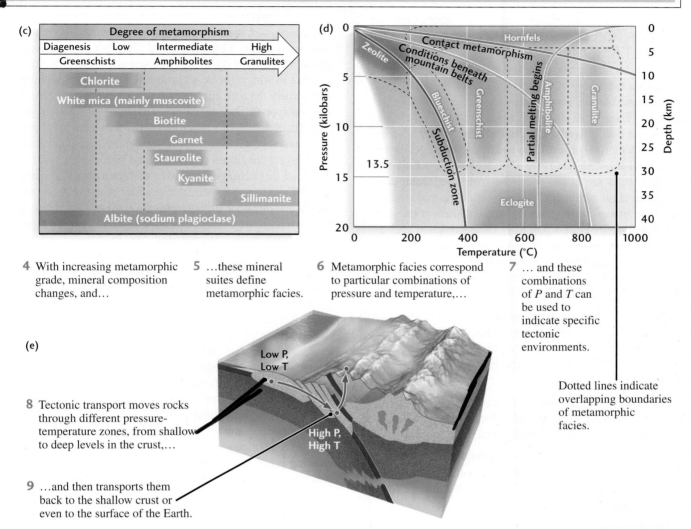

(c)

| Degree of metamorphism |
| Diagenesis | Low | Intermediate | High |
| Greenschists | Amphibolites | Granulites |

- Chlorite
- White mica (mainly muscovite)
- Biotite
- Garnet
- Staurolite
- Kyanite
- Sillimanite
- Albite (sodium plagioclase)

(d)

Pressure (kilobars) / Depth (km) vs Temperature (°C)

Zeolite, Hornfels, Contact metamorphism, Conditions beneath mountain belts, Blueschist, Greenschist, Amphibolite, Granulite, Subduction zone, Partial melting begins, Eclogite

13.5

[(slate, phyllite, schist, gneiss): John Grotzinger/Ramón Rivera-Moret/Harvard Mineralogical Museum. (blueschist): courtesy of Mark Cloos. (migmatite): Kip Hodges.]

4 With increasing metamorphic grade, mineral composition changes, and...

5 ...these mineral suites define metamorphic facies.

6 Metamorphic facies correspond to particular combinations of pressure and temperature,...

7 ... and these combinations of P and T can be used to indicate specific tectonic environments.

Dotted lines indicate overlapping boundaries of metamorphic facies.

(e)

8 Tectonic transport moves rocks through different pressure-temperature zones, from shallow to deep levels in the crust,...

9 ...and then transports them back to the shallow crust or even to the surface of the Earth.

Low P, Low T

High P, High T

Table 6.2   Major Minerals of Metamorphic Facies Produced from Parent Rocks of Different Composition

Facies	Minerals Produced from Shale Parent	Minerals Produced from Basalt Parent
Greenschist	Muscovite, chlorite, quartz, albite	Albite, epidote, chlorite
Amphibolite	Muscovite, biotite, garnet, quartz, albite, staurolite, kyanite, sillimanite	Amphibole, plagioclase feldspar
Granulite	Garnet, sillimanite, albite, orthoclase, quartz, biotite	Calcium-rich pyroxene, calcium-rich plagioclase feldspar
Eclogite	Garnet, sodium-rich pyroxene, quartz/coesite, kyanite	Sodium-rich pyroxene, garnet

facies is that they give us clues to the tectonic processes responsible for metamorphism (see Figure 6.7e).

PLATE TECTONICS AND METAMORPHISM

Soon after the theory of plate tectonics was proposed, geologists started to see how patterns of metamorphism fit into the larger framework of plate tectonic movements that cause volcanism and **orogeny.** Orogeny means "mountain making," particularly by the folding and thrusting of rock layers, often with accompanying magmatic activity. Regional metamorphic belts are often associated with continental collisions that build mountains. In the cores of the major mountain belts of the world, from the Appalachians to the Alps, we find long belts of regionally metamorphosed and deformed sedimentary and volcanic rocks that parallel the lines of folds and faults in the mountains.

Different types of metamorphism are likely to occur in different tectonic settings (see Figure 6.3):

• *Plate interiors.* Contact metamorphism, burial metamorphism, and perhaps regional metamorphism occur at the base of the crust. Shock metamorphism is likely to be best preserved in this setting because plate interiors are large exposed areas.

• *Divergent plate margins.* Seafloor metamorphism and contact metamorphism around intruding plutons in the ocean crust are found at divergent plate margins.

• *Convergent plate margins.* Regional metamorphism, high-pressure and ultra-high-pressure metamorphism, and contact metamorphism around intruding plutons are found at convergent plate boundaries.

• *Transform plate margins.* In oceanic settings, seafloor metamorphism may occur. In both oceanic and continental settings, we find extensive shearing along the plate boundary.

Metamorphic Pressure-Temperature Paths

The concept of metamorphic grade, introduced above, is completely static. This means that the grade of metamorphism can inform us of the maximum pressure or temperature to which a rock was subjected, but it says nothing about where the rock encountered these conditions or how it was transported back to Earth's surface. It is important to understand that most metamorphism is a dynamic process, not a static event. Metamorphism generally is characterized by changing conditions of pressure and temperature, and the history of these changes is called a **metamorphic P-T path.** The P-T path can be a sensitive recorder of many important factors that influence metamorphism—such as the sources of heat, which change temperatures, and the rates of tectonic transport, which change pressures.

To obtain a P-T path, geologists must analyze specific metamorphic minerals in the laboratory. One of the most widely used minerals is garnet, which serves as a sort of recording device (**Figure 6.9**). During metamorphism, garnet grows steadily, and as the pressure and temperature of the environment change, the composition of the garnet changes. The oldest part of the garnet is its core and the youngest is its outer edge, so the variation in composition from core to edge will yield the history of metamorphic conditions. From a measured value of garnet composition in the lab, the corresponding values for pressure and temperature can be obtained and then plotted as a P-T path. P-T paths have two segments. The *prograde* segment indicates increasing pressure and temperature, and the *retrograde* segment indicates decreasing pressure and temperature.

Ocean-Continent Convergence

The rock assemblages that form when a plate carrying a continent on its leading edge converges with a subducting

1 During metamorphism, a garnet crystal grows, and its composition changes as the temperature and pressure around it change.

2 Composition of crystal can be plotted on the P-T path as it grows from ❶ in its center to ❷ at its edge.

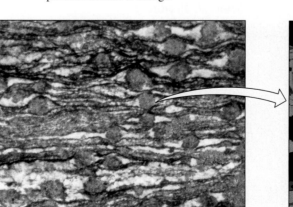

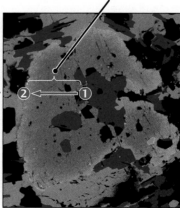

Figure 6.9 Metamorphic pressure-temperature paths. The path that a metamorphic rock typically follows begins with an increase in pressure and temperature, the prograde path, followed by a decrease in pressure and temperature, the retrograde path.
[Photos courtesy of Kip Hodges.]

Thin section of garnet gneiss

Growth zoning in garnet

oceanic plate are shown in **Figure 6.10**. Thick sediments eroded from the continent rapidly fill the adjacent depressions in the seafloor around the subduction zone. As it descends, the cold oceanic slab stuffs the region below the inner wall of the trench (the wall closer to land) with these sediments and with deep-sea sediments and ophiolite shreds scraped off the descending plate. Regions of this sort, located between the magmatic arc on the continent and the trench offshore, are enormously complex and variable. The deposits are all highly folded, intricately sliced, and metamorphosed. They are difficult to map in detail but are recognizable by their distinctive combination of materials and structural features. Such a chaotic mix is called a **mélange** (French for "mixture"). The metamorphism is the kind characteristic of high pressure and low temperature, because the material may be carried relatively rapidly to depths as great as 30 km, where recrystallization occurs in the environment of the still-cold subducting slab.

Subduction-Related Metamorphism *Blueschists*—The metamorphosed volcanic and sedimentary rocks whose minerals (see Figure 6.10) indicate that they were produced under very high pressures but at relatively low temperatures—form in the forearc region of a subduction zone, the area between the seafloor trench and the volcanic arc. Here sediments are carried down the subduction zone along the surface of a cool subducting lithospheric slab. The subducted plate moves down so quickly that there is little time for it to heat up, whereas the pressure increases rapidly.

Eventually, as part of the subduction process, the material rises back to the surface. This **exhumation** occurs because of two effects: buoyancy and circulation. Imagine trying to push a basketball below the surface of a swimming pool. The air-filled basketball has a lower density than the surrounding water, so it tends to rise back to the surface. In a similar way, the subducted metamorphic rocks are driven upward by their inherent buoyancy relative to the surrounding crust. But what "pushes" the material down to begin

with? A natural circulation sets up in the subduction zone. You can think of a subduction zone as an eggbeater. As the eggbeater rotates, it moves the froth in a circular direction. What moves in one direction eventually moves in the opposite direction because of the circular motion. In an analogous way, the sinking slab in a subduction zone sets up a circular motion of material above the slab, first pulling material down to great depths, then returning it to the surface.

Figure 6.10 shows the typical P-T path for rocks subjected to blueschist-grade metamorphism during subduction and exhumation. The P-T path is superimposed on the metamorphic facies diagram. Note that the P-T path forms a loop on this diagram. The prograde part of the path represents subduction, as shown by a rapid increase in pressure for only a relatively small increase in temperature. During exhumation, the path loops back around because temperature is still slowly increasing, but now pressure is rapidly decreasing. The retrograde part of the P-T path represents the exhumation process described above.

Evidence of Ancient Ocean-Continent Convergence
The essential elements of these collisional rock assemblages have been found at many places in the geologic record. One can see mélange in the Franciscan formation of the California Coast Ranges and in the parallel belt of arc magmatism in the Sierra Nevada to the east. These rocks mark the Mesozoic collision between the North American Plate and the Farallon Plate, which has disappeared by subduction (see Figure 20.6). The location of mélange on the west and magmatism on the east shows that the now-absent Farallon Plate was the subducted one, overridden by the North American Plate on the east. Analysis of the P-T paths for metamorphic minerals in the blueschist-grade Franciscan mélange reveals a loop similar to that shown in Figure 6.10, indicating rapid descent to high pressures, which is a diagnostic attribute of subduction.

Other examples of mélange-arc pairs can be found along the continental margins framing the Pacific Basin—in

3 As rock is carried deeper in Earth's crust and is subjected to higher temperatures and pressures (the **prograde** path), the garnet crystal initially grows in a schist but ends up growing in a gneiss as metamorphism progresses.

4 The **retrograde** path indicates decreasing temperatures and pressures as rocks are carried toward Earth's surface.

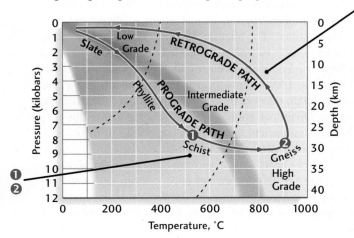

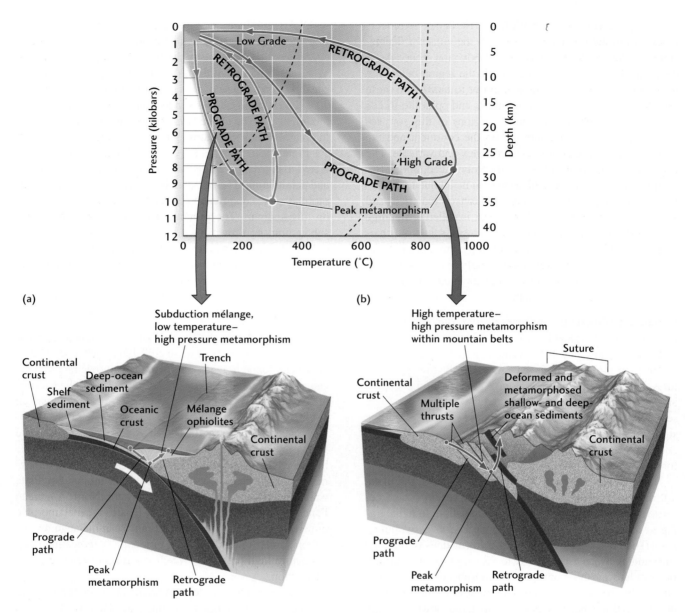

Figure 6.10 P-T paths and rock assemblages associated with (a) ocean-continent plate convergence and (b) continent-continent plate convergence. The P-T paths differ in illustrating the lower geothermal gradients present in subduction zones. Rocks transported to similar depths—and pressures—beneath mountain belts become much hotter at an equivalent depth.

Japan, for instance. The central Alps were uplifted by the convergence of a Mediterranean plate with the European continent. The Andes Mountains near the western coast of South America are products of a collision between ocean and continental plates. Here the Nazca Plate collides with and is subducted under the South American Plate.

Continent-Continent Collision

Plates may have continents embedded in them, and a continent can collide with another continent, as shown in Figure 6.10b. Because continental crust is buoyant, both continents may resist subduction and stay afloat. As a result, they collide, and a wide zone of intense deformation develops at the boundary where the continents grind together. The remnant of such a boundary left behind in the geologic record is called a suture. The intense deformation that occurs during orogeny results in a much-thickened continental crust in the collision zone, often producing high mountains such as the Himalaya. Belts of magmatism characteristically form at depth within the core of the mountain range adjacent to the suture. Ophiolites are often found near the suture; they are relics of an ancient ocean that disappeared in the convergence of two plates (see Chapter 4).

As continents collide and the lithosphere thickens, the deep parts of the continental crust heat up and metamorphose to different grades. In deeper zones, melting may begin at the same time. In this way, a complex mixture of metamorphic and igneous rocks forms the cores of the orogenic belts that evolve during mountain building. Millions of years afterward, when erosion has stripped off the surface layers, the cores are exposed at the surface, providing a rock record of the metamorphic processes that formed the schists, gneisses, and other metamorphic rocks.

P-T paths for metamorphic rocks produced by continental collision have a different shape from those produced by subduction alone. Continental collision generates higher temperatures than subduction. Therefore, as a rock is pushed to greater depths during collision, the temperature that corresponds to a given pressure will be higher (see Figure 6.10b). The P-T path begins at the same place as the path for subduction but shows a more rapid increase in temperature as greater pressures and depths are reached. Geologists generally interpret the prograde segment of a collisional P-T path as indicating the burial of rocks beneath high mountains during orogeny. The retrograde segment represents uplift and exhumation of the buried rocks during the collapse of mountains, either by erosion or by postcollision stretching and thinning of the continental crust.

The prime example of a collision of continents is the Himalaya, which began to form some 50 million years ago when the Indian continent collided with the Asian continent. The collision continues today: India is moving into Asia at a rate of a few centimeters per year, and the uplift is still going on, together with faulting and very rapid erosion rates caused by rivers and glaciers.

Exhumation: Links Between the Plate Tectonic and Climate Geosystems

Forty years ago, plate tectonics theory provided a ready explanation for how metamorphic rocks could be produced by seafloor spreading, plate subduction, and continental collision. By the mid-1980s, the study of P-T paths provided a more highly resolved picture of the specific tectonic mechanisms involved in the deep burial of rocks. At the same time, however, it surprised geologists by providing an equally well resolved image of the subsequent, and often very rapid, uplift and exhumation of these deeply buried rocks. Since the time of this discovery, geologists have been searching for exclusively tectonic mechanisms that could bring these rocks back to Earth's surface so quickly. One popular idea is that mountains, having been built to great elevations during collision-induced crustal thickening, suddenly fail by gravitational collapse. The old saying, "what goes up must go down," applies here, but with surprisingly fast results. So fast, in fact, that some geologists don't believe this is the only important effect—other forces must also be at work.

As we will learn in Chapter 22, geologists who study landscapes have discovered that extremely high erosion rates can be produced by glaciers and rivers in tectonically active mountainous regions. Over the past decade, these geologists have presented a new hypothesis that links rapid rates of uplift and exhumation to rapid erosion rates. The idea is that *climate,* not tectonics alone, drives the flow of rocks from the deep crust to the shallow crust through the process of rapid erosion. Thus, tectonics—which acts through orogeny and mountain building—and climate—which acts through weathering and erosion—*interact* to control the flow of metamorphic rocks to Earth's surface. After decades of emphasis on solely tectonic explanations for regional and global Earth processes, it now seems that two apparently unrelated disciplines of geology—metamorphism and surface processes—may be linked in a very elegant way. As one geologist exclaimed: "Savory the irony should the metamorphic muscles that push mountains to the sky be driven by the pitter patter of tiny raindrops."

SUMMARY

What factors cause metamorphism? Metamorphism—alteration in the solid state of preexisting rocks—is caused by increases in pressure and temperature and by reactions with chemical components introduced by migrating fluids. As pressures and temperatures deep within the crust increase as a result of tectonic or igneous activity, the chemical components of the parent rock rearrange themselves into a new set of minerals that are stable under the new conditions. Rocks metamorphosed at relatively low pressures and temperatures are referred to as low-grade rocks. Those metamorphosed at high temperatures and pressures are called high-grade rocks. Chemical components of a rock may be added or removed during metamorphism, usually by the influence of fluids migrating from nearby intrusions.

What are the various types of metamorphism? The three major types of metamorphism are (1) regional metamorphism, during which large areas are metamorphosed by high pressures and temperatures generated during orogenies; (2) contact metamorphism, during which rocks surrounding magmas are metamorphosed primarily by the heat of the igneous body; and (3) seafloor metamorphism, during which hot fluids percolate through and metamorphose various crustal rocks. Three additional types are (1) low-grade, or burial, metamorphism, during which deeply buried sedimentary rocks are altered by the more or less normal increases in pressure and temperature with depth in the crust; (2) high-pressure and ultra-high-pressure metamorphism, in which rocks may be subjected to pressures as great as 40 kbar, equivalent to depths greater than 120 km; and (3) shock metamorphism, which results from the impact of meteorites. Country rock is shattered by propagating shock waves and, in the process, quartz can be transformed to its denser, high-pressure forms, coesite and stishovite.

What are the chief kinds of metamorphic rocks? Metamorphic rocks fall into two major textural classes: the foliated (displaying fracture cleavage, schistosity, or other forms of preferred orientation of minerals) and the granoblastic, or nonfoliated. The kinds of rocks produced by metamorphism depend on the composition of the parent rock and the grade of metamorphism. The regional metamorphism of a shale leads to zones of foliated rocks of progressively higher grade, from slate to phyllite, schist, gneiss, and migmatite. These zones are marked by isograds defined by the first appearance of an index mineral. Regional metamorphism of mafic volcanic rocks progresses from zeolite facies to greenschist facies and then to amphibolite and pyroxene granulite facies. Among granoblastic rocks, marble is derived from the metamorphism of limestone, quartzite from quartz-rich sandstone, and greenstone from basalt. Hornfels is the product of contact metamorphism of fine-grained sedimentary rocks and other types of rock containing an abundance of silicate minerals. According to the concept of metamorphic facies, rocks of the same grade may differ because of variations in the chemical composition of the parent rocks, whereas rocks of the same composition may vary because of different grades of metamorphism.

How do metamorphic rocks relate to plate tectonic processes? During both subduction and continental collision, preexisting rocks and sediments are pushed to great depths in the Earth, where they are subjected to increasing pressures and temperatures that result in metamorphic mineral reactions. The shapes of metamorphic P-T paths provide insight into how these rocks are metamorphosed. In convergent margin settings, P-T paths indicate rapid subduction of rocks and sediments to sites with high pressures and relatively low temperatures. In settings where subduction leads to continental collision, rocks are pushed down to depths where pressure and temperature are both high. In both settings, the P-T paths form loops. The loops show that after the rocks experienced the maximum pressures and temperatures, they were pushed back up to shallow depths. This process of exhumation may be driven by the collapse of mountain belts either through enhanced weathering and erosion at Earth's surface or through tectonic stretching and thinning of the continental crust.

KEY TERMS AND CONCEPTS

amphibolite (p. 139)

blueschist (p. 141)

contact metamorphism (p. 134)

eclogite (p. 136)

exhumation (p. 145)

foliated rock (p. 138)

foliation (p. 138)

gneiss (p. 138)

granoblastic rock (p. 139)

granulite (p. 139)

greenschist (p. 141)

greenstone (p. 139)

high-pressure (and ultra-high-pressure) metamorphism (p. 136)

hornfels (p. 139)

low-grade (burial) metamorphism (p. 136)

marble (p. 139)

mélange (p. 145)

metamorphic facies (p. 141)

metamorphic P-T path (p. 144)

metamorphic rock (p. 131)

metasomatism (p. 134)

migmatite (p. 139)

orogeny (p. 144)

phyllite (p. 138)

porphyroblast (p. 140)

quartzite (p. 139)

regional metamorphism (p. 134)

schist (p. 138)

seafloor metamorphism (p. 136)

shock metamorphism (p. 136)

slate (p. 138)

stress (p. 134)

zeolite (p. 141)

EXERCISES

1. What types of metamorphism are related to igneous intrusions?

2. What does preferred orientation refer to in a metamorphic rock?

3. What is a porphyroblast?

4. Contrast a schist and a gneiss.

5. How do isograds help determine facies?

6. What is the difference between a granite and a slate?

7. How are metamorphic facies related to temperatures and pressures?

8. In which plate tectonic settings would you expect to find regional metamorphism?

9. What controls exhumation of metamorphic rocks?

10. What is the significance of eclogites at Earth's surface?

THOUGHT QUESTIONS

1. At what depths in the Earth do metamorphic rocks form? What happens if temperatures get too high?

2. Why are there no metamorphic rocks formed under natural conditions of very low pressure and temperature, as shown in Figure 6.2?

3. How is slaty cleavage related to deformation? What forces cause minerals to align with one another?

4. Are cataclastic rocks more likely to be found in a continental rift valley or in a volcanic arc?

5. Would you choose to rely on chemical composition or type of foliation to determine metamorphic grade? Why?

6. You have mapped an area of metamorphic rocks and have observed north-south isograd lines running from kyanite in the east to chlorite in the west. Were metamorphic temperatures higher in the east or in the west?

7. Draw a P-T path for shock metamorphism of country rock during impact of a meteorite.

8. Which kind of pluton would produce the highest grade of metamorphism, a granite intrusion 20 km deep or a gabbro intrusion at a depth of 5 km?

9. Convection of seawater at mid-ocean ridges is driven by heating of seawater as it flows through hot, fractured rock. Draw a sketch of how the metamorphism of seafloor basalts might take place.

10. Subduction zones are generally characterized by high pressure-low temperature metamorphism. In contrast, continental collision zones are marked by moderate pressure-high temperature metamorphism. Which region has a higher geothermal gradient? Explain.

CHAPTER SEVEN

DEFORMATION
Modification of Rocks by Folding and Fracturing

E arly geologists understood that most sedimentary rocks were originally deposited as soft horizontal layers at the bottom of the sea and hardened over time. But they were puzzled by the many hard rocks that were *deformed*—tilted, bent, or fractured—sometimes into crazy-looking patterns. What forces could have deformed them in this way? Why were particular patterns of deformation repeated time and time again throughout geologic history? Plate tectonics, discovered in the 1960s, provided the answers.

This chapter examines the processes that deform continental crust near plate boundaries and shows how geologists collect and interpret field observations to reconstruct the history of continental deformation.

According to the theory of plate tectonics (described in Chapter 2), the steady relative motion between two rigid plates of lithosphere causes deformation at the plate boundary. There are three basic types of plate boundary: spreading centers, where plates diverge and plate area increases; subduction zones, where plates converge and plate area decreases; and transform-fault boundaries, where two plates slide past each other with no change in plate area. If plates were perfectly rigid, the plate boundaries would be sharp lineations, and points on either side of these boundaries would move at the relative plate velocity. This idealization is often a good approximation in the oceans, where mid-ocean rift valleys, deep ocean trenches, and near-vertical transform faults form narrow plate boundary zones, often just a few kilometers wide.

In continents, however, the relative plate motions can be "smeared out" across plate boundary zones hundreds or even thousands of kilometers wide. Within these broad zones, the

Folded sedimentary rocks, northwest Canada. The folds have a wavelength of about 1 km. [John Grotzinger.]

(a)

(b)

Figure 7.1 (a) An outcrop of originally horizontal rock layers bent into folds by compressive tectonic forces. (b) An outcrop of once-continuous rock layers displaced on small faults by tensional tectonic forces. [(a) Phil Dombrowski. (b) Tom Bean.]

continental crust does not behave rigidly, and rocks at the surface are deformed by **folding** and **faulting.** Folds in rocks are like folds in clothing. Just as cloth pushed together from opposite sides bunches up in folds, layers of rock slowly compressed by forces in the crust can be pushed into folds (**Figure 7.1a**). The tectonic forces between plates can also cause a rock formation to break and slip on both sides of a fracture, forming a fault (Figure 7.1b). When a fault breaks suddenly, it causes an earthquake. Active zones of continental deformation are marked by frequent earthquakes.

Geologic folds and faults can range in size from centimeters to tens of kilometers or more. Many mountain ranges are actually a series of large folds and faults that have been weathered and eroded. From the geologic record of deformation, geologists can deduce the motions across ancient plate boundaries and reconstruct the tectonic history of the continental crust.

MAPPING GEOLOGIC STRUCTURE

To understand crustal deformation, geologists need information about the geometry of rock formations. A basic source of this information is the *outcrop*, where the bedrock that underlies the surface is exposed (not obscured by soil or loose boulders). Figure 7.1a is a picture of an outcrop showing a sedimentary bed bent into a fold. Often, however, folded rocks are only partly exposed in an outcrop and can be seen only as an inclined layer (**Figure 7.2**). The orientation of the layer is an important clue the geologist can use to piece together a picture of the overall geologic structure. Two measurements describe the orientation of a rock layer exposed at an outcrop: the strike and the dip of the layer surface.

Figure 7.2 Dipping limestone and shale beds on the coast of Somerset, England. Children are walking along the strike of beds that dip to the left at an angle of about 15°. [Chris Pellant.]

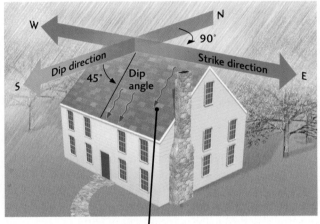

Water trickles down slope parallel to dip.

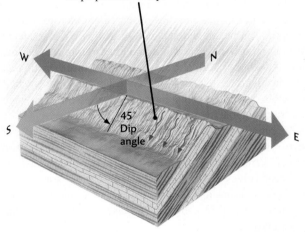

Figure 7.3 Geologists use the strike and dip of a formation to define its orientation at a particular place. Dip is the angle of steepest descent of the bed from the horizontal; strike is at right angles to the dip direction. Here the strike is east-west and the dip angle is 45°. [After A. Maltman, *Geological Maps: An Introduction*. New York: Van Nostrand Reinhold, 1990, p. 37.]

Measuring Strike and Dip

The **strike** is the compass direction of a rock layer as it intersects with a horizontal surface. The **dip,** which is measured at right angles to the strike, is simply the amount of tilting—the angle at which the bed inclines from the horizontal. **Figure 7.3** shows how the strike and the dip are measured in the field. A geologist might describe the outcrop in this figure as "a bed of coarse-grained sandstone striking west and dipping 45 degrees south." Strike and dip can be used to map the orientation of other types of geologic surfaces, such as fault planes or ancient erosional surfaces.

Geologic Maps

Geologic maps represent the rock formations exposed at Earth's surface (**Figure 7.4**). When preparing a map, a geologist must choose an appropriate *scale*—the ratio of distance on the map to true surface distance. A common scale for geologic field mapping is 1:24,000 (pronounced "one to twenty-four thousand"), which specifies that 1 inch on the map corresponds to 24,000 inches (2000 feet) on Earth's surface. To depict the geology of an entire state, a geologist would choose a smaller scale, say 1:1,000,000, where 1 centimeter represents 10 kilometers and 1 inch almost 16 miles. Less detail can be depicted on a map with a smaller scale.

Geologists keep track of different rock formations by assigning each formation a particular color on the map, usually keyed to the rock type and age (see Figure 7.4). Many rock formations may be exposed in highly deformed regions, so geologic maps can be very colorful! Softer rocks, such as mudstones and other poorly consolidated sediments, are more easily eroded than harder formations of limestone or metamorphic rocks. Consequently, the rock types can exert a strong influence on the topography of the land surface and the exposure of rock formations. The important

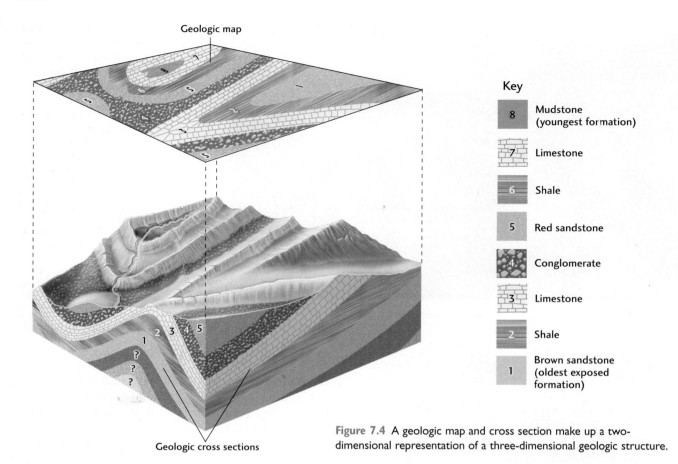

Geologic map

Key

8	Mudstone (youngest formation)
7	Limestone
6	Shale
5	Red sandstone
4	Conglomerate
3	Limestone
2	Shale
1	Brown sandstone (oldest exposed formation)

Geologic cross sections

Figure 7.4 A geologic map and cross section make up a two-dimensional representation of a three-dimensional geologic structure.

relationships between geology and topography can be made clear by plotting the geologic structure on a base map that shows topographic contours.

Maps are annotated with special symbols to indicate the local strike and dip of the formations and with special types of lines to mark faults and other significant boundaries. Because geologic maps can give a huge amount of information, they have been called "textbooks on a piece of paper."

Of course, not every detail of the surface geology can be put on a map, so geologists must simplify the structures they see, perhaps by representing a complex zone of faulting as a single fault trace or ignoring folds too small to show at the map scale they have chosen. They may also "dust off" their maps by ignoring thin layers of soil and loose rock that cover up the geologic structure, portraying the structure as if outcrop existed everywhere. You should therefore think of a geologic map as a *scientific model* of the surface geology.

Geologic Cross Sections

Once a region is mapped, the two-dimensional map must be interpreted in terms of the three-dimensional geology. How can the deformed shapes of the rock layers be reconstructed, even when erosion has removed parts of a formation? The process is like putting together a three-dimensional jigsaw puzzle with some of the pieces missing. Common sense and intuition play important roles, as do basic geologic principles.

To help them piece together the puzzle, geologists construct **geologic cross sections**—diagrams showing the features that would be visible if vertical slices were made through part of the crust. Small cross sections can sometimes be observed in the vertical faces of cliffs, quarries, and road cuts. Cross sections spanning much larger regions can be constructed from the information on a geologic map, including the strikes and dips observed at outcrops. Surface mapping can be supplemented with geologic data collected from boreholes or by seismic imaging to improve the location of deep rock formations on cross sections. (Drilling and seismic data are expensive to collect, so they are usually available only in areas that have been explored for oil, water, or other valuable natural resources.)

Figure 7.4 shows the geologic map and a geologic cross section of an area where sedimentary rocks, originally horizontal, were bent into a series of folds and eroded into a set of linear ridges and valleys. We will explore some of the geologic relationships seen on the map later in this chapter. But first we will investigate the basic processes by which rocks deform.

HOW ROCKS DEFORM

Rocks deform in response to the forces acting on them. Determining whether they respond by folding, faulting, or some combination of the two can be a complex business, depending on the orientation of the forces, the rock type, and the physical conditions during deformation. We will focus primarily on the big picture: how the forces of plate tectonics cause continental crust to deform.

Plate Tectonic Forces

In the upper crust near Earth's surface, the main tectonic forces are produced by horizontal movements of the plates. The tectonic forces that act on rocks in plate boundary zones are thus *horizontally directed,* and the type of force depends on the relative plate motion:

- **Tensional forces,** which stretch and pull formations apart, dominate at divergent boundaries, where plates move away from each other.

- **Compressive forces,** which squeeze and shorten rock formations, dominate at convergent boundaries, where plates move toward each other.

- **Shearing forces,** which push two sides of a formation in opposite directions, dominate at transform-fault boundaries, where plates slide past each other.

We will study the ways that continental crust deforms when subjected to horizontally directed compression, tension, and shear.

Brittle and Ductile Behavior of Rocks in the Laboratory

Early geologists were baffled by the problem of how rocks, which seem strong and rigid, could be distorted into folds or broken along faults by tectonic forces. The problem was solved in the mid-1900s when geologists used big hydraulic rams to bend and break small samples of rock. Engineers had invented such machines to measure the strength of concrete and other building materials, but geologists modified them to track the details of how rocks deform at pressures and temperatures high enough to simulate physical conditions deep in the crust.

In one such experiment, a compressive force was applied by pushing down with a hydraulic ram on one end of a small cylinder of marble, while at the same time maintaining the force of the surrounding pressure on the cylinder. (Surrounding pressure is the squeezing your body feels as you dive deeper underwater; it increases steadily with depth in proportion to the weight of the overlying mass.) Under low surrounding pressures, the marble sample deformed only a small amount until the compressive force on its end was increased to the point that the entire sample suddenly broke by brittle fracturing (**Figure 7.5**). This experiment showed that marble behaves as a **brittle material** at the low surrounding pressures found in the shallow crust. Repeating the experiment under high surrounding pressures produced a different result: the marble sample slowly and steadily deformed into a shortened, bulging shape without fracturing.

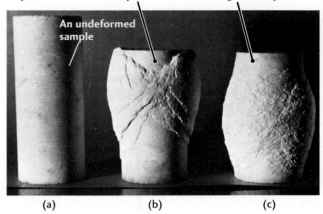

This sample was compressed under conditions representative of the shallow crust. The fracture indicates that the marble is brittle at the laboratory equivalent of shallow depth.

This sample was compressed under conditions representative of the deeper crust. It has deformed smoothly, indicating that marble is ductile at greater depth.

An undeformed sample

(a) (b) (c)

Figure 7.5 Results of laboratory experiments conducted to discover how a rock—in this case, marble—is deformed by compressive forces. [M. S. Patterson, Australian National University.]

Marble thus behaves as a pliable or **ductile material** at the high surrounding pressures found deep in the crust.

In other experiments, geologists demonstrated that when the sample was heated, the marble acted as a ductile material at a lower surrounding pressure—just as heating wax changes it from a hard material that can break into a soft one that flows. They concluded that a layer of this particular marble would deform by faulting at depths shallower than a few kilometers but by folding at greater crustal depths.

Brittle and Ductile Behavior of the Crust

Natural conditions cannot be reproduced exactly in the laboratory. Tectonic forces are applied over millions of years, whereas laboratory experiments are rarely conducted for more than a few hours or perhaps a few weeks. Nevertheless, laboratory experiments provide important clues for interpreting field observations. Geologists keep the following points in mind as they map crustal folds and faults:

- Some rocks are brittle, others are ductile, and the same rock can be brittle at shallow depths and ductile deep in the crust.

- Rock type affects the way rocks deform. In particular, the old, hard igneous and metamorphic rocks that form the crystalline basement act as brittle materials, fracturing along fault planes in earthquakes, while the younger, softer sediments that overlie them fold gradually as ductile materials.

- A rock formation that would flow as a ductile material if deformed slowly may break as a brittle material if deformed more rapidly. (Think of Silly Putty, which deforms as a ductile clay when you squeeze it slowly but breaks into pieces when you smash it quickly onto a hard surface.)

• Rocks break more easily in tension than in compression. Sedimentary formations that will deform by folding during compression will often break along faults when subjected to tensional forces.

BASIC DEFORMATION STRUCTURES

Faults and folds are examples of the basic features geologists observe and map to understand crustal deformation. They use simple geometrical concepts (and a rich vocabulary) to classify these features into different types of deformation structures.

Faults

A fault is a surface across which rock formations have been displaced. We can measure the orientation of the fault plane by a strike and a dip, just as we do for other geologic surfaces (see Figure 7.3). The movement of one side of the fault with respect to the other can be described by a *slip direction* and a total displacement or *offset*. For a small fault, such as the one pictured in Figure 7.1b, the offset might be only a couple of meters, whereas the offset along a major transform fault such as the San Andreas can amount to hundreds of kilometers (**Figure 7.6**).

Faults are classified by their slip direction (**Figure 7.7**). A **dip-slip fault** is one on which there has been relative movement of the blocks up or down the dip of the fault plane. A **strike-slip fault** is one on which the movement has been horizontal, parallel to the strike of the fault plane. Movement along the strike and simultaneously up or down the dip produces an *oblique-slip fault.* Dip-slip faults are caused by compressive or tensional forces, whereas strike-slip faults are the work of shearing forces. An oblique-slip fault results from shear in combination with either compression or tension.

Faults require further classification, because the movement can be up or down, or right or left. A dip-slip fault is called a **normal fault** if the rocks above the fault plane move down relative to the rocks below the fault plane, extending the structure horizontally (see Figure 7.7a). On a *reverse fault,* the rocks above the fault plane move upward in relation to the rocks below, causing a shortening of the structure (see Figure 7.7b)—the reverse of what geologists have (somewhat arbitrarily) chosen as "normal." A **thrust fault** is a low-angle reverse fault; that is, one with a dip less than 45°, so that the overlying block moves mainly horizontally (see Figure 7.7c). When subjected to lateral compression, brittle rocks of the continental crust usually break along thrust faults with dip angles of about 30° or less, rather than along more steeply dipping reverse faults.

A strike-slip fault (see Figure 7.7d) is a *right-lateral fault* if an observer on one side of the fault sees the block on the opposite side move to the right. It is a *left-lateral fault* if the

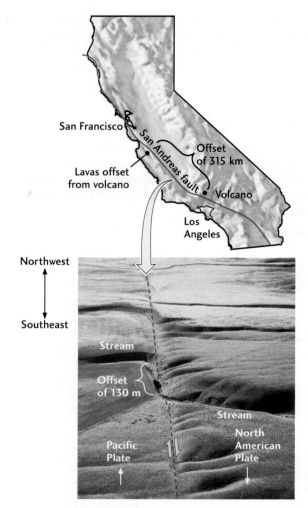

Figure 7.6 View of the San Andreas fault showing the northwestward movement of the Pacific Plate with respect to the North American Plate. The fault runs from top to bottom (dashed line) near the middle of the photograph. Note the offset of the stream (Wallace Creek) by 130 m as it crosses the fault. Elsewhere in California, a formation of volcanic rocks 23 million years old has been displaced by 315 km. [John S. Shelton.]

block on the opposite side moves to the left. As you can tell from the stream offset in Figure 7.6, the San Andreas fault is a right-lateral transform fault.

Geologists recognize faults in the field in several ways. A fault may form a *scarp* (small cliff) that marks the trace of the fault across the ground surface (**Figure 7.8**). If the offset has been large, as it is for transform faults such as the San Andreas, the formations currently facing each other across the fault usually differ in lithology and age. When movements are smaller, offset features can be observed and measured. (As an exercise, try to match up the beds offset by the small-scale fault in Figure 7.1b.) In establishing the time of faulting, geologists apply a simple rule: a fault must be younger than the youngest rocks that it cuts (the rocks had to be there before they could break!) and older than the oldest undisrupted formation that covers it.

Key Figure 7.7 Tectonic forces determine the style of faulting.

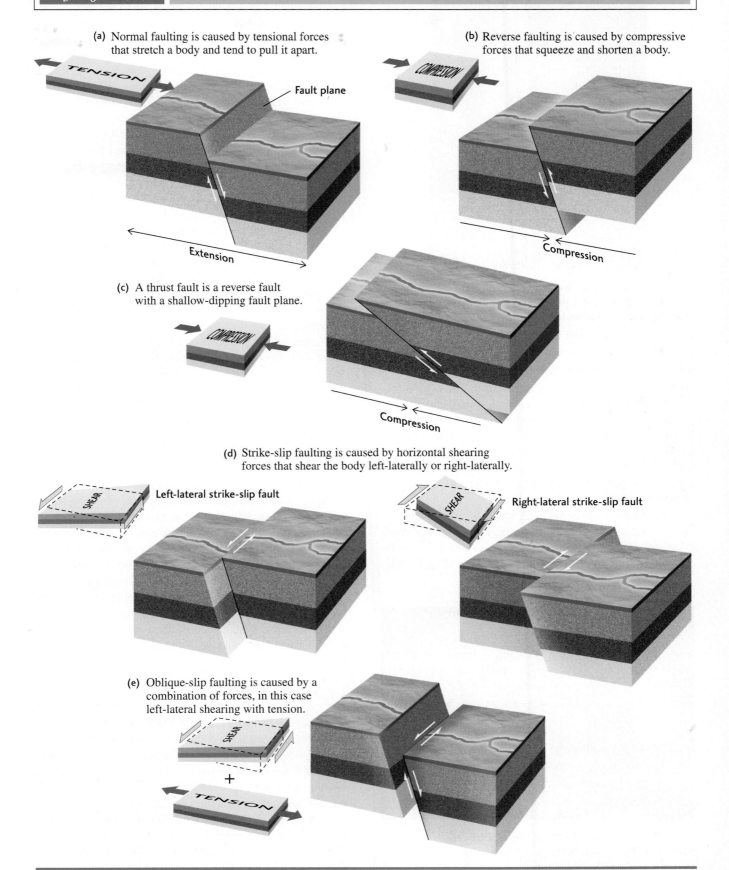

(a) Normal faulting is caused by tensional forces that stretch a body and tend to pull it apart.

(b) Reverse faulting is caused by compressive forces that squeeze and shorten a body.

(c) A thrust fault is a reverse fault with a shallow-dipping fault plane.

(d) Strike-slip faulting is caused by horizontal shearing forces that shear the body left-laterally or right-laterally.

(e) Oblique-slip faulting is caused by a combination of forces, in this case left-lateral shearing with tension.

Figure 7.8 This scarp is a fresh surface feature that formed by normal faulting during the 1954 Dixie Valley earthquake in Nevada. [Karl V. Steinbrugge Collection, Earthquake Engineering Research Center.]

Folds

Folding is a common form of deformation observed in layered rocks. Indeed, the term *fold* implies that an originally planar structure, such as a sedimentary bed, has been warped into a curved structure. The bending can be produced either by horizontally directed forces or by vertically directed forces in the crust, just as pushing together opposite edges of a piece of paper or pushing up or down on one side or the other can fold it.

Like faults, folds come in all sizes. In many mountain systems, majestic, sweeping folds can be traced over many kilometers. On a much smaller scale, very thin beds can be crumpled into folds a few centimeters long (**Figure 7.9**).

Figure 7.9 Small-scale folds in sedimentary anhydrite (light) and shale (dark) layers, west Texas. [John Grotzinger/Ramón Rivera-Moret/Harvard Mineralogical Museum.]

The bending can be gentle or severe, depending on the magnitude of the applied forces, the length of time that they were applied, and the ability of the beds to resist deformation.

Layered rocks that upfold into arches are called **anticlines;** those that downfold into troughs are called **synclines** (**Figure 7.10**). The two sides of a fold are its *limbs*. The **axial plane** is an imaginary surface that divides a fold as symmetrically as possible, with one limb on either side of the plane. The line made by the lengthwise intersection of the axial plane with the beds is the *fold axis*. A symmetrical horizontal fold has a horizontal fold axis and a vertical axial plane with limbs dipping symmetrically away from the axis. If its axis is not horizontal, it is called a *plunging fold.*

Folds rarely stay horizontal. Follow the axis of any fold in the field and sooner or later the fold dies out or appears to plunge into the ground. **Figure 7.11** diagrams the geometry of plunging anticlines and plunging synclines. In eroded mountain belts, a zigzag pattern of outcrops may appear in the field after erosion has removed much of the surface rock. The geologic map of Figure 7.4 shows this characteristic pattern.

With increasing amounts of deformation, the folds can be pushed into asymmetrical shapes, with one limb dipping more steeply than the other (see Figure 7.10). This can also occur if the direction of the deformational force is oblique to the layering of the beds. Such *asymmetrical folds* are common. When the deformation is intense and one limb has been tilted beyond the vertical, the fold is called an *overturned fold*. Both limbs of an overturned fold dip in the same direction, but the order of the layers in the bottom limb is precisely the reverse of their original sequence—that is, older rocks are on top of younger rocks.

Observations in the field seldom provide geologists with complete information. Bedrock may be obscured by overlying soils or erosion may have removed much of the evidence of former structures. So geologists search for clues they can use to work out the relationship of one bed to another. For example, in the field or on a map, an eroded anticline would be recognized by a strip of older rocks forming a core bordered on both sides by younger rocks dipping away from the core. These relationships are illustrated in Figures 7.4 and 7.11. An eroded syncline would show as a core of younger rocks bordered on both sides by older rocks dipping toward the core.

Circular Structures

The deformation along plate boundaries is usually expressed in linear faults and folds oriented nearly parallel to the plate boundary by horizontally directed forces. However, some types of deformation are more symmetrical, forming nearly circular structures called domes and basins. In many (though not all) cases, these structures result from the upward force of rising material or the downward force of sinking material, rather than the horizontally directed forces of plate tectonics. In fact, circular structures tend to be more common in the interior of plates, far away from active plate boundaries.

Key Figure 7.10 | Rock folding is described by the orientation of the fold axis and the axial plane.

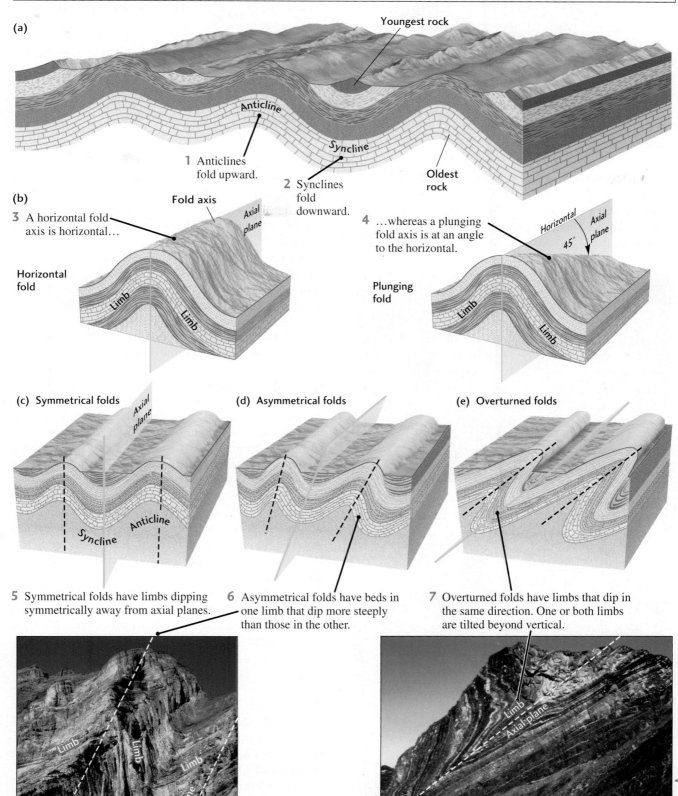

(a)

Youngest rock

Anticline

Syncline

1 Anticlines fold upward.

2 Synclines fold downward.

Oldest rock

(b)

Fold axis

3 A horizontal fold axis is horizontal…

Axial plane

Horizontal fold

Limb

Limb

4 …whereas a plunging fold axis is at an angle to the horizontal.

Horizontal

Axial plane

45°

Plunging fold

Limb

Limb

(c) Symmetrical folds

Axial plane

Syncline

Anticline

(d) Asymmetrical folds

(e) Overturned folds

5 Symmetrical folds have limbs dipping symmetrically away from axial planes.

6 Asymmetrical folds have beds in one limb that dip more steeply than those in the other.

7 Overturned folds have limbs that dip in the same direction. One or both limbs are tilted beyond vertical.

Limb

Limb

Limb

Axial plane

Limb

Axial plane

[(left photo) Courtesy of Mark McNaught. (right photo) John Grotzinger.]

Plunging anticline

Plunging syncline

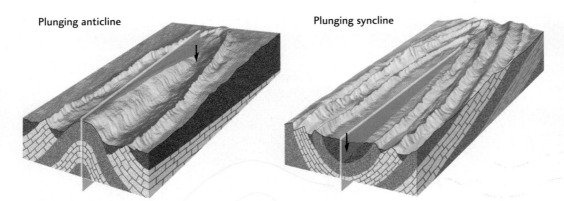

Figure 7.11 The geometry of plunging folds. Note the converging pattern of the layers where they intersect the surface, which you can also see in Figure 7.4.

A **dome** is an anticlinal structure, a broad circular or oval upward bulge of rock layers. The flanking beds of a dome encircle a central point and dip radially away from it (**Figure 7.12**). Domes are very important in oil geology because oil is buoyant and tends to migrate upward through permeable rocks. If the rocks at the high point of a dome are not easily penetrated, the oil becomes trapped against them.

A **basin** is a synclinal structure, a bowl-shaped depression of rock layers in which the beds dip radially toward a central point.

Domes and basins are typically many kilometers in diameter. They are recognized in the field by outcrops that outline their characteristic circular or oval shapes. At these outcrops, the formations dip downward toward the center of the basin or upward toward the top of the dome.

Several types of deformation can produce domes and basins. Some domes are formed by rising bodies of buoyant material—magma, hot igneous rock, or salt—that push the overlying sediments upward. Others are caused by multiple episodes of deformation, for instance, when rocks are compressed in one direction and then again in a direction nearly perpendicular to the original direction. As we saw in Chapter 5, some sedimentary basins form when a heated portion of the crust cools and contracts, causing the overlying sediments to subside (thermal subsidence basins). Others result when tectonic forces stretch and thin the crust (rift basins) or compress it downward (flexural basins). The weight of sediments deposited by a river delta can depress the crust into a sedimentary basin, such as the very large basin now forming at the mouth of the Mississippi in the Gulf of Mexico.

There are many domes and basins in the central portion of the United States. The Black Hills of South Dakota are an eroded dome; almost all of the lower peninsula of Michigan is a very large sedimentary basin.

Oldest formation exposed on the surface

Youngest formation

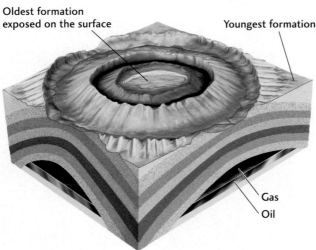

Gas

Oil

Figure 7.12 Sinclair dome, an eroded dome in strata 10 km east of Rawlings, Wyoming. The highway and railroad at the lower right suggest the dome's dimensions. The characteristic circular outcrop pattern of a dome and the extension to the subsurface are shown in the diagram. The oldest bed is in the core; the flanking formations are successively younger and dip away from the core. [Photo by John S. Shelton.]

Joints

We have seen how rock deformation depends on tectonic forces and the conditions under which the forces are applied. Some layers crumple into folds, and some fracture. A fracture that has displaced the geologic formations on either side is called a fault. A second type of fracture is a **joint**—a crack along which there has been no appreciable movement.

Joints are found in almost every outcrop. Some are caused by tectonic forces. Like any other easily broken material, brittle rocks fracture more easily at flaws or weak spots when they are subjected to pressure. These flaws can be tiny cracks,

through the crust and probably down into the mantle. At depths up to about 20 km, the fault zone is thought to be very narrow and characterized by cataclastic textures, indicating brittle deformation. Earthquakes are generated in this zone. Below 20 km, however, earthquakes do not occur, and the fault is thought to be characterized by a broad zone of ductile deformation that produces mylonites.

STYLES OF CONTINENTAL DEFORMATION

If we look closely enough, we can find all the basic deformation structures—faults, folds, domes, basins, joints—in any zone of continental deformation. But when we view continental deformation at the scale of the plate boundary, we see distinctive patterns of faulting and folding that relate directly to the tectonic forces causing the deformation. **Figure 7.15** depicts the deformation styles typical of the three main types of tectonic forces.

Tensional Tectonics

In brittle crust, the tensional forces that produce normal faulting may split a plate apart, resulting in a *rift valley*—a long, narrow trough formed by a block that has dropped down relative to its two flanking blocks along nearly parallel, steeply dipping normal faults (**Figure 7.16**). The rift valleys of East Africa, the rifts of mid-ocean ridges, the Rhine River valley, and the Red Sea Rift are well-known examples. These structures form basins that fill with sediments eroded from the mountains of the rift walls and with volcanic rocks extruded from tensional cracks in the crust.

In the Basin and Range province, which is centered on the Great Basin of Nevada and Utah, a region more than 800 km wide has been stretched in a northwest-southeast direction by a factor of two during the last 15 million years. Here normal faulting has created an immense landscape of eroded, rugged mountains and smooth, sediment-filled valleys, some covered with recent volcanics. The extensional deformation, which appears to be caused by upwelling convection currents beneath the Basin and Range, still continues.

Extension of the upper continental crust usually involves normal faults with high dip angles, typically 60° or more. However, below a depth of about 20 km, crustal rocks are hot enough to behave as ductile materials, and the deformation occurs by stretching rather than by fracturing. This change in rock behavior causes the fault dip to flatten with increasing depth, which results in normal faults with curved surfaces, as shown in Figure 7.15a. The crustal blocks moving along these curved faults are tilted back as the stretching continues. Geologists can explain the tilting of the fault-block mountains and valleys observed in the Basin and Range by this simple geometry.

Compressive Tectonics

In subduction zones, an oceanic plate slips beneath an overriding oceanic or continental plate along a huge thrust fault, or *megathrust* (see Figure 2.6). The world's largest earthquakes are caused by sudden slips on megathrusts, such as the great Sumatran earthquake of December 26, 2004, which generated a disastrous tsunami that killed more than 300,000 people. Thrusting is also the most common type of faulting in continental regions undergoing tectonic compression. During

(a)

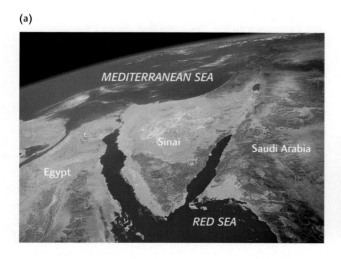

(b)

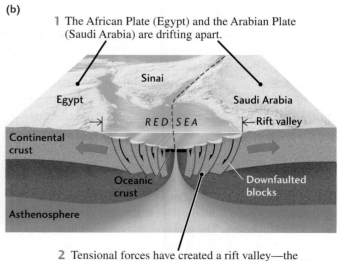

1 The African Plate (Egypt) and the Arabian Plate (Saudi Arabia) are drifting apart.

2 Tensional forces have created a rift valley—the result of downfaulted blocks—filled by the Red Sea.

Figure 7.16 A rift valley results from tensional forces and normal faulting. The African Plate, on which Egypt rides, and the Arabian Plate, bearing Saudi Arabia, are drifting apart. The tensional forces

have created a rift valley, filled by the Red Sea. [NASA/TSADO/ Tom Stack.] The diagram shows parallel normal faults bounding the rift valley in the crust beneath the sea.

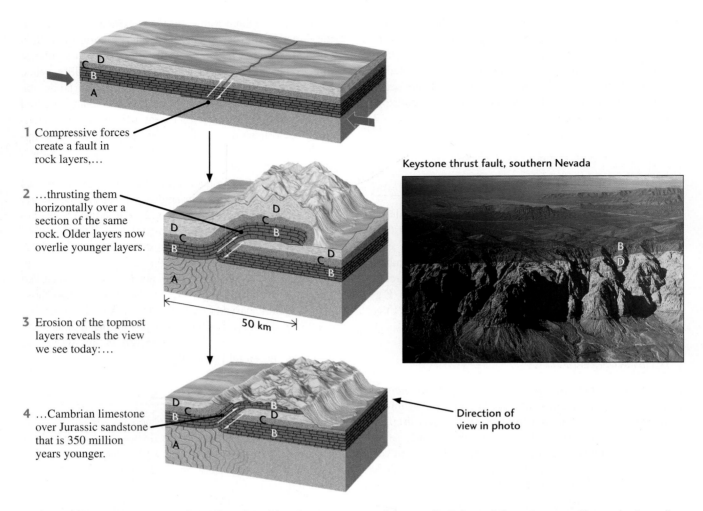

1 Compressive forces create a fault in rock layers,...

2 ...thrusting them horizontally over a section of the same rock. Older layers now overlie younger layers.

3 Erosion of the topmost layers reveals the view we see today:...

50 km

4 ...Cambrian limestone over Jurassic sandstone that is 350 million years younger.

Keystone thrust fault, southern Nevada

Direction of view in photo

Figure 7.17 The Keystone thrust fault of southern Nevada is a large-scale overthrust sheet of a kind found in California and southern Nevada. Compressive forces have detached a sheet of rock layers—D, C, B—and thrust it a great distance horizontally over the section D, C, B, A. [Photo by John S. Shelton.]

mountain building, sheets of crust may glide over one another for tens of kilometers along nearly horizontal thrust faults, forming *overthrust* structures (**Figure 7.17**).

When two continental plates collide, the crust can be compressed across a wide zone, resulting in spectacular episodes of mountain building. During such collisions, the brittle basement rocks ride over one another by thrust faulting while the more ductile overlying sediments compress into a series of great folds, forming what geologists call a *fold and thrust belt* (see Figure 7.15b).

The ongoing collisions of Africa, Arabia, and India with the southern margin of the Eurasian continent have created fold and thrust belts from the Alps to the Himalaya, many of which are still active. The great oil reservoirs of the Middle East are located in anticlines and other structural traps formed by this deformation. Compression across western North America, caused by its westward motion during the opening of the Atlantic Ocean, created the fold and thrust belt of the Canadian Rockies. The Valley and Ridge province of the Appalachians is an ancient fold and thrust belt

that dates back to the collisions that created the supercontinent of Pangaea (**Figure 7.18**).

Shearing Tectonics

A transform fault is a strike-slip fault that forms a plate boundary. Transform faults such as the San Andreas can offset geologic formations for long distances (see Figure 7.6), but as long as they stay aligned with the direction of relative plate motion, the blocks on either side can slide past each other without much internal deformation. Long transform faults are rarely straight, however, and the deformation can be much more complicated. The faults may have bends and jogs that change the tectonic forces acting across portions of the plate boundary from shearing to compression or tension. These forces, in turn, cause secondary faulting and folding (see Figure 7.15c).

A good example of how this works can be found in southern California, where the right-lateral San Andreas fault bends first to the left and then to the right as one moves

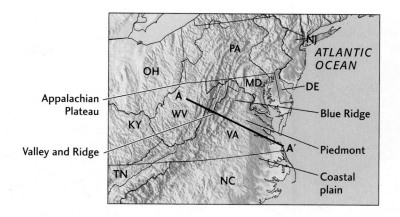

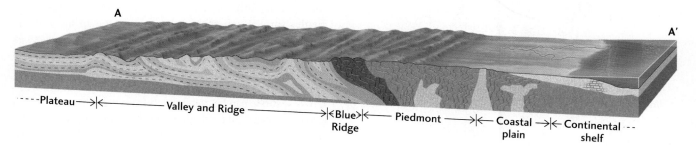

Figure 7.18 Map and cross section of the Valley and Ridge province of the Appalachian Mountains, the eroded remnant of a folded mountain belt. [Cross section after D. Johnson, *Stream Sculpture on the Atlantic Slope.* New York: Columbia University Press, 1931.]

along its trace (in either direction; see Figure 7.6). The segments of the fault on both sides of this "Big Bend" are aligned with the plate motion, so the blocks slip past each other in simple strike-slip faulting. Within the Big Bend, however, the change in the fault geometry compresses the blocks together, which causes thrust faulting south of the San Andreas. The thrusting along the fault has raised mountains to elevations exceeding 3000 m and, during the last half century, has produced a series of destructive ruptures, including the 1994 Northridge earthquake, which caused more than $40 billion in damage to Los Angeles (see Chapter 13).

In the Salton Trough at the southern end of the San Andreas, the Pacific–North American plate boundary jogs to the right in a series of steps. Within these steps, the plate boundary is subjected to tensional forces, and normal faulting has formed rift valleys that are volcanically active, rapidly subsiding, and filling with sediments. The extension of the Salton Trough occurs within 200 km of the Big Bend compression, demonstrating how variable the tectonics along continental transform faults can be!

UNRAVELING GEOLOGIC HISTORY

The geologic history of a region is a succession of episodes of deformation and other geologic processes. Let's see how some of the concepts introduced in this chapter lead to a simple interpretation of a complicated geologic structure. The cross sections in **Figure 7.19** represent a few tens of kilometers of a geologic province that underwent a succession of events. First, horizontal layers of sediment were deposited, then they were tilted and folded by horizontal forces of compression. After that, they were uplifted above sea level. There, erosion gave them a new horizontal surface, which was covered by lava when forces deep in Earth's interior caused a volcanic eruption. In the final stage, tensional forces resulted in normal faulting, which broke the crust into blocks. The geologist sees only the last stage but visualizes the entire sequence. When the sedimentary beds have been identified, the geologist starts with the knowledge that the beds must originally have been horizontal and undeformed at the bottom of an ancient ocean. The succeeding events can then be reconstructed.

Present-day surface relief in young mountain ranges—such as the Alps, the Rocky Mountains, the Pacific Coast Ranges, and the Himalaya—can be traced in large part to deformation that occurred over the past few tens of millions of years. These young systems still contain much of the information that the geologist needs to piece together the history of deformation. Deformation that occurred hundreds of millions of years ago no longer shows as the rugged mountains that once existed, however. Erosion has left behind only the remnants of folds and faults expressed as low ridges and shallow valleys (see Figure 7.18). As we will see in Chapter 10, even older episodes of mountain building are

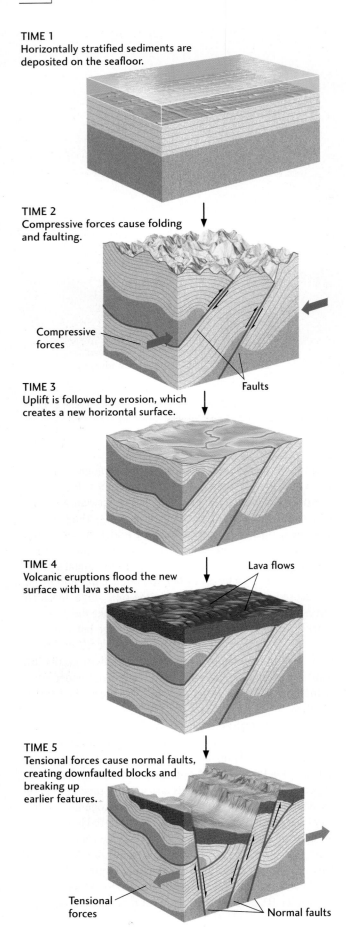

TIME 1
Horizontally stratified sediments are deposited on the seafloor.

TIME 2
Compressive forces cause folding and faulting.

Compressive forces

TIME 3
Uplift is followed by erosion, which creates a new horizontal surface.

Faults

TIME 4
Volcanic eruptions flood the new surface with lava sheets.

Lava flows

TIME 5
Tensional forces cause normal faults, creating downfaulted blocks and breaking up earlier features.

Tensional forces

Normal faults

Figure 7.19 Stages in the development of a geologic province. A geologist sees only the last stage and attempts to reconstruct from the structural features all the earlier stages in the history of a region.

evident from the twisted, highly metamorphosed formations that constitute the basement rocks of the interiors of continents.

SUMMARY

What are geologic maps and cross sections? A geologic map is a scientific model of the rock formations exposed at Earth's surface, presented in the form of a map showing the outcrops of various rock formations as well as fault traces and other significant boundaries. A geologic cross section is a diagram representing the geologic features that would be visible if a vertical slice were made through part of the crust. Cross sections can be constructed from the information on a geologic map, although they can be improved by supplementing the surface mapping with subsurface data collected from boreholes or by seismic imaging.

What do laboratory experiments tell us about the way rocks deform when they are subjected to crustal forces? Laboratory studies show that some rocks deform as brittle materials, others as ductile materials. These behaviors depend on the kind of rock, the temperature, the surrounding pressure, the magnitude of the force, and the speed with which it is applied. The same rock can be brittle at shallow depths and ductile deep in the crust.

What are the basic deformation structures that geologists observe in the field? Among the geologic structures in rock formations that result from deformation are folds, fractures (faults and joints), circular structures (domes and basins), and deformation textures caused by the shearing of rock formations.

What kinds of forces produce these deformation structures? Horizontally directed forces at plate boundaries primarily produce linear structures such as faults and folds. Horizontal tensional forces at divergent boundaries produce normal faults; horizontal compressive forces at convergent boundaries produce thrust faults; and horizontal shearing forces at transform-fault boundaries produce strike-slip faults. Folds are usually formed in layered sedimentary formations by compressive forces, especially in regions where continental plates collide. Circular structures, such as domes and basins, can be produced by vertically directed forces far from plate boundaries. Some domes are caused by the rise of buoyant materials, such as magma or salt. Basins can form when tensional forces stretch the crust or when a heated portion of the crust cools and contracts. Joints can be caused by tectonic stresses or by the cooling and contraction of rock formations.

How do we reconstruct the geologic history of a region? Geologists see the end result of a succession of events: deposition, deformation, erosion, volcanism, and so forth. They deduce the deformational history of a region by identifying and fixing the ages of the rock layers, recording the geometric orientation of the beds on maps, mapping folds and faults, and reconstructing cross sections of the subsurface consistent with the surface observations. They can determine the relative age of a deformation by finding a younger undeformed formation lying unconformably on an older deformed bed.

KEY TERMS AND CONCEPTS

anticline (p. 158)

axial plane (p. 158)

basin (p. 160)

brittle material (p. 155)

compressive force (p. 155)

dip (p. 153)

dip-slip fault (p. 156)

dome (p. 160)

ductile material (p. 155)

faulting (p. 152)

folding (p. 152)

geologic map (p. 153)

geologic cross section (p. 154)

joint (p. 160)

normal fault (p. 156)

shearing force (p. 155)

strike (p. 153)

strike-slip fault (p. 156)

syncline (p. 158)

tensional force (p. 155)

thrust fault (p. 156)

EXERCISES

1. What type of fold is shown in Figure 7.1a? Is the small fault on the left side of Figure 7.1b a normal fault or a thrust fault? Estimate the fault's offset, expressing your answer in meters.

2. On a geologic map of 1:250,000 scale, how many centimeters would represent an actual distance of 2.5 km? What is the actual distance in miles of 1 inch on the same map?

3. The motion of the San Andreas fault has offset the stream channel in Figure 7.6 by 130 m. Geologists have determined that this channel is 3800 years old. What is the slip rate along the San Andreas fault at this site, expressed in millimeters per year?

4. What was the direction of the crustal forces that deformed the Appalachian block depicted in Figure 7.18?

5. Show that a left jog in a right-lateral strike-slip fault will produce compression, whereas a right jog in a right-lateral strike-slip fault will produce extension. Write a similar rule for left-lateral strike-slip faults.

6. Draw a geological cross section that tells the following story. A series of marine sediments was deposited and subsequently deformed by compressive forces into a fold and thrust belt. The mountains of the fold and thrust belt eroded to sea level and new sediments were deposited. The region then began to extend, and lava intruded the new sediments to create a sill. In the latest stage, tensional forces broke the crust to form a rift valley bounded by steeply dipping normal faults.

THOUGHT QUESTIONS

1. In what sense is a geologic map a scientific model of the surface geology? Is it fair to say that geologic cross sections in combination with a geologic map describe a scientific model of three-dimensional geologic structure? (In formulating your answers, you may want to refer to the discussion of scientific models in Chapter 1.)

2. Why is it correct to say that "large-scale geologic structures should be represented on small-scale geologic maps"? How big a piece of paper would be required to make a map of the entire U.S. Rocky Mountains at 1:24,000 scale?

3. A continental margin has a thick layer of sediments overlying metamorphic basement rocks. It collides with another continental mass, and the compressive forces deform it into a fold and thrust belt. During the deformation, which of the following geologic formations would be likely to behave as brittle materials and which as ductile materials? (a) The sedimentary formations in the upper few kilometers; (b) the metamorphic basement rocks at depths of 5–15 km; (c) lower crustal rocks at depths below 20 km. In which of these layers would you expect earthquakes?

4. It was the writer John McPhee who called geologic maps "textbooks on a piece of paper" in his epic narrative about a geologic traverse across North America, *Annals of the Former World* (p. 378). Can you locate a passage in this book that describes a geologic structure and sketch a geologic map consistent with McPhee's description?

5. Can you explain the geologic story in Exercise 6 of this chapter in terms of plate tectonic events? Where in the United States do geologists think this sequence of events has taken place?

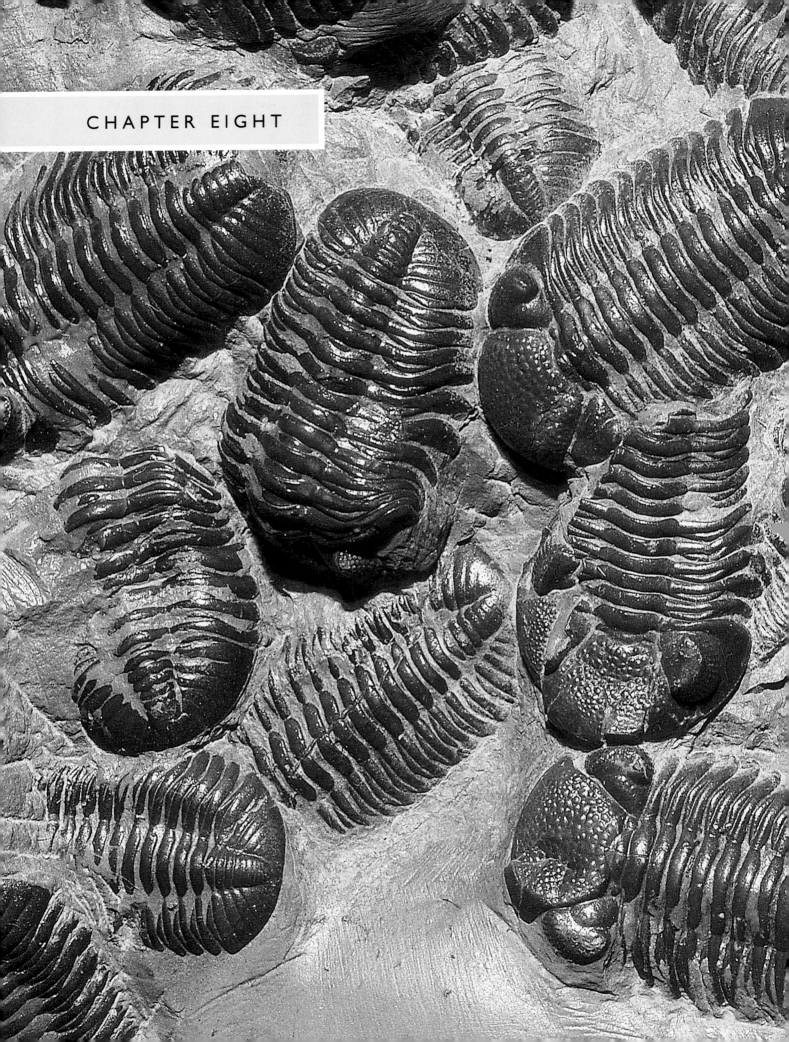

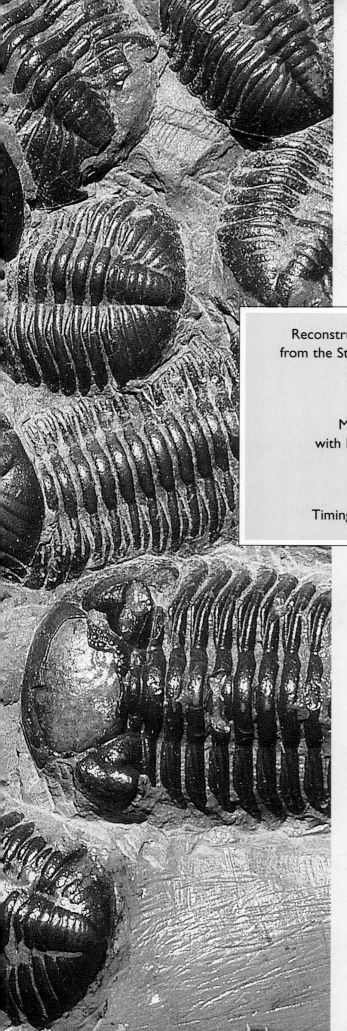

CLOCKS IN ROCKS
Timing the Geologic Record

The immensity of time—"deep time" measured in billions of years—was a great geological discovery. Knowledge of geologic time helped Darwin formulate his theory of evolution and has led to many other insights about the workings of the Earth system, the solar system, and the universe as a whole.

In this chapter, we will learn how geologists uncovered deep time by using the "clocks in rocks" to date geologic events. We will investigate how these clocks can measure the duration of geologic processes and cycles of the Earth system.

Geologic processes occur on time scales that range from seconds (meteorite impacts, volcanic explosions, earthquakes) to tens of millions of years (life cycle of oceanic lithosphere) and even billions of years (tectonic evolution of continents). We can directly measure short-term processes, such as beach erosion or the seasonal variations in the transport of sediment by rivers, and we can now monitor the slow movements of glaciers (meters per year) and even slower movements of the plates (centimeters per year). Historical documents can extend certain types of geologic data back for hundreds or, in some cases, thousands of years (**Figure 8.1**).

However, the record of human observation is far too short for the study of many slow geologic processes (**Figure 8.2**). In fact, it's not even long enough to capture some types of rapid but infrequent events; for example, we have never witnessed a meteorite impact as big as the one shown in Figure 1.2. We must rely instead on the geologic record—the information preserved in rocks that have survived erosion and subduction. Almost all oceanic crust older than 200 million years has been subducted back into the mantle, so most of Earth's history is documented only in the older rocks of the continents. Geologists can reconstruct subsidence from the record of

Trilobites preserved as fossils in rocks about 365 million years old. Ontario, Canada. [William E. Ferguson.]

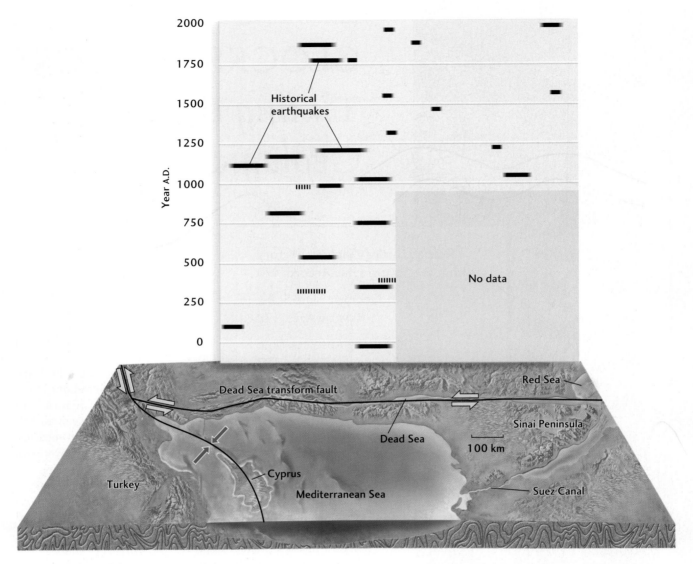

Figure 8.1 Map of earthquakes along the Dead Sea transform fault in the eastern Mediterranean, from historical records over the last 2000 years. Each bar represents a different earthquake, showing the length of the fault rupture on the transform fault (horizontal axis) and the date of its occurrence (vertical axis). [From research by Shmulik Marco, Amotz Agnon, Ronnie Ellenblum, and Tom Rockwell.]

sedimentation; uplift from the erosion of rock layers; and plate boundary deformation from faults, folds, and metamorphism. But to measure the pace of these processes and understand their common causes, we must be able to date the geologic record.

Geologists speak carefully about time. To them, dating refers not to a popular social activity but to measuring the **absolute age** of an event in the geologic record—the number of years elapsed from the event until now. Before the twentieth century, no one knew much about absolute ages; geologists could determine only how old one event was in relation to another—a **relative age.** They could say, for instance, that fish bones were first deposited in marine sediments before mammal bones appeared in land sediments, but they couldn't tell you how many millions of years ago

the first fish or mammals appeared. As we will see in this chapter, the physics and chemistry of atomic processes, particularly radioactivity, gave us powerful methods for estimating absolute ages and developing a precise and detailed geologic time scale.

RECONSTRUCTING GEOLOGIC HISTORY FROM THE STRATIGRAPHIC RECORD

Philosophers have struggled with the notion of time throughout human history, but until fairly recently they had very little data to constrain their speculations. The first geologic

1871

1968

Figure 8.2 Two photographs of Bowknot Bend on the Green River in Utah taken nearly 100 years apart show that little has changed in the configuration of rocks and formations in that time interval. (*left:* E. O. Beaman/USGS; *right:* H. G. Stevens/USGS.]

observations pertaining to the question of deep time came in the mid-seventeenth century from the study of fossils. A fossil is an artifact of life preserved in the geologic record (**Figure 8.3**). However, few people living in the seventeenth century would have understood this definition. Most thought that the seashells and other lifelike forms found in rocks dated from Earth's beginnings or grew there spontaneously.

In 1667, the Danish scientist Nicolaus Steno, who was working for the royal court in Florence, Italy, demonstrated that the peculiar "tongue stones" found in certain Mediterranean sediments were essentially identical to the teeth of

(a)

(b)

Figure 8.3 (a) Ammonite fossils, ancient examples of a large group of invertebrate organisms that are now largely extinct. Their sole representative in the modern world is the chambered nautilus. [Chip Clark.] (b) Petrified Forest, Arizona. These ancient logs are millions of years old. Their substance was completely replaced by silica, which preserved all the original details of form. [Tom Bean.]

modern sharks. He concluded that tongue stones *really were* ancient shark teeth preserved in the rocks and, more generally, that fossils were the remains of ancient life deposited with the sediments. To convince people of his ideas, Steno wrote a short but brilliant book about the geology of Tuscany in which he laid the foundation for the modern science of **stratigraphy**—the study of strata (layers) in rocks.

Principles of Stratigraphy

Geologists still use Steno's principles to interpret sedimentary strata. Two of his basic rules are so simple they seem obvious to us today:

1. The **principle of original horizontality** states that sediments are deposited under the influence of gravity as nearly horizontal beds. Observations in a wide variety of sedimentary environments support this principle. If we find folded or faulted strata, we know that the layers were deformed by tectonic forces after the sediments were deposited.

2. The **principle of superposition** states that each sedimentary layer of an undisturbed sequence is younger than the one beneath it and older than the one above it. A younger layer cannot be deposited beneath an existing layer. Thus, strata can be vertically ordered in time from the lowermost (oldest) bed to the uppermost (youngest) bed (**Figure 8.4**).

A chronologically ordered set of strata is a **stratigraphic succession.** We can apply Steno's principles in the field to determine whether one sedimentary layer is older than another. At each outcrop, there are distinct **formations,** groups of rock layers that can be identified throughout a region by their physical properties. Some formations consist of a single rock type, such as limestone. Others are made up of thin, interlayered beds of different kinds of sediments, such as sandstone and shale. However they vary, each formation comprises a distinctive set of rock layers that can be recognized and mapped as a unit. By piecing together the relative ages of formations exposed in different outcrops, we can sort them into chronological order and thus construct the stratigraphic succession of a region—at least in principle.

In practice, there were two problems. First, there were almost always gaps in a region's stratigraphic succession, indicating time intervals that had gone entirely unrecorded. Some were short, such as periods of drought between floods; others lasted for millions of years—for example, periods of regional tectonic uplift when thick sequences of sedimentary rocks were removed by erosion. Second, it was difficult to determine the relative ages of two widely separated formations; stratigraphy alone couldn't determine whether a sequence of mudstones in, say, Tuscany was older, younger, or the same age as a similar sequence in England. It was necessary to expand Steno's ideas about the biological origin of fossils to solve these problems.

Fossils as Recorders of Geologic Time

In 1793, William Smith, a surveyor working on the construction of canals in southern England, recognized that fossils can help us to order the relative ages of sedimentary rocks. Smith was fascinated by the variety of fossils, and he collected them from the rock strata exposed along the canal cuts. He observed that different layers contained different

1 Sediments are deposited in horizontal layers and slowly change into sedimentary rock.

2 If left undisturbed by tectonic processes, the youngest layers remain on the top and the oldest on the bottom.

Figure 8.4 Layers of sedimentary rock in Marble Canyon, part of the Grand Canyon. The Grand Canyon was cut by the Colorado River through what is now northern Arizona. These layers record millions of years of geologic history. Stratigraphy is the study of sedimentary sequences such as this one. [Fletcher and Baylis/Photo Researchers.]

sets of fossils, and he was able to tell one layer from another by the characteristic fossils in each. He established a general order for the sequence of fossils and strata, from lowermost (oldest) to uppermost (youngest) rock layers. Regardless of the location, Smith could predict the stratigraphic position of any particular layer or set of layers in any outcrop in southern England, basing his prediction on its fossil assemblages. This stratigraphic ordering of the fossils from animal species (fauna) is known as a *faunal succession.*

Smith's **principle of faunal succession** states that the layers of sedimentary rocks in an outcrop contain fossils in a definite sequence. The same sequence can be found in rocks at other locations, and so strata from one location can be matched to strata in another location.

Using faunal successions, Smith was able to identify formations of the same age found in different outcrops. By noting the vertical order in which the formations were found in each place, he compiled a composite stratigraphic succession for the entire region. His composite series showed how the complete succession would have looked if the formations at different levels in all the various outcrops could have been brought together at a single spot. **Figure 8.5** shows such a composite series for two formations.

Smith kept track of his work on stratigraphy by mapping surface exposures using colors assigned to specific formations, thus inventing the geologic map (described in Chapter 7). In 1815, he summarized his lifelong research by publishing his "General Map of Strata in England and Wales," a hand-colored masterpiece eight feet tall and six feet wide—the first geologic map of an entire country. The original still hangs in the Geological Society of London.

The geologists who followed in Steno and Smith's footsteps described and catalogued hundreds of fossils and their relationships to modern organisms, establishing the new science of *paleontology*—the historical study of ancient life forms. The most common fossils they found were the shells of invertebrate animals. Some were similar to clams, oysters, and other living shellfish; others represented strange species with no living examples, such as the trilobites shown in the photograph at the beginning of this chapter. Less common were the bones of vertebrates, such as mammals, birds, and the huge extinct reptiles they called dinosaurs. Plant fossils were found to be abundant in some rocks, particularly coal beds, where leaves, twigs, branches, and even whole tree trunks could be recognized. Geologists did not find fossils in intrusive igneous rocks—no surprise, since any biological material would have been destroyed in the hot

---

Key Figure 8.5 | Fossils can be used to correlate rock layers in different outcrops.

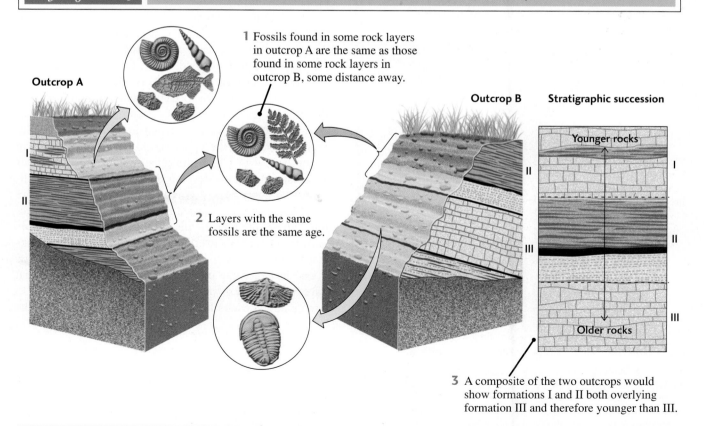

1 Fossils found in some rock layers in outcrop A are the same as those found in some rock layers in outcrop B, some distance away.

Outcrop A

Outcrop B   **Stratigraphic succession**

Younger rocks

2 Layers with the same fossils are the same age.

Older rocks

3 A composite of the two outcrops would show formations I and II both overlying formation III and therefore younger than III.

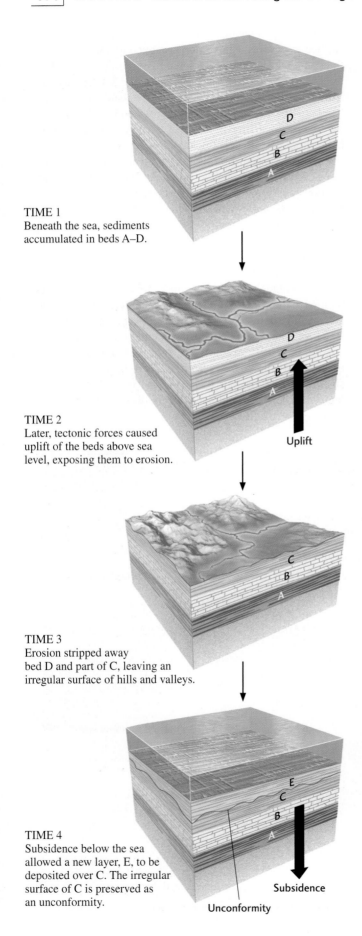

TIME 1
Beneath the sea, sediments accumulated in beds A–D.

TIME 2
Later, tectonic forces caused uplift of the beds above sea level, exposing them to erosion.

Uplift

TIME 3
Erosion stripped away bed D and part of C, leaving an irregular surface of hills and valleys.

TIME 4
Subsidence below the sea allowed a new layer, E, to be deposited over C. The irregular surface of C is preserved as an unconformity.

Subsidence

Unconformity

melt—nor did they find fossils in highly metamorphosed rocks, where any remains of organisms would have been distorted beyond recognition.

By the beginning of the nineteenth century, paleontology had become the single most important source of information about geologic history. The systematic study of fossils affected science far beyond geology. Charles Darwin studied paleontology as a young scientist, and he collected many unusual fossils on his famous voyage aboard the *Beagle* (1831–1836). During this world-circling tour, he also had the opportunity to study many unfamiliar animal and plant species in their native habitats. Darwin pondered what he had seen for a couple of decades and in 1859 proposed his theory of evolution, which revolutionized the science of biology and provided a sound theoretical framework for paleontology.

Unconformities: Gaps in the Geologic Record

In putting together the stratigraphic succession of a region, geologists often find places in the geologic record where a formation is missing. Either it was never deposited or it was eroded away before the next strata were laid down. The boundary along which the two existing formations meet is called an **unconformity**—a surface between two layers that were not laid down in an unbroken sequence (**Figure 8.6**). An unconformity, like a sedimentary sequence, represents the passage of time.

An unconformity may imply that tectonic forces raised the land above sea level, where it became eroded. Alternatively, it could have been produced by the erosion of land as the sea level fell. As we will see in Chapter 21, global sea level can be lowered by hundreds of meters during glacial ages when water from the oceans is withdrawn to form continental ice caps near the poles.

Unconformities are classified according to the relationships between the upper and lower sets of layers. An unconformity in which the upper set of layers overlies an erosional surface developed on an undeformed, still-horizontal lower set of beds is a *disconformity* (see Figure 8.6). Sea level drops caused by glaciations and broad tectonic uplifts often create disconformities. In a *nonconformity,* the upper beds overlie metamorphic or igneous rocks (see Feature 8.1 for an example). An *angular unconformity* is one in which the upper layers overlie lower beds that have been folded by tectonic processes and then eroded to a more or less even plane. In an angular unconformity, the

Figure 8.6 An unconformity is a surface between two layers that were not laid down in an unbroken sequence. In the series of events outlined here, the type of unconformity called a disconformity is created through uplift and erosion, followed by subsidence and another round of sedimentation.

Section through
Grand Canyon strata

Angular
unconformity

Figure 8.7 The Great Unconformity in the Grand Canyon, Arizona, is an angular unconformity between the horizontal Tapeats sandstones above and the steeply folded Precambrian Wapatai shales below. [GeoScience Features Picture Library.]

two sets of layers have bedding planes that are not parallel. **Figure 8.7** depicts a dramatic angular unconformity found near the bottom of the Grand Canyon. The formation of an angular unconformity by tectonic processes is illustrated in **Figure 8.8**.

Cross-Cutting Relationships

Other disturbances of the layering of sedimentary rocks also provide clues for determining relative age. Recall that dikes can cut through sedimentary layers; sills can be intruded parallel to bedding planes (Chapter 4); and faults can displace bedding planes, dikes, and sills as they shift blocks of rock (Chapter 7). These cross-cutting relationships can be used to establish the relative ages of igneous bodies or faults within the stratigraphic succession. Because the deformation or intrusive events must have taken place after the affected sedimentary layers were deposited,

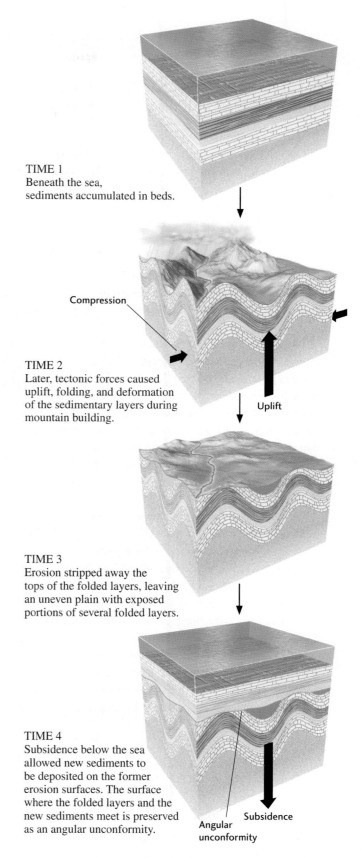

TIME 1
Beneath the sea,
sediments accumulated in beds.

Compression

TIME 2
Later, tectonic forces caused uplift, folding, and deformation of the sedimentary layers during mountain building.

Uplift

TIME 3
Erosion stripped away the tops of the folded layers, leaving an uneven plain with exposed portions of several folded layers.

TIME 4
Subsidence below the sea allowed new sediments to be deposited on the former erosion surfaces. The surface where the folded layers and the new sediments meet is preserved as an angular unconformity.

Subsidence

Angular
unconformity

Figure 8.8 An angular unconformity is a surface of erosion that separates two sets of layers having bedding planes that are not parallel. This sequence shows how such a surface can form.

EARTH ISSUES

8.1 The Grand Canyon Sequence and Regional Correlation of Strata

The rocks of the Grand Canyon and other parts of the Colorado Plateau have many stories to tell. They record a long history of sedimentation in a variety of environments, sometimes on land and sometimes under the sea. From correlation of the rock sequences exposed at different localities, geologists can reconstruct a geologic history over a billion years long.

The lowermost—and therefore oldest—rocks exposed at the Grand Canyon are dark igneous and metamorphic rocks forming the Vishnu group, known to be about 1.8 billion years old based on isotopic dating techniques.

Above the Vishnu are the younger Precambrian Grand Canyon beds. These beds contain fossils of millimeter-scale single-cell microorganisms. A nonconformity separates the Vishnu and Grand Canyon beds, signifying a period of structural deformation accompanying metamorphism of the Vishnu and then erosion before the deposition of the Grand Canyon beds. The tilting of the Grand Canyon beds from their originally horizontal position shows that they, too, were folded after deposition and burial.

An angular unconformity divides the Grand Canyon beds from the overlying horizontal Tapeats Sandstone. This unconformity indicates a long period of erosion after the lower rocks had been tilted. The Tapeats Sandstone and Bright Angel Shale can be dated as Cambrian by their fossils, many of which are trilobites.

Above the Bright Angel Shale is a group of horizontal limestone and shale formations (Muav Limestone, Temple Butte Limestone, Redwall Limestone) that represent about 200 million years from the late Cambrian period to the end of the Mississippian period. There are so many time gaps represented by disconformities in these rocks that less than 40 percent of the Paleozoic is actually represented by rock strata.

The next set of strata, high up on the canyon wall, is the Supai group of formations (Pennsylvanian and Permian), which contains fossils of land plants like those found in coal beds of North America and other continents. Overlying the Supai is the Hermit, a sandy red shale.

Continuing up the canyon wall, we find another continental deposit, the Coconino Sandstone, which contains vertebrate animal tracks. The animal tracks suggest that the Coconino was formed in a terrestrial environment during Permian times. At the top of the cliffs at the canyon rim are two more formations of Permian age: the Toroweap, made mostly of limestone, overlain by the Kaibab, a massive layer of sandy and cherty limestone. These two formations record subsidence below sea level and the deposition of marine sediments.

The succession of strata at the Grand Canyon, though picturesque and informative, represents an incomplete picture of Earth's history. Younger periods of geologic time are not preserved, and we must travel to locations in Utah such as Zion and Bryce Canyon national parks to fill in this younger history. At Zion, we find equivalents of the Kaibab and Moenkopi, which allow us to correlate back to the Grand Canyon area and establish a link. In contrast to the Grand Canyon area, however, the Zion rocks extend upward in age to Jurassic time, including ancient sand dunes represented by sandstones of the Navajo formation. If we travel a bit farther, we find that the Navajo formation also occurs in Bryce Canyon but that at this location the strata stack upward to the Wasatch formation of the Paleogene period.

The correlation of strata among these three areas of the Colorado Plateau shows how widely separated localities—each with an incomplete record of geologic time—can be pieced together to build a composite record of Earth's history.

Generalized stratigraphic section of the rock units in the Grand Canyon, Zion Canyon, and Bryce Canyon sequences. [*Grand Canyon:* John Wang/Photo Disc/Getty Images. *Zion Canyon:* David Muench/Corbis. *Bryce Canyon:* Tim Davis/Photo Researchers.]

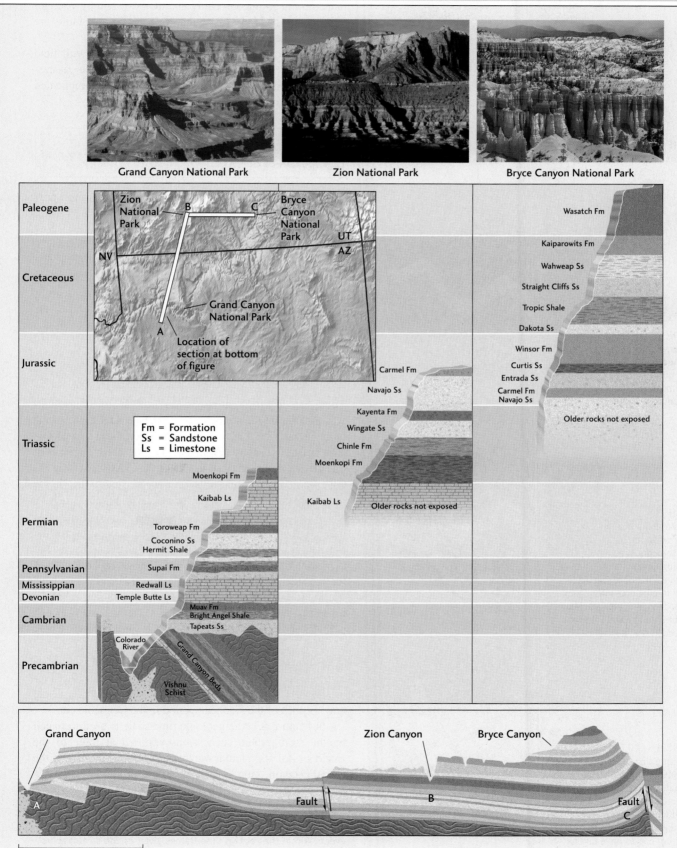

Grand Canyon National Park

Zion National Park

Bryce Canyon National Park

Paleogene		Wasatch Fm
		Kaiparowits Fm
Cretaceous		Wahweap Ss
		Straight Cliffs Ss
		Tropic Shale
		Dakota Ss
Jurassic	Carmel Fm	Winsor Fm
	Navajo Ss	Curtis Ss
		Entrada Ss
		Carmel Fm
		Navajo Ss
	Kayenta Fm	Older rocks not exposed
Triassic	Wingate Ss	
	Chinle Fm	
	Moenkopi Fm	
	Moenkopi Fm	
	Kaibab Ls	
Permian	Kaibab Ls — Older rocks not exposed	
	Toroweap Fm	
	Coconino Ss	
	Hermit Shale	
Pennsylvanian	Supai Fm	
Mississippian	Redwall Ls	
Devonian	Temple Butte Ls	
Cambrian	Muav Fm	
	Bright Angel Shale	
	Tapeats Ss	
	Colorado River	
Precambrian	Grand Canyon Beds	
	Vishnu Schist	

Zion National Park — B — C — Bryce Canyon National Park

NV — UT — AZ

Grand Canyon National Park

A — Location of section at bottom of figure

Fm = Formation
Ss = Sandstone
Ls = Limestone

Grand Canyon

Zion Canyon — Bryce Canyon

A — Fault — B — Fault — C

30 km

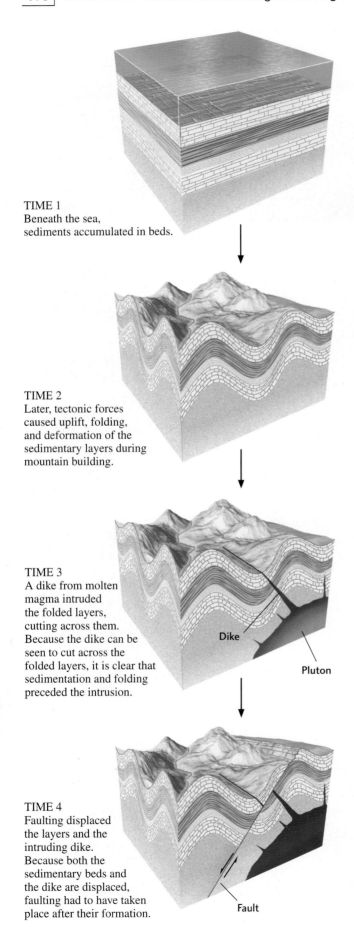

TIME 1
Beneath the sea, sediments accumulated in beds.

TIME 2
Later, tectonic forces caused uplift, folding, and deformation of the sedimentary layers during mountain building.

TIME 3
A dike from molten magma intruded the folded layers, cutting across them. Because the dike can be seen to cut across the folded layers, it is clear that sedimentation and folding preceded the intrusion.

Dike

Pluton

TIME 4
Faulting displaced the layers and the intruding dike. Because both the sedimentary beds and the dike are displaced, faulting had to have taken place after their formation.

Fault

they must be younger than the rocks that they cut (**Figure 8.9**). If the intrusions or fault displacements are eroded and planed off at an unconformity and then overlain by younger sedimentary beds, we know that the intrusions or faults are older than the younger strata.

We can combine cross-cutting relationships with field observations of unconformities and stratigraphic successions to decipher the history of geologically complicated regions (**Figure 8.10**).

GEOLOGIC TIME SCALE: RELATIVE AGES

Early in the nineteenth century, geologists began to apply Steno and Smith's stratigraphic principles to outcrops all over the world. The same distinctive fossils were discovered in similar formations on many continents. Moreover, faunal successions from different continents often displayed the same changes in fossil type. Thus, by matching up faunal successions, the relative ages of rock formations could be determined on a global basis. By the end of the century, geologists had pieced together a worldwide history of geologic events.

Divisions of Geologic Time

The **geologic time scale** divides Earth's history into intervals marked by a distinctive set of fossils, and it places the boundaries of the intervals at times when these sets of fossils changed abruptly (**Figure 8.11**). The basic intervals of this relative time scale are the **eras:** the Paleozoic (from the Greek *paleo,* meaning "old," and *zoi,* meaning "life"), the Mesozoic ("middle life"), and the Cenozoic ("new life").

The eras are subdivided into **periods,** usually named for the geographic locality in which the formations were first or best described, or for some distinguishing characteristic of the formations. The Jurassic period, for example, is named for the Jura Mountains of France and Switzerland and the Carboniferous period for the coal-bearing sedimentary rocks of Europe and North America. The Paleogene and Neogene periods of the Cenozoic are two exceptions; these Greek names mean "old origin" and "new origin," respectively.

Periods are further subdivided into **epochs,** such as the Miocene and Pliocene epochs of the Neogene period (see Figure 8.11). Today we are living in the Holocene ("completely new") epoch of the Neogene period in the Cenozoic era.

In Feature 8.1, the geologic time scale is used to match up the stratigraphic successions exposed at three of the world's most spectacular outcrops, the colorful cliffs of the Grand Canyon, Zion, and Bryce Canyon national parks.

Figure 8.9 Cross-cutting relationships allow us to place geologic events within the relative time frames given by the stratigraphic succession.

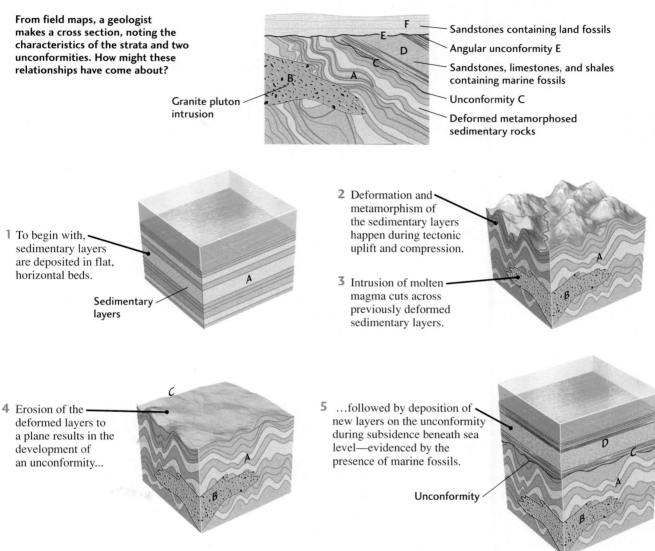

Earth System Figure 8.10 — Geologists use stratigraphic principles and cross-cutting relationships to establish a relative chronology.

From field maps, a geologist makes a cross section, noting the characteristics of the strata and two unconformities. How might these relationships have come about?

Sandstones containing land fossils

Angular unconformity E

Sandstones, limestones, and shales containing marine fossils

Unconformity C

Deformed metamorphosed sedimentary rocks

Granite pluton intrusion

1 To begin with, sedimentary layers are deposited in flat, horizontal beds.

Sedimentary layers

2 Deformation and metamorphism of the sedimentary layers happen during tectonic uplift and compression.

3 Intrusion of molten magma cuts across previously deformed sedimentary layers.

4 Erosion of the deformed layers to a plane results in the development of an unconformity...

5 ...followed by deposition of new layers on the unconformity during subsidence beneath sea level—evidenced by the presence of marine fossils.

Unconformity

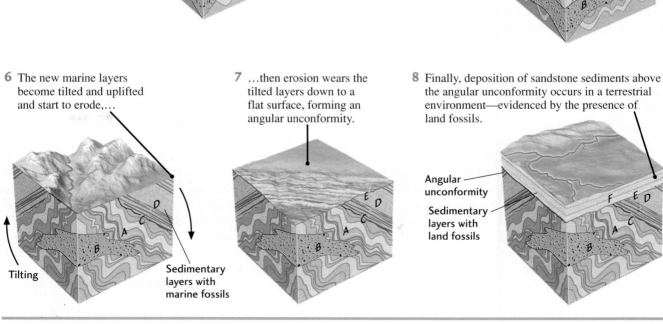

6 The new marine layers become tilted and uplifted and start to erode,...

Tilting

Sedimentary layers with marine fossils

7 ...then erosion wears the tilted layers down to a flat surface, forming an angular unconformity.

8 Finally, deposition of sandstone sediments above the angular unconformity occurs in a terrestrial environment—evidenced by the presence of land fossils.

Angular unconformity

Sedimentary layers with land fossils

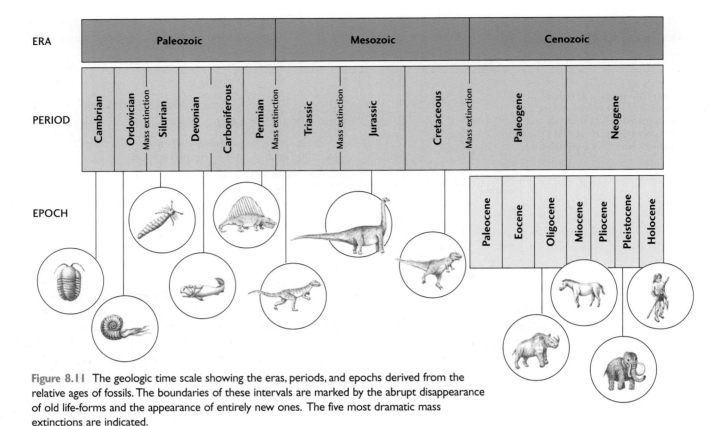

Figure 8.11 The geologic time scale showing the eras, periods, and epochs derived from the relative ages of fossils. The boundaries of these intervals are marked by the abrupt disappearance of old life-forms and the appearance of entirely new ones. The five most dramatic mass extinctions are indicated.

Interval Boundaries as Extreme Events

Many of the major boundaries in the geologic time scale represent *mass extinctions*—short intervals when many species simply disappeared from the geologic record, followed by the blossoming of many new species. These abrupt changes in the faunal succession were a big mystery to the geologists who discovered them. Darwin explained how new species could evolve, but what caused the mass extinctions?

In some cases, we think we know. The mass extinction at the end of the Cretaceous, which killed off 75 percent of the living species including all the dinosaurs, was almost certainly the result of a large meteorite impact that darkened the atmosphere and plunged Earth's climate into many years of bitter cold. This disaster marks the end of the Mesozoic era and the beginning of the Cenozoic.

In other cases, we are still not sure. The largest mass extinction, at the end of the Permian period, which defines the Paleozoic-Mesozoic boundary, eliminated nearly 95 percent of all species, but the cause of this event is still the subject of debate. Possibilities include the formation of the supercontinent Pangaea, abrupt climate changes, a meteorite impact, or a huge volcanic eruption in Siberia—or perhaps some combination of these causes. The extreme events that separate the geologic periods remain active areas of geologic research. We will return to this subject in Chapter 11.

MEASURING ABSOLUTE TIME WITH RADIOACTIVE CLOCKS

The geologic time scale based on studies of stratigraphy and fossils is a relative scale. It tells us if one formation is older than another, but not how long the eras, periods, and epochs were in actual years. Estimates of how long it takes mountains to erode and sediments to accumulate suggested that most geologic periods had lasted for millions of years, but geologists of the nineteenth century did not know whether the duration of a specific period was 10 million years, 100 million years, or even longer.

They did know that the geologic time scale was incomplete. The earliest period of geologic history recorded by faunal successions was the Cambrian, when animal life in the form of shelly fossils suddenly appeared in the geologic record. Many formations were clearly older than the Cambrian period, because they occurred below Cambrian rocks in the stratigraphic succession. But these formations contained no recognizable fossils, so there was no way to determine their relative ages. All geologists could do was lump such rocks into the general category of *Precambrian*. What fraction of Earth's history was locked up in these cryptic rocks? How old was the oldest Precambrian rock?

These questions sparked a huge debate in the latter half of the nineteenth century. Physicists and astronomers used theoretical arguments (now known to be incorrect) to deduce a maximum age of less than 100 million years, but most geologists regarded this age as much too young even though there were no precise data to back them up.

Discovery of Radioactivity

In 1896, a major advance in physics paved the way for reliable and accurate measurements of absolute age. Henri Becquerel, a French physicist, discovered radioactivity in uranium. Within a year, the French chemist Marie Sklodowska-Curie discovered and isolated a new and highly radioactive element, radium.

In 1905, the physicist Ernest Rutherford suggested that the absolute age of a rock could be determined by measuring the decay of radioactive elements found in rocks. He calculated the age of one rock by measuring its uranium content. This was the start of **isotopic dating,** the use of naturally occurring radioactive elements to determine the ages of rocks. The dating methods were refined over the next few years as more radioactive elements were found and the processes of radioactive decay became better understood. Within a decade of Rutherford's first attempt, geologists were able to show that some Precambrian rocks were billions of years old.

In 1956, the geologist Clare Paterson measured the decay of uranium in meteorites to determine that the solar system, and by implication planet Earth, was formed 4.56 billion years ago. This age has been modified by less than 10 million years since Paterson's original measurement, so we might say that he completed the discovery of geologic time.

Radioactive Atoms: The Clocks in Rocks

How do geologists use radioactivity to determine the age of a rock? As we saw in Chapter 3, the nucleus of an atom consists of protons and neutrons. For a given element, the number of protons is constant, but the number of neutrons can vary, forming different *isotopes* of the same element. Most isotopes are stable, but the nucleus of a radioactive isotope will spontaneously disintegrate, emitting particles and transforming to an atom of a different element. We call the original atom the *parent* and the decay product its *daughter.*

A useful element for isotopic age dating is rubidium, which has 37 protons and two naturally occurring isotopes: rubidium-85, which has 38 neutrons and is stable, and rubidium-87, which has 40 neutrons and is radioactive. A neutron in the nucleus of a rubidium-87 atom can spontaneously decompose, ejecting an electron and producing a new proton that stays in the nucleus. The former rubidium atom thus becomes a strontium atom, with 38 protons and 39 neutrons (**Figure 8.12**). In other words, the radioactive parent isotope rubidium-87 transforms into the stable daughter isotope strontium-87 by radioactive decay.

A parent isotope decays into a daughter isotope at a constant rate. The rate of radioactive decay is measured by

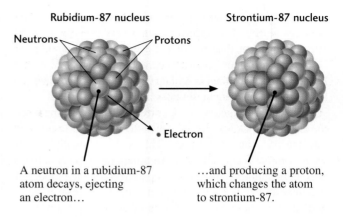

A neutron in a rubidium-87 atom decays, ejecting an electron…

…and producing a proton, which changes the atom to strontium-87.

Figure 8.12 The radioactive decay of rubidium to strontium.

the isotope's **half-life**—the time required for one-half of the original number of parent atoms to transform into daughter atoms. For instance, the half-life of rubidium-87 is 47 billion years, meaning that this is the time it takes for half of the rubidium in a rock sample to decay to strontium. Of course, we must know the initial amount that was present in a rock to calculate the isotopic age. In the case of rubidium-strontium dating, the initial amount of strontium-87 can be estimated from the amount of strontium-86, a stable isotope that is not the product of radioactive decay and therefore doesn't change as the mineral ages.

Radioactive isotopes make good clocks because the half-life does not vary with temperature, chemistry, pressure, or other changes that can accompany geologic processes on Earth or other planets. So when atoms of a radioactive isotope are created anywhere in the universe, they start to act like a ticking clock, steadily altering from one type of atom to another at a fixed rate. At the end of the first half-life, the number of parent atoms has decreased by a factor of two; at the end of a second half-life, by a factor of four; at the end of the third half-life, by a factor of eight, and so forth (**Figure 8.13**).

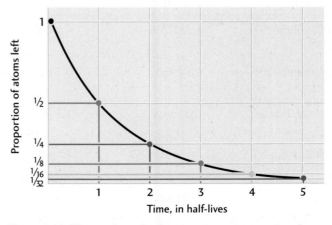

Figure 8.13 The number of radioactive atoms in any mineral declines at a fixed rate over time. This rate of decay is given by the half-life of the isotope.

Geologists can measure the ratio of parent and daughter isotopes with a mass spectrometer, a very precise and sensitive instrument that can detect even minute quantities of isotopes. From these measurements and knowledge of the half-life, they can then calculate the time elapsed since the radioactive clock began to tick.

The isotopic age of a rock corresponds to the time since the isotopic clock was "reset" when the isotopes were locked into the minerals of the rock. This locking usually occurs when a mineral crystallizes from a magma or recrystallizes during metamorphism. During crystallization, the number of daughter atoms in a mineral is not necessarily reset to zero, and so the initial amount must be taken into account when calculating the isotopic age.

Many other complications make isotopic age dating a tricky business. A mineral can lose daughter isotopes by weathering or can be contaminated by fluids circulating in the rock. For igneous rocks, metamorphism can reset the isotopic composition of the mineral to a date much younger than the crystallization age.

Isotopic Dating Methods

To determine the ages of rocks, geologists use a number of naturally occurring radioactive elements, each of which has its own half-life (Table 8.1). Isotopes that decay slowly over billions of years, such as rubidium-87, are most useful in measuring the ages of older rocks, whereas those that decay rapidly, such as carbon-14, can only be used to date younger rocks. Isotopic dating is possible only if a measurable number of parent and daughter atoms remain in the rock. For example, if a rock is very old and the decay rate is fast,

almost all the parent atoms will already have been transformed. In that case, we could determine that the isotopic clock has run down, but we would have no way of knowing how long ago it stopped.

Carbon-14 is especially useful for dating fossil bone, shell, wood, and other organic materials in sediments less than a few tens of thousands of years old. Carbon is an essential element in the living cells of all organisms. As green plants grow, they continuously incorporate carbon into their tissues from carbon dioxide in the atmosphere. When a plant dies, it stops absorbing carbon dioxide. At that moment, the amount of carbon-14 in relation to the stable carbon isotopes in the plant is identical to that in the atmosphere. Afterward, the amount of carbon-14 in the dead tissue decreases with a half-life of about 5700 years. Nitrogen-14, the daughter isotope of carbon-14, is a gas and thus leaks from the fossil organic material buried in the sediments, so it cannot be measured to determine the time that has elapsed since the plant died. We can, however, estimate this absolute age by comparing the amount of carbon-14 left in the plant material with the amount in the atmosphere at the time the plant died.

The accuracy of isotopic dating depends on precise measurements of the often minute amounts of parent and daughter atoms found in rocks. Techniques have now advanced to the point that mass spectrometers can count a very small number of atoms, providing accurate dates for very old rocks.

One of the most precise dating methods for old rocks is based on two related isotopes: the decay of uranium-238 to lead-206 and the decay of uranium-235 to lead-207. Both isotopes of uranium behave similarly in the chemical reactions that alter rocks, because the chemistry of an element

Table 8.1		Major Radioactive Elements Used in Radiometric Dating		
Isotopes		**Half-Life of Parent (years)**	**Effective Dating Range (years)**	**Minerals and Materials That Can Be Dated**
Parent	**Daughter**			
Uranium-238	Lead-206	4.4 billion	10 million– 4.6 billion	Zircon Apatite
Uranium-235	Lead-207	0.7 billion	10 million– 4.6 billion	Zircon Apatite
Potassium-40	Argon-40	1.3 billion	50,000– 4.6 billion	Muscovite, Biotite Hornblende
Rubidium-87	Strontium-87	47 billion	10 million– 4.6 billion	Muscovite, Biotite Potassium feldspar
Carbon-14	Nitrogen-14	5730	100–70,000	Wood, charcoal, peat Bone and tissue Shells and other calcium carbonates

depends mainly on its atomic number, not its atomic mass. However, the uranium isotopes have different half-lives (see Table 8.1), so together they provide a consistency check that helps account for the problems of weathering, contamination, and metamorphism discussed above. These days, the lead isotopes from single crystals of zircon—a crustal mineral with a relatively high concentration of uranium—can be used to date the oldest rocks on Earth with a precision of less than 1 percent. These formations turn out to be about 4 billion years old.

GEOLOGIC TIME SCALE: ABSOLUTE AGES

Armed with isotopic dating techniques, geologists of the twentieth century were able to nail down the absolute ages of the key events on which their predecessors had based the geologic time scale. More important, they were able explore the early history of the planet recorded in Precambrian rocks. **Figure 8.14** presents the results of this century-long effort as a ribbon of geologic time.

Calibrating the geologic time scale with absolute ages reveals differences in the time intervals spanned by the geologic periods. The Cretaceous period (80 million years) turns out to be more than three time longer than the Neogene (only 23 million years), and the Paleozoic era (291 million years) lasted longer than the Mesozoic and Cenozoic combined. The biggest surprise is the Precambrian, which had a duration of over 4000 million years—almost nine-tenths of Earth's history!

To handle this extension of the time scale, a division of geologic history longer than the era, called the **eon,** was introduced. Based on the isotopic ages of terrestrial rocks and meteorites, four eons are recognized.

Hadean Eon The earliest eon, whose name comes from *Hades* (the Greek word for "hell"), began with the formation of Earth 4.56 billion years ago and ended about 3.9 billion years ago. During its first 660 million years, Earth was

| Key Figure 8.14 | The ribbon of geologic time shows the complete geologic time scale. |

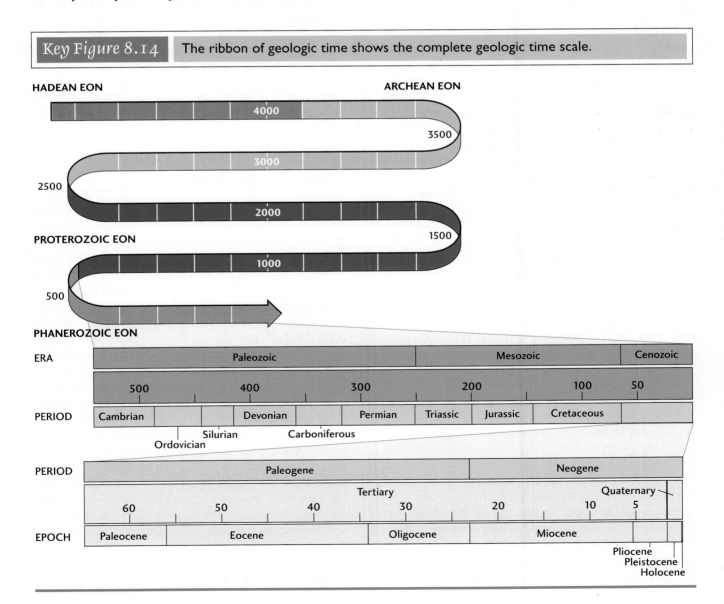

bombarded by chunks of material from the early solar system. Although very few rock formations survived this violent period, individual mineral grains 4.4 billion years old have been discovered. There is also evidence that some liquid water existed on Earth's surface at about this time, suggesting that the planet cooled rapidly. In Chapter 9, we will explore this early phase of Earth's history in more detail.

Archean Eon The name is from the Greek *archaios,* meaning "ancient." Rocks of Archean age range from 3.9 billion to 2.5 billion years old. The geodynamo and the climate systems were established during the Archean eon, and felsic crust accumulated to form the first stable continental masses (described in Chapter 10). The processes of plate tectonics were probably operating, although perhaps substantially differently from the way they did later in Earth's history. Life in the form of primitive unicellular microorganisms was established, as indicated by the fossils found in sedimentary rocks of this age (see Chapter 11).

Proterozoic Eon The last part of the Precambrian is called the Proterozoic eon (from the Greek *proteros* and *zoi,* meaning "earlier life"), which spans the time interval from 2.5 billion to 542 million years ago. By the beginning of this eon, the plate tectonic and climate systems were operating pretty much as they do today. Throughout the Proterozoic, organisms that produced oxygen as a waste product (as plants do today) increased the amount of oxygen in Earth's atmosphere, precipitating tremendous quantities of iron oxide from seawater to form distinctive red beds. Although life remained soft-bodied, some organisms evolved into sophisticated creatures with cells containing nuclei. The increase in oxygen to nearly present-day levels toward the end of the Proterozoic may have encouraged single-celled organisms to evolve into multicellular algae and animals, which are preserved in the late Proterozoic fossil record. A more complete story of life will be presented in Chapter 11.

Phanerozoic Eon The start of the Phanerozoic eon is marked by the first appearance of shelly fossils at the beginning of the Cambrian period, now dated at 542 million years. The name—from the Greek *phaneros* and *zoi* ("visible life")—certainly fits, because it comprises all three eras recognized in the fossil record: the Paleozoic (542 to 251 million years ago), the Mesozoic (251 to 65 million years ago), and the Cenozoic (65 million years ago to the present).

One way to appreciate the extraordinary length of geologic history is to think of Earth's age as one calendar year. On January 1, Earth formed. Within the first week, Earth was organized into core, mantle, and crust. The first primitive organisms appeared in mid-March. By mid-June, stable continents had developed, and throughout the summer and fall, biological activity increased oxygen in the atmosphere. On November 18, at the beginning of the Cambrian period, complex organisms, including those with shells, arrived. On December 11 reptiles evolved, and late on Christmas Day the dinosaurs became extinct. Modern humans, *Homo sapiens,* first appeared on the scene at 11:42 P.M. on New Year's Eve, and the last glacial age ended at 11:58 P.M. Three and a half seconds before midnight, Columbus landed on a West Indian island, and a couple of tenths of a second ago, you were born.

TIMING THE EARTH SYSTEM

The discovery of deep time changed our thinking about how Earth operates as a system. Two fathers of modern geology, James Hutton and Charles Lyell, led us to understand that the planet was not shaped by a series of catastrophic events over a mere few thousand years, as many people had believed. Rather, what we see today is the product of ordinary geologic processes operating over much longer time intervals. Hutton stated this understanding in the *principle of uniformitarianism,* described in Chapter 1.

We know that the time scales of geologic processes are not uniform, however, but vary from seconds to billions of years. We must therefore use a variety of methods for timing the Earth system—some to determine the age of very old rocks, others to measure rapid changes. New methods for determining relative and absolute ages continue to be developed, and these tools have steadily improved our understanding of how the Earth system works. To cap off our story of the geologic time scale, we will describe a few of the recent advances.

Sequence Stratigraphy

Until a few decades ago, geologists had to rely on rocks exposed at outcrops, in mines, and by drilling to map stratigraphic successions. Technological innovations in the field of seismology (described in Chapter 14) now allow us to see below Earth's surface without actually going there. From recordings of seismic waves generated by controlled explosions, as well as natural earthquakes, we can construct three-dimensional images of deeply buried structures (**Figure 8.15**). The basic geologic unit observed by the seismic imaging of sediments is a series of beds bounded above and below by unconformities—a sedimentary sequence. This type of detailed geologic mapping is called *sequence stratigraphy.*

Sedimentary sequences form on the edges of continents when sediment deposition by rivers is modified by fluctuations in sea level. In the example shown in Figure 8.15, sediment is laid down in a delta where a river enters the sea. As the sediment builds up the seafloor to sea level, the delta advances toward the sea. When the sea level falls (say, during a period of continental glaciation), the deltaic deposits are exposed to erosion. The shoreline shifts inland when sea level rises again, and a new deltaic sequence begins to cover the old one, creating an unconformity. Over millions of years, this cycle may be repeated many times, producing a complex set of sedimentary sequences.

1 Seismic technology can be used to create seismic profiles to reveal sequence stratigraphy,...

2 ...which allows geologists to see individual beds in a sequence.

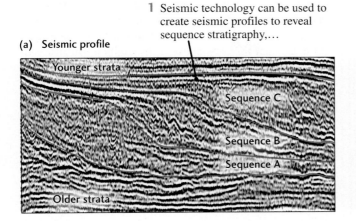

(a) Seismic profile

Younger strata

Sequence C

Sequence B

Sequence A

Older strata

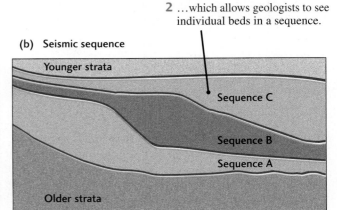

(b) Seismic sequence

Younger strata

Sequence C

Sequence B

Sequence A

Older strata

(c)

3 The seismic sequence reveals changes in sedimentation such as those that occur at a river delta.

Delta

B

A

Sediment

(d)

C

B

A

4 A sequence of delta sediments, B, accumulates over previous sediments, A.

5 The sea level rises, and the shoreline recedes inland.

6 Another sedimentary sequence, C, accumulates over sequence B.

Figure 8.15 A comparison of seismic profiles (a) with seismic sequences (b) reveals the depositional process that creates bedding patterns. When tectonic subsidence or events such as global climate change have caused the sea level to rise, two deltaic sequences are found (c and d).

Because sea level fluctuations are worldwide, we can match sedimentary sequences of the same age over wide areas. These relative ages can then be used to reconstruct a region's geologic history, accounting for any regional tectonic uplift or subsidence in addition to the global changes in sea level. Sequence stratigraphy has been especially effective in finding oil and gas deeply buried on continental margins, such as the Gulf of Mexico and the Atlantic margin.

Chemical Stratigraphy

The layers of sedimentary rocks often contain minerals and chemicals that identify them as distinctive units. For example, the amount of iron or manganese may vary from layer to layer in carbonate sediments because the composition of seawater changed during precipitation of the carbonate minerals. When the sediments get buried and converted to sedimentary rocks, these chemical variations can be preserved, "fingerprinting" the formations. Often these chemical fin-

gerprints extend regionally or even globally, allowing us to match up sedimentary rocks in a *chemical stratigraphy* where no other features, such as fossils, are available.

Paleomagnetic Stratigraphy

Another technique for fingerprinting rock units is *paleomagnetic stratigraphy*. As discussed in Chapter 1, Earth's magnetic field reverses itself irregularly, so that the north pole become the south pole and vice versa. These reversals are recorded in the orientation of magnetic minerals in volcanic rocks, which can be dated by isotopic methods. The resulting chronology of magnetic reversals—the magnetic time scale—allows us to "replay the magnetic tape" of seafloor spreading and determine the rates of plate motions, as we saw in Chapter 2. Even more detailed patterns of magnetic reversals can be observed in sediment cores, and these fingerprints can be dated as the stratigraphic age of the fossils in the sediments. Paleomagnetic stratigraphy has recently become one of the main methods for measuring sedimentation rates along the continental margins and in the deep oceans.

Clocking the Climate System

Sequence stratigraphy and other data indicate that the last two epochs of geologic time, the Pleistocene and Holocene (see Figure 8.14), have been especially active in terms of global climate change. As we will see in Chapter 15, repeated cycles of glaciation have occurred with periods ranging from 40,000 to 100,000 years, and shorter-term cycles of hundreds to thousands of years are also evident. There are a number of methods to measure these climate cycles.

For example, we can use the stable isotopes of oxygen in shelly fossils to estimate temperatures and carbon-14 to determine when in the recent past the shells were formed. By carefully measuring temperatures and ages along the length of sediment cores taken from various spots on the seafloor, we can see how temperature changed over time. This procedure has provided us with a precise record of glacial cycles since the beginning of the Pleistocene epoch, 1.8 million years ago (see Chapter 21).

We can also measure the amount of carbon dioxide in the atmosphere by analyzing the gases trapped in another type of core—ice samples drilled out of the thick Antarctic and Greenland ice caps, which have accumulated from countless snowstorms over hundreds of thousands of years. Carbon dioxide plays a very important role in global climate because it is a "greenhouse gas" that helps to trap solar heat in Earth's atmosphere. Unfortunately (and somewhat ironically), the carbon dioxide itself cannot be dated by the carbon-14 method; the small gas bubbles just don't contain enough atoms of carbon-14. However, the core can be accurately dated by other methods, including isotopic dating of the volcanic ash that occurs as thin layers in the ice cores and counting the annual layers in the ice cores. (The latter is much like counting tree rings, which have also been used to date climate changes.)

Combining the sediment-core records of temperature with the ice-core records of atmospheric composition shows that, throughout the last 400,000 years, the global temperatures were higher when the atmosphere contained more carbon dioxide and lower when it contained less. This is one of the reasons most geologists are concerned about our burning of fossil fuels, which is causing carbon dioxide in the atmosphere to increase very rapidly and will almost certainly result in substantial global warming (see Chapter 23).

SUMMARY

How do we know whether one rock is older than another? We determine the order in which rocks formed by studying the stratigraphy, fossils, and cross-cutting relation-

ships of rock formations observed at outcrops. An undeformed sequence of sedimentary rock layers will be horizontal, with each layer younger than the layers beneath it and older than the ones above it. In addition, because animals and plants have evolved progressively over time, the fossils found in each layer reflect the organisms that were present when that layer was deposited. Knowing the faunal succession makes it easier to spot unconformities, which indicate intervals of time missing in the sedimentary record.

How was a global geologic time scale created? Using fossils to correlate rocks of the same geologic age and piecing together the sequences exposed in hundreds of thousands of outcrops around the world, geologists compiled a stratigraphic sequence applicable everywhere on Earth. The composite sequence represents the geologic time scale. The use of isotopic dating allowed scientists to assign absolute dates to the eons, eras, periods, and epochs that constitute the time scale. Isotopic dating is based on the decay of radioactive elements, in which unstable parent atoms are transformed into stable daughter isotopes at a constant rate. When radioactive elements are locked into minerals as igneous rocks crystallize or metamorphic rocks recrystallize, the number of daughter atoms increases as the number of parent atoms decreases. By counting parent and daughter atoms, we can calculate absolute ages.

Why is the geologic time scale important to geology? The geologic time scale enables us to reconstruct the chronology of events that have shaped the planet. The time scale has been instrumental in validating and studying plate tectonics and in estimating the rates of geologic processes too slow to be monitored directly, such as the opening of an ocean over millions to hundreds of millions of years. The development of the time scale revealed that Earth is much older than anyone had imagined and that it has undergone almost constant change throughout its history. The creation of the geologic time scale paralleled the development of paleontology and the theory of evolution, one of the most revolutionary and powerful ideas in science.

KEY TERMS AND CONCEPTS

absolute age (p. 170)
eon (p. 183)
epoch (p. 178)
era (p. 178)
formation (p. 172)
geologic time scale (p. 178)
half-life (p. 181)
isotopic dating (p. 181)
period (p. 178)
principle of faunal succession (p. 173)
principle of original horizontality (p. 172)
principle of superposition (p. 172)
relative age (p. 170)
stratigraphic succession (p. 172)
stratigraphy (p. 172)
unconformity (p. 174)

EXERCISES

1. Many fine-grained muds are deposited at a rate of about 1 cm per 1000 years. At this rate, how long would it take to accumulate a stratigraphic sequence half a kilometer thick?

2. Construct a cross section similar to the one at the top of Figure 8.10 to show the following sequence of geologic events: (a) sedimentation of a limestone formation; (b) uplift and folding of the limestone; (c) erosion of the folded terrain; (d) subsidence of the terrain and sedimentation of a sandstone formation.

3. How many formations can you count in the geologic cross section of the Grand Canyon in Feature 8.1? How many are the same formations observed in Zion Canyon? Are any of the formations observed in both the Grand Canyon and Bryce Canyon cross sections?

4. What type of unconformity would a broad tectonic uplift followed by subsidence be likely to produce? What type of unconformity might separate young flat-lying sediments from older metamorphosed sediments?

5. Mass extinctions have been dated at 444 million years, 416 million years, and 359 million years. How are these events expressed in the geologic time scale of Figure 8.14?

6. A geologist discovers a distinctive set of fish fossils that dates from the Devonian period within a low-grade metamorphic rock. The rubidium-strontium isotopic age of the rock is determined to be only 70 million years. Give a possible explanation for the discrepancy.

7. Explain why the last eon of geologic history is named the Phanerozoic.

8. At the present rate of seafloor spreading, the entire seafloor is recycled every 200 million years. Assuming that this rate has been constant, calculate how many times the seafloor has been recycled since the end of the Archean eon.

THOUGHT QUESTIONS

1. As you pass by an excavation in the street, you see a cross section showing paving at the top, soil below the paving, and bedrock at the base. You also notice that a vertical water pipe extends through a hole in the street into a sewer in the soil. What can you say about the relative ages of the various layers and the water pipe?

2. Why did nineteenth-century geologists constructing the geologic time scale find sedimentary strata deposited in the sea more useful than strata deposited on land?

3. In studying an area of tectonic compression, a geologist discovers a sequence of older, more deformed sedimentary rocks on top of a younger, less deformed sequence, separated by an angular unconformity. What tectonic process might have created the angular unconformity?

4. A geologist documents a distinctive chemical signature caused by organisms of the Proterozoic eon that has been preserved in a sedimentary rock. Would you consider this chemical signature to be a fossil?

5. Is carbon-14 a suitable isotope for dating geologic events in the Pliocene epoch?

6. How does determining the ages of igneous rocks help to date fossils?

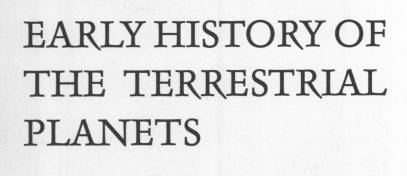

EARLY HISTORY OF THE TERRESTRIAL PLANETS

In a series of six *Apollo* landings from 1969 through 1972, astronauts trained in geology explored the lunar surface, taking photographs, mapping outcrops, conducting experiments, and collecting dust and rock samples for analysis back on Earth (see chapter opening photo). This unprecedented achievement was possible only through the close collaboration of engineers and scientists and the funding agencies that recognized the importance of basic research in developing new technologies. Perhaps the most important ingredient of all was the irrepressible drive, inherent in all human beings, to explore the unknown. The desire to explore our universe has existed for as long as humans have been able to think. Edwin Powell Hubble, a famous astronomer, best captured the spirit of space exploration when he modestly noted that "Equipped with his five senses, man explores the universe around him and calls the adventure science."

The modern era of space exploration began in the early 1900s when a handful of scientists with a yearning to escape the confines of Earth's gravity began to develop the first generation of rockets (**Figure 9.1a**). By the late 1920s, these backyard rockets powered by liquid propellants were ready for use. Early success fueled enthusiasm, and developments occurred rapidly over the next few decades, culminating in the fevered cold-war race between the United States and the Soviet Union to put the first rocket into space, the first satellite into Earth orbit, the first human on the Moon, and the first robot on Mars. By the mid-1970s—50 years after the first liquid-fueled rockets were invented—all of these goals had been achieved.

As we will see in this chapter, the scientific dividends of the space race were tremendous. The age of the solar system, the evidence for liquid water on early Mars, and the thick atmosphere of Venus were all revealed by the mid-1970s. Since that time, we have carried on our exploration of the solar system and beyond. Using instruments on spacecraft sent to the far limits of our solar system, we have obtained

Geologist-astronaut Harrison Schmitt, *Apollo 17* lunar module pilot, retrieves samples of a lunar rock for their return to Earth. The adjustable sampling scoop, used to obtain samples too small for tongs, is another geological tool used by astronauts. [NASA.]

(a)

(b)

Figure 9.1 Humans have always dreamed of exploring space. (a) Robert H. Goddard, one of the fathers of rocketry, fired this liquid oxygen–gasoline rocket on March 16, 1926, at Auburn, Massachusetts. [NASA.] (b) Seventy years later, on November 2, 1995, the Hubble Space Telescope (in orbit around Earth) took this stunning photograph of the Eagle Nebula. The dark, pillar-like structures are actually columns of cool hydrogen gas and dust that give birth to new stars. [NASA/ESA/STSci.]

a much better view of what, literally, is *way* out there! Of all these instruments, none has produced such visually spectacular images of deep space as the Hubble Space Telescope (see Figure 9.1b). Not since Galileo turned his telescope toward the heavens in 1610 has any instrument so changed our understanding of the universe.

In this chapter, we will explore the solar system not only in the vast reaches of interplanetary space but also backward in time to its earliest history. We will see how Earth and the other planets formed around the Sun and how they differentiated into layered bodies. We will compare the geologic processes that have shaped Earth with those of Mercury, Mars, Venus, and the Moon, and we will see how exploration of the solar system by spacecraft might answer fundamental questions about the evolution of our planet and the life it contains.

ORIGIN OF THE SOLAR SYSTEM

The search for the origins of the universe and our own small part of it goes back to the earliest recorded mythologies. Today, the generally accepted scientific explanation is the Big Bang theory, which holds that our universe began about 13.7 billion years ago with a cosmic "explosion." Before that moment, all matter and energy were compacted into a

single, inconceivably dense point. Although we know little of what happened in the first fraction of a second after time began, astronomers have a general understanding of the billions of years that followed. In a process that still continues, the universe has expanded and thinned out to form the galaxies and stars. Geology explores the latter third of that time: the past 4.5 billion years, during which our solar system—the star that we call the Sun and the planets that orbit it—formed and evolved. Specifically, geologists look to the formation of the solar system to understand the formation of Earth and the Earthlike planets.

The Nebular Hypothesis

In 1755, the German philosopher Immanuel Kant suggested that the origin of the solar system could be traced to a rotating cloud of gas and fine dust. Discoveries made in the past half century have led astronomers back to this old idea, now called the **nebular hypothesis.** Equipped with modern telescopes, they have found that outer space beyond our solar system is not as empty as we once thought. Astronomers have recorded many clouds of the type that Kant surmised, and they have named them *nebulae* (plural of the Latin word for "fog" or "cloud"). They have also identified the materials that form these clouds. The gases are mostly hydrogen and helium, the two elements that make up all but a small

Key Figure 9.2 | The nebular hypothesis explains the evolution of the solar system.

A diffuse, roughly spherical, slowly rotating nebula begins to contract.

As a result of contraction and rotation, a flat, rapidly rotating disk forms with the matter concentrated at the center that will become the proto-Sun.

The enveloping disk of gas and dust forms grains that collide and clump together into small chunks or planetesimals.

Planetesimals   Protostar

Planetesimal

~ 1 km

The terrestrial planets build up by multiple collisions and accretion of planetesimals by gravitational attraction. Giant outer planets grow by gas accretion.

Terrestrial planets

Giant outer planets

Planetesimals

Gas

Sun   Planets

Solar system

fraction of our Sun. The dust-sized particles are chemically similar to materials found on Earth.

How could our solar system take shape from such a cloud? This diffuse, slowly rotating cloud contracted under the force of gravity (**Figure 9.2**). Contraction, in turn, accelerated the rotation of the particles (just as ice skaters spin more rapidly when they pull in their arms), and the faster rotation flattened the cloud into a disk.

The Sun Forms

Under the pull of gravity, matter began to drift toward the center, accumulating into a protostar, the precursor of our present Sun. Compressed under its own weight, the material in the proto-Sun became dense and hot. The internal temperature of the proto-Sun rose to millions of degrees, at which point nuclear fusion began. The Sun's nuclear fusion, which continues today, is the same nuclear reaction that occurs in a hydrogen bomb. In both cases, hydrogen atoms, under intense pressure and at high temperature, combine (fuse) to form helium. Some mass is converted into energy in the process. The Sun releases some of that energy as sunshine; an H-bomb releases it as a great explosion.

The Planets Form

Although most of the matter in the original nebula was concentrated in the proto-Sun, a disk of gas and dust, called the **solar nebula,** remained to envelop it. The solar nebula grew hot as it flattened into a disk. It became hotter in the inner region, where more of the matter accumulated, than in the less dense outer regions. Once formed, the disk began to cool, and many of the gases condensed. That is, they changed into their liquid or solid form, just as water vapor condenses into droplets on the outside of a cold glass and water solidifies into ice when it cools below the freezing point. Gravitational attraction caused the dust and condensing material to clump together in "sticky" collisions as small, kilometer-sized chunks, or **planetesimals.** In turn, these planetesimals collided and stuck together, forming larger, Moon-sized bodies. In a final stage of cataclysmic impacts, a few of these larger bodies—with their larger gravitational attraction—swept up the others to form our nine planets in their present orbits.

As the planets formed, those in orbits close to the Sun and those in orbits farther from the Sun developed in markedly different ways. The composition of the inner planets is quite different from that of the outer planets.

Inner Planets The four inner planets, in order of closeness to the Sun, are Mercury, Venus, Earth, and Mars (**Figure 9.3**). They are also known as the terrestrial ("Earthlike") planets. In contrast with the outer planets, the four inner planets are small and are made up of rocks and metals. They grew close to the Sun, where conditions were so hot that most of the volatile materials—those that become gases— boiled away. Radiation and matter streaming from the Sun

The inner planets are small and rocky.

The four giant outer planets and their moons are gaseous, with rocky cores.

Pluto is a snowball of methane, water, and rock.

Figure 9.3 The solar system. The figure shows the relative sizes of the planets and the asteroid belt separating the inner and outer planets.

blew away most of the hydrogen, helium, water, and other light gases and liquids on these planets. The inner planets formed from the dense matter that was left behind, which included the rock-forming silicates as well as metals such as iron and nickel. From the age of meteorites that occasionally strike Earth and are believed to be remnants of preplanetary time, we know that the inner planets began to come together (accrete) at about 4.56 Ga (see Chapter 8). Theoretical calculations indicate that they would have grown to planetary size in a remarkably short time—perhaps as quickly as 10 million years or less.

Giant Outer Planets Most of the volatile materials swept from the region of the terrestrial planets were carried to the cold outer reaches of the solar system to form the giant outer planets made up of ices and gases—Jupiter, Saturn, Uranus, and Neptune—and their satellites. The giant planets were big enough and their gravitational attraction strong enough to enable them to hold onto the lighter nebular constituents. Thus, although they have rocky and metal-rich cores, they (like the Sun) are composed mostly of hydrogen and helium and the other light constituents of the original nebula.

Small Bodies of the Solar System

Not all of the material from the solar nebula ended up in the major planets. Some of the planetesimals collected between the orbits of Mars and Jupiter to form the *asteroid belt* (see Figure 9.3). This region now contains about 300 **asteroids** with diameters larger than 100 km and more than 10,000 with diameters larger than 10 km. The biggest is *Ceres,* which has a diameter of 930 km. Most **meteorites** that strike Earth are tiny pieces of asteroids ejected during collisions with one another. Astronomers originally thought the asteroids were the remains of a large planet that had broken apart early in the history of the solar system, but it now appears they are pieces that never coalesced into a planet, probably owing to the gravitational influence of Jupiter.

Another important group of small, solid bodies is the *comets,* aggregations of dust and ice that condensed in the cooler, outer reaches of the solar nebula. Most of the comets —there are probably many millions of them with diameters larger than 10 km—orbit the Sun far beyond the outer planets, forming concentric "halos" around the solar system. Occasionally, collisions or near misses will throw one of them into an orbit that penetrates the inner solar system. We can then observe it as a bright object with a tail of gases blown away from the Sun by the solar wind. Comets are intriguing to geologists because they provide clues about the more volatile components of the solar nebula, including water and carbon-rich compounds, which they contain in abundance.

EARLY EARTH: FORMATION OF A LAYERED PLANET

We know Earth is a layered planet with a core, mantle, and crust surrounded by its fluid ocean and atmosphere (see Chapter 1). How did Earth evolve from a hot, rocky mass to a living planet with continents, oceans, and a pleasant climate? The answer lies in **differentiation:** the transformation of random chunks of primordial matter into a body whose interior is divided into concentric layers that differ from one another both physically and chemically. Differentiation occurred early in Earth's history, when the planet got hot enough to melt.

Earth Heats Up and Melts

Although Earth probably accreted as a mixture of planetesimals and other remnants of the solar nebula, it did not retain this form for long. To understand Earth's present layered structure, we must return to the time when Earth was still subject to violent impacts by planetesimals and larger bodies. When these objects crashed into the primitive Earth, most of this energy of motion was converted into heat, another form of energy. The heat caused melting. A planetesimal colliding with Earth at a typical velocity of 15 to 20 km/s would deliver as much energy as 100 times its weight in TNT. The impact energy of a body the size of Mars colliding with Earth would be equivalent to exploding several trillion 1-megaton nuclear bombs (a single one of these terrible weapons would destroy a large city), enough to eject a vast amount of debris into space and to generate enough heat to melt most of what remained of Earth.

Many scientists now think that such a cataclysm did indeed occur during the middle to late stages of Earth's accretion. A giant impact by a Mars-sized body created a shower of debris from both Earth and the impacting body and propelled it into space. The Moon aggregated from this debris (**Figure 9.4**). According to this theory, Earth re-formed as a body with an outer molten layer hundreds of kilometers thick—a **magma ocean.** The huge impact sped up Earth's rotation and changed its spin axis, knocking it from vertical with respect to Earth's orbital plane to its present 23° inclination. All this occurred about 4.5 Ga, between the beginning of Earth's accretion (4.56 Ga) and the age of the oldest Moon rocks brought back by the *Apollo* astronauts (4.47 billion years).

Another source of heat that contributed to melting early in Earth's history was radioactivity. When radioactive elements such as uranium disintegrate spontaneously, they emit heat. Though present only in small amounts, radioactive elements have had an enormous effect on Earth's evolution and continue to keep the interior hot.

Differentiation of Earth's Core, Mantle, and Crust

As a result of the tremendous energy released during Earth's formation, its entire interior was heated to a "soft" state in which its components could move around. Heavy material sank to the interior to become the core, and lighter material floated to the surface and formed the crust. The rising lighter matter brought interior heat to the surface, where it could radiate into space. In this way, Earth cooled and mostly solidified and was transformed into a differentiated or zoned planet with three main layers: a central core and an outer crust separated by a mantle (**Figure 9.5**). We review the results, which were first presented in Chapter 1.

Earth's Core Iron, which is denser than most of the other elements, accounted for about a third of the primitive planet's material (see Figure 1.8). The iron and other heavy elements such as nickel sank to form a central *core,* which begins at a depth of about 2900 km. By probing the core with seismic waves, scientists have found that it is molten on the outside but solid in a region called the *inner core,* which extends from a depth of about 5200 km to Earth's center at about 6400 km. The inner core is solid because the pressures at the center are too high for iron to melt.

Earth's Crust Other molten materials were less dense, so they floated toward the surface of the magma ocean. There they cooled to form Earth's solid *crust,* which today

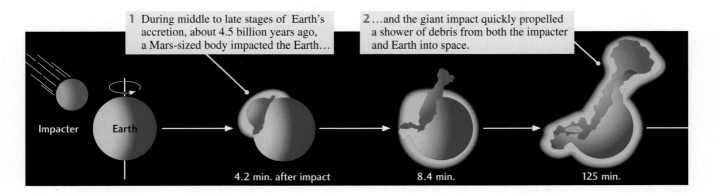

1 During middle to late stages of Earth's accretion, about 4.5 billion years ago, a Mars-sized body impacted the Earth…

2 …and the giant impact quickly propelled a shower of debris from both the impacter and Earth into space.

Impacter   Earth

4.2 min. after impact   8.4 min.   125 min.

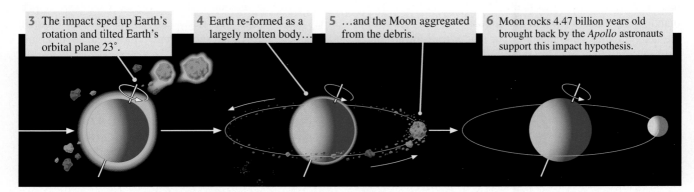

3 The impact sped up Earth's rotation and tilted Earth's orbital plane 23˚.

4 Earth re-formed as a largely molten body…

5 …and the Moon aggregated from the debris.

6 Moon rocks 4.47 billion years old brought back by the *Apollo* astronauts support this impact hypothesis.

Figure 9.4 Computer simulation of the origin of the Moon by the impact of a Mars-sized body on Earth. [*Solid Earth Sciences and Society,* National Research Council, 1993.]

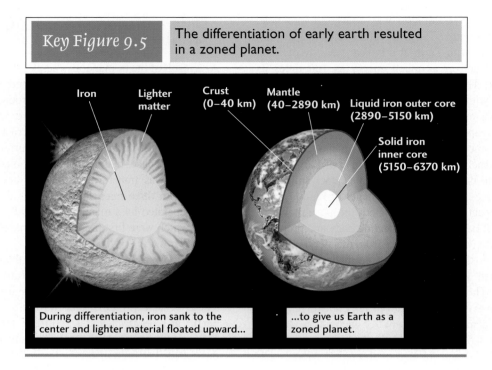

Key Figure 9.5 — The differentiation of early earth resulted in a zoned planet.

Iron · Lighter matter · Crust (0–40 km) · Mantle (40–2890 km) · Liquid iron outer core (2890–5150 km) · Solid iron inner core (5150–6370 km)

During differentiation, iron sank to the center and lighter material floated upward... ...to give us Earth as a zoned planet.

ranges in thickness from about 7 km in the oceans to about 40 km in the continents. We know that oceanic crust is constantly generated by seafloor spreading and recycled back into the mantle by subduction. In contrast, the continental crust began to accumulate early in Earth's history from silicates of relatively low density that have felsic compositions (that is, rich in sodium and potassium) and low melting temperatures.

Recently, in Western Australia, grains of the mineral zircon were found that have been dated, using uranium and lead isotopes, at about 4.4 billion years old, which would make them the oldest terrestrial material yet discovered. Chemical analysis indicates that the sample formed near the surface in the presence of water under relatively cool conditions. This finding suggests that Earth had cooled enough for a crust to exist only 100 million years after the planet reformed following the giant impact.

Earth's Mantle Between the core and the crust lies the *mantle,* a region that forms the bulk of the solid Earth. The mantle is the material left in the middle zone after most of the heavy matter sank and the light matter rose toward the surface. It is about 2900 km thick and consists of ultramafic silicate rocks containing more magnesium and iron than crustal silicates. Plate tectonics theory indicates that the mantle removes heat from Earth's interior by convection (see Chapter 2).

Because the mantle was hotter early in Earth's history, it was probably convecting more vigorously than it does today. Some form of plate tectonics may have been operating even then, although the "plates" would likely have been much smaller and weaker, and the tectonic features were probably far different from the linear mountain belts and long mid-ocean ridges we see on Earth's surface today. Some scientists think that Venus today provides an analog for these long-vanished processes on Earth. We will compare tectonics on Earth and Venus shortly.

Earth's Oceans and Atmosphere Form

Early melting led to the formation of Earth's crust and eventually the continents. It brought lighter materials to Earth's outer layers and allowed even lighter gases to escape from the interior. These gases formed most of the atmosphere and oceans. Even today, trapped remnants of the original solar nebula continue to be emitted as primitive gases in volcanic eruptions.

Some geologists think that most of the air and water on Earth today came from volatile-rich matter of the outer solar system that impacted the planet after it formed. For example, the comets we see are composed largely of water ice plus frozen carbon dioxide and other gases. Countless comets may have bombarded Earth early in its history, bringing water and gases that subsequently gave rise to the early oceans and atmosphere.

Many other geologists believe that the oceans and atmosphere can be traced back to the "wet birth" of Earth itself. According to this hypothesis, the planetesimals that aggregated into our planet contained ice, water, and other volatiles. Originally, the water was locked up in certain minerals carried by the aggregating planetesimals. Similarly, nitrogen and carbon were chemically bound in minerals. As Earth heated and its materials partially melted, water vapor and other gases were freed, carried to the surface by magmas, and released through volcanic activity.

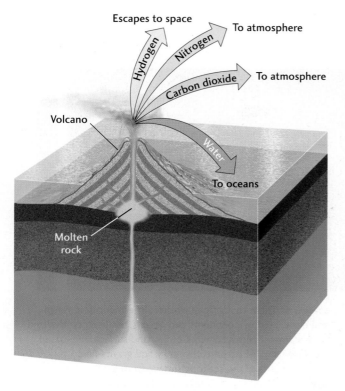

Escapes to space

To atmosphere

Hydrogen

Nitrogen

Carbon dioxide

To atmosphere

Volcano

Water

To oceans

Molten rock

Figure 9.6 Early volcanic activity contributed enormous amounts of water vapor, carbon dioxide, and other gases to the atmosphere and oceans and solid materials to the continents. Photosynthesis by microorganisms removed carbon dioxide and added oxygen to the primitive atmosphere. Hydrogen, because it is light, escaped into space.

The gases released from volcanoes around 4 Ga probably consisted of the same substances that are expelled from present-day volcanoes (though not necessarily in the same relative abundances): primarily hydrogen, carbon dioxide, nitrogen, water vapor, and a few other gases (**Figure 9.6**). Almost all of the hydrogen escaped to outer space, while the heavier gases enveloped the planet. This early atmosphere lacked the oxygen that makes up 21 percent of the atmosphere today. Oxygen did not enter the atmosphere until oxygen-producing organisms evolved, as described in Chapter 11.

DIVERSITY OF THE PLANETS

By about 4.4 Ga, in less than 200 million years since its origin, Earth had become a fully differentiated planet. The core was still hot and mostly molten, but the mantle was fairly well solidified, and a primitive crust and continents had developed. The oceans and atmosphere had formed, probably from substances released from Earth's interior, and the geologic processes that we observe today were set in motion.

But what of the other planets? Did they go through the same early history? Information transmitted from our spacecraft indicates that all the terrestrial planets have undergone differentiation, but their evolutionary paths have varied.

Mercury has a thin atmosphere consisting mostly of helium. The atmospheric pressure at the surface is less than a trillionth of Earth's atmospheric pressure. There is no surface wind or water to erode and smooth the ancient surface of this innermost planet. It looks like the Moon: intensely cratered and covered by a layer of rock debris, the fractured remnants of billions of years of meteorite impacts. Because it has essentially no atmosphere to protect it and is located close to the Sun, the planet warms to a surface temperature of 470°C during the day and cools to −170°C at night. This is the largest temperature range for any planet.

Even though it is a much smaller planet, Mercury's mean density is nearly as great as Earth's (Table 9.1). Accounting for the differences in interior pressure (remember, higher pressures increase density), scientists conclude that Mercury's iron-nickel core must make up about 70 percent of its mass, also a record for solar system planets (Earth's core is only one-third of its mass). Perhaps Mercury lost part of its silicate mantle in a giant impact. Alternatively, the Sun could have vaporized part of its mantle during an early phase of intense radiation. Scientists are still debating these theories.

Venus evolved into a planet with surface conditions surpassing most descriptions of hell. It is wrapped in a heavy, poisonous, incredibly hot (475°C) atmosphere composed

Table 9.1	Characteristics of the Terrestrial Planets and Earth's Moon				
	Mercury	**Venus**	**Earth**	**Mars**	**Earth's Moon**
Radius (km)	2440	6052	6378	3388	1737
Mass (Earth = 1)	0.06 (3.3×10^{23} kg)	0.81 (4.9×10^{24} kg)	1.00 (6.0×10^{24} kg)	0.11 (6.4×10^{23} kg)	0.01 (7.2×10^{22} kg)
Mean density (g/cm³)	5.43	5.24	5.52	3.94	3.34
Orbit period (Earth days)	88	224	365	687	27
Distance from Sun (millions of km)	57	108	148	228	
Moons	0	0	1	2	0

Figure 9.7 A comparison of the solid surfaces of Earth, Mars, and Venus, all at the same scale. Mars topography, which shows the greatest range, was measured in 1998 and 1999 by a laser altimeter aboard the orbiting *Mars Global Surveyor* spacecraft. Venus topography, which shows the lowest range, was measured in 1990–1993 by a radar altimeter aboard the orbiting *Magellan* spacecraft. Earth topography, which is intermediate in range and dominated by continents and oceans, has been synthesized from altimeter measurements of the land surface, ship-based measurements of ocean bathymetry, and gravity-field measurements of the seafloor surface from Earth-orbiting spacecraft. [Courtesy of Greg Neumann/MIT/GSFC/NASA.]

mostly of carbon dioxide and clouds of corrosive sulfuric acid droplets. A human standing on its surface would be crushed by the pressure, boiled by the heat, and eaten away by the sulfuric acid. Radar images that see through the thick cloud cover show that at least 85 percent of Venus is covered by lava flows. The remaining surface is mostly mountainous—evidence that the planet has been geologically active (**Figure 9.7**). Venus is close to Earth in mass and size (see Table 9.1), and its core also seems to be about the same size as Earth's. How it could evolve into a planet so different from Earth is a question that intrigues planetary geologists.

Mars has undergone many of the same geologic processes that have shaped Earth (see Figure 9.7). The Red Planet is considerably smaller than Earth, only about one-tenth of Earth's mass (see Table 9.1). However, the Martian core appears to have a radius of about half its surface radius, similar to both Earth and Venus, and it may also have a liquid outer part and solid inner part.

Mars has a thin atmosphere composed almost entirely of carbon dioxide. No liquid water is present on the surface today—the planet is too cold and its atmosphere is too thin, so water would either freeze or evaporate. Networks of valleys and dry river channels, however, indicate that liquid water was abundant on the surface of Mars before 3.5 Ga. Some of the rocks observed by *Spirit* and *Opportunity,* the *Mars Exploration Rovers,* have proved that some minerals on Mars must have formed in water and that aqueous processes were once widespread. Spacecraft circling Mars have recently found evidence that large amounts of water ice may be stored below the surface and in polar ice caps. Life might have formed on the wet Mars of billions of years ago and may exist today as microbes below the surface. NASA is designing a mission—*Mars Science Laboratory*—that could answer the question of life on Mars in a few years.

Most of the surface of Mars is older than 3 billion years. On Earth, in contrast, most surfaces older than about 500 million years have been obliterated through the combined activities of the plate tectonic and climate systems. Later in this chapter, we will compare the surface processes on Earth and Mars in more detail.

Other than Earth, the *Moon* is the best-known body in the solar system because of its proximity and the programs of manned and unmanned exploration. As stated earlier, we believe that the Moon formed from remnants of Earth's matter after a giant impact on Earth (see Figure 9.4). In bulk, the Moon's materials are lighter than Earth's because the heavier matter of the giant impactor and its primeval target remained embedded in Earth. The Moon has no atmosphere and, like Venus, is mostly bone dry, having lost its water in the heat generated by the giant impact. There is some new evidence from spacecraft observations that water ice may be present in small amounts deep within sunless craters at the Moon's north and south poles. The heavily cratered surface we see today is that of a very old, geologically dead body, dating back to a time early in the history of the solar system known as the **Heavy Bombardment,** when crater-forming impacts were very frequent (**Figure 9.8**). Once topography is created on any planetary body, tectonic and climate processes will work to "resurface" it, as they have on Venus and Mars. However, in the absence of these processes, the planet will remain pretty much the way it was since the time of Heavy Bombardment. Thus, the presence of heavily cratered terrains on little-studied planetary bodies, such as Mercury, indicate that they lack both a convecting mantle and an atmosphere.

The giant gaseous outer planets—*Jupiter, Saturn, Uranus,* and *Neptune*—will remain a puzzle for a long time. These huge gas balls are so chemically distinct and so large that they must have followed an evolutionary course

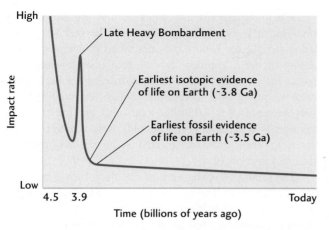

Figure 9.8 The number of impacts varies over time. The planets were formed by a chain of collisions that tapered off over the first 500 million years of their evolution. However, there was a later phase, known as the Heavy Bombardment, that peaked around 3.9 Ga.

entirely different from that of the much smaller terrestrial planets. All four of the giant planets are thought to have rocky, silica-rich and iron-rich cores surrounded by thick shells of liquid hydrogen and helium. Inside Jupiter and Saturn, the pressures become so high that we believe the hydrogen turns into a metal.

Exactly what lies beyond the orbit of Neptune, the most distant giant planet, remains a mystery. The ninth planet, tiny *Pluto,* is a strange frozen mixture of gas, ice, and rock with an unusual orbit that sometimes brings it closer to the Sun than Neptune. It is the only major planet not yet visited by our spacecraft, so we have observed it only from afar.

Pluto lies within a belt of icy bodies that is the source region for periodic comets (such as Halley's comet, which has a period of 76 years and was last seen in 1986). In July 2005, Caltech astronomer Mike Brown discovered that one of these bodies, named "2003 UB313," is actually larger than Pluto and may therefore qualify as the long-sought "tenth planet." Like Pluto, its surface shows a mixture of ice and rock. Other planet-sized objects are likely to be found as we explore the outer regions of the solar system.

WHAT'S IN A FACE? THE AGE AND COMPLEXION OF PLANETARY SURFACES

Like members of a family, the four terrestrial planets all bear a certain resemblance to one another. They are all differentiated and have iron-nickel cores, silicate mantles, and an outer crust. But, as we have just seen, there are no twins. Important distinctions relate to their different sizes and masses and their variable distances from the Sun, which all affect the surfaces of these planets.

Like human faces, planetary faces also reveal their age. Instead of forming wrinkles as they get older, the terrestrial planets are marked by craters. The surfaces of Mercury,

Mars, and the Moon are heavily cratered and therefore old. In contrast, Venus and Earth have very few craters, because their surfaces are much younger. In this section, we will study planetary faces to learn about the tectonic and climatic processes that have shaped their surfaces. Earth is excluded because it is the subject of this book, and Mars will be mentioned only briefly because its surface is more thoroughly described in the next section.

The Old Man in the Moon: A Planetary Time Scale

If you look at the face of the Moon through binoculars on a clear night, you will see two distinct types of terrain: rough areas that appear light-colored with lots of big craters, and smooth, dark areas, usually circular in shape, where craters are small or nearly absent (**Figure 9.9**). The light-colored regions are the mountainous **lunar highlands,** which cover about 80 percent of the surface, whereas the dark regions are low-lying plains called **lunar maria,** which is Latin for "seas," because they looked like seas to early Earth-bound observers.

In preparation for the *Apollo* landings, geologists used these features to develop a relative time scale for the formation of lunar surfaces based on the following simple principles:

• Craters are absent on a new geologic surface; older surfaces have more craters than younger surfaces.

• Impacts by small bodies are more frequent than impacts by large bodies; older surfaces have larger craters.

• More recent impact craters cross-cut or cover older craters.

Figure 9.9 The Moon has two terrains: the lunar highlands, with many craters, and the lunar lowlands, with very few craters. The lowlands appear darker due to the presence of widespread basalt that flowed across the surface over 3 Ga. The highlands are lighter because the abundant craters reflect sunlight better. In the center is the Orientale Basin, almost 1000 km in diameter, formed about 3.8 Ga by the impact of an asteroid-sized body. To the right are the lunar lowlands, which expose part of the great, dark Oceanus Procellarum. [NASA/USGS.]

The last statement is just the principle of superposition, which we described in Chapter 8. By applying these principles, geologists were able to show that the lunar highlands are older than the maria. They interpreted the maria to be basins formed by the impacts of asteroids or comets that were subsequently flooded with basalts, which "repaved" the basins. Based on detailed mapping of the numbers and sizes of craters—a procedure known as *crater counting*—they were able to group different parts of the Moon into geologic time periods, analogous to the relative time scale worked out by nineteenth-century geologists on Earth.

In the pre-*Apollo* days, no one knew the absolute ages of either the maria or the highlands, but the smart money held that both were very old. The intense cratering evident in the highlands and the big impacts that formed the maria are consistent with theoretical models of the early solar system. These models predict the Heavy Bombardment, during which the planets collided with the residual matter that still cluttered the solar system after the planets had been assembled. According to the models, the number and sizes of impacting objects should be greatest just after the planets formed and quickly decrease as the material was swept up by the planets.

When the *Apollo* astronauts visited the Moon, they brought back rock samples that could be precisely dated using radioactive isotopes, as described in Chapter 8. The dating of these samples allowed geologists to assign absolute ages to the Moon's surface, thereby calibrating the crater-counting time scale. Sure enough, the highlands are very ancient (4.4 to 4.0 billion years old), and the maria are younger (4.0 to 3.2 billion years old), as indicated by crater counting. These events are plotted on the ribbon of geologic time in **Figure 9.10**.

The relatively young ages of the maria turned out to be a puzzle, however. The best theoretical models indicated that the Heavy Bombardment should have been over rather quickly, perhaps in a few hundred million years or even less. Why then did some of the biggest impacts observed on the Moon—those that formed the maria—occur so late in lunar history?

The theoretical models missed an important event. The rate at which large objects impacted the Moon did decrease quickly, as the models predicted, but then spiked up again in a period known as the *Late* Heavy Bombardment, which occurred between about 4.0 and 3.8 Ga (see Figure 9.10).

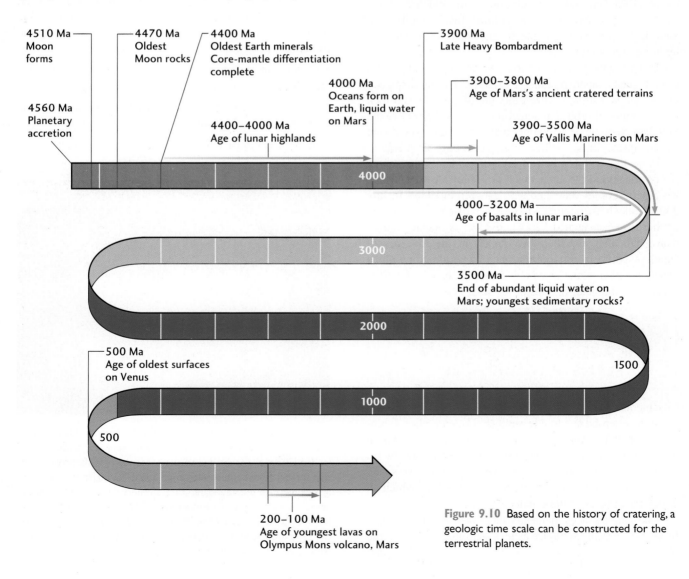

Figure 9.10 Based on the history of cratering, a geologic time scale can be constructed for the terrestrial planets.

The explanation of this event is still controversial, but it looks as though small but abrupt changes in the orbits of Jupiter and Saturn about 4 Ga (caused by their mutual gravitational interactions as their orbits "settled down" to the way they are today) perturbed the orbits of the asteroids. Some of the asteroids were sent into the inner solar system, where they collided with the Moon and terrestrial planets, including Earth. The Late Heavy Bombardment explains why so few rocks on Earth have ages greater than 3.9 billion years. It is this event that marks the end of the Hadean eon and the beginning of the Archean eon (see Figure 9.10).

We have been able to extend the crater-counting time scale first developed for the Moon to other planets by taking into account the differences in impact rates resulting from the planet's mass and position in the solar system.

Mercury: The Ancient Planet

The topography of Mercury is poorly understood, with only limited coverage obtained during a single mission. *Mariner 10* was the first and only spacecraft to visit Mercury, when

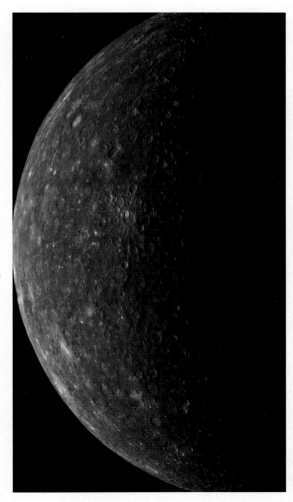

Figure 9.11 Mercury has a dormant, very ancient surface, similar to that of Earth's Moon. [NASA/JPL/Northwestern University.]

Figure 9.12 The prominent scarp that snakes across the image is thought to have formed as Mercury was compressed, possibly during cooling of the planet following its formation. Note that the scarp must be younger than the craters it offsets. [NASA/JPL/Northwestern University.]

it flew by the planet in March 1974. It mapped less than half the planet, and we have little idea of what is on the other side.

The *Mariner 10* mission confirmed that Mercury has a dormant, Moonlike surface (**Figure 9.11**). Its surface is the oldest of all the terrestrial planets, and it has no atmosphere. It is heavily cratered and very ancient, like the surface of Earth's Moon. Plains, which are slightly younger, sit between the largest old craters. Many scientists think these plains are volcanic, and *Mariner 10* images show a difference in color between the ancient craters and the plains, which supports this hypothesis. Unlike Earth and Venus, very few features seen on Mercury are clearly due to tectonic forces having reshaped the surface.

In many respects, the face of Mercury seems very similar to that of Earth's Moon, perhaps because they are similar in size and mass. Like the Moon, most of the geologic activity on Mercury took place within the first billion years of the planet's history. But there is one striking difference. Mercury's face has several scars marked by cliffs nearly 2 km high and up to 500 km long (**Figure 9.12**). These features are common on Mercury but rare on Mars and absent on Earth's Moon. The cliffs are thought to have resulted from compression, or squeezing, of Mercury's brittle crust, which formed enormous thrust faults (see Chapter 7). Some scientists think they formed during cooling of the planet's crust, immediately after its formation.

On August 3, 2004, the first new mission to Mercury in 30 years was launched successfully. If all goes well, *Messenger* will arrive and enter orbit in September 2009. *Messenger* will provide information about Mercury's surface composition, geologic history, and core and mantle and will search for evidence of water ice and other frozen gases such as carbon dioxide at the poles.

Venus: The Volcanic Planet

Venus is the planet most familiar to people, often highly visible just before sunset. Yet in the early decades of space exploration, Venus frustrated scientists because it is shrouded in fog—which makes it impenetrable using ordinary cameras. As mentioned earlier, the fog is created by a mixture of carbon dioxide, water vapor, and sulfuric acid that envelops the entire planet. Although many spacecraft were sent to Venus, only a few were able to penetrate this acid fog, The first that tried to land on its surface were crushed under the tremendous weight of the atmosphere. By the 1980s, we had some idea of the largest features on the Venusian surface, such as those resembling large volcanoes.

It was not until August 10, 1990, after traveling 1.3 billion kilometers, that the *Magellan* spacecraft arrived at Venus and took the first high-resolution pictures of its surface (**Figure 9.13**). *Magellan* did this using *radar* (shorthand for *radio detection and ranging*), similar to the cameras that police officers use to enforce speed limits (they "see" at night, and through the fog, to clock your speed). Radar cameras bounce radio waves off stationary surfaces (like those of planets) or moving surfaces (like those of cars).

The images that *Magellan* returned to Earth show clearly that beneath its fog, Venus is a surprisingly diverse and geologically active planet. It has mountains, plains, volcanoes, and rift valleys. As we have seen, one way to tell the age of Venus's surface is to count the number of impact craters. The plains of Venus have far fewer craters than the Moon's youngest lava plains, indicating they must be even younger. Lava must have covered many of the craters, suggesting that Venus has been geologically active relatively recently. Most of the planet is covered by these lavas, which form the lowlands shown in Figure 9.13. The lavas are

thought to range in age between 1600 and 300 million years. Because it doesn't rain on Venus, there is very little erosion, and so the features we see now have been "locked in" for all that time.

Venus has many features consistent with widespread volcanism, including hundreds of thousands of small domes 2 to 3 km across and perhaps 100 m or so high that dot the younger plains. These features are similar to *shield volcanoes* such as Earth's Hawaiian Islands (see Chapter 12), and they form over places where Venus's crust got very hot. There are larger, isolated volcanoes as well, up to 3 km high and 500 km across (**Figure 9.14a**). *Magellan* also observed surface features that have never been seen before on any other planet. These are broad, circular features called *coronae* that appear to be blobs of hot lava that rose, created a large bulge or dome in the surface, and then sank, collapsing the dome and leaving a ring that looks like a fallen soufflé (see Figure 9.14b)

Because Venus has so much evidence of widespread volcanism, it has been called the Volcanic Planet. Earth's crust is formed of a set of lighter, relatively thick plates that float on a heavier, fluid mantle. Where the plates collide, mountain ranges can form; and where they separate or slide beneath each other, linear belts of volcanoes occur. We use the term *plate tectonics* to describe these motions and patterns of deformation and volcanism on Earth's surface. But Venus,

Elevation (km)

0 2 4 6 8 10 12 14

Figure 9.13 This map shows the topography of Venus based on more than a decade of mapping culminating in the 1990–1994 *Magellan* mission. Regional variations in height are illustrated by the highlands (tan colors), the uplands (green colors), and the lowlands (blue colors). Vast lava plains are found in the lowlands. [NASA/USGS.]

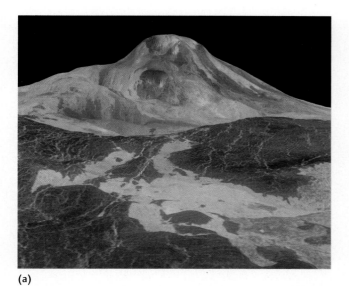

(a)

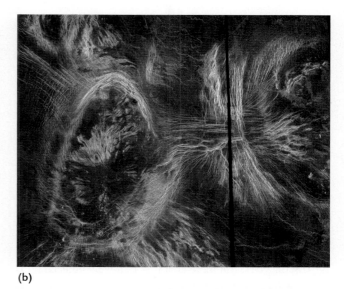

(b)

Figure 9.14 (a) Venus has mountains made of volcanoes, such as Maat Mons, which is up to 3 km high and 500 km across. (b) Volcanic features called coronae are not observed on any other planet except Venus. The visible lines that define the coronae are fractures, faults, and folds produced when a large blob of hot lava collapsed like a fallen soufflé. Each corona is a few hundred kilometers across. [Images from NASA/USGS.]

unlike Earth, does not appear to have thick plates. Instead, the geology of Venus seems dominated by vigorous convection currents in the mantle beneath a crust that is formed of a very thin layer of frozen lava (**Figure 9.15**). As these convection currents push and stretch the surface, the crust breaks up into flakes or crumples like a rug. As the mantle of Venus moves around, blobs of hot lava bubble up to form large landmasses, mountains, and the volcanic deposits mentioned earlier. Scientists have called the unique geology of Venus **flake tectonics.** When the Earth was younger and hotter, it is possible that flakes, rather than plates, were the expression of global tectonics.

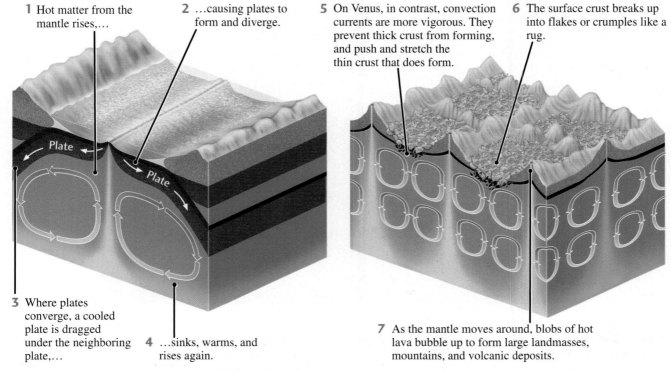

1 Hot matter from the mantle rises,…

2 …causing plates to form and diverge.

3 Where plates converge, a cooled plate is dragged under the neighboring plate,…

4 …sinks, warms, and rises again.

5 On Venus, in contrast, convection currents are more vigorous. They prevent thick crust from forming, and push and stretch the thin crust that does form.

6 The surface crust breaks up into flakes or crumples like a rug.

7 As the mantle moves around, blobs of hot lava bubble up to form large landmasses, mountains, and volcanic deposits.

Figure 9.15 Plate tectonics on Earth versus flake tectonics on Venus.

Mars: The Red Planet

Of all the planets, Mars is the most similar to Earth. Although Venus is closer in size and mass (see Table 9.1), Mars has surface features that indicate liquid water once flowed across its surface. Liquid water may still be stable in the deep subsurface. And where there is water, there may be life. No other planet has as much chance of harboring extra-terrestrial life forms as does Mars.

The abundance of iron oxide minerals on the surface of Mars make it red, giving rise to the name Red Planet. Iron oxide minerals are common on Earth and tend to form where weathering of iron-bearing silicates occurs (see Chapter 3). We now know that many other minerals common on Earth, such as olivine and pyroxene, which form in basalt, are also present on Mars. But there are other relatively unusual minerals, such as sulfates, that record an earlier, wetter phase on Mars, when liquid water may have been stable. The discovery of sulfates on Mars paints a picture of past aqueous processes that unite the early histories of Mars and Earth.

The surface of Mars is morphologically one of the most dramatic in the solar system, with several entries in the solar system book of records. Olympus Mons, at 25 km high, is a giant volcano that is also the tallest mountain in the solar system (**Figure 9.16**). The Vallis Marineris canyon, at 4000 km long and averaging 8 km deep, is as long as the distance from New York to Los Angeles and five times deeper than the Grand Canyon (**Figure 9.17**). Recently, geologists have

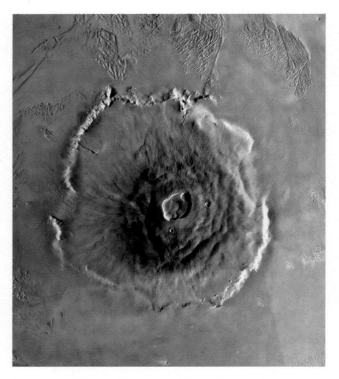

Figure 9.16 Olympus Mons is the tallest volcano in the solar system, with a summit caldera almost 25 km above the surrounding plains. Encircling the volcano is an outward-facing scarp 550 km in diameter and several kilometers high. Beyond the scarp is a moat filled with lava, most likely derived from Olympus Mons. [NASA/USGS.]

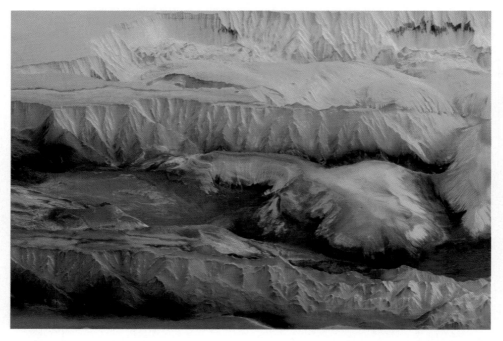

Figure 9.17 Vallis Marineris is the longest (4000 km) and deepest (up to 10 km) canyon in the solar system. It is five times deeper than the Grand Canyon. In this image, the canyon is exposed as series of fault basins whose sides have partially collapsed (such as at upper left), leaving piles of rock debris. The walls of the canyon are 6 km high here! Also, the layering of the canyon walls suggests deposition of sedimentary or volcanic rocks prior to faulting. [ESA/DLR/FU Berlin.]

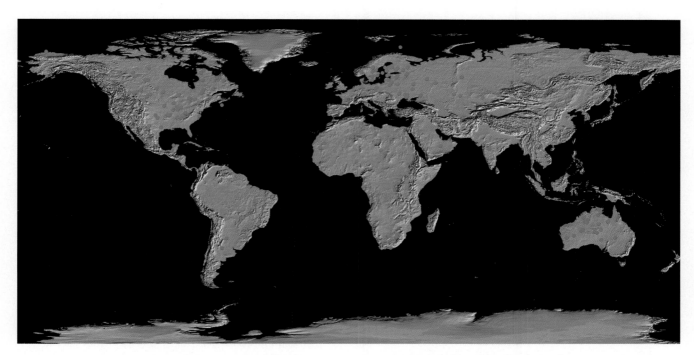

Figure 9.18 Impact craters formed by meteorites and asteroids are rare on Earth compared to the other terrestrial planets. Recycling of Earth's crust by plate tectonics has erased almost all of the evidence. We see impact craters preserved only on the continents. [NASA/JPL/ASU.]

discovered evidence of past glacial processes, when ice sheets similar to the ones that covered North America during the last ice age flowed across the surface of Mars. Finally, like the Moon, Mercury, and Venus, Mars has both heavily cratered ancient highlands and younger lowlands. However, unlike the other planets and the Moon, the lowlands of Mars are created not just by lava flows but also by the accumulation of windblown dust and possibly sediments and sedimentary rocks.

The face of Mars is sophisticated but has not always been easy to read, despite being visited and viewed more than any other planet except Earth. But, as we will see shortly, Mars's secrets are finally being revealed.

Earth: No Place Like Home

Every view of Earth underscores the unique beauty created by the overwhelming influence of plate tectonics, liquid water, and life. From its blue skies and oceans and its green vegetation to its rugged ice-covered mountains and moving continents, there truly is no place like home. Earth's remarkable appearance is maintained by the delicate balance of conditions necessary to support and sustain life.

The features that define the face of our planet are discussed throughout this book, but one process that is appropriate to review here is collision and cratering by meteorites and asteroids. Impacting is preserved in the record of the terrestrial planets, but Earth has a unique expression of this process because the record of Heavy Bombardment is al-

most completely erased. In contrast to the other terrestrial planets, whose surfaces are mostly frozen in time, Earth preserves very few vestiges of its beginning. The reason is that recycling by plate tectonics on Earth, which is even more efficient than flake tectonics on Venus, has almost completely resurfaced our planet. What craters do remain are much younger than the end of the period of Late Heavy Bombardment and are preserved entirely on continents, which resist subduction (**Figure 9.18**).

Nevertheless, Earth still accumulates a lot of junk from space (Table 9.2). At present, some 40,000 tons of extraterrestrial material fall on Earth each year, mostly as dust and unnoticed small objects. Although the rate of impacts is now orders of magnitude smaller than it was in the Heavy Bombardment period, a large chunk of matter 1 to 2 km in size still collides with Earth every few million years or so. Although such collisions have become rare, telescopes are being assigned to search space and warn us in advance of sizable bodies that might slam into Earth. NASA astronomers recently predicted "with nonnegligible probability" (1 chance in 300) that an asteroid 1 km in diameter will collide with Earth in March 2880. Such an event would threaten civilization.

We already know that collisions with the Earth can greatly upset the conditions that support life. As we will see in Chapter 11, a major impact did occur 65 million years ago. The asteroid, about 10 km in diameter, caused the extinction of 75 percent of Earth's species, including all dinosaurs. This event may have made it possible for mammals to become

Table 9.2	Impacts by Asteroids and Meteors and Their Effects on Life on Earth			
	Example(s)	**Most Recent**	**Planetary Effects**	**Effects on Life**
Supercolossal radius (R) > 2000 km	Moon-forming event	4.45×10^9 years ago	Melts planet	Drives off volatiles; wipes out life on Earth
Colossal R > 700 km	Pluto	More than 4.3×10^9 years ago	Melts crust	Wipes out life on Earth
Huge R > 200 km	4 Vesta (large asteroid)	About 4.0×10^9 years ago	Vaporizes oceans	Life may survive below surface
Extra large R > 70 km	Chiron (largest active comet)	3.8×10^9 years ago	Vaporizes upper 100 m of oceans	May wipe out photosynthesis
Large R > 30 km	Comet Hale-Bopp	About 2×10^9 years ago	Heats atmosphere and surface to about 1000 K	Burns continents
Medium R > 10 km	K/T impactor; 433 Eros (largest near-Earth asteroid)	65×10^6 years ago	Causes fires, dust, darkness; chemical changes in ocean and atmosphere; large temperature swings	K/T impact caused extinction of 75% of species and all dinosaurs
Small R > 1 km	About size of near-Earth asteroids	About 300,000 years ago	Causes global dusty atmosphere for months	Interrupts photosynthesis; individuals die but few species extinct; threatens civilization
Very small R > 100 m	Tunguska event (Siberia)	1908	Knocked over trees tens of kilometers away; caused minor hemispheric effects; dusty atmosphere	Newspaper headlines; romantic sunsets; increased birth rate

Modified from J. D. Lissauer, *Nature* 402 (1998): C11–C14.

the dominant species and paved the way for humankind's emergence. Table 9.2 describes the effects of impacts of various sizes on our planet and its life.

MARS ROCKS!

In June 2003, two golf-cart-sized rovers destined to land on the Martian surface were launched from Cape Canaveral, Florida, and began their 300-million-kilometer journey to the Red Planet. A third spacecraft equipped with geologic remote sensing tools for use while orbiting Mars was also launched in June 2003. These missions succeeded beyond anyone's expectations, making 2004 and 2005 two of the greatest years in the history of space exploration.

The *Mars Exploration Rovers* (**Figure 9.19**) were designed to survive 3 months under the hostile Martian surface conditions and drive no farther than 300 m. At the time of publication of this book, the rovers had operated for more than two years and had driven a combined distance of more

Figure 9.19 *Spirit,* one of the *Mars Exploration Rovers,* is about the size of a golf cart. *Spirit* is standing next to a twin of the *Sojourner* rover that was sent to Mars in 1997. [NASA/JPL.]

than 12 km! The rovers have had to survive nighttime temperatures below −90°C, dust devils that could have tipped them over, global dust storms that diminished their solar power, and drives along rocky slopes of almost 30° and through piles of treacherous windblown dust. The rovers have also discovered a treasure trove of geologic wonders. These discoveries include compelling evidence for water on the ancient Martian surface—a necessary condition for life (see Chapter 11).

The *Mars Express* orbiter has been equally successful in mapping the rocks and minerals of Mars at an unprecedented level of detail. Whereas the rovers are limited to several kilometers in their investigations of the Martian surface, the *Mars Express* orbiter can map anywhere on the planet. It is equipped with several instruments, including a high-resolution stereo color camera capable of resolving objects on the surface of Mars as small as 10 m across. Another important device looks at the sunlight reflected from the Martian surface to reveal the presence of water-bearing minerals. One of the remarkable observations by the *Mars Express* orbiter is the discovery of what look like icebergs embedded within a now-frozen sea or lake (**Figure 9.20**).

Mission to Mars: Flybys, Orbiters, Landers, and Rovers

Earlier missions to Mars helped lay the groundwork for the success of the current missions. All spacecraft sent to Mars since the early 1960s work in one of four ways. First, the early pioneers of Mars exploration, such as *Mariner 4,* flew by Mars while quickly acquiring all the data they could before disappearing into deep space.

The second, and most common, mode of operation is to orbit Mars in the same way that satellites orbit Earth. *Mariner 9,* launched in May 1971, was the first spacecraft to orbit another planet. Since that time, eight other orbiters have helped map the surface of Mars. *Mars Global Surveyor, Mars Odyssey,* and *Mars Express* are still active today.

The third method of observing Mars involves landing a spacecraft on the Martian surface. *Viking 1* touched down on the surface of Mars on July 20, 1976, and became the first spacecraft to land on another planet and transmit useful data back to Earth. The *Viking* mission, which featured a second lander as well, gave us our first look at the surface of another planet from the ground. *Viking* also provided our first chemical analyses of Martian rocks, which showed a great abundance of basalt, and the first life-detection experiments.

The fourth method of exploring Mars is the use of a rover—a robotic vehicle that can move about on the surface of the planet. As exciting as the *Viking* mission was, it was two decades until another spacecraft landed safely on the surface of Mars. This time it was *Pathfinder,* which arrived on the 4th of July, 1997. However, the *Pathfinder* lander also included a shoe box–sized rover—called *Sojourner*—that

Figure 9.20 An image recently acquired (February 2005) by the *Mars Express* orbiter shows what appears to be a dust-covered frozen sea near the Martian equator. The dark areas may be icebergs locked in this frozen sea. The scene is a few tens of kilometers across. [ESA/DLR/FU Berlin.]

was able to ramble around on the surface analyzing rocks and soils within a few meters of the *Pathfinder* lander. The *Sojourner* rover was the first mobile vehicle to operate successfully on another planet and became the prototype of the much larger and more capable *Mars Exploration Rovers* that landed in 2004 (see Figure 9.19).

Early Missions: *Mariner* (1965–1971) and *Viking* (1976–1980) These early missions returned the first detailed images of Mars. We saw a cratered Moonlike terrain over part of its surface. In other areas, we saw spectacular features including enormous volcanoes, huge canyons as long as the width of North America, ice caps at both poles, and the Martian moons Phobos and Deimos. Early images also confirmed global dust storms that had previously been observed from Earth. Orbiting spacecraft continue to monitor dust storms (**Figure 9.21**). Vast dune fields were also revealed.

In addition, extensive networks of river channels were discovered, providing the first evidence that liquids—possibly water—may once have flowed across the surface of Mars (**Figure 9.22**). Collectively, these data also reveal something that had not been appreciated before: the planet is divisible into two main regions, northern low plains and southern cratered highlands.

The *Viking* landers provided high-resolution views of the terrain. Both landing sites were strewn with rocks, somewhat rounded by the effects of wind-related sandblasting. Chemical sensors showed that rocks and soils were basaltic in composition. But all rocks were loose, and there was no evidence of any exposed bedrock. An onboard biology experiment found no evidence of life at either site. It was shown that the Red Planet is red because of the presence of

Dusty

Clear

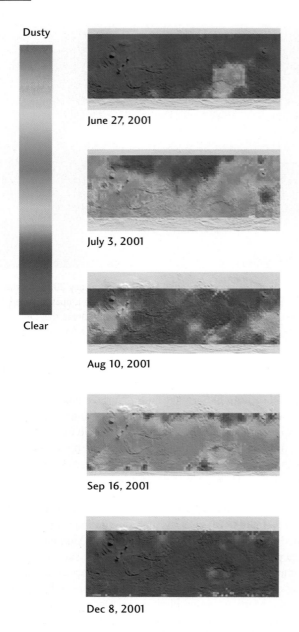

June 27, 2001

July 3, 2001

Aug 10, 2001

Sep 16, 2001

Dec 8, 2001

Figure 9.21 Global dust storms occur on Mars. The storms begin locally and gradually expand to envelop the entire planet, as seen in these images. [NASA/JPL/ASU.]

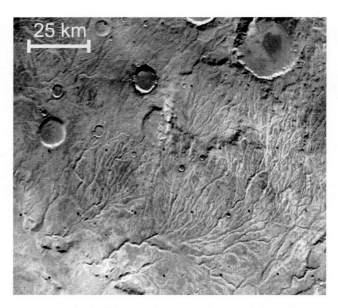

Figure 9.22 Channel networks carved into the surface of Mars are revealed by the *Viking* orbiter. The complexity of these channels suggests that liquid water was probably the main force of erosion. [NASA/Washington University.]

iron oxides in the soils. The color of the Martian sky is not blue but pink because of the high concentration of suspended iron oxide dust particles.

Pathfinder (1997) The *Pathfinder* camera returned images very similar to those of the *Viking* cameras: the landing sites were rocky, with windblown sand forming tails behind some rocks, and there was no evidence of any exposed bedrock. However, in addition to evidence of basalts, the *Sojourner* rover detected evidence of andesites. Recall from Chapter 4 that andesites are more evolved than basalts, which can be produced simply through partial melting of the mantle. The presence of andesites on Mars indicates that at least some parts of the Martian crust were formed by partial melting of previously formed basalts, suggesting a more complex history of crustal evolution.

The *Pathfinder* instrument suite also included a magnet that collected dust from the atmosphere. The dust was thought to contain magnetic minerals that form only in environments low in oxygen.

***Mars Global Surveyor* and *Mars Odyssey* (1996–today)** These two spacecraft provided a vastly improved global mapping capability, resulting in a number of significant discoveries. A laser-based altimeter surveyed Martian topography with unprecedented resolution. New images provided the strongest evidence yet for liquid water, this time expressed as meandering river deposits (see Chapter 18) formed of loose sediment (**Figure 9.23**). The channels carved into bedrock observed by *Viking* suggested flowing water; however, the presence of meandering river deposits is even stronger evidence for flowing water on the surface of Mars. But it was not until 2004 that the *Mars Exploration Rovers* first confirmed the presence of minerals that *require* liquid water to have been present.

Mars Global Surveyor and *Mars Odyssey* also showed that permafrost (ice-rich soil; see Chapter 21) composed of water ice underlies the soil from the midlatitudes all the way to the poles. Widespread glaciers were also shown to have been present in the relatively recent past, suggesting that Mars—like Earth—may have experienced ice ages driven by changes in global climate. Finally, *Mars Global Surveyor* discovered rare patches of hematite (Fe_2O_3) scattered over the surface of Mars. As we will see, this discovery contributed to the success of the *Mars Exploration Rovers* mission.

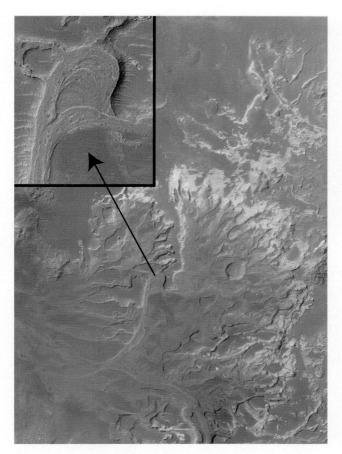

Figure 9.23 This image acquired by the *Mars Global Surveyor* orbiter shows clear evidence of meandering patterns within sediments deposited inside Holden Crater. Liquid water appears to have flowed across the Martian surface and entered the crater, where it deposited sediments in meandering channels similar to those seen in the Mississippi River on Earth today (see Chapter 18). [NASA/JPL/MSSS.]

Mars Exploration Rovers: Spirit and Opportunity

The *Mars Exploration Rovers*—named *Spirit* and *Opportunity*—are the first spacecraft sent to Mars that function almost as well as a human geologist would. Unlike orbiters, which look from afar, and landers, which can't move, *Spirit* and *Opportunity* can move around from rock to rock, picking and choosing which ones to study in more detail (see Figure 9.19). And when the right rock is found, the rover can look at with its hand lens—just as geologists do here on Earth in the classroom and in the field. But unlike geologists on Earth, these rovers provide a mobile laboratory so that the rocks can be analyzed on the spot without having to pay the enormous costs of flying them back to Earth. Because of this remarkable capability, *Spirit* and *Opportunity* have been dubbed the first *robotic geologists* on Mars.

To be sure, these rovers have some significant limitations. They can't climb walls (but they can crawl up hills), they can't get in a car or airplane and move to an entirely different location (but they can move around pretty well in the ones they're at), and their mineral analyzers provide only a very limited sense of what minerals are present in the rocks (but the chemical analyzers give us a pretty good idea of what elements are in the rocks). And these rovers were not equipped to do life-detection experiments or search for fossils.

Because of the 10-minute delay of radio signals between Earth and Mars, the rovers must have some self-controlled navigation capability and method for avoiding hazards (bumping into rocks, driving off cliffs, and so forth). However, almost every other decision is made by a team of humans back on Earth. This ensures that the rovers "think" as geologists do.

What's Under the Hood? *Spirit* and *Opportunity* both come equipped with 6-wheel drive, a color stereo camera with human vision, front-and-back hazard avoidance cameras, a magnifying "hand lens" for close-up inspection of rocks and soils, and instruments to detect the chemical and mineral composition of rocks and soils. The rovers are powered by solar energy and controlled by scientists on Earth, who send daily command sequences to each rover via radio signals. Onboard computers receive these commands, which control each rover activity including driving; taking pictures of the terrain, rocks, and soils; analyzing rocks and minerals; and studying the atmosphere and moons of Mars. After two years of operation, the rovers have sent over 125,000 images back to Earth and analyzed hundreds of rock and soil samples.

Rover Landing Sites The *Mars Exploration Rovers* mission was motivated by the search for evidence of liquid water on Mars in its geologic past. The rovers were built with this idea in mind and sent to two locations where orbiter data from *Mars Global Survey* and *Mars Odyssey* suggested the chances of success would be high. (However, some of the best places were eliminated from consideration because of the extreme risk of landing in a rocky terrain.) Two sites out of several hundred were chosen, both near the Martian equator but on opposite sides of the planet. The equatorial position provides the rovers' solar panels with maximum energy throughout the year.

Spirit was sent to Gusev Crater, a large crater about 160 km in diameter that is thought to have been filled up with water to form a large lake (**Figure 9.24a**). *Opportunity* was sent to the Meridiani Planum ("Plains of Meridiani") where the mineral hematite (Fe_2O_3) had been detected by *Mars Global Surveyor* (see Figure 9.24b). Since landing, *Spirit* has trekked across a volcanic plain and ascended the Columbia Hills. *Opportunity* landed in a small crater (Eagle Crater, about 20 m in diameter), where it spent 60 days studying the first sedimentary rocks ever found on another planet and showing that they must have formed in water. *Opportunity* then moved on to another, larger crater (Endurance Crater, about 180 m in diameter), where it spent the

(a)

100 km

0 %   Hematite Abundance   20 %

(b)

Figure 9.24 *Mars Exploration Rovers* landing sites. (a) *Spirit* has been exploring Gusev Crater, about 160 km in diameter, which is thought to have been filled with water forming an ancient lake. A channel that might have supplied water to the crater is visible at lower right. [NASA/JPL/ASU/MSSS.] (b) *Opportunity* was sent to an area where hematite—a mineral that often forms in water on Earth—is abundant. The image shows the concentration of hematite, and the ellipse outlines the permissible landing area. [NASA/ASU.]

next 6 months putting those sedimentary rocks into a broader context of environmental evolution. *Opportunity* is now on its way to another, much larger crater (Victoria Crater, about 1 km in diameter) to explore for even more expansive outcrops of sedimentary rocks.

Recent Discoveries: The Environmental Evolution of Mars

The recent rover and orbiter missions to Mars have transformed our understanding of its early evolution. As on the Moon and other terrestrial planets, Mars has ancient cratered terrains that preserve the record of Late Heavy Bombardment. Therefore, these ancient cratered terrains are likely all made of rocks older than 3.8 to 3.9 billion years (see Figure 9.10). Younger surfaces, which formed after the time of Heavy Bombardment, are also widespread on Mars. Until recently, these younger surfaces were thought to be largely volcanic, as on Venus. However, data from the *Mars Exploration Rovers* and *Mars Express* show us that at least some— and perhaps many—of these younger surfaces are underlain by sedimentary rocks.

Some of these sedimentary rocks are composed of silicate minerals derived from erosion of older basaltic lavas

and the pulverized basaltic breccias of the ancient cratered terrains. For example, the delta deposits preserved in the crater of Figure 9.22 may have formed largely by the accumulation of basaltic detritus. However, most if not all of the sedimentary rocks beneath Meridiani Planum, where *Opportunity* has been exploring, are composed of sulfate minerals mixed with silicate minerals. These sedimentary rocks are chemical in origin. Their sulfate minerals were precipitated when water evaporated, probably in shallow lakes or seas.

The water must have been very salty to precipitate these minerals and included common minerals such as gypsum ($CaSO_4$). In addition, the presence of unusual sulfate minerals such as *jarosite* (**Figure 9.25**)—an iron-rich sulfate mineral—tells us that the water would also have been very acidic. On Mars, jarosite probably formed when abundant basaltic rocks interacted with water and were weathered, releasing their sulfur, which then made sulfuric acid. The acid-rich water then flowed through rocks, heavily fractured from impact, and over the surface to accumulate in lakes or shallow seas.

As discussed in Chapters 5 and 8, sedimentary rocks are very valuable tools in recording Earth's history. The vertical succession of sedimentary rocks—their *stratigraphy*—tells

Figure 9.25 This image shows the first outcrop studied on another planet (Mars). The outcrop is made of sedimentary rocks formed in part of sulfate minerals, including jarosite. Jarosite is important because it can form only in water—and only in acid-rich water. The photograph is about 50 cm in width. [NASA/JPL/Cornell.]

us how environments change over time. One of the most exciting findings of the *Mars Exploration Rovers* mission so far has been the discovery of a stratigraphic record at Endurance Crater. Because the crater is large, there is more outcrop to observe, and it is mostly unaffected by the crater-forming impact. **Figure 9.26a** shows the outcrop that contains all the stratigraphic clues. By using *Opportunity* to measure each layer, we were able to create a high-resolution stratigraphy (see Figure 9.26b), the first of its kind generated for another planet. Remarkably, this interpretive drawing—from a planet 300 million miles away from Earth—provides the same level of understanding that is typically obtained here on Earth, such as that illustrated in Figure 5.11.

Perhaps one day we will have enough understanding of the stratigraphy of Mars to be able to correlate its sedimentary and volcanic rocks from one part of the planet to another, as is illustrated for Earth in Feature 8.1. To do this, we'll need to link observations on the ground with observations provided by orbiters overhead. The *Mars Express* orbiter has shown that sulfate minerals are abundant in several places on Mars. They have been found in the Vallis Marineris, where they may form deposits up to several kilometers thick. This leads us to believe that their formation was related to a process that occurred globally, possibly over a significant interval of time.

The evidence is now compelling that at some point in Mars's history, there was liquid water on its surface and underground. The planet must have been warmer than it is today, unless the water was very short-lived, gushing to the surface briefly, then evaporating quickly or sinking back underground before being frozen, as would happen today. There are many questions left to answer. How much water was there? How long did it last? Did it ever rain, or was it all groundwater leaking to the surface? Did the water last

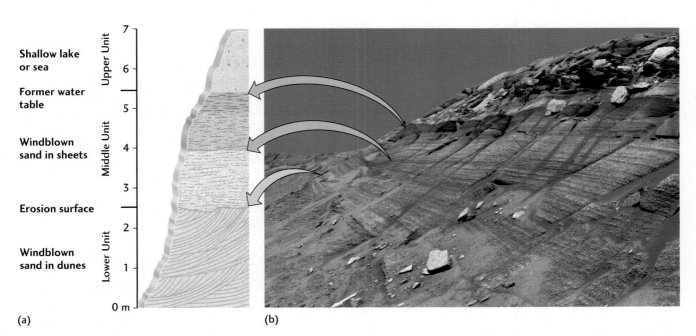

(a)
(b)

Figure 9.26 Stratigraphy exposed along the flank of Endurance Crater. (a) The record can be read in the outcrop and depicted as an interpretive drawing that shows each stage in the history of the environment. (b) The vertical succession of layers in this outcrop preserves an excellent record of early Martian environments. [Both parts from NASA/JPL/Cornell.]

long enough and have the right composition to allow life to get started? Only one thing is certain at this point. More missions are required to answer these questions.

EXPLORING THE SOLAR SYSTEM AND BEYOND

An astronomer staring through a telescope is the first image that comes to mind when most people think of how the solar system is explored. But most modern telescopes have no eyepiece at all and instead record their images with digital cameras. Many telescopes, such as the Hubble Space Telescope, are not even located on Earth but are positioned in space.

Regardless of how the telescope takes its photographs or where it is stationed, the principle is still the same: to gather more light than we can see with the naked eye. The photographs can be processed to increase their brightness further or enhance their contrast; such photographs reveal important surface features such as craters and canyons. All the geologic surface features of the planets discussed so far in this chapter are studied in this way.

However, the light gathered by telescopes and other digital cameras such as those on *Spirit* and *Opportunity* can be studied using a second technique. Once we have a record of the light coming from an object of interest—say a planet, star, or outcrop—we can study its **spectrum.** We are all familiar with how light, when passed through a prism, splits into a rainbow of colors called its spectrum (plural *spectra*). In a similar way, the light generated by stars, or reflected off the surface of a planet or outcrop, also produces spectra. And the colors of these spectra reveal the chemical composition of the light-producing or light-reflecting materials. Geologists can look at the spectra of the planets and know which gases are in their atmospheres and which chemicals and minerals are in their rocks and soils.

Astronomers use this same principle to look at the light coming from other faraway stars and galaxies. The spectra they see show the age of stars and galaxies, reveal how they evolved, and even provide mind-boggling insights into the origin and evolution of the universe.

Space Missions

Most observations of our solar system and beyond are still done from Earth. Over the past 50 years, we have increasingly sent all kinds of machines, robots, and even humans into space in our quest to explore the unknown. Missions to space are a costly business requiring a tremendous effort by hundreds and sometimes thousands of people at a cost of hundreds of millions to billions of dollars. The *Mars Exploration Rovers* project cost on the order of $800 million for both rovers. And space missions are also a risky business; fewer than half the missions ever sent to Mars have succeeded. As the Space Shuttle program reminds us, space exploration is risky for humans, too.

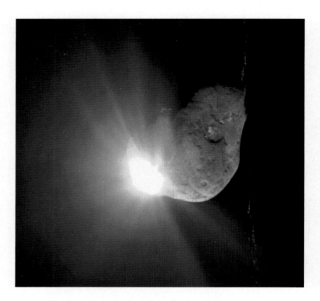

Figure 9.27 The image shows the first moments after *Deep Impact*'s probe collided with comet Tempel 1. The debris from the interior of the comet is expanding from the impact site. [NASA/JPL-Caltech/UMD.]

Are these efforts, costs, and risks worth it? For thousands of years, people have looked up at the skies and pondered the universe. What are the stars and planets made of? How did the universe form? Is there any life out there? To answer these questions, we have to look for clues, and most of these will only ever be provided by missions to space. The issue is not so much whether or not to explore space, but how. Most debates focus on whether it is essential to send humans into space or whether the *Mars Exploration Rovers* have demonstrated the adequacy of robots.

We actively explore space in many different ways. Spacecraft have been sent to orbit other planets, their moons, and asteroids or to fly by planets and comets in the outer solar system and beyond. On other occasions, we have instructed landers, rovers, and other probes to descend to planetary surfaces and make direct measurements of rocks, minerals, gases, and fluids. Some successes are awesome. On July 3, 2005, a probe was released from the *Deep Impact* spacecraft to deliberately collide with comet Tempel 1. The resulting depth of the crater and the light emitted at the time of collision (**Figure 9.27**) will show what the interior of the comet is made of and thus contribute to our understanding of comets in general.

The *Cassini* Mission to Saturn

An even more remarkable story of deep space exploration involves the *Cassini-Huygens* mission. Earlier in 2005, the *Huygens* lander became the farthest-traveled explorer to reach another land and live to tell about it.

Cassini-Huygens is one of the most ambitious missions ever launched into space. The mission includes two spacecraft: the *Cassini* orbiter and the *Huygens* lander. The spacecraft was launched from Earth on October 15, 1997. After traveling over a billion kilometers across deep space in almost 7 years, *Cassini-Huygens* sailed through the rings of Saturn on July 1, 2004. Saturn's beautiful rings are what set it apart from the other planets in our solar system (**Figure 9.28a**). It is the most extensive and complex ring system in our solar system, extending hundreds of thousands of kilometers from the planet. Made up of billions of particles of ice and rock—ranging in size from grains of sand to houses—the rings orbit Saturn at varying speeds. Understanding the nature and origin of these rings is a major goal of *Cassini* scientists.

On December 24, 2004, the *Huygens* lander was released from the orbiter and traveled over 5 million kilometers to reach Titan, one of Saturn's 18 moons. On January 14, 2005, it reached Titan's upper atmosphere, where it deployed a parachute and then plunged to the surface, where it landed successfully. The surface appears to have drainage networks, similar to those seen on Earth and Mars, and volcanoes. The landing site was strewn with rocks up to 10 to 15 cm in diameter (see Figure 9.28b). However, these "rocks" are likely ices made of methane (CH_4) and other organic compounds.

Bigger than the planets Mercury and Pluto, Titan is of particular interest to scientists because it is one of the few moons in our solar system with its own atmosphere. The moon is cloaked in a thick, smoglike haze that scientists believe may be very similar to Earth's atmosphere before life began more than 3.8 Ga (see Chapter 11). Organic compounds, including gases made of methane, are rich on Titan. Further study of this moon promises to reveal much about planetary formation and, perhaps, about the early days of Earth as well.

Other Solar Systems

For ages, scientists and philosophers have speculated that there may be planets around stars other than our Sun. In the 1990s, astronomers discovered planets orbiting nearby Sun-like stars. In 1999, the first family of **exoplanets**—planets that orbit another star—was found. These planets are too dim to be seen directly by telescopes, but their existence can be inferred from their slight gravitational pull on the stars they orbit, causing to-and-fro movements of the star that can be measured. We see these motions recorded in the spectra of their light. In about 10 years or so, spacecraft above Earth's atmosphere should be able to search for the dimming of the parent star's light as an orbiting planet passes in front of it along the line of sight to Earth.

By mid-2004, astronomers had discovered over 150 new planets, organized in 108 solar systems. Most planets found in this way are Jupiter-sized or larger and are close to their parent stars—many within scorching distance. Earth-sized

(a)

(b)

Figure 9.28 (a) Saturn and its rings completely fill the field of view in this natural color image taken by the *Cassini* spacecraft on March 27, 2004. Color variations in the rings reflect differences in the composition of materials that make them up, such as ice and rock. *Cassini* scientists will investigate the nature and origin of the rings as the mission progresses. (b) The surface of Titan is strewn with rocks of ice composed of frozen methane and other carbon-containing compounds. [Images from NASA/JPL/SSI/ESA/University of Arizona.]

planets have been discovered in recent years using other methods.

We are fascinated by planetary systems around other stars because of what they might teach us about our own origins. Our overriding interest, though, is in the profound scientific and philosophical implications posed by the question: "Is anyone else out there?" In about 15 years, a spacecraft named *Life Finder* could be carrying instruments to analyze the atmospheres of exoplanets in our galaxy for signs of the presence of some kind of life. Based on what we know about biological processes, life on an exoplanet would probably be carbon-based and require liquid water. The benign temperatures we enjoy on Earth—not too far outside the range between the freezing and boiling points of water—appear to be essential (see Chapter 11). An atmosphere is needed to filter harmful radiation from the parent star, and so the planet must be large enough for its gravitational field to keep the atmosphere from escaping into space. For a habitable planet with advanced life *as we know it* to exist would require conditions even more limiting. For example, if the planet were too massive, delicate organisms such as humans would be too weak to withstand its larger gravitational force. Are these requirements too restrictive for life to exist elsewhere? Many scientists think not, considering the billions of Sunlike stars in our own galaxy.

SUMMARY

How did our solar system originate? The Sun and its family of planets probably formed when a primeval cloud of gas and dust condensed about 4.5 Ga. The planets vary in chemical composition in accordance with their distance from the Sun and with their size.

How did Earth form and evolve over time? Earth probably grew by the accretion of colliding chunks of matter. Soon after it formed, it was struck by a giant asteroid. Matter ejected into space from both Earth and the impactor reassembled to form the Moon. The impact melted much of the Earth. Radioactivity also contributed to early heating and melting. Heavy matter, rich in iron, sank toward Earth's center, and lighter matter floated up to form the outer layers that became the crust and continents. Outgassing contributed to the growth of the oceans and a primitive atmosphere. In this way, Earth was transformed into a differentiated planet with chemically distinct zones: an iron core; a mantle that is mostly magnesium, iron, silicon, and oxygen; and a crust rich in radioactive elements and the light elements oxygen, silicon, aluminum, calcium, potassium, and sodium.

What are some major events in the early history of the solar system? The age of the solar system as determined from meteorites is about 4.56 billion years. The major planets formed within about 10 million years, and they differentiated into a core-mantle-crust layering in less than 100 million years. The Moon formed from a giant impact at about 4.5 Ga. Minerals as old as 4.4 billion years have survived in Earth's crust. The lunar maria were formed by impact in the Late Heavy Bombardment, which peaked around 3.9 Ga and marks the end of the Hadean eon on Earth.

How can planetary surfaces be dated? The surface of the Moon has been dated by isotopic methods from rocks returned by the *Apollo* missions. The lunar highlands show ages from 4.47 to about 4.0 billion years. The lunar maria show ages from 4.0 to 3.2 billion years. These isotopic ages have allowed us to calibrate the crater-counting time scale, which can be used to estimate the ages of other planetary surfaces.

Do other planets have plate tectonics? Venus is the only planet, other than Earth, that has active tectonics controlled by global convection of its mantle. But Venus does not appear to have thick plates. Instead, it has a crust formed of a very thin layer of frozen lava that breaks up into flakes or crumples like a rug. Geologists refer to this as "flake tectonics." Flake tectonics may have occurred on Earth when it was younger and hotter.

How have Mars and the other planets been explored using spacecraft? Four types of spacecraft are used. During a flyby, a spacecraft comes close to a planet, makes observations, and then continues out into deep space or another planet. An orbiter circles a planet, making remote observations of its surface and interior. A lander can actually touch down on the surface of a planet to make local observations. A rover can leave the landing site and travel up to several kilometers to investigate new terrains.

Does water exist on Mars? Today, water is present only as ice at its poles and in the shallow subsurface. In the past, it may have been present as a liquid, when it ran across the surface to carve channels and deposit sediments in rivers and deltas. It also accumulated in shallow lakes or seas where it dried to precipitate a variety of salts, including sulfate minerals.

How do we use light in exploring stars and our solar system? Sometimes light reveals shapes and textures that tell us about the geology of distant objects and how they may have formed. In other cases, we use the spectrum of light, which varies depending on the composition of materials that produce the light or reflect it toward us.

Is our solar system unique? Up until just 10 years ago, we didn't know. But now we have evidence for more than 150 planets that circle other stars. In several cases, there is more than one planet in these solar systems. Because these new planets lie outside our solar system, they are called exoplanets.

KEY TERMS AND CONCEPTS

asteroid (p. 192)

differentiation (p. 192)

exoplanet (p. 211)

flake tectonics (p. 201)

Heavy Bombardment (p. 196)

lunar highlands (p. 197)

lunar maria (p. 197)

magma ocean (p. 193)

meteorite (p. 192)

nebular hypothesis (p. 190)

planetesimal (p. 191)

solar nebula (p. 191)

spectrum (p. 210)

EXERCISES

1. What factors made Earth a particularly congenial place for life to develop?

2. How and why do the inner planets differ from the giant outer planets?

3. What caused Earth to differentiate, and what was the result?

4. Mercury's average density is less than Earth's, but its core is denser than Earth's. How can you explain this?

5. What aspects of Earth's history are consistent with high impact rates during the Late Heavy Bombardment?

6. What surface features would you look for on Mars if you were searching for evidence of liquid water in the geologic past?

7. What are the advantages and disadvantages of spacecraft that fly by and orbit a celestial body versus those that land and move around on its surface?

THOUGHT QUESTIONS

1. If a giant impact such as the one that formed the Moon had occurred after life had arisen on Earth, what would have been the consequences?

2. If you were an astronaut landing on an unexplored planet, how would you decide whether the planet was differentiated and whether it was still geologically active?

3. Knowing how the Moon formed, what might you expect as a result if you are told that a large meteorite has collided with a planet twice its size? What could be the effect of this collision on the interior composition of this planet? How would the result of the impact differ if the meteorite were significantly smaller than the planet?

4. During a dust storm on Mars, sediments fill the atmosphere with dust. But Mars has an atmosphere much thinner than Earth's. To move sand, would the wind have to blow faster on Mars to compensate for this difference?

5. Basalt is the most common rock on Mars. If Mars once had liquid water, which minerals in basalt would weather first? Which new minerals might form as weathering proceeded? (*Hint:* See Chapter 16 for help on weathering.)

6. Many scientists think that water is present on Mars. Today it is frozen, but at 4 Ga water may have been liquid. What happened? Describe all the possible mechanisms for this change. What evidence would you search for to help decide among these possibilities?

7. How does the discovery of planets orbiting other stars contribute to the debate about the possibility of life elsewhere in the cosmos? What are the scientific and philosophical implications of the existence of life on the planets of other stars?

8. What are the advantages and disadvantages of living on a differentiated planet? On a geologically active planet?

SHORT-TERM TEAM PROJECT

Robotic Geologist

Missions to other planets require many people with many different talents to work together in teams to help accomplish their goals. Most missions begin with the seed of an idea posed by a scientist—usually an important problem that could be solved with just the right kind of data. Once the question is posed, other scientists and engineers are consulted to help design a spacecraft equipped with the right capabilities and instruments to collect the data necessary to solve the problem.

You are faced with a difficult mission: to look for evidence of water in the cliffs of the Vallis Marineris on Mars. Features you might want to examine include evidence of modern water seeping out between layers of rocks or evidence of past water in the form of mineral deposits, plastered to the cliffs, which precipitated as past water emerged and then dried up. The problem is that a rover with wheels just won't do the trick; the cliffs are too steep. What other types of robots could examine these cliffs—which are up to several kilometers high—without risk of crashing? Think about what instruments you would want to include on the robot to solve the science problem and what kind of mobility design the robot would need to get the job done.

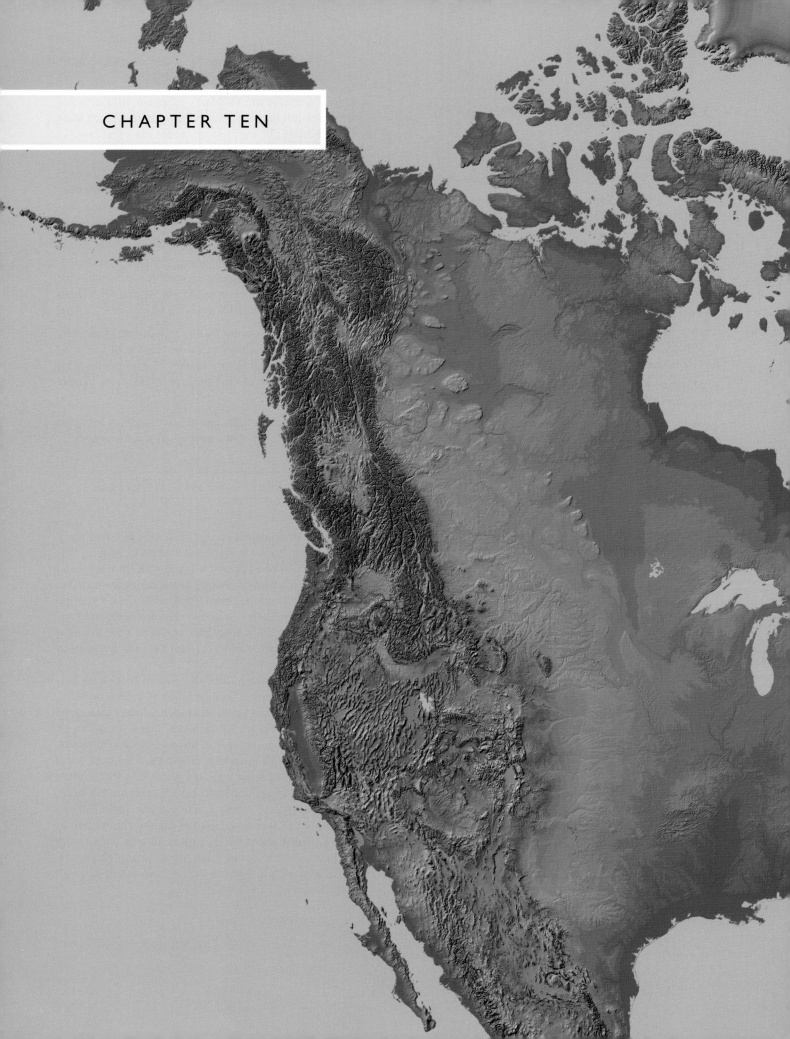

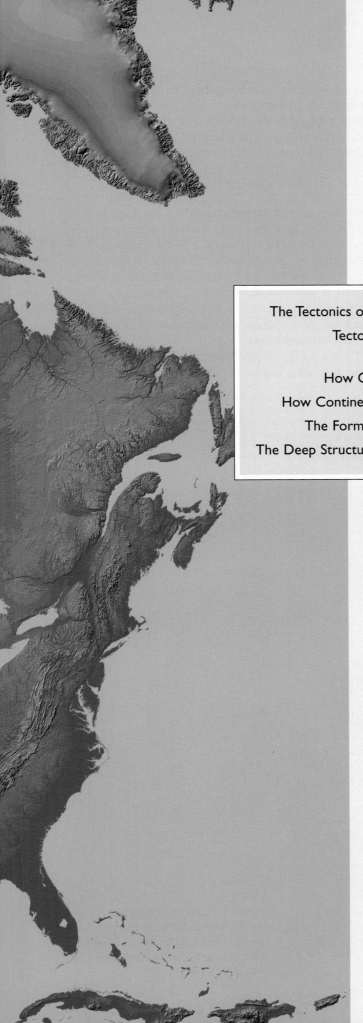

EVOLUTION OF THE CONTINENTS

Nearly two-thirds of Earth's surface—all of the oceanic crust—was created by sea-floor spreading over the past 200 million years, an interval that spans a mere 4 percent of Earth's history. To understand how Earth has evolved since its fiery beginnings, we must look to the continents, which contain rocks as old as 4 billion years.

The geologic record of the continental crust is very complex, but our ability to read this record has improved immensely in just the last few years. Earth scientists use the concepts of plate tectonics to interpret eroded mountain belts and ancient rock assemblages in terms of closing ocean basins and colliding continents. New geochemical tools, such as isotopic dating, help decipher the history of continental rocks. We can now image the structure of the continents far below Earth's surface using networks of seismographs and other sensors. One discovery, unanticipated by the theory of plate tectonics, is that in areas of the oldest continental crust, the plates have deep "keels" that are much thicker than the lithosphere beneath even the oldest ocean basins.

This chapter will examine the 4-billion-year history of the continents and what it tells us about how continental lithosphere formed and how it has evolved over geologic time.

We will see how plate tectonic processes have added new material to the continental crust, how plate convergence has thickened this crust into mountain belts, and how these mountains have been eroded to expose the metamorphic basement rocks found in many older regions of the continents. We also reach back into the earliest period of continental evolution, the Archean eon (4.0 to 2.5 billion years ago), to ponder two of the great puzzles of Earth's history: How did continents and their deep keels form, and how have they survived through billions of years of plate tectonics and continental drift?

Continents, like people, show a great variety of features that reflect their parentage and experience over time. Yet, also like people, continents share many similarities in their basic structure and

Shaded relief map of the North American continent.
[U.S. Geological Survey.]

growth. Before considering continents in general, we will discuss the major features of one continent we know very well: North America.

THE TECTONICS OF NORTH AMERICA

The large-scale geologic features of North America reflect the continent's long-term tectonic evolution (**Figure 10.1**). Parts of the crust built during the most ancient episodes of deformation tend to be found in the northern interior of the continent. This region, which includes most of Canada and the closely connected landmass of Greenland, is *tectonically stable*. In other words, it has remained largely undisturbed in recent episodes of continental rifting, drift, and collision, and it has been eroded nearly flat. On the edges of these older terrains are the younger deformation belts, where most of the present-day mountain chains are found. They form elongated

topographic features near the margins of the continent. The two main examples are the Cordillera, which runs down the western edge of North America and includes the Rocky Mountains, and the Appalachian fold belt, which trends southwest to northeast on the continent's eastern margin.

The Stable Interior

Much of central and eastern Canada is a landscape of very old crystalline basement rocks—a huge tectonic domain (8 million km^2) called the *Canadian Shield* (**Figure 10.2**). It is dominated by granitic and metamorphic rocks, such as gneisses, together with highly deformed and metamorphosed sedimentary and volcanic rocks. This primitive region represents one of the oldest records of Earth's history, much of it of Archean age. It contains major deposits of iron, gold, copper, diamonds, and nickel.

The nineteenth-century Austrian geologist Eduard Suess named these areas continental **shields** because they emerge

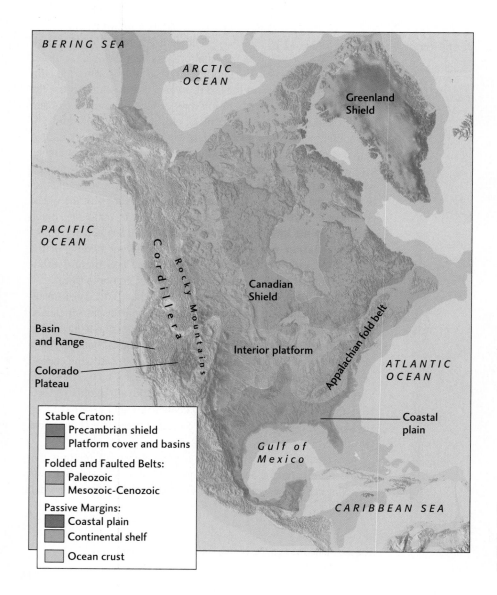

Figure 10.1 Major tectonic features of North America: Canadian Shield; interior platform; Cordilleran orogenic belt; Colorado Plateau; Appalachian fold belt; coastal plain.

Figure 10.2 An aerial view of the ancient, eroded metamorphic rocks exposed on the surface of the Canadian Shield. Nunavut, Canada. [Roy Tanami/Ursus Photography.]

from the surrounding sediments like a shield partially buried in the dirt of a battlefield. In North America, flat-lying platform sediments occur around the periphery of the Canadian Shield and also near its center, beneath Hudson Bay (see Figure 10.1). South of the Canadian Shield is a vast sediment-covered region of low topography called the *interior platform* that comprises the Great Plains of Canada and the United States. The Precambrian basement rocks beneath the interior platform are a subsurface continuation of the Canadian Shield, although here they are under layers of Paleozoic sedimentary rocks typically less than about 2 km thick.

The North American platform sediments were laid down on the deformed and eroded Precambrian basement under a variety of conditions. Some rock assemblages (marine sandstones, limestones, shales, deltaic deposits, evaporites) indicate sedimentation in extensive shallow inland seas. Other assemblages (nonmarine sediments, coal deposits) indicate deposition on alluvial plains or in lakes or swamps. Many of the continent's deposits of uranium, coal, oil, and gas are found in the sedimentary cover of the interior platform.

Within the platform are broad sedimentary basins, roughly circular or oval depressions where the sediments are thicker than the surrounding platform sediments. The Michigan Basin, a circular area of about 200,000 km^2 that covers most of the Lower Peninsula of Michigan, subsided throughout much of Paleozoic time and received sediments more than 5 km thick in its central, deepest part (**Figure 10.3**). The sandstones and other sedimentary rocks of the basins, laid down under quiet conditions, have remained

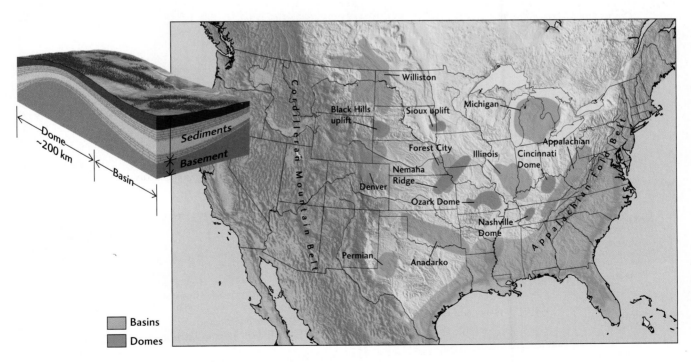

Basins
Domes

Figure 10.3 A map of the interior platform of North America, showing its basin and dome structure. The basins are nearly circular regions of thick sediments, such as the Michigan and Illinois basins, whereas the domes are regions where the sediments are anomalously thin. Crystalline basement rocks are exposed on the tops of some domes, such as the Black Hills uplift and the Ozark Dome.

unmetamorphosed and only slightly deformed to this day. Most of these basins subsided when heated portions of the lithosphere cooled and contracted.

The Appalachian Fold Belt

Along the eastern side of North America's stable interior are the old, eroded Appalachian Mountains. This classic fold-and-thrust belt, which we first examined in Chapter 7, extends along eastern North America from Newfoundland to Alabama. The rock assemblages and structures found there today resulted from plate tectonic events that took place from late Precambrian time to the present. The western side of the Appalachians is bounded by the Allegheny Plateau, a region of slightly uplifted, mildly deformed sediments. Moving eastward, we encounter regions of increasing deformation (**Figure 10.4**):

• *Valley and Ridge province.* Thick Paleozoic sedimentary rocks laid down on an ancient continental shelf were folded and thrust to the northwest by compressive forces from the southeast. The rocks show that deformation occurred in three mountain-building episodes, one beginning in the middle Ordovician period (about 470 million years ago), one in the middle to late Devonian period (380 to 350 million years ago), and one in the late Carboniferous and early Permian periods (320 to 270 million years ago).

• *Blue Ridge province.* These eroded mountains are composed largely of Precambrian and Cambrian highly metamorphosed crystalline rock. The Blue Ridge rocks were not intruded and metamorphosed in place but were thrust as sheets over the sedimentary rocks of the Valley and Ridge province near the end of the Paleozoic era, about 300 million years ago.

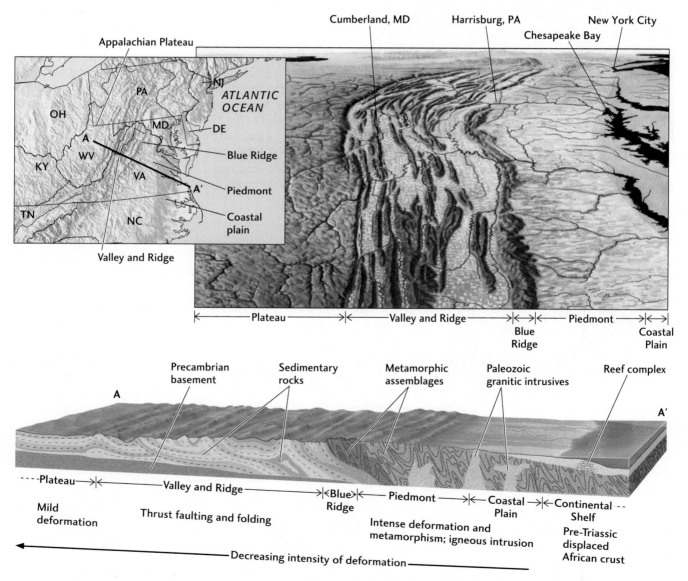

Figure 10.4 The Appalachian Mountain region of the south and central eastern United States, with an aerial view to the northeast and an idealized cross section. The major physiologic subregions extend from west to east in order of increasing intensity of deformation. [After S. M. Stanley, *Earth System History.* New York: W. H. Freeman, 2005. Aerial view from NASA.]

• *Piedmont.* This hilly region contains Precambrian and Paleozoic metamorphosed sedimentary and volcanic rocks intruded by granite, all now eroded to low relief. Volcanism began in the late Precambrian and continued into the Cambrian. The Piedmont was thrust over Blue Ridge rocks along a major thrust fault, overriding them to the northwest. At least two episodes of deformation are evident, one in the middle to late Devonian (380 to 350 million years ago), the other in the late Carboniferous and early Permian (320 to 270 million years ago).

• *Coastal plain.* The modern coastal plain developed after the Triassic-Jurassic splitting of North America from Africa and the opening of the modern North Atlantic Ocean, about 180 million years ago. Here, relatively undisturbed sediments of Jurassic age and younger are underlain by rocks similar to those of the Piedmont.

Coastal Plain and Continental Shelf

The Atlantic coastal plain and the continental shelf, its offshore extension (see Figure 10.1), began to develop in the Triassic period with the rifting that preceded the opening of the modern Atlantic Ocean. The rift valleys formed basins that trapped a thick series of nonmarine sediments. As these deposits were accumulating, they were intruded by basaltic sills and dikes. The Connecticut River valley and the Bay of Fundy are such sediment-filled rift valleys.

In the early Cretaceous period, the deeply eroded sloping surface of the Atlantic coastal plain and continental shelf began to subside and to receive sediments from the continent. Cretaceous and Tertiary sediments as much as 5 km thick filled the slowly subsiding trough, and even more material was dumped into the deeper water of the continental rise. This still-active offshore sedimentary basin continues to receive sediments. If the present stage of opening of the Atlantic is reversed some millions of years from now, the sediments in this basin will be folded and faulted in the same kind of process that produced the Appalachians.

The coastal plain and shelf of the Gulf of Mexico are continuous extensions of the Atlantic coastal plain and shelf, interrupted only briefly by the Florida Peninsula, a large carbonate platform. The Mississippi, Rio Grande, and other rivers that drain the interior of the North American continent have delivered enough sediments to fill a trough some 10 to 15 km deep running parallel to the coast. The Gulf coastal plain and shelf are rich reservoirs of petroleum and natural gas.

The North American Cordillera

The stable interior platform of North America is bounded on the west by a younger complex of mountain ranges and deformation belts (**Figure 10.5**). This region is part of the North American Cordillera, a mountain belt extending

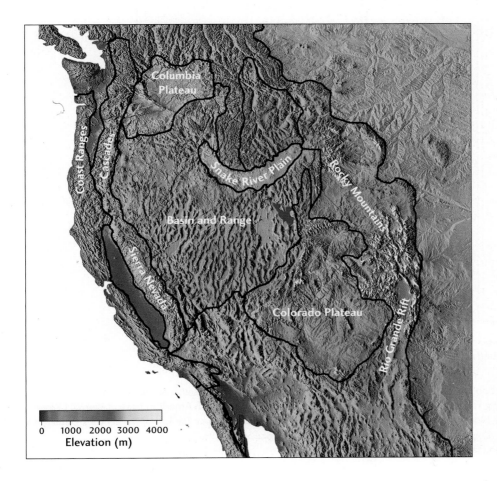

Figure 10.5 Color shaded relief map of the western United States. Computer manipulation of digitized elevation data produced an image in which the major structural provinces and tectonic history of the area are clearly visible, as if illuminated by a light source low in the west.

from Alaska to Guatemala, and it contains some of the highest peaks on the continent. Across its middle section, between San Francisco and Denver, the Cordilleran system is about 1600 km wide and includes several different geographic provinces: the Coast Ranges along the Pacific Ocean; the lofty Sierra Nevada; the Basin and Range province; the high tableland of the Colorado Plateau; and the rugged Rocky Mountains, which end abruptly at the edge of the Great Plains on the stable interior.

The history of the Cordillera is a complicated one, with details that vary along its length. It is a story of the interaction of the Pacific, Farallon, and North American plates over the past 200 million years. Before the breakup of Pangaea, the Farallon Plate occupied most of the eastern Pacific Ocean. As North America moved westward, most of this oceanic lithosphere was subducted eastward under the continent. Today, all that is left of the Farallon Plate are small remnants, which include the Juan de Fuca and Cocos plates (**Figure 10.6**). During this period of subduction, the western margin of North America was modified by the accretion of

continental and oceanic fragments, thrusting and folding of deep-water and shelf deposits, volcanism and the intrusion of granitic plutons, and metamorphism. Together, these processes swept together deformed rock assemblages that vary in age and origin and led to the reworking of older tectonic features that characterize the Cordillera.

The main phase of Cordilleran mountain building occurred in the last half of the Mesozoic era and in the early Paleocene period (150 to 50 million year ago), more recently than the Appalachian orogenies. The Cordilleran system is topographically higher than the Appalachians, which is not surprising—there has been less time for erosion to wear it down. The form and height of the Cordillera that we see today resulted from even more recent events in the Neogene period, over the past 15 or 20 million years. During these periods, the mountains underwent **rejuvenation;** that is, they were raised again and brought back to a more youthful stage. At that time, the central and southern Rockies attained much of their present height as a result of a broad regional upwarp. The Rockies were raised 1500 to

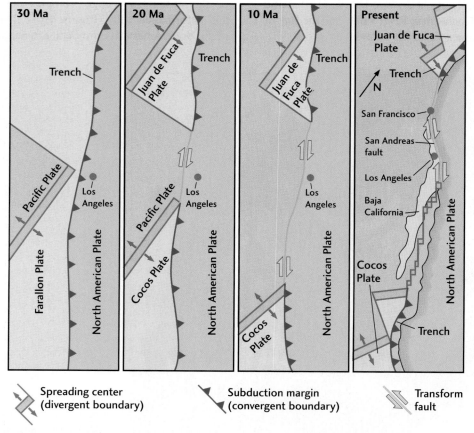

Figure 10.6 Maps illustrating the interaction of the western coast of North America with the shrinking Farallon Plate, as it was progressively consumed beneath the North American Plate, leaving the present-day Juan de Fuca and Cocos plates as small remnants of the previously larger Farallon Plate. Large solid arrows show the present-day sense of relative movement between the Pacific and North American plates. [After W. J. Kious and R. I. Trilling, *This Dynamic Earth: The Story of Plate Tectonics.* Washington, D.C.: U.S. Geological Survey, 1996.]

Figure 10.7 Owens Valley and Mount Whitney, on the east side of the Sierra Nevada, illustrate the uplift of the Sierra Nevada along a major normal fault. [Galen Rowell/Corbis.]

2000 m as Precambrian basement rocks and their veneer of later-deformed sediments were pushed above the level of their surroundings. Stream erosion accelerated, the mountain topography sharpened, and the canyons deepened. Other examples of rejuvenation by upwarping include the Adirondacks of New York and the Labrador Highlands of Canada. You will learn in Chapter 22 that rejuvenation is driven not only by tectonic processes but also by an interaction between climate change and tectonics. For example, some increase in relief of Cordilleran mountain chains may have occurred as a result of the onset of the ice ages in North America.

The Basin and Range province developed by the uplift and stretching of the crust in a northwest-southeast direction. This extension began with the heating of the lithosphere about 15 million years ago and continues to the present day. It resulted in a wide zone of normal faulting extending from southern Oregon to Mexico and encompassing Nevada, western Utah, and parts of eastern California, Arizona, New Mexico, and western Texas (**Figure 10.7**). Thousands of steeply dipping normal faults have sliced the crust into a pattern of upheaved and downdropped blocks, forming scores of rugged and nearly parallel mountain

ranges separated by alluvium-filled rift valleys. The Wasatch Range of Utah and the Teton Range of Wyoming are being uplifted on the eastern edge of the Basin and Range province, while the Sierra Nevada of California are being uplifted and tilted on the province's westward edge.

The Colorado Plateau seems to be an island of stability that has experienced no major extension or compression since the Precambrian. The broad uplift of the plateau has allowed the Colorado River to cut through flat-lying rock formations, creating the Grand Canyon (see Feature 8.1). Geologists believe this uplift was caused by the same type of lithospheric heating that led to stretching of the crust.

TECTONIC PROVINCES AROUND THE WORLD

We will now expand our view from North America to Earth's other continents (**Figure 10.8**). From a global perspective, the geologic structure of continents exhibits a general, though highly irregular, pattern. The continental shields and platforms are stable parts of the crust, called **cratons,** that contain the eroded remnants of ancient deformed rocks.

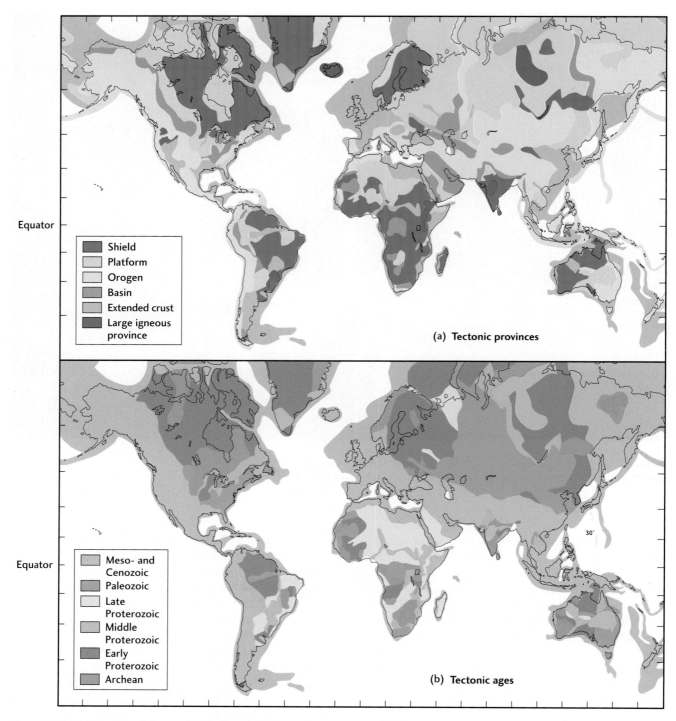

Figure 10.8 Global maps of the continents showing (a) tectonic provinces and (b) tectonic ages. [W. Mooney/USGS.]

Around these cratons are arrayed elongated mountain belts or **orogens** (from the Greek *oros*, meaning "mountain," and *gen*, "be produced") that were formed by later episodes of compressive deformation. The youngest orogenic (mountain-building) systems are found along the active margins of the continents, where plate tectonic motions continue to deform the weak continental crust.

Types of Tectonic Provinces

The general pattern of cratons bounded by orogens can be seen in Figure 10.8a, which summarizes the major tectonic provinces of the continents. The classifications portrayed on this map are closely related to the regions we used to describe the tectonics of North America:

- *Shield.* A region of uplifted and exposed crystalline basement rocks of Precambrian age, which have remained undeformed throughout the Phanerozoic eon (542 million years ago to the present). Example: Canadian Shield.

- *Platform.* A region where Precambrian basement rocks are overlain by less than a few kilometers of relatively flat-lying sediments. Examples: interior platform of central North America, Hudson Bay.

- *Continental basin.* A region of prolonged subsidence where thick sediments have accumulated during the Phanerozoic, with layers dipping into the margins of the basin. Examples: Michigan and Illinois basins.

- *Phanerozoic orogen.* A region where mountain building has been active during the Phanerozoic. Examples: Appalachian fold belt, Cordillera.

- *Extended crust.* A region where the most recent deformation has involved large-scale crustal extension. Examples: Basin and Range province, Atlantic margin.

Tectonic Ages

The tectonic ages shown in Figure 10.8b are grouped according to the geologic eras defined in Chapter 8. The **tectonic age** of a rock is the time of the last major episode of crustal deformation of that rock. Most continental basement rocks have survived a long and complex history of repeated deformation, melting, and metamorphism. We can often use isotopic dating techniques and other age indicators (see Chapter 8) to assign more than one age to any particular rock. The tectonic age indicates the *last* time the radiometric "clock" within a rock was reset by tectonic activity.

For example, many of the igneous rocks in the southwestern United States were originally derived from the melting of crust and mantle in the middle Proterozoic (1.9 to 1.6 billion years ago) (**Figure 10.9**). However, these rocks were substantially metamorphosed during subsequent periods, including several episodes of compressive deformation in the Mesozoic and rifting in the Cenozoic. Geologists thus assign this region to the youngest age category (Mesozoic-Cenozoic).

A Global Puzzle

The current distribution of continental provinces and ages is like a giant puzzle in which the original pieces have been rearranged and reshaped by continental rifting, drift, and collision over billions of years of plate tectonics. Only the past 200 million years of these motions are reliably known

Figure 10.9 The Vishnu schist, part of the middle Proterozoic (1.8 billion years old) basement found at the bottom of the Grand Canyon. [Stephen Trimble.]

from existing oceanic crust. Earlier plate motions must be inferred from the indirect evidence found in continental rocks. In Chapter 2, we saw that geologists have made amazing progress in reconstructing earlier configurations of the continents from paleomagnetic and paleoclimate data and from the signatures of deformation exposed in ancient mountain belts.

We now continue the story of the continents back into earlier geologic times. We focus on three key questions of continental evolution: What geologic processes built the continents we see today? How do these processes fit into the theory of plate tectonics? Can plate tectonics explain the formation of the oldest continents, the Archean cratons? As we will see, these questions have been answered only partially by geologic research.

HOW CONTINENTS GROW

Over their 4-billion-year history, the continents have grown at an average rate of about 2 km³ per year. Earth scientists continue to debate whether the growth of continental crust occurs gradually over geologic time or was concentrated early in Earth's history. In this section, we examine some of the basic growth processes that are currently operating in the plate tectonic system.

Magmatic Addition

In the modern plate tectonic system, the most buoyant, silica-rich crust is born in subduction zones, where the water squeezed out of downgoing slabs melts the mantle wedge above the slabs. The magmas, which are of basaltic to andesitic composition, migrate toward the surface, pooling in magma chambers near the base of the crust. Here they incorporate crustal materials and further differentiate into silica-rich magmas that migrate into the upper crust, forming dioritic and granodioritic plutons capped by andesitic volcanoes.

This process can add new crustal material directly to active continental margins. Subduction of the Farallon Plate beneath North America during the Cretaceous period, for example, created the batholiths along the western edge of the continent, including the rocks now exposed in Baja California and the Sierra Nevada. Subduction of the remnant Juan de Fuca Plate continues to add new material to the crust in the volcanically active Cascade Range of the Pacific Northwest, just as subduction of the Nazca Plate is building up the crust in the Andes Mountains of South America.

Buoyant crust is also produced far away from continents, in island arcs where one oceanic plate subducts beneath another. Over time, island arcs can merge into thick sections of silica-rich crust, such as those found in the Philippines and other island groups of the southwestern

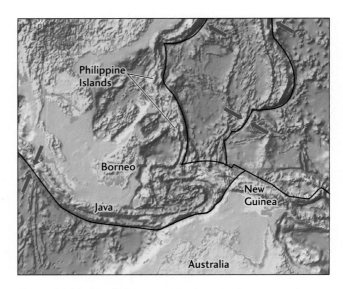

Figure 10.10 The Philippines and other island groups in the southwestern Pacific illustrate how island arcs amalgamate to form protocontinental crust.

Pacific (**Figure 10.10**). Plate motions eventually attach this crust to active continental margins through accretion.

Continental Accretion

Accretion is a process of continental growth in which buoyant fragments of crust are attached (accreted) to continents during plate motions. The geologic evidence for accretion can be found on the active margins of continents such as North America. In the Pacific Northwest and Alaska, the crust consists of a mix of odd pieces—island arcs, seamounts and remnants of thickened oceanic plateaus, old mountain ranges and other slivers of continental crust—that were plastered onto the leading edge of the continent as it moved across Earth's surface (**Figure 10.11**). The individual fragments of crust are called **accreted terranes.** Geologists use the term *terrane* (a variation of the more familiar term *terrain,* meaning a tract of land and its topography) to define a large piece of crust, tens to hundreds of kilometers in geographic extent, with shared characteristics and a common origin, usually transported great distances by plate movements.

The geologic arrangement of accreted terranes can be very chaotic. Adjacent provinces can contrast sharply in the assemblages of rock types, the nature of folding and faulting, and the history of magmatism and metamorphism. Geologists often find fossils indicating that these blocks originated in different environments and at different times from those of the surrounding area. For example, a terrane comprising ophiolites (rocks characteristic of oceanic crust) that contain deep-water fossils might be surrounded by remnants of island arcs and continental fragments containing shallow-water fossils of a completely different age.

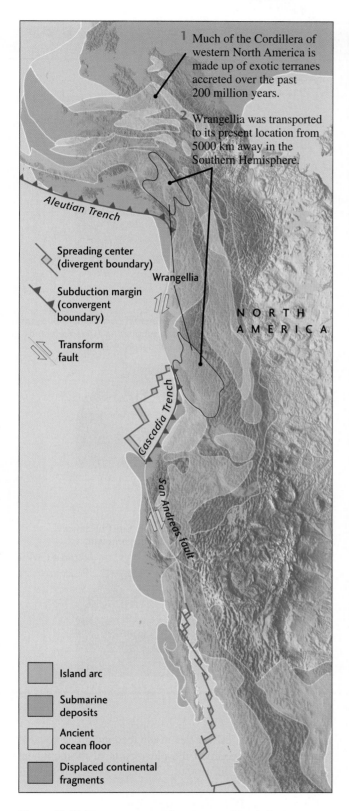

1 Much of the Cordillera of western North America is made up of exotic terranes accreted over the past 200 million years.

2 Wrangellia was transported to its present location from 5000 km away in the Southern Hemisphere.

Aleutian Trench

↯ Spreading center (divergent boundary)

◤ Subduction margin (convergent boundary)

⟰ Transform fault

Wrangellia

NORTH AMERICA

Cascadia Trench

San Andreas fault

☐ Island arc

☐ Submarine deposits

☐ Ancient ocean floor

☐ Displaced continental fragments

Figure 10.11 Terranes accreted to western North America in the past 200 million years. They are made up of island arcs, ancient seafloor crust, continental fragments, and submarine deposits. [After D. R. Hutchison, "Continental Margins." *Oceanus* 35 (Winter 1992–1993): 34–44; modified from work of D. G. Howell, G. W. Moore, and T. J. Wiley.]

The boundaries between terranes are almost always major faults that have undergone substantial slip (although the nature of the faulting is often difficult to discern). Fragments that seem completely out of place are called *exotic terranes*.

Before plate tectonics theory, exotic terranes were the subject of a fierce debate among geologists, who had difficulty coming up with reasonable explanations for their origins. Now terrane analysis is a specialized field within plate tectonics research. Well over 100 areas of the Cordillera of western North America have been identified as exotic terranes accreted during the last 200 million years (many more than depicted in Figure 10.11). One such terrane, formerly a large oceanic plateau now called Wrangellia, appears to have been transported over 5000 km from the Southern Hemisphere to its current location in Alaska and western Canada (see Figure 10.11). Extensive accreted terranes have also been mapped in Japan, Southeast Asia, China, and Siberia.

In only a few cases do we know precisely where these terranes originated. For the others, we can begin to decipher how these pieces came together by considering four distinct tectonic processes (**Figure 10.12**):

1. The transfer of a crustal fragment from a subducting plate to the overriding plate, which can occur when the fragment is too buoyant to be subducted. The fragments can be small pieces of continent ("microcontinents") or thickened sections of oceanic crust (large seamounts, oceanic plateaus).

2. The closure of a marginal sea that separates an island arc from a continent. A collision with the advancing edge of the continent can attach the thickened island-arc crust to the continent.

3. The transport of terranes laterally along continental margins by strike-slip faulting. Today, the southwestern part of California attached to the Pacific Plate is moving northwestward relative to the North American Plate along the San Andreas transform fault. Strike-slip faulting landward of the trench in oblique subduction zones can also transport terranes hundreds of kilometers.

4. The suturing of two continental margins by continent-continent collision and the subsequent breakup of this zone by continental rifting.

The fourth process explains some of the accreted terranes found on the passive eastern margin of North America. The Appalachian fold belt, ranging from Newfoundland to the southeastern United States, contains slices of ancient Europe and Africa, as well as a variety of exotic terranes. Florida's oldest rocks and fossils are more like those in Africa than like those found in the rest of the United States, indicating that most of this peninsula was probably a piece of Africa left behind when North America and Africa split apart about 200 million years ago.

Key Figure 10.12 Four distinct processes explain the accretion of exotic terranes.

1 ACCRETION OF A BUOYANT FRAGMENT TO A CONTINENT
Fragment
A buoyant oceanic or continental fragment is carried into a plate collision zone.
Continental crust
Lithosphere
Asthenosphere
The fragment is more buoyant than the subducting lithosphere and is not subducted.
Accreted terrane
The fragment becomes welded to the overriding plate.
2 ACCRETION OF AN ISLAND ARC TO A CONTINENT
A plate carrying a continent subducts beneath an oceanic island arc.
Island arc
Continental crust
The continental crust is more buoyant than the subducting lithosphere and is not subducted with it.
The island arc crust becomes welded to the continent.
Accreted terrane

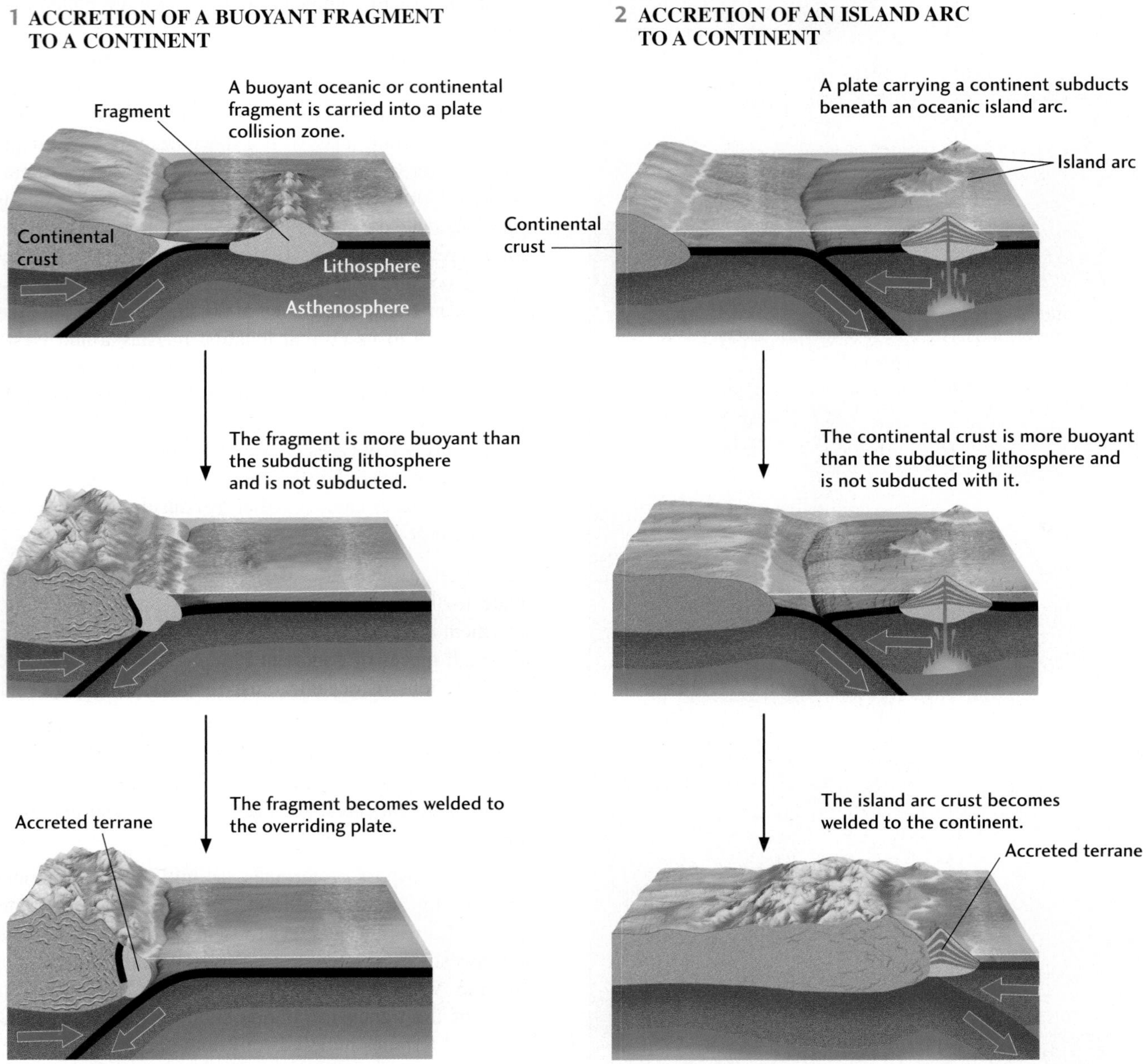

HOW CONTINENTS ARE MODIFIED

The Cordilleran region of western Canada includes many exotic terranes accreted during the drift of North America since the breakup of Pangaea. The geology of this youthful part of the continent looks nothing like that of the ancient Canadian Shield, which lies directly east of the Cordillera. In particular, the accreted terranes do not show the high degree of melting or the high-grade metamorphism that characterize the Precambrian crust of the shield. Why such a difference? The answer lies in the tectonic processes that have repeatedly modified the older parts of the continent throughout its long history.

Orogeny: Modification by Plate Collision

The continental crust is profoundly altered by **orogeny**—the mountain-building processes of folding, faulting, magmatism, and metamorphism. Most orogenies (periods of mountain building) involve plate convergence. When one or both plates are made of oceanic lithosphere, the conver-

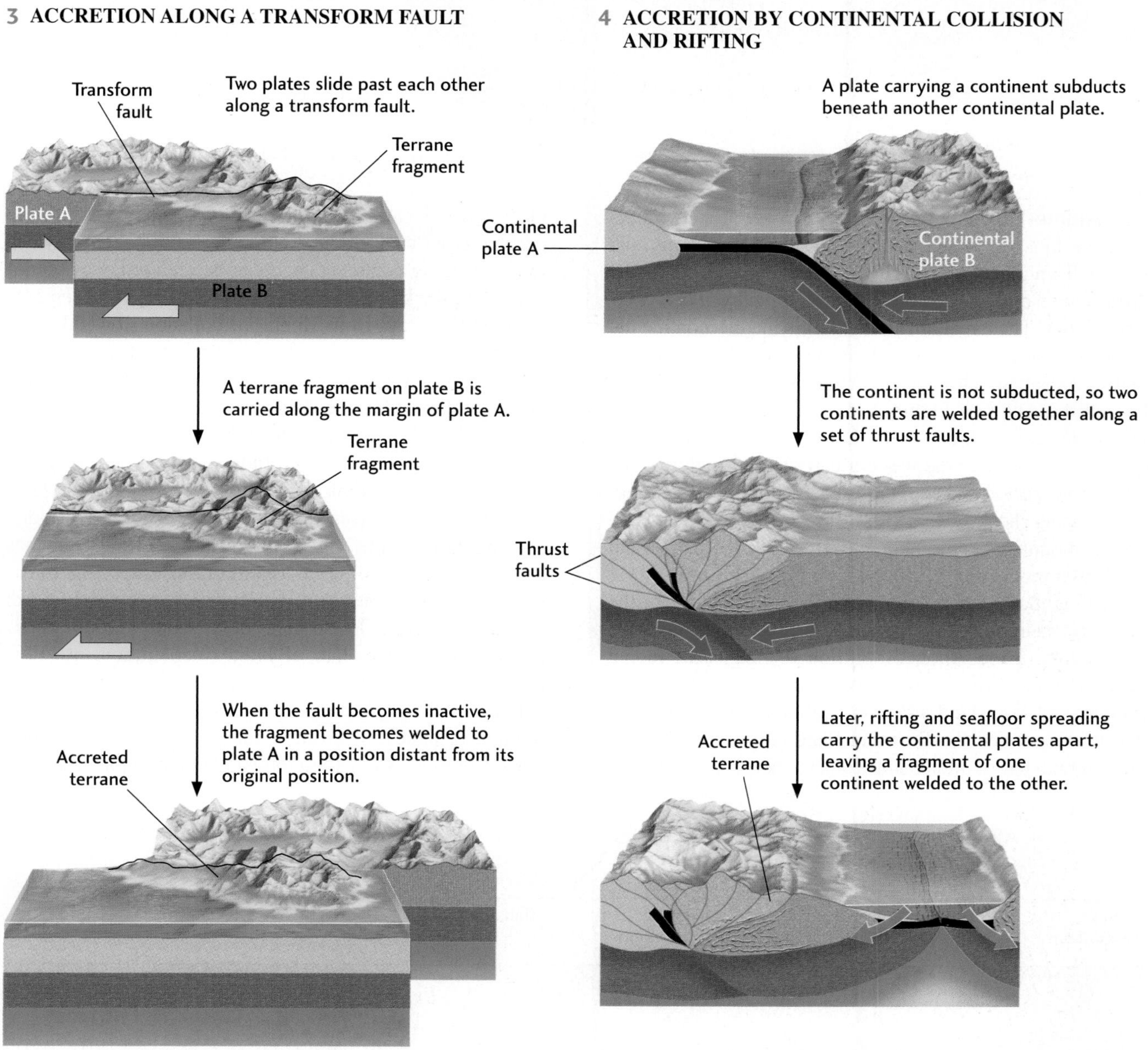

3 ACCRETION ALONG A TRANSFORM FAULT

Two plates slide past each other along a transform fault.

Transform fault

Terrane fragment

Plate A

Plate B

A terrane fragment on plate B is carried along the margin of plate A.

Terrane fragment

When the fault becomes inactive, the fragment becomes welded to plate A in a position distant from its original position.

Accreted terrane

4 ACCRETION BY CONTINENTAL COLLISION AND RIFTING

A plate carrying a continent subducts beneath another continental plate.

Continental plate A

Continental plate B

The continent is not subducted, so two continents are welded together along a set of thrust faults.

Thrust faults

Later, rifting and seafloor spreading carry the continental plates apart, leaving a fragment of one continent welded to the other.

Accreted terrane

gence is taken up primarily by subduction. Orogenies can result when a continent rides forcefully over a subducting oceanic plate, as in the Andean orogeny now under way in South America, but the most intense orogenies are caused by the convergence of two or more continents. As we observed in Chapter 2, when two continental plates collide, a basic tenet of plate tectonics—the rigidity of plates—must be modified.

Continental crust is much more buoyant than the mantle, so colliding continents resist being subducted with the plates that carry them. Instead, the continental crust deforms and breaks in a combination of intense folding and faulting that can extend hundreds of kilometers from the collision (see Chapter 7). Low-angle thrust faulting caused by the convergence can stack the upper part of the crust into multiple thrust sheets tens of kilometers thick, deforming and metamorphosing the rocks they contain (**Figure 10.13**). Wedges of continental-shelf sediments can be detached from the basement on which they were deposited and be thrust inland. Compression throughout the crust can double its thickness, causing the rocks in the lower crust to melt. This melting can generate huge amounts of granitic magma, which rises to form extensive batholiths in the upper crust.

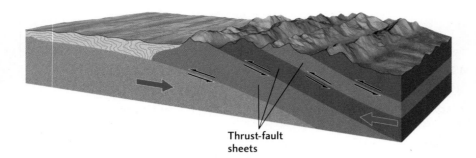

Figure 10.13 When plates bearing continents collide, the continental crust can break into multiple thrust-fault sheets stacked one above the other.

Thrust-fault sheets

The Alpine-Himalayan Orogeny To see orogeny in action, we look to the great chains of high mountains that stretch from Europe through the Middle East and across Asia, known collectively as the *Alpine-Himalayan belt* (**Figure 10.14**). The breakup of Pangaea sent the continental crust of Africa, Arabia, and India northward, causing the Tethys Ocean to close as its lithosphere was subducted beneath Eurasia (see Figure 2.15). These former pieces of Gondwanaland collided with Eurasia in a complex sequence, beginning in the western part of Eurasia during the Cretaceous period and continuing eastward through the Tertiary, raising the Alps in central Europe, the Caucasus and Zagros mountains in the Middle East, and the Himalaya and other high mountain chains across central Asia.

The Himalaya, the world's highest mountains, are the most spectacular result of this modern episode of continent-continent collision. About 50 million years ago, the Indian subcontinent, riding on the subducting Indian Plate, first encountered the island arcs and continental volcanic belts that then bounded the Eurasian Plate (**Figure 10.15**). As the landmasses of India and Eurasia merged, the Tethys

Ocean disappeared through subduction. Pieces of the oceanic crust were trapped along the suture zone between the converging continents and can be seen today as ophiolites along the Indus and Tsangpo river valleys that separate the high Himalaya from Tibet. The collision slowed India's advance, but the plate continued to drive northward. So far, India has penetrated over 2000 km into Eurasia, causing the largest and most intense orogeny of the Cenozoic era.

The Himalaya were formed from overthrust slices of the old northern portion of India, stacked one atop the other. This process took up some of the compression. Horizontal compression also thickened the crust north of India, causing uplift of the huge Tibetan Plateau, which now has a crustal thickness of 60 to 70 km (almost twice the thickness of normal continental crust) and stands nearly 5 km above sea level. These and other compression zones account for perhaps half of India's penetration into Eurasia. The other half has been accommodated by pushing China and Mongolia eastward, out of India's way, like toothpaste squeezed from a tube. The movement took place along the Altyn Tagh fault and other major strike-slip

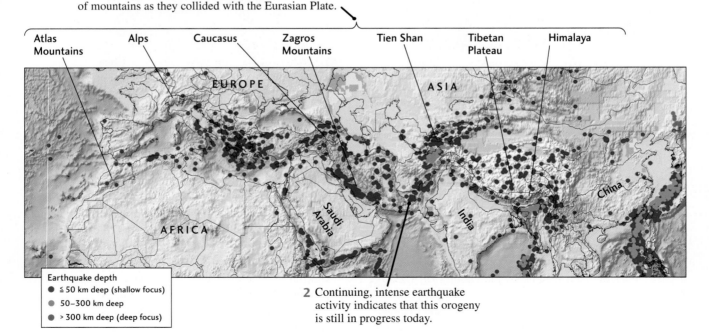

1 The African, Arabian, and Indian plates raised chains of mountains as they collided with the Eurasian Plate.

Atlas Mountains | Alps | Caucasus | Zagros Mountains | Tien Shan | Tibetan Plateau | Himalaya

EUROPE   ASIA

AFRICA   Saudi Arabia   India   China

Earthquake depth
● ≤ 50 km deep (shallow focus)
● 50–300 km deep
● > 300 km deep (deep focus)

2 Continuing, intense earthquake activity indicates that this orogeny is still in progress today.

Figure 10.14 The Alpine-Himalayan belt showing the chains of high mountains built by the ongoing collision of the African, Arabian, and Indian plates with the Eurasian Plate. This orogenesis is marked by intense earthquake activity.

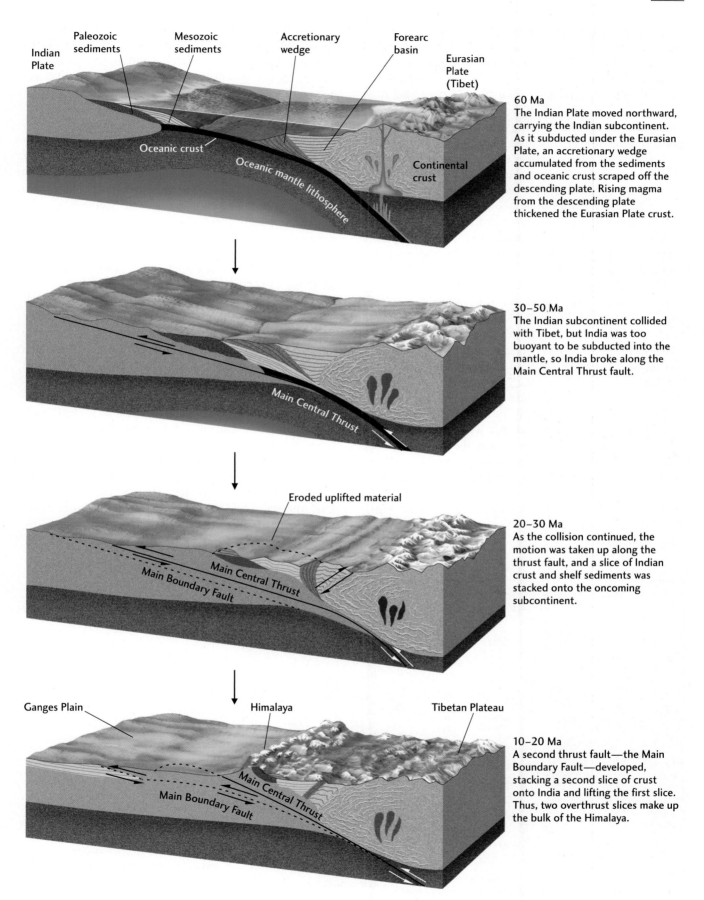

Indian Plate

Paleozoic sediments

Mesozoic sediments

Accretionary wedge

Forearc basin

Eurasian Plate (Tibet)

Oceanic crust

Oceanic mantle lithosphere

Continental crust

60 Ma
The Indian Plate moved northward, carrying the Indian subcontinent. As it subducted under the Eurasian Plate, an accretionary wedge accumulated from the sediments and oceanic crust scraped off the descending plate. Rising magma from the descending plate thickened the Eurasian Plate crust.

Main Central Thrust

30–50 Ma
The Indian subcontinent collided with Tibet, but India was too buoyant to be subducted into the mantle, so India broke along the Main Central Thrust fault.

Eroded uplifted material

Main Central Thrust

Main Boundary Fault

20–30 Ma
As the collision continued, the motion was taken up along the thrust fault, and a slice of Indian crust and shelf sediments was stacked onto the oncoming subcontinent.

Ganges Plain

Himalaya

Tibetan Plateau

Main Central Thrust

Main Boundary Fault

10–20 Ma
A second thrust fault—the Main Boundary Fault—developed, stacking a second slice of crust onto India and lifting the first slice. Thus, two overthrust slices make up the bulk of the Himalaya.

Figure 10.15 Cross sections showing the sequence of events that have caused the Himalayan orogeny, simplified and vertically exaggerated. [After P. Molnar, "The Structure of Mountain Ranges." *Scientific American* (July 1986): 70.]

faults shown on the map in **Figure 10.16**. The mountains, plateaus, faults, and great earthquakes of Asia, thousands of kilometers from the Indian-Eurasian suture, are thus affected by the Himalayan orogeny, which continues as India ploughs into Asia at a rate of 40 to 50 mm/year.

Paleozoic Orogenies During the Assembly of Pangaea If we go further back in geologic time, we find abundant evidence of older orogenies caused by earlier episodes of plate convergence. We have already mentioned, for example, that at least three distinct orogenies were responsible for the Paleozoic deformation now exposed in the eroded Appalachian fold belt of the eastern United States. These three periods of mountain building were caused by plate convergence that led to the assembly of the Pangaean supercontinent near the end of the Paleozoic.

The supercontinent of Rodinia began to break up toward the end of the Proterozoic eon, spawning several paleocontinents (see Figure 2.13). One was the large continent of *Gondwana*. Two of the others were *Laurentia*, which included the North American craton and Greenland, and *Baltica*, comprising the lands around the Baltic Sea (Scandinavia, Finland, and the European part of Russia). In the Cambrian period, Laurentia was rotated almost 90° from its present orientation and straddled the equator; its southern (today, eastern) side was a passive continental margin. To its immediate south was the proto-Atlantic or *Iapetus Ocean* (in Greek mythology, Iapetus was the father of Atlantis), which was being subducted beneath a distant island arc. Baltica lay off to the southeast, and Gondwana was thousands of kilometers to the south. **Figure 10.17** shows the sequence of events as the three continents converged.

The island arc built up by the southward-directed subduction of Iapetus lithosphere collided with Laurentia in the middle to late Ordovician (470 to 440 million years ago), causing the first episode of mountain building—the *Taconic orogeny*. (You can see some of the rocks accreted and deformed during this period when you drive the Taconic Parkway, which runs east of the Hudson River for about 160 km north of New York City.) The second episode began when Baltica and a connected set of island arcs began to collide in the early Devonian (about 400 million years ago). The convergence of Baltica with Laurentia deformed southeastern Greenland, northwestern Norway, and Scotland in what European geologists refer to as the *Caledonian orogeny*. The deformation continued into present-day North America as the *Acadian orogeny*, when island arcs added the terranes of maritime Canada and New England to Laurentia in the middle to late Devonian (380 to 350 million years ago).

The grand finale in the assembly of Pangaea was the collision of the behemoth landmass of Gondwana with Laurasia and Baltica, by then joined into a continent named *Laurussia* (see Figure 10.17). The collision began around 340 million years ago with the *Variscan orogeny* in what is now central Europe and continued along the margin of the North American craton with the *Alleghenian orogeny* (320 to 270 million years ago). This latter phase of assembly pushed Gondwanan crust over Laurentia, lifting the Blue Ridge into a mountain chain perhaps as high as the modern Himalaya and causing much of the deformation now seen in the Appalachian fold belt. Also during this phase, Siberia and other Asian terranes converged with Laurussia, forming the Ural Mountains (and thus *Laurasia*), while extensive deformation created new mountain belts across Europe and northern Africa (the *Hercynian orogeny*).

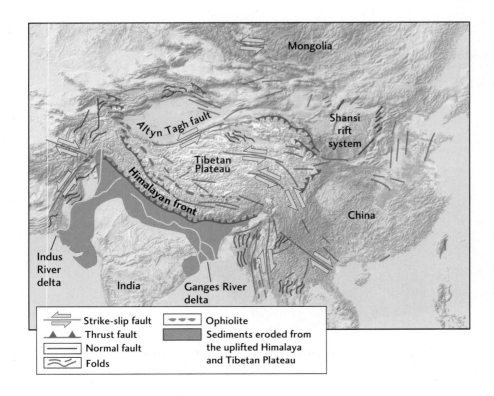

Figure 10.16 Tectonic features caused by the collision between India and Eurasia, including large-scale faulting and uplift. [After P. Molnar and P. Tapponier, "The Collision Between India and Eurasia." *Scientific American* (April 1977): 30.]

Middle Cambrian (510 Ma)
After the breakup of Rodinia, the continent of Laurentia straddled the equator, and its southern side was a passive continental margin, bounded on the south by the Iapetus Ocean.

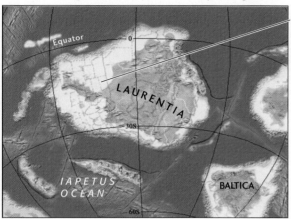

Late Ordovician (450 Ma)
The island arc built up by the southward-directed subduction of Iapetus lithosphere collided with Laurentia in the middle to late Ordovician, causing the Taconic orogeny.

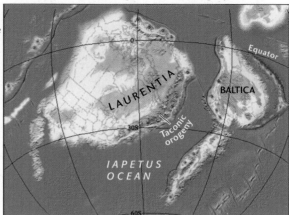

Early Devonian (400 Ma)
The collision of Laurentia with the continent of Baltica caused the Caledonian orogeny and formed Laurussia. The southward continuation of the convergence caused the Acadian orogeny.

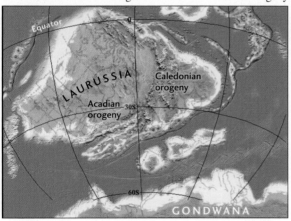

Late Mississippian (340 Ma)
The collision of Gondwana with Laurussia began with the Variscan orogeny in what is now central Europe…

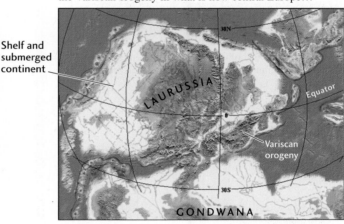

Upper Pennsylvanian (300 Ma)
… and continued along the margin of the North American craton with the Appalachian orogeny. During this terminal cataclysm, Siberia converged with Laurussia in the Ural orogeny to form Laurasia, while the Hercynian orogeny created new mountain belts across Europe and northern Africa.

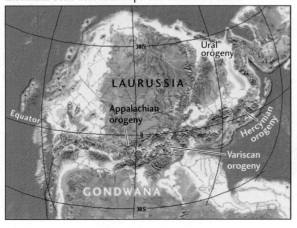

Early Permian (270 Ma)
The end product of these episodes of continental convergence was the supercontinent of Pangaea.

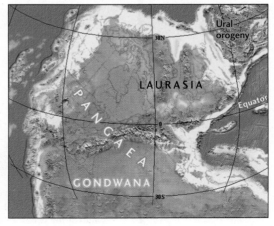

Figure 10.17 Paleogeographic reconstructions of the North Atlantic region, showing the sequence of orogenic events that resulted from the assembly of Pangaea. [Ronald C. Blakey, Northern Arizona University, Flagstaff.]

Earth System Figure 10.18 The Wilson cycle describes the plate tectonic movements that open and close ocean basins and their effects on the continents.

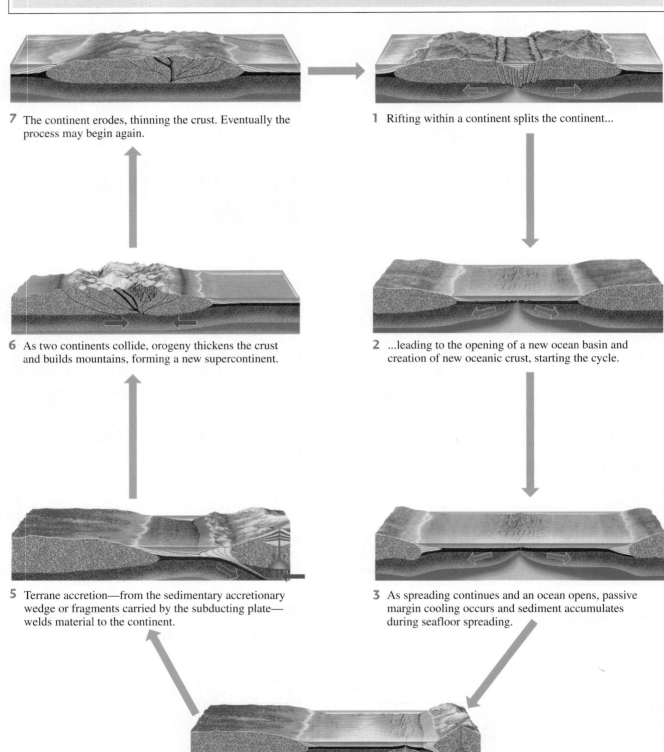

7 The continent erodes, thinning the crust. Eventually the process may begin again.

1 Rifting within a continent splits the continent...

6 As two continents collide, orogeny thickens the crust and builds mountains, forming a new supercontinent.

2 ...leading to the opening of a new ocean basin and creation of new oceanic crust, starting the cycle.

5 Terrane accretion—from the sedimentary accretionary wedge or fragments carried by the subducting plate—welds material to the continent.

3 As spreading continues and an ocean opens, passive margin cooling occurs and sediment accumulates during seafloor spreading.

4 Convergence begins; an oceanic plate subducts beneath a continental plate, creating a volcanic chain at the active margin.

The crunching together of all these continental masses profoundly altered the structure of the crust. The rigid cratons were little affected, but the younger accreted terranes caught in between were consolidated, thickened, and metamorphosed. The lower parts of this juvenile crust were partially melted, producing granitic magmas that rose to form batholiths in the upper crust and volcanoes at the surface. Uplifted mountains and plateaus were eroded, exposing high-grade metamorphic rocks that were once many kilometers deep and depositing thick sedimentary sequences. Sediments laid down following the first orogeny were deformed and metamorphosed by later mountain-building episodes.

Earlier Orogenies So far, we have investigated two major periods of mountain building: the Paleozoic orogenies associated with the assembly of Pangaea and the Cenozoic continent-continent collisions across the Alpine-Himalayan belt. In Chapter 2, we discussed the late Proterozoic supercontinent of Rodinia. By now, it should not surprise you to learn that major orogenies accompanied the formation of this earlier supercontinent.

Some of the best evidence comes from the eastern and southern margins of the Canadian Shield in a broad belt known as the Grenville province, where new crustal material was added to the continent about 1.1 to 1.0 billion years ago (see Figure 10.8). Geologists believe that these rocks, which are now highly metamorphosed, originally consisted of continental volcanic belts and island-arc terrains that were accreted and compressed by the collision of Laurentia with the western part of Gondwana. They have drawn analogies between what happened during this *Grenville orogeny* and what is happening today in the Himalayan orogeny. A Tibet-like plateau was formed by compressive thickening of the crust through folding and thrust faulting, which metamorphosed the upper crust and partially melted large parts of the lower crust. Once the orogeny ceased, erosion of the plateau thinned the crust and exposed crystalline rocks of high metamorphic grade. Geologists have found orogenic belts of similar age on continents worldwide, and although many of the details remain uncertain, they have reconstructed from this evidence (which includes paleomagnetic data) a general picture of how Rodinia came together between 1.3 and 0.9 billion years ago.

From our brief look at the history of eastern North America, we can infer that the edges of many cratons have experienced multiple episodes in a general plate tectonic cycle that comprises four main phases (**Figure 10.18**):

1. Rifting during the breakup of a supercontinent

2. Passive-margin cooling and sediment accumulation during seafloor spreading and ocean opening

3. Active-margin volcanism and terrane accretion during subduction and ocean closure

4. Orogeny during the continent-continent collision that forms the next supercontinent

This idealized sequence of events in the opening and closing of ocean basins is called the **Wilson cycle** after the Canadian pioneer of plate tectonics, J. Tuzo Wilson, who first recognized its importance in the evolution of continents.

The geologic data suggest that the Wilson cycle has operated throughout the Proterozoic and Phanerozoic eons (**Figure 10.19**). Based on the geochronology of ancient rock formations, geologists have postulated the existence of at

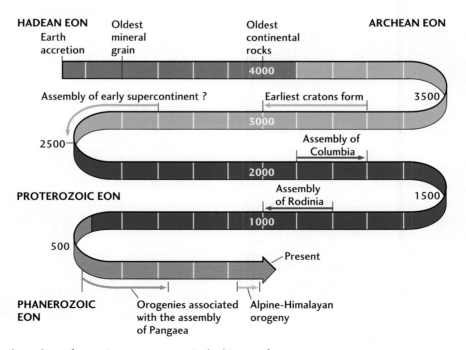

Figure 10.19 The geochronology of some important events in the history of the continents.

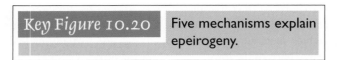

Key Figure 10.20 Five mechanisms explain epeirogeny.

(a) 1 GLACIAL REBOUND

A glacial ice load downwarps the continental lithosphere,…   …which rebounds once the ice is removed.

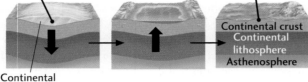

Continental crust
Continental lithosphere
Asthenosphere

Continental glacier

(b) 2 HEATING OF LITHOSPHERE

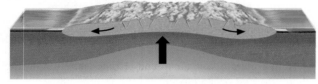

Upwarping and thinning of continental lithosphere are a result of heating and rifting.

(c) 3 COOLING OF LITHOSPHERE IN CONTINENTAL INTERIOR

Sedimentary basin

As the lithosphere cools and contracts, it subsides to form a basin within the continent.

(d) 4 COOLING OF LITHOSPHERE ON CONTINENTAL MARGIN

Sedimentary wedges

When a new episode of seafloor spreading splits a continent apart, the edges of the continent subside as they cool, accumulating thick sedimentary wedges.

(e) 5 HEATING OF DEEP MANTLE

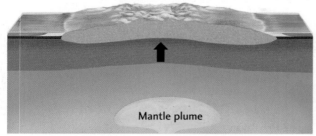

Mantle plume

A plume rising from the mantle heats the continent and raises the base of the lithosphere, upwarping the surface over a broad area.

least two episodes of supercontinent formation prior to Rodinia, one at 1.9 to 1.7 billion years ago (a supercontinent named *Columbia*) and an even earlier assembly at 2.7 to 2.5 billion years ago, which marks the transition from the Archean eon to the Proterozoic eon. Did the Wilson cycle also operate in the Archean eon? We will return to the question of Archean continental evolution shortly.

Epeirogeny: Modification by Vertical Motions

So far, our consideration of continental evolution has emphasized accretion and orogeny, processes that involve horizontal plate motions and are usually accompanied by deformation in the form of folding and faulting. Throughout the world, however, sedimentary rock sequences record another kind of motion that has modified the continents: gradual downward and upward movements of broad regions of crust without significant folding or faulting. These vertical motions involve a set of processes called **epeirogeny,** a term coined in 1890 by the American geologist Clarence Dutton (from the Greek *epeiros,* meaning "mainland").

Epeirogenic downward movements usually result in a sequence of relatively flat-lying sediments, such as those found in the stable interior platform of North America. Upward movements cause erosion and gaps in the sedimentary record seen as unconformities. Erosion can lead to the exposure of crystalline basement rocks, such as those found in the Canadian Shield.

Geologists have identified several causes of epeirogenic movements. One example is glacial rebound (**Figure 10.20a**). The weight of large glaciers depresses the continental crust during ice ages. When they melt, the crust rebounds upward for tens of millennia. Glacial rebound explains the uplift of Finland and Scandinavia and the raised beaches of northern Canada (**Figure 10.21**) following the Wisconsin glaciation, which ended about 17,000 years ago. Although rebound

Figure 10.21 Raised beaches, evidence of upward recovery of the crust after removal of glacial load. Baffin Island, Nunavut, Canada. [L. M. Cumming, Geological Survey of Canada.]

seems slow by human standards, it is a rapid process, geologically speaking.

Cooling and heating of the continental lithosphere are important causes of epeirogenic movements on longer time scales. Heating causes rocks to expand, decreasing their density and thus raising the surface (Figure 10.20b). A good example is the Colorado Plateau, which has been uplifted to about 2 km during the last 10 million years or so. Geologists think this heating results from an active mantle upwelling (which is also stretching the crust in the Basin and Range province on the western and southern sides of the plateau). The uplift has caused the Colorado River to cut deeply into the plateau, forming the Grand Canyon.

Conversely, the cooling of the lithosphere after heating and rifting increases the density of the lithosphere, making it sink under its own weight and creating a sedimentary basin (Figure 10.20c). Cooling of a once-hot area may explain the Michigan and Illinois basins and other deep basins in central North America (see Figure 10.3). When a new episode of seafloor spreading splits a continent apart, the uplifted edges are eroded and eventually subside as they cool, allowing sediments to be deposited and carbonate platforms to accumulate on the passive margins (Figure 10.20d). This process has led to the formation of a thick sedimentary wedge along the eastern coast of the United States.

One intriguing puzzle is the South African Plateau, where a craton has been uplifted during the Cenozoic to heights of almost 2 km above sea level—more than twice the elevation of most cratons. However, the lithosphere in this part of the continent does not appear to be unusually hot. One explanation might be that forces originating deep within the mantle may be responsible for the uplift of the southern African craton and other broad upwarps. A hot and buoyant solid plume rising a few centimeters per year from deep within the mantle (see Chapter 2) applies upward forces at the base of the lithosphere, sufficient to raise the surface by about a kilometer (Figure 10.20e).

None of the proposed mechanisms for epeirogeny explains a central feature of continental cratons: the existence of raised continental shields and subsided platforms. These regions are too vast and have persisted too long to be explained by the plate tectonic processes we have discussed so far.

THE FORMATION OF CRATONS

From the map in Figure 10.8b, you can see that every continental craton contains regions of ancient crust that were last deformed during the Archean eon (4.0 to 2.5 billion years ago). This crust formed the stable landmasses around which new crust was accreted and deformed during the subsequent Wilson cycles of plate tectonics. But how were these central parts of the cratons created in the first place?

Earth was a hotter planet 4 billion years ago, owing to the heat generated by the decay of radioactive elements,

which were more abundant, as well as the energy released by differentiation and during the Heavy Bombardment period (see Chapter 9). Evidence for a hotter mantle comes from a peculiar type of ultramafic volcanic rock found only in Archean crust, called *komatiite* (named after the Komati River in southeastern Africa, where it was first discovered). Komatiites contain a very high percentage (up to 33 percent) of magnesium oxide, and their formation involved a much higher degree of mantle melting than found anywhere on Earth today.

Mantle convection during the Archean was certainly very active, but we don't know to what extent the motions of rocks on Earth's surface resembled modern plate tectonics. The plates may have been smaller and might have moved more rapidly. Volcanism was widespread, and the crust formed at spreading centers was probably thicker. Although lithosphere must surely have been recycled back into the mantle, some geologists believe that the plates formed at this time were too thin and light to be subducted in the same way that oceanic plates are consumed in modern subduction zones.

We do know that a silica-rich continental crust existed at this early stage in Earth's history. The Acasta gneiss, in the northwestern part of the Canadian Shield, has been dated at 4.0 billion years, and these rocks look very similar to modern gneisses (**Figure 10.22**). Formations as old as 3.8 billion

Figure 10.22 The Acasta gneiss in the Archean Slave province of northwestern Canada, dated at 4.0 billion years, is the oldest rock formation so far discovered. It demonstrates that continental crust existed on Earth's surface at the beginning of the Archean eon. [Courtesy of Sam Bowring, Massachusetts Institute of Technology.]

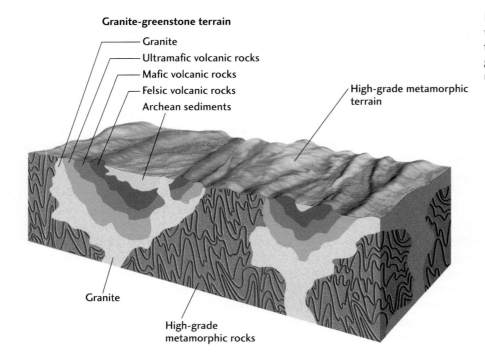

Granite-greenstone terrain
— Granite
— Ultramafic volcanic rocks
— Mafic volcanic rocks
— Felsic volcanic rocks
— Archean sediments

High-grade metamorphic terrain

Granite

High-grade metamorphic rocks

Figure 10.23 Schematic diagram of the two major types of rock formations found in the Archean cratons: granite-greenstone terrains and high-grade metamorphic terrains.

years have been found on many continents; most are metamorphic rocks evidently derived from even older continental crust. In Australia, single grains of zircon (a very hard mineral that survives erosion) have been dated at 4.1 to 4.4 billion years. The composition of oxygen isotopes from these grains suggests that liquid water may have been present on Earth's surface.

In the early part of the Archean eon, the continental crust that had differentiated from the mantle was very mobile, perhaps organized into small rafts being rapidly pushed together and torn apart by the intense tectonic activity. The first cratons with long-term stability began to form around 3.3 to 3.0 billion years ago. In North America, the oldest surviving example is the central Slave craton in northwestern Canada, where the Acasta gneiss is found, which stabilized around 3 billion years ago. The rock assemblages in these Archean cratons fall into two major groups (**Figure 10.23**):

1. **Granite-greenstone terrains,** areas of massive granite intrusions that surround smaller pockets of deformed, metamorphosed volcanic rocks (greenstones), primarily of mafic composition, capped with sediments. The origin of the greenstones is controversial, but many geologists think they were pieces of oceanic crust formed at small spreading centers behind island arcs, incorporated into the continents during compressional episodes, and later engulfed by the granite intrusions.

2. **High-grade metamorphic terrains,** areas of high-grade (granulite facies) metamorphic rocks derived primarily from the compression, burial, and subsequent erosion of granitic crust.

The high-grade metamorphic terrains found in the Archean cratons look similar to the deeply eroded parts of

modern orogenic belts, but the geometry of the deformation was different. Modern orogenies typically produce linear mountain belts where the edges of large cratons converge. In the Archean, the style of deformation was less lineated (that is, more circular or sinusoidal in planform), reflecting the fact that the cratons were much smaller with boundaries that were more curved.

By the end of the Archean, 2.5 billion years ago, enough continental crust had been stabilized in the cratons to allow the formation of larger and larger continents. Plate tectonics was probably operating much as it does today. It is at about this time that we see the first evidence of major continent-continent collisions and the assembly of supercontinents. From this point onward in Earth's history, the evolution of the continents was governed by the plate tectonic processes of the Wilson cycle.

THE DEEP STRUCTURE OF CONTINENTS

In this chapter, we have surveyed the most important processes in the development of the continental crust. However, we have not yet explained one very basic aspect of continental behavior: the long-term stability of the cratons. How have the cratons survived being knocked around by plate tectonics for billions of years? The answer to this question lies not in the crust but in the subcontinental mantle.

Cratonic Keels

Using earthquake waves to "see" into Earth's interior (a technique described in Chapter 14), we have discovered a remarkable fact: the continental cratons are underlain by a thick layer of strong mantle that moves with the plates during continental drift. These thickened sections of lithosphere extend to depths of more than 200 km—more than twice the thickness of the oldest oceanic plates.

At 100 to 200 km beneath the oceans (as well as beneath most younger regions of the continents), the mantle rocks are hot and weak. They are part of the convecting asthenosphere that flows relatively easily, allowing the plates to slide across Earth's surface. The lithosphere beneath cratons extends into this region like the hull of a boat into water, so we refer to these mantle structures as **cratonic keels** (**Figure 10.24**). All cratons on every continent appear to have such keels.

Deep continental keels present many puzzles that scientists are still trying to solve. Less heat is coming from the mantle beneath the cratons than from the mantle beneath the oceanic crust. This indicates that the keels are strong because they are several hundred degrees colder than the surrounding asthenosphere. If the rocks in the mantle beneath the cratons are so cold, however, why don't they sink into the mantle under their own weight, just as cold, heavy slabs of oceanic lithosphere sink in subduction zones?

Composition of the Keels

The cratonic keels would indeed sink if their chemical compositions were the same as those of ordinary mantle peridotites. To get around this problem, geologists have hypothesized that the cratonic keels are made of rocks with different, less dense compositions. The lower density of the rocks that make up the keels counteracts the increased density resulting from colder temperatures. Evidence in support of this hypothesis has come from mantle samples found in kimberlite pipes—the same types of volcanic deposits that produce diamonds.

Kimberlite pipes are the eroded necks of volcanoes that have erupted explosively from tremendous depths. Almost all kimberlites that contain diamonds are located within the Archean cratons. A diamond will revert to graphite at depths shallower than 150 km unless its temperature drops quickly. Therefore, the presence of diamonds in these pipes means that the kimberlite magmas come from deeper than 150 km and that they erupted through the keels when magma fractured the lithosphere very rapidly.

Key Figure 10.24 Rock composition counterbalances temperature effects to form cratonic keels.

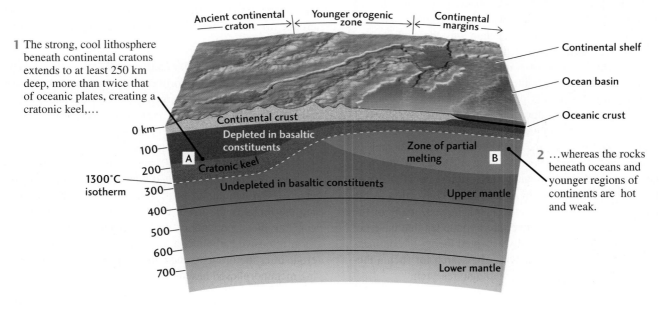

1 The strong, cool lithosphere beneath continental cratons extends to at least 250 km deep, more than twice that of oceanic plates, creating a cratonic keel,...

2 ...whereas the rocks beneath oceans and younger regions of continents are hot and weak.

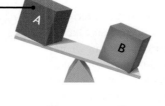

3 The rocks of continental keels are peridotites that are depleted of heavier constituents, such as the element iron and the mineral garnet, so they are lighter under standard conditions than normal continental mantle.

Density at standard (surface) conditions

4 However, at 150-km depth, the continental rocks are colder than the oceanic rocks so that their densities are approximately equal.

Density at mantle conditions

[After T. H. Jordan, "The Deep Structure of Continents," *Scientific American* (January 1979): 92.]

During a violent kimberlite eruption, fragments of the keel, some containing diamonds, are ripped off and brought to the surface in the magma as mantle *xenoliths* (Greek for "foreign rocks"). The majority of the mantle xenoliths turn out to be peridotites with less iron (a heavy element) and less garnet (a heavy mineral) than ordinary mantle rocks. Such rocks can be produced by extracting a basaltic (or komatiitic) magma from the convecting asthenosphere. In other words, the mantle beneath the cratons is the depleted residue left over from melting sometime earlier in Earth's history. A cratonic keel made of these depleted rocks can still float atop the mantle, despite the fact that the keel is cold (see Figure 10.24).

Age of the Keels

By analyzing the xenoliths from kimberlites and the diamonds they contain, we have learned that the cratonic keels are about the same age as the Archean crust above them. (The diamond in your ring or necklace is likely to be several billion years old!) Therefore, the rocks now in the cratonic keels were depleted by the extraction of a basaltic melt very early in Earth's history, and they were subsequently positioned beneath the continental crust about the time the Archean cratons were stabilized.

In fact, keel formation was probably responsible for the tectonic stabilization of the cratons. The existence of a cold, mechanically strong keel can explain why the Archean cratons have managed to survive through many continental collisions, including at least four episodes of supercontinent formation, without much internal deformation.

Many aspects of this process are still not understood. How did the keels cool down? How did they achieve the balance between temperature and composition illustrated in Figure 10.24? Why are the cratons with the thickest keels of Archean age? Some scientists believe the continents may play a major role in the mantle convection system that drives plate tectonics, but how the keels affect convection in the mantle is not completely understood. Indeed, many of the ideas presented in this chapter are hypotheses that have not yet been established as a fully accepted theory of continental evolution and deep structure. The search for such a theory remains a central focus of geologic research.

SUMMARY

What are the major geologic features of North America? The most ancient crust is exposed in the Canadian Shield. South of the Canadian Shield is the interior platform, where Precambrian basement rocks are covered by layers of Paleozoic sedimentary rocks. These older regions form the continental craton. Around the edges of the craton are the elongated mountain chains of the younger orogenic belts. The main orogenic belts are the Cordillera, which runs down the western edge of North America, and the Appalachian fold belt, which trends southwest to northeast on its eastern margin. The coastal plains and shelf of the Atlantic Ocean and Gulf of Mexico are parts of a passive continental margin that subsided after rifting during the breakup of Pangaea.

How do continents grow? Buoyant silica-rich rocks are produced by differentiation of magmas, primarily in subduction zones. Plate motions accrete this material to continental margins through four main processes: the transfer of buoyant crustal fragments from a subducting plate to an overriding continental plate; the closure of marginal basins, which adds thickened island-arc crust to the continent; the transport of crust laterally along continental margins by strike-slip faulting; and the suturing of two continental margins by continent-continent collision and the subsequent breakup of this zone by continental rifting.

What are epeirogeny and orogeny? Forces within the lithosphere can deform large regions of the continents. Some regional movements are simple up-and-down displacements without severe deformation of the rock formations (epeirogeny); examples are the Colorado Plateau and the postglacial uplift of Scandinavia and central Canada. In other cases, horizontal forces arising mainly from plate convergence can produce mountains by extensive and complex folding and faulting (orogeny), as in the Cordillera and Appalachians of North America, the Alps of Europe, and the Himalaya of Asia. The Rockies were eroded to low relief, then rejuvenated by recent broad regional uplift.

How does orogeny modify continents? Orogeny caused by plate convergence can deform continental crust hundreds of kilometers from the convergence zone. Low-angle thrust faulting can stack the upper part of the crust into multiple thrust sheets tens of kilometers thick, deforming and metamorphosing the rocks they contain. Wedges of continental-shelf sediments can be detached from the basement on which they were deposited and then be thrust inland. Compression throughout the crust can double its thickness, causing the rocks in the lower crust to melt. This melting can generate huge amounts of granitic magma, which rises to form extensive batholiths in the upper crust. The mountains erode after the orogeny ends, thinning the crust and exposing metamorphosed basement rocks.

What is the Wilson cycle? The Wilson cycle describes four main phases in the opening and closing of an Atlantic-type ocean basin: rifting during the breakup of a supercontinent; passive-margin cooling and sediment accumulation during seafloor spreading and ocean opening; active-margin volcanism and terrane accretion during subduction and ocean closure; and orogeny during continent-continent collision.

How have the Archean cratons survived billions of years of plate tectonics? The Archean cratons are underlain by a layer of cold, strong mantle more than 200 km thick that moves with the plates during continental drift. These keels appear to be as old as the cratons themselves. They are formed from mantle peridotites that have been depleted by the extraction of mafic magmas, which lowers their density and stabilizes the keel against disruption by mantle convection and plate tectonic processes.

KEY TERMS AND CONCEPTS

accreted terrane (p. 224)

accretion (p. 224)

craton (p. 221)

cratonic keel (p. 237)

epeirogeny (p. 234)

granite-greenstone terrains (p. 236)

high-grade metamorphic terrains (p. 236)

orogen (p. 222)

orogeny (p. 226)

rejuvenation (p. 220)

shield (p. 216)

tectonic age (p. 223)

Wilson cycle (p. 233)

EXERCISES

1. Draw a rough topographic profile of the United States from San Francisco to Washington, D.C., and label the major tectonic features.

2. Why is the topography of the North American Cordillera higher than that of the Appalachian Mountains? How long ago were the Appalachians at their highest elevation?

3. Describe the tectonic province in which you live.

4. Are the interiors of continents usually younger or older than their margins? Explain your answer using the concept of the Wilson cycle.

5. Four basic processes of continental accretion are described in Figure 10.12. Illustrate two of them with examples of accreted terranes in North America.

6. Two continents collide, thickening the crust from 35 km to 70 km and forming a high, Tibet-like plateau. After hundreds of millions of years, the plateau is eroded down to sea level. (a) What kinds of rocks might be exposed at the surface by this erosion? (b) Estimate the crustal thickness after the erosion has occurred. (c) Where in North America has this sequence of events been recorded in surface geology?

7. How many times have the continents been joined in a supercontinent since the end of the Archean eon? Use this number to estimate the typical duration of a Wilson cycle and how fast plate tectonics moves continents.

8. How was orogenesis in the Archean eon different from orogenesis during the Proterozoic and Phanerozoic eons? What factors might explain these differences?

THOUGHT QUESTIONS

1. How would you recognize an accreted terrane? How could you tell if it originated far away or nearby?

2. How would you identify a region where active orogeny is taking place today? Give an example.

3. Would you prefer to live on a planet with orogenies or without them? Why?

4. Figure 10.8b shows more continental crust of Mesozoic-Cenozoic tectonic age than continental crust of any other geologic period. Does this observation contradict the hypothesis that most of the continental crust was differentiated from the mantle in the first half of Earth's history?

5. Why are the ocean basins just about the right size to contain all the water on Earth's surface?

6. What would happen at the surface if the cold keel beneath a craton suddenly heated up? How might this effect be related to the formation of the Colorado Plateau?

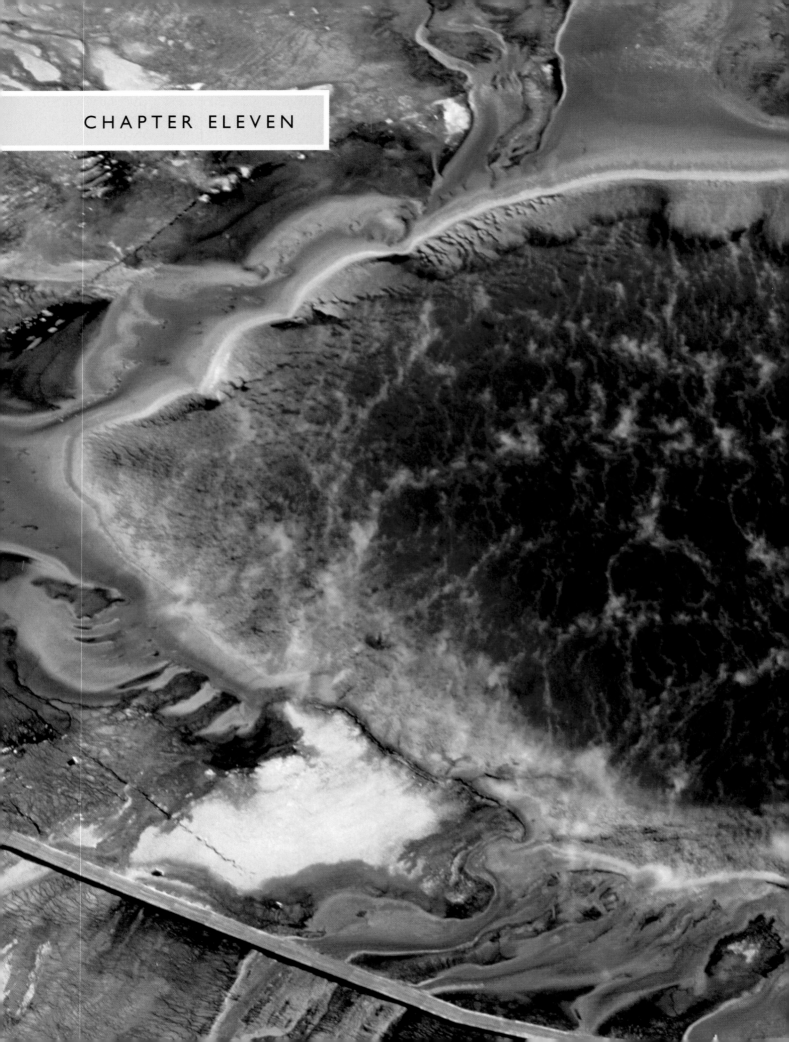

CHAPTER ELEVEN

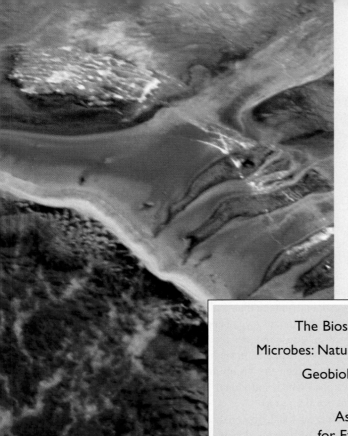

GEOBIOLOGY
Life Interacts with the Earth

Geology is the study of the physical and chemical processes that control the Earth, today and in the past. Biology is the study of life and living organisms, including their structure, function, growth, and origin. As separate as geology and biology may seem, significant interactions occur among rocks, organisms, and Earth's environment. We have long recognized that biology and geology are intimately related, but until recently, we have not known exactly how. Fortunately, technological advances in both the Earth and life sciences now allow us to ask and answer questions that were previously beyond our scope. Over the past decade, scientists working at the frontier have begun to understand how several important *geobiological* processes work.

Geobiology can be defined as the study of how organisms have influenced and been influenced by Earth's environment. Organisms can change the Earth. For example, organisms contribute to weathering of rocks by releasing gases that help break down minerals. Through this process, they obtain nutrients essential to their growth. Our atmosphere is distinct from that of every other planet in having a significant amount of oxygen—the result of the evolution of oxygen-producing microorganisms billions of years ago. Similarly, geologic processes can change life, as when an asteroid struck the Earth 65 million years ago, causing a mass extinction that killed off all the dinosaurs.

The goal of this chapter is to explore the links between organisms and Earth's environment. We'll describe how the biosphere works as a system and what gives Earth its ability to support life. Next, we'll explore the remarkable ways in which tiny microorganisms play an important role in Earth processes, including mineral and rock formation (and destruction). We'll also discuss geobiological interactions over Earth's history and some of the

Grand Prismatic Hot Spring, Yellowstone National Park, Wyoming. The striking array of colors reflects different communities of microorganisms that are very sensitive to water temperature. Water flowing away from the center of the spring (blue) cools down, causing a given community of microorganisms to be replaced by a different community that grows best at the new, lower temperature. The boardwalk visible in the lower part of the photo allows tourists to peer into its depths and provides a sense of scale. [Douglas Faulkner/Photo Researchers.]

major events that changed our planet. Finally, we will consider the key ingredients for sustaining life and ponder the eternal question posed by *astrobiologists:* Is there life out there?

THE BIOSPHERE AS A SYSTEM

Life is everywhere on Earth. The Earth's **biosphere** is the part of our planet that contains all of its living organisms. It includes all the plants and animals as well as the nearly invisible microorganisms that live in some of the most extreme environments on Earth. These organisms live on Earth's surface, in its atmosphere and oceans, and within the upper crust. Because the biosphere intersects with the lithosphere, hydrosphere, and atmosphere, it can influence or even control basic geologic and climatic processes. The biosphere maintains Earth's oxygen-rich atmosphere and produces and consumes carbon dioxide, which helps to control Earth's global temperature. Green plants help control the flow of water on Earth, which modifies and moderates local and regional climate.

The biosphere works as a system because it has inputs and important processes that control outputs. Inputs include energy, such as sunlight, and matter, such as carbon, nutrients, and water. Organisms use these inputs to function and grow. In the process, they create an amazing variety of outputs, some of which can be very significant geologically. At a local scale, perhaps that of a pore within loose sediment, a small group of organisms may have a geological effect that is limited to a particular sedimentary environment. At larger scales, the activities of organisms may help influence the composition of gases in the atmosphere or the flow of certain elements through Earth's crust.

Ecosystems

Think of a class project in which each member of a team might have special skills that allow the team as a whole to exceed the capabilities of individuals by themselves. Groups of organisms act in similar ways: individual organisms play particular roles that allow the group as a whole to achieve some special advantage that is also useful to the individuals. In the case of humans, we accomplish this group behavior as a result of conscious decisions. For all other organisms, group behavior happens through trial and error and involves feedbacks between the group and individuals. These feedbacks expose which individual behaviors result in the best outcome for the group.

Whether at local, regional, or global scales, the group activities of organisms define organizational units known as **ecosystems.** Ecosystems are composed of organisms and geologic components that function in a balanced, interrelated fashion. Different ecosystems occur at many different scales (**Figure 11.1**). They are often separated by geologic barriers such as mountains, deserts, or oceans at the largest scale, or by different water temperatures within a single hot spring at a much smaller scale.

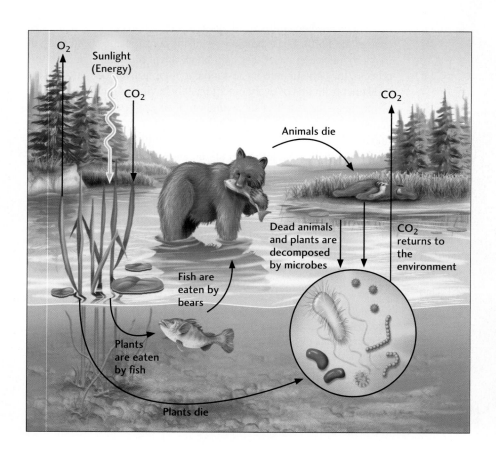

Figure 11.1 Organisms interact with one another and the Earth in community groups called ecosystems. Ecosystems can be small or large, but there is always a flow of energy and matter through them. Sunlight may be used as an energy source by plants, which are then eaten by fish, which are then eaten by bears. The plants, fish, and bears will die and be decomposed by microbes. Their matter will then be returned to the environment, where it can be used again. Thus, energy and matter cycle between organisms and their environment.

(a)

(b)

(c)

Figure 11.2 When humans or natural events interfere with ecosystems, dramatic imbalances can occur. These organisms all dominate their environments. (a) Kudzu, introduced to stop highway erosion. [Kerry Britton/Forest Service, USDA.] (b) Purple loosestrife, introduced from Europe as a garden flower. [Reimar Gaertner/INSADCO Photography/Alamy.] (c) Zebra mussels aggressively colonize and overwhelm ordinary mussels. [Courtesy of U.S. Fish and Wildlife Service/Washington, DC Library.]

A typical ecosystem might involve, say, a river and its adjacent environment, where different groups of organisms are adapted to live in the water (fish), in the sediment (worms, snails), on the banks (grass, trees, muskrats), and in the sky above it (birds, insects). In one sense, the river controls where the organisms live by supplying the ecosystem with water, sediment, and dissolved mineral nutrients. Conversely, the organisms influence how the river behaves; for example, grass and trees stabilize the river banks against the destructive effects of floods. The balance between such biologically controlled and geologically controlled processes ensures the long-term stability of the ecosystem.

Earth's history shows us that ecosystems respond sensitively to geologic processes or the introduction of new groups of organisms. When severe imbalances in ecosystems occur, responses are often dramatic. Consider the effects of introducing a new organism into your neighborhood environment. People sometimes bring in plants to control hillslope erosion along highways, a new sport fish for their favorite fishing spots, or a pretty new plant for their gardens. In all too many cases, the new organism is better suited for its new environment than the current inhabitants. In cases of successful invasions, the new organism often comes from a place where the physical environment is similar but where biological competition is more intense, and so it is more likely to win the battle for nutrients and space in its new home, squeezing out the previous inhabitants. In some cases, the organism that is squeezed out may become *extinct* if it is outcompeted by the invader in all the regions it formerly occupied.

Kudzu, Eurasian water milfoil, purple loosestrife, zebra mussels, and the snakehead fish create problems in different parts of the United States and Canada by completely dominating their local ecosystems (**Figure 11.2**). Extinction of overwhelmed organisms is not the only concern. For example, in 2002, the U.S. Fish and Wildlife Service estimated that $5 billion was spent by electrical utility companies just to unclog water intake pipes where zebra mussels grow. Over the course of Earth's history competition from new types of organisms, impacts by meteorites, superabundant volcanism, rapid global warming (or freezing), and loss of oxygen in the oceans are just a few of the processes that contributed to the demise or extinction of major groups of organisms. We will explore some of these effects later in this chapter.

Inputs: The Stuff Life Is Made Of

The organisms of any ecosystem can be subdivided into producers or consumers according to the way they obtain their food, which is their source of energy (**Figure 11.3** and Table 11.1). Producers, or **autotrophs,** are organisms that make their own food. They manufacture organic compounds such as carbohydrates that they use as sources of energy and nutrients. Consumers, or **heterotrophs,** get their food by feeding directly or indirectly on producers.

It is often said that you are what you eat. This is true not only for humans but for all organisms. Our foods are all made up of more or less the same materials: molecules composed of carbon, hydrogen, oxygen, nitrogen, phosphorus, and sulfur. So it doesn't matter whether the organism is an autotroph or a heterotroph, it still ingests mostly the same

Figure 11.3 Autotrophs are organisms, such as plants, that make their own food as a source of energy. Heterotrophs are organisms, such as koalas, that get their energy by feeding on autotrophs. [*left:* Frans Lanting/Minden Pictures; *right:* Juniors Bildarchiv/Alamy.]

six elements. What differs is what form (that is, molecular structure) the food comes in. When heterotrophs such as humans eat bread, we are feeding ourselves on *carbohydrates*—large molecules formed of carbon, hydrogen, and oxygen. Even the lowliest microorganisms dine on carbon-bearing molecules such as carbon dioxide (CO_2) or methane (CH_4). The difference is that what is food for a microbe may not seem like food to us!

Carbon   The fundamental building block of all life on Earth is carbon. Where life is involved, carbon must be present. Carbon acts as the template around which all organic molecules, such as carbohydrates and *proteins,* are built. These molecules provide the organism with energy and allow it to grow. Part of the reason that carbon is so important is that it forms four covalent bonds with itself and other elements (see Chapter 3), which allows for a wide variety of structures. If water is re-

moved, the composition of all organisms, including humans, is dominated by the element carbon. The availability of carbon is critically important to life because it goes into manufacturing everything from genes to body structures.

The biosphere largely controls the flow of carbon through the Earth system. Marine organisms extract carbon—present as carbon dioxide—from seawater to form carbonate shells and other skeletons. When the organisms die, their shells settle to the floor of the ocean where they accumulate as sediment, effectively removing carbon from the system. Over geologically significant intervals of time, the accumulation of the organic remains of organisms in freshwater swamps and on the ocean floor has also removed carbon from the system to become oil, gas, and coal deposits. Today, when we extract and burn these deposits, we are returning carbon to the system in the form of carbon dioxide.

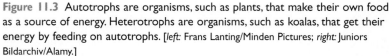

Table 11.1	Organisms as Producers and Consumers		
Type	**Energy Source**	**Carbon Source**	**Example**
Photoautotroph	Sun	CO_2	Cyanobacteria
Photoheterotroph	Sun	Organic compounds	Purple bacteria
Chemoautotroph	Chemicals	CO_2	H, S, Fe bacteria
Chemoheterotroph	Chemicals	Organic compounds	Most bacteria, fungi, and animals, including humans

Nutrients Nutrients are elements or chemical compounds that contribute to or enhance an organism's metabolism. Common plant nutrients, such as garden fertilizer, contain the elements phosphorus, nitrogen, and potassium. Other organisms also depend on iron and calcium. Some organisms can manufacture their own nutrients, but others must obtain them in their diets from materials in the environment. Microorganisms have a special ability to derive nutrients from the dissolution of minerals.

Water Life as we know it requires water (H_2O). We humans, and all life on Earth, are aqueous beings. Water is important because organisms are composed primarily of water, typically 50 to 95 percent. No organism can live without water, and it is well known that organisms such as humans can live for weeks without food, but most will perish in a few days without water. Even microorganisms that live in the atmosphere obtain water from tiny droplets that condense around dust particles, and viruses must obtain water from their hosts.

Water's chemical properties and the way it responds to changes in temperature make it an ideal medium for biological activity. The cells of all organisms are made mostly of a water solution that promotes the chemical reactions that enable an organism to grow and reproduce. Water is the habitat in which life first emerged and in which much of it still thrives. Water helps moderate Earth's climate, which has supported life for at least 3.5 billion years.

Water is such an important ingredient for life that the search for extraterrestrial life must first begin with the search for water. The successful Mars Exploration Rovers mission of 2004–2005 was designed to search for evidence of water on Mars. The two rovers, *Spirit* and *Opportunity,* achieved this goal, and a follow-up rover mission is now planned to search for evidence of life itself, with a launch date perhaps as soon as 2009. Had the rovers failed to detect any evidence of water on the surface of Mars, any future plans to search for evidence of life on Mars would have been postponed or possibly even canceled.

Energy All organisms need energy to live and grow. Some of the simplest organisms, such as diatoms and algae, obtain energy from sunlight. Others acquire energy from the breakdown of minerals into dissolved chemicals. Some organisms get energy by feeding on other organisms—including plants, animals, and microorganisms. Energy is important because it allows simple molecules such as carbon dioxide and water to be combined into larger molecules such as carbohydrates and protein, which are essential for life.

Processes and Outputs: How Organisms Live and Grow

Metabolism is the process that all organisms use to convert inputs to outputs. During some types of metabolism, organisms take in small molecules such as CO_2, H_2O, and CH_4 and use energy to create larger molecules, such as certain types of carbohydrates, that enable the organism to function and grow (**Figure 11.4**). Other carbohydrates—for example,

Figure 11.4 During the metabolic process of photosynthesis, organisms use carbon dioxide and water from the environment and the energy of sunlight to make carbohydrates such as glucose.

Table 11.2	Comparison of Photosynthesis and Respiration

Photosynthesis	Respiration
Stores energy as sugar	Releases energy from sugar
Uses CO_2 and H_2O	Releases CO_2 and H_2O
Increases weight	Decreases weight
Produces oxygen	Consumes oxygen

a particular type of sugar called glucose—are later used as an energy source, that is, food.

One particularly well-known metabolic process is **photosynthesis.** In photosynthesis, organisms such as green plants and algae use energy from sunlight to convert water and carbon dioxide to carbohydrates, such as sugar, and oxygen (Table 11.2). This reaction proceeds as follows:

water + carbon dioxide + sunlight →
$6H_2O$ + $6CO_2$ + energy →

glucose (sugar) + oxygen
$C_6H_{12}O_6$ + $6O_2$

Oxygen is released to the atmosphere, and the sugar is stored as an energy source for future use by the organism. An important group of microbes, known as the *cyanobacteria,* also use photosynthesis to create carbohydrates.

Another important metabolic process is **respiration,** which allows the organism to release the energy stored in carbohydrates (see Table 11.2). All organisms use oxygen to burn, or respire, sugars to create energy, but different organisms respire in different ways. For example, humans, and many other organisms, consume oxygen gas (O_2) present in the atmosphere to metabolize carbohydrates and release CO_2 and H_2O. In this case, the reaction is the reverse of photosynthesis:

glucose (sugar) + oxygen →
$C_6H_{12}O_6$ + $6O_2$ →

water + carbon dioxide + energy
$6H_2O$ + $6CO_2$ + energy

(for the organism to do other things)

But other organisms, such as microbes that live in environments where oxygen is absent, have a more difficult task. They must break down oxygen-containing compounds dissolved in water, such as sulfate (SO_4^{-2}), to obtain oxygen. During the course of these reactions, different gases—such as hydrogen (H_2), hydrogen sulfide (H_2S), and methane (CH_4)—may be produced.

When organisms produce oxygen, it is released to the atmosphere and causes iron-bearing silicate minerals such as pyroxene and amphibole to oxidize during weathering and form iron-bearing oxide minerals such as hematite (see Chapter 16). When organisms produce CO_2 and CH_4, which are both heat-trapping gases, they escape to the atmosphere and contribute to global warming. Conversely, when organisms consume these gases, they contribute to global cooling.

Biogeochemical Cycles

In the course of living and dying, organisms continuously exchange energy and matter with their environment. This exchange occurs at the scale of the individual organism, the ecosystem of which it is a part, and the global biosphere. The metabolic control of gases such as CO_2 and CH_4 is a good example of how organisms may exert a global control on Earth's climate. CO_2 and CH_4 are **greenhouse gases**—gases that absorb heat emitted by the Earth and trap it in the atmosphere. When organisms produce more CO_2 and CH_4 than they consume, the climate will warm; when they consume more CO_2 and CH_4 than they produce, the climate will cool. This is not the only control on global climate, as we will learn in Chapter 15, but it is an important one that directly involves the biosphere.

Geobiologists keep track of these exchanges by studying **biogeochemical cycles.** A biogeochemical cycle is a pathway through which a chemical element or molecule moves between the biologic ("bio") and environmental ("geo") components of an ecosystem.

Because ecosystems vary in scale, so do biogeochemical cycles. For example, phosphorus, which is an essential nutrient, might cycle back and forth between the water and the microorganisms in the pores of sediments, or it might cycle back and forth between uplifted rocks and sediments deposited along the margins of ocean basins (**Figure 11.5**). In either case, when the organisms die, the phosphorus—before being recycled—may accumulate for a variable period of time in a temporary place. Sediments and sedimentary rocks are an important repository for this element.

Biogeochemical cycles are very important for understanding the mechanisms associated with major evolutionary events throughout Earth's history, as we will see later in this chapter. They are also critical to understanding how elements and compounds that humans emit to the atmosphere and oceans will interact with the biosphere, to be discussed in Chapters 15 and 23.

MICROBES: NATURE'S TINY CHEMISTS

Single-celled organisms including bacteria, some fungi and algae, and protozoa are known as **microbes.** Where there is water, there are microbes. Microbes, like other organisms, need water to live and reproduce. Microbes are as small as

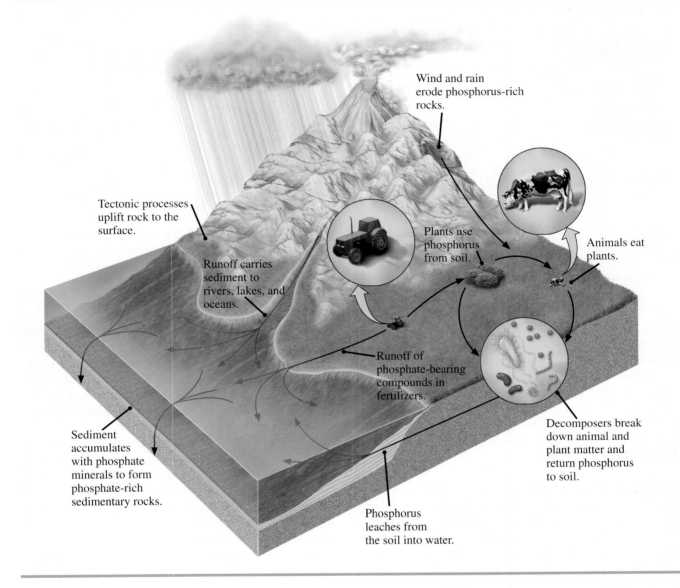

Sedimentation of phosphate minerals is a key component in the phosphorus cycle.

Wind and rain erode phosphorus-rich rocks.

Tectonic processes uplift rock to the surface.

Plants use phosphorus from soil.

Animals eat plants.

Runoff carries sediment to rivers, lakes, and oceans.

Runoff of phosphate-bearing compounds in fertilizers.

Decomposers break down animal and plant matter and return phosphorus to soil.

Sediment accumulates with phosphate minerals to form phosphate-rich sedimentary rocks.

Phosphorus leaches from the soil into water.

1/1000 of a millimeter and can inhabit almost every nook and cranny you can think of, from at least 5 km beneath Earth's surface to more than 10 km high in the atmosphere. They live in the air, in soil, on and in rocks, inside roots, in piles of toxic waste, on frozen snowfields, and in water bodies of every type, including boiling hot springs. They live at temperatures that range from lower than –20°C to higher than the boiling point of water (100°C).

People have exploited the useful effects of microbial metabolism for thousands of years to produce bread, wine, and cheese. Today, people also use microbes to help create antibiotics and other valuable drugs. Geobiologists study microbes to understand their important roles in dissolving and precipitating minerals and to understand the early evo-

lution of Earth's biosphere before the advent of higher organisms.

Abundance and Diversity of Microbes

Microbes dominate the Earth in terms of the number of individuals that live on it. Concentrations ranging from 10^3 to 10^9 microbes/cm^3 have been reported from soils, sediments, and natural waters. Every time you walk on the ground, you step on billions of microbes! More important, microbes are the most genetically diverse group of organisms on Earth. Under a given set of environmental conditions, **genes** are large molecules within the cells of every organism that

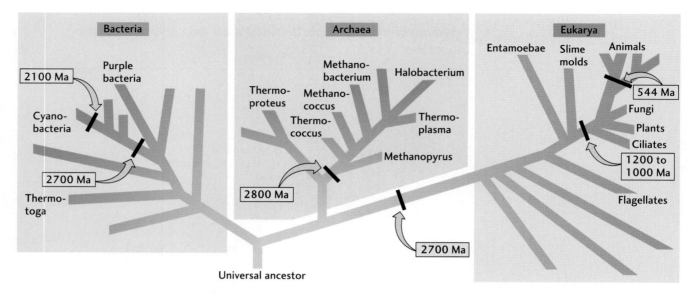

Figure 11.6 The universal tree of life shows how all organisms relate to one another. Three great domains subdivide all organisms: the Bacteria, Archaea, and Eukarya. These domains are united by a universal common ancestor. All domains are dominated by microorganisms. Each branch point represents a significant evolutionary episode during which a common ancestor gave rise to two descendants. Note that animals, including humans, appear at the end of the Eukarya domain.

encode all of the information within that organism—be it plant, animal, or microbe—that determines what it will look like, how it will live and reproduce, and how it differs from all other organisms. Diversity is important because it allows microbes to colonize and thrive in environments that would be lethal to other organisms. Diversity is also important because it allows microbial ecosystems to cycle important materials in a broad—even extreme—range of geological environments.

The Universal Tree of Life Biologists compare the genes of different organisms within a particular group to establish the hierarchy of ancestors and descendants and construct a family tree, called the **universal tree of life (Figure 11.6)**. Genes are the basic hereditary units passed on from generation to generation. About 30 years ago, a startling discovery was made when the first family trees for microbes were constructed. When the genes for *all microbes* were compared, it was shown that despite their similar sizes (tiny) and shapes (simple rods and ellipses), there were enormous differences in their genetic content (see Figure 11.6). Furthermore, when the genes of *all organisms* were compared, including plants and animals, it was revealed that the differences among groups of microbes were much greater than the differences between plants and animals, including humans (see Figure 11.6).

The Three Domains of Life The single root of the universal tree of life shown in Figure 11.6 is called the **universal ancestor.** This universal ancestor led to three major groups, or *domains,* of descendants: the Bacteria, the Archaea, and the Eukarya (see Figure 11.6). The *Bacteria*

and *Archaea* probably evolved first and all are single-celled microorganisms. The **Eukarya** appear to have evolved later and are distinguished by cells with more complicated internal structure, such as the presence of a nucleus. **Eukaryotes** have a more advanced cellular structure that makes it possible to form larger, multicelled organisms, an essential step in the evolution of animals.

For geobiologists, the universal tree of life is a map that reveals how the tiny and featureless microbes relate to one another and interact with the Earth. Names such as halobacterium, thermococcus, and methanopyrus give us some clue that these organisms can live in extreme environments that are very salty (*halo,* "halite"), or hot (*thermo*), or high in methane (*methano*). Microbes that live in extreme environments are almost exclusively Archaea and Bacteria (see Figure 11.6).

Extremophiles: Microbes That Live on the Edge

Extremophiles are microbes that live in environments that would kill any other organisms (Table 11.3). Extremophiles live on all kinds of foods including oil and toxic wastes. Some "breathe" substances other than oxygen, such as nitric acid, sulfuric acid, iron, arsenic, or uranium, which they use to produce energy.

The suffix *-phile* is derived from the Latin word *philus,* which means "to have a strong affinity or preference for." *Acidophiles* are microbes that can live in acidic environments such as the Rio Tinto in Spain (see Feature 3.1 in Chapter 3). The water that flows from old mines where sulfide minerals were mined is usually very acidic because sul-

Table 11.3	Characteristics of Extremophiles		
Type	**Tolerance**	**Environment**	**Example**
Halophile	High salinity	Playa lake Marine evaporite	Great Salt Lake
Acidophile	High acidity	Mine drainage Water near volcano	Rio Tinto (see Chapter 3)
Thermophile	High temperature	Thermal spring Mid-ocean ridge vents	Yellowstone Hot Springs
Anaerobe	No oxygen	Pores of wet sediments Groundwater Microbial mats Mid-ocean ridge vents	Cape Cod Bay sediment

fide reacts with oxygen and water to form sulfuric acid. Acidophiles can tolerate acid levels high enough to kill other organisms. They are able to survive in this acid habitat because they have developed a way to pump out the acid that accumulates inside their cells.

Thermophiles are microbes that live and grow in extremely hot environments that would kill most other microorganisms. They grow best in temperatures that are between 50° and 70°C and can tolerate temperatures up to 120°C. They will not grow if the temperature drops to 20°C. Thermophiles live both in geothermal habitats such as hot springs and mid-ocean ridge vent systems and in environments that create their own heat such as compost piles and garbage landfills. The microbes that cover the bottom of Grand Prismatic Hot Spring (see the chapter opening photo) are dominated by thermophiles. In the universal tree of life shown in Figure 11.6, the Eukarya (includes humans) are generally the least tolerant of high temperatures (60°C

seems to be the upper limit), the Bacteria are more tolerant (upper limit close to 90°C), and the Archaea are the most tolerant, able to withstand temperatures of up to 120°C. The microbes that can stand temperatures above 80°C are called *hyperthermophiles.*

Halophiles are microbes that live and grow in highly saline environments. The salt content in halophilic environments is usually 10 times the salt content of normal ocean water. Halophiles live in playa lakes such as the Great Salt Lake and the Dead Sea and in some parts of the ocean such as the southern end of San Francisco Bay where seawater is commercially evaporated to form salt. The pink color shown in the evaporation pools of **Figure 11.7** is due to a pigment in halophilic bacteria. These microbes keep the salt concentration inside their cells normal because they are able to expel extra salt from their cells into the environment.

Anaerobes are another group of extremophiles that live in environments completely devoid of oxygen. At the

Figure 11.7 Humans have dammed off parts of the sea to create pools where seawater can evaporate to precipitate halite for table salt and other uses. Halophilic bacteria thrive in these hypersaline environments and produce a distinctive pigment that turns the ponds pink. [Yann Arthus-Bertrand/Corbis.]

Figure 11.8 Microbes can form layered deposits called microbial mats. The top part of the mat is exposed to the Sun, and photosynthetic autotrophic microbes are revealed by the green color. Farther down in the mat are nonphotosynthetic autotrophs, revealed by the purple color. These autotrophs occur within the aerobic zone. Deeper in the mat, the color turns gray, revealing the anaerobic zone where heterotrophs live. [John Grotzinger.]

bottoms of ponds, lakes, rivers, and seas, the pore fluids of sediments just a few millimeters or centimeters below the sediment-water interface are starved of oxygen. Microbes that live right at the air-water interface use up all the oxygen during respiration, creating what is known as an **anaerobic** condition farther down in the sediment. The oxygen-rich upper zone of most sediment layers is known as the **aerobic zone.** Many microorganisms that live in the aerobic zone would die in the anaerobic zone, and vice versa. This boundary is often very sharp, as shown in **Figure 11.8.**

Microbe-Mineral Interactions

Microbes play a critical role in many geologic processes. These include mineral precipitation, mineral dissolution, and the flow of important elements through Earth's crust. As we will learn later in the chapter, microbial processes were also crucial in helping to regulate the evolutionary history of higher organisms.

Mineral Precipitation Microorganisms precipitate minerals in two distinct ways, *indirectly* during growth and *directly* as a result of metabolic activity (**Figure 11.9**). Indirect precipitation occurs when dissolved minerals in an oversaturated solution precipitate on the surface of individual microbes. This happens because the surface of a microbe has sites that bind dissolved mineral-forming elements. Mineral precipitation often leads to the complete encrustation of the microbes, which are effectively buried alive. Microbial precipitation of carbonate minerals and silica in hot springs are good examples of this type of microbial biomineralization. Thermophiles may become completely overgrown by the mineral deposits they help precipitate.

Minerals are directly precipitated by the metabolic activities of microorganisms. For example, iron and manganese

oxides are deposited by microbes that use these substances to generate energy for growth.

Microbial precipitation of pyrite (FeS_2) (**Figure 11.10**) occurs in the anaerobic zone of sediments that contain iron-bearing minerals and water with sulfate (SO_4). Precipitation of pyrite is caused by the biological process of respiration. As we have learned, all organisms—even microbes—need oxygen to create energy by respiring sugars. However, in the anaerobic zone, O_2 is not available, and heterotrophic microbes (see Table 11.1) have adapted to this harsh, but very common, condition by learning how to obtain oxygen from other sources. Heterotrophic microbes can remove the oxygen contained in sulfate (SO_4), which is abundant in most pore fluids, to respire the organic compounds they feed on. In the process, they make hydrogen sulfide gas (H_2S), which has the unpleasant odor of rotten eggs that is released when you dig into sandy or muddy sediments during low tide. In the final step of the process, hydrogen is replaced by iron to create pyrite. Pyrite is remarkably abundant in sedimentary rocks that contain organic matter, such as black shales. Many black shales become buried, and the associated organic matter may be broken down to create oil and gas (see Chapter 23).

Another example of metabolically controlled direct mineral precipitation is the formation of tiny magnetite particles inside microbes (see Figure 11.9), which use these crystals to help them navigate. You may recall from Chapter 1 that one of Earth's three global geosystems is the geodynamo, which creates Earth's magnetic field. Magnetite crystals within microbes and other organisms, such as honeybees, help these organisms find their way by sensing Earth's magnetic field. Thus, geologic phenomena deep within the Earth are capable of influencing the biosphere.

Mineral Dissolution Some elements that are essential for metabolism, such as sulfur and nitrogen, are readily avail-

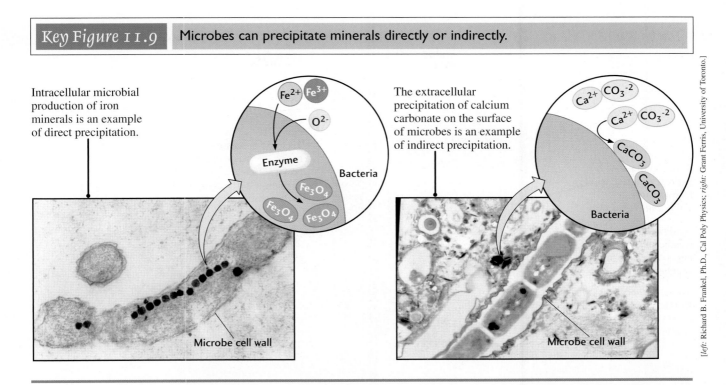

Intracellular microbial production of iron minerals is an example of direct precipitation.

The extracellular precipitation of calcium carbonate on the surface of microbes is an example of indirect precipitation.

Microbe cell wall

Microbe cell wall

able from natural waters; but others, such as iron and phosphorus, must be actively scavenged from minerals by the microbe. All microbes need iron, but iron concentrations in near-surface waters are generally so low that the microbe must obtain iron by dissolving nearby minerals. In a similar way, phosphorus—required for construction of biologically important molecules—is available by dissolution of minerals such as apatite (calcium phosphate). Some autotrophs derive their energy from the chemicals produced when minerals are dissolved. These organisms are known as chemoautotrophs (see Table 11.1). For example Mn^{2+}, Fe^{2+}, S, NH_4^+,

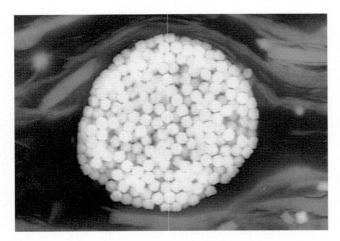

Figure 11.10 Pyrite commonly forms small globules in sediments that have anaerobic pore waters. Microbes use the sulfur in seawater and the iron dissolved from nearby minerals to precipitate pyrite. [Courtesy of Juergen Schieber.]

and H_2 all supply microbes with energy when they are released from minerals.

Microbes can dissolve minerals because the interactions among their cells, organic molecules, and minerals liberate ions from mineral surfaces. Rates of mineral dissolution are normally slow but may be enhanced where minerals containing desirable metals such as Fe and Mn are coated by microbial biofilms. Biofilms have tremendous concentrations, as high as 10^8 microbes/cm^2 of surface area. Acidophiles thrive in sites where mineral dissolution results in prolific acid formation.

Microbes and Global Biogeochemical Cycles Microbial pyrite precipitation is an example of an important global biogeochemical process (**Figure 11.11**). Iron and sulfur precipitate as pyrite, which accumulates abundantly within sediments. As layers of sediment are deposited, the pyrite becomes buried and encapsulated in sedimentary rocks. The pyrite will then stay buried until the rocks are returned to Earth's surface during the uplift phase of the rock cycle. During weathering of the sedimentary rocks, the iron and sulfur are recycled by dissolving as ions in water or become incorporated in new minerals that accumulate in sediments, starting the biogeochemical cycle over again.

On a global scale, microbes play a major role in several other important biogeochemical cycles. Microbial precipitation of phosphate minerals contributes to the flow of phosphorus into sediments, particularly along the western coasts of South America and Africa, where phosphorus-rich deep ocean water rises to the surface and is made available to microbes that live in shallower water. The chemical weathering

Earth System Figure 11.11 — Microbial pyrite precipitation is a key component of the sulfur cycle.

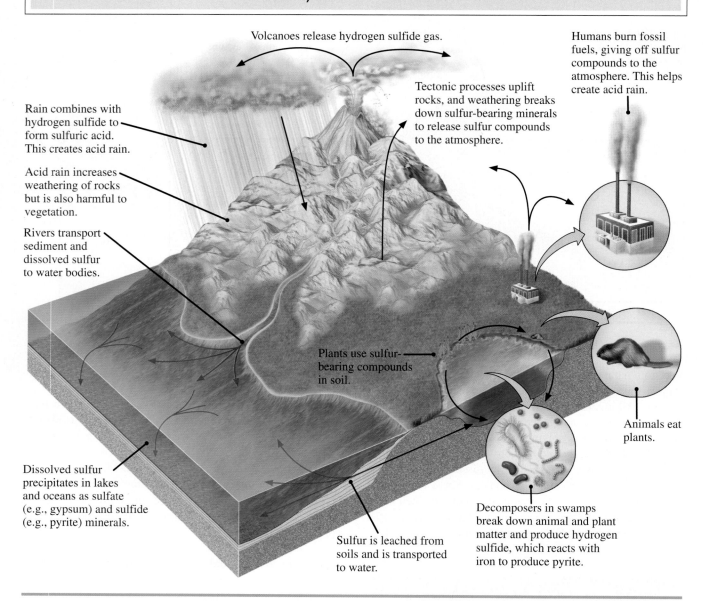

Volcanoes release hydrogen sulfide gas.

Humans burn fossil fuels, giving off sulfur compounds to the atmosphere. This helps create acid rain.

Tectonic processes uplift rocks, and weathering breaks down sulfur-bearing minerals to release sulfur compounds to the atmosphere.

Rain combines with hydrogen sulfide to form sulfuric acid. This creates acid rain.

Acid rain increases weathering of rocks but is also harmful to vegetation.

Rivers transport sediment and dissolved sulfur to water bodies.

Plants use sulfur-bearing compounds in soil.

Animals eat plants.

Dissolved sulfur precipitates in lakes and oceans as sulfate (e.g., gypsum) and sulfide (e.g., pyrite) minerals.

Sulfur is leached from soils and is transported to water.

Decomposers in swamps break down animal and plant matter and produce hydrogen sulfide, which reacts with iron to produce pyrite.

of continental rocks is influenced by microbes that can increase the acidity of soils, leading to faster weathering rates. And finally, as we discussed in Chapter 5, the precipitation of carbonate minerals in marine environments is also stimulated by microbial processes. This last example is especially important because carbonate minerals serve as a sink for atmospheric carbon dioxide and for cations such as Ca^{2+} and Mg^{2+} released during weathering of silicate minerals, such as feldspar, amphibole, pyroxene, and olivine.

Microbial Mats

Microbial mats are layered microbial communities. The microbial mats you are most likely to see are those that are exposed to the Sun (see Figure 11.8). They commonly occur in tidal flats, hypersaline lagoons, and thermal springs. On the top, we usually find a layer of oxygen-producing cyanobacteria that obtain energy through photosynthesis. This uppermost layer is green because it has the same light-absorbing pigment that green plants have. This layer might be as thin as 1 mm, yet it can be as effective in producing energy from the Sun as hardwood forests or grasslands. This uppermost green layer defines the aerobic zone of the mat. The anaerobic zone occurs below the cyanobacterial layer and is often a dark gray color. Although this anaerobic part of the mat contains no oxygen, it still can be very active. Anaerobic microbes are heterotrophic because they derive energy from feeding on the organic matter produced by the cyanobacteria. This activity often results in the formation of sedimentary pyrite, as discussed earlier.

Microbial mats are miniature models of the same biogeochemical cycles that occur at regional or even global scales. In a microbial mat, photosynthetic autotrophs obtain energy from sunlight to convert carbon in atmospheric CO_2 into carbon in larger molecules such as carbohydrates, which the microbe uses to grow. After the microbes die, heterotrophs use the carbon in the dead microbes as an energy source. The heterotrophs then convert some of this carbon to CO_2, which is returned to the atmosphere, where it can be used by the next generation of photosynthetic autotrophs, and so on. In the case of microbes, this cycle is confined to the very small size of a layer of sediment, but it is directly analogous to the process by which rain forests—at a global scale—extract CO_2 during photosynthesis. Although individual trees do the actual work, one can think of a rain forest as a giant photosynthesis machine that removes enormous quantities of CO_2 and produces enormous quantities of carbohydrates. When the trees die, their organic matter is used by heterotrophs on the forest floor to produce energy and returns enormous amounts of carbon—in the familiar form of CO_2—back to the atmosphere.

Stromatolites Microbial mats are restricted to places on Earth where plants and animals can't interfere with their growth. However, before the existence of plants and animals, microbial mats were widespread and are one of the most common features preserved in Precambrian sedimentary rocks formed in marine and lake environments. **Stromatolites**—rocks with distinctive lamination—are believed to have been formed by ancient microbial mats. Stromatolites range in shape from flat sheets to domal structures with complex branching patterns (**Figure 11.12**). They are one of the most ancient types of fossils on Earth and give us a glimpse of a world once ruled by microorganisms.

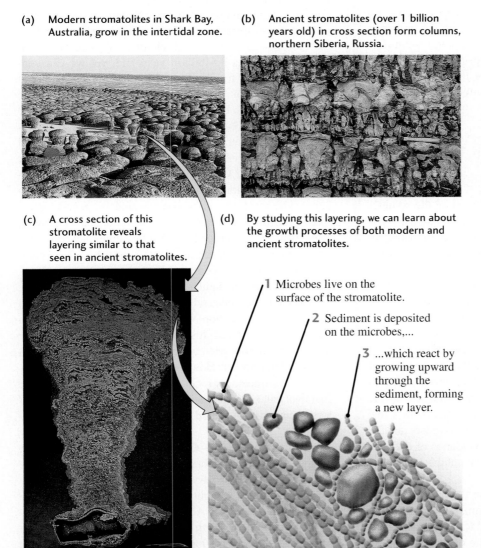

(a) Modern stromatolites in Shark Bay, Australia, grow in the intertidal zone.

(b) Ancient stromatolites (over 1 billion years old) in cross section form columns, northern Siberia, Russia.

(c) A cross section of this stromatolite reveals layering similar to that seen in ancient stromatolites.

(d) By studying this layering, we can learn about the growth processes of both modern and ancient stromatolites.

1 Microbes live on the surface of the stromatolite.

2 Sediment is deposited on the microbes,...

3 ...which react by growing upward through the sediment, forming a new layer.

Figure 11.12 Stromatolites are sedimentary features that result from the interaction of microbes with their environment. [Images from John Grotzinger.]

Most stromatolites probably formed when sediment raining down on microbial mats was trapped and bound by microbes that colonized the surface of the mat (see Figure 11.12). Once covered by sediment, the microbes grew upward between the sediment particles and spread laterally to bind the particles in place. Each stromatolite lamina corresponds to deposition of a sediment layer followed by trapping and binding of that layer.

In some cases, however, stromatolites form by mineral precipitation rather than by trapping and binding. Mineral precipitation is indirectly controlled by microbes or is simply the result of oversaturation of minerals in the surrounding fluids. As discussed in Chapter 5, the ocean is oversaturated with the carbonate minerals calcite and aragonite. These minerals are important for the growth of stromatolites formed by mineral precipitation.

The potential role of microbes in stromatolite formation is important because these laminated, dome-shaped structures are used as evidence for life on the early Earth and have a record back to approximately 3.5 billion years ago (Ga) (see below). But if stromatolites can be built by non-microbial mineral precipitation, their use as evidence for life is uncertain. Only by carefully studying the processes by which microbes interact with minerals and sediments, and the chemical and textural fingerprints of these interactions, will we be able to determine if the formation of stromatolites on the early Earth required the presence of microorganisms.

GEOBIOLOGICAL EVENTS IN EARTH'S HISTORY

The geologic time scale divides time based on the comings and goings of now-extinct fossil groups (see Chapter 8). The patterns themselves provide a convenient ruler for subdividing Earth's history, but the events they represent were almost always associated with global environmental change. In most cases, Earth may have experienced a one-time event, usually related to evolution of some new group of organisms. For instance, the rise of oxygen was related to the rise of photosynthetic microbes during Archean time. The evolution of land plants during late Cambrian time made Earth green.

In most cases, the boundaries of the Phanerozoic part of the geologic time scale are marked by the demise, or *extinction,* of a particular group of organisms, followed by the rise, or *radiation,* of a new group of organisms. In a few cases—for example, the Permian-Triassic boundary and the Cretaceous-Tertiary boundary—the boundaries of the geologic time scale are marked by environmental catastrophes of truly global magnitude.

We will now study a few of these dramatic events in Earth's history—events in which the link between life and environment is clearly visible. **Figure 11.13** shows the great antiquity of life on Earth and the timing of several of these major events.

Origin of Life and the Oldest Fossils

When Earth formed some 4.5 billion years ago, it was lifeless and inhospitable. A billion years later, it was teeming with microbes. How did life begin? Along with other grand puzzles such as the origin of the universe, this question remains one of science's greatest mysteries.

The question of *how* life may have originated is very different from the question of *why* life originated. Science offers an approach only to understanding the "how" part of this mystery because, as you may recall from Chapter 1, we can use both observations and experiments to create testable hypotheses. These hypotheses may explain the series of steps involved in the origin and evolution of life, and they can be tested by searching for evidence in the fossil and geologic records. However, observations and experiments do not provide a testable approach to the question of why life evolved.

The fossil record tells us that microbes originated first and that they evolved into all the multicelled organisms that constitute the younger part of the geologic record. The fossil record also shows us that most of life's history involved the evolution of single-celled microbes. We can find fossil microbes in rocks 3.5 billion years old, yet we can conclusively identify fossils of multicellular organisms only in rocks younger than 1 billion years. Thus, microbes were the only organisms on Earth for at least 2.5 billion years!

The theory of evolution predicts that these first microbes—and all that came after—evolved from life's universal ancestor (see Figure 11.6). What did this universal ancestor actually look like? We really don't know, but most geobiologists agree that it must have had several important characteristics. The most crucial of these would be genetic information: instructions for growth and reproduction. Otherwise, there would be no descendants. Another important characteristic of the universal ancestor is that it must have been composed of carbon-rich compounds. As we have seen, all organic substances, including life, are made of carbon.

How did the universal ancestor arise? One approach to answering this question would be to search for clues in rocks. Well-preserved fossils occur only in sedimentary rocks that have not been affected by metamorphism or significant deformation. However, because there are no well-preserved sedimentary rocks from the time during which life first evolved, scientists must use other approaches. Laboratory chemists have played a very important role here.

Prebiotic Soup: The Original Experiment on the Origin of Life In laboratory experiments that probe the origin of life, scientists try to recreate some of the environmental conditions thought to have existed on the primitive Earth, before life arose. Although we cannot hope to reproduce all of primitive Earth's conditions, we do hope to learn some of the basic conditions that would have allowed life to emerge. Therefore, the experimenter must think carefully

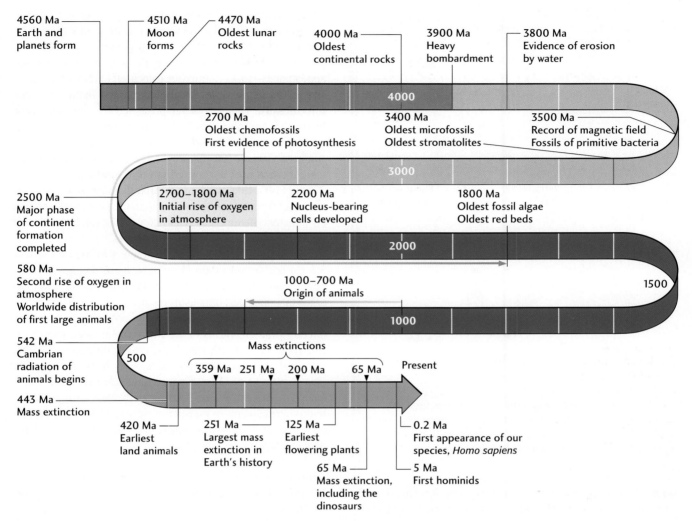

4560 Ma
Earth and
planets form

4510 Ma
Moon
forms

4470 Ma
Oldest lunar
rocks

4000 Ma
Oldest
continental rocks

3900 Ma
Heavy
bombardment

3800 Ma
Evidence of erosion
by water

4000

2700 Ma
Oldest chemofossils
First evidence of photosynthesis

3400 Ma
Oldest microfossils
Oldest stromatolites

3500 Ma
Record of magnetic field
Fossils of primitive bacteria

3000

2500 Ma
Major phase
of continent
formation
completed

2700–1800 Ma
Initial rise of oxygen
in atmosphere

2200 Ma
Nucleus-bearing
cells developed

1800 Ma
Oldest fossil algae
Oldest red beds

2000

580 Ma
Second rise of oxygen in
atmosphere
Worldwide distribution
of first large animals

1000–700 Ma
Origin of animals

1500

542 Ma
Cambrian
radiation of
animals begins

500

Mass extinctions

1000

359 Ma   251 Ma   200 Ma   65 Ma   Present

443 Ma
Mass extinction

420 Ma
Earliest
land animals

251 Ma
Largest mass
extinction in
Earth's history

125 Ma
Earliest
flowering plants

65 Ma
Mass extinction,
including the
dinosaurs

0.2 Ma
First appearance of our
species, *Homo sapiens*

5 Ma
First hominids

Figure 11.13 The timing of several major geobiological events in Earth's history.

about the particular set of conditions and components the experiments must begin with.

In the early 1950s, Stanley Miller, a graduate student at the University of Chicago, did the first experiment designed to explore life-building chemical reactions on the primitive Earth. The experiment was amazingly simple (**Figure 11.14**). At the bottom of a flask, he created an "ocean" of water that he then heated to create water vapor. The water vapor was mixed with other gases to create an "atmosphere" that was thought to contain some of the more important compounds in Earth's primitive atmosphere. The experimental atmosphere was composed of methane (CH_4), ammonia (NH_3), hydrogen (H_2), and the water vapor. Oxygen—an important gas in Earth's atmosphere today—was absent. In the next step, Miller exposed this atmosphere to electrical sparks ("lightning"), which caused the gases to react with one another and with the water in the "ocean."

The results were impressive. The experiment yielded many compounds called *amino acids,* in addition to other carbon-bearing compounds. Amino acids are the fundamental building blocks of all proteins, including those essential

for an organism's development, survival, and reproduction. The discovery was exciting because it showed that amino acids would have been abundant on the primitive Earth, leading to the concept that Earth's oceans and atmosphere formed a sort of prebiotic soup in which life originated. If you want to build an organism, creating amino acids is a really good place to start. Miller's experiment, and others since then, suggests that our universal ancestor contained genetic material that enabled amino acids to create proteins, which it then relied on for self-perpetuation.

Miller's experiment predicted that early planetary materials might contain amino acids. This prediction was realized years later when, in 1969, a meteorite hit the Earth near Murchison, Australia. When geologists analyzed it, they discovered that the *Murchison meteorite* had many (about 20) of the amino acids created in the lab! In fact, the Murchison meteorite even had similar relative amounts of these amino acids.

The message of all these discoveries is the same: amino acids can form on a planet without oxygen. But the opposite is also true: in a world where oxygen is present, amino acids

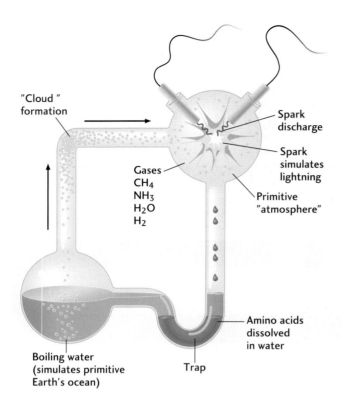

"Cloud" formation

Spark discharge

Spark simulates lightning

Gases
CH₄
NH₃
H₂O
H₂

Primitive "atmosphere"

Amino acids dissolved in water

Boiling water (simulates primitive Earth's ocean)

Trap

Figure 11.14 This simple experimental design was used to convert ammonia (NH_3), hydrogen (H_2), water vapor (H_2O), and small carbon-bearing molecules such as methane (CH_4) into amino acids—a key component of living organisms.

do not form or are present only in tiny amounts. This is one of several reasons that Earth scientists think the early Earth was a planet without oxygen.

The Oldest Fossils and Early Life Whatever the processes by which life originated, the oldest possible fossils on Earth suggest that life had originated by 3.5 billion years ago (see Figure 11.6). Stromatolites, shaped like small cones, provide some of the best evidence for life at this time (**Figure 11.15a**). The oldest fossils that preserve possible morphologic evidence for life are tiny threads similar in size and appearance to modern microbes, encased in chert. The small size of these possible fossils has earned them the name **microfossils.** Younger, better preserved microfossils occur in the 3.2-billion-year-old Fig Tree formation of South Africa and in the 2.1-billion-year-old Gunflint formation of southern Canada (see Figure 11.15b). The Gunflint fossils were the first ever discovered in a Precambrian rock and in 1954 set off a tidal wave of research that lasts to this day. In the past 50 years, we have seen, in many new localities, just how ancient life on Earth is and how well it can be preserved under the right geological circumstances.

Most geobiologists agree that there was life on Earth at 3.5 Ga but are uncertain about how these organisms functioned and obtained energy and nutrients. Some scientists argue that the oldest organisms on the universal tree of life appear to have been chemoautotrophic, obtaining their energy directly from chemicals in the environment (see Figure 11.6 and Table 11.2). Furthermore, the closest descendants of the universal ancestor may also have been hyperthermophilic (see Figure 11.6). This suggests that life may have originated in very hot water, such as in thermal springs or the hydrothermal vents of mid-ocean ridges, where sunlight was unavailable as an energy source but chemicals were abundant (**Figure 11.16**).

Chemofossils and Eukaryotes Form and size alone are not enough to deduce the function of microbes, and so microfossils are ultimately limited in the information they provide. Additional information is required. In recent years, geobiologists have discovered **chemofossils,** the chemical remains of organic compounds made by the organism while it was alive. When an organism is alive, it makes all kinds of organic compounds including carbohydrates and proteins.

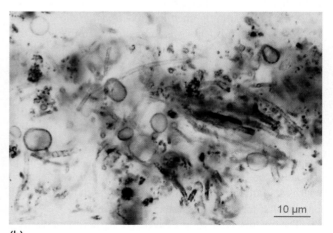

(a)

(b)

10 µm

Figure 11.15 (a) Early Archean (3.4 billion years old) stromatolites in the Warrawoona Group, Western Australia. Their conical shape suggests that microbes may have grown toward the

sunlight. (b) By 2.1 Ga, microfossils were well preserved and abundant in the Gunflint formation, southern Ontario, Canada. [Images courtesy of H. J. Hofmann.]

Figure 11.16 Hydrothermal vents along mid-ocean spreading ridges are full of nutrients from which autotrophic microbes obtain their energy. Life may have originated in such environments. [Dudley B. Foster, Woods Hole Oceanographic Institution.]

After it dies, most of these compounds are very quickly broken down into much smaller molecules, usually by heterotrophs. However, some of these molecules are very stable and resist recycling. For example, *cholestane* is a remarkably durable substance that is very similar to the well-known compound cholesterol. Cholestane has been identified in 2.7-billion-year-old rocks from Western Australia. Cholestane chemofossils are diagnostic of the Eukarya (see Figure 11.6), which are a more sophisticated group of organisms. The presence of cholestane chemofossils at 2.7 Ga tells us that single-celled eukaryotic microorganisms had emerged by then (see Figure 11.13). However, multicellular eukaryotes, including animals, did not evolve until much later.

Origin of Earth's Oxygenated Atmosphere

The rise of oxygen—the stuff that we breathe—is another important milepost in the history of interactions between life and environment. As we have just noted, evolution within the Eukarya led to the development (much later) of multicellular animals. Because these organisms require oxygen to create energy, their development would have been possible only after an increase in oxygen levels on Earth. Interestingly, it was the evolution of an entirely different domain of organisms—the Bacteria (see Figure 11.6)—that led to the origin of photosynthesis and thus the ability to produce the oxygen required by multicellular animals.

Photosynthesis originated in a group of microorganisms called the **cyanobacteria.** Remarkably, the same Australian rocks that preserve chemofossil evidence for eukaryotes also preserve chemofossil evidence for cyanobacteria. Because of this evidence, we believe that photosynthesis became an important metabolic process by 2.7 Ga (see Figure 11.13).

Thus, one group of organisms (cyanobacteria) influenced the Earth's environment by forever changing the composition of its atmosphere, while another group of organisms (eukaryotes) was influenced by this change in environment to evolve in new directions (animals).

It is widely accepted that, early in Earth's history, the atmosphere was dominated by CO_2 and that oxygen concentrations increased with the evolution of the first cyanobacteria. However, once these organisms began to produce oxygen, it was not abundant enough to accumulate as gas in the atmosphere, a necessary step toward evolving oxygen-breathing animals and humans. As it turns out, oxygenation of the Earth likely occurred in two steps, separated by perhaps more than a billion years (see Figure 11.13). The first step begins with the origin of cyanobacteria at 2.7 Ga and the occurrence of sedimentary rocks known as banded iron formations and ends with the origin of eukaryotic algae and the appearance of red beds at 2.1 to 1.8 Ga. The second step occurred near the end of Precambrian time.

Banded iron formations (see Chapter 5) are sedimentary rocks composed of alternating thin layers formed of iron oxide minerals, such as magnetite and hematite, and silica-rich minerals, such as chert and iron silicates (**Figure 11.17**). Most geologists agree that banded iron formations were precipitated from seawater when oxygen, produced by cyanobacteria, reacted with iron dissolved in seawater. Iron is soluble when oxygen concentrations are low, as would have been the case on Earth before the advent of oxygen-producing cyanobacteria. However, iron is highly insoluble when oxygen concentrations are high. Therefore, oxygen produced by cyanobacteria would have immediately caused the iron in seawater to precipitate and fall to the bottom of the ocean, where it accumulated in thin layers of iron oxide minerals such as those shown in Figure 11.17. This process

(a)

(b)

(c)

Figure 11.17 Unusual sedimentary rocks and new, larger organisms mark the rise of oxygen in the atmosphere between 2.7 and 2.1 Ga. (a) Banded iron formation. [Pan Terra.] (b) *Grypania,* a fossil of eukaryotic algae visible with the naked eye. [Courtesy of H. J. Hofmann.] (c) Red beds are mostly sandstones and shales partly cemented by iron oxide minerals (red). [John Grotzinger.]

would have continued until all the iron was used up and then, finally, oxygen would have been free to accumulate in the atmosphere.

Oxygen in the atmosphere began to build around 2.4 Ga and reached an initial peak around 2.1 to 1.8 Ga when we see evidence for Earth's first eukaryotes—fossil algae—that are visible to the naked eye (see Figure 11.17). The large size of the fossil shown in this illustration—at least 10 times larger than anything that came before it—is thought to be a consequence of this increase in oxygen. Geologists also have noted the first appearance of **red beds,** which are usually sandstones and shales that are red because of the presence of iron oxide cement that binds the grains together (see Figure 11.17). Most of these red beds were deposited on land as ancient river deposits, indicating that oxygen must have been present in the atmosphere to precipitate the iron oxide cements.

Once eukaryotic algae and red beds entered the scene, not much else happened for a very long time, perhaps a billion years or more. But then, about 600 million years ago, the first really large animals suddenly appeared—and the rock record indicates that oxygen may have risen dramatically, close to its modern level (see Figure 11.13). One idea to explain this long period of inactivity is that the oceans were largely devoid of oxygen, except for a thin layer at its top, for about a billion years after oxygen first appeared in the atmosphere. Something was required to enrich the deep ocean in oxygen, as it is today. In a mechanism somewhat similar to the process that produces banded iron formations, oxygen reacts easily with organic matter, usually helped by microbes. As long as there is organic matter around, oxygen will be consumed. One way to remove organic matter from the system is to bury it in sediments so that it cannot react with available oxygen. Perhaps the second step in the rise of atmospheric oxygen was related to an increase in sediment production and burial.

The causes of this rise are still not understood, but one point remains clear: the evolution of all modern groups of animals occurred shortly after this, ushering in the Phanerozoic eon and a world full of big, complex, multicellular organisms.

Radiation of Life: The Cambrian Explosion of Animals

Perhaps the most remarkable geobiological event in Earth's history—aside from the origin of life itself—is the sudden appearance of large animals with skeletons at the end of Precambrian time (see Figure 11.13). This event coincides with the Precambrian-Cambrian boundary of the geologic time scale at 542 Ma (see Figure 8.11). It also coincides with the base of the Paleozoic era and the Phanerozoic eon.

As noted earlier, after life originated, Earth was dominated mostly by microorganisms for billions of years. Curiously, after almost 3 billion years of very slow evolution, animals suddenly appeared in a dramatic radiation of life (**Figure 11.18**). A **radiation** is the relatively rapid develop-

Figure 11.23 An artist's rendition of the Cretaceous-Tertiary scene. [Alfred Kamajian.]

The high concentration of debris in the atmosphere blocked out the Sun, vastly reducing the light energy needed for photosynthesis. In addition to the solid debris, poisonous sulfur- and nitrogen-bearing gases were also injected into the atmosphere, where they reacted with water vapor to form toxic sulfuric and nitric acids that would have rained down on Earth. The combination of these two effects, and others, would have been devastating to both marine and terrestrial ecosystems, which depended on photosynthetic autotrophs to create the base of the food chain. Heterotrophs, including the dinosaurs, were next. Once their food sources died off, they died off as well.

Global Warming Disaster Leads to Radiation of Mammals

Global warming is a worrisome climatic condition that many scientists feel may threaten society in the coming decades (see Chapter 23). Geologists are concerned because there is firm evidence that global temperatures have been increasing and because they understand the potentially disastrous effects associated with global warming events in the past. In some cases, these past events were extreme, resulting in mass extinction of many species. When we face a threatening situation, public policy decisions often are best informed by consideration of past events so that history does not repeat itself.

The mass extinction at the Paleocene-Eocene boundary (~55 Ma; see Figure 8.11) was directly related to abrupt global warming. This extinction is significant in the evolution of life because it paved the way for mammals, including primates, to finally radiate as an important group. But unlike the mass extinction that wiped out the dinosaurs, this mass extinction had no extraterrestrial cause. Instead, the Earth system was pushed into a severe imbalance (unusually high temperatures) when one of its inputs (greenhouse gases) accumulated in an uncontrolled fashion.

We now understand that this global warming event occurred when the ocean suddenly burped an enormous amount of methane—a very potent greenhouse gas—into the atmosphere. Global warming was the primary cause of death that led to the mass extinction. However, the extreme imbalance in the Earth system was caused by the dramatic influx of methane. But what produced the methane? And why was it suddenly expelled? And how long did it take for the Earth system to recover its balance? Geologists have recently found many clues that provide the answers to these questions. Unraveling this mystery involves weaving together many of the concepts we have learned in this chapter, including microbial processes, biogeochemical cycles, and global behavior of the biosphere.

Microbes Sow the Seeds of Disaster The story begins with Earth's cycle of carbon, which will be discussed in more detail in Chapter 15. Normally, carbon is removed from the atmosphere by photosynthetic autotrophs—both algae and microbes—mostly in the world's oceans. After these marine organisms die, they slowly settle to the bottom of the ocean, where they accumulate as organic debris. Some of this organic debris is buried in sediments, but some is also consumed by heterotrophic microbes as food. As you may recall, during respiration this food is converted to energy. Some microbes produce carbon dioxide (CO_2) as a by-product. This is released back into the ocean, where it will eventually escape to the atmosphere, its point of origin.

However, other microbes may produce methane (CH_4) during respiration. This methane accumulates in the pores of sediments, where it may be transformed into a solid. Unlike

EARTH ISSUES

11.1 The Mother of All Mass Extinctions: Whodunit?

The Cretaceous-Tertiary and Paleocene-Eocene extinctions are clear-cut examples of how a dramatic perturbation in Earth's environment can cause the catastrophic collapse of ecosystems, leading to extinction. They were big, but not the biggest.

The extinction at the end of Permian time was even worse—95 percent of all species became extinct (see Figure 11.21)! Here, there is no smoking gun, and it seems unlikely that something as simple as a colliding asteroid could explain how almost every species on Earth was killed. Not surprisingly, the absence of clear-cut evidence has resulted in a long list of hypotheses. Some of these are based on extraterrestrial effects, such as impact by a comet or an increase in the solar wind. Others argue for effects that come from the Earth itself, such as an increase in volcanism, the depletion of oxygen in the oceans, or the sudden release of carbon dioxide from the oceans. As in the Paleocene-Eocene extinction, the sudden release of methane from the oceans has also been proposed.

Recently, it has been shown that the extinction is exactly 251 million years old. Perhaps it is no coincidence that the age of *flood basalts* in Siberia is also 251 million years. Flood basalts, as you will learn in Chapter 12, are huge volumes of volcanic rock that pour out across the surface of the Earth in a relatively short period of time. The Siberian flood basalts spewed out some 2 million cubic kilometers of basaltic lava covering an area of 1.6 million square kilometers, about the size of Europe. Geochronologic data also show that *all* of this lava poured out in 1 million years or less. It is hard to escape the impression that the mass extinction at the end of the Permian was somehow related, at least in part, to this catastrophic eruption in Siberia. Potential negative effects associated with these tremendous volcanic emissions would be caused by the injection of CO_2 and SO_2 gases into the atmosphere. CO_2 contributes to global warming, and SO_2 is the principal source of acid rain. Both are bad for life if levels get too high.

More work is required to test all these hypotheses. Massive outpourings of basaltic lava have occurred at other times in Earth's history, but without such apparently devastating effects. Whatever the cause of the Permian mass extinction, one point is clear: just as in the Cretaceous-Tertiary and Paleocene-Eocene extinctions, the ultimate cause was the collapse of ecosystems. We know that this collapse occurred, but we don't know exactly how. The message that we should take away from this history lesson is that the past may repeat itself. The decisions that humans make today inevitably affect the environment—we just don't know exactly how, at least not yet.

---

carbon dioxide, methane is highly susceptible to small changes in temperature and pressure. At current deep ocean temperatures (average is about 3°C), microbially generated methane will combine with water to form a frozen solid and can be stored within sediments as long as the ocean stays cold. Today, frozen methane occurs abundantly in the upper 1500 m of sediments along continental margins. But if temperatures rise by even a few degrees, it can be transformed—almost instantly—into a gas. And that's when the trouble may start.

The Oceans Bubble Methane, a Greenhouse Gas At the end of Paleocene time, average temperatures in the deep ocean may have risen by as much as 6°C. Once the first methane solids thawed and were transformed back into gas, they bubbled up through the oceans and entered the atmosphere, where they reinforced the global warming trend. In turn, this raised temperatures further in the deep ocean, which accelerated the rate of thawing. These positive feedbacks eventually resulted in a sudden—and catastrophic—release of methane that caused average global temperatures to rise dramatically. As much as 2 *trillion* tons of carbon as methane may have escaped to the atmosphere over a period as short as 10,000 years or less!

Because methane easily reacts with oxygen to produce carbon dioxide, it also caused oxygen levels to decrease dramatically. The oxygen decrease and temperature rise was devastating in the deep ocean, with up to 80 percent of bottom-feeders, such as clams, becoming extinct. Marine life was essentially suffocated when oxygen levels dropped below a critical level.

Recovery and the Evolution of Modern Mammals Following the catastrophe, it took about 100,000 years for the Earth to return to its previous, normal state. During this time, temperatures remained unusually high until Earth was able to absorb all the extra carbon that had been released to the atmosphere. The warmer temperatures allowed rapid expansion of forests into higher latitudes: redwoods—related to the giant sequoias of California—grew as far north as 80°, rain forests were widespread in Montana and the Dakotas, and tropical palms flourished near London, England. Primitive mammals rapidly evolved into the ancestors of today's modern mammals, which adapted to cope with the high heat at that time. One particular group of mammals—the *Primates*—evolved as the ancestors to humans.

Methane Deposits on the Modern Earth: A Ticking Time Bomb? Whether driven by anthropogenic processes or not, the global average temperature of Earth has been rising for over 100 years. In the frozen tundra of northern Canada and other arctic regions of the world, there may be as much as half a trillion tons of frozen methane. Deep-sea sediments around the world contain much more; the global inventory of all methane deposits is estimated to be 10 to 20 trillion tons of carbon present as methane.

Are these deposits a ticking time bomb? Once these methane deposits start to thaw, the runaway cycle of extreme warming could begin all over again. It is difficult to predict what may happen in the future, but it is clear that we must be mindful of greenhouse gas emissions. Today, humans are adding carbon dioxide to the atmosphere, through burning fossil fuels, at a rate comparable to that at which methane was added to the atmosphere when the Paleocene-Eocene boundary event occurred. If Earth continues to warm and the oceans heat up, it is possible that the current methane deposits could thaw. Humans would be wise to pay attention to the lessons of history.

Figure 11.24 The Allende meteorite is full of organic compounds, some of which may have helped in the origin of life. This meteorite fell to Earth near Allende, Mexico, in 1969. [John Grotzinger/Ramón Rivera-Moret/Harvard Mineralogical Museum.]

ASTROBIOLOGY: THE SEARCH FOR EXTRATERRESTRIAL LIFE

Looking up at the stars on a clear night, it's hard not to wonder if we are alone in the universe. As we have learned, the activities of life on our planet create distinctive biogeochemical signatures. Some of these signatures could be detected remotely, such as the presence of oxygen in the atmosphere of another planet in a different solar system. In other cases, we might be able to land a spacecraft equipped with instruments to detect diagnostic organic compounds or morphologic fossils preserved in rocks.

In the past few decades, **astrobiologists** have begun to search systematically for evidence of life on other worlds. Although no organisms have yet been discovered beyond the Earth, we should be encouraged to pursue this quest. Life may have gotten started even if it didn't flourish. In our own solar system, Mars and Europa (a moon of Jupiter) are tantalizing targets because they are similar to Earth in several important ways. In addition, new discoveries of planets orbiting other stars allow us to extend this search to other solar systems.

The search for life on other worlds requires a systematic, scientific approach. One must also have patience, because any victories will not be easily won. The most widely accepted approach has been to recognize that life as we know it, here on Earth, is based on liquid water and carbon-bearing organic compounds. Therefore, a sensible search strategy might begin with looking for liquid water and

organic compounds. Earlier in this chapter, we noted that no other chemical element can match carbon in the variety and complexity of compounds it can form. We also noted that liquid water provides a stable medium in which carbon-bearing molecules on the primitive Earth—and other planets—could have interacted to form larger organic molecules such as amino acids. Moreover, compounds made of carbon are already common in the universe. Astronomers find evidence for them everywhere, from interstellar gas and dust particles to the meteorites that land on Earth (**Figure 11.24**).

Of course, there is some risk in the "life as we know it" approach to searching for extraterrestrial life. We might miss forms of life we know nothing about, and one could imagine a whole host of other elements and compounds that life could be based on. In general, however, these alternative schemes mostly provide fuel for science fiction writers. At least for the time being, carbon and water are regarded as the key substances for all life in the universe.

Habitable Zones Around Stars

At the broadest scale, life must be restricted to surfaces of planets and moons that orbit stars (**Figure 11.25**). The trick is to keep water stable as a liquid for a long enough period of time that life can originate. This may take hundreds of millions of years, based on our experience on Earth. If a planetary surface is too close to its star, water will boil off and become a gas. This is what happened on Venus, which is 30 percent closer than Earth to the Sun and whose surface temperature is almost 500°C. If the planetary surface is too

Too close: Habitable
Temperature   zone
above boiling
point of water

Too far:
Temperature
below freezing
point of water

Figure 11.25 Stars have habitable zones where life on an orbiting planet could exist if the temperature were not too warm (too close to the star) or too cold (too far away from the star). The width of the zone is the distance over which water won't boil away or freeze.

far from its star, the water will freeze and become a solid. This is the case on Mars today, which is 50 percent farther from the Sun than Earth and may reach temperatures of below –150°C. Earth is in the middle zone where water is stable as a liquid and temperatures are just right for life. For every star, there is a **habitable zone,** marked by the distance away from the star to the point where water is stable as a liquid. If a planet is within the habitable zone, there a chance that life might have originated there.

Greenhouse gases such as carbon dioxide also play an important role in controlling the distance of the habitable zone. In the case of Mars, its atmosphere may have had high levels of greenhouse gases such as carbon dioxide and methane early in its history. Thus, even though Mars is farther from the Sun than Earth, it would have been warmed through the greenhouse effect, as Earth is today. New data suggest that liquid water was present on the surface of Mars, but we don't know how long it may have been stable. Thus, it might have been possible for life to originate during the history of Mars. But once the greenhouse gases were lost, Mars was transformed into the icy desert it is today.

Habitable Environments on Mars

People have long wondered about life on Mars. Mars is the planet most resembling Earth and is therefore the most likely planet in our solar system to host or have hosted life. As we discussed in Chapter 9, there is clear evidence for liquid water on its surface at some point in its past. Based on crater counts to estimate the age of surface features (see Chapter 9), geologists estimate that water on Mars was stable at 3 Ga,

when it carved canyons deep across the planet's surface, dissolved rocks and minerals, and then precipitated them in a variety of basins where the water evaporated.

Water is also present today, but only as ice or vapor. If life had evolved early on, it would have had to seek refuge deep beneath the surface because of the frigid modern climate. Any organisms that had been present would now be thoroughly frozen. However, the interior of Mars is warm because of the decay of radioactive elements, and so at some point within Mars the ice present at or just below its surface must turn into liquid water. It is therefore possible that organisms—perhaps microbial extremophiles—live within what might be a watery upper crust of Mars.

Unfortunately, the restriction of liquid water is not the only challenge that modern or ancient life would have to face on Mars. The sedimentary rocks discovered by the *Opportunity* rover are full of sulfate minerals, including an unusual iron sulfate mineral called *jarosite* (**Figure 11.26**). In Chapter 3, we discussed how jarosite forms by alteration of sulfide minerals to form sulfuric acid, from which jarosite then precipitates. On Earth, jarosite accumulates in some of the most acidic waters ever observed in natural environments. Our skin would dissolve if we went swimming in these waters.

It seems that the evolution of life on Mars would require it to wrestle with limited and possibly very acidic water. The good news is that extremophiles on Earth can live in these environments. But the really important question is whether or not life can originate in such environments. The bad news is that experiments on the origin of life suggest that this might be difficult to do. Some of the simple reactions that

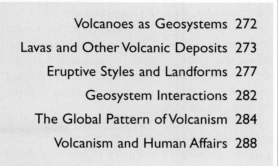

VOLCANOES

The northwest corner of Wyoming is a geologic wonderland of geysers, hot springs, and steam vents—the visible signs of a vast active volcano that stretches across the wilderness of Yellowstone National Park. Every day, this volcano expels more energy in the form of heat than all the electric power consumed in the three surrounding states of Wyoming, Idaho, and Montana combined. The energy is not released steadily; some of it builds up in hot magma chambers until the volcano blows its top. A cataclysmic eruption of the Yellowstone volcano 630,000 years ago ejected 1000 km^3 of rock into the air, covering regions as far away as Texas and California with a layer of volcanic ash.

The geologic record shows that volcanic explosions nearly this big or even bigger have occurred in the western United States at least six times during the last 2 million years, so we can be fairly certain that such an eruption will happen again. We can only imagine what it might do to civilization. Hot ash will snuff out all life within 100 km or more, and cooler but choking ash will blanket the ground more than 1000 km away. Dust thrown high into the stratosphere will dim the Sun for several years, dropping temperatures and plunging the Northern Hemisphere into an extended volcanic winter.

The hazards volcanoes pose to our society, as well as the energy and mineral resources they can provide, are certainly good reasons to study them. Volcanoes are also fascinating because they are windows through which we can see into Earth's deep interior to understand the igneous processes of plate tectonics that have generated Earth's oceanic and continental crust.

In this chapter, we examine how magma rises through the crust, emerges onto the surface as lava, and cools into hard volcanic rock. We will see how plate tectonics and mantle convection can explain volcanism at plate boundaries and at "hot spots" within plates. We will give examples of how volcanoes interact with other components of the Earth system, particularly the atmosphere and hydrosphere. Finally, we will consider the destructive power of volcanoes and the potential benefits of their chemical riches and heat energy.

Ancient philosophers were awed by volcanoes and their fearsome eruptions of molten rock. In their efforts to explain volcanoes, they spun myths about a hot, hellish underworld below Earth's surface. Basically, they had the right idea. Modern

Grand Canyon of the Yellowstone, where the Yellowstone River cuts down 250 m through brightly colored rhyolitic lavas. The lavas were deposited by a huge volcanic eruption less than a million years ago. [Kerry L. Thalmann.]

researchers, using science rather than mythology, also see evidence of Earth's internal heat in volcanoes.

VOLCANOES AS GEOSYSTEMS

Temperature readings of rocks as far down as humans have drilled (about 10 km) show that the Earth does indeed get hotter with depth. We now believe that temperatures at depths of 100 km and more—within the asthenosphere—reach at least 1300°C, high enough for the rocks there to begin to melt. For this reason, we identify the asthenosphere as a main source of *magma*, the molten rock below Earth's surface that we call *lava* after it erupts. Sections of the solid lithosphere that ride above the asthenosphere may also melt to form magma.

Because magmas are liquids, they are less dense than the rocks that produced them. Therefore, as magma accumulates, it begins to rise buoyantly (float upward) through the litho-sphere. In some places, the melt may find a path to the surface by fracturing the lithosphere along zones of weakness. In other places, the rising magma melts its way toward the surface. Some of the magma eventually reaches the surface and erupts as lava. A **volcano** is a hill or mountain constructed from the accumulation of lava and other erupted materials.

Taken together, the rocks, magmas, and processes needed to describe the entire sequence of events from melting to eruption constitute a **volcanic geosystem.** This type of geosystem can be viewed as a chemical factory that processes the input material (magmas from the asthenosphere) and transports the end product (lava) to the surface through an internal plumbing system.

Volcanic Plumbing Systems

Figure 12.1 is a simplified diagram of a volcano's plumbing system, through which melts journey to the surface. Mag-

Earth System Figure 12.1 Volcanoes transport magma from Earth's interior to form rocks at its surface and inject gases into its atmosphere.

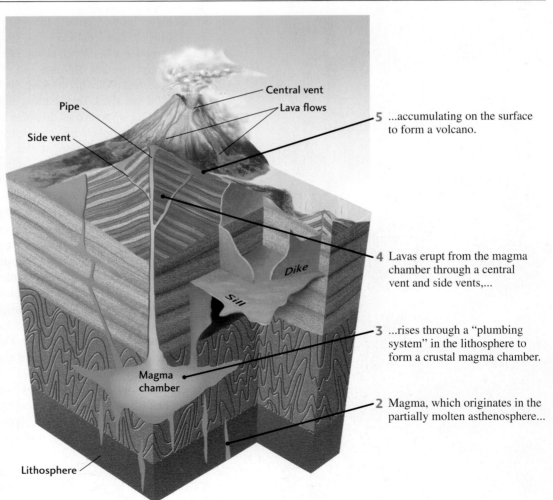

1 A volcanic geosystem includes interactions between the lithosphere and asthenosphere and the flux of gases into the atmosphere (land volcano) or hydrosphere (underwater volcano).

Pipe

Side vent

Central vent

Lava flows

Dike

Sill

Magma chamber

Lithosphere

5 ...accumulating on the surface to form a volcano.

4 Lavas erupt from the magma chamber through a central vent and side vents,...

3 ...rises through a "plumbing system" in the lithosphere to form a crustal magma chamber.

2 Magma, which originates in the partially molten asthenosphere...

mas rising through the lithosphere pool in a magma chamber, usually at shallow depths in the crust. This reservoir periodically empties to the surface through a pipelike conduit in repeated cycles of *central eruptions*. Lava can also erupt from vertical cracks (fissures) and other vents on the flanks of a volcano. These *fissure eruptions* become more important as the size of the volcano increases.

Volcanoes as Chemical Factories

Volcanic geosystems are chemical factories that process magmas to produce lavas. As we saw in Chapter 4, only a very small fraction of the asthenosphere melts in the first place. A magma gains chemical components as it melts the surrounding rocks while rising through the lithosphere. It loses other components when crystals settle out along the way or in shallow magma chambers. And its gaseous constituents escape to the atmosphere or ocean as it erupts at the surface. By accounting for these changes, we can extract clues to the chemical composition and physical state of the upper mantle where the magmas originate. We can also learn about eruptions millions or even billions of years ago using isotopic dating (Chapter 8) to determine the age of the lavas.

LAVAS AND OTHER VOLCANIC DEPOSITS

The several types of lava leave behind different landforms: volcanic mountains that vary in shape and solidified lava flows that vary in character. The differences depend on the chemical composition, gas content, and temperature of lavas. The higher the silica content and the lower the temperature, for example, the more viscous (more resistant to flow) the lava is and the more slowly it moves. The more gas a lava contains, the more violent its eruption is likely to be.

Types of Lava

Erupted lavas, the end products of volcanic geosystems, usually solidify into one of three major types of igneous rock (see Chapter 4): basalts, andesites, and rhyolites.

Basaltic Lavas Basalt is an extrusive rock of mafic composition (high in iron, magnesium, and calcium) with a relatively low silica content; its intrusive equivalent is gabbro. Basaltic magmas are the most common type of melt. They are produced along mid-ocean spreading centers and continental rift valleys, as well as from hot spots in the mantle beneath oceanic plates. The volcanic island of Hawaii lies above a mantle hot spot.

When cool, **basaltic lava** is black or dark gray, but at its high eruption temperature (1000° to 1200°C), it glows in reds and yellows (**Figure 12.2**). Because its temperature is high and silica content low, basaltic lava is extremely fluid and can flow downhill fast and far. Lava streams flowing as fast as 100 km/hour have been observed, although velocities of a few kilometers per hour are more common. In 1938, two daring Russian volcanologists measured temperatures and collected gas samples while floating down a river of molten basalt on a raft of colder solidified lava. The surface temperature of the raft was 300°C, and the river temperature was 870°C. Lava streams have been observed to travel more than 50 km from their sources.

Basaltic eruptions are rarely explosive. Basalt erupts when hot, fluid magmas fill up the volcano's plumbing system and overflow, sending lava down the flanks of the volcano in great streams that can engulf everything in their path. Basaltic lavas take on different forms depending on

Figure 12.2 The central vent eruption from Kilauea shield volcano on the island of Hawaii, showing a river of hot, fast-flowing basaltic lava. [J. D. Griggs/USGS.]

Aa lava

Pahoehoe lava

~1 m

Figure 12.3 Two types of basaltic lava, ropy pahoehoe (*bottom*) and jagged blocks of aa (*top*). Mauna Loa volcano, Hawaii. [Kim Heacox/DRK.]

how they cool. On land, they solidify as pahoehoe (pronounced pa-ho'-ee-ho'u-ee) or aa (ah-ah) (**Figure 12.3**).

Pahoehoe (the word is Hawaiian for "ropy") forms when a highly fluid lava spreads in sheets and a thin, glassy elastic skin congeals on its surface as it cools. As the molten liquid continues to flow below the surface, the skin is dragged and twisted into coiled folds that resemble rope.

"*Aa*" is what the unwary exclaim after venturing barefoot onto lava that looks like clumps of moist, freshly plowed earth. Aa is lava that has lost its gases and consequently flows more slowly than pahoehoe, allowing a thick skin to form. As the flow continues to move, the thick skin breaks into rough, jagged blocks. The blocks pile up in a steep front of angular boulders that advances like a tractor tread. Aa is truly treacherous to cross. A good pair of boots may last about a week on it, and the traveler or geologist can count on cut knees and elbows.

A single downhill basaltic flow commonly has the features of pahoehoe near its source, where the lava is still fluid and hot, and of aa farther downstream, where the flow's surface—having been exposed to cool air longer—has developed a thicker outer layer.

Basaltic magma that cools under water forms *pillow lavas*—piles of ellipsoidal, pillowlike blocks of basalt about a meter wide (see Figure 4.13). Pillow lavas are an important indicator that a region on dry land was once under water. Scuba-diving geologists have actually observed pillow lavas forming on the ocean floor off Hawaii. Tongues of

Figure 12.4 Mount St. Helens, an andesitic volcanic cone in southwestern Washington State, before, during, and after its cataclysmic eruptions in May 1980, which ejected about 1 km³ of pyroclastic material. [*Before:* Emil Muench/Photo Researchers. *Erupting:* U.S. Geological Survey. *After:* David Weintraub/Photo Researchers.]

Figure 12.5 Phreatic explosion on Nisino-sima, a new volcano that rose above the sea in 1973 after a submarine eruption in the Pacific Ocean about 900 km south of Tokyo. [Hydrographic Department, Maritime Safety Agency, Japan.]

molten basaltic lava develop a tough, plastic skin on contact with the cold ocean water. Because lava inside the skin cools more slowly, the pillow's interior develops a crystalline texture, whereas the quickly chilled skin solidifies to a crystal-less glass.

Andesitic Lavas Andesite is an extrusive rock with an intermediate silica content; its intrusive equivalent is diorite. Andesitic magmas are erupted mainly in the volcanic belts along active continental or oceanic margins above subduction zones. The name comes from a prime example—the Andes mountains of South America.

The temperatures of **andesitic lavas** are lower than those of basalts, and their silica contents are higher, so they flow more slowly and lump up in sticky masses. If one of these sticky masses plugs up the throat of the volcano, gases can build up beneath the plug and eventually blow off the top of the volcano. The explosive eruption of Mount St. Helens in 1980 (**Figure 12.4**) is a famous example.

Some of the most destructive volcanic eruptions in history are *phreatic,* or steam, explosions, which occur when hot, gas-charged magma encounters groundwater or seawater, generating vast quantities of superheated steam (**Figure 12.5**). The island of Krakatoa, an andesitic volcano in Indonesia, was destroyed by a phreatic eruption in 1883. This legendary explosion was heard thousands of kilometers away, and it generated a tsunami (sea wave) that killed more than 40,000 people.

Rhyolitic Lavas Rhyolite is an extrusive rock of felsic composition (high in sodium and potassium) with silica content greater than 68 percent; its intrusive equivalent is granite. Rhyolitic magmas are produced in zones where heat from the mantle has melted large volumes of continental crust. Today, the Yellowstone volcano is producing huge amounts of rhyolitic magmas that are building up in shallow magma chambers for the next eruption.

Rhyolite is light in color, often a pretty pink. It has a lower melting point than andesite, erupting at temperatures of only 600° to 800°C. Because **rhyolitic lava** is richer in silica than any other lava type, it is the stickiest and least fluid. A rhyolite flow typically moves more than 10 times slower than a basaltic flow, and it tends to pile up in thick, bulbous deposits. Gases are easily pent up beneath rhyolite lavas, and large rhyolite volcanoes such as Yellowstone produce the most explosive of all volcanic eruptions.

Texture of Volcanic Rocks

The textures of volcanic rocks, like the surfaces of frozen lavas, reflect the conditions under which they solidified. Coarse-grained textures with visible crystals can result if lavas cool slowly beneath the surface. Lavas that cool quickly tend to have fine-grained textures. If they are silica-rich, rapidly cooled lavas can form obsidian, a volcanic glass.

Volcanic rock often contains little bubbles, created as gas is released during an eruption. Lava is typically charged with gas, like soda in an unopened bottle. When lava rises, the pressure on it decreases, just as the pressure on the soda drops when its bottle cap is removed. And just as the carbon dioxide in the soda forms bubbles when the pressure is released, the water vapor and other dissolved gases escaping from lava create gas cavities, or *vesicles* (**Figure 12.6**). *Pumice* is an extremely vesicular volcanic rock, usually rhyolitic in composition. Some pumice has so much void space that it is light enough to float on water.

~0.25 m

Figure 12.6 Vesicular basalt sample. [Glenn Oliver/Visuals Unlimited.]

Pyroclastic Deposits

Water and gases in magmas can have even more dramatic effects. Before a magma erupts, the confining pressure of the overlying rock keeps these volatiles from escaping. When the magma rises close to the surface and the pressure drops, the volatiles may be released with explosive force, shattering the lava and any overlying solidified rock into fragments of various sizes, shapes, and textures (**Figure 12.7**). Explosive eruptions are particularly likely with gas-rich, viscous rhyolitic and andesitic lavas.

Volcanic Ejecta Fragmentary volcanic rocks ejected into the air are called **pyroclasts.** These rocks, minerals, and glasses are classified according to their size. The finest frag-

Figure 12.8 Volcanologist Katia Krafft examines a volcanic bomb ejected from Asama volcano, Japan. [Science Source/Photo Researchers.]

ments, less than 2 mm in diameter, are called *volcanic ash.* Fragments ejected as blobs of lava that cool in flight and become rounded or chunks torn loose from previously solidified volcanic rock can be much larger. These are called *volcanic bombs* (**Figure 12.8**). Volcanic bombs as large as houses have been thrown more than 10 km in violent eruptions. Volcanic ash fine enough to stay aloft can be carried great distances. Within two weeks of the 1991 eruption of Mount Pinatubo in the Philippines, its volcanic dust was traced all the way around the world by Earth-orbiting satellites.

Sooner or later pyroclasts fall to Earth, building the largest deposits near their source. As they cool, hot sticky fragments become welded together (lithified). The rocks created from smaller fragments are called *volcanic tuffs;* those formed from larger fragments are called *volcanic breccias* (**Figure 12.9**).

Pyroclastic Flows One particularly spectacular and often devastating form of eruption occurs when hot ash, dust, and gases are ejected in a glowing cloud that rolls downhill at high speeds. The solid particles are buoyed up by the hot gases, so there is little frictional resistance in this incandescent *pyroclastic flow* (**Figure 12.10**).

In 1902, with very little warning, a pyroclastic flow with an internal temperature of 800°C exploded from the side of Mont Pelée, on the Caribbean island of Martinique. The avalanche of choking hot gas and glowing volcanic ash plunged down the slopes at a hurricane speed of 160 km/hour. In 1 minute and with hardly a sound, the searing emulsion of gas, ash, and dust enveloped the town of St. Pierre and killed 29,000 people. It is sobering to recall the statement of one Professor Landes, issued the day before the cataclysm: "The Montagne Pelée presents no more danger to the inhabitants of St. Pierre than does Vesuvius to those of Naples." Professor Landes perished with the others. In 1991, French volcanologists Maurice and Katia Krafft (who is pictured in Fig-

Figure 12.7 Pyroclastic eruption at Arenal volcano, Costa Rica. [Gregory G. Dimijian/Photo Researchers.]

~0.3 m

(a)

(b)

Figure 12.9 (a) Welded tuff from an ash-flow deposit in the Great Basin of northern Nevada. (b) Volcanic breccia. [(a) John Grotzinger. (b) Doug Sokell/Visuals Unlimited.]

Figure 12.10 A pyroclastic flow plunged down the slopes of Mount Unzen, in Japan, in June 1991. Note the fireman and fire engine in the foreground, trying to outrun the hot ash cloud descending on them. Three scientists who were studying this volcano died when they were engulfed by a similar flow. [AP/Wide World Photos.]

ure 12.8) were killed by the pyroclastic flow on Mount Unzen, Japan, shown in Figure 12.10.

ERUPTIVE STYLES AND LANDFORMS

The shape a volcano takes as it ejects material varies with the properties of the magma, especially its chemical composition and gas content, as well as with the environmental conditions under which the lava erupts, such as on land or under the sea. Volcanic landforms also depend on the rate at which lava is manufactured and the plumbing system that gets it to the surface (**Figure 12.11**).

Central Eruptions

Central eruptions discharge lava or pyroclastic materials from a *central vent*. The vent is an opening atop a pipelike feeder channel rising from the magma chamber. The material ascends through this channel to erupt at Earth's surface. Central eruptions create the most familiar of all volcanic features—the volcanic mountain, shaped like a cone.

Shield Volcanoes A lava cone is built by successive flows of lava from a central vent. If the lava is basaltic, it flows easily and spreads widely. If flows are copious and frequent, they create a broad, shield-shaped volcano two or more kilometers high and many tens of kilometers in circumference. The slopes are relatively gentle. Mauna Loa, on the Big Island of Hawaii, is the classic example of a **shield volcano** (see Figure 12.11a). Although it rises only 4 km above sea level, it is actually the world's tallest structure: measured from its base on the seafloor, Mauna Loa is 10 km high, higher than Mount Everest! It has a base diameter of 120 km, covering three times the area of Rhode Island. It grew to this enormous size by the accumulation of thousands of lava

Key Figure 12.11 — Magma types determine eruptive styles and landforms.

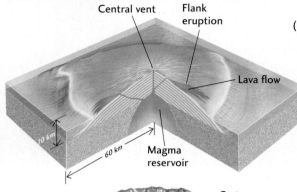

Central vent · Flank eruption · Lava flow · 10 km · 60 km · Magma reservoir

(a) **Shield volcanoes** are built up by the accumulation of thousands of thin basaltic flows that spread as gently sloping sheets. Each layer in the diagram represents many hundreds of thin flows. Magma can erupt on the flanks of a volcano as well as from the central vent.

Mauna Loa (Hawaii)

Lava dome · Crater

(b) **Volcanic domes** are bulbous masses of felsic lava, which are so viscous that instead of flowing, they pile up over the vent. The photo shows a growing dome within the crater of Mount St. Helens after its 1980 eruption.

Mount St. Helens (Washington)

Central vent filled with rock fragments · Successive layers of ejected material

(c) **Cinder-cone volcanoes** are formed when ejected material is deposited as layers that dip away from the crater at the summit. The vent beneath the crater is filled with fragmental debris. The photo is of Cerro Negro, shown erupting in 1968, a cinder cone built on an older terrain of lava flows.

Cerro Negro (Nicaragua)

Central vent filled from previous eruption · Radiating dikes · Pyroclastic layers · Lava flows

(d) **Stratovolcanoes** are built from alternating layers of pyroclastic material and lava flows. Lava that has solidified in fissures forms riblike dikes that strengthen the cone.

Mount Fuji (Japan)

(e) **Craters** are found at the summits of most volcanoes. After an eruption, lava often sinks back into the vent and solidifies, to be blasted out by a later pyroclastic explosion.

Mt. Etna (Sicily, Italy)

(f) **Calderas** result when a violent eruption empties a volcano's magma chamber, which then cannot support the overlying rock. It collapses, leaving a large, steep-walled basin.

Crater Lake (Oregon)

[(a) U.S. Geological Survey. (b) Lyn Topinka/USGS Cascades Volcano Observatory. (c) Mark Hurd Aerial Surveys. (d) Corbis. (e) Fabrizio Villa/AP/Wide World Photos. (f) Greg Vaughn/Tom Stack & Associates.]

Figure 12.17 A fumarole becomes encrusted with sulfur deposits on the Merapi volcano in Indonesia. [R. L. Christiansen/USGS.]

steam through small vents called *fumaroles* (**Figure 12.17**). These emanations contain dissolved materials that precipitate onto surrounding surfaces as the water evaporates or cools, forming various encrusting deposits (such as travertine). Some precipitates can contain valuable minerals.

Fumaroles are a surface manifestation of **hydrothermal activity**—the circulation of water through hot volcanic rocks and magmas. Circulating groundwater that reaches buried magma (which stays hot for hundreds of thousands of years) is heated and returned to the surface as hot springs and geysers. A *geyser* is a hot-water fountain that spouts intermittently with great force, frequently accompanied by a thunderous roar. The best-known geyser in the United States is Old Faithful in Yellowstone National Park, which erupts about every 65 minutes, sending a jet of hot water as high as 60 m into the air (**Figure 12.18**).

The steam and hot water formed by hydrothermal activity can be tapped as a source of geothermal energy. Hydrothermal activity is also responsible for the deposition of unusual minerals that concentrate relatively rare elements, particularly metals, into ore deposits of great economic value. As noted in Chapter 4, hydrothermal interactions are especially intense at the spreading centers of mid-ocean ridges, where huge volumes of water and magma come into contact.

Volcanism and the Atmosphere

The relationship between volcanic eruptions and changes in weather and climate is another good example of Earth system interactions. The 1982 eruption of El Chichón in southern Mexico and the 1991 eruption of Mount Pinatubo in the Philippines injected sulfurous gases into the atmosphere tens of kilometers above the Earth. Through various chemical reactions, the gases formed an aerosol (a fine airborne mist) containing tens of millions of metric tons of sulfuric acid. This aerosol blocked enough of the Sun's radiation from reaching Earth's surface to lower global temperatures for a year or two. The eruption of Mount Pinatubo, one of the largest explosive eruptions of the century, led to a global

cooling of at least 0.5°C in 1992. (Chlorine emissions from Pinatubo also hastened the loss of ozone in the atmosphere, nature's shield that protects the biosphere from the Sun's ultraviolet radiation.)

The debris lofted into the atmosphere during the 1815 eruption of Mount Tambora in Indonesia resulted in even greater cooling. The next year, the Northern Hemisphere suffered a very cold summer; in New England, there were snowstorms in July! The drop in temperature and the ash fallout caused widespread crop failures. More than 90,000 people perished in that "year without a summer," inspiring Lord Byron to write a gloomy poem, "Darkness":

I had a dream, which was not all a dream.
The bright sun was extinguish'd, and the stars
Did wander darkling in the eternal space,
Rayless, and pathless, and the icy earth
Swung blind and blackening in the moonless air;
Morn came and went—and came, and brought no day.
And men forgot their passions in the dread
Of this their desolation; and all hearts
Were chill'd into a selfish prayer for light.

Figure 12.18 Old Faithful geyser, in Yellowstone National Park, erupts 60 m high about every 65 minutes. [Simon Fraser/SPL/Photo Researchers.]

THE GLOBAL PATTERN OF VOLCANISM

Before the advent of the plate tectonics theory, geologists noted a concentration of volcanoes around the rim of the Pacific Ocean and nicknamed it the Ring of Fire. The explanation of the Ring of Fire in terms of plate subduction was one of the great successes of the new theory. We have already examined how lava compositions vary with plate tectonic setting (see Figure 4.11). In this section, we show how plate tectonics can explain essentially all major features in the global pattern of volcanism.

Figure 12.19 shows the locations of the world's active volcanoes that occur on land or above the ocean's surface. About 80 percent are found at boundaries where plates converge, 15 percent where plates separate, and the remaining few within the plate interiors. There are many more active volcanoes than shown on this figure, however. Most of the lava erupted on Earth's surface comes from vents beneath the sea, located on the spreading centers of the mid-ocean ridges.

Basalt Production at Spreading Centers

Each year, approximately 3 km³ of basaltic lava erupts along the mid-ocean ridges in the process of seafloor spreading.

This is a truly enormous volume. In comparison, all the active volcanoes along the convergent boundaries (about 400) generate volcanic rock at a rate of less than 1 km³/year. Enough magma has erupted in seafloor spreading over the past 200 million years to create all of the present-day seafloor, which covers nearly two-thirds of Earth's surface. This "crustal factory" lies beneath a rift valley a few kilometers wide, and it extends along the thousands of kilometers of mid-ocean ridge (**Figure 12.20**). The magma and volcanic rocks are formed during decompression melting of mantle peridotite, as discussed in Chapter 4.

Hydrothermal Activity at Spreading Centers

Tensional fissures at the mid-ocean ridge crests allow seawater to circulate throughout the new oceanic crust. Heat from the hot volcanic rocks and deeper magmas drives a vigorous convection that pulls cold seawater into the crust, heats it through contact with magma, and expels the hot water back into the overlying ocean (see Figure 4.13).

Given the common occurrence of hot springs and geysers in volcanic geosystems on land, the evidence for pervasive hydrothermal activity at spreading centers immersed in deep water should come as no surprise. Nevertheless, geologists were amazed once they recognized the intensity of the convection and discovered some of its chemical and

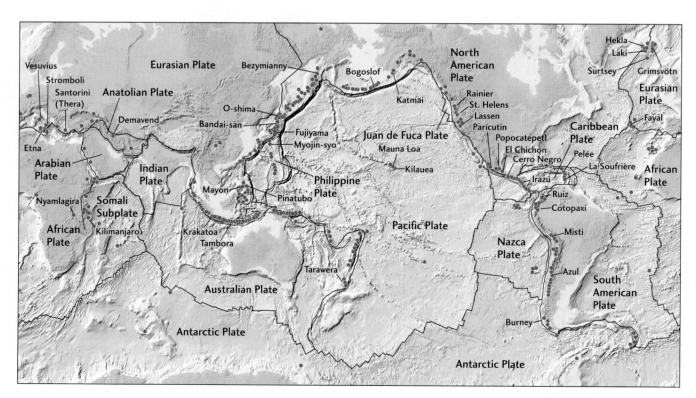

Figure 12.19 The active volcanoes of the world with vents on land or above the ocean surface (red dots). About 80 percent are found at boundaries where plates collide, 15 percent where plates separate, and the remaining few at intraplate hot spots. Black lines are plate boundaries. Not shown on this map are the numerous axial volcanoes of the mid-ocean ridge system below the water's surface.

Key Figure 12.20   Plate tectonics explains the global pattern of volcanism.

At ocean-ocean convergent boundaries, magmas originating from partial melting of the mantle give rise to volcanic island arcs erupting mostly basaltic lavas.

Magmas formed at ocean-continent convergences are mixtures of basalts from the mantle, remelted felsic continental crust, and materials melted off the top of the subducted plate. They give rise to volcanoes erupting andesitic lavas.

Island arc

Active volcano over hot spot

Continental volcanic belt

Extinct volcano

Mid-ocean ridge

Ocean plate

Hot spot

Plate separation at a mid-ocean ridge and magma drawn from a broad region of the asthenosphere result in basaltic volcanism and the creation of new oceanic crust and lithosphere.

Rising magma

Continental crust

Plate motion over hot spots creates midplate chain of basaltic volcanic islands.

Mantle plume

Continental mantle lithosphere

---

biological consequences. The most spectacular manifestations of this process were first found in the eastern Pacific Ocean in 1977. Plumes of hot, mineral-laden water with temperatures as high as 350°C spout as "black smokers" through hydrothermal vents on the mid-ocean ridge crest (see Figure 11.16). The rates of fluid flow turned out to be very high. Marine geologists have estimated that the entire volume of the ocean's water is circulated through the cracks and vents of the spreading centers in only 10 million years.

Scientists have come to realize that the interactions between hydrosphere and lithosphere at spreading centers profoundly affect the geology, chemistry, and biology of the oceans.

• The creation of new lithosphere accounts for almost 60 percent of the energy flowing out of Earth's interior. Circulating seawater cools the new lithosphere very efficiently and therefore plays a major role in the outward transport of Earth's internal heat.

• Hydrothermal activity leaches metals and other elements from the new crust, injecting them into the oceans. These elements contribute as much to seawater chemistry as the mineral components dumped into the oceans by all the world's rivers.

• Metal-rich minerals precipitate out of the circulating seawater and form ores of zinc, copper, and iron in shallow parts of the oceanic crust. The ores form when seawater sinks through porous volcanic rocks, is heated, and leaches these elements from the new crust. When the heated seawater, enriched with dissolved minerals, rises and reenters the cold ocean, the ore-forming minerals precipitate.

The hydrothermal energy and nutrients from circulating seawater feed unusual colonies of strange organisms whose energy comes from Earth's interior rather than from sunlight. Complex ecosystems at spreading centers include microorganisms that provide food for giant clams and for tube worms up to several meters long. These types of microorganisms also populate thermal springs with temperatures above the boiling point of water. Some scientists have speculated that life on Earth may have begun in the energetic, chemically rich environments at ocean spreading centers (see Chapter 11).

Volcanism in Subduction Zones

One of the most striking features of subduction zones is the chain of volcanoes that parallels the convergent boundary above the sinking slab of oceanic lithosphere, regardless of whether the overriding plate is oceanic or continental (see Figure 12.20). The magmas that feed subduction-zone volcanoes are produced by fluid-induced melting (see Chapter 4) and are more varied than the basalts of mid-ocean ridge volcanism. The lavas range from mafic to felsic—that is, from basaltic to rhyolitic—though intermediate (andesitic) compositions often dominate.

Where the overriding plate is oceanic lithosphere, these volcanoes and their products form the islands of ocean volcanic arcs, such as the Aleutian Islands of Alaska and the Mariana Islands of the western Pacific. Where subduction takes place beneath a continent, the volcanoes and volcanic rocks coalesce to form a mountainous volcanic belt on land, such as the Andes mountain range that marks the subduction of the oceanic Nazca Plate beneath continental South America. Similarly, subduction of the small Juan de Fuca Plate beneath western North America has generated the Cascade Range with its active volcanoes in northern California, Oregon, and Washington.

The terrain of the Japanese islands is a prime example of the complex of intrusives and extrusives that evolves over many millions of years at a subduction zone. Everywhere in this small country are all kinds of extrusive igneous rocks of various ages, mixed in with mafic and intermediate intrusives, metamorphosed volcanic rocks, and sedimentary rocks derived from erosion of the igneous rocks. The erosion of these various rocks has contributed to the distinctive landscapes portrayed in so many classical and modern Japanese paintings.

Intraplate Volcanism: The Mantle Plume Hypothesis

Decompression melting explains volcanism at spreading centers, and fluid-induced melting can account for the volcanism above subduction zones, but what mechanism produces *intraplate volcanism*—that is, volcanoes far from plate boundaries?

Hot Spots and Mantle Plumes Consider the Hawaiian Islands, which stretch across the middle of the Pacific Plate (see Figure 12.19). This island chain begins with the active volcanoes on the Big Island of Hawaii and continues to the northwest as a string of progressively older, extinct, eroded,

and submerged volcanic ridges and mountains. In contrast to the seismically active mid-ocean ridges, the Hawaiian chain is not marked by frequent large earthquakes (except near its volcanic center). It is essentially aseismic (without earthquakes) and is therefore called an *aseismic ridge*. Volcanically active **hot spots** at the beginnings of progressively older aseismic ridges can be found elsewhere in the Pacific and in other large ocean basins. The active volcanoes of Tahiti, at the southeastern end of the Society Islands, and the Galápagos Islands, at the western end of the aseismic Nazca Ridge, are two examples.

Once the general pattern of plate motions had been worked out, geologists were able to show that these aseismic ridges approximated the volcanic tracks that the plates would make over a set of hot spots fixed relative to one another, as if they were blowtorches anchored in Earth's mantle (**Figure 12.21**). Based on this evidence, they hypothesized that hot spots were the volcanic manifestations of hot, solid material rising in narrow, cylindrical jets from deep within the mantle (perhaps as deep as the core-mantle boundary), called **mantle plumes.** When the peridotites brought up in a mantle plume reach lower pressures at shallow depths, they begin to melt, producing basaltic magma. The magma penetrates the lithosphere and erupts at the surface. The current position of the plate over the hot spot is marked by an active volcano, which becomes inactive as the plate moves it away from the hot spot. The plate motion thus generates a trail of extinct, progressively older volcanoes. As shown in Figure 12.21, the Hawaiian Islands fit this pattern well, yielding a rate of movement of the Pacific Plate over the Hawaiian hot spot of about 100 mm/year.

Some aspects of intraplate volcanism within continents have also been explained using the plume hypothesis. Yellowstone is an example. The Yellowstone Caldera in northwestern Wyoming, only 600,000 years old, is still volcanically active with geysers, boiling springs, uplift, and earthquakes. It is the youngest member of a chain of sequentially older and now extinct calderas that supposedly mark the movement of the North American Plate over the Yellowstone hot spot (see Figure 12.21). The oldest member of the chain, a volcanic area in Oregon, erupted about 16 million years ago, producing the flood basalts of the Columbia Plateau. A simple calculation indicates that the North American Plate moved over the Yellowstone hot spot to the southwest at a rate of about 25 mm/year during the past 16 million years. Accounting for the relative motion of the Pacific and North American plates, this rate and direction are consistent with the plate motions inferred from Hawaii.

Assuming that hot spots are anchored by plumes coming from the deep mantle, geologists can use the worldwide distribution of their volcanic tracks to compute how the global system of plates is moving with respect to the deep mantle. The results are sometimes called "absolute plate motions" to distinguish them from the relative motions between plates. The absolute plate motions derived from the hot-spot tracks

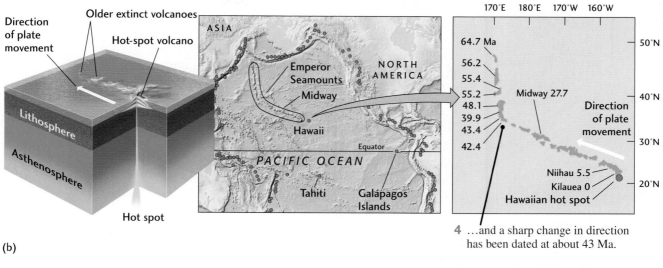

(a)

1 The Pacific Plate has moved northwest over the Hawaiian hot spot...

2 ...resulting in a chain of volcanic islands and seamounts.

3 The ages of the mountains are consistent with plate movement of about 100 mm/yr...

4 ...and a sharp change in direction has been dated at about 43 Ma.

(b)

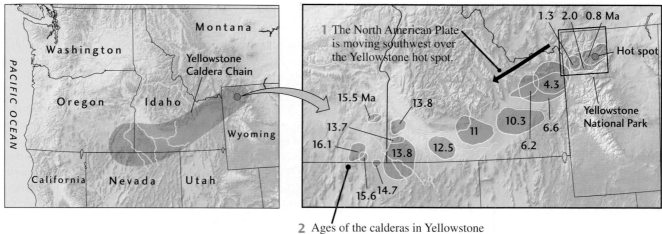

1 The North American Plate is moving southwest over the Yellowstone hot spot.

2 Ages of the calderas in Yellowstone chain can be dated up to 16 Ma.

Figure 12.21 Plate motion generates a trail of progressively older volcanoes. (a) The Hawaiian Island chain and its extension into the northwestern Pacific, showing the northwestward trend toward progressively older ages. (b) The Yellowstone volcanic track marks the movement of the North American Plate over a hot spot during the past 16 million years. [Wheeling Jesuit University/NASA Classroom of the Future.]

have helped geologists understand the forces driving the plates. Plates that are being subducted along large fractions of their boundaries—such as the Pacific, Nazca, Cocos, and Indian-Australian plates—are moving fast with respect to the hot spots, whereas plates without much subducting slab—such as the Eurasian and African plates—are moving slowly. This observation supports the hypothesis that the gravitational pull of the dense, sinking slabs is an important plate-driving force (see Chapter 2).

Hot-spot tracks can be used to reconstruct the history of plate motions relative to the deep mantle, just as the magnetic isochrons have allowed geologists to reconstruct how the plates have moved relative to one another. As we have

seen, this idea works fairly well for recent plate motions. Over longer periods of time, however, a number of problems arise. For instance, according to the fixed-hot-spot hypothesis, the sharp bend in the Hawaiian aseismic ridge 43 million years ago (where it becomes the north-trending Emperor seamount chain; see Figure 12.21) should coincide with an abrupt shift in Pacific Plate motion. However, no sign of a shift is evident in the magnetic isochrons, leading some geologists to question the fixed-hot-spot hypothesis. Others have pointed out that, in a convecting mantle, plumes would not necessarily remain fixed relative to one another but might be moved about by the shifting convection currents.

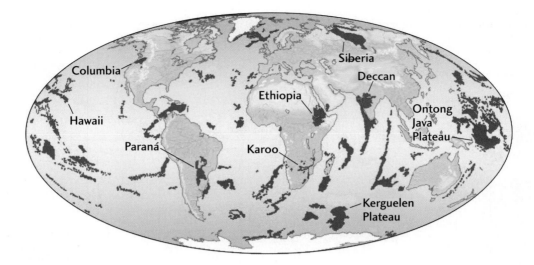

Figure 12.22 The global distribution of large igneous provinces on continents and in the ocean basins. These provinces are marked by unusually large outpourings of basaltic magmas through hot spots and fissure eruptions. They are hypothesized to be the result of major melting events caused by the arrival of plume heads at Earth's surface. [After M. Coffin and O. Eldholm, *Rev. Geophys.* 32 (1994):1–36, Figure 1.]

Almost all geologists accept the notion that hot-spot volcanism is caused by some type of upwelling in the mantle beneath the plates. However, the mantle plume hypothesis—that these upwellings are narrow conduits of material rising from the deep mantle—remains controversial. Even more controversial is the idea that plumes are responsible for the great outpourings of flood basalts and other large igneous provinces.

Large Igneous Provinces The origin of fissure eruptions of basalt on continents—such as those that formed the Columbia Plateau and even larger lava plateaus in Brazil-Paraguay, India, and Siberia—is a major puzzle. The geologic record shows that immense amounts of lava, up to several million cubic kilometers, can be released in a period as short as a million years. During these events, the lava eruption rates at a single spot on Earth's surface can equal the eruption rate of the entire mid-ocean ridge system!

Flood basalts are not limited to continents; they also create large oceanic plateaus, such as the Ontong-Java Plateau on the northern side of the island of New Guinea and major parts of the Kerguelen Plateau in the southern Indian Ocean (**Figure 12.22**). These features are all examples of what geologists call **large igneous provinces** (LIPs). LIPs are large volumes of predominantly mafic extrusive and intrusive rocks whose origins lie in processes other than normal seafloor spreading. LIPs include continental flood basalts and associated intrusive rocks, ocean basin flood basalts, and the aseismic ridges of hot spots. A global map of these provinces is given in Figure 12.22.

The volcanism that covered much of Siberia with lava is of special interest because it happened at the same time as the greatest extinction of species in the geologic record, which occurred at the end of the Permian period, about 251 million years ago. Some geologists think that the eruption caused the extinction, perhaps by polluting the atmosphere with volcanic gases that triggered a major climate change.

Many geologists believe that almost all LIPs were created at hot spots by mantle plumes. However, the amount of lava erupting from the most active hot spot on Earth today, Hawaii, is paltry compared to the enormous outpourings during flood basalt episodes. What explains these unusual bursts of basaltic magma from the mantle? Some geologists have speculated that they result when a new plume rises from the core-mantle boundary. According to this hypothesis, a large, turbulent blob of hot material—a "plume head"—leads the way. When this plume head reaches the top of the mantle, it generates a huge quantity of magma by decompression melting, which erupts in massive flood basalts (**Figure 12.23**). Others dispute this hypothesis, pointing out that continental flood basalts often seem to be associated with preexisting zones of weakness in the continental plates—suggesting that the magmas are generated by convective processes localized in the upper mantle. Sorting out the origins of LIPs is one of the most exciting areas of current geological research.

VOLCANISM AND HUMAN AFFAIRS

Large volcanic eruptions are not just of academic interest to geologists. They are a significant natural hazard to our society, which we must understand to reduce the risks they pose. In a world of growing human consumption, we also need to appreciate the benefits volcanoes bring in the form of geothermal energy and mineral resources.

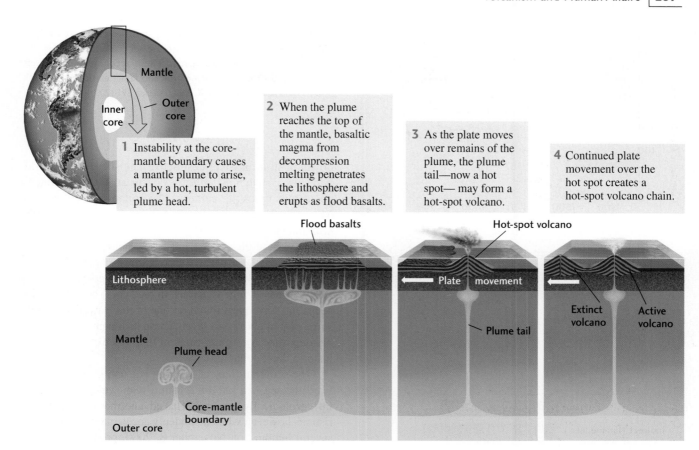

1 Instability at the core-mantle boundary causes a mantle plume to arise, led by a hot, turbulent plume head.

2 When the plume reaches the top of the mantle, basaltic magma from decompression melting penetrates the lithosphere and erupts as flood basalts.

3 As the plate moves over remains of the plume, the plume tail—now a hot spot— may form a hot-spot volcano.

4 Continued plate movement over the hot spot creates a hot-spot volcano chain.

Figure 12.23 A speculative model for the formation of flood basalts and other large igneous provinces. A new plume rises from the core-mantle boundary, led by a hot, turbulent plume head. When the plume head reaches the top of the mantle, it flattens, generating a huge volume of basaltic magma, which erupts as flood basalts.

Volcanic Hazards

Volcanic eruptions have a prominent place in human history and mythology. The original source of the lost continent of Atlantis myth may be the explosion of Thera, a volcanic island in the Aegean Sea (also known as Santorini). The eruption has been dated at 1623 B.C., and it formed a caldera 7 km by 10 km in diameter, visible today as a lagoon as much as 500 m deep, with two small active volcanoes in the center. The volcanic debris and tsunami resulting from this ancient catastrophe destroyed dozens of coastal settlements over a large part of the eastern Mediterranean. Some scientists have attributed the mysterious demise of the Minoan civilization to this cataclysm.

Of Earth's 500 to 600 active volcanoes, at least one in six is known to have claimed human lives. In the past 500 years alone, more than 250,000 people have been killed by volcanic eruptions. The statistics of these fatalities and their causes are shown in **Figure 12.24**.

Volcanoes can kill people and damage property in many ways, some of which are depicted in **Figure 12.25**. Several aspects of volcanic hazards are of special concern.

Lahars Among the most dangerous volcanic events are the torrential mudflows of wet volcanic debris called **lahars.** They can occur when a pyroclastic flow meets a river or a snowbank; when the wall of a crater lake breaks, suddenly releasing water; when a lava flow melts glacial ice; or when heavy rainfall transforms new ash deposits into mud. One extensive layer of volcanic debris in the Sierra Nevada of California contains 8000 km³ of material of lahar origin, enough to cover all of Delaware with a deposit more than a kilometer thick. Lahars have been known to carry huge boulders for tens of kilometers. When Nevado del Ruiz in the Colombian Andes erupted in 1985, lahars triggered by the melting of glacial ice near the summit plunged down the slopes and buried the town of Armero 50 km away, killing more than 25,000 people.

Flank Collapse A volcano is constructed from thousands of deposits of lava or ash or both, which is not the best way to build a stable structure. A volcano's sides can become too steep and break or slip off. In recent years, volcanologists have discovered many prehistoric examples of catastrophic structural failures in which a big piece of a volcano broke off, perhaps triggered by an earthquake, and slid

(a)

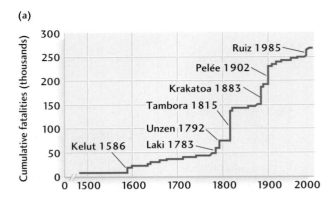

(b)

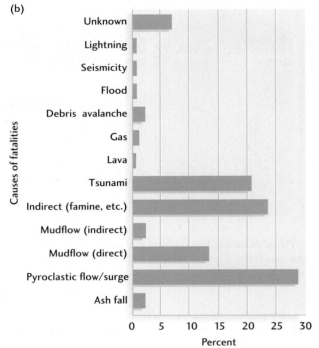

Figure 12.24 (a) Cumulative volcano statistics since A.D. 1500. The seven eruptions that dominate the record, each claiming 10,000 or more victims, are named. These account for two-thirds of the total deaths. (b) Cause of volcano fatalities since A.D. 1500. [After T. Simkin, L. Siebert, and R. Blong. *Science* 291 (2001): 255.]

downhill in a massive, destructive landslide. On a worldwide basis, such *flank collapses* occur at an average rate of about four times per century. The collapse of one side of Mount St. Helens was the most damaging part of its 1980 eruption (see Figure 12.4).

Surveys of the seafloor off Hawaii have revealed many giant landslides on the underwater flanks of the Hawaiian ridge. When they occurred, these massive movements most likely triggered huge tsunamis. In fact, coral-bearing marine sediments have been found on one of the Hawaiian Islands, some 300 m above sea level. These sediments were probably deposited by a giant tsunami that was excited by a prehistoric flank collapse.

The southern flank of Kilauea volcano is advancing toward the sea at a rate of 250 mm/year, which is relatively fast, geologically speaking. This advance became more worrisome, however, when it suddenly accelerated by a factor of several hundred on November 8, 2000, probably triggered by heavy rainfall a few days earlier. A network of motion sensors detected an ominous surge in velocity of about 50 mm/day lasting for 36 hours, after which the normal motion was reestablished. Someday, maybe thousands of years from now but perhaps sooner, the flank will probably break off and slide into the ocean. This catastrophic event would trigger a tsunami that could prove disastrous for Hawaii, California, and other Pacific coastal areas.

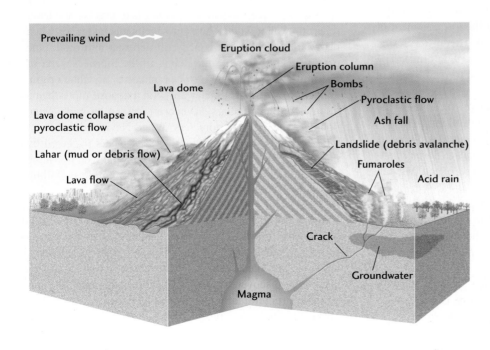

Figure 12.25 Some of the volcanic hazards that can kill people and destroy property. [B. Meyers et al./USGS.]

Caldera Collapse Though infrequent, the collapse of a large caldera is one of the most destructive natural phenomena on Earth. Monitoring caldera unrest is very important because of the long-term potential for widespread destruction. Fortunately, no catastrophic collapses have occurred in North America during recorded history, but geologists are wary of an increase in small earthquakes in Yellowstone and Long Valley calderas and other indications of activity in their underlying magma chambers. For example, carbon dioxide leaking into soil from magma deeper in the crust has been killing trees since 1992 on Mammoth Mountain, a volcano on the boundary of Long Valley Caldera. Other indications of resurgence of the Long Valley Caldera are uplift of the center of the caldera by more than half a meter in the past 20 years and the occurrence of nearly continuous swarms of small earthquakes. More than a thousand occurred in a single day in 1997.

Eruption Clouds With the growth in air travel, a volcanic hazard that is attracting increased attention is the encounter by commercial jet passenger planes of volcanic ash lofted into the air traffic lanes from erupting volcanoes. Over a period of 25 years, more than 60 airplanes have been damaged by such accidents. One Boeing 747 temporarily lost all four engines when ash from an erupting volcano in Alaska was sucked into the engines and caused them to flame out. Fortunately, the pilot was able to make an emergency landing. Warnings of eruptions ejecting volcanic ash near air traffic lanes are now being issued by several countries.

Intraplate Volcanism: The Mantle Plume Hypothesis

Volcanic eruptions cannot be prevented, but their catastrophic effects can be significantly reduced by a combination of science and enlightened public policy. Volcanology has progressed to the point that we can identify the world's dangerous volcanoes and characterize their potential hazards from deposits laid down in earlier eruptions. Some potentially dangerous volcanoes in the United States and Canada are identified in **Figure 12.26**. Assessments of their hazards can be used to guide zoning regulations to restrict land use—the most effective measure to reduce property losses and casualties.

Such studies indicate that Mount Rainier, because of its proximity to the heavily populated cities of Seattle and Tacoma, Washington, probably poses the greatest volcanic risk in the United States. Some 150,000 people live in areas where the geological record shows evidence of floods and lahars that have swept down from the volcano over the past 6000 years. An eruption that was not predicted could kill thousands of people and cripple the economy of the Pacific Northwest.

Predicting Eruptions These potential risks raise the question: Can volcanic eruptions be predicted? In many cases, the

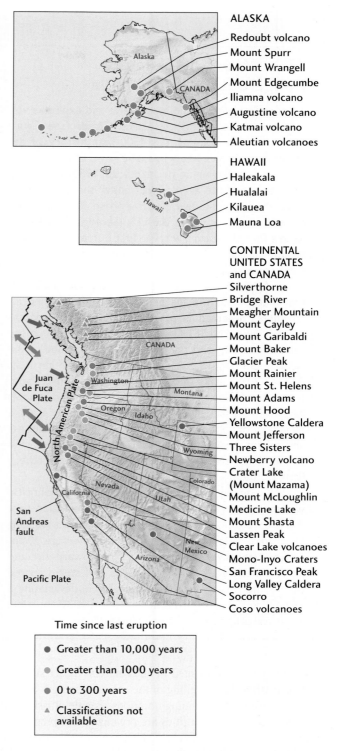

Figure 12.26 Locations of potentially hazardous volcanoes in the United States and Canada. Volcanoes within each U.S. group are color-coded blue, green, and red in the order of increasing probable cause for concern, subject to revision as studies progress (danger classifications are not available for Canadian volcanoes). Note the relationship between the volcanoes, which extend from northern California to British Columbia, and the subduction plate boundary between the North American Plate and the Juan de Fuca Plate. [After R. A. Bailey, P. R. Beauchemin, F. P. Kapinos, and D. W. Klick/USGS.]

EARTH POLICY

12.1 Mount St. Helens: Dangerous but Predictable

Mount St. Helens is the most active and explosive volcano in the contiguous United States (see Figure 12.4). It has a documented 4500-year history of destructive lava flows, hot pyroclastic flows, lahars, and distant ash falls. Beginning on March 20, 1980, a series of small to moderate earthquakes under the volcano signaled the start of a new eruptive phase after 123 years of dormancy, motivating the U.S. Geological Survey to issue a formal hazard alert. The first outburst of ash and steam erupted from a newly opened crater on the summit one week later.

In April, the seismic tremors increased, indicating that magma was moving beneath the summit, and instruments detected an ominous swelling of the northeastern flank. The U.S. Geological Survey issued a more serious warning, and people were ordered out of the vicinity. On May 18, the main eruption began abruptly. A large earthquake apparently triggered the collapse of the north side of the mountain, loosening a massive landslide, the largest ever recorded anywhere. As a huge flow of debris plummeted down the mountain, gas and steam under high pressure were released in a tremendous lateral blast that blew out the northern flank of the mountain.

U.S. Geological Survey geologist David A. Johnston was monitoring the volcano from his observation post 8 km to the north. He must have seen the advancing blast wave before he radioed his last message: "Vancouver, Vancouver, this is it!" A northward-directed jet of superheated (500°C) ash, gas, and steam roared out of the breach with hurricane force, devastating a zone 20 km outward from the volcano and 30 km wide. A vertical eruption sent an ash plume 25 km into the sky,

twice as high as a commercial jet flies. The ash cloud drifted to the east and northeast with the prevailing winds, bringing darkness at noon to an area 250 km to the east and depositing ash as deep as 10 cm on much of Washington, northern Idaho, and western Montana. The energy of the blast was equivalent to about 25 million tons of TNT. The volcano's summit was destroyed, its elevation was reduced by 400 m; and its northern flank disappeared. In effect, the mountain was hollowed out.

Local devastation was spectacular. Within an inner blast zone extending 10 km from the volcano, the thick forest was denuded and buried under several meters of pyroclastic debris. Beyond this zone, out to 20 km, trees were stripped of their branches and blown over like broken matchsticks aligned radially away from the volcano. As far as 26 km away, the hot blast was so intense that it overturned a truck and melted its plastic parts. Some fishermen were severely burned and survived only by jumping into a river. More than 60 other people were killed by the blast and its effects. A lahar formed when the landslide and pyroclastic debris—fluidized by groundwater, melted snow, and glacial ice—flowed 28 km down the valley of the Toutle River. The valley bottom was filled to a depth of 60 m. Beyond this debris pile, muddy water flowed into the Columbia River, where sediments clogged the ship channel and stranded many vessels in Portland.

Earthquakes and magmatic activity have continued off and on for the last 25 years. After more than a decade of relative quiescence, the volcano reawoke in September 2004, with a series of minor steam and ash eruptions that have continued into 2005. Growth of the central volcanic dome (see Figure 12.4) suggests that the current phase of eruptive activity may persist for some time into the future.

answer is yes. Instrumented monitoring can detect signals such as earthquakes, swelling of the volcano, and gas emissions that warn of impending eruptions. People at risk can be evacuated if the authorities are organized and prepared. Scientists monitoring Mount St. Helens were able to warn people of its eruption in 1980, as described in Feature 12.1. Government infrastructures were in place to evaluate the warnings and to enforce evacuation orders, so very few people were killed.

Another successful warning was issued a few days before the cataclysmic eruption of Mount Pinatubo in the Philippines on June 15, 1991. A quarter of a million people were evacuated, including some 16,000 residents of the nearby U.S. Clark Air Force Base (since permanently aban-

doned). Tens of thousands of lives were saved from the lahars that destroyed everything in their paths. Casualties were limited to the few who disregarded the evacuation order. In 1994, 30,000 residents of Rabaul, Papua New Guinea, were successfully evacuated by land and sea hours before two volcanoes on either side of the town erupted, destroying or damaging most of it. Many owe their lives to the government for conducting evacuation drills and to scientists at the local volcano observatory who issued a warning when their seismographs recorded the ground tremor that signaled magma moving toward the surface.

Improving our ability to predict eruptions is important because there are about 100 high-risk volcanoes in the world, and some 50 volcanic eruptions per year.

Figure 12.27 The Geysers, one of the world's largest supplies of natural steam. The geothermal energy is converted into electricity for San Francisco, 120 km to the south. [Pacific Gas and Electric.]

Controlling Eruptions Can we go further by actually controlling volcanic eruptions? Not likely, because large volcanoes release energy on a scale that dwarfs our capabilities for control. However, under special circumstances and on a small scale, the damage can be reduced. Perhaps the most successful attempt to manage volcanic activity was made on the Icelandic island of Heimaey in January 1973. By spraying advancing lava with seawater, Icelanders cooled and slowed the flow, preventing the lava from blocking the port entrance and saving some homes from destruction.

In the years ahead, the best policy for protecting the public will be the establishment of more warning and evacuation systems and more rigorous restriction of settlements in potentially dangerous locations.

Natural Resources from Volcanoes

We have seen something of the beauty of volcanoes and something of their destructiveness. Volcanoes contribute to our well-being in many, though often indirect, ways. Soils derived from volcanic materials are exceptionally fertile because of the mineral nutrients they contain. Volcanic rock, gases, and steam are also sources of important industrial materials and chemicals, such as pumice, boric acid, ammonia, sulfur, carbon dioxide, and some metals. Seawater circulating through fissures in the ocean-ridge volcanic system is a major factor in the formation of ores and in the maintenance of the chemical balance in the oceans.

Heat energy from volcanism—a form of *geothermal energy*—is being harnessed in more and more places. Underground water heated to temperatures of 80° to 180°C by magmas near the surface is being used to warm residential, commercial, and industrial spaces. Reykjavík, the capital of Iceland, is almost entirely heated by geothermal energy derived from the on-shore extension of the Mid-Atlantic Ridge.

Geothermal reservoirs with temperatures above 180°C are useful for generating electricity. They are present primarily in regions of recent volcanism as hot dry rock, natural hot water, or natural steam. One of the world's largest supplies of natural steam is The Geysers, located 120 km north of San Francisco (**Figure 12.27**). Generators there currently produce more than 600 megawatts of electricity. Some 70 geothermal electricity-generating plants operate in California, Utah, Nevada, and Hawaii, producing 2800 megawatts of power—enough to supply about a million people. Geothermal energy will never replace petroleum as a major source of power, but it may help to meet our future energy needs.

SUMMARY

Why study volcanoes? Volcanoes are important geosystems for three reasons: volcanism is a fundamental tectonic process in constructing Earth's crust; volcanoes can be major natural hazards as well as sources of energy and minerals; and lavas from volcanoes provide geologists with samples from which they can infer properties of Earth's interior.

Why does volcanism occur? The basic cause of volcanism is Earth's internal heat. Magma (molten rock inside the Earth) rises buoyantly to the surface because, as a fluid, it is less dense than surrounding rock. Volcanism occurs when magmas are extruded from pipes or cracks as lava or explode to produce pyroclastic materials.

What are the three major categories of lava? Lavas are classified as felsic (rhyolite), intermediate (andesite), or mafic (basalt) on the basis of the decreasing amounts of silica and the increasing amounts of magnesium and iron they contain.

How are the structure and terrain of a volcano related to the kind of lava it emits and the style of its eruption? The chemical composition and gas content of lava are important factors in the form an eruption takes. Basalt can be highly fluid. On continents, it can erupt from fissures and flow out in thin sheets to build a lava plateau. A shield volcano grows from repeated eruptions of basalt from vents. Silica-rich magma is more viscous and, when charged with gas, tends to erupt explosively. The resulting pyroclastic debris may pile up into a cinder cone or cover an extensive area with ash-flow sheets. A stratovolcano is built of alternating layers of lava flows and pyroclastic deposits. The rapid ejection of magma from a magma chamber a few kilometers below the surface, followed by collapse of the chamber's roof, results in a large surface depression, or caldera. Giant resurgent calderas are among the most destructive natural cataclysms.

How is volcanism related to plate tectonics? The ocean crust forms from basaltic magma that rises from the asthenosphere (as a result of decompression melting) into fissures of the ocean ridge-rift system where plates separate. All three major types of lava—basaltic, andesitic, and rhyolitic—can erupt in convergent zones. Andesitic lavas are common in the volcanic belts of ocean-continent convergent plate boundaries. Rhyolitic lavas are produced by the melting of felsic continental crust. Within plates, basaltic volcanism occurs above hot spots, which are manifestations of plumes of hot material that rise through the mantle.

What are some beneficial and hazardous effects of volcanism? Volcanoes can kill people and damage property by ash falls, mudflows (lahars), lavas, gases, pyroclastic flows, and tsunamis. Volcanic eruptions have killed about 250,000 people in the past 500 years. In Earth's evolution, volcanic eruptions released much of the water and gases that formed the oceans and atmosphere. Geothermal heat drawn from areas of recent volcanism is a useful source of energy. An important ore-forming process takes place when groundwater circulates around buried magma or seawater circulates through ocean-floor rifts.

KEY TERMS AND CONCEPTS

andesitic lava (p. 275)
ash-flow deposit (p. 282)
basaltic lava (p. 273)
caldera (p. 279)
crater (p. 279)
diatreme (p. 280)
fissure eruption (p. 281)
flood basalt (p. 282)
hot spot (p. 286)
hydrothermal activity (p. 283)
lahar (p. 289)
large igneous province (p. 288)
mantle plume (p. 286)
pyroclast (p. 276)
rhyolitic lava (p. 275)
shield volcano (p. 277)
stratovolcano (p. 279)
volcanic geosystem (p. 272)
volcano (p. 272)

EXERCISES

1. On Earth's surface as a whole, what process generates the greater volume of volcanic rock, decompression melting or fluid-induced melting? Which of these processes creates the more dangerous volcanoes?

2. What is the difference between magma and lava? Describe a geologic situation in which a magma does not form a lava.

3. What are the three major types of volcanic rocks and their intrusive counterparts? Is kimberlite one of these three types?

4. What type of volcano is the Arenal volcano, shown in Figure 12.7?

5. Most volcanism occurs near plate boundaries. What type of plate boundary can produce large amounts of rhyolitic lavas?

6. What evidence suggests that Yellowstone Caldera was produced by a mantle hot spot?

7. How do scientists predict volcanic eruptions?

THOUGHT QUESTIONS

1. What might be the effects on civilization of a Yellowstone-type caldera eruption, such as the one described at the beginning of the chapter?

2. Give a few examples of what geologists have learned about Earth's interior by studying volcanoes and volcanic rocks.

3. Why are eruptions of stratovolcanoes generally more explosive than those of shield volcanoes?

4. While on a field trip, you come across a volcanic formation that resembles a field of sandbags. The individual ellipsoidal forms have a smooth, glassy surface texture. What type of lava is this, and what information does this give you about its history?

5. Why are the volcanoes on the northwest side of the Hawaiian Islands dormant whereas those on the southeast side are more active?

6. How do interactions between volcanic geosystems and the climate system increase volcanic hazards?

EARTHQUAKES

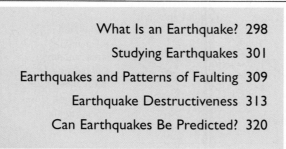

arthquakes rival all other natural disasters in the threat they pose to human life and property. We inhabit houses, office buildings, and transportation systems anchored in Earth's active crust, a fragile "built environment" extremely vulnerable to seismic movements and the ground ruptures, landslides, and tsunamis they can produce. Some events of the past century are sobering illustrations of this fact.

On a fine April morning in 1906, the citizens of northern California were awakened by the roar and violent shaking caused by the breaking of the San Andreas fault. This rupture extended over 400 km from the mission town of San Juan Batista all the way to Cape Mendocino, creating the largest earthquake the nation had yet experienced. The ensuing fires destroyed the city of San Francisco; by the time the flames died away, nearly 3000 of its inhabitants were dead. Almost a century later, on December, 26, 2004, a much larger fault broke west of the Indonesian island of Sumatra, lifting up the seafloor and sending a huge tsunami across the Indian Ocean. This planetary monster drowned more than 300,000 people living on coastlines from Thailand to Africa. In the time between these two memorable events, earthquakes killed an average of 10,000 people per year around the world.

To cope with the all-too-frequent destruction caused by earthquakes, we have long sought to improve our knowledge about where and when such events might occur and what happens when they do. Science has shown that seismic activity can be understood in terms of a basic machinery of deformation that shapes the face of the planet. As a result, attempts to reduce earthquake risk have become increasingly fused with the quest for a more fundamental understanding of the geologically active Earth.

This chapter will examine what happens during an earthquake and how Earth scientists locate, measure, and try to predict earthquakes. We will also consider what steps people can take to reduce the risk of losses from earthquakes.

In Chapter 2, we discussed how the observation that earthquakes are concentrated in narrow zones contributed to the development of plate tectonics theory. Here we will examine earthquake patterns in more detail. On a global scale, most earthquakes, such as the two just described, do indeed occur at the boundaries where plates converge, split apart, or slide past each other. However, in many earthquake-prone areas, such as California and Japan, the pattern of tectonic faulting that

Workers of a rescue operation stand before a building completely demolished in the September 1985 earthquake in Mexico City, Mexico. About 10,000 people were killed in Mexico City by this earthquake. [Owen Franken/Corbis.]

| Key Figure 13.1 | Elastic rebound theory explains the earthquake cycle. |

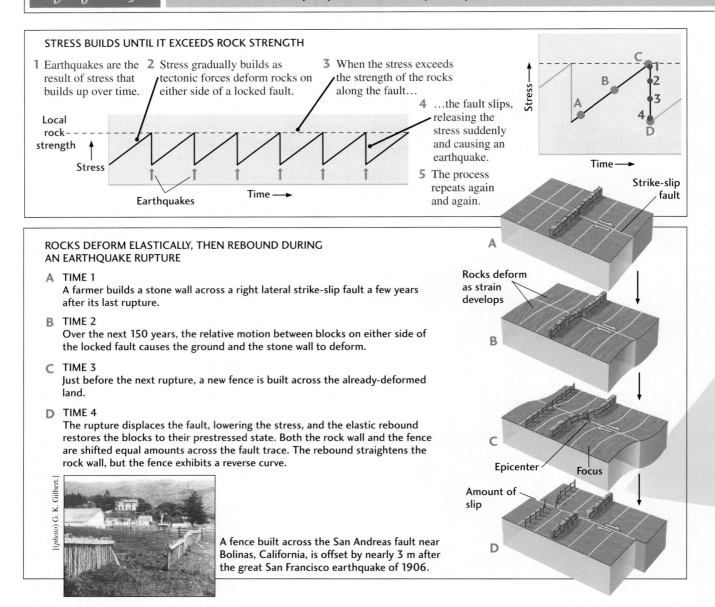

STRESS BUILDS UNTIL IT EXCEEDS ROCK STRENGTH

1 Earthquakes are the result of stress that builds up over time.

2 Stress gradually builds as tectonic forces deform rocks on either side of a locked fault.

3 When the stress exceeds the strength of the rocks along the fault…

4 …the fault slips, releasing the stress suddenly and causing an earthquake.

5 The process repeats again and again.

Local rock strength

Stress

Earthquakes

Time →

Stress ↑

Time →

Strike-slip fault

ROCKS DEFORM ELASTICALLY, THEN REBOUND DURING AN EARTHQUAKE RUPTURE

A TIME 1
A farmer builds a stone wall across a right lateral strike-slip fault a few years after its last rupture.

B TIME 2
Over the next 150 years, the relative motion between blocks on either side of the locked fault causes the ground and the stone wall to deform.

C TIME 3
Just before the next rupture, a new fence is built across the already-deformed land.

D TIME 4
The rupture displaces the fault, lowering the stress, and the elastic rebound restores the blocks to their prestressed state. Both the rock wall and the fence are shifted equal amounts across the fault trace. The rebound straightens the rock wall, but the fence exhibits a reverse curve.

Rocks deform as strain develops

Epicenter   Focus

Amount of slip

[(photo) G. K. Gilbert.]

A fence built across the San Andreas fault near Bolinas, California, is offset by nearly 3 m after the great San Francisco earthquake of 1906.

causes earthquakes is much more complex than the simplest models of plate tectonics might suggest. These complexities contribute to the difficulty of predicting earthquakes.

We cannot yet predict earthquakes reliably, although their destructiveness can be reduced considerably. To lessen earthquake damage, we must use our geological knowledge of where large earthquakes are likely to occur in designing buildings, dams, bridges, and other structures to withstand earthquake shaking.

WHAT IS AN EARTHQUAKE?

We have seen that plate movements generate forces in narrow zones at the boundaries between tectonic plates. These

global forces act on crustal rocks in ways that can be described by the concepts of stress, strain, and strength. *Stress* is the local force per unit area that causes rocks to deform. *Strain* is the relative amount of deformation, expressed as the percentage of distortion (for example, compression of a rock by 1 percent of its length). Rocks fail—that is, they lose cohesion and break into two or more parts—when they are stressed beyond a critical value called the *strength*. As we saw in Chapter 7, brittle rock formations on faults commonly fail when they are stressed beyond their strength. An **earthquake** occurs when rocks being stressed suddenly break along a new or preexisting fault. The two blocks of rock on either side of the fault slip suddenly, setting off ground vibrations, or **seismic waves** (from the Greek *seis-*

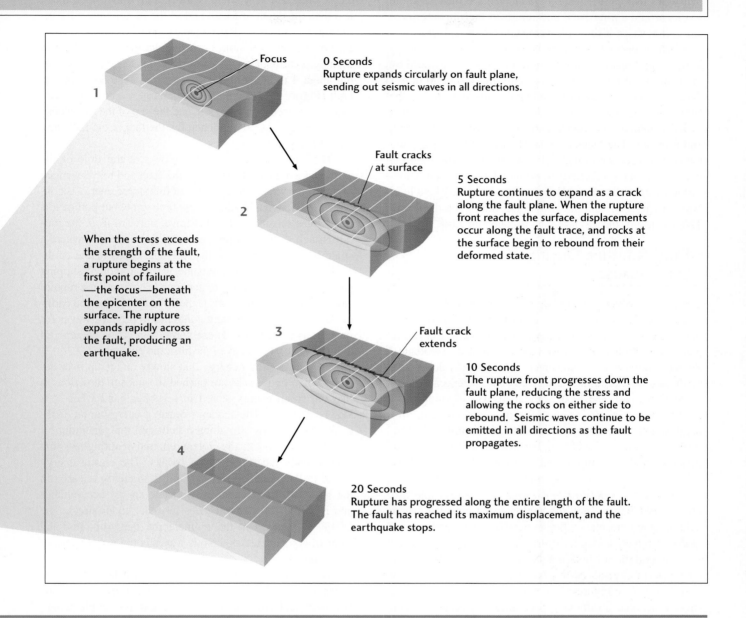

1 When the stress exceeds the strength of the fault, a rupture begins at the first point of failure —the focus—beneath the epicenter on the surface. The rupture expands rapidly across the fault, producing an earthquake.

Focus

0 Seconds
Rupture expands circularly on fault plane, sending out seismic waves in all directions.

2 Fault cracks at surface

5 Seconds
Rupture continues to expand as a crack along the fault plane. When the rupture front reaches the surface, displacements occur along the fault trace, and rocks at the surface begin to rebound from their deformed state.

3 Fault crack extends

10 Seconds
The rupture front progresses down the fault plane, reducing the stress and allowing the rocks on either side to rebound. Seismic waves continue to be emitted in all directions as the fault propagates.

4

20 Seconds
Rupture has progressed along the entire length of the fault. The fault has reached its maximum displacement, and the earthquake stops.

---

mos, meaning "shock" or "earthquake"), that are often destructive. When the fault slips, the stress is reduced, dropping to a level below the rock strength. After the earthquake, the stress begins to increase again, and the cycle is repeated (**Figure 13.1**). Earthquakes occur most often at plate boundaries, where stresses are concentrated and straining of the crust is intense.

Elastic Rebound Explains Why Earthquakes Occur

The earthquake on the San Andreas fault that devastated San Francisco in 1906 was studied more than any earthquake up to that time. In 1910, a geologist who investigated that catas-

trophe, Henry Fielding Reid of Johns Hopkins University, advanced the **elastic rebound theory** to explain why earthquakes occur.

To visualize what happens in an earthquake, imagine the following experiment carried out across a strike-slip fault between two crustal blocks. Suppose that surveyors had painted straight lines on the ground, running perpendicular to a fault and extending from one block to the other, as in Figure 13.1. The two blocks are being pushed in opposite directions by plate motions. The weight of the overlying rock presses them together, however, so friction locks them in place along the fault. They do not move, just as a car does not move when the emergency brake is engaged. Instead of slipping along the fault as stress builds up, the blocks are strained elastically near the fault, as shown by the bent lines in Figure 13.1. By *elastically,* we mean that the blocks would spring back and return to their undeformed, stress-free shape if the fault were suddenly to unlock.

As the slow plate movements continue to push the blocks in opposite directions, the strain in the rocks—evidenced by the bending of the survey lines—continues to build up for decades, centuries, or even millennia. At some point, the strength of the rocks is exceeded. The frictional bond that locks the fault can no longer hold somewhere along the fault, and it breaks. The blocks slip suddenly, and the rupture extends over a section of the fault. Figure 13.1 shows how the two blocks have rebounded—sprung back to their undeformed state—after the earthquake. The imaginary bent lines have straightened, and the two blocks have been displaced. The distance of the displacement is called the **fault slip.**

Fault Rupture During Earthquakes

The point at which the slip begins is the **focus** of the earthquake (see Figure 13.1). The **epicenter** is the geographic point on Earth's surface directly above the focus. For example, you might hear in a news report: "The U.S. Geological Survey reports that the epicenter of last night's destructive earthquake in California was located 6 kilometers east of Los Angeles city hall. The depth of the focus was 10 kilometers."

For most earthquakes occurring in continental crust, the focal depths range from about 2 to 20 km. Continental earthquakes are rare below 20 km, because under the high temperatures and pressures found at greater depths, the crust deforms as a ductile material and cannot support brittle fracture (just as hot wax flows when stressed, whereas cold wax breaks; see Chapter 7). In subduction zones, however, where cold oceanic lithosphere plunges back into the mantle, earthquakes originate at depths as great as 690 km.

The fault rupture does not happen all at once. It begins at the focus and expands outward on the fault plane, typically at 2 to 3 km/s (see Figure 13.1). The rupture stops where the stresses become insufficient to continue breaking the fault (such as where the rocks are stronger) or where the rupture enters ductile material in which it can no longer propagate as a fracture. As we will see later in this chapter, the size of an earthquake is related to the total area of fault rupture. Most earthquakes are very small, with rupture dimensions much less than the depth of focus, and so the rupture never breaks the surface. In large, destructive earthquakes, however, surface breaks are common. For example, the great 1906 San Francisco earthquake caused surface displacements averaging about 4 m along a 400-km section of the San Andreas fault (**Figure 13.2**). Faulting in the largest earthquakes can extend for more than 1000 km, and the slip of the two blocks can be as large as 20 m. Generally, the longer the fault rupture, the larger the fault slip.

The sudden slipping of the blocks at the time of the earthquake reduces the stress on the fault and releases much of the stored strain energy. Most of this stored energy is converted to frictional heating of the fault zone, but part of it is released as seismic waves that travel outward from the rupture, much as waves ripple outward from the spot where a stone is dropped into a still pond. The focus of the earthquake generates the first seismic waves, although slipping parts of the fault continue to generate waves until the rupture stops. In a large event such as the 1906 San Francisco earthquake, the propagating fault continues to produce waves for many tens of seconds. These waves can cause damage all along the fault break, even far from the epicenter.

The elastic strain energy that slowly builds up over decades when two blocks are pushed in opposite directions is like the strain energy stored in a rubber band when it is slowly stretched. The sudden release of energy in an earthquake, signaled by slip along a fault and the generation of seismic waves, is like the violent rebound or springback that occurs when the rubber band breaks. The elastic energy stored in the stretched rubber band is suddenly released in the backlash. In the same way, elastic energy accumulates and is stored for many decades in rocks under stress. The energy is released at the moment the fault ruptures, and some of it is radiated as seismic waves in the few minutes of an earthquake.

The elastic rebound model implies that there should be a periodic buildup and release of strain energy at faults. However, real earthquakes rarely exhibit this simple behavior, which is one reason earthquakes are so difficult to predict. For instance, all of the strain accumulated since the last earthquake may not be released in the next—that is, the

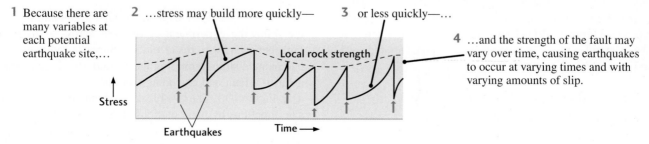

1 Because there are many variables at each potential earthquake site,...

2 ...stress may build more quickly—

3 or less quickly—...

Local rock strength

4 ...and the strength of the fault may vary over time, causing earthquakes to occur at varying times and with varying amounts of slip.

Stress

Earthquakes   Time →

Figure 13.2 Irregularities in the earthquake cycle can be caused by variations in rock strength and stress accumulation. [H. Kanamori and E. E. Brosky, *Physics Today* (June 2001): 34–39.]

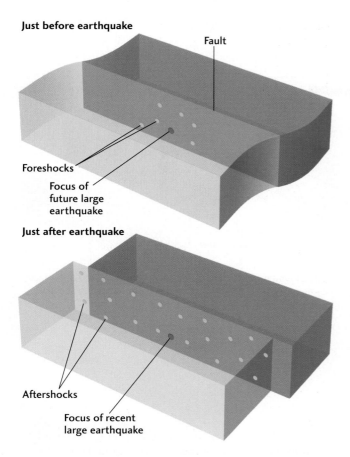

Just before earthquake

Fault

Foreshocks

Focus of
future large
earthquake

Just after earthquake

Aftershocks

Focus of recent
large earthquake

Figure 13.3 Foreshocks occur near, but before, the main shock. Aftershocks are smaller shocks caused by the main shock.

rebound may be incomplete—or the stress may not build up as steadily (see Figure 13.2).

Foreshocks and Aftershocks

An aftershock is an example of a complexity not described by simple elastic rebound. An **aftershock** is an earthquake that occurs as a consequence of a previous earthquake of larger magnitude. Aftershocks follow the main earthquake in sequences, and their foci are distributed in and around the rupture plane of the main shock (**Figure 13.3**).

The number and sizes of aftershocks depend on the main shock's magnitude, and their frequencies decrease inversely with time after the main shock. The aftershock sequence of a magnitude 5 earthquake might last for only a few weeks, whereas one for a magnitude 7 earthquake could persist for several years. The size of the largest aftershock is usually about one magnitude unit smaller than the main shock. In other words, a magnitude 7 earthquake might have an aftershock as big as magnitude 6. In populated regions, the shaking from large aftershocks can be very dangerous, compounding the damage caused by the main shock.

A **foreshock** (see Figure 13.3) is a small earthquake that occurs near, but before, a main shock. One or more foreshocks have preceded many large earthquakes, so scientists

have tried to use them to predict when and where large earthquakes might happen. Unfortunately, it is usually very hard to distinguish foreshocks from other small earthquakes that occur randomly on active faults, so this method has only rarely proved successful.

STUDYING EARTHQUAKES

As in any experimental science, instruments and field observations provide the basic data used to study earthquakes. These data enable investigators to locate earthquakes, determine their sizes and numbers, and understand their relationships to faults.

Seismographs

The **seismograph,** which records the seismic waves that earthquakes generate (**Figure 13.4**), is to the Earth scientist what the telescope is to the astronomer—a tool for peering into inaccessible regions. The ideal seismograph would be a device affixed to a stationary frame not attached to the Earth. When the ground shook, the seismograph could measure the changing distance between the frame, which did not move, and the vibrating ground, which did. As yet, we have no way to position a seismograph that is not attached to the Earth—although modern space technology is beginning to remove this limitation. So we compromise. A mass is attached to the Earth so loosely that the ground can vibrate up and down or side to side without causing much motion of the mass.

One way to achieve this loose attachment is to suspend the mass from a spring (see Figure 13.4a). When seismic waves move the ground up and down, the mass tends to remain stationary because of its inertia (an object at rest tends to stay at rest), but the mass and the ground move relative to each other because the spring can compress or stretch. In this way, the vertical displacement of the Earth caused by seismic waves can be recorded by a pen on chart paper or, nowadays, digitally on a computer.

Loose attachment of the mass can also be achieved with a hinge. A seismograph that has its mass suspended on hinges like a swinging gate (see Figure 13.4b) can record the horizontal motions of the ground. In modern seismographs, advanced electronic technology is used to amplify ground motion before it is recorded. These instruments can detect ground displacements of less than a billionth of a meter—an astounding feat, considering that such small displacements are of atomic size!

Seismic Waves

Install a seismograph anywhere, and within a few hours it will record the passage of seismic waves generated by an earthquake somewhere on Earth. The waves will have traveled from the earthquake focus through the Earth and arrived at the seismograph in three distinct groups. The first

(a)   Seismograph designed to detect vertical movement

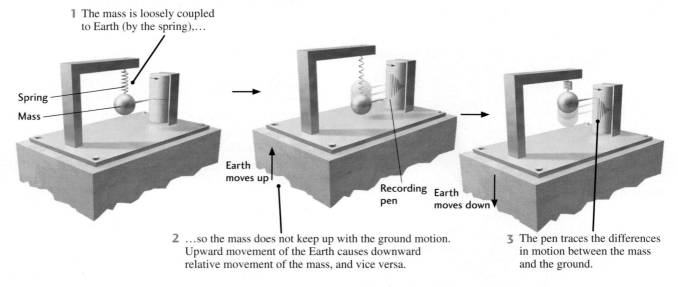

1 The mass is loosely coupled to Earth (by the spring),…

Spring

Mass

Earth moves up

Recording pen

Earth moves down

2 …so the mass does not keep up with the ground motion. Upward movement of the Earth causes downward relative movement of the mass, and vice versa.

3 The pen traces the differences in motion between the mass and the ground.

(b)   Seismograph designed to detect horizontal movement

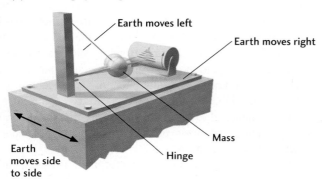

Earth moves left

Earth moves right

Earth moves side to side

Mass

Hinge

Figure 13.4 Seismographs record (a) vertical or (b) horizontal motion. Because of its loose coupling to the Earth through the spring (a) or hinge (b) and its inertia, the mass does not keep up with the motion of the ground. A typical observatory has instruments to measure three components of ground motion: up-down, horizontal east-west, and horizontal north-south.

waves to arrive are called primary waves or **P waves.** The secondary or **S waves** follow. Both P and S waves travel through Earth's interior. Finally come the **surface waves,** which travel around Earth's surface (**Figure 13.5**).

P waves in rock are analogous to sound waves in air, except that P waves travel through the solid rock of Earth's crust at about 6 km/s, which is about 20 times faster than sound waves travel through air. Like sound waves, P waves are *compressional waves,* so called because they travel through solid, liquid, or gaseous materials as a succession of compressions and expansions. P waves can be thought of as push-pull waves: they push or pull particles of matter in the direction of their path of travel.

S waves travel through solid rock at a little more than half the speed of P waves. They are called *shear waves* because they displace material at right angles to their path of travel. Shear waves do not exist in liquids or gases.

Surface waves are confined to Earth's surface and outer layers, like waves on the ocean. Their speed is slightly less than that of S waves. One type of surface wave sets up a rolling motion in the ground; another type shakes the ground sideways (see Figure 13.5).

People have felt seismic waves and witnessed their destructiveness throughout history, but not until the close of the nineteenth century were seismologists (scientists who study seismic waves and their sources) able to devise instruments to record them accurately. Seismic waves enable us to locate earthquakes and determine the nature of the faulting that produces them, and they provide our most important tool for probing Earth's deep interior.

Locating the Epicenter

Locating a quake's epicenter is like deducing the distance to a lightning bolt on the basis of the time interval between the flash of light and the sound of thunder—the greater the distance to the bolt, the larger the time interval. Light travels faster than sound, so the lightning flash may be likened to the P waves of earthquakes and the thunder to the slower S waves.

The time interval between the arrival of P and S waves depends on the distance the waves have traveled from the focus. This relationship is established by recording seismic waves from an earthquake or underground nuclear explosion that is at a known distance from the seismograph. To

Key Figure 13.5 Three different types of seismic waves are recorded by seismographs.

THE THREE DIFFERENT TYPES OF SEISMIC WAVES MOVE AT DIFFERENT SPEEDS

1 Seismic waves generated at an earthquake's focus travel through Earth and over its surface, arriving at a seismograph far from the earthquake.

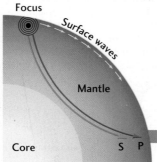

2 Primary, secondary, and surface waves travel at different speeds and arrive at the seismograph at different times.

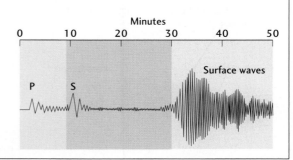

SEISMIC WAVES ARE CHARACTERIZED BY DISTINCT KINDS OF MOTION

P-wave motion

1 P waves (primary waves) are compressional waves—like sound waves—that travel quickly through rock.

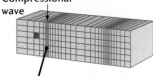

2 P waves travel as a series of contractions and expansions, pushing and pulling particles in the direction of their path of travel.

3 The red square charts the contraction and expansion of a section of rock.

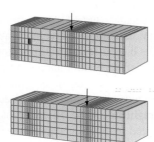

S-wave motion

4 S waves (secondary waves) travel at about half the speed of P waves.

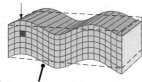

5 S waves are shear waves that push material at right angles to their path of travel.

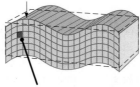

6 The red square shows how a section of rock shears from a square to a parallelogram as the S wave passes.

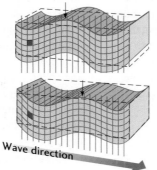

Surface-wave motion

7 Surface waves ripple across Earth's surface, where air above the surface allows free movement. There are two types of surface waves.

8 In one type, the ground surface moves in a rolling, elliptical motion that dies down with depth beneath the surface.

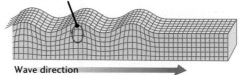

9 In the second type, the ground shakes sideways, with no vertical motion.

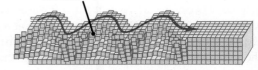

Key Figure 13.6 Readings at different seismographic stations reveal the location of the earthquake epicenter.

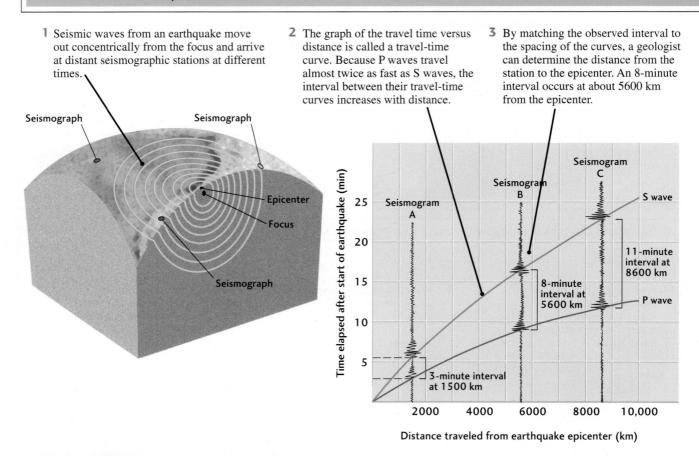

1 Seismic waves from an earthquake move out concentrically from the focus and arrive at distant seismographic stations at different times.

2 The graph of the travel time versus distance is called a travel-time curve. Because P waves travel almost twice as fast as S waves, the interval between their travel-time curves increases with distance.

3 By matching the observed interval to the spacing of the curves, a geologist can determine the distance from the station to the epicenter. An 8-minute interval occurs at about 5600 km from the epicenter.

determine the approximate distance to an epicenter, seismologists read from a seismogram the amount of time that elapsed between the arrival of the first P waves and the later arrival of the S waves. Then they use a table or a graph like the one shown in **Figure 13.6** to determine the distance from the seismograph to the epicenter. If they know the distances from three or more stations, they can locate the epicenter. They can also deduce the time of the shock at the epicenter, because the arrival time of the P waves at each station is known, and it is possible to determine from a graph or table how long the waves took to reach the station. Today, this entire process is carried out repeatedly by a computer until the data from a large number of seismographic stations agree on where the epicenter is, the time at which the earthquake began, and the depth of the focus below the surface.

Measuring the Size of an Earthquake

Locating earthquakes is only one step on the way to understanding them. The seismologist must also determine their sizes, or magnitudes. Other variables (such as the distance to the focus and the regional geology) being equal, an earthquake's magnitude is the main factor that determines the intensity of the seismic waves and thus the earthquake's potential destructiveness.

Richter Magnitude In 1935, Charles Richter, a California seismologist, devised a simple procedure to determine the size of an earthquake. Richter studied astronomy as a young man and learned that astronomers give each star a magnitude—a measure of its brightness. Adapting this idea to earthquakes, Richter assigned each earthquake a number, now called the _Richter magnitude_. Just as the brightness of stars varies over a huge range, so do the sizes of earthquakes. To compress his **magnitude scale,** Richter took the logarithm of the largest ground motion registered by a seismograph as his measure of earthquake size. He accounted for the distance between the seismograph and the fault rupture by correcting for the weakening of the seismic waves as they spread away from the focus (**Figure 13.7**).

4 If the geologist then draws a circle with a radius calculated from the travel-time curves around each seismographic station,…

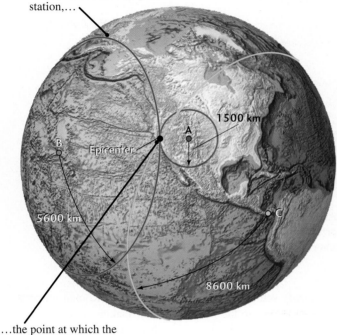

5 …the point at which the circles intersect will locate the earthquake's epicenter.

On Richter's scale, two earthquakes at the same distance from a seismograph differ by one magnitude if the size of their ground motions differs by a factor of 10. The ground motion of an earthquake of magnitude 3, therefore, is 10 times that of an earthquake of magnitude 2. Similarly, a magnitude 6 earthquake produces ground motions that are 100 times greater than those of a magnitude 4 earthquake. The energy released as seismic waves increases even more with earthquake magnitude, by a factor of 33 for each Richter unit. A magnitude 8 earthquake releases 33×33 or 1000 times the energy of a magnitude 6 earthquake.

Using Richter's procedure, seismologists anywhere could study their records and in a few minutes come up with nearly the same value for the magnitude of an earthquake, no matter how close or far away their instruments were from the focus. His method came to be used throughout the world.

Moment Magnitude Although the Richter magnitude is a household term, seismologists now prefer a measure of earthquake size more directly related to the physical properties of the faulting that causes the earthquake. The *moment magnitude* of an earthquake is defined as the product of the area and the average slip across the fault break; it increases by about one unit for every 10-fold increase in the area of faulting. Although both Richter's method and the moment method produce roughly the same numerical values, the moment magnitude can be measured more accurately from seismograms, and it can also be determined directly from field measurements of the fault.

Large earthquakes occur much less often than small ones. This observation can be expressed by a very simple relationship between earthquake frequency and magnitude

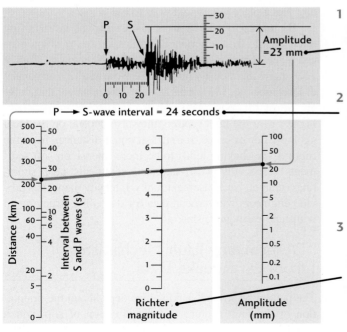

1 A geologist measures the amplitude of the largest seismic wave (23 mm)…

2 …and the time interval between the P-wave and S-wave arrivals (24 s) to determine the distance from the epicenter to the station (210 km).

3 By plotting the two measurements on these graphs and connecting the points, the geologist determines the Richter magnitude of the earthquake (5.0).

Figure 13.7 The maximum amplitude of the ground shaking and the P-S wave interval, indicated on the seismographic record, is used to assign a Richter magnitude to an earthquake. [California Institute of Technology.]

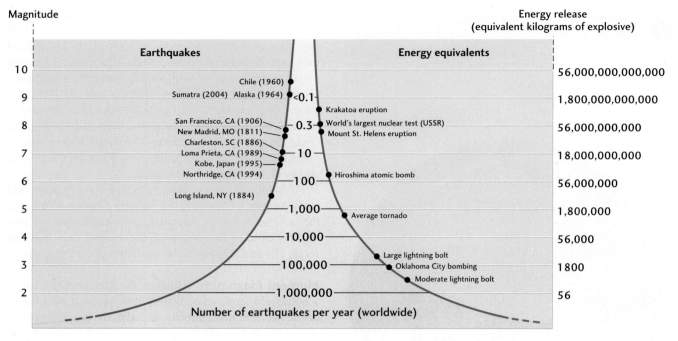

Magnitude

Energy release
(equivalent kilograms of explosive)

Figure 13.8 Relationship between moment magnitude (scale on left), earthquake energy release (scale on right), number of earthquakes per year worldwide (scale on horizontal lines in colored area), and other large sources of sudden energy release. [IRIS Consortium, http://www.iris.edu.]

(**Figure 13.8**). Worldwide, approximately 1,000,000 earthquakes with magnitudes greater than 2.0 take place each year, and this number decreases by a factor of 10 for each magnitude unit. Hence, there are about 100,000 earthquakes with magnitudes greater than 3, about 1000 with magnitudes greater than 5, and about 10 with magnitudes greater than 7. According to this statistical rule, there should be, on average, about 1 earthquake of magnitude 8 per year. In fact, magnitude 8 earthquakes occur on average only about once every 3 years, which implies that there is an upper limit to how big fault ruptures can be. Truly huge earthquakes, like the ones that occurred on thrust faults in the subduction zones off Sumatra in 2004 (moment magnitude 9.2), Alaska in 1964 (moment magnitude 9.2) and Chile in 1960 (moment magnitude 9.5), are rare.

Shaking Intensity Earthquake magnitude does not by itself describe the earthquake hazard, since the amplitude of the shaking also depends on distance from the fault rupture. A magnitude 8 earthquake in a remote area far from the nearest city might cause no human or economic losses, whereas a magnitude 6 quake immediately beneath a city is likely to cause serious damage.

In the late nineteenth century, before Richter invented his magnitude scale, seismologists and earthquake engineers developed methods for estimating the intensity of earthquake shaking directly from an event's destructive effects. Table 13.1 shows the scale that remains in most common use today, called the *modified Mercalli intensity* after Giuseppe Mercalli, the Italian scientist who proposed it in 1902. This

intensity scale assigns a value, given as a Roman numeral from I to XII, to the intensity of the shaking at a particular site. For example, a location where an earthquake is just barely felt by a few people is assigned an intensity of II, whereas one that was felt by nearly everyone is given an intensity of V. Numbers at the upper end of the scale describe increasing amounts of damage. The narrative attached to the highest value, XII, is tersely apocalyptic: "Damage total. Lines of sight and level are distorted. Objects thrown into the air." Not a place you would want to be!

By making observations at many sites and interviewing many people who experienced an earthquake, seismologists can make maps showing contours of equal intensity. **Figure 13.9** shows an intensity map for the New Madrid earthquake of December 16, 1811, a magnitude 7.5 event near the southern tip of Missouri. Although the intensities are generally highest near the fault rupture, they also depend on the local geology. For example, for sites at equal distances from the rupture, the shaking tends to be more intense on soft sediments (especially water-saturated sediments near shorelines) than on hard basement rocks. Intensity maps thus provide engineers with crucial data for designing structures to withstand earthquake shaking.

Determining Fault Mechanisms from Earthquake Data

The pattern of ground shaking also depends on the orientation of the fault rupture and the direction of slip, which together specify the **fault mechanism** of an earthquake. The

Table 13.1	Modified Mercalli Intensity Scale

Intensity Level	Description
I	Not felt.
II	Felt only by a few people at rest. Suspended objects may swing.
III	Felt noticeably indoors. Many people do not recognize it as an earthquake. Parked cars may rock slightly.
IV	Felt indoors by many, outdoors by few. Dishes, windows, doors rattle. Parked cars rock noticeably.
V	Felt by most; many awakened. Some dishes, windows broken. Unstable objects overturned.
VI	Felt by all. Some heavy furniture moves. Damage slight.
VII	Slight to moderate damage in well-built structures; considerable damage in poorly built structures; some chimneys broken.
VIII	Considerable damage in well-built structures. Damage great in poorly built structures. Fall of chimneys, factory stacks, columns, monuments, walls.
IX	Damage great in well-built structures, with partial collapse. Buildings shifted off foundations.
X	Some well-built wooden structures destroyed; most masonry and frame structures destroyed. Rails bent.
XI	Few if any masonry structures remain standing. Bridges destroyed. Rails bent greatly.
XII	Damage total. Lines of sight and level are distorted. Objects thrown into the air.

fault mechanism tells us whether the rupture was on a *normal, reverse,* or *strike-slip* fault. If the rupture was on a strike-slip fault, the fault mechanism also tells us whether the sense of motion was *right-lateral* or *left-lateral* (see Chapter 7 for the definition of these terms). We can then use this information to infer the regional pattern of tectonic forces (**Figure 13.10**).

For shallow ruptures that break the surface, we can sometimes determine the fault mechanism from field observations of the fault scarp. As we have seen, however, most ruptures are too deep to break the surface, so we must deduce the fault mechanism from the motions measured on seismographs.

For large earthquakes at any depth, this turns out to be easy to do, because there are enough seismographs around the world to surround the earthquake's focus. In some directions from the focus, the very first movement of the ground recorded by a seismograph—the P wave—is a *push away* from the focus, causing upward motion on a vertical seismograph. In other directions, the initial ground movement is a *pull toward* the focus, causing downward motion on a vertical seismograph. In other words, the slip on a fault looks like a push if you view it from one direction but like a pull if you view it from another. The pushes and pulls can be divided

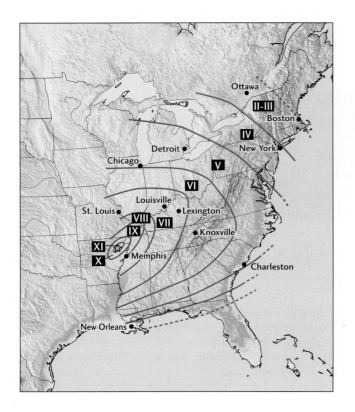

Figure 13.9 Modified Mercalli intensities measured for the New Madrid earthquake of December 16, 1811, a magnitude 7.5 event near the juncture of Missouri, Arkansas, and Tennessee. Regions near the fault rupture show intensities greater than IX, and intensities as high as VI were observed 200 km from the epicenter (see Table 13.1). [Carl W. Stover and Jerry L. Cossman, USGS Professional Paper 1527, 1993.]

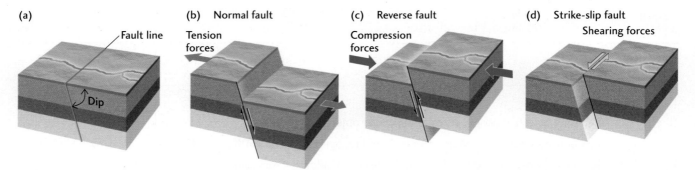

(a)

Fault line

Dip

(b) Normal fault

Tension forces

(c) Reverse fault

Compression forces

(d) Strike-slip fault

Shearing forces

Figure 13.10 The three main types of fault movements that initiate earthquakes and the stresses that cause them. (a) The situation before movement takes place; (b) normal faulting due to tensile stress; (c) reverse faulting due to compressive stress; (d) strike-slip faulting due to shearing stress (in this case, left-lateral).

into four sections based on the positions of the seismographic stations, as shown in **Figure 13.11**. One of the two boundaries is the fault orientation; the other is a plane perpendicular to the fault. The slip direction is determined from the arrangement of pushes and pulls. In this manner, without surface evidence, seismologists can deduce whether the horizontal crustal forces that triggered an earthquake were tensional, compressive, or shear.

GPS Measurements and "Silent" Earthquakes

As discussed in Chapter 2, GPS stations can record the slow movements of plates. These instruments can also measure the strain that builds up from such movements, as well as the sudden slip on a fault when it ruptures in an earthquake. Seismologists are now using GPS observations to study another kind of movement along active faults. It has been known for many years that a section of the San Andreas fault in central California creeps continuously rather than rupturing suddenly. This creep slowly deforms structures and cracks pavements that cross the fault trace. Recently, new networks of GPS stations have found surface movements at convergent plate boundaries that reflect transient (short-lived) creep events. These events may last a few weeks at a time. They have been named _silent earthquakes_ because the gradual movements do not trigger destructive seismic waves. Nevertheless, they can release large amounts of stored strain energy.

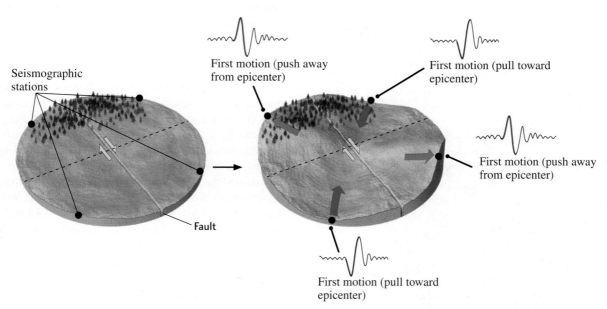

Seismographic stations

First motion (push away from epicenter)

First motion (pull toward epicenter)

First motion (push away from epicenter)

First motion (pull toward epicenter)

Fault

Figure 13.11 The first motion of P waves arriving at seismographic stations is used to determine the orientation of the fault plane and the direction of slip. The case shown here is for the rupture of a right-lateral strike-slip fault. Note that the alternating pattern of pushes and pulls would remain the same if the plane perpendicular to the fault ruptured with left-lateral displacement. Seismologists can usually choose between the two possibilities using additional information, such as field mapping of the fault trace or alignments of aftershocks along the fault plane.

These observations raise many questions that geologists are now trying to answer. Why gradual creep and not sudden breaking? What causes faults to stick and slip catastrophically in some places and creep in others? Will the release of strain energy by slow events make destructive earthquakes in these regions less frequent or less severe? Can these slow events be used to predict earthquakes? Scientific research is under way to address all these questions.

EARTHQUAKES AND PATTERNS OF FAULTING

Networks of sensitive seismographs locate earthquakes around the globe, measure their magnitudes, deduce their fault mechanisms, and reveal new information about tectonic processes on scales much smaller than the plates themselves. In this section, we summarize the global pattern of earthquake occurrence from the perspective of plate tectonics and show how regional studies of active fault networks are improving our understanding of earthquakes along plate boundaries and within plate interiors.

The Big Picture: Earthquakes and Plate Tectonics

The **seismicity map** in **Figure 13.12** shows the epicenters of earthquakes recorded around the world in a 27-year period. The most obvious features of this map, known to seismologists for many decades, are the earthquake belts that mark the major plate boundaries.

Divergent Boundaries The narrow belts of earthquakes that run through the ocean basins coincide with mid-ocean ridge crests and their offsets on transform faults, as shown in Figure 13.12. The fault mechanisms of the ridge-crest earthquakes, revealed by analysis of the first P-wave motion, correspond to normal faulting. The faults strike parallel to the ridges and dip toward the mid-ocean rift valley. Normal faulting indicates that tensional forces are at work as the plates are pulled apart during seafloor spreading, which explains why rift valleys develop at the ridge crests. Earthquakes also have normal fault mechanisms in zones where continental crust is being pulled apart, such as in the East African rift valleys and in the Basin and Range province of western North America.

Transform-Fault Boundaries Earthquake activity is even greater along the transform-fault boundaries that offset the ridge segments. These earthquakes show strike-slip fault mechanisms—just as one would expect where plates slide past each other in opposite directions. Moreover, for earthquakes along transform faults between ridge segments, the slip indicated by the fault mechanisms is left-lateral where the ridge crest steps right and right-lateral where it steps left

(see Figure 13.12). These directions are the opposite of what would be needed to create the offsets of the ridge crest but are consistent with the direction of slip predicted by seafloor spreading. In the mid-1960s, seismologists used this diagnostic property of transform faults to support the hypothesis of seafloor spreading. The motions on transform faults that run through continental crust, such as California's San Andreas fault and New Zealand's Alpine fault (both right-lateral), also agree with the predictions of plate tectonics.

Convergent Boundaries The world's largest earthquakes occur at convergent plate boundaries. Examples include the great earthquakes in Sumatra (2004), Alaska (1964) and Chile (1960). During the Chile earthquake—the biggest ever recorded (moment magnitude 9.5)—the crust of the Nazca Plate slipped an average of 15 m beneath the crust of the South American Plate on a fault rupture with an area larger than Kansas! The fault mechanisms show that these great earthquakes occur by horizontal compression along huge thrust faults, called _megathrusts_, that form the boundaries where one plate subducts beneath another (see Figure 13.12). Large ruptures on megathrusts can displace the seafloor, generating devastating tsunamis such as the one produced by the 2004 Sumatra megathrust earthquake (see Feature 13.1).

Earth's deepest earthquakes also occur at convergent boundaries. Almost all earthquakes originating below 100 km rupture the descending oceanic plate in subduction zones (see Figure 13.12). The fault mechanisms of these deep earthquakes show a variety of orientations, but they are consistent with the deformation expected within the descending plate as gravity pulls it back into the convecting mantle. The deepest earthquakes, located at depths of 600 to 700 km, take place in the oldest—and therefore coldest—descending plates, such as those beneath South America, Japan, and the island arcs of the western Pacific Ocean.

Intraplate Earthquakes Although most earthquakes occur at plate boundaries, the map in Figure 13.12 shows that a small percentage of global seismicity originates within plate interiors ("intraplate"). The foci of these earthquakes are relatively shallow, and most occur on continents. Among these earthquakes are some of the most famous in American history: a sequence of three large events near New Madrid, Missouri, in 1811–1812; the Charleston, South Carolina, earthquake of 1886; and the Cape Ann earthquake, near Boston, Massachusetts, in 1755. Many of these _intraplate earthquakes_ occur on old faults that were once parts of ancient plate boundaries. The faults no longer form plate boundaries but remain zones of crustal weakness that concentrate and release intraplate stresses.

One of the deadliest intraplate earthquakes (magnitude 7.6) occurred near Bhuj in Gujarat State, western India, in 2001. It is estimated that some 20,000 lives were lost. The epicenter was 1000 km south of the boundary between the Indian Plate and the Eurasian Plate. Indian geologists believe

Earth System Figure 13.12 | Earthquakes indicate how tectonic plates interact at their boundaries.

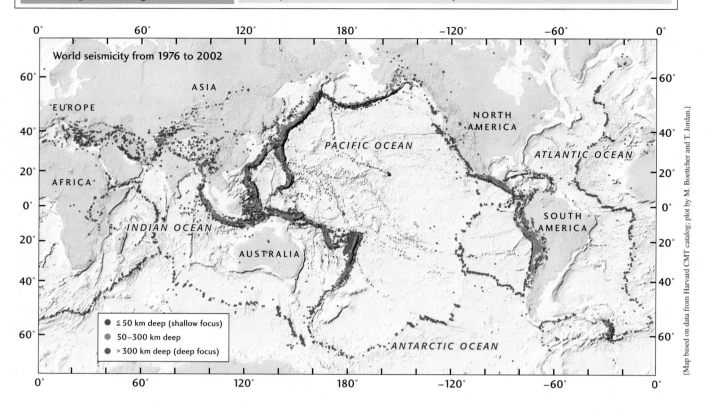

World seismicity from 1976 to 2002

● ≤ 50 km deep (shallow focus)
● 50–300 km deep
● > 300 km deep (deep focus)

[Map based on data from Harvard CMT catalog; plot by M. Boettcher and T. Jordan.]

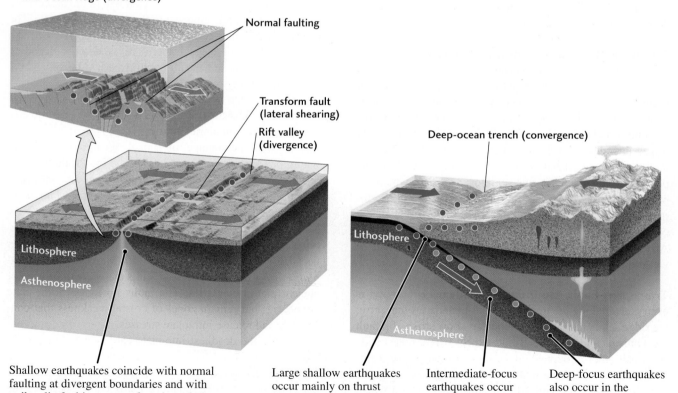

Mid-ocean ridge (divergence)

Normal faulting

Transform fault (lateral shearing)

Rift valley (divergence)

Deep-ocean trench (convergence)

Lithosphere

Asthenosphere

Lithosphere

Asthenosphere

Shallow earthquakes coincide with normal faulting at divergent boundaries and with strike-slip faulting at transform boundaries.

Large shallow earthquakes occur mainly on thrust faults at the plate boundary.

Intermediate-focus earthquakes occur in the descending slab.

Deep-focus earthquakes also occur in the descending slab.

that the compressive stresses responsible for the Bhuj earthquake originated in the continuing northward collision of India with Eurasia. Apparently, strong crustal forces can still develop and cause faulting within a lithospheric plate far from modern plate boundaries—in this case triggering a previously unknown thrust fault at a depth of about 20 km.

Regional Fault Systems

Although most major earthquakes conform to the types of faulting predicted by plate tectonics, a plate boundary can rarely be described as just one fault, particularly when the boundary involves continental crust. Rather, the zone of deformation between two moving plates usually comprises a network of interacting faults—a *fault system*. The fault system in southern California provides an interesting example (**Figure 13.13**).

The "master fault" of this system is our old nemesis, the San Andreas—a right-lateral strike-slip fault that runs northwestward through California from the Salton Sea near the Mexican border until it goes offshore in the northern part of the state (see Figure 7.7). There are a number of

(a) Southern California fault traces

In the Carrizo Plain of central California, the San Andreas fault is parallel to the Pacific–North American plate motion, and the faulting is right-lateral strike slip.

The westward "Big Bend" in the San Andreas fault causes the Pacific Plate to compress against the North American Plate, causing thrust faulting in the Los Angeles region south of the fault. This convergence raises the San Gabriel Mountains.

(b) Southern California earthquakes (July 1970–June 1995)

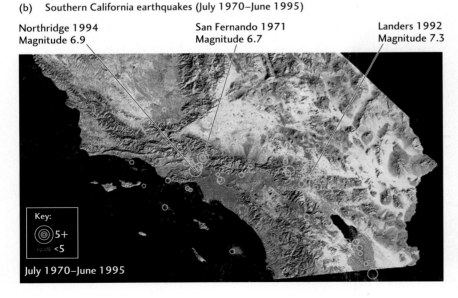

Figure 13.13 (a) Fault system of southern California, showing the surface traces of the San Andreas and other active faults. (b) Locations of earthquakes during the period July 1970–June 1995. [Southern California Earthquake Center.]

EARTH ISSUES

13.1 The Great Indian Ocean Tsunami of 2004

The disastrous Indian Ocean tsunami was produced by the Sumatra earthquake of December 26, 2004. This magnitude 9.2 event occurred on the megathrust marked by the Sunda Trench, where the Indian and Australian plates are subducted beneath the island arcs of Southeast Asia. The focus of the earthquake was offshore northern Sumatra at a depth of about 30 km. The rupture continued for 9 minutes, moving northward 1300 km along the shallow-dipping thrust fault—the longest single fault break ever recorded. During the earthquake, the overriding plate shifted toward the trench by 10 to 20 m, uplifting the seafloor. The energy of the displaced water mass above the rupture spread away in a series of sea waves. Owing to the fault geometry, more energy was directed east and west than north and south.

Within 15 minutes, the first wave ran up the Sumatran coastline southwest of the city of Banda Ache. Few eyewitnesses survived, but geologic investigations after the tsunami indicate that the maximum wave height on the beaches of the west-facing coast was about 15 m (see lower photograph), and the run-up attained heights of 25 to 35 m, reaching inland up to 2 km and wiping out most built structures, vegetation, and human life in its path. It is believed that more than 200,000 people perished along the Sumatran coastline, though no one will ever be sure because many bodies were washed out to sea.

The tsunami moved faster in the deep ocean to the west and slower in the shallow waters to the east, so that the waves arrived at the heavily populated coastlines of Sri Lanka and Thailand at nearly the same time, some

The tsunami struck without warning on a beach of Phuket, Thailand. [David Rydevik.]

2 hours after the earthquake. In Sri Lanka and Thailand, more than 40,000 and 15,000 were killed, respectively. The tsunami struck India at 3 hours (15,000 deaths), the Maldive Islands at 4 hours (108 deaths), and Somalia on the African coast at 9 hours (298 deaths). Enough time was available to forewarn the coastal residents of the impending tsunami, but no system for this purpose existed in the Indian Ocean, so the waves appeared without any warning (see upper photograph).

A small headland near Banda Ache on the west coast of Sumatra previously covered by dense jungle to the waterline but now stripped clean to height of about 15 m by the tsunami. [José Borrero, University of Southern California/Tsunami Research Group.]

subsidiary faults on either side of the San Andreas, however, that generate large earthquakes. In fact, most of the damaging earthquakes in southern California during the last century have occurred on these subsidiary faults.

Why is the San Andreas fault system so complex? Part of the explanation has to do with the geometry of the San Andreas fault itself. Notice in Figure 13.13a that although the strike of the San Andreas is approximately in the direction of the relative motion between the Pacific and North American plates, as expected for a transform-fault boundary, the fault is not straight but curves westward just north of Los Angeles. This "Big Bend" causes compression in the region, which is taken up by thrust faulting south of the San Andreas. The thrust faulting was responsible for two recent deadly earthquakes, the San Fernando earthquake of 1971 (magnitude 6.6, 65 people killed) and the Northridge earthquake of 1994 (magnitude 6.9, 58 people killed). Over the past several million years, thrust faulting associated with the Big Bend has raised the San Gabriel Mountains to elevations of 1800 to 3000 m.

Another complication is the extensional deformation taking place east of California in the Basin and Range province, which spans the states of Nevada and much of Utah and Arizona (see Chapter 7). This broad zone of divergence connects up with the San Andreas system through a series of faults that run along the eastern side of the Sierra Nevada and through the Mojave Desert. Faults of this system were responsible for the 1992 Landers earthquake (magnitude 7.3) and the 1999

Hector Mine earthquake (magnitude 7.1), as well as the 1872 Owens Valley earthquake (magnitude 7.6).

EARTHQUAKE DESTRUCTIVENESS

Over the last century, earthquakes worldwide have caused an average of 13,000 deaths per year and hundreds of billions of dollars in economic losses. Two California earthquakes—the 1989 Loma Prieta earthquake (magnitude 7.1), which occurred on the San Andreas fault some 80 km south of San Francisco, and the 1994 Northridge earthquake—were among the costliest disasters in U.S. history. Damage amounted to more than $10 billion in the Loma Prieta earthquake and $40 billion in the Northridge quake because of their proximity to urban centers. About 60 people died in each event, but the death toll would have been many times higher if stringent earthquake-resistant building codes had not been in place (**Figure 13.14**).

Destructive earthquakes are even more frequent in Japan than in California. The recorded history of destructive earthquakes in Japan, going back 2000 years, has left an indelible impression on the Japanese people. Perhaps that is why Japan is the best prepared of any nation in the world to deal with earthquakes. It has impressive public education campaigns, building codes, and warning systems. Despite this preparedness, more than 5600 people were killed in a devastating (magnitude 6.9) earthquake that

Figure 13.14 Sixteen people died in the Northridge Meadows apartment building in Los Angeles during the 1994 Northridge earthquake. The victims lived on the first floor and were crushed when the upper levels collapsed. Many more buildings would have collapsed if the newer buildings in the area had not been constructed according to stringent codes for earthquake resistance. [Nick Ut, Files/AP Photo.]

Figure 13.15 This elevated expressway, in Kobe, Japan, was overturned during an earthquake in 1995. [Tom Wagner/Corbis/SABA.]

struck Kobe on January 16, 1995 (**Figure 13.15**). The casualties and the enormous failure of structures (50,000 buildings destroyed) resulted partly from the less stringent building codes that were in effect before 1980, when much of the city was built, and from the location of the earthquake rupture so close to the city.

How Earthquakes Cause Damage

Earthquakes can damage our environment in several ways. They proceed as chain reactions in which the primary effects of faulting and ground shaking induce secondary effects, such as landslides and tsunamis, as well as destructive processes within our built environment, particularly fires.

Faulting and Shaking The *primary hazards* are the breaks in the ground surface that occur when faults rupture the surface, the permanent subsidence and uplift of the ground surface caused by faulting, and the ground shaking caused by seismic waves radiated during the rupture. Seismic waves can shake structures so hard that they collapse. The ground accelerations near the epicenter of a large earthquake can

approach and even exceed the acceleration of gravity, so an object lying on the surface can literally be thrown into the air. Very few structures built by human hands can survive such severe shaking, and those that do are severely damaged. The collapse of buildings and other structures is the leading cause of economic damage and casualties during earthquakes, including the high death tolls in the Tangshan, China, earthquake of 1976 (more than 240,000 killed); the Spitak, Armenia, earthquake of 1988 (25,000 killed); the Izmit, Turkey, earthquake of 1999 (15,600 killed); the Bhuj, India, earthquake of 2001 (20,000 killed); the Bam, Iran, earthquake of 2003 (30,000 killed); and the Pakistan earthquake of 2005 (86,000 killed).

Landslides and Other Types of Ground Failure The primary effects of faulting and ground shaking generate *secondary hazards* such as slumps, landslides, and other forms of ground failure. When seismic waves shake water-saturated soils, the soils can behave like a liquid and become unstable. The ground simply flows away, taking buildings, bridges, and everything else with it. Soil *liquefaction* destroyed the residential area of Turnagain Heights near Anchorage, Alaska, in the 1964 earthquake (see Figure 16.16); the Nimitz Freeway near San Francisco in the 1989 Loma Prieta earthquake; and areas of Kobe in the 1995 earthquake.

In some instances, ground failures can cause more damage than the ground shaking itself. A great earthquake in China's Kansu Province in 1920 triggered an extensive debris flow that covered a region larger than 100 km² and resulted in roughly 200,000 deaths. An immense rock and snow landslide (over 80 million cubic meters) triggered by a 1970 earthquake in Peru destroyed the mountain towns of Yungay and Ranrahirca, killing 66,000 (**Figure 13.16**). Many of the 844 people killed in the 2001 El Salvador earthquake were buried by a muddy landslide loosened from a slope in the suburbs of the country's capital, San Salvador.

Tsunamis A large earthquake that occurs beneath the ocean can generate a destructive sea wave, sometimes called a tidal wave but more accurately named a **tsunami,** the Japanese term for "harbor wave." Tsunamis are by far the deadliest and most destructive hazards associated with the world's largest earthquakes—the megathrust events that occur in subduction zones. When a megathrust ruptures, it can push the seafloor landward of the oceanic trench upward by as much as 10 m, displacing a large mass of the overlying ocean water. This disturbance flows outward in a wave that travels across the ocean at speeds of up to 800 km/hour (about as fast as a commercial jetliner). In the deep ocean, a tsunami is hardly noticeable, but when it approaches shallow coastal waters, the waves slow down and pile up, inundating the shoreline in walls of water that can reach heights of tens of meters (**Figure 13.17**).

The destructive power of a great tsunami was brought home by the terrifying video images captured on December 26, 2004, as the tsunami unleashed by the magnitude 9.2 Sumatra earthquake swept over low-lying coastal areas of the Indian Ocean from Indonesia and Thailand to Sri Lanka,

Figure 13.16 The mountain towns of Yungay and Ranrahirca, Peru, were buried when the magnitude 8 earthquake of 1970 triggered a landslide. [Courtesy of Servicio Aerofotografico Nacional de Peru, June 13, 1970/George Plafker/USGS.]

India, and the east coast of Africa. This event, one of the worst calamities in recorded history, is described in Feature 13.1. Tsunamis from megathrusts are more common in the Pacific Ocean, which is ringed with very active subduction zones. For example, the 1960 Chile earthquake (magnitude 9.5) and 1964 Alaska earthquake (magnitude 9.2) generated ocean-crossing tsunamis that caused death and destruction thousands of kilometers from the epicenters. At one location, near Valdez, Alaska, the 1964 tsunami ran up a mountainside to a height of 67 m!

Disturbances of the seafloor caused by landslides or volcanic eruptions can also produce tsunamis. The 1883 explosion of the Krakatoa volcano in Indonesia generated a tsunami that reached 40 m in height and drowned 36,000 people on nearby coasts.

Fires Secondary hazards also include the fires ignited by ruptured gas lines or downed electrical power lines. Damage to water mains in an earthquake can make firefighting all but impossible—a circumstance that contributed to the burning of San Francisco after the 1906 earthquake. Most of the 140,000 fatalities in the 1923 Kanto earthquake, one of Japan's greatest disasters, resulted from fires in the cities of Tokyo and Yokohama.

Key Figure 13.17 Megathrust earthquakes generate tsunamis that can propagate across ocean basins.

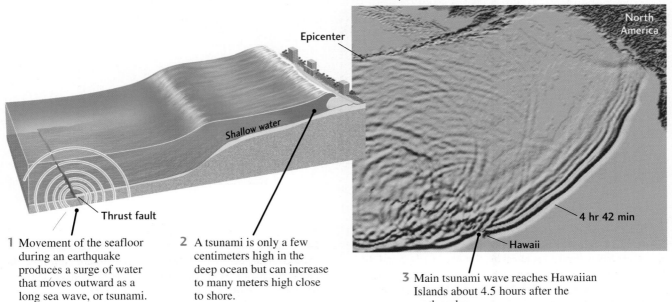

Tsunami generation

Computer simulation of tsunami radiation caused by a magnitude 7.7 earthquake in the Aleutian Islands

1 Movement of the seafloor during an earthquake produces a surge of water that moves outward as a long sea wave, or tsunami.

2 A tsunami is only a few centimeters high in the deep ocean but can increase to many meters high close to shore.

3 Main tsunami wave reaches Hawaiian Islands about 4.5 hours after the earthquake.

[NOAA. Pacific Marine Environmental Laboratory.]

Reducing Earthquake Risk

In assessing the possibility of damage from earthquakes, it is important to distinguish between hazard and risk. **Seismic hazard** describes the *intensity* of seismic shaking and ground disruption that can be expected over the long term at some specified location. The hazard depends on the proximity of the site to active faults that might generate earthquakes, and it can be expressed in the form of a seismic hazard map. **Figure 13.18** displays the national seismic hazard map produced by the U.S. Geological Survey.

In contrast, **seismic risk** describes the *damage* that can be expected over the long term for a specified region, such as a county or state, usually measured in terms of average dollar loss per year. The risk depends not only on the seismic hazard but also on the region's exposure to seismic damage (its population, number of buildings, and other infrastructure) and its *fragility* (the vulnerability of its built structures to seismic shaking). Because so many geologic

and economic variables must be considered, estimating seismic risk is a complex job. The results of the first comprehensive national study, published by the Federal Emergency Management Agency in 2001, are presented in **Figure 13.19**.

The differences between seismic hazard and risk can be appreciated by comparing the two types of national maps. For instance, although the seismic hazard levels in Alaska and California are both high (see Figure 13.18), California's exposure is much greater, yielding a much larger total risk (see Figure 13.19). California leads the nation in seismic risk, with about 75 percent of the national total; in fact, a single county, Los Angeles, accounts for 25 percent. Nonetheless, the problem is truly national: 46 million people in metropolitan areas outside of California face substantial earthquake risks. Those areas include Hilo, Honolulu, Anchorage, Seattle, Tacoma, Portland, Salt Lake City, Reno, Las Vegas, Albuquerque, Charleston, Memphis, Atlanta, St. Louis, New York, Boston, and Philadelphia.

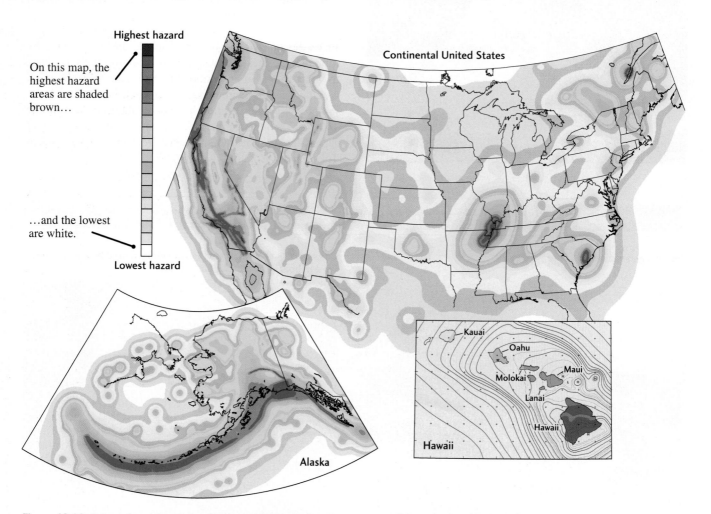

Figure 13.18 Seismic hazard map for the United States. The region of highest hazard lies along the San Andreas fault and the Transverse Ranges in California, with a branch extending into eastern California and western Nevada. High hazards are also found along the coast of the Pacific Northwest and in a zone

following the intermountain seismic belt. In the central and eastern United States, the highest hazard areas are New Madrid, Missouri; Charleston, South Carolina; eastern Tennessee; and portions of the Northeast. [U.S. Geological Survey, http://geohazards.cr.usgs.gov/eq/.]

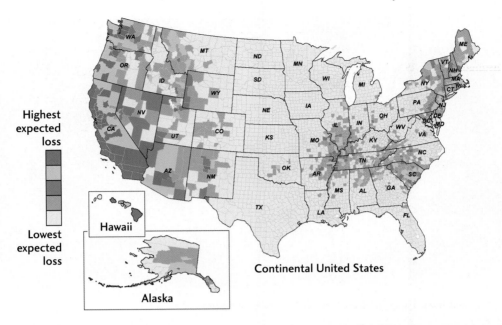

Highest
expected
loss

Lowest
expected
loss

Hawaii

Alaska

Continental United States

Figure 13.19 Seismic risk map for the United States. The map shows current annualized earthquake losses (AEL) on a county-by-county basis. Twenty-four states have an AEL greater than $10 million. The total AEL estimated for the entire United States is about $4.4 billion. [Federal Emergency Management Agency, Report 366, Washington, D.C., 2001.]

Not much can be done about seismic hazard because we have no way to prevent or control earthquakes. However, there are many important steps that society can take to reduce seismic risk.

Hazard Characterization The first step is to follow the old proverb, "Know thy enemy." We still have much to learn about the sizes and frequencies of ruptures on active faults. For example, it is only in the past decade that we have come to appreciate that an earthquake in the Cascadia subduction zone, which stretches from northern California through Oregon and Washington to British Columbia, could produce a tsunami as large as the one that devastated the Indian Ocean region in 2004. These dangers became apparent when geologists found evidence of a magnitude 9 earthquake that occurred in 1700, before any written historical accounts of the area existed. This monstrous rupture caused major ground subsidence along the Cascadia coastline and left a record of flooded, dead coastal forests. A tsunami at least 5 m high hit Japan, where historical records pin down its exact date (January 26, 1700). The recurrence of this event is consistent with plate tectonics. The Juan de Fuca Plate is being subducted under the North American Plate at a rate of about 45 mm/year along the coast of the Pacific Northwest. Geologists have debated whether this motion occurs seismically or is taken up by continuous creep or perhaps by silent earthquakes, but current opinion pegs the average time between magnitude 9 earthquakes in this subduction zone at 500 to 600 years.

Although we have a good understanding of seismic hazards in some parts of the world—the United States and Japan, in particular—we know much less about others. During the 1990s, the United Nations sponsored an effort to map seismic hazards worldwide as part of the International Decade of Natural Disaster Reduction. This effort resulted in the first global seismic hazard map, shown in **Figure 13.20**. The map is based primarily on historical earthquakes, so it may underestimate the hazard in some regions where the historical record is short. Much more needs to be done to characterize seismic hazards on a global scale.

Land-Use Policies The exposure of buildings and other structures to earthquake risk can be reduced by policies that restrict land use. This approach works well where the hazard is localized, as in the case of fault rupture. Erecting buildings on known active faults, as was done in the residential developments pictured in **Figure 13.21**, is clearly unwise, because few buildings can withstand the deformation to which they might be subjected when the fault ruptures. In the 1971 San Fernando earthquake, a fault ruptured under a densely populated area of Los Angeles, destroying almost 100 structures. The state of California responded in 1972 with a law that restricts the construction of new buildings across an active fault. For existing residences on a fault, real estate agents are required to disclose the information to potential buyers. A notable omission is that the act does not cover publicly owned or industrial facilities.

Earthquake Engineering Although land-use policies help to reduce the risk from localized hazards such as ground ruptures and liquefaction, they are less effective where seismic

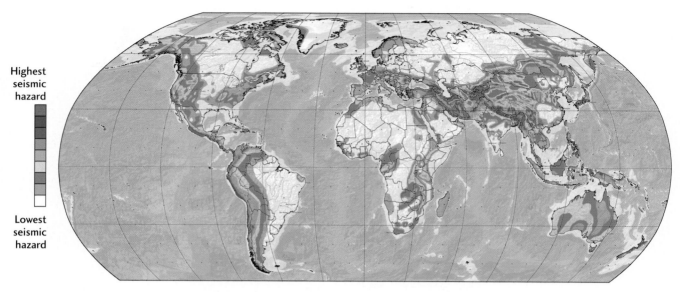

Figure 13.20 World seismic hazard map. [K. M. Shedlock et al., *Seismological Research Letters* 71(2000): 679–686.]

Highest seismic hazard

Lowest seismic hazard

Figure 13.21 Housing tracts constructed within the San Andreas fault zone, San Francisco Peninsula, before the state passed legislation restricting this practice. The white line indicates the approximate fault trace, along which the ground ruptured and slipped about 2 m during the earthquake of 1906. [R. E. Wallace/USGS.]

shaking is distributed across large regions—for example, Los Angeles and San Francisco. The risks from seismic shaking can best be reduced by good engineering and construction. Standards for the design and construction of new buildings are regulated by building codes enacted by state and local governments. A **building code** specifies the forces a structure must be able to withstand, based on the maximum intensity of shaking expected from the seismic hazard. In the aftermath of an earthquake, engineers study buildings that were damaged and make recommendations about modifications to the building codes that could reduce future damage from similar earthquakes.

As a result of this procedure, U.S. building codes have been largely successful in preventing loss of life during earthquakes. For example, from 1983 to 2004, 131 people died in nine severe earthquakes in the western United States, whereas more than 460,000 people were killed by earthquakes worldwide. Nevertheless, more can be done to improve earthquake engineering. Damage to structures can be reduced by using specialized construction materials and advanced engineering methods, such as putting entire buildings on movable supports to isolate them from the shaking.

For individuals, earthquake preparedness begins at home. Feature 13.2 summarizes some of the steps you can take to protect yourself and your family in an earthquake.

Emergency Response Once an earthquake happens, networks of seismographs can transmit signals automatically to central processing facilities. In a fraction of a minute, computers can pinpoint the earthquake focus, measure its magnitude, and determine its fault mechanism. If equipped with strong-motion sensors that accurately record the most violent

Use a map of the United States and time-travel curves (see Figure 13.6) to obtain a rough epicenter.

7. At a place along a boundary fault between the Nazca Plate and the South American Plate, the relative plate motion is 80 mm/year. The last great earthquake, in 1880, showed a fault slip of 12 m. When should local residents begin to worry about another great earthquake?

THOUGHT QUESTIONS

1. The belts of shallow-focus earthquakes, shown by the blue dots in Figure 13.12, are wider and more diffuse in the continents than in the oceans. Why? (In answering this question, you might want to review Chapter 7.)

2. Why are earthquakes with focal depths greater than 20 km infrequent in continental lithosphere?

3. Why do the largest earthquakes occur on subduction megathrusts and not, say, on continental strike-slip faults?

4. Why are great tsunamis, such as the Indian Ocean tsunami of December 26, 2004, so rare?

5. In Figure 13.1, the right-lateral fault offsets the fence line to the right. In Figure 13.12, the mid-ocean ridge crest is also offset to the right. Why then is the transform fault in Figure 13.12 left-lateral?

6. Would you support legislation to prevent owners from building structures close to active faults?

7. Taking into account the possibility of false alarms, reduction of casualties, mass hysteria, economic depression, and other possible consequences of earthquake prediction, do you think the objective of predicting earthquakes should have a high priority?

CHAPTER FOURTEEN

EXPLORING EARTH'S INTERIOR

Humans have burrowed in mines to depths as great as 4 km to extract gold and other minerals, and they have drilled to almost 10 km in the search for petroleum. But these efforts, though heroic, have barely scratched the surface of our massive planet. The crushing pressures and red-hot temperatures of Earth's deeper layers make the planet's interior inaccessible for the foreseeable future. Nevertheless, we have learned much about the structure and composition of Earth's interior by making measurements near and above its surface.

Some of the best information comes from seismology. We saw in Chapter 13 how the terrible shaking of earthquakes can wreak destruction. Yet this same energy can be harnessed to illuminate Earth's deepest regions, allowing us to construct three-dimensional images of geologic features in the lower crust, the rising and falling of convection currents throughout the mantle, and the workings of the central core.

In this chapter, we will explore Earth's interior down to its center, nearly 6400 km beneath our feet. In addition to the techniques of seismology, we will examine the data from rocks brought up in volcanoes and the information contained in Earth's gravity and magnetic fields.

We will investigate the high temperatures deep inside the Earth and delve further into the internal machinery of the two great heat engines first introduced in Chapter 1: the geodynamo in the liquid iron outer core, which generates the magnetic field, and the convection of the solid mantle, which drives plate tectonics.

EXPLORING THE INTERIOR
WITH SEISMIC WAVES

Different types of waves—light, sound, and seismic—have a common characteristic: the velocity at which they travel depends on the material

Exploring the Interior with Seismic Waves 325

Layering and Composition of the Interior 328

Earth's Internal Heat and Temperature 332

The Three-Dimensional Structure of the Mantle 335

Earth's Magnetic Field and the Geodynamo 337

This open-pit excavation of a kimberlite volcano at Jwaneng, Botswana, is the world's richest diamond mine, producing 12.4 million carats worth more than $1.3 billion in 2001. In addition to diamonds, kimberlite volcanoes bring up rocks from more than 200 km deep, providing critical information about Earth's deep interior. [Peter Essick/Aurora Photos.]

through which they are passing. Light waves travel quickest through a vacuum, more slowly through air, and even more slowly through water. Sound waves, on the other hand, travel faster through water than through air and not at all through a vacuum. Why? Sound waves are simply propagating variations in pressure. Without something to compress, such as air or water, they cannot exist. The more force it takes to compress a material, the faster sound will travel through it. The speed of sound in air—Mach 1 in the jargon of jet pilots—is typically 0.3 km/s, or about 670 miles per hour. Water resists compression much more than air, so the speed of sound waves in water is correspondingly higher, about 1.5 km/s. Solid materials are even more resistant to compression, so sound waves travel through them at even higher speeds. In granite rock, sound travels at about 6 km/s —nearly 13,500 miles per hour!

Basic Types of Waves

The push-pull motions of sound waves in a solid are called **compressional waves** to distinguish them from the side-to-side motions of **shear waves** (see Figure 13.5). It's harder to compress solids than to shear them, so compressional waves always travel faster than shear waves. This physical principle explains a relationship we discussed in Chapter 13: compressional waves are always the P (primary) arrivals, and shear waves are the S (secondary) arrivals. It also explains why the speed of shear waves in gases and liquids must be zero—they have no resistance to shear. Shear waves cannot propagate through any fluid—air, water, or the liquid iron in Earth's outer core.

From seismograms, geologists can calculate the speed of a P or S wave by dividing the distance traveled by the travel time. The measurements of these wave speeds can then be used to infer which materials the waves encountered along their paths. For example, P and S waves travel about 20 percent more rapidly through rock typical of oceanic crust (gabbro) than through typical upper continental crust (granite), and they travel about 30 percent more rapidly through the upper mantle (peridotite).

The concepts of travel times and wave paths sound simple enough, but complications arise when waves pass through more than one type of material. At the boundary between two different materials, some of the waves bounce off—that is, they are *reflected*—and others are transmitted into the second material, just as light is partly reflected and partly transmitted when it strikes a windowpane. The waves that cross the boundary between two materials are bent, or *refracted,* as their velocity changes from that in the first material to that in the second. **Figure 14.1** shows a laser light beam whose path bends as it goes from air into water, much as a P or an S wave bends as it travels from one material to another. By studying how fast seismic waves travel and how they are refracted and reflected at Earth's internal boundaries, seismologists have been able to measure the layering of Earth's crust, mantle, and core with great accuracy.

Figure 14.1 In this experiment, two beams of laser light enter a bowl of water from the top. Both beams are reflected from a mirror on the bottom of the bowl. One is then reflected at the water-air interface and passes through the bowl to make a bright spot on the table. Most of the energy from the other beam is bent downward (refracted) as it passes from the water to the air, although a small amount is reflected to form a second spot on the table. You can also trace the paths of other beams reflected by the interfaces. [Susan Schwartzenberg/The Exploratorium.]

Paths of Seismic Waves in the Earth

If Earth were made of a single material with constant properties from the surface to the center, P and S waves would travel from the focus of an earthquake to a distant seismograph along straight lines through the interior. When the first global networks of seismographs were installed about a century ago, however, seismologists discovered that the structure of Earth's interior was much more complicated.

Waves Refracted Through Earth's Interior The first observations of long-distance seismic waves showed that the paths of the P and S waves curved upward through the mantle, as illustrated in **Figure 14.2**. From the travel times and the amount of upward bending, seismologists were able to demonstrate that P waves travel much faster through rocks at great depths than they do through rocks found on the surface. This was hardly surprising, because rocks subjected to great pressures in Earth's interior are squeezed into tighter crystal structures. The atoms in these tighter structures are more resistant to further compression, which causes P waves to travel through them more quickly.

Key Figure 14.2   Earth's core creates P-wave and S-wave shadows.

(a) The pattern of P-wave paths through Earth's interior (solid blue lines). The dashed blue lines show the progress of wave fronts through the interior at 2-minute intervals. Distances are measured in angular distance from the earthquake focus. The P-wave shadow zone extends from 105° to 142°. P waves cannot reach the surface within this zone because of the way they are bent when they enter and leave the core.

(b) S-wave paths through Earth's interior (solid green lines). The larger S-wave shadow zone extends from 105° to 180°. Although S waves strike the core, they cannot travel through its fluid outer region and therefore never emerge beyond 105° from the focus.

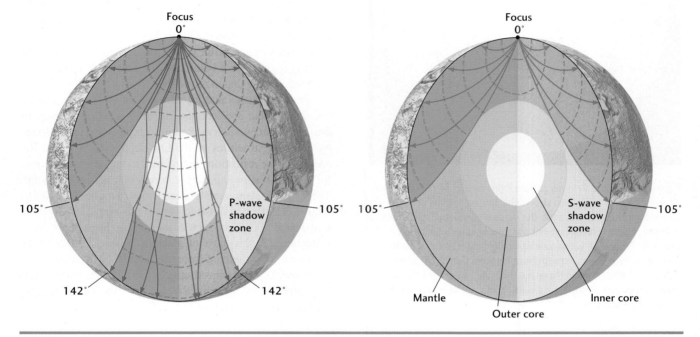

Seismologists were very surprised, however, by what they found at progressively greater distances from an earthquake focus. After the P waves and S waves had traveled beyond about 11,600 km from the earthquake focus, they suddenly disappeared! (Like airplane pilots and ship captains, seismologists prefer to measure distances traveled on Earth's surface in angular degrees, from 0° at the earthquake focus to 180° at a point on the opposite side of the Earth. Each degree measures 111 km at the surface, so 11,600 km corresponds to an angular distance of 105°, as shown in Figure 14.2.) When they looked at seismograms recorded beyond this distance, they did not see the distinct P- and S-wave arrivals that were so clear on seismograms recorded at shorter distances. Then, beyond about 15,800 km from the focus (142°), the P waves suddenly reappeared as big arrivals, but they were much delayed compared to their expected travel times. The S waves never reappeared.

In 1906, the British seismologist R. D. Oldham put these observations together to provide the first evidence that Earth has a liquid outer core. S waves cannot travel through the outer core because it is liquid, and there is thus an S-wave **shadow zone** beyond 105° from the earthquake focus (see

Figure 14.2b). The propagation of P waves is more complicated (see Figure 14.2a). At 105°, their paths just miss the core, whereas waves that would have traveled to greater distances encounter the core-mantle boundary. At the core-mantle boundary, the P-wave speed drops by almost a factor of two. Therefore, the waves are refracted downward into the core and emerge at greater distances after the delay caused by their detour through the core. This refraction effect forms a P-wave shadow zone at angular distances between 105° and 142°.

Waves Reflected at Earth's Internal Boundaries The core-mantle boundary turned out to be very sharp, so when seismologists looked at records of earthquake waves made at angular distances of less than 105°, they found arrivals corresponding to waves reflected from this boundary. They called the compressional wave that reflects from the top of the outer core PcP and the shear wave ScS. (The lowercase *c* indicates a reflection from the core.) In 1914, a German seismologist, Beno Gutenberg, used the travel times of these core reflections to determine an accurate depth for the core-mantle boundary—just under 2900 km.

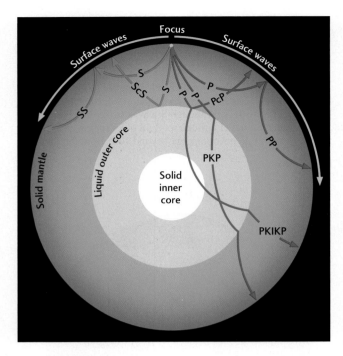

Figure 14.3 P and S waves radiate from an earthquake focus in all directions. This diagram shows the simple labeling scheme seismologists use to describe the various paths the waves take. PcP and ScS are compressional and shear waves that bounce off the core. PP and SS waves are internally reflected from Earth's surface. A PKP wave is transmitted through the liquid outer core, and a PKIKP wave traverses the solid inner core. Surface waves propagate along Earth's outer surface, like waves on the surface of a pond.

Figure 14.3 shows examples of the paths taken by these core-reflected waves, as well as the paths and symbolic names that have been attached to some other prominent arrivals seen on seismograms. For example, a compressional wave reflected once at Earth's surface is called PP, and a shear wave with a similar path is called SS. **Fig-**

ure 14.4 shows these internal reflections on seismograms recorded at different distances from an earthquake focus.

The path of a compressional wave through the outer core is labeled with a *K* (from the German word for "core"), so PKP describes a compressional wave that propagates from an earthquake through the crust and mantle, into the outer core, and back through the mantle and crust to a surface receiver. In 1936, Danish seismologist Inge Lehmann discovered Earth's inner core by observing the compressional waves refracted from its outer boundary, which she determined to be at a depth of about 5150 km. Paths through the inner core are labeled with an *I*, so the refracted waves she used are called PKIKP. Others have since observed compressional waves (PKiKP) reflected from the top side of the inner core–outer core boundary (the lowercase *i* indicates a reflection rather than a refraction).

Probing Earth's interior with seismic waves has a number of practical applications. Seismic waves generated by artificial sources, such as dynamite explosions, are reflected by geologic structures at shallow depths in the crust. Recording these reflections has proved to be the most successful method for finding deeply buried oil and gas reservoirs. This type of seismic exploration is now a multibillion-dollar industry. Reflected seismic waves are employed in other practical applications, such as measuring the depth of water tables and the thickness of glaciers. At sea, compressional waves can be generated from mechanical sources similar to a loudspeaker, and ships routinely use the underwater sound they produce to measure the depth of the ocean and the thickness of sediments on the seafloor.

LAYERING AND COMPOSITION OF THE INTERIOR

Thousands of sensitive seismographs and highly accurate clocks have enabled seismologists throughout the world to measure the travel times of many types of seismic waves very

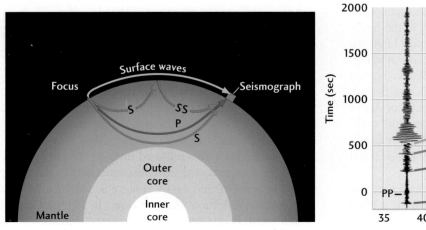

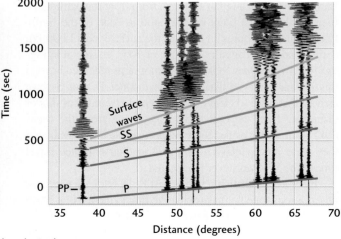

Figure 14.4 Seismograms recorded at various distances from an earthquake in the Aleutian Islands, Alaska, showing P, S, SS, and surface waves.

precisely. Nuclear explosions set off underground at nuclear test sites also excite seismic waves and add valuable data to those derived from earthquakes. From these measurements, seismologists can plot travel-time curves of the sort shown in Figure 13.6 for the various kinds of seismic waves.

The travel times of compressional and shear waves depend on their speeds as they pass through the materials in Earth's interior. The key to making travel times a useful geological tool is to learn how to convert them into a graph or table that shows how the speeds of seismic waves change with depth in the Earth. Solving this problem is something like guessing which of several possible routes a driver took on a trip from New York to Chicago and how fast he traveled along the route, knowing only the speed limits along the way and that the trip took 12.1 hours.

The model that seismologists have devised is displayed in **Figure 14.5**. The illustration shows how the speeds of both compressional and shear waves change with depth and how these changes are related to Earth's major layers. To understand the structure and composition of these layers, scientists must combine the seismological information on this diagram with information from many other types of

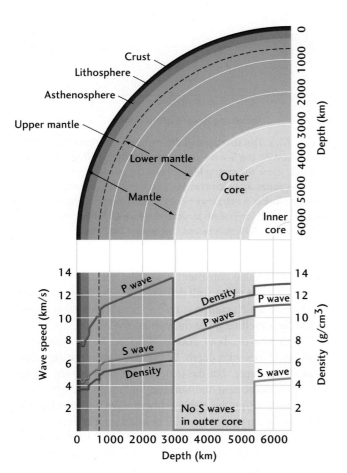

Figure 14.5 Earth's layering revealed by seismology. The lower diagram shows the changes in the compressional and shear speeds and density with depth in the Earth. The upper diagram is a cross section through Earth on the same depth scale, showing how these changes are related to the major layers.

geologic studies. We will explore what they have found by taking a downward journey through Earth's interior, from its outer crust to its inner core.

The Crust

Earth's outermost layer, the crust, varies in thickness. It is thin (about 7 km) under oceans, thicker (about 40 km) under continents, and thickest (as much as 70 km) under high mountains. P waves move through crustal rocks at about 6 to 7 km/s. By sampling various materials typical of the crust and mantle and by measuring the velocities of waves passing through these materials in the laboratory, we can compile a library of seismic wave speeds through all sorts of Earth materials. Rough values for the P-wave speeds in igneous rocks, for example, are as follows:

- Felsic rocks typical of the upper continental crust (granite): 6 km/s
- Mafic rocks typical of oceanic crust or the lower continental crust (gabbro): 7 km/s
- Ultramafic rocks typical of the upper mantle (peridotite): 8 km/s

The P-wave speeds differ because they depend on a rock's density and its resistance to compression and shear, all of which change with composition and crystal structure. In general, higher densities correspond to higher P-wave velocities; typical densities for granite, gabbro, and peridotite are 2.6 g/cm^3, 2.9 g/cm^3, and 3.3 g/cm^3, respectively.

We know from the correlations of wave velocity with rock type that the upper part of the continental crust is made up mostly of low-density granitic rocks and that no granite exists on the floor of the deep ocean. The crust there consists entirely of basalt and gabbro overlain by sediments. Below the crust, the velocity of P waves increases abruptly to 8 km/s. This increase marks a sharp boundary between crustal rocks and underlying mantle known as the **Mohorovičić discontinuity** (*Moho*, for short), after the European seismologist who discovered it in 1909 (see Chapter 1). The velocity of 8 km/s indicates that the mantle below the Moho is made primarily of dense peridotites.

This crustal structure is consistent with the theory that lighter materials floated up from the mantle to form the crust. As discussed in Chapter 1, continental surfaces typically have elevations of 0 to 1 km above sea level, whereas deep ocean basins typically have elevations of 4 to 5 km below sea level. The higher elevations of continents relative to the deep seafloor can be explained by the **principle of isostasy,** which states that the buoyancy force that pushes a low-density continent upward must be balanced by the gravitational force that pulls it downward.

To understand *isostasy* (Greek for "equal standing"), consider a block of wood floating in water. The buoyancy force pushing the wood upward depends on its density—the lower the density, the greater the buoyancy—whereas the gravitational force pulling it downward depends on how much of the wood rises above the water surface. If the density of the wood is half that of water (as in the case of balsa wood), it will float high, with half of its volume out of the water. However, if its density is only slightly less than that of water (as in the case of oak), it will float low, barely sticking up at all.

In a similar way, the large volume of continental crust that projects into the denser mantle provides the buoyancy that allows the continents to float 5 km above the oceanic crust. In regions of high, mountainous topography, the additional weight of the mountains is usually supported by the buoyancy from a thicker root of low-density crustal rocks, as shown in Figure 1.7. The principle of isostasy thus explains why Tibet, the highest plateau in the world, also has the world's thickest crust.

The principle of isostasy also implies that, over long periods, the mantle has little strength and flows as a viscous solid when it is forced to support the weight of mountains or continental ice sheets (see Feature 14.1).

The Mantle

The primary rock type of the upper mantle, peridotite, is made up mainly of olivine and pyroxene, two silicates that contain magnesium and iron (see Chapter 4). Laboratory experiments show that changes in pressure and temperature alter the properties and forms of olivine and pyroxene. These minerals begin to melt under the conditions found in the upper mantle. At greater depths, pressure forces the atoms of these minerals closer together into more compact crystal structures. The major changes in the mineralogy of mantle peridotites with depth are marked by increases or decreases in S-wave speed (**Figure 14.6**).

In the outermost zone of the mantle, the S-wave speed is typical of the mantle peridotites that constitute the subcrustal parts of the rigid lithosphere. The average thickness of this layer is about 100 km, but it is highly variable geographically, ranging from almost no thickness near the spreading centers where new oceanic lithosphere is forming from hot, rising mantle material to over 200 km beneath the cold, stable continental cratons.

Near the base of the lithosphere, the S-wave speed abruptly decreases to form a **low-velocity zone.** The decrease in speed occurs because the temperature of the mantle increases with depth. At about 100 km, it approaches the melting temperature of the mantle rocks, partially melting some minerals in the peridotite. The small amount of melt (in most places, less than 1 percent) decreases the rigidity of the rock, which slows the S waves passing through the rock. Because partial melting also allows the rocks to flow more easily, geologists identify the low-velocity zone with the top part of the asthenosphere—the weak layer across which the rigid lithospheric plates slide. This idea fits nicely with evidence that the asthenosphere is the source of most basaltic magma (see Chapters 4 and 12).

The base of the low-velocity zone occurs at about 200 to 250 km below ocean plates, where the S-wave speed in-

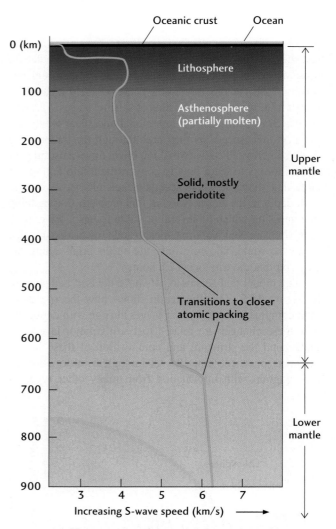

Figure 14.6 The structure of the mantle beneath an old ocean basin, showing the S-wave speed to a depth of 900 km. Changes in speed mark the strong lithosphere, the weak asthenosphere, and two zones in which changes occur because increasing pressure forces a rearrangement of the atoms into denser and more compact crystalline structures. [After D. P. McKenzie, "The Earth's Mantle." *Scientific American* (September 1983): 66.]

creases to a value consistent with solid peridotite. The low-velocity zone is not as well defined under the stable continental cratons, which comprise colder mantle rocks that remain below their melting point throughout this region.

At depths of about 200 to 400 km, the S-wave speed again increases with depth. Within this layer, pressure continues to increase, but the temperature is not rising as rapidly as it does near the surface, owing to the effects of convection within the asthenosphere. (We will discuss why this is so in the next section.) The combined effects of pressure and temperature cause the amount of melting to decrease with depth and cause the rock rigidity—and thus the S-wave speed—to rise.

About 400 km below the surface, S-wave speed increases by about 10 percent within a narrow zone less than 20 km thick. This jump in S-wave speed can be explained by a **phase change** in olivine, the major mineral constituent of the upper mantle, which transforms from its ordinary

EARTH ISSUES

14.1 The Uplift of Scandinavia: Nature's Experiment with Isostasy

If you depress a cork floating in water with your finger and then release it, the cork pops up almost instantly. A cork floating in molasses would rise more slowly, because the drag of the viscous fluid would slow the process. If we could perform a similar experiment on the Earth, we could learn much about how isostasy works—in particular, about the viscosity of the mantle and how it affects rates of uplift and subsidence. How convenient it would be if we could push the crust down somewhere, remove the depressive force, and then sit back and watch the depressed area rise.

Nature has been good enough to perform this experiment for us. The force is the weight of a continental glacier—an ice sheet 2 to 3 km thick. With the onset of an ice age, ice sheets can appear within the geologically short period of a few thousand years. The crust is depressed by the ice load, and a downward bulge develops under the glacier to the extent needed to provide buoyant support. At the onset of a warming trend, the glacier melts rapidly. With the removal of the weight, the depressed crust begins to rebound. We can discover the rate of uplift by dating ancient beaches that are now above sea level. Raised beaches can tell us how long ago a particular stretch of land was at sea level.

Such depression and uplift have occurred in Norway, Sweden, Finland, Canada, and elsewhere in glaciated regions. The ice cap retreated from these areas some 10,000 years ago, and the land has been rising ever since. Figure 10.21 shows a series of raised beaches that have allowed geologists to measure the speed of this glacial rebound and infer the viscosity of the mantle.

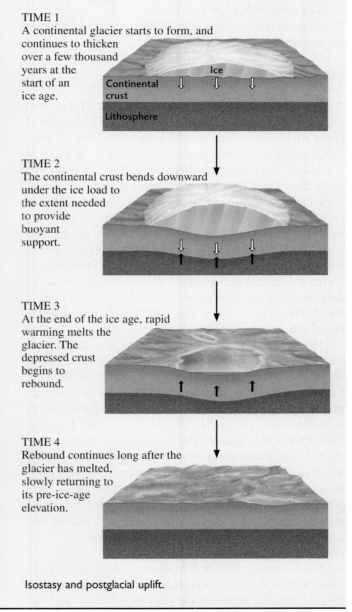

TIME 1
A continental glacier starts to form, and continues to thicken over a few thousand years at the start of an ice age.

TIME 2
The continental crust bends downward under the ice load to the extent needed to provide buoyant support.

TIME 3
At the end of the ice age, rapid warming melts the glacier. The depressed crust begins to rebound.

TIME 4
Rebound continues long after the glacier has melted, slowly returning to its pre-ice-age elevation.

Isostasy and postglacial uplift.

---

crystal structure to a denser, more closely packed structure at high pressure. This hypothesis has been confirmed by laboratory experiments. When olivine is subjected to high pressures in the laboratory, the atoms that form its crystal structure shift into a more compact arrangement at the temperatures and pressures corresponding to depths of about 400 km. Moreover, the jumps in the P- and S-wave speeds measured in the laboratory match the increase observed for seismic waves at this depth.

In the region from 420 to 650 km below the surface, the mantle properties change little as depth increases. Near 660 km, however, the S-wave speed abruptly increases again, indicating a second major phase change to an even more closely packed crystal structure. Laboratory experi-

ments on olivine have also confirmed the existence of another major mineralogical phase change at this depth.

Phase changes involve a transition in a rock's mineralogy but not in its chemical composition. Some geologists have believed that the increase in seismic wave speeds at a depth of 660 km comes in part from a change in chemical composition. The issue is critical to understanding the plate tectonic system, because a chemical change would imply that the convective overturning that drives plate tectonics does not penetrate much beyond this depth; convection in the mantle would be layered, as shown in Figure 2.14. However, the evidence from detailed studies of mantle structure now indicates very little if any chemical change in this region of the mantle.

Below the transition at a depth of 660 km, the seismic wave speeds increase gradually and do not show any unusual features, such as major phase changes, until close to the core-mantle boundary (see Figure 14.5). This relatively homogeneous region, more than 2000 km thick, is called the **lower mantle.**

The Core-Mantle Boundary

At the **core-mantle boundary,** about 2890 km below the surface, we encounter the most extreme change in properties found anywhere in Earth's interior. From the way seismic waves reflect from this boundary, seismologists can tell that it is a very sharp interface. Here the material changes abruptly from a solid silicate rock to a liquid iron alloy. Because of the complete loss of rigidity, the S-wave speed drops from about 7.5 km/s to zero, and the P-wave speed drops from more than 13 km/s to about 8 km/s, causing the core shadow zone. Density, on the other hand, increases by about 4.5 g/cm^3 (see Figure 14.5). This large density jump, which is even greater than the increase in density at Earth's surface, keeps the core-mantle boundary very flat (you could probably skateboard on it!) and prevents any large-scale mixing of the mantle and core.

The core-mantle boundary appears to be a very active place. Heat conducted out of the core increases the temperatures at the base of the mantle by as much as 1000°C. Indeed, the seismic waves that pass near the base of the mantle show peculiar complications, suggesting a region of exceptional geologic activity. In a thin layer above the core-mantle boundary, seismologists have recently discovered a steep (10 percent or more) decrease in seismic wave speeds, which may be an indication that the mantle in contact with the core is partially molten, at least in some places. We noted in Chapter 12 that some geologists believe this hot region to be the source of mantle plumes that rise all the way to Earth's surface, creating volcanic hot spots such as Hawaii and Yellowstone.

The lowermost boundary layer of the mantle, a region about 300 km thick, may also be the ultimate graveyard of some lithospheric material subducted at the surface, such as the denser, iron-rich parts of the oceanic crust. It is possible that the region above the core-mantle boundary might involve an upside-down version of the tectonics we see at Earth's surface. For example, accumulations of heavy, iron-rich material might form chemically distinct "anticontinents" that are constantly pushed to-and-fro across the core-mantle boundary by convection currents. Seismologists are teaming with other geologists who study mantle and core convection to learn more about whatever geologic processes might be active in this strange place.

The Core

We know quite a bit about the composition of the core, but not from direct observation. Our understanding has been derived from years of research using a combination of astronomical data, laboratory experiments, and seismological data. An iron-nickel composition is consistent with many lines of evidence. These metals are abundant in the cosmos. They are dense enough to explain the mass of the core (about one-third of Earth's total mass) and to be consistent with the theory that the core formed by gravitational differentiation. This hypothesis, first proposed by Emil Wiechert in the late nineteenth century, was buttressed by discoveries of meteorites made almost entirely of iron and nickel, which presumably came from the breakup of a planetary body that also had an iron-nickel core (see Figure 1.6).

Laboratory measurements at appropriately high pressures and temperatures have led to a slight revision of this hypothesis. A pure iron-nickel alloy turns out to be about 10 percent too dense to match the data for the outer core. Therefore, it has been proposed that the core includes minor amounts of some lighter element. Oxygen and sulfur are leading candidates, although the precise composition remains the subject of research and debate.

Seismology tells us that the core below the mantle is a fluid, but it is not fluid to the very center of the Earth. As Lehmann first discovered, P waves that penetrate to depths of 5150 km suddenly speed up, indicating the presence of an inner core, a metallic sphere two-thirds the size of the Moon. Seismologists have recently shown that the inner core does transmit shear waves, confirming early speculations that it is solid. In fact, some calculations suggest that the inner core spins at a slightly faster rate than the mantle, acting like a "planet within a planet."

The very center of the planet is not a place you would like to be. The pressures are immense, over 4 million times the atmospheric pressure at Earth's surface. And it's also very hot, as we will see.

EARTH'S INTERNAL HEAT AND TEMPERATURE

The evidence of Earth's internal heat is everywhere: volcanoes, hot springs, and the elevated temperatures in mines and boreholes. The internal heat fuels convection in the mantle, which drives the plate tectonic system, as well as the geodynamo in the core, which produces Earth's magnetic field.

Heat in the interior comes from several sources. During the planet's violent origin, kinetic energy released by infalling chunks of matter heated its outer regions, while gravitational energy released by differentiation of the core heated its deep interior (see Chapter 9). The disintegration of the radioactive elements uranium, thorium, and potassium continues to generate heat.

After Earth formed, it began to cool, and it is cooling to this day as heat flows from the hot interior to the cool surface. The temperatures of the planet's interior result from a balance between the heat gained and that lost through these processes.

Heat Flow Through Earth's Interior

The Earth cools in two main ways: through the slow transport of heat by conduction and the more rapid transport of heat by convection. **Conduction** dominates in the lithosphere, whereas **convection** is more important throughout most of Earth's interior.

Conduction Through the Lithosphere Heat energy exists in a material as the vibration of atoms; the higher the temperature, the more intense the vibrations. The conduction of heat occurs when thermally agitated atoms and molecules jostle one another, mechanically transferring the vibrational motion from a hot region to a cool one. Heat is conducted from regions of high temperature to regions of low temperature by this process.

Materials vary in their ability to conduct heat. Metal is a better conductor than plastic (think of how rapidly the metal handle of a frying pan heats up in comparison with one made of plastic). Rock and soil are very poor heat conductors, which is why underground pipes are less susceptible to freezing than those above ground and why wine cellars and other underground vaults have a nearly constant temperature despite large seasonal temperature changes at the surface. Rock conducts heat so poorly that a lava flow 100 m thick takes about 300 years to cool from 1000°C to surface temperatures.

The conduction of heat through the outer surface of the lithosphere causes it to cool slowly over time. As it cools, the thickness of the lithosphere increases, just as the cold crust on a bowl of hot wax thickens over time. Rock, like wax, contracts and becomes denser with decreasing temperature, so the average density of the lithosphere must increase over time, and, by the principle of isostasy, its surface must sink to lower levels. Thus, the mid-ocean ridges stand high because the lithosphere there is young, thin, and hot, whereas the abyssal plains are deep because the lithosphere is old, cold, and thick.

From these considerations, geologists have constructed a simple but precise theory of seafloor topography that can explain the large-scale features of the ocean basins. The theory predicts that the depth of the oceans should depend primarily on the age of the seafloor. According to this theory, the ocean depth should increase as the square root of age. In other words, seafloor that is 40 million years old should have subsided twice as much as seafloor that is only 10 million years old (because $\sqrt{40/10} = \sqrt{4} = 2$). This simple mathematical relationship matches seafloor topography near the mid-ocean ridge crests amazingly well, as demonstrated in **Figure 14.7**.

Conductive cooling of the lithosphere accounts for a wide variety of other geologic phenomena, including the subsidence of passive continental margins and the growth of many sedimentary basins. It explains why the heat flowing out of the oceanic lithosphere is high near spreading centers and decreases as the oceanic lithosphere gets older, and it

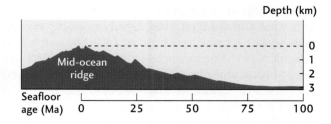

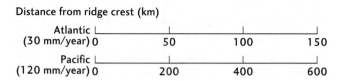

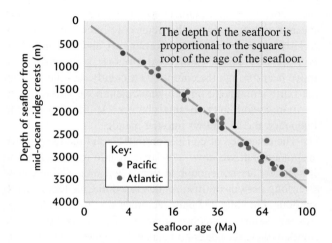

Figure 14.7 Topography of mid-ocean ridges in the Atlantic and Pacific oceans, showing how the water depth increases in proportion to the square root of lithosphere age as the plate moves away from the spreading centers. The same theoretical curve, which is derived by assuming that the lithosphere cools by conduction, matches the data for both ocean basins, even though seafloor spreading is much faster in the Pacific than in the Atlantic.

tells us why the average thickness of the oceanic lithosphere is about 100 km. The establishment of this theory was one of the great successes of plate tectonics.

Conductive cooling does not explain all aspects of heat flow through Earth's outer surface, however. Marine geologists have found that seafloor older than about 100 million years does not continue to subside as the simple theory would predict. Moreover, simple conductive cooling is far too inefficient to account for the cooling of the Earth over its entire history. It can be shown that if the 4.5-billion-year-old Earth cooled by conduction alone, very little of the heat from depths greater than about 500 km would have reached the surface. The mantle, which was molten in Earth's early history, would be far hotter than it is now. To understand these facts, we must consider the second mode of heat transport, convection, which is more efficient than conduction in getting the heat out of Earth's interior.

Convection in the Mantle and Core Convection occurs when a heated fluid, either liquid or gas, expands and rises because it has become less dense than the surrounding material. Convection moves heat more efficiently than conduction because the heated material itself moves, carrying its heat with it. Colder material flows in to take the place of the hot rising fluid, is itself heated, and then rises to continue the cycle. This is the process by which water is heated in a kettle (see Figure 1.11a). Liquids conduct heat poorly, so a kettle of water would take a long time to heat to the boiling point if convection did not distribute the heat rapidly. Convection moves heat when a chimney draws, when warm tobacco smoke rises, or when clouds form on a hot day.

We have already seen how seismic waves reveal that the outer core is a liquid. Other types of data demonstrate that the iron-rich material in the outer core has a low viscosity (resistance to permanent deformation) and can therefore convect very easily. Convective motions in the outer core move heat through the core very efficiently, and they generate Earth's magnetic field, a phenomenon we will examine in more detail later in this chapter. At the core-mantle boundary, the heat flows into the mantle.

The existence of convection in the solid mantle is more surprising, but we now know that mantle rocks below the lithosphere are hot enough to flow when subjected to forces over long periods of time. As discussed in Chapters 1 and 2, seafloor spreading and plate tectonics are direct evidence of this solid-state convection at work. The rising hot matter under mid-ocean ridges builds new lithosphere, which cools as it spreads away. In time, it sinks back into the mantle, where it is eventually resorbed and reheated. This cyclical process is a form of convection; heat is carried from the interior to the surface by the motion of matter.

Geologists are still debating many aspects of mantle convection. Some believe that only the upper mantle is involved in the convection that drives plates, which would imply that the upper and lower mantles do not mix. Many others think that the plate tectonic system of convection extends throughout the mantle, including most of the lower mantle. New ways of exploring Earth's interior using seismic waves, as well as evidence from other methods, are beginning to resolve these issues, as we will see shortly.

Regardless of the specifics, almost all geologists now agree that the convective movement of heat and matter from the interior to the surface is the dominant mechanism for Earth's cooling over geologic time.

Temperatures in the Earth

Geologists want to know how hot Earth gets deeper in its interior. Temperature and pressure determine whether matter is solid or molten, how atoms are packed together in crystals, and the resistance of solid matter to flow (its viscosity). The higher the temperature at depth, the lower the viscosity and the more rapidly convecting matter will move.

The curve that describes how temperature increases with depth is called the **geotherm.** One possible geotherm is

illustrated in **Figure 14.8**. Near the surface, geologists can directly measure temperature to depths of 4 km in mines and almost 10 km in boreholes. They find that the *geothermal gradient* (the change of temperature with depth) is 20° to 30°C per kilometer in normal continental crust. Conditions below the crust can be inferred from lavas and rocks erupted in volcanoes. The data indicate that temperatures near the base of the lithosphere range from 1300° to 1400°C. As Figure 14.8 shows, these temperatures are very near the melting point of mantle rocks, consistent with widespread observations that basaltic magmas are produced by partial melting in the upper part of the asthenosphere.

The steep geothermal gradient near Earth's surface tells us that heat is transported through the lithosphere by conduction. Below this depth, the temperature does not rise as rapidly. If it did, the temperatures in the deeper parts of the mantle would be so high (tens of thousands of degrees) that the lower mantle would be molten, which is inconsistent with seismological observations. Instead, the change in temperature with depth drops to about 0.5°C per kilometer, which is the geothermal gradient expected in a convecting mantle. This drop occurs because convection mixes cooler material near the top of the mantle with warmer material at greater depths, averaging out the temperature differences (just as temperatures are evened out when you stir your bathwater).

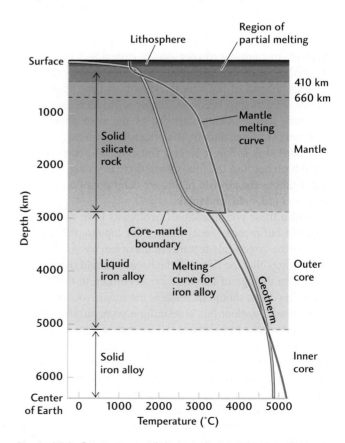

Figure 14.8 One estimate of the geotherm, which describes how temperatures increase with depth in the Earth. The geotherm lies above the temperature at which Earth materials first begin to melt (red line) in the upper mantle, forming the partially molten low-velocity zone, and in the outer core.

In the mid-mantle, phase changes—observed as steep increases in seismic wave speeds—occur at depths of 410 km and 660 km. Seismology can accurately determine the depths (and thus the pressures) of these phase changes, so the temperatures at which the phase changes take place can be calibrated using high-pressure laboratory experiments. The values obtained from the laboratory data are consistent with the geotherm shown in Figure 14.8.

We have more limited information about the temperatures at greater depths. Most geologists agree that convection extends throughout the mantle, vertically mixing material and keeping the geothermal gradient low. Near the base of the mantle, however, we expect temperatures to increase more rapidly, because the core-mantle boundary restricts vertical mixing. Motions near the core-mantle boundary, like motions near the surface, are primarily sideways rather than vertical. Close to this boundary, heat is transported from the core into the mantle mainly by conduction, and the geothermal gradients should therefore be high, like they are in the lithosphere.

Seismology tells us that the outer core is liquid, which means it is hot enough to melt the iron alloy that constitutes it. Laboratory data indicate that this temperature is probably greater than 3000°C, consistent with the high geothermal gradients at the base of the mantle predicted by convection models. The inner core, on the other hand, is solid. Because its iron-nickel composition is nearly the same as that of the outer core, the boundary between the inner core and outer core should correspond to the depth where the geotherm crosses the core's melting curve. According to Figure 14.8, this hypothesis implies that the temperature at Earth's center is slightly less than 5000°C.

Many aspects of this story can be debated, however, especially in regard to the deeper parts of the geotherm. For example, some geologists believe that the temperature at Earth's center may be as high as 6000° to 8000°C. More laboratory experiments and better calculations are required to reconcile these differences.

THE THREE-DIMENSIONAL STRUCTURE OF THE MANTLE

So far, we have investigated how the properties of Earth's materials vary with depth. Such a one-dimensional description would suffice if our planet were a perfect sphere, but of course it is not so symmetric. At the surface, we can see *lateral variations* (geographic differences) in Earth's structure associated with oceans and continents and with the basic features of plate tectonics: spreading centers at mid-ocean ridges, subduction zones at deep-sea trenches, and mountain belts lifted up by continent-continent collisions. Below the crust, we can expect that convection will cause changes in temperature from one part of the mantle to another. Downwelling currents, such as those associated with subducted lithospheric plates, will be relatively cold; whereas upwelling currents, such as those associated with mantle plumes, will be relatively hot.

Computer models tell us that the lateral variations in temperature due to mantle convection should be on the order of several hundred degrees. From laboratory experiments on rocks, we know that a temperature increase of 100°C reduces the speed of an S wave traveling through a mantle peridotite by about 1 percent (or even more if the rock is close to its melting temperature). If the mantle is indeed convecting, the seismic wave speeds should vary by several percentage points from place to place. Seismologists can make three-dimensional maps of these small lateral variations in wave speeds—and therefore three-dimensional maps of mantle convection—using the techniques of seismic tomography.

Seismic Tomography

Seismic tomography is an adaptation of a method commonly used in medicine to map the human body, called computerized axial tomography (CAT). CAT scanners construct three-dimensional images of organs by measuring small differences in X rays that sweep the body in many directions. Similarly, we can use the seismic waves from earthquakes recorded on thousands of seismographs all over the world to sweep Earth's interior in many different directions and construct a three-dimensional image of what's inside. We can find places where seismic waves speed up or slow down. The reasonable assumption, consistent with laboratory experiments, is that regions where seismic waves speed up are composed of relatively cool, dense rock (for example, subducted ocean plates), whereas regions where seismic waves slow down indicate relatively hot, buoyant matter (for example, rising convection plumes).

Seismic tomography has revealed features in the mantle clearly associated with mantle convection. Researchers at Harvard University constructed a tomographic model of S-wave speed variations in the mantle (**Figure 14.9**). The model is displayed as a series of global maps at depths ranging from just below the crust down to the core-mantle boundary. Near the surface, you can clearly see the structure of plate tectonics. The low S-wave speeds caused by the upwelling of hot asthenosphere along the mid-ocean ridges are shown in warm colors; the high S-wave speeds from cold lithosphere in the old ocean basins and beneath the continental cratons are shown in cool colors. At greater depths, the features become more variable and less coherent with the surface plates, reflecting what is inferred to be a complex pattern of mantle convection. Some large-scale features stand out, however. For example, you will notice that, just above the core-mantle boundary, there is a red region of relatively low S-wave speeds beneath the central Pacific Ocean, surrounded by a broad blue ring of higher S-wave speeds. Seismologists have speculated that the high speeds represent a "graveyard" of oceanic lithosphere subducted beneath the Pacific's volcanic arcs—the Ring of Fire—during the last 100 million years or so.

Seismic tomography shows that cold lithospheric slabs plunge all the way through the mantle, demonstrating that the plate tectonic system involves convection extending

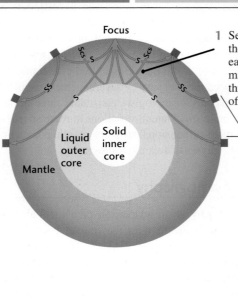

Focus

Mantle

Liquid outer core

Solid inner core

1 Seismic tomography uses the travel times from many earthquakes recorded at many receivers to create three-dimensional images of Earth's interior.

Seismographic stations

2 Regions with faster S-wave speeds indicate relatively colder, denser rock. Regions with slower S-wave speeds (red and yellow) indicate relatively hotter, less dense rock. A tomographic section through Earth reveals hot rocks, such as a mantle plume rising from Earth's core beneath South Africa,…

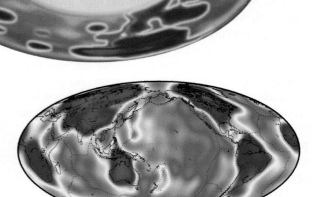

North America

Africa

3 …and colder rocks, such as the descending remnants of the Farallon Plate under the North American Plate.

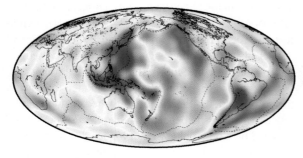

70 km

4 Near Earth's surface, hot rocks in the asthenosphere slow S waves, as revealed by the warm colors (red and yellow) along oceanic spreading centers.

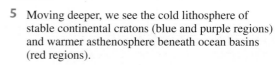

200 km

5 Moving deeper, we see the cold lithosphere of stable continental cratons (blue and purple regions) and warmer asthenosphere beneath ocean basins (red regions).

500 km

6 Deeper in the mantle, the features no longer match the continental positions.

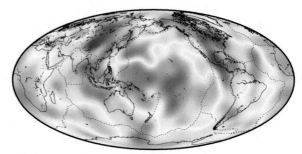

2800 km (near mantle-core boundary)

7 Near the mantle-core boundary, the S-wave patterns reveal colder regions around the Pacific that may be the "graveyards" of sinking lithospheric slabs.

[S-wave speeds courtesy of G. Ekström and A. Dziewonski, Harvard University. Cross section from M. Gurnis, *Scientific American* (March 2001): 40. Maps courtesy of L. Chen and T. Jordan, University of Southern California.]

almost as deep as the core-mantle boundary. An example is given in the cross section through the mantle shown in Figure 14.9. The image clearly reveals material from the once-large Farallon Plate, which has been mostly subducted under North America (see Chapter 10). The obliquely sinking slab material (in blue) appears to have penetrated the entire mantle. The image also indicates sinking colder rock beneath Indonesia. In addition, a large yellow blob of hotter rock, thought to be a "superplume," can be seen rising at an angle from the core-mantle boundary to a position beneath southern Africa. This hot buoyant mass pushing up the cooler material above it may explain the uplifted, mile-high plateaus of South Africa. The other blobs of hotter and cooler materials may be evidence of material exchanges among the lithosphere, the mantle, and the core-mantle boundary layer.

Earth's Gravity Field

The same temperature variations that speed up and slow down seismic waves also change the densities of mantle rocks. Laboratory experiments show that the expansion of a rock due to a 300°C increase in temperature will reduce its density by about 1 percent. This might seem to be a small effect, but the mass of the mantle is enormous (about 4 billion trillion tons!), so even small changes in the distribution of mass can lead to observable variations in the pull of Earth's gravity.

Geologists can determine the gravity field and features of Earth's mass distribution by observing bulges and dimples in the shape of the planet. Through careful analysis, they have been able to show that the shape measured by Earth-orbiting satellites matches the pattern of mantle convection imaged by seismic tomography (see Feature 14.2). This agreement has allowed us to refine our models of the mantle convection system.

EARTH'S MAGNETIC FIELD AND THE GEODYNAMO

Like the mantle, Earth's outer core transports most of its heat by convection. But the same techniques that have revealed so much about mantle convection—seismic tomography and the study of Earth's gravity field—have provided almost no information about convection in the core. Why not?

The problem has to do with the fluidity of the outer core. The mantle is a viscous solid that flows very slowly. As a result, convection creates regions where the temperatures are significantly higher or lower than the average mantle geotherm. We see these regions in Figure 14.9 as places where the density and seismic wave speeds are lower or higher than they are in the average mantle. The outer core, in contrast, has a very low viscosity—it can flow as easily as water or liquid mercury. Even small density variations caused by convection are quickly smoothed out by

the rapid flow of core fluid under the force of gravity. Thus, temperature fluctuations of just a few degrees cannot be sustained in the outer core. Any lateral variations in density and seismic-wave speeds caused by convection in the core are therefore much too small for us to see using seismic tomography, and they do not cause measurable distortions in the shape of the planet.

However, we can investigate convection in the outer core through observations of Earth's magnetic field. In Chapter 1, we briefly described the magnetic field and its generation by the core's geodynamo system. In Chapter 2, we discussed magnetic-field reversals and the paleomagnetism of volcanic rocks used to measure seafloor spreading. Here we will further explore the nature of the magnetic field and its origin in the geodynamo of the outer core.

Dipole Field

The most basic instrument for sensing Earth's magnetic field is the magnetic compass, invented by the Chinese more than 22 centuries ago. For hundreds of years, explorers and ship captains used compasses to navigate, but they had little understanding of how this ancient device actually worked. In 1600, William Gilbert, physician to Queen Elizabeth I, provided a scientific explanation. He proposed that "the whole Earth is itself a great magnet" whose field acts on the small magnet of a compass needle to align it in the direction of the north magnetic pole.

Scientists of Gilbert's day had begun to visualize a magnetic field as lines of force, such as those revealed by the alignment of iron filings on a piece of paper above a bar magnet. Gilbert showed that Earth's magnetic lines of force point into the ground at the north magnetic pole and outward at the south magnetic pole, as if a powerful bar magnet were located at Earth's center and oriented along an axis inclined about 11° from Earth's axis of rotation (see Figure 1.12). In other words, the lines of force look like a **dipole** (two-pole) magnetic field.

Complexity of the Magnetic Field

Gilbert solved an important problem for a seafaring nation dependent on the compass for good navigation, but his explanation was only partially correct. We now know that the source of the magnetic field is a geodynamo powered by core convection rather than a permanent magnet at Earth's center (which would be quickly destroyed by high temperatures in the core). The magnetic field produced by the geodynamo is considerably more complex than a simple dipole, and it is constantly changing with time owing to fluid motions in the outer core.

Within a few decades after Gilbert's famous pronouncement, careful observers had realized that the magnetic field varies with time. Not surprisingly, some of the best evidence for these changes came from the compass measurements systematically recorded by the British navy. Navigators had

14.2 The Geoid: Shape of Planet Earth

The surface of the ocean is warped upward in places where the gravitational pull is stronger and downward where the pull is weaker. The shape of the ocean's surface can be accurately measured by radar altimeters mounted on satellites. By averaging out wave motions and other fluctuations, oceanographers can map the small-scale variations in gravity caused by geologic features on the seafloor, such as faults and seamounts (see Chapter 20). These variations in gravity are also produced by the much larger features caused by mantle convection currents.

A perfectly still ocean has an upper boundary that conforms to what geologists call the *geoid*. The surface of a still body of water is perfectly "flat" in the sense that the pull of gravity is perpendicular to this surface—otherwise the water would flow "downhill" to make the surface flatter. The geoid is defined as an imaginary surface at some reference height above Earth adjusted to be everywhere perpendicular to the local gravitational force. Because the ocean surface approximates the geoid, we usually take the reference height to be sea level. When we measure the height of a mountain relative to sea level, we are actually measuring its height above the geoid at that point. In this sense, the geoid is just the "shape of the Earth." Geologists can use the geoid shape to calculate the size and direction of the gravitational force at any point on the planet's surface and infer how the rock density varies in Earth's interior.

Radar altimeters can easily map the geoid over the oceans, but how can we get this information on dry land? It turns out that the geoid can be measured for the entire Earth by tracking orbiting satellites. Three-dimensional mass variations in the mantle exert a small gravitational pull on the satellites, shifting their orbits slightly. By moni-

toring these shifts for long periods of time, scientists can create a two-dimensional map of the geoid over continents as well as oceans.

A smoothed version of the observed geoid is shown in the upper panel of the accompanying figure. The highs and lows on this map tell us about the large-scale features in Earth's gravity field. Relative to what sea level would be on an Earth without any lateral variations in mass, the height of the geoid varies from a low of about −110 m at a point near the coast of Antarctica to a high of just over +100 m on the island of New Guinea in the western Pacific.

The geoid shows some similarities to the large-scale features in the deeper parts of the mantle, which you can see by comparing the geoid map with Figure 14.9. The agreement suggests that the three-dimensional variations in the rock density and S-wave speed are both related to temperature differences arising from the large-scale convective flow in the mantle.

This hypothesis can be tested in the following way. Using laboratory data for calibration, geologists first calculate the three-dimensional density differences from the seismic wave speed variations mapped by tomography. They then construct a computer model of convective flow by assuming that the heavier parts of the mantle are sinking while the lighter parts are rising. Finally, they calculate what the geoid shape should be according to this convection model. The lower panel of the accompanying figure gives the results obtained more than a decade ago by the geophysicists Brad Hager and Mark Richards. You can see that the match to the observed geoid is quite good, especially for the largest features. This agreement has given geologists confidence that temperature variations within the mantle convection system can explain what we see in both the seismic images and the gravity field.

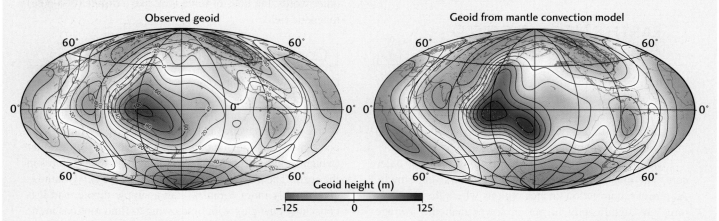

Observed geoid   Geoid from mantle convection model

Geoid height (m)

−125   0   125

The left panel of this figure is a smoothed map of the geoid or "shape of the Earth" derived from satellite observations. The contours, given here in meters, show how sea level deviates from an ideal Earth without any lateral variations in rock density. The panel on the right is the same map computed from a model of mantle convection that is consistent with the temperature structure of the mantle derived from seismic tomography. By matching the observations with a theoretical model, geologists have improved their understanding of the mantle convection system. [Observed geoid from NASA; geoid from model from B. Hager, Massachusetts Institute of Technology; maps courtesy of L. Chen and T. Jordan, University of Southern California.]

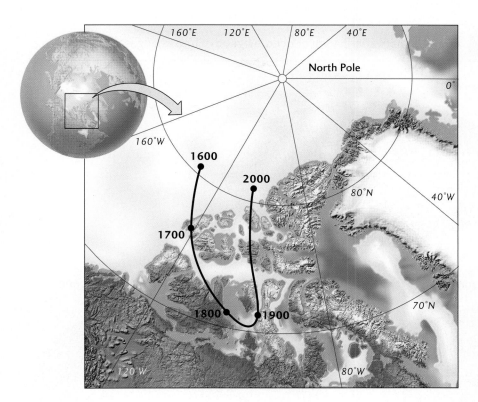

Figure 14.10 Change in location of the north magnetic pole from 1600 to 2000.

to correct their compass bearings to account for the displacement of the north magnetic pole (magnetic north) from the north rotational pole (true north), and these corrections showed that the north magnetic pole was moving at rates of 5° to 10° per century (**Figure 14.10**). Little did the British sailors know that these changes were coming from convective motions deep in Earth's core!

Nondipole Field Measurements at Earth's surface have revealed that only about 90 percent of the magnetic field can be described by the simple dipole illustrated in Figure 1.12. The remaining 10 percent, which geologists refer to as the *nondipole field,* has a more complex structure. This structure can be seen by comparing the strength of the field calculated for a simple dipole (**Figure 14.11a**) with the observed field (Figure 14.11b). If we extrapolate the field lines down to the core-mantle boundary, the size of the nondipole field actually increases relative to the size of the dipole field (Figure 14.11c). The poorly conducting mantle tends to smooth out complexities in the magnetic field, making the dipole field seem bigger than it really is.

Secular Variation Magnetic records for the last 300 years (many from the British navy) show that both the dipole and nondipole parts of the field are changing over time, but this *secular* (time-related) *variation* is fastest for the nondipole part. The secular variation is very evident in the comparison of today's magnetic field at the core-mantle boundary with maps reconstructed for previous centuries (see Figure 14.11e,f). Changes in field strength occur on time scales of decades and indicate that fluid motions

within the geodynamo system are on the order of millimeters per second.

Scientists can use the secular variation to help them understand convection in the outer core. With high-performance computers, they have been able to simulate the complex convective motions and electromagnetic interactions in the outer core that might be creating the geodynamo. The magnetic field lines from one such simulation are shown in Figure 14.11d. Away from the core, the field lines can be approximated by a dipole, but they become more complicated near the core-mantle boundary. Within the core itself, they are hopelessly entangled by the strong convective motions.

Magnetic Reversals Computer simulations also allow us to understand a remarkable behavior of the geodynamo system: the spontaneous reversals of the magnetic field. As we learned in Chapter 2, the magnetic field reverses its direction at irregular intervals (ranging from tens of thousands to millions of years), exchanging the north and south magnetic poles as if the magnet depicted in Figure 1.12 were flipped 180°. Recent computer simulations of the geodynamo can reproduce these sporadic reversals in the absence of any other external triggers (**Figure 14.12**). In other words, Earth's magnetic field spontaneously reverses, purely through internal interactions.

This behavior illustrates a fundamental difference between the geodynamo and the dynamos used in power plants. A steam-powered dynamo is an artificial system engineered by humans to do a particular job. The geodynamo, in contrast, exemplifies a *self-organized natural system*—one whose behavior is not predetermined by external constraints but

Earth System Figure 14.11   Earth's magnetic field produced by geodynamo changes with time.

(a)   Map of ideal dipole tilted by 11°

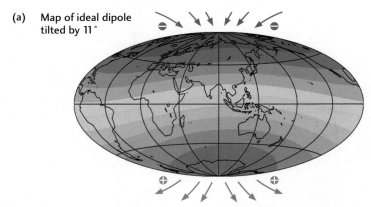

(b)   Magnetic field mapped at surface in 2000

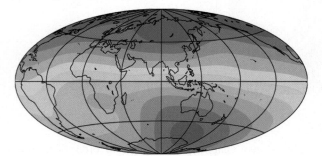

(c)   Magnetic field mapped at core-mantle boundary in 2000

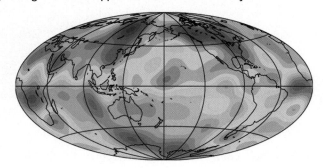

(d)   Computer model of magnetic field lines

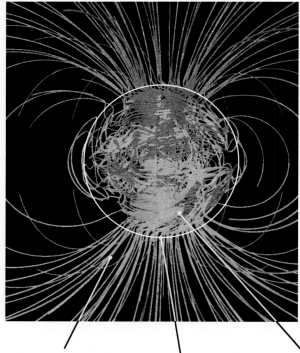

(e)   Magnetic field mapped at core-mantle boundary in 1900

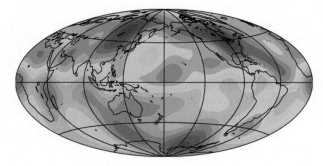

(f)   Magnetic field mapped at core-mantle boundary in 1800

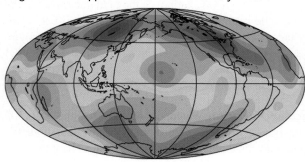

Magnetic field lines in mantle approximate those of a dipole.

Field mapped at core-mantle boundary reveals complexities in the core.

Field lines in core are entangled owing to convective motions that create the geodynamo.

[Maps courtesy of J. Bloxham, Harvard University. Computer model by G. Glatzmaier, University of California, Santa Cruz.]

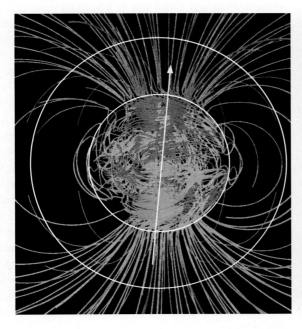

Time 1

Magnetic field lines with normal orientation prior to reversal. The magnetic field lines in the mantle approximate those of a magnetic dipole.

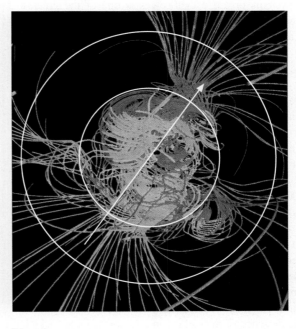

Time 2

Beginning of magnetic reversal. Geodynamo spontaneously begins to reorganize its magnetic field, increasing the complexity of the field lines within the outer core and decreasing the strength of the dipole component of the magnetic field.

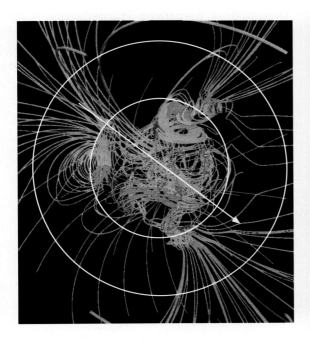

Time 3

Reversal continues with rapid changes in the structure of the magnetic field, which continues to have a weak dipole component.

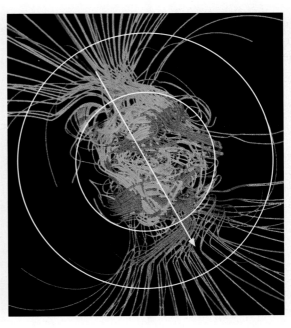

Time 4

Reversal nearly complete. Dipole field restrengthens with its north pole now pointing south.

[Computer model courtesy of G. Glatzmaier, University of California, Santa Cruz.]

emerges from internal interactions. The other two global geosystems, plate tectonics and climate, also display a wide variety of self-organized behaviors. Understanding how these natural systems organize themselves is one of the greatest challenges to geoscience. We will return to this subject when we discuss the climate system in Chapter 15.

Paleomagnetism

We have seen repeatedly how the geologic record of ancient magnetism, or **paleomagnetism,** has become a crucial source of information for understanding Earth's history. Magnetic stripes mapped on oceanic crust confirmed the existence of seafloor spreading and still provide the best data to explain how plate motions have evolved since the breakup of Pangaea 200 million years ago (see Chapter 2). The paleomagnetism of old continental rocks has been essential for establishing the existence of earlier supercontinents, such as Rodinia (see Chapter 10).

Scientists have also used paleomagnetism to reconstruct the history of Earth's magnetic field. The oldest magnetized rocks found so far formed around 3.5 Ga and indicate that Earth had a magnetic field at that time similar to the present one. The presence of magnetism in the most ancient rocks is consistent with the ideas of Earth's differentiation discussed in Chapter 1, which imply that a convecting fluid core must have been established very early in Earth's 4.5-billion-year history.

Let's delve a little more deeply into the rock-forming processes that have allowed geologists to draw these remarkable conclusions.

Thermoremanent Magnetization In the early 1960s, an Australian graduate student found a fireplace in an ancient campsite where the Aborigines had cooked their meals. The stones were magnetized. He carefully removed several stones that had been baked by the fires, first noting their physical orientation. Then he measured the direction of the stones' magnetization and found that it was exactly the reverse of Earth's present magnetic field. He proposed to his disbelieving professor that, as recently as 30,000 years ago, when the campsite was occupied, the magnetic field was the reverse of the present one—that is, a compass needle would have pointed south rather than north.

Recall that high temperatures destroy magnetism. An important property of many magnetizable materials is that, as they cool below about 500°C, they become magnetized in the direction of the surrounding magnetic field. This happens because groups of atoms of the material align themselves in the direction of the magnetic field when the material is hot. When the material has cooled, these atoms are locked in place and therefore are always magnetized in the same direction. This process is called **thermoremanent magnetization,** because the magnetization is "remembered" by the rock long after the magnetizing field has disappeared. Thus, the Australian student was able to determine the direction of the field when the stones cooled after the last fire and took on the magnetization of Earth's magnetic field of that time (**Figure 14.13**).

The discovery of magnetic reversals and the means to decipher them was a key ingredient in formulating the theory of plate tectonics. You should review the material in Figure 2.11 and its accompanying text, which discusses thermoremanent magnetization and magnetic stratigraphy.

Depositional Remanent Magnetization Some sedimentary rocks can take on a different type of remanent magnetization. Recall that marine sedimentary rocks form when particles of sediment that have settled through the ocean

30,000 years ago

Today

Figure 14.13 Earth's magnetic field 30,000 years ago was the reverse of today's, as evidenced by the discovery of reversely magnetized rocks found in the fireplace of an ancient campsite.

The rocks, cooling after the last fire, became magnetized in the direction of the ancient magnetic field, leaving a permanent record of it, just as a fossil leaves a record of ancient life.

1 Magnetic mineral grains transported to the ocean with other sediments become aligned with the Earth's magnetic field while settling through the water.

Direction of magnetic field

Ocean

2 This orientation is preserved in the lithified sediments, which thus "remember" the field that existed at the time of deposition.

Magnetic particles in ocean sediment

Figure 14.14 Newly formed sedimentary deposits can become magnetized in the same direction as the contemporaneous magnetic field of the Earth.

to the seafloor become lithified. Magnetic grains among the particles—chips of the mineral magnetite, for example—become aligned in the direction of Earth's magnetic field as they fall through the water, and this orientation can be incorporated into the rock when the particles becomes lithified. The **depo-**

sitional remanent magnetization of a sedimentary rock results from the parallel alignment of all these tiny magnets, as if they were compasses pointing in the direction of the field prevailing at the time of deposition (**Figure 14.14**).

Magnetic Stratigraphy Reversals of Earth's magnetic field are clearly indicated in the fossil magnetic record of layered lava flows. Each layer of rocks from the top down represents a progressively earlier period of geologic time, whose age can be determined by isotopic dating methods. The direction of remanent magnetism can be obtained for each layer, and in this way the time sequence of flip-flops of the field—that is, the **magnetic stratigraphy**—can be deduced. Geologists can also get data on Earth's magnetic reversal history by mapping magnetic stripes on the seafloor. From a combination of these data, they have worked out a detailed history of reversals for the last 200 million years (**Figure 14.15**). This information is of use to archaeologists and anthropologists as well as geologists. For example, the magnetic stratigraphy of continental sediments has been used to date sediments containing the remains of predecessors of our own species.

About half of all rocks studied are found to be magnetized in a direction opposite that of Earth's present magnetic field. Apparently, then, the field has flipped frequently over geologic time, and normal (same as now) and reversed fields are equally likely. Normal and reversed periods, which are called *magnetic chrons* (from the Greek word for "time"),

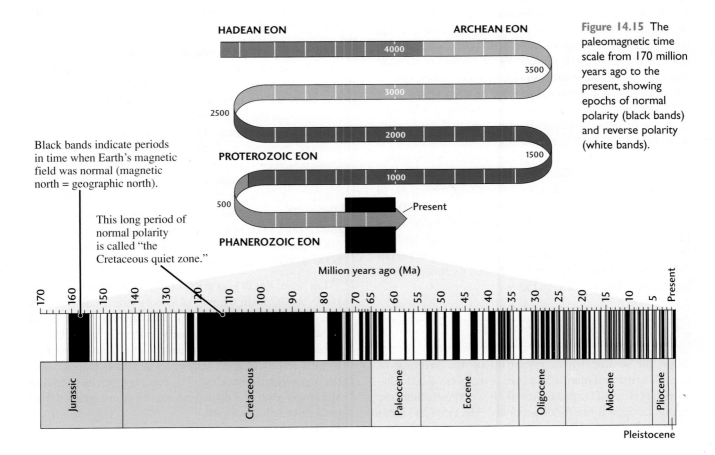

HADEAN EON   ARCHEAN EON

4000

3500

3000

2500

2000

1500

PROTEROZOIC EON

1000

Black bands indicate periods in time when Earth's magnetic field was normal (magnetic north = geographic north).

500

Present

This long period of normal polarity is called "the Cretaceous quiet zone."

PHANEROZOIC EON

Million years ago (Ma)

Figure 14.15 The paleomagnetic time scale from 170 million years ago to the present, showing epochs of normal polarity (black bands) and reverse polarity (white bands).

170 160 150 140 130 120 110 100 90 80 70 65 60 55 50 45 40 35 30 25 20 15 10 5 Present

Jurassic | Cretaceous | Paleocene | Eocene | Oligocene | Miocene | Pliocene

Pleistocene

have irregular lengths, but on average they last about a half-million years. Superimposed on the chrons are transient, short-lived reversals of the field, known as *subchrons,* which may last anywhere from several thousand years to tens of millions of years. The Australian graduate student apparently found a new reversed subchron within the present normal magnetic chron.

The Magnetic Field and the Biosphere

From the rock record, we know that the geodynamo began to operate early in Earth's history, and life therefore evolved within a strong magnetic field. The consequences turn out to be rather surprising. For example, many types of organisms—pigeons, sea turtles, whales, and even bacteria—have evolved sensory systems that use the magnetic field for precise navigation. The basic sensors are small crystals of the mineral magnetite (Fe_3O_4) that become magnetized by the Earth's background field as they are biologically precipitated within an organism. These crystals subsequently act as tiny compasses to orient the organism in the background field. Geobiologists have recently discovered that some animals can even use arrays of magnetite crystals to sense the strength of the magnetic field, which provides them with additional information for navigation.

The magnetic field is not just a convenient frame of reference for the flying and swimming species. It constitutes a part of the Earth system that is essential for sustaining a rich and delicate biosphere at the planet's surface. Although the machinery of the geodynamo operates deep within the core, its magnetic lines of force reach far into outer space, forming a barrier that shields Earth's surface from the damaging radiation of the solar wind. Without the protection of a strong magnetic field, this intense stream of high-energy, electrically charged particles would be lethal to many organisms.

Moreover, if the geodynamo were to stop producing a magnetic field, bombardment by the solar wind would gradually strip away Earth's atmosphere, further degrading the terrestrial environment. This appears actually to have happened in the case of Mars. Paleomagnetism of the ancient Martian crust has been detected by orbiting spacecraft, so we know the Red Planet once had an active geodynamo that generated a strong magnetic field. Sometime early in the planet's history, its geodynamo ceased to operate, perhaps because the Martian core cooled enough to freeze, and exposure to the solar wind subsequently eroded its atmosphere to the tenuous state we observe today.

SUMMARY

What do seismic waves reveal about the layering of Earth's crust and mantle? Seismic waves were used to discover that beneath Earth's felsic crust lies a denser ultramafic mantle composed mostly of peridotite. The crust and outer part of the mantle make up the rigid lithosphere. Beneath the lithosphere lies the asthenosphere, the weak layer of the mantle across which the lithosphere slides in plate tectonics. At the top of the asthenosphere, the temperature is high enough to partially melt peridotite, forming a zone where S-wave speeds decrease with depth. Below about 200 km, seismic wave speeds increase with depth down to the Earth's core. However, at two places, 410 km and 660 km below the surface, they show jumps caused by phase changes that transform mantle minerals to denser, more closely packed crystal structures stable at higher pressures.

What do seismic waves tell us about the layering of Earth's core? Seismic waves reflected from the core-mantle boundary locate this chemical transition at a depth of 2890 km. The failure of S waves to penetrate below the core-mantle boundary indicates that the outer core is a fluid. Seismic wave speeds indicate that the fluid outer core becomes a solid inner core at a depth of 5150 km. Several lines of evidence show that the core is composed mostly of iron and nickel, with minor amounts of lighter elements such as oxygen and sulfur.

What has seismic tomography revealed about structures in the mantle? Geologists use seismic tomography to make three-dimensional images of Earth's interior. The images show how plates vary from very thin under mid-ocean ridges to very thick under continental cratons. They also reveal many features of mantle convection, such as lithospheric slabs sinking into the lower mantle (some all the way to the core-mantle boundary) and superplumes rising from deep within the mantle.

How hot does it get in Earth's interior? Earth's interior is hot because it still retains much of the heat of its violent formation as well as the heat generated by the decay of radioactive elements. It has cooled over geologic time primarily by convection in the mantle and core and by conduction of heat through the lithosphere. The geotherm describes how temperature increases with depth. Within normal continental crust, it increases at a rate of 20° to 30°C per kilometer. Temperatures near the base of the lithosphere reach 1300° to 1400°C, which is hot enough to begin to melt mantle peridotites. The temperature in the fluid core is probably greater than 3000°C. The temperature at the Earth's center is about 5000°C according to some geologists, or it may be as high as 6000° to 8000°C according to others.

What does Earth's gravity field tell us about the interior? Variations in the pull of gravity over Earth's surface and corresponding distortions in its shape can be measured by satellites. These variations arise primarily from the temperature variations caused by mantle convection, which affect the density of rocks (higher temperatures reduce densities and seismic velocities). The observed gravity field is in agreement with the pattern of mantle convection inferred from seismic tomography.

What does Earth's magnetic field tell us about the fluid outer core? Convective motions in the outer core stir the electrically conducting iron-rich fluid, forming a geodynamo that produces the magnetic field. The magnetic field at the surface is primarily a dipole, but it has a small nondipole part. Maps of the magnetic field derived from compass readings show that it has changed over the last several centuries, which tells us about the type of fluid motions that drive the geodynamo.

What is paleomagnetism and what is its importance? Geologists have discovered that rocks can become magnetized in the direction of Earth's magnetic field at the time they form. This remanent magnetization can be preserved in rocks for millions of years. Paleomagnetism tells us that Earth's magnetic field has reversed (flipped back and forth) over geologic time. The chronology of reversals has been worked out so that the direction of remanent magnetization of a rock formation is often an indicator of stratigraphic age.

KEY TERMS AND CONCEPTS

compressional wave (p. 326)

conduction (p. 333)

convection (p. 333)

core-mantle boundary (p. 332)

depositional remanent magnetization (p. 343)

dipole (p. 337)

geotherm (p. 334)

lower mantle (p. 332)

low-velocity zone (p. 330)

magnetic stratigraphy (p. 343)

Mohorovičić discontinuity (Moho) (p. 329)

paleomagnetism (p. 342)

phase change (p. 330)

principle of isostasy (p. 329)

seismic tomography (p. 335)

shadow zone (p. 327)

shear wave (p. 326)

thermoremanent magnetization (p. 342)

EXERCISES

1. The speed of compressional waves in the lower part of the oceanic crust averages about 7 km/s. What rock type is most consistent with this observation and other knowledge of the oceanic crust?

2. What evidence suggests that the asthenosphere is probably partially molten?

3. What evidence indicates that Earth's outer core is molten and composed mostly of iron and nickel?

4. What is the depth to the core, and how do we know it?

5. What is the difference between heat conduction and convection? Which process is more efficient in transporting heat through the mantle?

6. Would the temperature at the Moho beneath a continental craton be hotter or cooler than at the Moho beneath an ocean basin?

7. Why can features of mantle convection, such as rising and descending convection currents, be seen by seismic tomography?

8. How can a mountain float on the mantle when both are composed of rock?

9. How do igneous rocks become magnetized when they form? How does the magnetization of sedimentary rocks differ from this process?

10. What evidence supports the hypothesis that Earth's magnetic field is generated by a geodynamo in its outer core?

11. Does the magnetic field change by an observable amount over the span of a human lifetime? What does this say about fluid motions in the outer core?

THOUGHT QUESTIONS

1. The Moon shows no evidence of tectonic plates or their motions, nor has it been volcanically active for billions of years. What does this observation imply about the state and temperature of the interior of this planetary body?

2. How does the existence of Earth's magnetic field, iron meteorites, and the abundance of iron in the cosmos support the idea that Earth's core is mostly iron and the outer core is liquid?

3. How would you use seismic waves to find a chamber of molten magma in the crust?

4. How would seismic tomography answer the question, "How deep do the subducted slabs go before they are assimilated?"

5. Where in the mantle might you look to find regions of anomalously low S-wave speeds?

THE CLIMATE SYSTEM

In the last several chapters, we descended into Earth's deep interior to explore the heat engines of the plate tectonic and geodynamo systems. We now rise back to its surface to examine a global geosystem powered not by Earth's internal heat but by external heat from the Sun—the climate system.

No aspect of Earth science is more important to our continued well-being than the study of climate. Throughout geologic time, the rise and fall of organisms have been closely connected to changes in climate. Even the short history of our own species is deeply imprinted by climate change: agricultural societies began to flourish only about 11,000 years ago, when the harsh climate of the last glacial age rapidly transformed into the mild and steady climate of the Holocene epoch. Now a globalized human society based on a petroleum-fueled economy is injecting greenhouse gases into the atmosphere at a phenomenal rate with potentially dire consequences—global warming, sea-level rise, and unfavorable changes in weather patterns. The climate system is a huge, incredibly complex machine, and, like or not, our hands are on the controls. We're in the driver's seat with pedal to the metal, so we had better understand how the contraption works!

In this chapter, we will examine the main components of the climate system and the way these components interact to produce the climate we live in today. We will investigate the geologic record of climate change and discuss the important role of the carbon cycle in regulating climate. Finally, we will look at the evidence for recent global warming caused by fossil-fuel emissions into the atmosphere.

Knowledge of the climate system will equip us to study a wide range of geologic processes that shape the face of our planet—weathering, erosion, sediment transport by wind and water, the interaction of tectonics and climate—which will be the topics of the next seven chapters. The material presented

A snapshot of the climate system taken by sensors on several spacecraft, showing cloud cover, sea-surface temperature, and land-surface properties, including vegetation. [R. B. Husar/NASA Visible Earth.]

here will also set up the final subject of this book: a geosystem perspective on the resource needs and environmental impacts of human society, discussed in Chapter 23.

COMPONENTS OF THE CLIMATE SYSTEM

At any point on Earth's surface, the energy received from the Sun changes daily, yearly, and on longer-term cycles associated with Earth's motions through the solar system. The cyclical variations in the input of solar energy force changes in the conditions of the surface environment; temperatures heat up in the day and cool off at night, heat up in summer and cool off in winter. **Climate** describes the average surface conditions and their variation during these cycles of solar forcing.

Climate is measured by the daily and seasonal statistics of atmospheric temperature, humidity, cloud cover, rate of rainfall, wind speed, and other weather conditions. Table 15.1 gives an example of temperature statistics in New York City, which include measures of temperature variability (record highs and lows) as well as the mean values. In addition to the common weather statistics, a full scientific description of climate involves the nonatmospheric components of the surface environment, such as soil moisture and streamflow on land, and sea-surface temperature and the velocity of currents in the ocean.

The **climate system** includes all parts of the Earth system and all of the interactions needed to explain how climate behaves in space and time (**Figure 15.1**). The main components of the climate system are the atmosphere, hydrosphere, cryosphere, lithosphere, and biosphere. Each component plays a different role in the climate system, depending on its ability to store and transport energy.

The Atmosphere

Earth's atmosphere is the most mobile and rapidly changing part of the climate system. Like Earth's interior, the atmosphere is layered (**Figure 15.2**). About three-quarters of its mass is concentrated in the lowest layer, the **troposphere,** which has an average thickness of about 11 km and convects vigorously owing to surface heating by the Sun, causing storms and other short-term disturbances in the weather (*tropos* is the Greek word for "turn" or "mix"). Above the troposphere is the **stratosphere,** a colder, dryer layer that extends to an altitude of about 50 km. The atmosphere has no abrupt cutoff; it slowly becomes thinner and fades away into outer space.

Because of Earth's rotation, the convection in the troposphere sets up a series of wind belts with a generally eastward flow at midlatitudes, which we will describe in Chapter 19. In temperate regions, the wind belts transport a typical parcel of air eastward around the globe in about a month (which is why it takes a few days for storms to blow across the country). The spiral-like circulation of air in the global wind belts (see Figure 19.1) also transports heat energy from the hotter equatorial regions to the colder polar regions.

The atmosphere is a mixture of gases, mostly molecular nitrogen (78 percent by volume in dry air) and oxygen (21 percent by volume in dry air). The remaining 1 percent consists of argon (0.93 percent), carbon dioxide (0.035 percent), and other minor gases (0.035 percent). Among the minor gases is water vapor, which is concentrated near Earth's surface in highly variable amounts (up to 3 percent, but typically about 1 percent). Water vapor (H_2O) and carbon dioxide (CO_2) are the principal greenhouse gases. If there were no greenhouse gases, heat generated by solar radiation would pass out easily through the atmosphere, and Earth's climate would be much colder. Greenhouse gases absorb solar radiation at certain wavelengths, raising the surface temperature. We will discuss the greenhouse effect in more detail shortly.

Another minor constituent of the atmosphere is ozone ($O_3{}^+$), a highly reactive gas produced primarily by the ionization of molecular oxygen by ultraviolet (UV) radiation from the Sun. In the lower part of the atmosphere, ozone exists in only tiny amounts, although it is a strong enough greenhouse gas to play a significant role in the atmospheric heat budget. Most of the ozone is found in the stratosphere, with a maximum concentration at an altitude of 25 to 30 km (see Figure 15.2). Absorption of ultraviolet energy by the stratospheric "ozone shield" filters out this potentially dam-

| Table 15.1 | Daily and Seasonal Temperatures (°F) in Central Park, New York City |

Data type[a]	January 1	April 1	July 1	October 1
Record high	62	83	100	88
Mean high	39	56	82	69
Mean low	28	39	67	55
Record low	−4	12	52	36

[a]Mean temperatures are for 1970–2004; record temperatures are for 1869–2004.

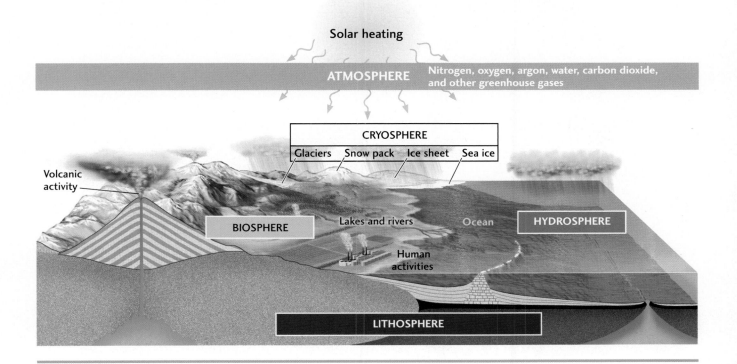

aging part of the solar radiation, protecting the biosphere at Earth's surface.

The Hydrosphere

The hydrosphere comprises all the liquid water on Earth's surface, including oceans, lakes, rivers, and groundwater. Almost all the liquid water is in the global ocean (1350 million cubic

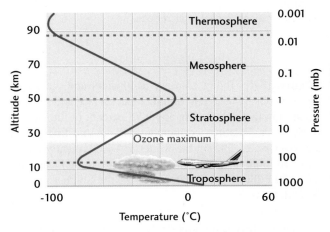

Figure 15.2 Layers of the atmosphere, showing the variation of temperature and pressure with altitude.

kilometers); lakes, rivers, and groundwater constitute a mere 1 percent (15 million cubic kilometers). However, these continental components play a vital role in the climate system as reservoirs for land moisture, as well as providing the transportation system for returning precipitation to the oceans and supplying the oceans with salt and other minerals.

Although water moves more slowly in the oceans than air moves in the atmosphere, water can store a much larger amount of heat. For this reason, ocean currents transport energy very effectively. Winds blowing across the ocean generate surface currents and large-scale circulation patterns within the ocean basins (**Figure 15.3a**). As in the atmosphere, the most important currents for regulating climate are those that transport heat from equatorial to polar regions. These currents involve vertical convection as well as horizontal motion. An example is the Gulf Stream, which flows along the western Atlantic margin, bringing waters from the Caribbean Sea that warm the climate of the North Atlantic and Europe.

In the North Atlantic, the water cools and becomes more saline (because less fresh water enters the sea from rivers at high latitudes). The cooler, saltier water sinks because it is has become denser. In this way, a subsurface cold current is created that flows southward as part of a general **thermohaline circulation**—so called because it is driven by differences in temperature and salinity. On a planetary scale, the

(a)

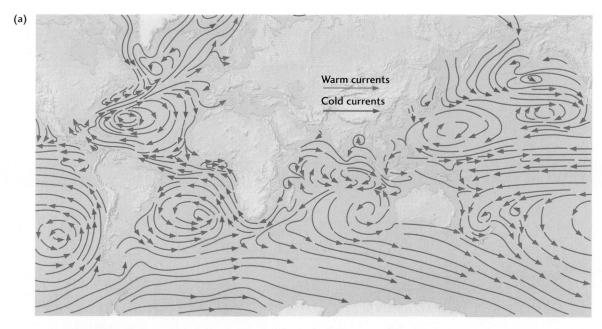

Warm currents

Cold currents

(b)

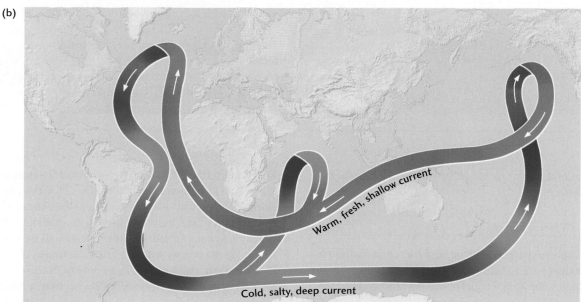

Warm, fresh, shallow current

Cold, salty, deep current

Figure 15.3 Two major current systems in the oceans.
(a) Currents at the surface of the oceans. [U.S. Naval Oceanographic Office.] (b) A schematic representation of thermohaline circulation, which acts as a conveyor belt to transport heat from warm equatorial regions to cool polar regions.

thermohaline circulation acts like an enormous conveyor belt running through the oceans that moves heat from the equatorial regions toward the poles (see Figure 15.3b). Changes in this circulation pattern can strongly influence the global climate.

The Cryosphere

The ice component of the climate system is called the *cryosphere*. It comprises 33 million cubic kilometers of ice, primarily in the ice caps and glaciers of the polar regions

(Figure 15.4). Today continental glaciers and ice sheets cover about 10 percent of the land surface (15 million square kilometers), storing about 75 percent of the world's fresh water. Floating ice includes sea ice and frozen lake and river water. During the winter, sea ice typically covers 14 to 16 million square kilometers of the Arctic Ocean and 17 to 20 million square kilometers of the Southern Ocean around Antarctica.

The seasonal exchange of water between the cryosphere and hydrosphere is an important process of the climate system. About 33 percent of the land surface is covered by sea-

Figure 15.4 Sea ice is an important part of the cryosphere. This satellite image shows sea ice flowing through the Bering Strait in May 2002. According to a recent study by the U.S. Navy, global warming since the 1950s has shrunk the polar ice cap by as much as 40 percent and reduced the ice extent by 20 percent, disrupting whale migrations and the habitats of seals and polar bears that live at the ice edge. [NASA MODIS satellite.]

sonal snows, almost entirely (all but 2 percent) in the Northern Hemisphere. Melting snow is the source of much of the fresh water in the hydrosphere. For example, snowfall accounts for 60 to 70 percent of annual precipitation in the U.S. Sierra Nevada and Rocky Mountains, which is later released as water during spring snow melt and river runoff.

The role of the cryosphere in the climate system differs from the role of the liquid hydrosphere, because ice is relatively immobile and white, reflecting almost all of the solar energy that falls on it. Large masses of water are exchanged between the cryosphere and the hydrosphere during glacial cycles. At the last glacial maximum 18,000 years ago, sea level was about 130 m lower than it is today, and the volume of the cryosphere was three times larger.

The Lithosphere

The part of the lithosphere most important to the climate system is the land surface, which makes up 30 percent of Earth's total area. The composition of the land surface affects how solar energy is absorbed and returned to the atmosphere. As the temperature rises, the land radiates more

energy as infrared waves back into the atmosphere, and more water evaporates from the land surface. Evaporation requires considerable energy, so soil moisture and other factors that influence evaporation—such as vegetation and the subsurface flow of water—are very important in controlling surface temperatures.

Topography has a direct effect on climate through its influence on atmospheric winds and circulation. Air masses that flow over large mountain ranges dump rain on the windward side, creating a rain shadow in the lee of the mountains (see Figure 17.3). In fact, the overall asymmetry of continents, a direct consequence of plate tectonics, induces hemispheric asymmetries in the global climate system.

Geologists have documented many long-term changes in the climate system associated with the rearrangement of land areas by plate tectonics. Changes in the shape of the seafloor due to seafloor spreading induce sea-level changes, and the drift of continents over the poles leads to the growth of continental glaciers. Tectonic movements can also block ocean currents or open gateways through which currents can flow. The emergence of the Isthmus of Panama, for example, closed a passage connecting the Atlantic and Pacific oceans about 5 million years ago, which may have initiated the Pleistocene glacial cycles (see Chapter 21). If future geologic activity were to close the narrow channel between the Bahamas and Florida through which the Gulf Stream flows, temperatures in western Europe might drop drastically.

Volcanism, which occurs in the lithosphere, affects climate by changing the composition of the atmosphere. Major volcanic eruptions can inject aerosols and other particulates into the stratosphere, blocking solar radiation from reaching the surface and temporarily lowering atmospheric temperature. After the massive April 1815 eruption of Mount Tambora in Indonesia, New England suffered through a "year without a summer" in 1816. According to a diarist in Vermont, "no month passed without a frost, nor one without a snow," and crop failures were common. Careful studies have shown that more recent large volcanic eruptions—including Krakatoa (1883), El Chichón (1982), and Mount Pinatubo (1991)—produced an average dip of 0.3°C in global temperatures about 14 months after the eruption (local temperature variations can be much larger, of course). Temperatures returned to normal in about 4 years. Although the effects of most eruptions on global climate are short-lived, they provide an important way to calibrate numerical climate models.

The Biosphere

The term *biosphere* was coined as early as 1875 by Eduard Suess, the same geologist who named Gondwanaland and first described the exposed continental cratons as "shields." The biosphere comprises all organisms living near Earth's surface: plants, animals, and microbes, both marine and terrestrial. Life is ubiquitous at Earth's surface, but the amount of life depends on local climate conditions, as we

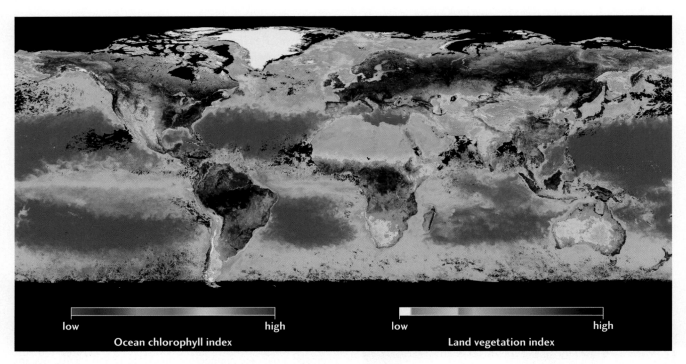

low — high
Ocean chlorophyll index

low — high
Land vegetation index

Figure 15.5 The global distribution of plant life as mapped by NASA's SeaWiFS satellite. [NASA Visible Earth.]

can see from the satellite images of plant concentrations in **Figure 15.5**.

The amount of energy contained and transported by living organisms is relatively small—less than 0.1 percent of the solar energy flux is used by plants in photosynthesis—but the biosphere is strongly coupled to the climate system through its interactions with the atmosphere, hydrosphere, and lithosphere. We have mentioned the effects of land vegetation on surface moisture and temperature. Life also regulates the composition of the atmosphere, including greenhouse gases such as carbon dioxide and methane. Through photosynthesis, marine and terrestrial plants extract carbon dioxide from the atmosphere. Some of it is precipitated as shells made of calcium carbonate or buried as organic matter in sediments. The biosphere thus plays a central role in the carbon cycle.

Of course, humans are part of the biosphere, though hardly an ordinary part. Our influence over the biosphere is growing rapidly, and we have become the most active agents of environmental change. As an organized society, we behave in fundamentally different ways from other species. For example, we can study climate change scientifically and modify our actions according to what we have learned.

THE GREENHOUSE EFFECT

The Sun is a yellow star that puts out about half its radiation in the visible part of the electromagnetic spectrum. The other half is split between infrared waves, which have longer wavelengths and lower energies, and ultraviolet waves, which

have shorter wavelengths and higher energies. (The fact that the Sun's radiation is most intense in the visible band is no coincidence, by the way; our eyes evolved to optimize their sensitivity to yellow sunlight.) At Earth's position in the solar system, the average amount of radiation its surface receives throughout the year is 342 W/m^2 (1 watt = 1 joule per second; a joule is a unit of energy or heat). In comparison, the average amount of heat flowing out of Earth's deep interior from mantle convection is minuscule, only 0.06 W/m^2. Essentially all the energy driving the climate system ultimately comes from the Sun (**Figure 15.6**).

For the global temperature averaged over daily and seasonal cycles to remain the same, Earth's surface must radiate energy back into space at a rate of precisely 342 W/m^2. Any less would cause the surface to heat up; any more would cause it to cool down. How is this *radiation balance* —the equilibrium between incoming and outgoing radiant energy—achieved?

A Planet Without Greenhouse Gases

Suppose Earth had no atmosphere at all but was a rocky sphere like the Moon. Some of the sunlight falling on the surface would be reflected back into space, and some would be absorbed by the rocks, depending on the color of the surface. A perfectly white body would reflect all the solar energy, whereas a perfectly black body would absorb it all. The fraction of the energy reflected is called the planet's **albedo** (from the Latin word *albus,* meaning "white").

1 Solar energy input to Earth's surface is, on average, 342 W/m².

2 Heat flowing out of Earth's deep interior is much smaller — only 0.06 W/m².

3 Therefore, heat radiating from Earth must balance solar input.

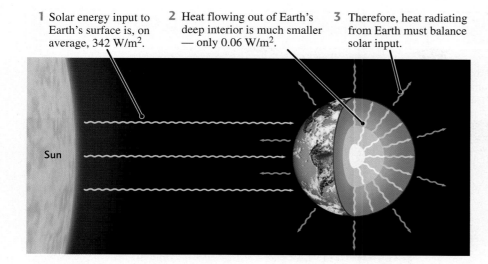

Figure 15.6 Earth's energy balance is achieved by radiation of the incoming solar energy back into space. Heat from Earth's interior is negligible in comparison to that from solar energy.

Although a full Moon looks bright to us, the rocks on its surface are mainly dark basalts, so its albedo is only about 7 percent. In other words, the Moon is very nearly black.

The energy radiated by a black body goes up rapidly with increasing temperature. A cold bar of iron is black and gives off little heat. If you heat the bar to 100°C, it gives off warmth in the form of infrared waves (like a steam radiator). If you heat the bar to 1000°C, it becomes bright orange, radiating heat at visible wavelengths (like the burner on an electric stove).

A black body exposed to the Sun heats up until its temperature is just the right value for it to radiate the incoming solar energy back into space. The same principle applies to a "gray body" like the Moon, except you must exclude the reflected energy from this balance. The Moon's daytime temperatures rise to 130°C and its nighttime temperatures drop to −170°C. Not a pleasant environment!

Earth rotates much faster than the Moon (once per day rather than once per month), which evens out the day-and-night extremes of temperature. Earth's albedo, about 31 percent, is much higher than the Moon's, because Earth's blue oceans, white clouds, and ice caps are more reflective than dark lunar basalts. If Earth's atmosphere did not contain greenhouse gases, the average surface temperature required to balance the nonreflected solar radiation would be about −19°C (−2°F), cold enough to freeze all the water on the planet. Instead, Earth's average surface temperature remains a balmy 14°C (57°F). The difference of 33°C is a result of the greenhouse effect.

Earth's Greenhouse Atmosphere

At a given temperature, a planet with an atmosphere containing greenhouse gases radiates solar energy back into space less efficiently than a planet with no atmosphere. Therefore, its surface temperature must rise to achieve the radiation balance. In this way, the atmosphere acts like the glass in a greenhouse, allowing solar energy to pass through but trapping heat. This is known as the **greenhouse effect.**

Greenhouse gases such as water vapor, carbon dioxide, and ozone absorb solar energy coming directly from the Sun or reflected from Earth's surface and reradiate it as infrared energy in all directions, including downward to the surface. The net effect is to trap heat within the atmosphere by increasing the temperature of the surface relative to the temperature at higher levels of the atmosphere.

How Earth's atmosphere balances incoming and outgoing radiation is illustrated in **Figure 15.7**. The incoming solar radiation not directly reflected is absorbed by the atmosphere and the surface. To achieve radiation balance, Earth radiates this same amount of energy back into space as infrared waves. Because of the heat trapped by the greenhouse gases, the amount of energy transported away from Earth's surface, both by radiation and by the flow of hot air and moisture from the surface, is significantly larger than the amount Earth receives as direct solar radiation. The excess is exactly the energy received as Earthward infrared radiation from the greenhouse gases. It is this "back radiation" that causes Earth's surface to be 33°C warmer than it would be if the atmosphere contained no greenhouse gases.

Balancing the System Through Feedbacks

How does the climate system actually achieve the radiation balance illustrated in Figure 15.7? Why does the greenhouse effect yield an overall warming of 33°C and not some other temperature, larger or smaller? The answers to these questions are not simple, because they depend on interactions among the many components of the climate system. The most important interactions involve feedbacks.

Feedbacks come in two basic types: **positive feedback,** in which a change in one component is *enhanced* by the

Key Figure 15.7 | Earth's greenhouse atmosphere balances incoming and outgoing radiation.

Earth's albedo (31%) is energy reflected by clouds (22%) and Earth's surface (9%). The remaining incoming solar radiation is absorbed by Earth's atmosphere (20%) and the surface (49%).

To achieve radiation balance, Earth radiates the sum of the radiation absorbed by the atmosphere and surface back into space.

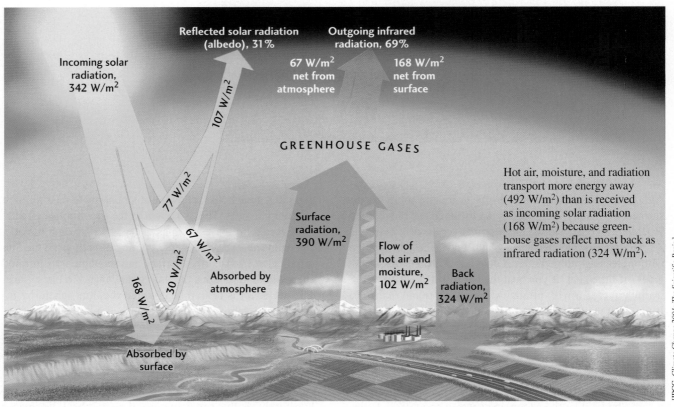

Reflected solar radiation (albedo), 31%

Outgoing infrared radiation, 69%

Incoming solar radiation, 342 W/m²

67 W/m² net from atmosphere

168 W/m² net from surface

107 W/m²

GREENHOUSE GASES

77 W/m²

67 W/m²

30 W/m²

168 W/m²

Surface radiation, 390 W/m²

Flow of hot air and moisture, 102 W/m²

Back radiation, 324 W/m²

Absorbed by atmosphere

Absorbed by surface

Hot air, moisture, and radiation transport more energy away (492 W/m²) than is received as incoming solar radiation (168 W/m²) because greenhouse gases reflect most back as infrared radiation (324 W/m²).

[IPCC, Climate Change 2001: The Scientific Basis.]

changes it induces in other components, and **negative feedback**, in which a change in one component is *reduced* by the changes it induces in other components. Positive feedbacks tend to amplify changes in the system, whereas negative feedbacks tend to stabilize the system against change.

Here are some of the feedbacks within the climate system that can significantly affect the temperatures achieved by radiation balance.

• *Water vapor feedback.* A rise in temperature increases the amount of water vapor in the atmosphere through evaporation. Water vapor is a greenhouse gas, so this increase enhances the greenhouse effect and the temperature rises—a positive feedback.

• *Albedo feedback.* A rise in temperature reduces the accumulation of ice and snow in the cryosphere, which decreases Earth's albedo and increases the energy its surface absorbs. The increased warming of the atmosphere enhances the temperature rise—another example of positive feedback.

• *Radiation feedback.* In "radiative damping," a rise in atmospheric temperature strongly increases the amount of infrared energy radiated back into space, which reduces the

temperature rise—a negative feedback. Radiative damping stabilizes Earth's climate against major changes, keeping the oceans from freezing up or boiling off and thus maintaining an equable habitat for water-loving life.

• *Plant growth feedback.* The stimulation of plant growth by carbon dioxide increases atmospheric CO_2, which causes temperatures to rise by enhancing the greenhouse effect. However, more plant growth also removes CO_2 from the atmosphere by converting it to carbon-rich organic matter, thus reducing the greenhouse temperature rise. This secondary effect is a negative feedback.

Feedbacks can involve much more complex interactions among components of the climate system. For example, an increase in atmospheric water vapor produces more clouds. Because clouds reflect solar energy, they increase the planetary albedo, which sets up a negative feedback between water vapor and temperature. On the other hand, clouds absorb infrared radiation efficiently, so increasing the cloud cover enhances the greenhouse effect, thus providing a positive feedback between water vapor and temperature. Does the net effect of clouds produce a positive or negative feedback?

Scientists have found it surprisingly difficult to answer such questions. Earth has only one climate system. The components of this system are joined through an amazingly complex web of interactions on a scale far beyond experimental control. Consequently, it is often impossible to gather data that isolate one type of feedback from all the others. Scientists must therefore turn to computer models to understand the inner workings of the climate system.

Climate Models and Their Limitations

Generally speaking, a **climate model** is any representation of the climate system that can reproduce one or more aspects of climate behavior. Some models are designed to study local or regional climate processes, such as the relationships between water vapor and clouds, but the most interesting representations are global models that describe how climate has changed in the past or predict how it might change in the future.

At the heart of such global models are schemes for computing the motions within the atmosphere and oceans based on the fundamental laws of physics. These *general circulation models* represent the currents of air and water driven by solar energy on scales ranging from small disturbances (storms in the atmosphere, eddies in the oceans) to global circulations (wind belts in the atmosphere, thermohaline convection in the oceans). Scientists represent the basic physical variables (temperature, pressure, density, velocity, and so forth) on three-dimensional grids comprising millions of geographic points. They use supercomputers to solve numerically the mathematical equations that describe how the variables change with time at each of these points (**Figure 15.8**).

You see the results of this type of calculation whenever you tune in the weather report on your favorite TV station. These days, most weather predictions are made by setting up a model that accurately describes the current conditions observed at thousands of weather stations and running it forward in time using an atmospheric general circulation model. Numerical weather predictions thus use the same basic computer programs that are used for climate modeling.

Climate modeling is more difficult, however. In predicting weather a few days from now, scientists can ignore such slow processes as changes in atmospheric greenhouse gases or ocean circulation. Climate predictions, on the other hand, require that we properly model these slow processes, including all important feedbacks, in addition to modeling the fast motions of air masses. Moreover, the simulation must be extended for years or decades into the future. Such enormous calculations require weeks of time on the world's largest supercomputers.

Because of their complexity, current models of the climate system must be viewed with some skepticism. Many questions remain about how the climate system works—for instance, how clouds affect atmospheric temperatures. Model predictions are subject to errors and have caused much debate among experts and governmental authorities

Figure 15.8 Climate can be simulated by general circulation models. This climate model, developed with support from the U.S. Department of Energy, portrays interactions among the atmosphere, hydrosphere, and cryosphere. [Warren Washington and Gary Strand/National Center for Atmospheric Research.]

who must deal with the regulation and consequences of human-induced climate change.

CLIMATE VARIABILITY

Earth's climate varies considerably from place to place—its poles are frigid and arid, its tropics sweltering and humid. Comparable variations in climate can also occur at different times. The geologic record shows us that periods of global warmth have alternated with epochs of glacial cold many times in the past. This climate variability appears to be highly irregular: dramatic changes can happen in just a few decades or evolve over time scales of many millions of years.

Knowing how the climate system has changed in the past will help us understand how it might change in the future. Some variations can be attributed to the external forcing of climate, such as fluctuations in the amount of radiation Earth receives from the Sun and changes in the distribution of land surface caused by continental drift. Others result from internal variations within the climate system itself, as mass and energy are exchanged among its components. Both types of variations, external and internal, can be amplified or suppressed by feedbacks. In this section, we will examine several types of climate variability and discuss their causes.

Ice Ages

An **ice age** occurs when Earth cools and water is transferred from the hydrosphere to the cryosphere. The amount of sea ice increases, and more snow falls on the continents in winter than melts in summer, increasing the volume of the continental ice sheets and decreasing sea level. As the polar ice caps expand, they reflect more solar energy back into space, and the surface temperatures fall further—an example of the albedo feedback.

Pleistocene Ice Ages During the last ice age, the *Wisconsin glaciation,* which peaked about 18,000 years ago, the volume of cryosphere exceeded 100 million cubic kilometers, more than three times the amount of ice on Earth today. Ice sheets with thicknesses of 2 to 3.5 km built up over North America, Europe, and Asia. In the Southern Hemisphere, the Antarctic ice expanded, and the southern tips of South America and Africa were covered with ice (**Figure 15.9**). The continents were slightly larger than they are today because the continental shelves surrounding them, some more than 100 km wide, were exposed by a drop in sea level that reached 130 m.

The Wisconsin glaciation is only the latest in a series of ice ages that occurred during the Pleistocene epoch, which began 1.8 million years ago. A precise record of these events can be obtained by measuring the proportion of two oxygen isotopes preserved in oceanic sediments. The lighter and more common isotope, oxygen-16, has a greater tendency to evaporate from the ocean surface than the heavier oxygen-18. Therefore, during glaciations, marine sediments become enriched in oxygen-18, because oxygen-16 is preferentially

evaporated from the oceans and trapped in glacial ice. Using this type of isotopic analysis, scientists have been able to trace the growth and shrinkage of continental glaciers and determine the temperatures of the sea surface over many glacial cycles.

The best records of climate variations during the last half-million years come from ice cores drilled in the East Antarctic ice sheet by Russian scientists at the Vostok Station and in the Greenland ice sheet by European and U.S. teams (see Feature 15.1). These cores have provided a detailed stratigraphic record of the annual layers produced by the conversion of snow into ice, and the oxygen isotopes of the ice layers have yielded records of the temperature variations. Additional information has come from bubbles of air trapped at the time the ice formed. **Figure 15.10** displays temperatures and the concentrations of two important greenhouse gases, carbon dioxide and methane (CH_4), recovered from the Vostok ice core.

The temperature record shows much variation, but the largest swings correspond to the sawtooth pattern of the glacial cycles—a gradual decline of about 6° to 8°C from a warm interglacial period to a cold ice age, followed by a rapid rise during a short interval of deglaciation. Temperatures began to drop in the last glacial cycle about 120,000 years ago, reaching their lowest values 25,000 to 18,000 years ago (the Wisconsin glacial maximum). Temperatures then rebounded to

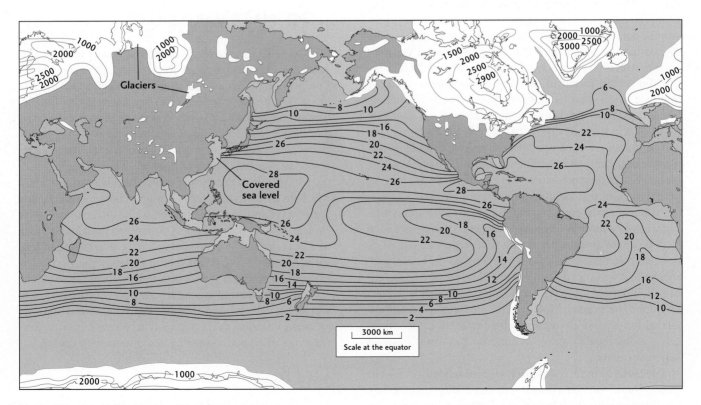

Figure 15.9 Ice extent during the peak of the Wisconsin glaciation 18,000 years ago, showing ice thickness in meters (blue contours) and sea-surface temperatures in degrees Celsius (black contours). Continental outlines reveal the lowering of sea level 85 m below the present level. [After CLIMAP, "The Surface of the Ice-Age Earth." *Science* 191 (1976): 1131–1137.]

1 There is a decline in both temperature and greenhouse gas concentrations during glacial periods...

3 Climate has been relatively warm and stable during the last 10,000 years—the Holocene interglacial period.

2 ...and a rapid rise during deglaciation.

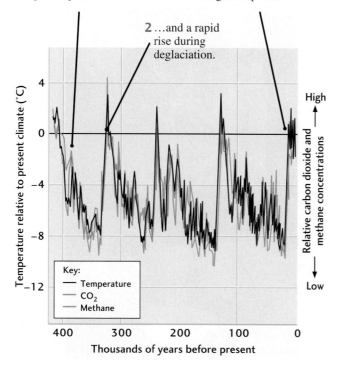

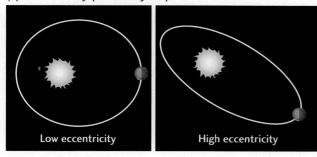

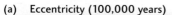

Figure 15.10 A graph showing three types of data recovered from the Vostok ice core in East Antarctica: temperatures inferred from oxygen isotopes (black line) and the concentrations of two important greenhouse gases, carbon dioxide (blue line) and methane (red line). The gas concentrations come from measurements of air samples trapped as tiny bubbles within the Antarctic ice. [IPCC, *Climate Change 2001: The Scientific Basis.*]

the warm interglacial climate of the last 11,000 years, corresponding to the Holocene epoch. Three other ice ages are visible in this record; they had minimum temperatures at about 140,000, 260,000, and 350,000 years ago.

Milankovitch Cycles The alternation between glacial and interglacial ages observed during the Pleistocene epoch is explained best by the **Milankovitch cycles,** named after a Yugoslavian geophysicist who first calculated them in the 1920s and 1930s. These cycles cause periodic variations in the amount of heat Earth receives from the Sun. The shape of Earth's orbit around the Sun changes cyclically, being more circular at some times and more elliptical at others. The degree of ellipticity of Earth's orbit around the Sun is known as *eccentricity.* A nearly circular orbit has low eccentricity, and a more elliptical orbit has high eccentricity (**Figure 15.11a**). The time interval of one cycle of variation from low to high eccentricity is about 100,000 years.

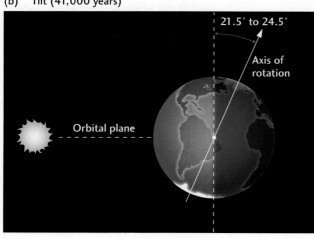

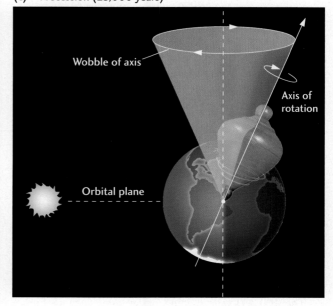

Figure 15.11 Three components of variations of Earth's orbit (here much exaggerated). (a) *Eccentricity* is the degree of ellipticity of Earth's orbit. (b) *Tilt* is the angle between Earth's axis of rotation and the vertical to the orbital plane. (c) *Precession* is the wobble of the axis of rotation. One can imagine this motion by thinking of the wobble of a spinning top.

EARTH ISSUES

15.1 Vostok and GRIP: Ice-Core Drilling in Antarctica and Greenland

At the Vostok science station in the frozen Antarctic, Russian scientists have been working year-round since 1960 to discover the climatological history of Earth hidden in the ice. In the 1970s, scientists at Vostok drilled boreholes 500 to 952 m deep in the East Antarctic ice sheet and brought up a set of ice cores showing layers produced by annual cycles of ice formation from snow. Careful counting of the layers, working from the top down, revealed the age of the ice, much as tree rings reveal the age of a tree. This time-stratigraphic record of the ice proved to correlate with the temperatures thought to prevail when the layers formed. A low ratio of oxygen-18 to oxygen-16 in an ice layer indicates that the ice formed when atmospheric temperature was relatively high. High concentrations of carbon dioxide, methane, and other greenhouse gases in an ice layer also suggest that it formed during a period of atmospheric greenhouse warming (see Figure 15.10).

By the 1990s, ice borers at Vostok had drilled to a depth of 2755 m, penetrating ice not only from the last glacial period but also from the interglacial period before it. They thereby accumulated a stratigraphic record for the past 160,000 years. These cores showed that eastern Antarctica, where the ice cores were drilled, was colder and drier during the last glaciation than during the 11,000 years of the current interglacial period. Variations in the oxygen isotope ratios of ice layers confirm other evidence that variations in Earth's orbit—the Milankovitch cycles—control the alternation of glacial and interglacial epochs. The carbon dioxide

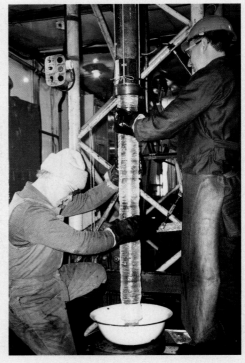

Russian scientists at Vostok science station in Antarctica carefully remove an ice core from a drill. The layers produced by annual cycles of ice formation are visible. [R. J. Delmas, Laboratoire de glaciologie et géophysique de l'environnement, Centre National de la Recherche Scientifique.]

content of ice layers formed during glacial intervals decreased markedly with climate cooling.

Meanwhile, in the Arctic, a group of scientists in 1992 completed 2 years of drilling the top 3 km of ice

---

Another astronomical cycle involves the angle of *tilt* of Earth's rotation axis. The angle cycles between 21.5° and 24.5° over a time period of about 41,000 years. This cycle also slightly changes the heat received from the Sun (see Figure 15.11b). In addition, Earth wobbles about its axis of rotation, giving rise to a *precession* with a time period of about 23,000 years (see Figure 15.11c). Precession, too, changes the amount of heat Earth receives from the Sun.

The 100,000-year spacing of ice ages seen in Figure 15.10 matches the times of high orbital eccentricity, when Earth received less radiation on average from the Sun. This correlation has led scientists to conclude that glacial cycles are externally forced by variations in the amount of solar radiation. The small changes in solar radiation during the Milankovitch cycles, however, cannot completely explain the large changes observed in Earth's surface temperature.

Some type of positive feedback must be operating within the climate system to amplify the external forcing. The data in Figure 15.10 clearly indicate that this feedback involves the greenhouse gases CO_2 and CH_4. Their concentrations precisely track the temperature variations throughout the glacial cycles—warm interglacial intervals are marked by high concentrations, cold glacial intervals by low concentrations. Exactly how this feedback works has not yet been fully explained, but it demonstrates the importance of the greenhouse effect in long-term climate variations.

Ancient Ice Ages In addition to the Pleistocene ice ages, there is good evidence in the geologic record for earlier episodes of continental glaciation during the Permian-Pennsylvanian and Ordovician periods and at least twice in the Proterozoic eon. Evidently, there are long-term changes in the Greenland ice cap. The Greenland Ice Core Project, better known as GRIP, is an outstanding example of international scientific cooperation in Europe. GRIP was sponsored by the European Science Foundation, which included the governments of Belgium, Denmark, France, Germany, Iceland, Italy, Switzerland, and the United Kingdom. The GRIP core penetrated ice layers of the last glacial period (11,000 to 115,000 years before the present) and the last interglacial period (the Eemian, 115,000 to 135,000 years before the present). It stopped in layers going back 235,000 years, near the end of the previous interglacial period. The accompanying diagram shows the relationship between depth of the ice core and times of glacial and interglacial periods for the past 135,000 years.

A parallel core was drilled 30 km to the west of the GRIP site by the U.S. Greenland Ice-Sheet Project (known as GISP2). The GRIP and GISP2 ice cores agreed very closely over the past 110,000 years, confirmed the glacial chronology at Vostok, and gave a good picture of the changes in climate and atmospheric composition of the past 250,000 years. Analysis of the cores also corrected earlier suggestions that the climates of the Eemian interglacial period were unstable and changed appreciably on a short time scale. The GRIP and GISP2 findings confirmed the remarkably rapid climate oscillations of the last glacial period. The GRIP and Vostok cores yielded precise oxygen-isotope profiles that correlated closely with those of deep-sea sediments, providing additional detail and further supporting the chronology of glacial-interglacial climate change.

These triumphs have not been won easily. The scientists had to take extreme care not to melt and contaminate the ice cores while drilling them, transporting them to the laboratories, and storing them until the ice and the air bubbles trapped in them could be analyzed. The scientists also had to guard against misleading results—caused, for example, by the reaction of carbon dioxide with impurities in the ice. It is a tribute to the patience, ingenuity, and care of these hardy bands of researchers that the worldwide scientific community has recognized the great importance of glacial ice-core research in understanding the history and predicting the future of global climate change.

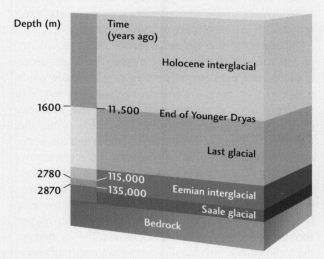

Comparison of ice depths and time before the present for recent glacial and interglacial intervals. [After J. W. C. White, "Don't Touch That Dial." *Nature* 364 (1993): 186.]

in the climate system that cause the formation of polar ice caps during certain periods of geologic history but not during others.

In most cases, the general cooling of polar regions can be well explained by rearrangements of the continents by plate tectonics, coupled with the albedo feedback and other feedbacks in the climate system. For most of Earth's history, there were no extensive land areas in the polar regions, and there were no ice caps. The oceans circulated through the open polar regions, transporting heat from the equatorial regions and helping the atmosphere to distribute temperatures fairly evenly over the Earth. When large land areas drifted to positions that obstructed the efficient transport of heat by the oceans, the differences in temperature between the poles and the equator increased. As the poles cooled, the ice caps formed.

An example is the Permian-Pennsylvanian glaciation, which is preserved in glacial deposits across the continents of the Southern Hemisphere. During this ice age, about 300 million years ago, the southern continents were joined together near the South Pole as part of the ancient continent of Gondwanaland (see Figure 21.26). We will return to the subject of ancient ice ages in Chapter 21.

Short-Term Variations

The temperatures shown in Figure 15.10 do not vary smoothly over time. Superposed on the 100,000-year glacial cycles are climate fluctuations of shorter duration, some nearly as large as the changes from glacial to interglacial periods. Geologists have combined the information from cores in polar ice sheets, mountain glaciers, lake sediments,

EARTH ISSUES

15.2 El Niño: The Wayward Child

El Niño is caused by an upset in the delicate balance between winds and currents of the tropical Pacific. The tropical Pacific absorbs an enormous amount of solar heat—more than any other ocean. Normally, the warmer waters of the western tropical Pacific off Indonesia cause lower pressure, towering thunderstorms, and heavy rainfall in that region. In contrast, air pressure is higher and rainfall is less over the cooler waters in the eastern tropical Pacific. These prevailing pressure patterns drive the trade winds that blow from east to west, pushing the heated tropical Pacific waters westward, where they pile up and maintain a large warm pool. In the eastern Pacific, colder deep water rises to replace the warm surface waters blown to the west, producing an equatorial "cold tongue" off the west coast of South America.

Sporadically, this system balance breaks down. Air pressure rises over the western tropical Pacific and drops in the central and eastern parts, causing the trade winds to weaken or occasionally even reverse direction. This recurring swing in air pressure is called the Southern Oscillation. With the collapse of the trade winds, the transport of warm water westward ceases and the warm pool migrates eastward, bringing with it thunderstorms and rainfall. The equatorial cold tongue fails to develop, and temperatures across the tropical Pacific that normally are warmer in the west and cooler in the east become equalized. These regional changes can trigger anomalous weather patterns worldwide. The jet stream, the atmospheric circulation, and the transport of moisture are changed in large parts of both the Northern Hemisphere and Southern Hemisphere. For example, cooler, wetter winters are more frequent in the southwestern United States, and milder winters tend to prevail in the Pacific Northwest and New England.

The 1997–1998 El Niño was the strongest on record, and its particularly severe climatic effects were felt throughout the world. Some examples of unusual weather attributed to this event were drought in Australia; record drought and forest fires in Indonesia; severe hurricanes and typhoons in the Pacific and milder hurricanes in the Caribbean; severe storms in California that caused landslides and floods; and heavy rains and flash floods in Peru, Ecuador, and Kenya. Crops failed and the fish catch was decimated in many areas. A single blizzard paralyzed Denver, and warmer than usual temperatures stole Christmas in New York City. In regions with increased rainfall, such as parts of South America and Africa, waterborne diseases (cholera, dysentery, typhoid, hepatitis, Rift Valley fever) increased. According to one estimate, the global disruption in weather patterns and ecosystems may have cost 23,000 lives and caused $33 billion in damage.

El Niño has a flip side, called La Niña ("the girl child"), that sometimes (but not always) follows El Niño by a year or so. La Niña is characterized by stronger trade winds, colder sea-surface temperatures in the eastern tropical Pacific, and warmer temperatures in the western tropical Pacific than are the norm. Global weather anomalies are generally the opposite of those that occur during El Niño. Some scientists attribute the severe droughts on the East Coast of the United States in the summer of 1999 to a severe La Niña event in the same year.

and deep-sea sediments to reconstruct a decade-by-decade — and in some cases, a year-by-year—history of short-term climate variations during the last glacial cycle. Here we summarize some of the basic features of this remarkable chronicle.

• During the last ice age, Earth's climate was highly variable, with shorter (1000-year) oscillations between warm and cold superposed on longer (10,000-year) cycles. The most extreme variations appear to have been in the North Atlantic region, where mean local temperatures rose and fell by as much as 15°C. Each 10,000-year cycle comprised a set of progressively cooler 1000-year oscillations and ended with an abrupt warming. The massive discharges of icebergs and fresh water resulting from the warming altered the thermohaline "conveyor belt" circulation in the oceans (see Figure 15.3b) and dumped large amounts of glacial drift into deep-sea sediments.

• The main phase of warming from the last ice age occurred in the interval between 14,500 and 10,000 years ago. It was not a smooth transition but involved two main stages, with a pause in deglaciation and a return to cold conditions during the Younger Dryas event 13,000 to 11,500 years ago (see Feature 15.1). The extremely abrupt increases in temperature at 14,500 and 11,500 years ago are perhaps the most astonishing aspect of this jerky transition. Broad regions of Earth experienced almost synchronous changes from glacial to interglacial temperatures during intervals as short as 30 to 50 years. Evidently, atmospheric circulation can reorganize very rapidly, flipping the entire climate system from one state (glacial cold) to another (interglacial warmth) in less than a human lifetime! This raises the possibility that human-induced global climate change could involve abrupt shifts to a new (and unknown) climate state, rather than just a gradual warming.

• The warm interglacial climate of the Holocene, since about 11,000 years ago, has been much more stable than the previous parts of the glacial cycle. The warmest tempera-

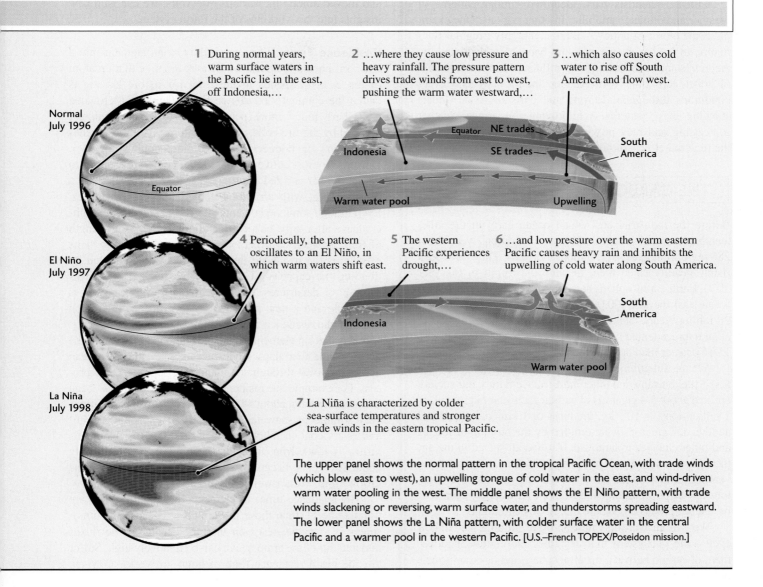

1 During normal years, warm surface waters in the Pacific lie in the east, off Indonesia,...

2 ...where they cause low pressure and heavy rainfall. The pressure pattern drives trade winds from east to west, pushing the warm water westward,...

3 ...which also causes cold water to rise off South America and flow west.

Normal
July 1996

Equator

Equator   NE trades

SE trades

Indonesia

South
America

Warm water pool

Upwelling

4 Periodically, the pattern oscillates to an El Niño, in which warm waters shift east.

5 The western Pacific experiences drought,...

6 ...and low pressure over the warm eastern Pacific causes heavy rain and inhibits the upwelling of cold water along South America.

El Niño
July 1997

Indonesia

South
America

Warm water pool

La Niña
July 1998

7 La Niña is characterized by colder sea-surface temperatures and stronger trade winds in the eastern tropical Pacific.

The upper panel shows the normal pattern in the tropical Pacific Ocean, with trade winds (which blow east to west), an upwelling tongue of cold water in the east, and wind-driven warm water pooling in the west. The middle panel shows the El Niño pattern, with trade winds slackening or reversing, warm surface water, and thunderstorms spreading eastward. The lower panel shows the La Niña pattern, with colder surface water in the central Pacific and a warmer pool in the western Pacific. [U.S.–French TOPEX/Poseidon mission.]

tures occurred during the beginning of this epoch. Geologists have documented regional variations of about 5°C on time scales of 1000 years or so, but the global changes during this period are much smaller, with a total range of only 2°C. It is interesting that the Holocene appears to be the longest stable warm period over the last 400,000 years. Equable Holocene conditions no doubt promoted the rapid rise of agriculture and civilization that followed the end of the last ice age.

Regional Patterns of Variability

Local and regional climates are much more variable than global climate, because averaging over large surface areas, like averaging over time, tends to smooth out small-scale fluctuations. Over periods of years to decades, the predominant regional variations result from interactions between atmospheric circulation and the sea and land surfaces. They generally occur in distinctive geographical patterns, although their timing and amplitudes can be highly irregular.

The most famous example is an anomalous warming of the eastern Pacific Ocean that occurs every 3 to 7 years and lasts for a year or so. Peruvian fishermen call such an event **El Niño** ("the Christ Child" in Spanish) because the warming typically reaches the surface waters off the coast of South America about Christmastime. El Niño events can be disastrous for local ecosystems, which depend on upwellings of cold water for their nutrient supply. Besides disrupting fishing in the eastern Pacific, they can trigger changes in wind and rain patterns over much of the globe, precipitating floods, prolonged droughts, forest fires, landslides, and the spread of waterborne diseases. Scientists have shown that El Niño and the complementary cooling events, known as La Niña, are part of a natural variation in the exchange of heat between the atmosphere and the tropical Pacific Ocean. This variation is known as ENSO, or El Niño–Southern Oscillation (see Feature 15.2).

Climate scientists have identified similar patterns of weather and climate variability in other regions. One example is the North Atlantic Oscillation, a highly irregular fluctuation in the barometric pressure between Iceland and the Azores Islands that has a strong influence on the movement of storms across the North Atlantic and thus affects weather conditions throughout Europe and parts of Asia. Understanding these patterns is improving long-range weather forecasting and may provide important information about the regional effects of human-induced climate change.

THE CARBON CYCLE

Before the industrial era, which began early in the nineteenth century, atmospheric CO_2 concentrations were about 280 ppm for at least several thousand years. They have been rising steadily since then, reaching 370 ppm in the year 2000. Earth's atmosphere has not contained this much CO_2 for at least the last 400,000 years and probably for the last 20 million years. The CO_2 concentration is now increasing at an unprecedented rate of 0.4 percent per year (see Figure 15.16), faster than at any time in recent geologic history.

Yet the situation could be worse. In the 1980s, through fossil-fuel burning and other industrial activities, our society emitted about 5.4 gigatons of carbon each year (a gigaton, or 1 billion tons, is 10^{12} kg, the mass of 1 km^3 of water), and another 1.7 gigatons were emitted by the burning of forests and other changes in land use. If it had all stayed in the air, the CO_2 increase would have been closer to 0.9 percent per year, more than twice the observed rate. Instead, 3.8 gigatons of carbon were removed from the atmosphere each year by natural processes. Where did all this carbon go?

We will answer this question by examining the **carbon cycle**—the geosystem that describes the continual movement of carbon between the atmosphere and its other principal reservoirs: the lithosphere, hydrosphere, and biosphere. We will then be able to consider an even deeper set of issues: What are the interactions between the carbon cycle and the climate system? Is the net feedback between these systems positive or negative—that is, will changes in the carbon cycle enhance or reduce global warming? Finally, how might humans intervene in the carbon cycle to stabilize the climate system against global warming?

Geochemical Cycles and How They Work

In discussing geochemical cycles, we view the components of the Earth system—atmosphere, hydrosphere, cryosphere, lithosphere, and biosphere—as **geochemical reservoirs** for holding terrestrial chemicals, linked by processes that transport chemicals among them. Geochemical cycles trace the flow, or *flux*, of chemicals from one reservoir to another. By quantifying the amounts of the chemicals that are stored in and moved among the various reservoirs, we can gain new insights into the workings of the Earth system.

Residence Time When the inflow of an element into a reservoir equals the outflow, the average time that an atom of the element spends in the reservoir before leaving is called the element's **residence time.** Think of a crowded bar where many more people want to get in than are allowed by the fire code. After the room fills up or reaches its *capacity,* the bouncer begins holding people at the door. During the most active hours, when people are waiting to get in, the bar is filled to capacity, or *saturated,* and is in a **steady state,** with arrivals exactly balancing departures. Even though some people come early and stay late and others leave after only a short time, there is an average length of time between each person's arrival and departure. This average is the residence time, which can be computed by dividing the capacity of the room by the rate of arrivals (*inflow*) or departures (*outflow*). If the room's capacity is 30 people and a new person is let in every 2 minutes on average, the residence time is 60 minutes.

Visualize an element's residence time in the ocean as the average time that elapses between its entry into the ocean and its removal through sedimentation or some other process. For example, the residence time of sodium in the ocean is extremely long, about 48 million years, because sodium is very soluble in seawater (the reservoir capacity is high) and rivers contain relatively small amounts (the element's inflow is low). Iron, in contrast, stays in the ocean only about 100 years, because its solubility is very low and the inflow from rivers is relatively high.

The residence times of elements in the atmosphere are usually shorter than those in the ocean, because the atmosphere is a smaller reservoir than the ocean and the fluxes into and out of the atmosphere can be relatively high. Sulfur dioxide has a residence time of hours to weeks. Oxygen, about 21 percent of the atmosphere, has a residence time of 6000 years. Nitrogen, about 78 percent of the atmosphere, has a residence time of 400 million years. A molecule of nitrogen that went into the atmosphere in the late Paleozoic era, about 300 million years ago, is still likely to be there.

Chemical Reactions Some chemicals are relatively inert, but most participate in chemical reactions. In fact, these reactions generally govern a chemical's residence time within a reservoir. A calcium ion (Ca^{2+}) can be removed from seawater by reacting with two bicarbonate ions ($2HCO_3^-$) to form carbonic acid (H_2CO_3) and calcite ($CaCO_3$). Carbonic acid can dissociate into water (H_2O) and carbon dioxide (CO_2), both of which can escape to the atmosphere, and calcite can precipitate out of the ocean in limestone sediments. How much calcium can be dissolved in seawater thus depends on the availability of bicarbonate ions, which in turn depends on the influx of carbon dioxide into the seawater. As this example illustrates, chemical reactions can couple

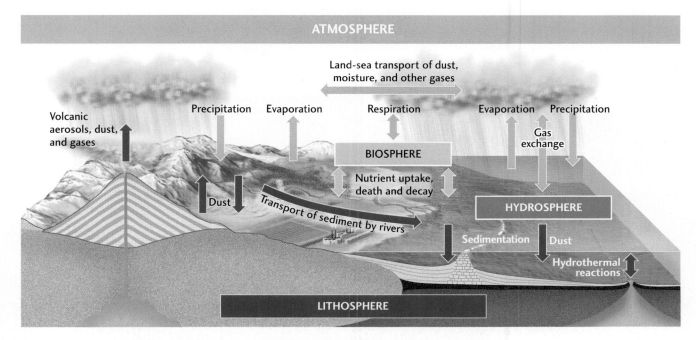

Figure 15.12 Transport processes between components of the climate system.

the geochemical cycle of one element (calcium) with that of another (carbon).

This coupling implies that geochemical cycles can rarely be considered in isolation but must be treated as interacting geosystems. Moreover, the chemical reactions depend on the physical conditions of the reservoirs, such as the local pressure and temperature. This dependence can couple geochemical cycles to the climate system itself.

Transport Across Interfaces Fluxes between reservoirs are governed by processes that transport chemicals into and out of them (**Figure 15.12**). For example, volcanic processes transport gases, aerosols, and dust from the lithosphere into the atmosphere (see Chapter 12). Wind lifts dust into the atmosphere, and gravity pulls it back to the surface. Windborne dust is an important mechanism for transferring minerals from the lithosphere to the hydrosphere, although by far the largest flux comes from dissolved or suspended minerals in the outflow from rivers.

Evaporation and precipitation transport huge amounts of water between the atmosphere and the surfaces of both land and ocean. At the sea surface, gas molecules escape from their dissolved state in the water and enter the atmosphere. Their escape is promoted by the evaporation of sea spray, which releases dissolved gases as well as dissolved salt in the form of tiny crystals. The gas exchange is balanced by the dissolution of atmospheric constituents into sea spray and rain falling on the ocean or by the dissolution of gases directly across the ocean surface.

Sedimentation is the great flux that keeps the ocean in a steady state, primarily by counterbalancing the influx of

river water. As sediments are buried, they become part of the oceanic crust. There they stay until they move into the mantle through subduction or become part of the continental crust through accretion. The uplift of continental regions by plate convergence and other mountain-building processes exposes crustal rocks to weathering and erosion, maintaining the balance of fluxes among all the reservoirs.

The biosphere is a unique reservoir, because each individual organism maintains an active interface with its environment, which often includes three of the other main surface reservoirs—atmosphere, hydrosphere, and lithosphere. The most important transport processes are the inflow and outflow of atmospheric gases by respiration, the inflow of nutrients from the lithosphere and hydrosphere, and the outflow of these nutrients through the death and decay of organisms. When the biosphere plays a key role in the flux of a chemical, geologists refer to the geosystem that describes the flux as a *biogeochemical cycle* (see Chapter 11). The carbon cycle, which depends critically on the organic pumping of carbon into and out of the atmosphere, is clearly a biogeochemical cycle.

Example: The Calcium Cycle The **calcium cycle** provides a fairly simple illustration of the concepts involved in geochemical cycles (**Figure 15.13**). The ocean contains about 560,000 gigatons of calcium, dissolved in a total ocean mass of about 1.4×10^9 gigatons. Calcium steadily enters this reservoir in large quantities through the rivers of the world, which transport dissolved and suspended calcium derived from the weathering of such minerals as calcite, gypsum, calcium feldspars, and other calcium silicates. A

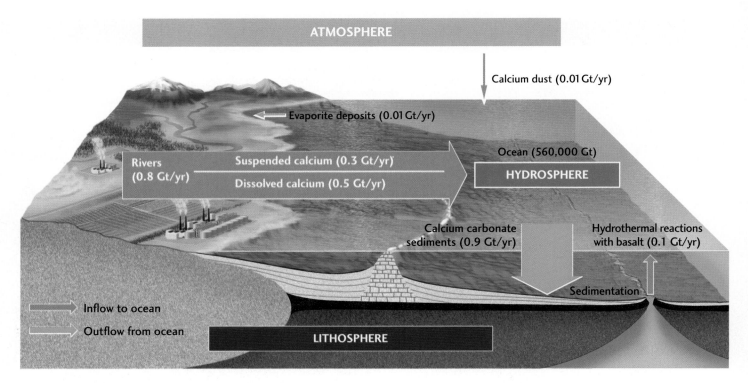

Figure 15.13 The calcium cycle, showing fluxes into and out of the ocean. Fluxes are in units of gigatons (10^{12} kg) per year. The inflow of calcium to the ocean is approximately equal to the outflow, resulting in a steady state.

much smaller amount enters the ocean via transport by windblown dust.

If the ocean continually received calcium liberated by weathering, without there being a way to remove excess calcium, it would quickly become supersaturated with this element. The flux that removes the majority of excess calcium from the ocean is the *sedimentation of calcium carbonate,* which we described in Chapter 5. A smaller amount of calcium is precipitated as gypsum in evaporite deposits.

The system diagram in Figure 15.13 shows this steady-state situation. The ocean's calcium capacity (the amount of calcium the ocean can hold) is much larger than the inflow and outflow of calcium, so calcium has a fairly long residence time in the ocean. By dividing the total annual influx (0.9 gigaton/ year) by the ocean's calcium capacity (560,000 gigatons), we obtain a residence time of about 600,000 years.

The Carbon Budget

The carbon cycle, depicted in **Figure 15.14**, involves four main reservoirs: the atmosphere; the global ocean, including marine organisms; the land surface, including terrestrial plants and soils; and the deeper lithosphere. We can describe the flux of carbon among these reservoirs in term of four basic subcycles. During times when Earth's climate is stable, each subcycle is in a steady state and can therefore be characterized by a constant flux.

Air-Sea Gas Exchange The exchange of CO_2 across the air-sea interface amounts to a carbon flux of about 90 gigatons/year. The process depends on many factors, including air and sea temperatures and the composition of the seawater, but it is particularly sensitive to the wind velocity, which increases gas transfer by stirring up the surface water and generating spray.

Photosynthesis and Respiration in the Terrestrial Biosphere The greatest carbon flux, 120 gigatons/year, comes from the cycling of atmospheric CO_2 by land plants and animals through photosynthesis, respiration, and decay. Plants take in this entire amount during photosynthesis and respire about half of it back into the atmosphere. The other half is incorporated into plant tissues—leaves, wood, and roots. Animals eat the plants, and microorganisms promote their decay; both processes oxidize plant tissues and respire CO_2. A small fraction (about 3 percent) is directly oxidized through forest fires and other combustion.

Dissolved Organic Carbon A small fraction of the CO_2 incorporated into plant tissues (0.4 gigaton/year) is dissolved into surface waters and transported by rivers to the ocean, where it is respired back into the atmosphere.

Carbonate Weathering and Precipitation The weathering of carbonate rocks removes about 0.2 gigaton of carbon per year from the lithosphere and an equal amount from the atmosphere. This inorganic carbon is dissolved into

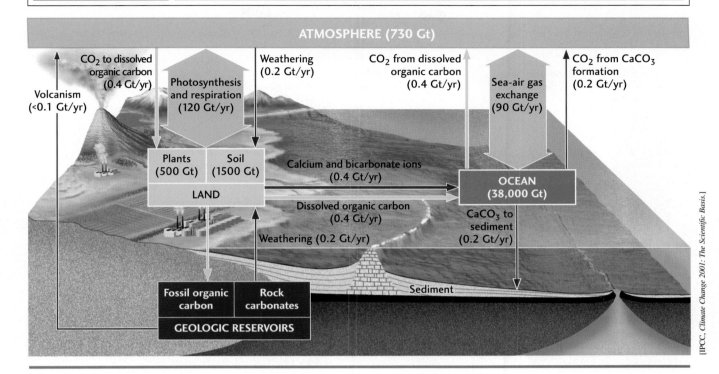

Key Figure 15.14   The carbon cycle regulates the flux of carbon in the climate system.

[IPCC, *Climate Change 2001: The Scientific Basis.*]

surface waters, primarily as bicarbonate ions, and is transported by rivers to the ocean. Here shell-forming marine organisms reverse the weathering reaction, precipitating calcium carbonate and releasing an equal amount of carbon back into the atmosphere as CO_2.

Other geological processes contribute to the carbon budget. Volcanism releases minor amounts of CO_2 into the atmosphere, and the weathering of silicate rocks and the burial of organic carbon by sedimentation consume CO_2. The flux of carbon by these processes is relatively small (less than 0.1 gigaton/year), so they are usually neglected in considerations of short-term climate change. Over the long term, however, their effects can be substantial. For example, regional uplift might cause changes in the carbon cycle by accelerating physical erosion and chemical weathering of the newly exposed silicate rocks in the mountains. In simplified form, the weathering reaction that removes atmospheric carbon dioxide is

silicate rock weathering

$$CaSiO_3 + CO_2 \rightarrow CaCO_3 + SiO_2$$

According to one controversial hypothesis, the uplift of the Himalaya and the Tibetan Plateau, which began about 40 million years ago, increased weathering rates enough to draw down CO_2 in the atmosphere. By weakening the greenhouse effect, this may have contributed to the subsequent climate cooling and growth of ice sheets that led to the Pleistocene glaciations.

Human Perturbations of the Carbon Cycle

With this background, we now return to the fate of human carbon emissions. Careful studies have provided good estimates of how human activities perturbed the carbon cycle during the 1980s. Over this decade, the average mass of carbon that human activities emitted into the atmosphere was 7.1 gigatons/year. **Figure 15.15** shows what happened to this carbon. Only 46 percent of the total (3.3 gigatons/year) stayed in the atmosphere as CO_2. The remainder was absorbed in nearly equal amounts by the oceans and the land surface. On the land surface, the uptake was primarily by plant growth at temperate latitudes in the Northern Hemisphere, caused in part by the "fertilization effect" of increasing CO_2 (plants love the stuff!) and in part by the rapid growth of new forests on land previously used for agriculture.

How will these numbers change as the atmospheric CO_2 concentration continues to rise? Scientists believe that the percentage of human carbon emissions taken up by the ocean and the land surface will decrease as the reservoirs become more saturated. Moreover, they also believe that the enhanced greenhouse warming expected from the rise in atmospheric CO_2 will further decrease the capacity of the oceans to absorb human carbon emissions. The magnitude of these effects is highly uncertain, but it does appear that changes in the carbon cycle caused by human activities may eventually enhance the greenhouse warming.

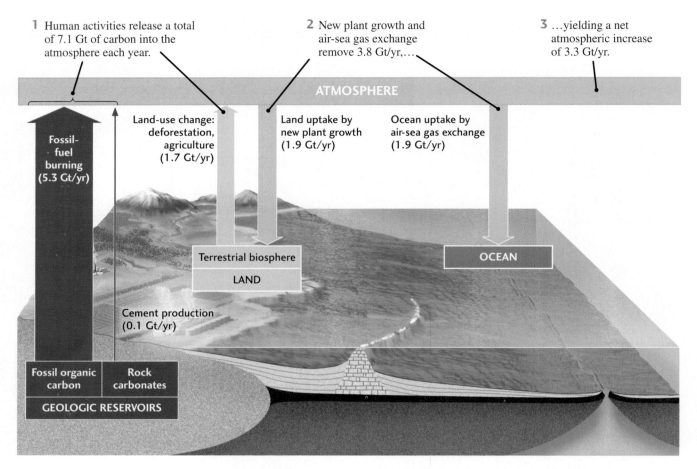

1 Human activities release a total of 7.1 Gt of carbon into the atmosphere each year.

2 New plant growth and air-sea gas exchange remove 3.8 Gt/yr,...

3 ...yielding a net atmospheric increase of 3.3 Gt/yr.

ATMOSPHERE

Fossil-fuel burning (5.3 Gt/yr)

Land-use change: deforestation, agriculture (1.7 Gt/yr)

Land uptake by new plant growth (1.9 Gt/yr)

Ocean uptake by air-sea gas exchange (1.9 Gt/yr)

Terrestrial biosphere
LAND

OCEAN

Cement production (0.1 Gt/yr)

Fossil organic carbon

Rock carbonates

GEOLOGIC RESERVOIRS

Figure 15.15 Humans add CO_2 to the atmosphere by burning fossil fuels; producing cement; and releasing carbon by deforestation, agriculture, and other land-use practices. The climate system responds by absorbing some of this carbon into the oceans and increasing plant production on land. The remainder stays in the atmosphere, increasing the concentration of CO_2. The annual amounts shown in this figure are for the 1980s. Since then, fossil-fuel burning has increased to more than 6 gigatons/year. [IPCC, *Climate Change 2001: The Scientific Basis.*]

TWENTIETH-CENTURY WARMING: HUMAN FINGERPRINTS OF GLOBAL CHANGE

Systematic measurements of weather and climate using thermometers and other scientific instruments began only a few hundred years ago. By the mid-nineteenth century, temperatures around the world were being reported by enough meteorological stations on land and on ships at sea to allow accurate estimation of Earth's average annual surface temperature. Between the end of the nineteenth century and the beginning of the twenty-first, the mean surface temperature rose by about 0.6°C, a climate trend called the **twentieth-century warming** (Figure 15.16).

Fossil-fuel burning and deforestation have substantially increased the amount of atmospheric CO_2 during the same interval, and human activities have also generated significant rises in other greenhouse gases, such as methane. We know that humans are responsible for the CO_2 increase be-cause the carbon isotopes of fossil fuels have a distinctive signature that precisely matches the changing isotopic composition of atmospheric carbon. But how certain can we be that the twentieth-century warming was a direct consequence of the CO_2 increase—that is, a result of the **enhanced greenhouse effect**—and not some fortuitous change associated with natural climate variability? This question lies at the heart of the global warming controversy.

Almost all experts on Earth's climate are now convinced that the twentieth-century warming was in part human-induced and that the warming will continue into the twenty-first century as the levels of atmospheric greenhouse gases continue to rise. They base this judgment on two principal lines of reasoning: the climate-change record and their understanding of how the climate system works.

The twentieth-century warming lies within the range of temperature variations that have been inferred for the Holocene. In fact, the average temperatures in many regions of the world were probably warmer 10,000 to 8000 years ago than they are today. The twentieth-century record is

(a)

1 A recent warming trend correlates with the increase in CO_2 from emissions since the industrial revolution of the 19th century.

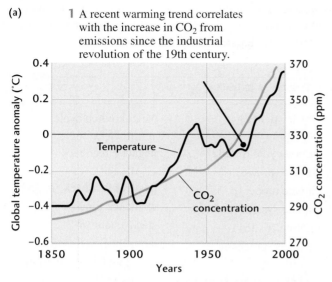

(b)

2 The 20th-century record is clearly anomalous when compared with climate change documented during the last millennium.

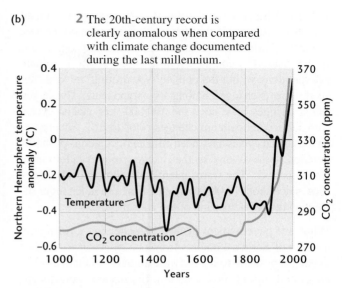

Figure 15.16 Earth heats up. A comparison of Earth's average annual surface temperature with CO_2 concentrations in the atmosphere, showing that the recent warming trend correlates clearly anomalous, however, when compared with the increase in CO_2 caused by human activities since the industrial revolution. [IPCC, *Climate Change 2001: The Scientific Basis.*]

clearly anomalous, however, when compared with the pattern and rate of climate change documented during the last millennium. Based on data from tree rings, corals, ice cores, and other climate indicators, scientists have drawn two major conclusions about climate during the 1000 years that preceded the twentieth century: (1) there was an irregular but steady global cooling of about 0.2°C over this entire interval, and (2) the maximum fluctuation in mean temperatures during any one of the nine previous centuries was probably less than 0.3°C. Against this background, the twentieth-century warming appears to be very abnormal (see Figure 15.16b).

The second argument, and to many scientists a more compelling one, comes from the agreement between the observed pattern of warming and the pattern predicted by the best climate-system models of the enhanced greenhouse effect. Models that include changes in atmospheric greenhouse gases not only satisfy the global temperature rise but also reproduce the pattern of temperature change both geographically and with altitude in the atmosphere—what some scientists have called the "fingerprints" of the enhanced greenhouse effect. For example, models of the enhanced greenhouse effect predict that as global warming occurs, nighttime low temperatures at the surface should increase more rapidly than daytime high temperatures, thus reducing the daily variation of temperature. Climate data for the last century confirm this prediction.

As we emphasized earlier in this chapter, poorly understood aspects of the climate system may introduce substantial errors in climate model predictions. Nevertheless, the consistency of the measured trends with the basic physics of the enhanced greenhouse effect lends powerful support to the hypothesis that we ourselves are the agents responsible

for the recent global warming. We will discuss global warming and the societal issues it poses in Chapter 23.

SUMMARY

What is the climate system? The climate system includes all parts of the Earth system and all the interactions among these components needed to describe how climate behaves in space and time. The main components of the climate system are the atmosphere, hydrosphere, cryosphere, lithosphere, and biosphere. Each component plays a different role in the climate system, depending on its ability to store and transport energy.

What is the greenhouse effect? For the global temperature averaged over daily and seasonal cycles to remain the same, Earth's surface must radiate as much energy back into space at it receives from the Sun. Much of the radiant energy from the Sun passes through the atmosphere and is absorbed by Earth's surface. The warmed surface radiates heat back to the atmosphere as infrared rays. Carbon dioxide and other trace gases absorb infrared rays selectively at certain wavelengths and radiate infrared energy at other wavelengths in all directions, including downward to Earth's surface. The net effect is to increase the temperature of the surface relative to the temperature at higher levels in the atmosphere. In this way, the atmosphere acts like the glass in a greenhouse, allowing solar radiant energy to pass through but trapping heat.

How has Earth's climate changed over time? Natural variations of climate occur on a wide range of scales in both time and space. Some occur by the external forcing of climate, such as fluctuations in the amount of solar radiation received at Earth's surface and changes in the distribution of land surface caused by continental drift. Others result from internal variations within the climate system itself, as mass and energy are exchanged among its components. The largest recent changes have been the 100,000-year glacial cycles, with surface temperature changes of 6° to 8°C. Superposed on the glacial cycles have been climate fluctuations of shorter duration, some as large as the changes from glacial to interglacial periods. The warm interglacial climate of the Holocene, since about 10,000 years ago, has been much more stable than the previous parts of the glacial cycle.

What are geochemical cycles? Geochemical cycles trace the flux of Earth's elements from one reservoir to another, making it possible to quantify the amounts of elements that are stored in and moved among the oceans, atmosphere, land surface, crust, and mantle. The calcium cycle shows how rivers transport calcium derived from rock weathering to the oceans and how sedimentation transports calcium back into the lithosphere. If the cycle is in steady state, inflow balances outflow, and the residence time can be calculated as the total amount of the element in the reservoir divided by the inflow.

What is the carbon cycle? The carbon cycle is the geosystem that describes the continual movement of carbon between the atmosphere and its other principal reservoirs—the lithosphere, hydrosphere, and biosphere. Major processes include gas exchange between the atmosphere and the ocean; photosynthesis, respiration, and the burning of organic materials; and the weathering and precipitation of calcium carbonate. Outflows of carbon dioxide from the atmosphere include the weathering of silicate rocks and the burial of carbon in the crust, which release oxygen to the atmosphere. Inflows of carbon into the atmosphere include volcanism.

How can internal geologic processes cause climatic change? Over the short term of a few years, sulfuric acid aerosols emanating from large volcanic eruptions can absorb solar radiation before it reaches the lower atmosphere and thus decrease global temperatures. Over the long term of millions of years, plate tectonic movements can cause continents to drift over the poles, stabilizing polar ice caps; block or open gateways to ocean currents; and cause uplift, which alters weather systems and rates of chemical erosion that draw down atmospheric CO_2.

Was the twentieth-century warming caused by human activities? Global warming of about 0.6°C during the twentieth century correlates with the significant rise in atmospheric CO_2 and other greenhouse gases caused by fossil-fuel burning, deforestation, and other human activities. Most experts on Earth's climate are now convinced that the twentieth-century warming was in part human-

induced and that the warming will continue into the twenty-first century as the levels of atmospheric greenhouse gases continue to rise.

KEY TERMS AND CONCEPTS

albedo (p. 352)
calcium cycle (p. 363)
carbon cycle (p. 362)
climate (p. 348)
climate model (p. 355)
climate system (p. 348)
El Niño (p. 361)
enhanced greenhouse effect (p. 366)
geochemical reservoir (p. 362)
greenhouse effect (p. 353)
ice age (p. 355)

Milankovitch cycle (p. 357)
negative feedback (p. 354)
positive feedback (p. 353)
residence time (p. 362)
steady state (p. 362)
stratosphere (p. 348)
thermohaline circulation (p. 349)
troposphere (p. 348)
twentieth-century warming (p. 366)

EXERCISES

1. What is the relationship between weather and climate?

2. What is a greenhouse gas, and how does it affect Earth's climate?

3. Why is it incorrect to assert that greenhouse gases prevent heat energy from escaping to outer space?

4. Give an example not discussed in this chapter of a positive feedback and a negative feedback in the climate system.

5. From the information given in Figure 15.14, estimate the residence time of carbon dioxide in the ocean.

6. List two inflows of chemical components into the ocean and two outflows of the same chemical components from the ocean.

7. List three causes of climate change that result from the interaction of Earth's internal and external systems.

THOUGHT QUESTIONS

1. Do Milankovitch cycles fully explain the warming and cooling of global climate during Pleistocene glacial cycles?

2. How would the calcium geochemical cycle be affected by a global increase in chemical weathering?

3. Devise a simple geochemical cycle for the element sodium, which is found in marine evaporites (halite) and clay minerals as well as dissolved in seawater.

4. What would the carbon cycle have looked like after life had originated but before photosynthesis had evolved?

5. Assume that we keep pumping carbon dioxide into the atmosphere at a steadily increasing rate and Earth warms significantly in the next 100 years. How might this affect the global carbon cycle?

6. Why are scientists reasonably sure that most of the twentieth-century warming was due to changes in the climate system caused by human activities?

SHORT-TERM TEAM PROJECT

Your Local Climatology

Investigate the climate at your locality by compiling data on the maximum and minimum annual temperatures and average rainfall during the last 40 years. A good place to start is the Web site of the National Climatic Data Center (http://www.ncdc.noaa.gov/oa/ncdc.html). Can you see evidence of global warming in your local climate record?

CHAPTER SIXTEEN

WEATHERING, EROSION, AND MASS WASTING
Interface Between Climate and Tectonics

Solid as the hardest rocks may seem, all rocks—like rusting old automobiles and yellowed old newspapers—eventually weaken and crumble when exposed to water and the gases of the atmosphere. Unlike those cars and newspapers, however, rocks may take thousands of years to disintegrate.

In this chapter, we will explore some of the interactions between the climate and plate tectonic systems. The systems interact to give us three geologic processes that break down and fragment rocks and transport these products over short distances: weathering, erosion, and mass wasting. These processes are the first steps in flattening mountains that have been uplifted by tectonic processes.

In Chapter 5, we discussed weathering as the general process by which rocks are broken down at Earth's surface. *Weathering* produces all the clays of the world, all soils, and the dissolved substances rivers carry to the ocean. *Chemical weathering* occurs when the minerals in a rock are chemically altered or dissolved. *Physical weathering* takes place when solid rock is fragmented by mechanical processes that do not change its chemical composition. Chemical and physical weathering reinforce each other. Chemical decay weakens fragments of rocks and makes them more susceptible to breakage. The smaller the pieces produced by physical weathering, the greater the surface area available for chemical weathering.

Once weathering reduces rocks to particles, they may accumulate as soil, or they may be removed by *erosion*. Erosion occurs along hillslopes; particles are transported to the starting points of channels,

On June 1, 2005, several homes were destroyed after sliding down a rain-soaked hill near Bluebird Canyon in Laguna Beach, California.
[Steven Georges/Long Beach Press-Telegram/Corbis.]

where they may be transported farther downslope. Sometimes large segments of hillslopes collapse, leading to what is known as *mass wasting*. The products of mass wasting—particles released during weathering as well as large masses of unweathered rock—are also transported to the starting points of river channels. Once in channels, streams and rivers can efficiently transport these weathered materials farther downslope, perhaps across continents and all the way to the world's oceans. We have also seen how rivers convey sediments from their source areas in mountains to their sinks in oceans. This topic will be covered in more detail in Chapter 18.

WEATHERING, EROSION, MASS WASTING, AND THE ROCK CYCLE

After tectonics and volcanism have made mountains, chemical decay and physical breakup join with rainfall, wind, ice, and snow to wear away those mountains. **Erosion** and **mass wasting** are the processes that loosen and transport soil and rock downhill or downwind. Erosion generally refers to processes that remove weathered Earth materials on a grain-by-grain basis, usually by moving currents. Mass wasting involves transport of weathered and unweathered Earth materials in larger amounts and as large single events. These events are usually driven by gravity and the tendency of Earth materials to move downhill. Both processes carry away weathered material on Earth's surface and deposit it elsewhere, exposing fresh, unaltered rock to weathering.

Weathering is one of the major geologic processes in the rock cycle that shape Earth's surface and alter rock materials, converting all kinds of rocks into sediment and forming soil. When rainwater weathers a rock such as pure limestone or rock salt, for example, all the material is completely dissolved and carried away in the water as ions in solution. Chemically weathered material contributes most of the dissolved matter in the oceans.

The early sections of this chapter emphasize chemical weathering, because it is in some ways the fundamental driving force of the whole process. As we saw in Chapter 15, chemical weathering is a major factor in regulating global climate. The effects of physical weathering, always important, depend largely on chemical decay. First, however, we examine the factors that control weathering.

CONTROLS ON WEATHERING

All rocks weather, but the manner and rate of their weathering vary. The four key factors that control the fragmentation and decay of rocks are the properties of the parent rock, the climate, the presence or absence of soil, and the length of time the rocks are exposed to the atmosphere. These four factors are summarized in Table 16.1.

The Properties of the Parent Rock

The nature of a parent rock affects weathering because (1) various minerals weather at different rates and (2) a rock's structure affects its susceptibility to cracking and fragmentation. Old inscriptions on gravestones are evidence of the varying rates at which rocks weather. The carved letters on a recently erected gravestone stand out clearly from the stone's

Table 16.1	Major Factors Controlling Rates of Weathering		
	Weathering Rate		
	Slow ————————————————————————————————————→		Fast
PROPERTIES OF PARENT ROCK			
Mineral solubility in water	Low (e.g., quartz)	Moderate (e.g., pyroxene, feldspar)	High (e.g., calcite)
Rock structure	Massive	Some zones of weakness	Very fractured or thinly bedded
CLIMATE			
Rainfall	Low	Moderate	High
Temperature	Cold	Temperate	Hot
PRESENCE OR ABSENCE OF SOIL AND VEGETATION			
Thickness of soil layer	None—bare rock	Thin to moderate	Thick
Organic content	Low	Moderate	High
LENGTH OF EXPOSURE	Short	Moderate	Long

Figure 16.1 Early-nineteenth-century gravestones at Wellfleet, Massachusetts. The stone on the right is limestone and is so weathered that it is unreadable. The stone on the left is slate, which retains its legibility under the same conditions. [Courtesy of Raymond Siever.]

polished surface. After a hundred years in a moderately rainy climate, however, the surface of a limestone will be dull and the letters inscribed on it will have almost melted away, much as the name on a bar of soap disappears after a few washes (**Figure 16.1**). Granite, on the other hand, will show only minor changes. The differences in the weathering of granite and limestone result from their different mineral compositions. Given enough time, however, even a resistant rock will ultimately decay. After several hundred years, the granite monument will also have weathered appreciably, and its surface and letters will be somewhat dulled and blurred.

Climate: Rainfall and Temperature

The rate of both chemical and physical weathering varies not only with the properties of the rock but also with the climate—the amount of rainfall and the temperature. High temperatures and heavy rainfall promote faster chemical weathering. Cold and dry climates slow the process. In cold climates, water can't dissolve minerals because it is frozen. In arid regions, water is relatively unavailable.

On the other hand, climates that minimize chemical weathering may enhance physical weathering. For example, freezing water may act as a wedge, widening cracks and pushing a rock apart.

The Presence or Absence of Soil

Soil, one of our most valuable natural resources, is composed of fragments of bedrock, clay minerals formed by the chemical alteration of bedrock minerals, and organic matter produced by organisms that live in the soil. Although soil is itself a product of weathering, its presence or absence affects the chemical and physical weathering of other materials. Soil production is a *positive feedback process*—that is, the product of the process advances the process itself. Once soil starts to form, it works as a geological agent to weather rock more rapidly. The soil retains rainwater, and it hosts a variety of vegetation, bacteria, and other organisms. These organisms create an acidic environment that, in combination with moisture, promotes chemical weathering, which alters or dissolves minerals. Plant roots and organisms tunneling through the soil promote physical weathering by helping to create fractures in a rock. Chemical and physical weathering, in turn, lead to the production of more soil.

The Length of Exposure

The longer a rock weathers, the greater its chemical alteration, dissolution, and physical breakup. Rocks that have been exposed at Earth's surface for many thousands of years form a *rind*—an external layer of weathered material ranging from several millimeters to several centimeters thick—that surrounds the fresh, unaltered rock. In dry climates, some rinds have grown as slowly as 0.006 mm per 1000 years.

We now consider the two types of weathering—chemical and physical—in more detail.

CHEMICAL WEATHERING

Chemical weathering occurs when minerals react with air and water. In these chemical reactions, some minerals dissolve. Others combine with water and atmospheric components such as oxygen and carbon dioxide to form new

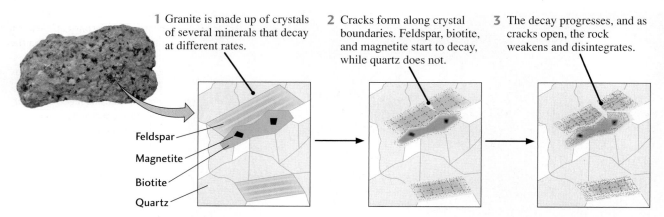

1 Granite is made up of crystals of several minerals that decay at different rates.

2 Cracks form along crystal boundaries. Feldspar, biotite, and magnetite start to decay, while quartz does not.

3 The decay progresses, and as cracks open, the rock weakens and disintegrates.

Feldspar
Magnetite
Biotite
Quartz

Figure 16.2 Diagrammatic microscopic views of stages in the disintegration of granite. [John Grotzinger/Ramón Rivera-Moret/Harvard Mineralogical Museum.]

minerals. We begin our investigation by examining the chemical weathering of feldspar, the most abundant mineral in Earth's crust.

The Role of Water in Weathering: Feldspar and Other Silicates

Feldspar is a key mineral in a great many igneous, sedimentary, and metamorphic rocks. Many other kinds of rock-forming silicate minerals also weather much as feldspar does. Feldspar is one of many silicates that are altered by chemical reactions to form the water-containing minerals known as clay minerals. Feldspar's behavior during weathering helps us understand the weathering process in general, for two reasons:

1. There is an overwhelming abundance of silicate minerals in the Earth.

2. The chemical processes of dissolution and alteration that characterize feldspar weathering also characterize weathering in other kinds of minerals.

In a sample of unweathered granite, the rock is hard and solid because the interlocking network of quartz, feldspar, and other crystals holds it tightly together. When the feldspar is altered to a loosely adhering clay, however, the network is weakened and the mineral grains are separated (**Figure 16.2**). In this instance, chemical weathering, by producing the clay, also promotes physical weathering because the rock now fragments easily along widening cracks at mineral boundaries.

The white to cream-colored clay produced by the weathering of feldspar is **kaolinite,** named for Gaoling, a hill in southwestern China where it was first obtained. Chinese artisans had used pure kaolinite as the raw material of pottery and china for centuries before Europeans borrowed the idea in the eighteenth century.

Only in the severely arid climates of some deserts and polar regions does feldspar remain relatively unweathered.

This observation points to water as an essential component of the chemical reaction by which feldspar becomes kaolinite. Kaolinite is a hydrous (that is, containing water in the crystal structure) aluminum silicate. In the reaction that produces kaolinite, the solid feldspar undergoes *hydrolysis* (*hydro* means "water" and *lysis* means "to loosen"). The feldspar is broken down and also loses several chemical components. The only part of a solid that reacts with a fluid is the solid's surface, so as we increase the surface area of the solid, we speed up the reaction. For example, as we grind coffee beans into finer and finer particles, we increase the ratio of their surface area to their volume. The finer the coffee beans are ground, the faster their reaction with water, and the stronger the brew becomes. Similarly, the smaller the fragments of minerals and rocks, the greater the surface area. The ratio of surface area to volume increases greatly as the average particle size decreases, as shown in **Figure 16.3**.

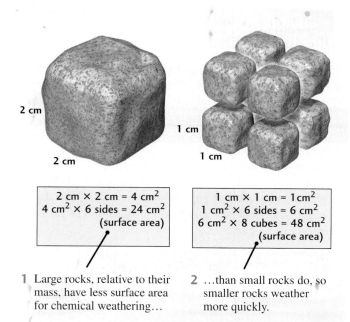

2 cm
2 cm
1 cm
1 cm

2 cm × 2 cm = 4 cm^2
4 cm^2 × 6 sides = 24 cm^2
(surface area)

1 cm × 1 cm = 1cm^2
1 cm^2 × 6 sides = 6 cm^2
6 cm^2 × 8 cubes = 48 cm^2
(surface area)

1 Large rocks, relative to their mass, have less surface area for chemical weathering…

2 …than small rocks do, so smaller rocks weather more quickly.

Figure 16.3 As a rock mass breaks into smaller pieces, more surface area becomes available for the chemical reactions of weathering.

Carbon Dioxide, Weathering, and the Climate System

Variability in the atmosphere's concentration of carbon dioxide leads to corresponding variability in the rate of weathering (**Figure 16.4**). Higher levels of carbon dioxide in the atmosphere lead to higher levels in the soil, which increases the rate of weathering. As discussed in Chapter 15, carbon dioxide, a greenhouse gas, makes Earth's climate warmer and thus promotes weathering. Weathering, in turn, converts carbon dioxide into bicarbonate ions and so decreases the amount of carbon dioxide in the atmosphere (see Figure 16.4). This decrease in carbon dioxide eventually results in a cooler climate. In this way, the weathering at the surface of a feldspar grain is linked to the causes of global climate change. As more and more carbon dioxide is used up through weathering and the climate cools, weathering decreases again. As weathering decreases, the amount of carbon dioxide in the atmosphere builds up again, and the climate warms, thus completing the cycle.

The Role of Carbon Dioxide in Weathering The reaction of feldspar with pure water in a laboratory is an extremely slow process that would take thousands of years to weather even a small amount of feldspar completely. If we wanted to, we could speed weathering by adding a strong acid (such as hydrochloric acid) to our solution and dissolve the feldspar in a few days. An acid is a substance that releases hydrogen ions (H^+) to a solution. A strong acid produces abundant hydrogen ions; a weak one, relatively few. The strong tendency of hydrogen ions to combine chemically with other substances makes acids excellent solvents.

On Earth's surface, the most common acid—and the one responsible for increasing weathering rates—is carbonic acid (H_2CO_3). This weak acid forms when carbon dioxide (CO_2) gas from the atmosphere dissolves in rainwater:

$$\text{carbon dioxide} + \text{water} \rightarrow \text{carbonic acid}$$
$$CO_2 \qquad\quad H_2O \qquad\qquad H_2CO_3$$

The amount of carbon dioxide dissolved in rainwater is small because the amount of carbon dioxide gas in the atmosphere is small. About 0.03 percent of the molecules in Earth's atmosphere are carbon dioxide. The amount of carbonic acid formed in rainwater is very small, only about 0.0006 g/l. As the burning of oil, gas, and coal increases the amount of carbon dioxide in the atmosphere, the amount of carbonic acid in rainwater increases slightly.

Most of the acidity of acid rain, however, comes not from carbon dioxide but from sulfur dioxide and nitrogen gases, which react with water to form strong sulfuric and nitric acids, respectively. These acids promote weathering to a greater degree than carbonic acid does. Volcanoes and coastal marshes emit gases of carbon, sulfur, and nitrogen into the atmosphere, but by far the largest source is industrial pollution. (See Chapter 23 for more information about acid rain.)

Although rainwater contains only a relatively small amount of dissolved carbon dioxide (carbonic acid), that amount is enough to weather feldspars and dissolve great quantities of rock over long periods of time. We can now write the chemical reaction for the weathering of feldspar with water:

$$\text{feldspar} + \text{carbonic acid} + \text{water} \rightarrow$$
$$2KAlSi_3O_8 \qquad 2H_2CO_3 \qquad\quad H_2O$$

dissolved +	dissolved +	dissolved +	dissolved bicarbonate
kaolinite	silica	potassium ions	ions
$Al_2Si_2O_5(OH)_4$	$4SiO_2$	$2K^+$	$2HCO_3^-$

This simple weathering reaction illustrates the three main effects of chemical weathering on silicates. It leaches, or dissolves away, cations and silica. It *hydrates*, or adds water to, the minerals. And it makes the solutions less acidic. Specifically, the carbonic acid in rainwater helps to weather feldspar in the following way (see Figure 16.4):

• A small proportion of carbonic acid molecules ionize, forming hydrogen ions (H^+) and bicarbonate ions (HCO_3^-) and thus making the water droplets slightly acidic.

• The slightly acidic water dissolves potassium ions and silica from feldspar, leaving a residue of kaolinite, a solid clay. The hydrogen ions from the acid combine with the oxygen atoms of the feldspar to form the water in the kaolinite structure. The kaolinite becomes part of the soil or is carried away as sediment.

• The dissolved silica, potassium ions, and bicarbonate ions are carried away by rain and river waters and are ultimately transported to the ocean.

In Nature: Feldspar in Outcrops and in Moist Soil

Now that we understand how acidic water weathers feldspar, we can better understand why feldspars on bare rock surfaces are much better preserved than those buried in damp soils. The chemical reaction for feldspar weathering gives us two separate but related clues: the amount of water and the amount of acid available for the chemical reaction. The feldspar on a bare rock weathers only while the rock is moist with rainwater. During all the dry periods, the only moisture that touches the bare rock is dew. The feldspar in moist soil is constantly in contact with the small amounts of water retained in spaces between grains in the soil. Thus feldspar weathers continuously in moist soil.

There is more acid in the water in the soil than there is in falling rain. Rainwater carries its original carbonic acid into the soil. As water filters through the soil, it picks up additional carbonic acid and other acids produced by the roots of plants, by the many insects and other animals that live in the soil, and by the bacteria that degrade plant and animal remains. Recently, it was discovered that some bacteria release organic acids, even in waters hundreds of meters deep in the ground. These organic acids then weather

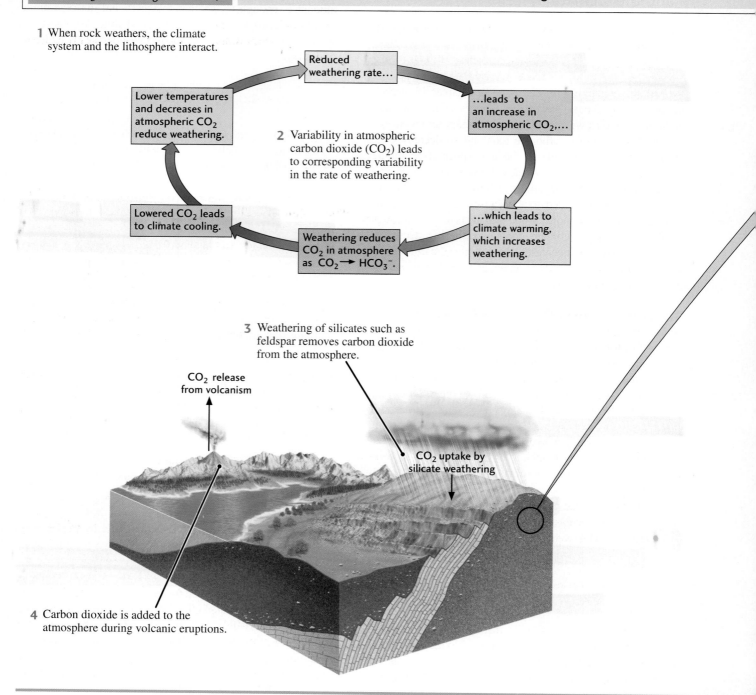

Earth System Figure 16.4 Atmospheric carbon dioxide influences weathering and climate.

1 When rock weathers, the climate system and the lithosphere interact.

Reduced weathering rate...

Lower temperatures and decreases in atmospheric CO₂ reduce weathering.

...leads to an increase in atmospheric CO₂,...

2 Variability in atmospheric carbon dioxide (CO₂) leads to corresponding variability in the rate of weathering.

Lowered CO₂ leads to climate cooling.

Weathering reduces CO₂ in atmosphere as CO₂ → HCO₃⁻.

...which leads to climate warming, which increases weathering.

3 Weathering of silicates such as feldspar removes carbon dioxide from the atmosphere.

CO₂ release from volcanism

CO₂ uptake by silicate weathering

4 Carbon dioxide is added to the atmosphere during volcanic eruptions.

feldspar and other minerals in rocks below the surface. Bacterial respiration in soils may increase the soils' carbon dioxide to as much as 100 times the atmospheric value!

Rock weathers more rapidly in the tropics than in temperate and cold climates mainly because plants and bacteria grow quickly in warm, humid climates, contributing the carbonic acid and other acids that promote weathering. Additionally, most chemical reactions, weathering included, speed up with an increase in temperature.

Chemical Stability: A Speed Control for Weathering

Why do weathering rates vary so widely among different minerals? Minerals weather at different rates because there are differences in their chemical stability in the presence of water at given surface temperatures.

Chemical stability is a measure of a substance's tendency to retain its chemical identity rather than reacting

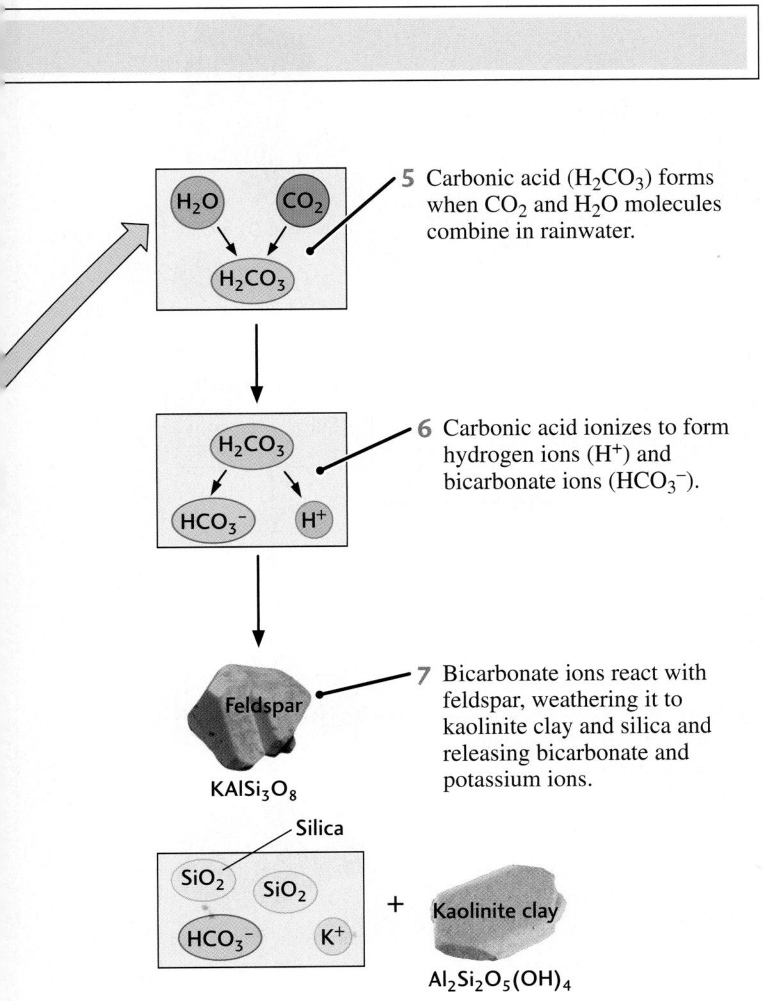

5 Carbonic acid (H_2CO_3) forms when CO_2 and H_2O molecules combine in rainwater.

6 Carbonic acid ionizes to form hydrogen ions (H^+) and bicarbonate ions (HCO_3^-).

7 Bicarbonate ions react with feldspar, weathering it to kaolinite clay and silica and releasing bicarbonate and potassium ions.

Feldspar
$KAlSi_3O_8$

Silica
SiO_2 SiO_2 HCO_3^- K^+

+ Kaolinite clay
$Al_2Si_2O_5(OH)_4$

[John Grotzinger/Ramón Rivera-Moret/Harvard Mineralogical Museum.]

Solubility The solubility of a specific mineral is measured by the amount of the mineral dissolved in water when the solution is saturated. Saturation is the point at which the water cannot hold any more of the dissolved substance. The higher a mineral's solubility, the lower its stability under weathering. Rock salt, for example, is unstable under weathering. It is highly soluble in water (about 350 g/l) and is leached from a soil by even small amounts of water. Quartz, in contrast, is stable under most weathering conditions. Its solubility in water is very low (only about 0.008 g/l), and it is not easily leached from a soil.

Rate of Dissolution A mineral's rate of dissolution is measured by the amount of the mineral that dissolves in an unsaturated solution in a given length of time. The faster the mineral dissolves, the less stable it is. Feldspar dissolves at a much faster rate than quartz, and, primarily for that reason, it is less stable than quartz under weathering.

Relative Stability of Common Rock-Forming Minerals Relative chemical stabilities of various minerals can be used to determine the intensity of weathering in a given area. In a tropical rain forest, only the most stable minerals will be left on an outcrop or in the soil, and so we know that the weathering there is intense. In an arid region such as the desert of North Africa, where weathering is minimal, alabaster (gypsum) monuments remain intact, as do many other unstable minerals. Table 16.2 shows the relative stabilities of all the common rock-forming minerals. Salt and carbonate minerals are the least stable, iron oxides the most stable.

Table 16.2 Relative Stabilities of Common Minerals Under Weathering

Stability of Minerals	Rate of Weathering
MOST STABLE	Slowest
Iron oxides (hematite)	
Aluminum hydroxides (gibbsite)	
Quartz	
Clay minerals	
Muscovite mica	
Potassium feldspar (orthoclase)	
Biotite mica	
Sodium-rich feldspar (albite)	
Amphiboles	
Pyroxene	
Calcium-rich feldspar (anorthite)	
Olivine	
Calcite	
Halite	
LEAST STABLE	Fastest

spontaneously to become a different chemical substance. Chemical substances are stable or unstable in relation to a specific environment or set of conditions. Feldspar, for example, is stable under the conditions found deep in Earth's crust (high temperatures and small amounts of water) but unstable under the conditions at Earth's surface (lower temperatures and abundant water). Two characteristics of a mineral—its solubility and its rate of dissolution—help determine its chemical stability.

The Role of Oxygen in Weathering: From Iron Silicates to Iron Oxides

Iron is one of the eight most abundant elements in Earth's crust, but iron metal, the chemical element in its pure form, is rarely found in nature. It is present only in certain kinds of meteorites that fall to Earth from other places in the solar system. Most of the iron ores used for the production of iron and steel are formed by weathering. These ores are composed of iron oxide minerals originally produced during the weathering of iron-rich silicate minerals, such as pyroxene and olivine. The iron released by dissolution of these minerals combines with oxygen from the atmosphere to form iron oxide minerals.

The iron in minerals may be present in one of three forms: metallic iron, ferrous iron, or ferric iron. In the metallic iron found only in meteorites, the iron atoms are uncharged: they have neither gained nor lost electrons by reaction with another element. In the *ferrous iron* (Fe^{2+}) found in silicate minerals such as pyroxene, the iron atoms have lost two of the electrons they have in the metallic form and thus have become ions. The iron in the most abundant iron oxide at Earth's surface, **hematite** (Fe_2O_3), is *ferric iron* (Fe^{3+}). The iron atoms in ferric iron have lost three electrons. Ferrous iron ions oxidize by losing an additional electron, going from 2^+ (ferrous) to 3^+ (ferric). In fact, all the iron oxides formed at Earth's surface are ferric. The electrons lost by the iron are gained by oxygen atoms in a process called *oxidation*. Thus, oxygen atoms from the atmosphere oxidize ferrous iron to ferric iron. Oxidation, like hydrolysis, is one of the important chemical weathering processes.

When an iron-rich mineral such as pyroxene is exposed to water, its silicate structure dissolves, releasing silica and ferrous iron to solution, where the ferrous iron is oxidized to the ferric form (**Figure 16.5**). The strength of the chemical bonds between ferric iron and oxygen make ferric iron insoluble in most natural surface waters. It therefore precipitates from the solution, forming a solid ferric iron oxide. We are familiar with ferric iron oxide in another form—rusting iron, which is produced when iron metal is exposed to the atmosphere.

We can show this overall weathering reaction of iron-rich minerals by the following equation:

$$\text{iron pyroxene} + \text{oxygen} \rightarrow \text{hematite} + \text{dissolved silica}$$
$$4FeSiO_3 \qquad O_2 \qquad 2Fe_2O_3 \qquad 4SiO_2$$

Although the equation does not show it explicitly, water is required for this reaction to proceed.

Iron minerals, which are widespread, weather to the characteristic red and brown colors of oxidized iron (**Figure 16.6**). Iron oxides are found as coatings and encrustations that color soils and weathered surfaces of iron-containing

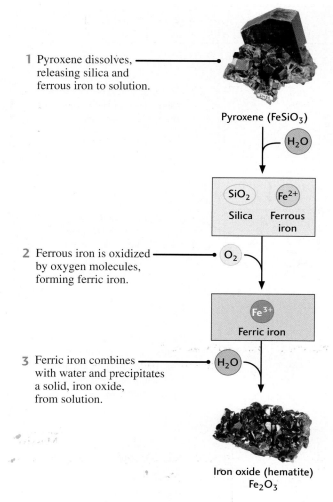

1 Pyroxene dissolves, releasing silica and ferrous iron to solution.

Pyroxene ($FeSiO_3$)

H_2O

SiO_2 Silica Fe^{2+} Ferrous iron

2 Ferrous iron is oxidized by oxygen molecules, forming ferric iron.

O_2

Fe^{3+} Ferric iron

3 Ferric iron combines with water and precipitates a solid, iron oxide, from solution.

H_2O

Iron oxide (hematite) Fe_2O_3

Figure 16.5 The general course of chemical reactions by which an iron-rich mineral, such as pyroxene, weathers in the presence of oxygen and water. [John Grotzinger/Ramón Rivera-Moret/Harvard Mineralogical Museum.]

rocks. The red soils of Georgia and other warm, humid regions are colored by iron oxides. Iron minerals weather so slowly in frigid regions that iron meteorites frozen in the ice of Antarctica are almost entirely unweathered.

PHYSICAL WEATHERING

Now that we have surveyed chemical weathering, we can turn to its partner, physical weathering. We can see the workings of physical weathering most clearly by examining its role in arid regions, where chemical weathering is minimal.

What Determines How Rocks Break?

Rocks can break for a variety of reasons, including stress along natural zones of weakness and biological and chemical activity.

Figure 16.6 Red and brown iron oxides color weathering rocks in Monument Valley, Arizona. [Betty Crowell.]

Natural Zones of Weakness Rocks have natural zones of weakness along which they tend to crack. In sedimentary rocks such as sandstone and shale, these zones are the bedding planes formed by the successive layers of solidified sediment. Metamorphic rocks such as slate form parallel planes of fractures that enable them to be split easily to form roofing tiles. Granites and other rocks are massive. Massive rocks tend to crack along regular fractures spaced one to several meters apart, called **joints** (Figure 16.7). These and less regular fractures form while rocks are still deeply buried in Earth's crust. Through uplift and erosion, the rocks rise slowly to Earth's surface. There, freed from the weight of overlying rock, the fractures open slightly. Once the fractures open a little, both chemical and physical weathering work to widen the cracks.

Activity of Organisms The activity of organisms affects both chemical and physical weathering. Bacteria and algae invade cracks, producing microfractures. These organisms, both those in cracks and those that may encrust the rock, produce acid, which then promotes chemical weathering. In some regions, acid-producing fungi are active in soils, contributing to chemical weathering. Animals burrowing or moving through cracks can break rock. Many of us have seen a crack in a rock that has been widened by a tree root. Physical weathering takes place when the force of the growing root system helps to pry cracks apart (Figure 16.8).

Frost Wedging One of the most efficient mechanisms for widening cracks is **frost wedging**—breakage resulting from the expansion of freezing water. As water freezes, it

Figure 16.7 Weathered, enlarged joint patterns developed in two directions in rocks at Point Lobos State Reserve, California. [Jeff Foott/DRK.]

Figure 16.8 Organisms such as these tree roots invade fractured rock, widening cracks and promoting further chemical and physical weathering. [Peter Kresan.]

Figure 16.9 Gneiss boulder, 3 m high, fractured by frost wedging. Taylor Valley, Victoria Land, Antarctica. [Michael Hambrey.]

Figure 16.10 Exfoliation on Half Dome, Yosemite National Park, California. [Tony Waltham.]

expands, exerting an outward force strong enough to wedge open a crack and split a rock (**Figure 16.9**). This is the same process that can crack open the engine block of a car that is not protected by antifreeze. Frost wedging is most important where water episodically freezes and thaws, such as in temperate climates and in mountainous regions.

Exfoliation There is a form of rock breakage not directly related to earlier fractures or joints caused by previous weathering. **Exfoliation** is a physical weathering process in which flat or curved sheets of rock fracture and are detached from an outcrop. These sheets may look like the layers peeled from a large onion (**Figure 16.10**). Even though

WEATHERING FACTORS

1. Duration of weathering		Less weathering, erosion, and soil formation over short periods of time	More weathering, erosion, and soil formation over long periods of time
2. Bedrock type		More stable minerals, (e.g., quartz), result in lower weathering	Less stable minerals, (e.g., feldspar), result in higher weathering
3. Climate	Lower temperatures	Less chemical weathering (dissolution, alteration to aid physical weathering, production of clay materials)	More physical weathering (thermal expansion and contraction, frost wedging, breakage of bedrock, fragmentation to smaller sizes)
	Higher temperatures	Less physical weathering	More chemical weathering
	Rainfall amount	Little rainfall (less dissolution of minerals, physical weathering, fragmentation, erosion)	Heavy rainfall (more dissolution of minerals, production of clay materials, production of small size particles, erosion)
	Rainfall acidity	Low acidity (less dissolution of minerals, less physical weathering)	High acidity (more dissolution of minerals, more production of clay materials)
4. Topography	Steep slopes	Less chemical weathering	More physical weathering, more erosion
	Gentle slopes	Less physical weathering, less erosion	More chemical weathering

Figure 16.11 Summary chart of weathering. Factors considered are the proportion of stable and unstable minerals in a rock, climate (including temperature and rainfall), and topography (including steep and gentle slopes).

exfoliation is common, no generally accepted explanation of its origin has yet emerged. Some geologists have suggested that exfoliation results from an uneven distribution of expansion and contraction caused by chemical weathering and temperature changes.

Physical Weathering and Erosion

Weathering and erosion are closely related, interacting processes. Physical weathering and erosion are tied to how wind, water, and ice work to transport weathered material. Physical weathering fractures a large rock into smaller pieces, which are more easily transported and therefore more easily eroded than the larger mass. The first steps in the erosion process are the downhill movements of masses of weathered rock, such as landslides, and the transportation of individual particles by flows of rainwater down slopes. The steepness of the slopes affects both physical and chemical weathering, which in turn affect erosion. Weathering and erosion are more intense on steep slopes, and their action makes slopes gentler. Wind may blow away the finer particles, and glacial ice can carry away large blocks torn from bedrock.

Chemical weathering rates are low at high altitudes, where temperatures are generally low, soil is thin or absent, and vegetation is sparse. Physical weathering is greater at high altitudes and in glacial terrains, where the ice tears apart the rock. We can see that the sizes of the materials formed by physical weathering are closely related to various erosional processes. As weathered material is transported, it may again change in size and shape, and its composition may change as a result of chemical weathering. When transportation stops, deposition of the sediment formed by weathering begins.

Figure 16.11 is a chart summarizing the processes of physical and chemical weathering.

SOIL: THE RESIDUE OF WEATHERING

On moderate and gentle slopes, plains, and lowlands, a layer of loose, heterogeneous weathered material remains overlying the bedrock. It may include particles of weathered and unweathered parent rock, clay minerals, iron and other metal oxides, and other products of weathering. Geologists use the term *soil* to describe layers of materials, initially created by fragmentation of rocks during weathering, that experience additions of new materials, losses of original materials, and modification through physical mixing and chemical reactions. In addition, soils often have the ability to support rooted plants. Organic matter, called **humus,** is often important in Earth's soil; it is the remains and waste products of the many plants, animals, and bacteria living in the soil. Leaf litter contributes significantly to the soil of forests. However, soils are not required to support life, and they occur in places such as Antarctica and Mars where life is limited or possibly absent altogether.

Soils vary in color, from the brilliant reds and browns of iron-rich soils to the black of soils rich in organic matter. Soils also vary in texture. Some are full of pebbles and sand; others are composed entirely of clay. Soils are easily eroded, and so they do not form on very steep slopes or where high altitude or frigid climate prevents plant growth. Soil scientists, agronomists, geologists, and engineers study the composition and origin of soils, their suitability for agriculture and construction, and their value as a guide to climate conditions in the past.

Soils form at the interface between the climatic and plate tectonic systems, and they are important to life on Earth. Soils are essential for the growth of plants that provide food, paper, lumber, and drugs, and they supply habitats for plants and animals. They help control the global climate by providing a way to both store and release carbon dioxide, which is an important greenhouse gas. Finally, soils filter water and recycle wastes and are also necessary for building and construction projects.

Soils as Geosystems

As we have seen, the approach to studying Earth as a set of interacting systems is of great value in understanding the behavior of Earth processes. Soils are no exception and can be described as a geosystem with inputs, processes, and outputs (**Figure 16.12**).

Input Material: Weathered Rocks, Organisms, and Dust Soils are formed of weathered rock, with additional inputs provided by organic matter and dust from the atmosphere. As discussed earlier, physical weathering breaks down rock into smaller pieces, and chemical weathering changes primary minerals, such as feldspar, into clays. Plants may colonize the soil, and when they die the plant tissue decomposes to enrich the soil in organic matter, or humus. The atmosphere also contributes matter to the soil, but this material is predominantly inorganic, such as dust made of carbonate or silicate minerals.

Processes: Transformations and Translocations As the soil ages and matures, the materials added or removed from it cause it to change through a set of *transformations*. For instance, the addition of decomposed plant matter provides a source of nutrients to encourage further plant growth. This process forms a positive feedback in the soil geosystem. Many transformations involve chemical weathering of primary minerals to form clay minerals.

Translocations are lateral and vertical movements of materials within the developing soil. Water is the main agent of translocation, usually transporting dissolved salts. Water commonly percolates down through the soil after rainfalls. However, it may also rise from below the surface when

Earth System Figure 16.12   The two basic soil-forming processes are translocation and transformation.

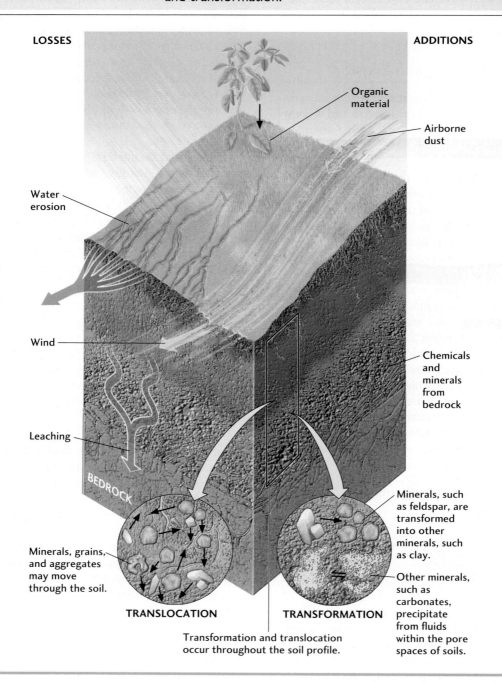

LOSSES

ADDITIONS

Organic material

Airborne dust

Water erosion

Wind

Chemicals and minerals from bedrock

Leaching

BEDROCK

Minerals, grains, and aggregates may move through the soil.

Minerals, such as feldspar, are transformed into other minerals, such as clay.

Other minerals, such as carbonates, precipitate from fluids within the pore spaces of soils.

TRANSLOCATION

TRANSFORMATION

Transformation and translocation occur throughout the soil profile.

temperatures increase and draw water to the surface as it evaporates. Organisms also play an important role in translocation, often moving the soil as they burrow through it.

Soils are dynamic and respond to changes in climate, interactions with organisms, and perturbations by humans. Five factors are important in their formation:

1. *Bedrock composition.* The solubility of minerals, the size of grains and crystals, and patterns of bedrock deformation, such as joints and cleavage.

2. *Climate.* The temperature and how it varies through the seasons and how much precipitation occurs.

3. *Topography.* Slope steepness and the direction the slope faces. Gentler slopes that face toward the Sun promote better soil development.

4. *Organisms.* The diversity and number of organisms.

5. *Time.* The amount of time that a soil has to form; the older a soil, the more likely it is to be well developed.

Output Material: Soil Profiles The final composition and appearance of a soil is known as its **soil profile** (see Figure 16.12). Soil profiles are revealed in vertical sections of exposed soils and consists of six *horizons*. These horizons are a set of distinct layers, usually parallel to the land surface.

The soil's topmost layer, called the *O-horizon,* is usually quite thin and consists of loose leaves and organic debris. Beneath this topmost layer is the *A-horizon,* typically not much more than a meter or two thick and usually the darkest layer because it contains the highest concentration of organic matter as a result of transformation processes. Next down is the *E-horizon,* which consists mostly of clay and insoluble minerals such as quartz. Soluble minerals have been leached from this layer by translocation processes. Beneath the E-horizon is the *B-horizon,* where organic matter is sparse. In this layer, soluble minerals and iron oxides have accumulated in small pods, lenses, and coatings. The climate influences the types of minerals that accumulate, with carbonate minerals and gypsum forming in more arid climates, for example. The next lowest layer, the *C-horizon,* is slightly altered bedrock, broken and decayed, mixed with clay from chemical weathering. Unaltered bedrock forms the lowest level, or *R-horizon.*

The five soil-forming factors interact to create 12 different soil profiles that are recognized by scientists who study soils (Table 16.3).

Table 16.3	Twelve Important Soil Types[a]	
Soil Name	**Description**	**Most Important Factors**
Alfisols	Soils of humid and subhumid climates with a subsurface horizon of clay accumulation, not strongly leached, common in forested areas	Climate, organisms
Andisols	Soils that formed in volcanic ash and contain compounds rich in organics and aluminum	Parent material
Aridisols	Soils formed in dry climates, low in organic matter and often having subsurface horizons with salt accumulations	Climate
Entisols	Soils lacking subsurface horizons because the parent material accumulated recently or because of constant erosion; common on floodplains, mountains, and badlands (highly eroded, rocky areas)	Time, topography
Gelisols	Weakly weathered soils formed in areas that contain permafrost (frozen soil in arctic and subarctic regions) within the soil profile	Climate
Histosols	Soils with a thick upper layer very rich in organic matter (>25%) and containing relatively little mineral material	Topography
Inceptisols	Soils with weakly developed subsurface horizons, little or no subsoil clay accumulation because the soil is young or the climate does not promote rapid weathering	Time, climate
Mollisols	Mineral soils of semiarid and subhumid midlatitude grasslands that have a dark, organic-rich A-horizon and are not strongly leached	Climate, organisms
Oxisols	Very old, extremely leached and weathered soils with subsurface accumulations of Fe and Al oxides, commonly found in humid tropical environments	Climate, time
Spodosols	Soils formed in cold, moist climates that have a well-developed B-horizon with accumulation of Al and Fe oxides, formed under pine vegetation in sandy parent material	Parent material, organisms, climate
Ultisols	Soils with a subsurface horizon of clay accumulation, strongly leached (but not as strongly as oxisols), commonly found in humid tropical and subtropical climates	Climate, time, organisms
Vertisols	Soils that develop deep, wide cracks when dry (shrink and swell) due to high clay content (>35%) and are not strongly leached	Parent material

[a]All five soil-forming factors combine to create these soils, but only the most important factors are listed.
Data adapted from E. C. Brevik, *Journal of Geoscience Education* 50 (2002): 541.

Paleosols: Working Backward from Soil to Climate

Recently, there has been much interest in ancient soils that have been preserved as rock in the geologic record. Some of these soils are more than a billion years old. These *paleosols,* as they are called, are being studied as guides to ancient climates and even to the amounts of carbon dioxide and oxygen in the atmosphere in former times. The mineralogy of paleosols billions of years old is evidence that there was no oxidation of soils at that early stage of Earth's history and therefore that oxygen had not yet evolved to become a major fraction of the atmosphere.

Soil formation is just one step in the evolution of a landscape. Weathering and rock fragmentation often destabilize the environment and lead to the more dramatic changes caused by mass wasting. This is an important part of the general erosion of the land, especially in hilly and mountainous regions.

MASS WASTING

On the morning of June 1, 2005, as residents of Laguna Beach, California, were waking and enjoying their morning coffee, the Earth broke loose beneath their feet and collapsed. Seven multimillion dollar homes were destroyed when a large mass of soil and weathered bedrock gave way and slid downhill. Twelve other homes were badly damaged, and hundreds more were evacuated as residents waited anxiously for geologists to evaluate their home sites and determine if it was safe to return home. Some houses completely collapsed; others literally broke in half; and others were left stranded at the top of the hill, where they jutted out over a large gash in the Earth formed where the sliding mass of earth broke away (see chapter opening photo). This event was triggered by the very high seasonal rainfall—the second highest ever recorded for this part of California—which saturated the soil and weathered bedrock and created the necessary conditions for an already unstable geological environment to tip the scales in favor of disaster. Earlier in the year, the same high rainfall had triggered similar events, which were also lethal, when homes slid downhill near Ventura, California.

These events—or **mass movements**—represent one of many kinds of downhill movements of masses of soil, rock, mud, or other unconsolidated (loose and uncemented) materials under the force of gravity. The masses are not pulled down primarily by the action of an erosional agent, such as wind, running water, or glacial ice. Instead, mass movements occur when the force of gravity exceeds the strength (resistance to deformation) of the slope materials. Earthquakes, floods, and other geological events trigger such movements. The materials then move down the slope, either at a slow or very slow rate or as a sudden, sometimes cata-

strophic, large movement. In various combinations of falling, sliding, and flowing, mass movements can displace small, almost imperceptible amounts of soil down a gentle hillside, or they can be huge landslides that dump tons of earth and rock on valley floors below steep mountain slopes.

Every year, mass movements take their toll of lives and property throughout the world. In late October and early November of 1998, one of the most catastrophic hurricanes of the century, Hurricane Mitch, dropped torrential rains on Central America, saturating the ground and causing raging floods and landslides. At least 9000 people were killed, and billions of dollars in damage was done as the floods and slides laid waste to once-fertile land and crops of corn, beans, coffee, and peanuts. One of the hardest-hit places was near the Nicaragua-Honduras border, where a series of landslides and mudflows buried at least 1500 people. Dozens of villages were simply obliterated, engulfed by a sea of mud. The flanks of a crater on Casita volcano collapsed and started a series of slides and flows that were described as a moving wall of mud more than 7 m high. Those in the direct path of the avalanche could not escape, and many were buried alive as they tried to outrun the fast-moving mud.

Because mass movements are responsible for so much destruction, we want to be able to predict them, and we certainly want to refrain from provoking them by unwise interference with natural processes. We cannot prevent most natural mass movements, but we can control construction and land development to minimize losses.

Mass wasting includes all the processes by which masses of rock and soil move downhill under the influence of gravity, eventually to be carried away by other transport agents. Mass movements change the landscape by scarring mountainsides as great masses of material fall or slide away from the slopes. The material that moves ends up as tongues or wedges of debris on the valley floor, sometimes piling up and damming a stream running through the valley. The scars and debris deposits, mapped in the field or from aerial photographs, are clues to past mass movements. By reading these clues, geologists may be able to predict and issue timely warnings about new movements likely to occur in the future.

In mass movements, as in many other kinds of geological processes, human interference can have severe effects. Although human engineering works seem small compared with natural processes, they are significant. In the United States alone, just one activity—excavation for houses and other buildings—breaks up and transports more than 700 million metric tons of surface materials each year, according to some calculations. This amount far exceeds the 550 million metric tons moved annually in the United States by natural processes.

Mass movements caused by natural processes are influenced by three primary factors (Table 16.4):

1. *The nature of the slope materials.* They may be solid masses of bedrock, the soil formed by weathering, or sedi-

Table 16.4	Factors That Influence Mass Movements		
Nature of Slope Material	**Water Content**	**Steepness of Slope**	**Stability of Slope**
UNCONSOLIDATED			
Loose sand or sandy silt	Dry	Angle of repose	High
	Wet		Moderate
Unconsolidated mixture of sand,	Dry	Moderate	High
silt, soil, and rock fragments	Wet		Low
	Dry	Steep	High
	Wet		Low
CONSOLIDATED			
Rock, jointed and deformed	Dry or wet	Moderate to steep	Moderate
Rock, massive	Dry or wet	Moderate	High
	Dry or wet	Steep	Moderate

ment. Slopes may be made up of **unconsolidated materials,** which are loose and uncemented, or **consolidated materials,** which are compacted and bound together by mineral cements.

2. *The amount of water in the materials.* This characteristic depends on how porous the materials are and on how much rain or other water they have been exposed to.

3. *The steepness and stability of slopes.* This factor contributes to the tendency of materials to fall, slide, or flow under various conditions.

All three factors operate in nature, but slope stability and water content are most strongly influenced by human activity, such as excavation for building and highway construction. All three produce the same result: they lower resistance to movement, and then the force of gravity takes over and the slope materials begin to fall, slide, or flow.

The Nature of Slope Materials

Slope materials vary greatly because they are so dependent on the physical properties of the local terrain. Thus, the metamorphic bedrock of one hillside may be badly fractured by foliation, whereas another slope only a few hundred meters away may be composed of massive granite. Slopes of unconsolidated material are the least stable of all.

Unconsolidated Sand and Silt The behavior of loose, dry sand and silt illustrates how the steepness and stability of slopes influence mass movements. Children's sandboxes have made nearly everyone familiar with the characteristic slope of a pile of dry sand. The angle between the slope of any pile of sand or silt and the horizontal is the same, whether the pile is a few centimeters or several meters high.

For most sands and silts, the angle is about 35°. If you scoop some sand from the base of the pile very slowly and carefully, you can increase the slope angle a little, and it will hold temporarily. If you then jump on the ground near the sandpile, however, the sand will cascade down the side of the pile, which will again assume its original slope of 35°.

The slope angle of the sandpile is its **angle of repose,** the maximum angle at which a slope of loose material will lie without cascading down. A slope that is steeper than the angle of repose is unstable and will tend to collapse to the stable angle. Sand or silt grains form piles with slopes at and below the angle of repose because of frictional forces between the individual sand grains. However, as more and more sand grains are placed on the pile and the slope steepens, the ability of the frictional forces to prevent sliding decreases, and the pile suddenly collapses.

The angle of repose varies significantly with a number of factors, one of which is the size and shape of the particles (**Figure 16.13a**). Larger, flatter, and more angular pieces of loose material remain stable on steeper slopes. The angle of repose also varies with the amount of moisture between particles. The angle of repose of wet sand can be higher than that of dry sand because the small amount of moisture between the grains tends to bind them together so that they resist movement. The source of this binding tendency is *surface tension*—the attractive force between molecules at a surface (see Figure 16.13b). Surface tension makes waterdrops spherical and allows a razor blade or paper clip to float on a smooth water surface. Too much water, on the other hand, separates the particles and allows them to move freely over one another. Saturated sand, in which all the pore space is occupied by water, runs like a fluid and collapses to a flat pancake shape (see Figure 16.13c). The surface

Key Figure 16.13 Mass movement depends on the nature of the materials, their water content, and the slope steepness.

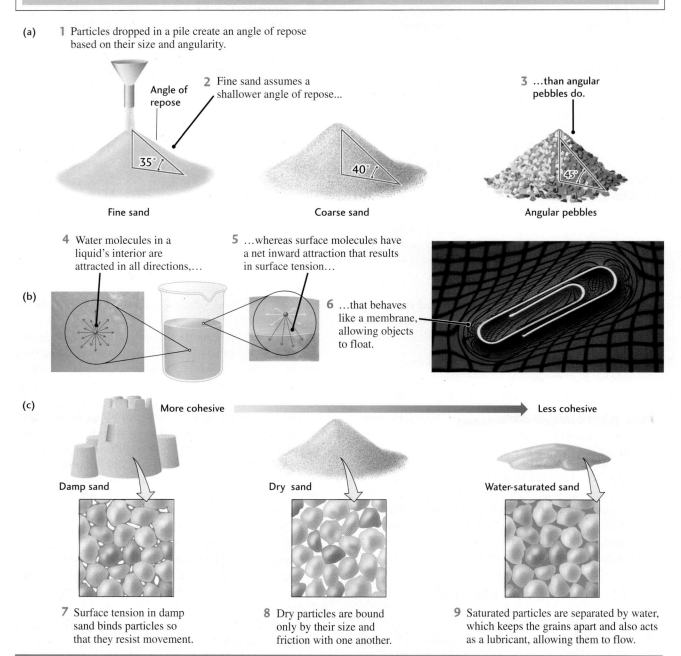

(a)

1 Particles dropped in a pile create an angle of repose based on their size and angularity.

Angle of repose

2 Fine sand assumes a shallower angle of repose...

3 ...than angular pebbles do.

35°

40°

45°

Fine sand   Coarse sand   Angular pebbles

4 Water molecules in a liquid's interior are attracted in all directions,...

5 ...whereas surface molecules have a net inward attraction that results in surface tension...

(b)

6 ...that behaves like a membrane, allowing objects to float.

(c)

More cohesive →→→ Less cohesive

Damp sand   Dry sand   Water-saturated sand

7 Surface tension in damp sand binds particles so that they resist movement.

8 Dry particles are bound only by their size and friction with one another.

9 Saturated particles are separated by water, which keeps the grains apart and also acts as a lubricant, allowing them to flow.

tension that binds moistened sand allows beach sculptors to create elaborate sand castles (**Figure 16.14**). When the tide comes in and saturates the sand, the structures collapse.

Unconsolidated Mixtures Slope materials composed of mixtures of unconsolidated sand, silt, clay, soil, and fragments of rock will form slopes with moderate to steep angles (see Table 16.4). The platy shape of clay minerals, the organic content of soils, and the rigidity of rock fragments are all key factors that change the ability of the material to form slopes with a specific angle.

Consolidated Materials Slopes of consolidated dry materials—such as rock, compacted and cemented sediments, and vegetated soils—may be steeper and less regular than slopes consisting of loose material. They can become unstable when they are oversteepened or denuded of vegetation. The particles of consolidated sediments such as dense clays are bound together by cohesive forces associated with tightly packed particles. *Cohesion* is an attractive force between particles of a solid material that are close together. The greater the cohesive forces in a material, the greater the resistance to movement.

Figure 16.14 Sand castles hold their shape because they are made of damp sand. The steepness of the slope is maintained by moisture between grains. [Kelly Mooney Photography/Corbis.]

Water Content

Mass movements of consolidated materials can usually be traced to the effects of moisture, often in combination with other factors such as the loss of vegetation or oversteepening of the slope. When the ground becomes saturated with water, the solid material is lubricated, the internal friction is lowered, and the particles or larger aggregates can move past one another more easily, so that the material may start to flow like a fluid. This process is called **liquefaction.**

Steepness and Stability of Slopes

Rock slopes range from relatively gentle inclines of easily weathered shales and volcanic ash beds to vertical cliffs of hard rocks such as granite. The stability of rock slopes depends on the weathering and fragmentation of the rock. Shales, for example, tend to weather and fragment into small pieces that form a thin layer of loose rubble covering the bedrock (**Figure 16.15**). The resulting slope angle of the bedrock is similar to the angle of repose of loose, coarse sand. The weathered rubble gradually builds up to an unstable slope, and eventually some of the loose material will slide down.

In contrast, limestones and hard, cemented sandstones in arid environments resist erosion and break into large blocks, forming steep, bare bedrock slopes above and gentler slopes covered with broken rock below. The bedrock cliffs are fairly stable, except for the occasional mass of rock that falls and rolls down to the rock-covered slope below. Where such limestones or sandstones are interbedded with shale, slopes may be stepped. (The rocks in the Grand Canyon are an example; see Feature 8.1.) As shale slides from under the sandstone beds, the harder beds are undercut, become less stable, and eventually fall as large blocks.

Figure 16.15 Weathered shale slope fragmented into a thin layer of rubble. Grand Canyon, Arizona. [Martin Miller.]

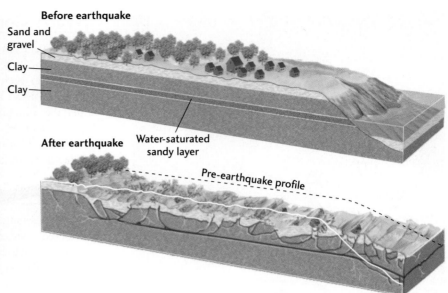

Before earthquake

Sand and gravel

Clay

Clay

Water-saturated sandy layer

After earthquake

Pre-earthquake profile

Figure 16.16 Landslide at Turnagain Heights, Alaska, triggered by the earthquake of 1964. [Steve McCutcheon/ Alaska Pictorial Service.]

Cross sections of the bluffs at Anchorage, Alaska, before and after the earthquake.

Triggers of Mass Movements

When the right combination of materials, moisture, and steepened slope angle makes a hillside unstable, a slide or flow is inevitable. All that is needed is a trigger. Sometimes a slide or debris flow, like the one at Laguna Beach illustrated at the beginning of this chapter, is provoked by a heavy rainstorm. Many slides are set off by vibrations, such as those that occur in earthquakes. Others may be precipitated just by gradual steepening that eventually results in sudden collapse of the slope.

Geological reports can help to minimize the human cost of mass movements, but only if city planners and individual home buyers heed the reports and avoid building or buying in unstable areas. The devastating mass movements in southern California during 2005 are clearly related to the unusually high seasonal rainfall during the winter of 2004–2005. This rainfall was related to the El Niño conditions that we learned about in Chapter 15 and that Earth scientists now understand to recur on a regular and therefore predictable basis.

Similarly, most of the damage in the great Alaskan earthquake of March 27, 1964, was caused by the slides it triggered. Mass movements of rock, earth, and snow wreaked havoc in residential areas of Anchorage, and there were major submarine slides along lakeshores and the seacoast. Huge landslides took place along the flat plains below the 30- to 35-m-high bluffs along the coast. The bluffs were composed of interbedded clays and silts. During the earthquake, the ground shook so hard that unstable, water-saturated sandy layers in the clay were transformed into fluid slurries. This is the process of liquefaction, which we discussed in the section on water content. Enormous blocks of clay and silt were shaken down from the bluffs and slid along the flat ground with the liquefied sediments, leaving a completely disrupted terrain of jumbled blocks and broken buildings (**Figure 16.16**). Houses and roads were carried along by the slides and destroyed. The whole process took only 5 minutes, beginning about 2 minutes after the first shock of the earthquake. At one locality, three people were killed and 75 homes were destroyed.

Studies of the stability of both the California and Alaska slopes and the likelihood of repeated high rainfall or earthquakes had indicated that both areas were prime candidates for landslides. A geological report issued more than a decade earlier had warned of the hazards of development in that part of Alaska, but the great scenic beauty of the area overwhelmed people's judgment. The same is true for southern California. In Alaska, people paid the price with their lives. Fortunately, at Laguna Beach, the cost was only the value of people's houses, but even this was steep in an area where the average price of a house was well over a million dollars.

CLASSIFICATION OF MASS MOVEMENTS

Although the popular press often refers to any mass movement as a landslide, there are many different kinds of mass movements, each with its own characteristics. We use the term *landslide* only in its popular sense, to refer to mass movements in general.

Geologists classify mass movements in accordance with three characteristics, as summarized in **Figure 16.17**:

1. The nature of the material (for example, whether it is rock or unconsolidated debris)

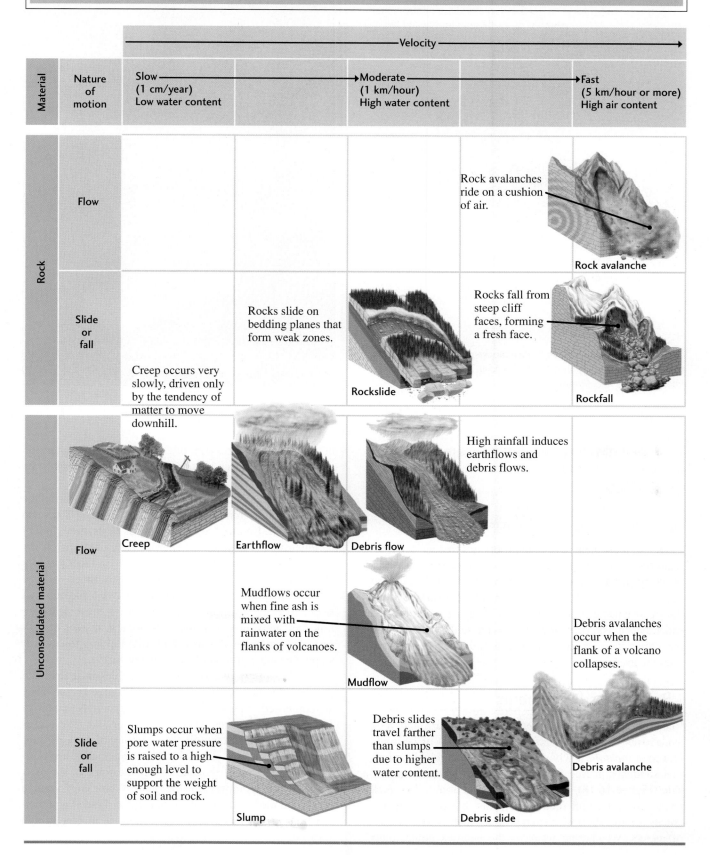

Key Figure 16.17 Mass movements are classified according to the dominant material, water or air content, and velocity of the movement.

Velocity

Material	Nature of motion	Slow (1 cm/year) Low water content		Moderate (1 km/hour) High water content		Fast (5 km/hour or more) High air content
Rock	Flow					Rock avalanches ride on a cushion of air. Rock avalanche
	Slide or fall		Rocks slide on bedding planes that form weak zones. Rockslide		Rocks fall from steep cliff faces, forming a fresh face. Rockfall	
Unconsolidated material	Flow	Creep occurs very slowly, driven only by the tendency of matter to move downhill. Creep	Earthflow	Debris flow	High rainfall induces earthflows and debris flows.	
			Mudflows occur when fine ash is mixed with rainwater on the flanks of volcanoes. Mudflow		Debris avalanches occur when the flank of a volcano collapses. Debris avalanche	
	Slide or fall	Slumps occur when pore water pressure is raised to a high enough level to support the weight of soil and rock. Slump	Debris slides travel farther than slumps due to higher water content. Debris slide			

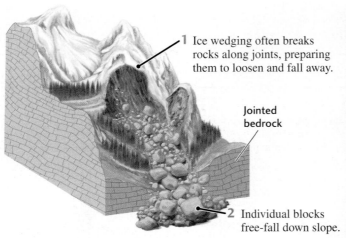

1 Ice wedging often breaks rocks along joints, preparing them to loosen and fall away.

Jointed bedrock

2 Individual blocks free-fall down slope.

Figure 16.18 In a rockfall, individual blocks plummet in a free fall from a cliff or steep mountainside. Rockfall, Zion National Park, Utah. [Sylvester Allred/Visuals Unlimited.]

2. The velocity of the movement (from a few centimeters per year to many kilometers per hour)

3. The nature of the movement: whether it is sliding (the bulk of the material moves more or less as a unit) or flowing (the material moves as if it were a fluid)

Some movements have characteristics that are intermediate between sliding and flowing. Most of the mass may move by sliding, for example, but parts of it along the base may move as a fluid. A movement is called a flow if that is the main type of motion. It is not always easy to tell the exact mechanism of a movement, because the nature of the movement must be reconstructed from the debris deposited after the event is over.

Rock Mass Movements

Rock movements include rockfalls, rockslides, and rock avalanches of small blocks or larger masses of bedrock. During a *rockfall,* newly detached individual blocks plummet suddenly in free fall from a cliff or steep mountainside (**Figure 16.18**). The velocities of the free-falling rock are the fastest of all rock movements, but the travel distances are the shortest, generally only meters to hundreds of meters. Weathering weakens the bedrock along joints

until the slightest pressure, often exerted by the expansion of water as it freezes in a crack, is enough to trigger the rockfall.

The evidence for the origin of rockfalls is clear from the blocks seen in a rocky accumulation at the foot of a steep bedrock cliff. Referred to collectively as **talus,** these blocks of fallen rock can be matched with rock outcrops on the cliff. Talus accumulates slowly, building up blocky slopes along the base of a cliff over long periods of time.

In many places, rocks do not fall freely but slide down a slope, forming *rockslides.* Although these movements are fast, they are slower than rockfalls because masses of bedrock slide more or less as a unit, often along downward-sloping bedding or joint planes (**Figure 16.19**).

Rock avalanches differ from rockslides in their much greater velocities and travel distances (**Figure 16.20**). They are composed of large masses of rocky materials that broke up into smaller pieces when they fell or slid. The pieces then flow farther downhill at velocities of tens to hundreds of kilometers per hour. Rock avalanches are typically triggered by earthquakes. They are some of the most destructive mass movements because of their large volume (many are more than a half-million cubic meters) and their fast transport of materials for thousands of meters at high velocities.

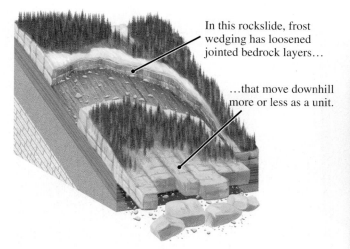

Figure 16.19 In a rockslide, large masses of bedrock move more or less as a unit in a fast, downward slide. Rockslide, Elephant Rock, Yosemite National Park, California. [Jeff Foott/DRK.]

Most rock mass movements occur in high mountainous regions; they are rare in low hilly areas. Rock masses tend to move where weathering and fragmentation have attacked rocks already predisposed to breakage by structural deformation, relatively weak bedding, or metamorphic cleavage planes. In many such regions, extensive talus accumulations have been built by infrequent but large-scale rockfalls and rockslides.

Unconsolidated Mass Movements

Unconsolidated material, often called *debris,* includes various mixtures of soil; broken-up bedrock; trees and shrubs; and materials of human construction, from fences to cars and houses. Most unconsolidated mass movements are slower than most rock movements, largely because of the lower slope angles at which these materials move. Although

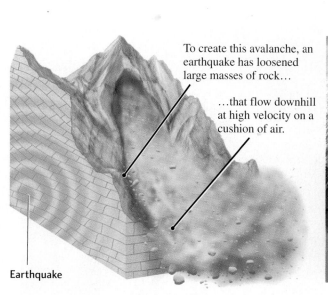

Earthquake

Figure 16.20 In a rock avalanche, large masses of broken rocky material flow, rather than slide, downward at high velocity. Two rock avalanches can be seen that were triggered by the November 3, 2002, earthquake along the Denali fault, Alaska.

The rock avalanches traveled down south-facing mountains, moved across the 1.5-mile-wide Black Rapids Glacier, and flowed partway up the opposite slope. [Photo by Dennis Trabant/USGS; mosaic by Rod March/USGS.]

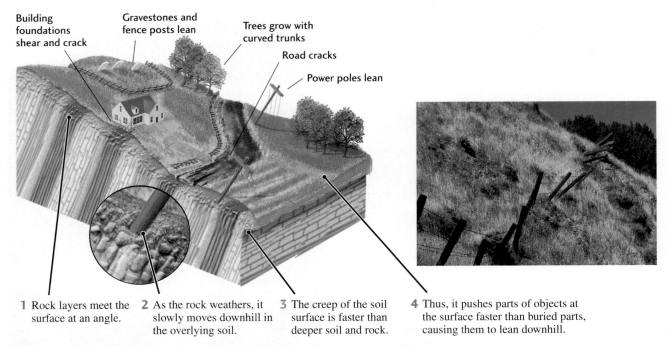

1 Rock layers meet the surface at an angle.

2 As the rock weathers, it slowly moves downhill in the overlying soil.

3 The creep of the soil surface is faster than deeper soil and rock.

4 Thus, it pushes parts of objects at the surface faster than buried parts, causing them to lean downhill.

Figure 16.21 Creep is the downhill movement of soil or other debris at a rate of about 1 to 10 mm/year. A fence offset by creep in Marin County, California. [Travis Amos.]

some unconsolidated materials move as coherent units, many flow like very viscous fluids, such as honey or syrup. (Viscosity, you will recall, is a measure of a fluid's resistance to flow.)

The slowest unconsolidated mass movement is **creep**—the downhill movement of soil or other debris at a rate rang-

ing from about 1 to 10 mm/year, depending on the kind of soil, the climate, the steepness of the slope, and the density of the vegetation (**Figure 16.21**). The movement is a very slow deformation of the soil, with the upper layers of soil moving down the slope faster than the lower layers. Such slow movements may cause trees, telephone poles, and fences

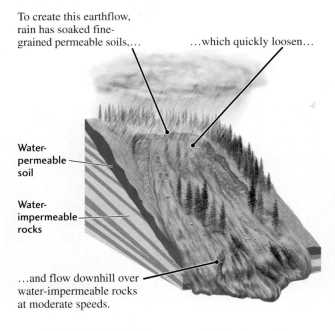

To create this earthflow, rain has soaked fine-grained permeable soils,…

…which quickly loosen…

Water-permeable soil

Water-impermeable rocks

…and flow downhill over water-impermeable rocks at moderate speeds.

Figure 16.22 An earthflow is a movement of relatively fine grained materials that travel as fast as a few kilometers per hour. Earthflow, Hogan Creek, Denali National Park, Alaska. [Steve McCutcheon/Visuals Unlimited.]

(a)

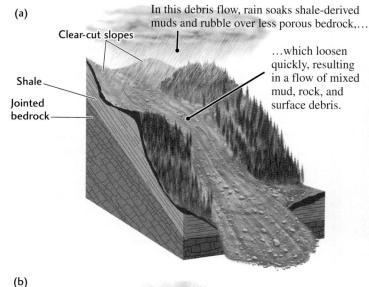

Clear-cut slopes

Shale

Jointed bedrock

In this debris flow, rain soaks shale-derived muds and rubble over less porous bedrock,...

...which loosen quickly, resulting in a flow of mixed mud, rock, and surface debris.

(b)

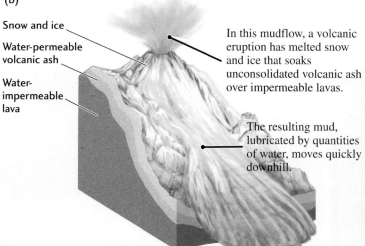

Snow and ice

Water-permeable volcanic ash

Water-impermeable lava

In this mudflow, a volcanic eruption has melted snow and ice that soaks unconsolidated volcanic ash over impermeable lavas.

The resulting mud, lubricated by quantities of water, moves quickly downhill.

Figure 16.23 (a) A debris flow contains material that is coarser than sand and travels at rates from a few kilometers per hour to many tens of kilometers per hour. Debris flow, Rocky Mountain National Park, Colorado. [E. R. Degginger.] (b) A mudflow contains large quantities of water, and many such flows move at a rate of several kilometers per hour. Mudflows tend to move faster than earthflows or debris flows. An earthquake in Tadzhikistan in January 1989 produced 15-m-high mudflows on slopes weakened by rain. [Vlastimir Shone/Getty Images.]

to lean or move slightly downslope. The heavy weight of the masses of soil creeping downhill can break poorly supported retaining walls and crack the walls and foundations of buildings. In icy regions where the ground is permanently frozen, *solifluction* is a type of movement that occurs when water in the surface layers of the soil alternately freezes and thaws, causing the soil to ooze downhill, carrying broken rocks and other debris with it.

Earthflows (**Figure 16.22**) and debris flows (**Figure 16.23a**) are fluid mass movements that usually travel faster than creep, as much as a few kilometers per hour, primarily because they have less resistance to flow. An *earthflow* is a fluid movement of relatively fine grained materials, such as soils, weathered shales, and clay. A *debris flow* is a fluid mass movements of rock fragments supported by a muddy matrix. Debris flows contain much material coarser than sand and tend to move more rapidly than earthflows. In some cases, debris flows may reach velocities of 100 km/hour.

Mudflows are flowing masses of material mostly finer than sand, along with some rock debris, containing large amounts of water (see Figure 16.23b). The mud offers little resistance to flow because of the high water content and thus tends to move faster than earth or debris. Many mudflows move at several kilometers per hour. Most common in hilly and semiarid regions, mudflows start after infrequent, sometimes prolonged, rains. The previously dry, cracked mud keeps absorbing water as the rain continues. In the process, its physical properties change; the internal friction decreases, and the mass becomes much less resistant to movement. The slopes, which are stable when dry, become unstable, and any disturbance triggers movement of waterlogged masses of mud. Mudflows travel down upper valley slopes and merge on the valley floor. Where mudflows exit from confined valleys into broader, lower valley slopes and flats, they may splay out to cover large areas with wet debris. Mudflows can carry huge boulders, trees, and even houses.

In this debris avalanche, unconsolidated ash and rock move downhill at high speeds, lubricated by high air or water content.

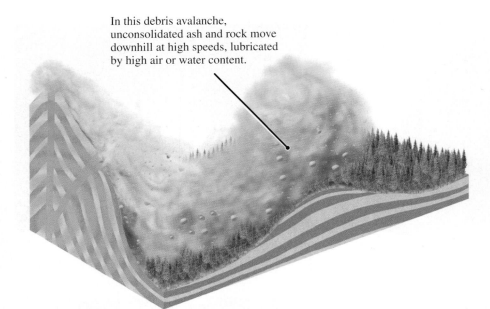

Figure 16.24 A debris avalanche is the fastest unconsolidated flow, owing to its high water content and movement down steep slopes. In 1970, an earthquake-induced debris avalanche on Mount Huascarán, Peru, buried the towns of Yungay and Ranrahirca. The avalanche traveled 17 km at a speed of up to 280 km/hour and is estimated to have consisted of up to 100 million cubic meters of water, mud, and rocks. The death toll from the earthquake and landslide was 66,700 persons. [Lloyd Cluff/Corbis.]

Towns of Yungay and Ranrahirca before an earthquake-induced debris avalanche on Mount Huascarán, Peru, buried these towns.

Aftermath of the avalanche.

Debris avalanches (**Figure 16.24**) are fast downhill movements of soil and rock that usually occur in humid mountainous regions. Their speed results from the combination of high water content and steep slopes. Water-saturated debris may move as fast as 70 km/hour, a speed comparable to that of water flowing down a moderately steep slope. A debris avalanche carries with it everything in its path. In 1962, a debris avalanche in the high Peruvian Andes traveled almost 15 km in about 7 minutes, engulfing most of eight towns and killing 3500 people. Eight years later, on May 31, 1970, an earthquake in the same active subduction zone toppled a large mass of glacial ice at the top of one of the highest mountains, Nevado de Huascarán. As the ice broke up, it mixed with the debris of the high slopes and became an ice-debris avalanche. The avalanche picked up debris as it raced downhill, increasing its speed to an almost unbelievable 200 to 435 km/hour. More than 50 million cubic meters of muddy debris roared down into the valleys, killing 17,000 people and wiping out scores of villages.

On May 30, 1990, an earthquake shook another mountainous area in northern Peru, again setting off mudflows and debris avalanches. It was the day before a memorial ceremony scheduled to commemorate the disaster that had occurred 20 years earlier. In regions where unstable slopes build up and earthquakes are frequent because of the subduction of an oceanic plate beneath a continental plate, there can be no doubt about the necessity of learning how to predict both earthquakes and the dangerous mass movements that follow.

A **slump** is a slow slide of unconsolidated material that travels as a unit (**Figure 16.25**). In most places, the slump slips along a basal surface that forms a concave-upward shape, like a spoon. Faster than slumps are *debris slides* (**Figure 16.26**), in which the rock material and soil move largely as one or more units along planes of weakness, such as a waterlogged clay zone either within or at the base of the debris. During the slide, some of the debris may behave as a chaotic, jumbled flow. Such a slide may become predomi-

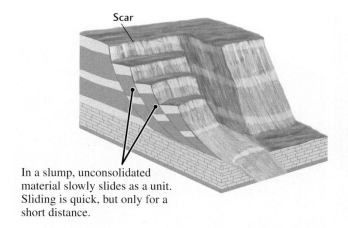

Scar

In a slump, unconsolidated material slowly slides as a unit. Sliding is quick, but only for a short distance.

Scar

Figure 16.25 A slump is a slow slide of unconsolidated material that travels as a unit. Soil slump, Sheridan, Wyoming. [E. R. Degginger.]

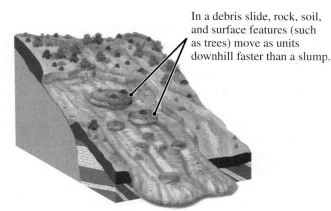

In a debris slide, rock, soil, and surface features (such as trees) move as units downhill faster than a slump.

nantly a flow as it moves rapidly downhill and most of the material mixes as if it were a fluid.

UNDERSTANDING THE ORIGINS OF MASS MOVEMENTS

To understand how slope steepness, the nature of the slope materials, and the materials' water content interact to create mass movements, geologists study both natural mass movements and those provoked by human activities. They investigate the causes of a modern slide by combining eyewitness reports with geological studies of the source of the slide and the distribution and nature of the debris dropped into the valley below. They can infer the causes of prehistoric slides from geological evidence alone where the debris is still present and can be analyzed for size, shape, and composition.

Natural Causes of Landslides

In April 1983, about 4 million cubic meters of unconsolidated mud and debris slid down from a wall of Spanish Fork Canyon in Utah. Near-record depths of snow had accumulated at higher elevations in the surrounding Wasatch Mountains. A warm spring melted the snow rapidly, and heavy rainstorms augmented the meltwaters, triggering debris flows and slides in unprecedented numbers in Spanish Fork and other canyons of the Wasatch Range. In this case, the nature of the materials—soil and rock that was structurally deformed, cracked, and weak—combined with the steep slopes of Spanish Fork to make the canyon susceptible to slides. The rain and melting snow saturated a mass of soil and weathered rock on the wall of the canyon, lubricating both the mass and the bedrock surface of the slope below it. Sooner or later, the mass had to give.

The 1925 landslide in the Gros Ventre River valley of western Wyoming illustrates again how water, the nature of slope materials, and slope stability interact to produce slides

Figure 16.26 A debris slide travels as one or more units and moves more quickly than a slump. A relatively recent debris slide has choked this narrow valley in the Tien Shan mountains in Kyrgyzstan. [Martin Miller/Visuals Unlimited.]

STEP 1
Heavy spring rain and snowmelt saturated the permeable sandstone dipping toward the Gros Ventre River.

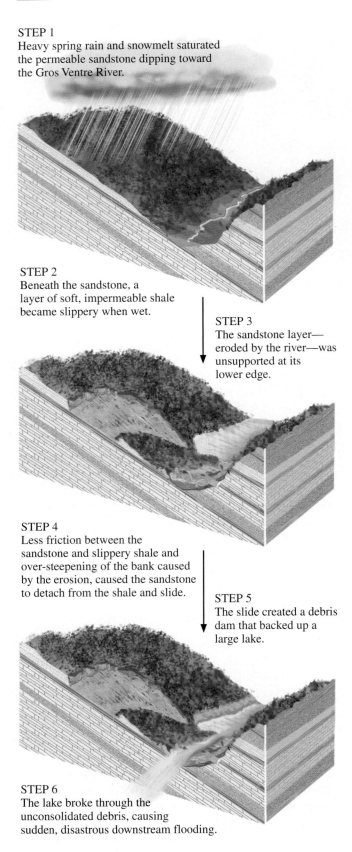

STEP 2
Beneath the sandstone, a layer of soft, impermeable shale became slippery when wet.

STEP 3
The sandstone layer—eroded by the river—was unsupported at its lower edge.

STEP 4
Less friction between the sandstone and slippery shale and over-steepening of the bank caused by the erosion, caused the sandstone to detach from the shale and slide.

STEP 5
The slide created a debris dam that backed up a large lake.

STEP 6
The lake broke through the unconsolidated debris, causing sudden, disastrous downstream flooding.

Figure 16.27 The 1925 Gros Ventre slide. [After W. C. Alden, "Landslide and Flood at Gros Ventre, Wyoming." *Transactions of the American Institute of Mining, Metallurgical, and Petroleum Engineers* (1928): 345–361.] The Gros Ventre landslide, Grand Teton National Park, Wyoming. [Steven Dutch.]

(Figure 16.27). In the spring of that year, melting snow and heavy rains swelled streams and saturated the ground in the valley. One local rancher, out on horseback, looked up to see the whole side of the valley racing toward his ranch. From the gate to his property, he watched the slide hurtle past him at about 80 km/hour and bury everything he owned.

About 37 million cubic meters of rock and soil slid down one side of the valley that day, then surged more than 30 m up the opposite side and fell back to the valley floor. Most of the slide was a confused mass of blocks of sandstone, shale, and soil, but one large section of the side of the valley, covered with soil and a forest of pine, slid down as a unit. The slide dammed the river, and a large lake grew over the next 2 years. Then the lake overflowed, breaking the dam and rapidly flooding the valley below.

The causes of the Gros Ventre slide were all natural. In fact, the stratigraphy and structure of the valley made a slide almost inevitable. On the side of the valley where the slide occurred, a permeable, erosion-resistant sandstone formation dipped about 20° toward the river, paralleling the slope of the valley wall. Under the sandstone were beds of soft, impermeable shale that became slippery when wet. The conditions became ideal for a slide when the river channel cut through most of the sandstone at the bottom of the valley wall and left it with virtually no support. Only friction along the bedding plane between the shale and sandstone kept the layer of sandstone from sliding. The river's removal of the sandstone's support was equivalent to scooping sand from the base of a sandpile—both cause oversteepening. The heavy rains and snow meltwater saturated the sandstone and the surface of the underlying shale, creating a slippery surface along the bedding planes at the top of the shale. No one knows what triggered the Gros Ventre slide, but at some point the force of gravity overcame friction and almost all of the sandstone slid down along the water-lubricated surface of the shale.

The formation of a dam on a river and the growth of a lake, as happened at both Spanish Fork Canyon and Gros

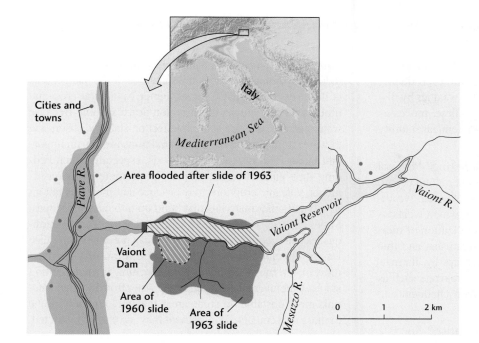

Figure 16.28 An ancient small slide on the north side of the walls of the Vaiont Dam reservoir, as well as a landslide in 1960 (shown by a dashed line in the brown area), warned of the danger of mass movement above the reservoir. In October 1963, a massive landslide that could have been predicted caused the water in the Vaiont reservoir to overflow the dam, flooding the downstream areas and killing 3000 people.

Ventre, are common consequences of a landslide. Because most slide materials are permeable and weak, such a dam is soon breached when the lake water reaches a high level or overflows. Then the lake drains suddenly, releasing a catastrophic torrent of water (see Figure 16.27).

Human Activities That Promote or Trigger Slides

Although the vast bulk of mass wasting is natural, humans may trigger landslides or make them more likely in vulnerable areas by activities that change natural slopes. Some geological settings are so susceptible to landslides that engineers may have to forgo construction projects in these areas entirely. A landslide in one such place ultimately killed 3000 people. The place was Vaiont, an Italian alpine valley, and the time was the night of October 9, 1963. A large reservoir in the valley was impounded by a concrete dam (the second highest in the world, at 265 m) and bordered by steep walls of interbedded limestone and shale. A great debris slide of 240 million cubic meters (2 km long, 1.6 km wide, and more than 150 m thick) plunged into the deep water of the reservoir behind the dam. The debris filled the reservoir for a distance of 2 km upstream of the dam and created a giant spillover. In the violent torrent that hurtled downstream as a 70-m-high flood wave, 3000 people died.

Engineers had underestimated three warning signs at Vaiont (**Figure 16.28**):

1. The weakness of the cracked and deformed layers of limestone and shale that made up the steep walls of the reservoir

2. The scar of an ancient slide on the valley walls above the reservoir

3. A forewarning of danger signaled by a small rockslide in 1960, just three years earlier

Although the 1963 landslide was natural and could not have been prevented, its consequences could have been much less severe. If the reservoir had been located in a geologically safer place, where the water was less likely to spill over its walls, damage might have been limited to a lesser loss of property and far fewer deaths. We cannot prevent most natural mass movements, but we can minimize our losses through more careful control of construction and land development.

SUMMARY

What is weathering and how is it geologically controlled? Rocks are broken down at Earth's surface by chemical weathering—the chemical alteration or dissolution of a mineral—and by physical weathering—the fragmentation of rocks by mechanical processes. Erosion wears away the land and transports the products of weathering, the raw material of sediment. The nature of the parent rock affects weathering because various minerals weather at different rates. Climate strongly affects weathering: warmth and heavy rainfall speed weathering; cold and dryness slow it down. The presence of soil accelerates weathering by providing moisture and an acidic environment, which promote chemical weathering and the growth of plant roots that aid physical weathering. All else being equal, the longer a rock weathers, the more completely it breaks down.

How does chemical weathering work? Potassium feldspar ($KAlSi_3O_8$), or orthoclase, weathers in much the same way as many other kinds of silicate minerals. Feldspar weathers in the presence of water by hydrolysis. In the chemical reaction of orthoclase with water, potassium (K) and silica (SiO_2) are lost to the water solution, and the solid feldspar changes into the clay mineral kaolinite, $Al_2Si_2O_5(OH)_4$. Carbon dioxide (CO_2) dissolved in water promotes chemical weathering by providing acid in the form of carbonic acid (H_2CO_3). Water in the ground and streams on the surface carry away dissolved ions and silica. Iron (Fe), which is found in ferrous form in many silicates, weathers by oxidation, producing ferric iron oxides in the process. Carbonates weather by dissolving completely, leaving no residue. These processes operate at varying rates, depending on the chemical stability of minerals under weathering conditions.

What are the processes of physical weathering? Physical weathering breaks rocks into fragments, either along crystal boundaries or along joints in rock masses. Physical weathering is promoted by chemical weathering, which weakens grain boundaries; by frost wedging, crystallization of minerals, and burrowing and tunneling by organisms and tree roots, all of which expand cracks; and perhaps by alternating extremes of heat and cold. Patterns of breakage such as exfoliation result from interactions between chemical and physical weathering processes.

What factors are important in soil formation? Soil is a mixture of clay minerals, weathered rock particles, and organic matter that forms as organisms interact with weathering rock and water. Weathering is controlled by climate and the activity of organisms, so soils form faster in warm, humid climates than in cold, dry ones. The five key factors that contribute to soil development are bedrock composition, climate, topography, organisms, and time.

What are mass movements, and what kinds of material do they involve? Mass wasting erodes and sculpts the land surface. Mass movements are slides, flows, or falls of large masses of material down slopes in response to the pull of gravity. The movements may be imperceptibly slow or far too fast to outrun. The masses consist of bedrock; consolidated material, including compacted sediment or soil; or unconsolidated material, such as loose, uncemented sediment or soil. Rock movements include rockfalls, rockslides, and rock avalanches. Unconsolidated material moves by creep, slump, debris slide, debris avalanche, earthflow, mudflow, and debris flow.

What factors are responsible for mass movements, and how are such movements triggered? The three factors that have the greatest bearing on the predisposition of material to move down a slope are (1) the steepness and stability of the slope, (2) the nature of the material, and (3) the water content of the material. Slopes become unstable when they are steeper than the angle of repose, the maximum slope angle that unconsolidated material will assume without cascading down. Slopes in consolidated material may also become unstable when they are oversteepened or denuded of vegetation. Water absorbed by the material contributes to instability in two ways: (1) by lowering internal friction (and thus resistance to flow) and (2) by lubricating planes of weakness in the material. Mass movements can be triggered by earthquakes or by sudden absorption of large quantities of water after a torrential rain. In many places, slopes build to a point of instability at which the slightest vibration will set off a slide, flow, or fall.

What factors are responsible for catastrophic mass movements, and how can such movements be prevented or min- imized? Analysis of both natural mass movements and those induced by human activity shows that one of the main causes is the oversteepening of slopes, either by natural erosional processes or by human construction or excavation. Because water content has such a strong effect on stability, absorption of water from prolonged or torrential rains is often an important factor. The structure of the beds, especially when bedding dips parallel to the slope, can promote mass movements. Volcanic eruptions may produce a tremendous fallout of ash and other materials that build up into unstable slopes. Slides or flows of the volcanic materials are triggered by earthquakes that accompany eruptions. We can prevent or minimize loss of life and damage to property from catastrophic mass movements by refraining from steepening or undercutting slopes. Careful engineering can keep water from making materials more unstable. In some areas that are extremely prone to mass movements, development may have to be restricted.

KEY TERMS AND CONCEPTS

angle of repose (p. 385)

chemical stability (p. 376)

chemical weathering (p. 373)

consolidated material (p. 385)

creep (p. 392)

erosion (p. 372)

exfoliation (p. 380)

frost wedging (p. 379)

hematite (p. 378)

humus (p. 381)

joint (p. 379)

kaolinite (p. 374)

liquefaction (p. 387)

mass movement (p. 384)

mass wasting (p. 372)

physical weathering (p. 381)

slump (p. 394)

soil (p. 373)

soil profile (p. 383)

talus (p. 390)

unconsolidated material (p. 385)

EXERCISES

1. What do the various kinds of rocks used for monuments tell us about weathering?

2. What rock-forming minerals found in igneous rocks weather to clay minerals?

3. How does abundant rainfall affect weathering?

4. Which weathers faster, a granite or a limestone?

5. How does physical weathering affect chemical weathering?

6. How do climates affect chemical weathering?

7. What role do earthquakes play in the occurrence of landslides?

8. What kinds of mass movements advance so rapidly that a person could not outrun them?

9. How does the absorption of water weaken unconsolidated material?

10. What is the angle of repose, and how does it vary with water content?

11. How does the steepness of a slope affect mass wasting?

12. What is a mudflow, and how is it produced?

THOUGHT QUESTIONS

1. In northern Illinois, you can find two soils developed on the same kind of bedrock: one is 10,000 years old and the other is 40,000 years old. What differences would you expect to find in their compositions or profiles?

2. Which igneous rock would you expect to weather faster, a granite or a basalt? What factors influenced your choice?

3. Assume that a granite with crystals about 4 mm across and a rectangular system of joints spaced about 0.5 to 1 m apart is weathering at Earth's surface. What size would you ordinarily expect the largest weathered particle to be?

4. Why do you think a road built of concrete, an artificial rock, tends to crack and develop a rough, uneven surface in a cold, wet region even when it is not subjected to heavy traffic?

5. Pyrite is a mineral in which ferrous iron is combined with sulfide ion. What major chemical process weathers pyrite?

6. Rank the following rocks in the order of the rapidity with which they weather in a warm, humid climate: a sandstone made of pure quartz, a limestone made of pure calcite, a granite, a deposit of rock salt (halite, $NaCl$).

7. What would a world look like if there were no weathering at the surface?

8. Would a prolonged drought affect the potential for landslides? How?

9. What geological conditions might you want to investigate before you bought a house at the base of a steep hill of bedrock covered by a thick mantle of soil?

10. What evidence would you look for to indicate that a mountainous area had undergone a great many prehistoric landslides?

11. What factors would make the potential for mass movements in a mountainous terrain in the rainy tropics greater or lesser than the potential in a similar terrain in a desert?

12. What kind(s) of mass movements would you expect from a steep hillside with a thick layer of soil overlying unconsolidated sands and muds after a prolonged period of heavy rain?

13. What factors weaken rock and enable gravity to start a mass movement?

SHORT-TERM PROJECTS

How Do Hoodoos Form?

As explained in this chapter, chemical and physical weathering processes reinforce each other. *Hoodoos* (a technical geologic term) are whimsically shaped pinnacles of rock and balanced boulders that are products of both physical and chemical weathering, commonly in dry climates. Bryce Canyon in Utah is a national park in large part because of its spectacular display of hoodoos (see the photo at the text's Web site). Hoodoos typically form in rock types such as sediments (as in Bryce), volcanic tuff, and granites. A set of vertical fractures (joints) is also a prerequisite for well-developed hoodoos.

At Bryce Canyon, soft siltstones and limestones from an ancient lake bed are exposed along the rim of the Paunsaugunt Plateau. Headward erosion into the rim of the plateau has created a maze of hoodoos. Snowmelt and heavy summer thunderstorms generate the runoff that readily erodes the soft sediments. Given the high elevation of the plateau, many days of freeze and thaw conditions also enhance physical weathering. The plateau rim is receding at a rate of 23 to 122 cm per 100 years, based on the study of growth rings of trees impacted by the erosion of the rim.

How do the hoodoos at Bryce Canyon form? Why are hoodoos and most weathered rocks typically rounded? To what degree does physical weathering enhance chemical weathering? Explore the answers to these questions at the text's Web site, **http://www.whfreeman.com/understandingearth**.

Environmental Changes and Land Use

While you are enrolled in your geology class, a variety of natural and human-induced environmental changes are likely to take place in your local area. There may be an earthquake, drought, flood, sinkhole collapse, or oil spill. Local officials may vote to open a new landfill, allow railroad transportation of hazardous materials through town, or develop the local waterfront.

In teams of four, prepare a series of news releases, approximately one per month, for your local media outlets in response to natural or human-induced environmental changes and to local land-use policy decisions. This project will require you to search for events pertinent to geology in local newspapers and broadcasts and to write responses from a geologist's perspective to the events that your team discovers.

Because you are writing for a nontechnical audience, you should make each news release brief (preferably only one page) and avoid the use of jargon. The title must grab the reader's attention. State the topic clearly in the first paragraph, describe the nature and significance of the event or issue, and emphasize its implications for the public.

Submit your news releases to your local newspapers and radio and television stations. Most local media outlets welcome newsworthy items about events in their communities.

CHAPTER SEVENTEEN

Porosity
store water

THE HYDROLOGIC CYCLE AND GROUNDWATER

Water is essential to a wide variety of geological processes. It is essential to weathering, both as a solvent of minerals in rock and soil and as a transport agent that carries away dissolved and weathered materials. Rivers and glacial ice are major agents of erosion that help to shape the landscape of the continents. Water that sinks into surface materials forms large reservoirs of groundwater. The study of the movements and characteristics of water on and under Earth's surface is known as **hydrology.**

We saw in Chapter 15 that water is exchanged between the oceans and atmosphere to form a critical link in regulating Earth's climate. Over the short term, El Niño transports water from the oceans to the atmosphere to the hillslopes of the western United States, where it lubricates the materials involved in catastrophic landslides. Over longer time scales, variations in Earth's orbit cause water to move from the oceans to the atmosphere to the ice sheets formed at Earth's poles. Climate scientists now recognize that understanding the *hydrologic cycle* of water on, over, and under Earth's surface is one of the most important steps in climate prediction.

Water is vital to all life on Earth. Humans cannot survive more than a few days without it, and even the hardiest desert plants and animals need some water. The amount of water that modern civilization requires is far greater than what humans need for simple physical survival. Water is used in immense quantities for industry, agriculture, and such urban needs as sewage systems. The United States, one of the heaviest users of water in the world, has been steadily increasing its consumption for over 100 years. In the 35 years between 1950 and 1985 alone, water use nearly tripled, from 34 billion gallons a day to about 90 billion gallons a day. By 1990, only 5 years later, that figure almost quadrupled, to 339 billion gallons a day. Some, but not all, of this increase is a result of population growth. In the United States, water consumption, corrected for population by using figures for consumption per person, actually fell by about 20 percent from 1980 to

Angel Falls, Venezuela, is the highest waterfall on Earth. The falls plunge 914 m from a flat-topped mountain composed of 1.7-billion-year-old sandstones. [Michael K. Nichols/National Geographic/Getty Images.]

1995. Developed countries have started to emphasize more efficient use of finite water resources.

Hydrology is becoming more important to all of us as the demand on limited water supplies increases and the need to understand Earth's climatic patterns heightens. To satisfy these concerns, we must understand not only where to find water but also how water supplies cycle through the Earth.

This chapter is a survey of water in and on the Earth.

FLOWS AND RESERVOIRS

We can see water in Earth's lakes, oceans, and polar ice caps, and we can see water moving over Earth's surface in rivers and glaciers. It is harder to see the massive amounts of water stored in the atmosphere and underground and the flows into and out of these storage places. As water evaporates, it vanishes into the atmosphere as vapor. As rain sinks into the ground, it becomes **groundwater**—the mass of water stored beneath Earth's surface.

Each place that stores water is a **reservoir.** Earth's main natural reservoirs are the oceans; glaciers and polar ice; groundwater; lakes and rivers; the atmosphere; and the biosphere. **Figure 17.1** shows the distribution of water among these reservoirs. The oceans are by far the largest repository of water on the planet. Although the total amount of water in rivers and lakes is relatively small, these reservoirs are important to human populations because they contain fresh water. The amount of groundwater is more than 100 times the amount in rivers and lakes, but much of it is unusable because it contains large quantities of dissolved material.

Reservoirs gain water from inflows, such as rain and river inflow, and lose water from outflows, such as evaporation and river outflow. If inflow equals outflow, the size of

the reservoir stays the same, even though water is constantly entering and leaving it. These flows mean that any given quantity of water spends a certain average time, called the *residence time,* in a reservoir.

How Much Water Is There?

The world's total water supply is enormous—about 1.46 billion cubic kilometers distributed among the various reservoirs. If it covered the land area of the United States, it would submerge the 50 states under a layer of water about 145 km deep. This total is constant, even though the flows from one reservoir to another may vary from day to day, year to year, and century to century. Over these geologically short time intervals, there is neither a net gain or loss of water to or from Earth's interior nor any significant loss of water from the atmosphere to outer space.

The Hydrologic Cycle

Water on or beneath Earth's surface cycles among the various reservoirs: the oceans, the atmosphere, and the land. The cyclical movement of water—from the ocean to the atmosphere by evaporation, to the surface through rain, to streams through runoff and groundwater, and back to the ocean—is the **hydrologic cycle.** **Figure 17.2** is a simplified illustration of the endless circulation of water and the amounts moved. Because organisms use water, it is also stored in the biosphere—for example, within the trees of rain forests.

Within the range of temperatures found at Earth's surface, water shifts among the three states of matter: liquid (water), gas (water vapor), and solid (ice). These transformations power some of the main flows from one reservoir to another in the hydrologic cycle. Earth's external heat engine, powered by the Sun, drives the hydrologic cycle, mainly by evaporating water from the oceans and transporting it as water vapor in the atmosphere. Under the right conditions of temperature and humidity, water vapor condenses to the tiny droplets of water that form clouds and eventually falls as rain or snow—together known as **precipitation**—over the oceans and continents. Some of the water that falls on land soaks into the ground by **infiltration,** the process by which water enters rock or soil through joints or small pore spaces between particles. Part of this groundwater evaporates through the soil surface. Another part is absorbed by the biosphere in plant roots, carried up to the leaves, and returned to the atmosphere by *transpiration*—the release of water vapor from plants. Other groundwater may return to the surface in springs that empty into rivers and lakes.

The rainwater that does not infiltrate the ground runs off the surface, gradually collecting into streams and rivers. The sum of all rainwater that flows over the surface, including the fraction that may temporarily infiltrate near-surface formations and then flow back to the surface, is called

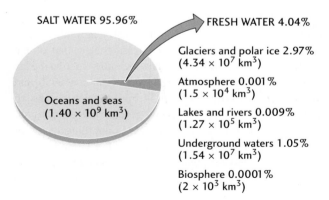

SALT WATER 95.96%

FRESH WATER 4.04%

Oceans and seas
$(1.40 \times 10^9 \ km^3)$

Glaciers and polar ice 2.97%
$(4.34 \times 10^7 \ km^3)$

Atmosphere 0.001%
$(1.5 \times 10^4 \ km^3)$

Lakes and rivers 0.009%
$(1.27 \times 10^5 \ km^3)$

Underground waters 1.05%
$(1.54 \times 10^7 \ km^3)$

Biosphere 0.0001%
$(2 \times 10^3 \ km^3)$

Figure 17.1 The distribution of water on Earth. [Revised from J. P. Peixoto and M. Ali Kettani, "The Control of the Water Cycle." *Scientific American* (April 1973): 46; E. K. Berner and R. A. Berner, *Global Environment.* Upper Saddle River, N.J.: Prentice Hall, 1996, pp. 2–4.]

Earth System Figure 17.2   Water cycles through Earth's crust, atmosphere, oceans, lakes, and rivers.

SEA	
36	Runoff from land
+ 398	Precipitation over sea
434	Evaporation

SEA		LAND	
434	Evaporation	107	Precipitation
− 398	Precipitation	− 71	Evaporation
36	Excess to land via precipitation	36	Runoff to ocean

LAND	
107	Precipitation
− 36	Runoff to ocean
71	Evaporation

1 Flux in and flux out over oceans is enormous and is almost balanced due to net evaporation and precipitation over water.

2 Excess is moved to land and precipitates (net precipitation over land).

3 The precipitation runs off into lakes, streams, and oceans…

4 …or filters into soil and rock, where it moves as groundwater.

Evaporation 434 — Precipitation 398 — Evaporation 71 — Precipitation 107 — Runoff 36 — Surface runoff — Groundwater table — Groundwater flow — Infiltration

runoff. Some runoff may later seep into the ground or evaporate from rivers and lakes, but most of it flows into the oceans.

Snowfall may be converted into ice in glaciers, which return water to the oceans by melting and runoff and to the atmosphere by *sublimation*, the transformation from a solid (ice) directly into a gas (water vapor). Most of the water that evaporates from the oceans returns to them as precipitation. The remainder falls over the land and either evaporates or returns to the ocean as runoff.

Figure 17.2 shows how the total flows among reservoirs balance one another in the hydrologic cycle. The land surface, for example, gains water from precipitation and loses the same amount of water by evaporation and runoff. The ocean gains water from runoff and precipitation and loses the same amount by evaporation. As you can see from Figure 17.2, more water evaporates from the oceans than falls on them as rain. This loss is balanced by the water returned as runoff from the continents. Thus, the size of each reservoir stays constant.

The hydrologic cycle is global in scale. The movement of water by evaporation and precipitation takes place within zones of latitude that circle the Earth. In Chapter 19, we will discuss how these zones control the location of the world's deserts—regions where precipitation is extremely limited—and global wind patterns. The global pattern of precipitation controls where the "green" parts of the Earth develop, such as the equatorial rain forests and midlatitude temperate forests.

How Much Water Can We Use?

As threats of water shortages loom, our use of water enters the arena of public policy debate (see Feature 17.1). The global hydrologic cycle ultimately controls water supplies. Almost all the water we use is fresh water—water that is not salty. Artificial desalination of seawater (the removal of salt from it) produces small but steadily growing amounts of fresh water in areas such as the arid Middle East. In the natural world, however, fresh water is supplied only by rain, rivers, lakes, some groundwaters, and water melted from snow or ice on land. All these waters are ultimately supplied by precipitation. Therefore, the practical limit to the amount of natural fresh water that we can ever envision using is the amount steadily supplied to the continents by precipitation.

EARTH ISSUES

17.1 Water Is a Precious Resource: Who Should Get It?

Until recently, most people in the United States have taken our water supply for granted. Analyses of available supplies and user needs, however, indicate that many areas of the country will experience water shortages more frequently. These shortages will create conflict among the several sectors of consumers—residential, industrial, agricultural, and recreational—over who has the greatest rights to the water supply.

In recent years, widely publicized droughts and mandatory restrictions on water use—such as have occurred in California, Florida, Colorado, and many other places—have alerted the public that the nation faces major water problems. Public concern waxes and wanes, however, as periods of droughts and abundant rainfall come and go, and governments are not pursuing long-term solutions with the urgency that they deserve. Here are some facts to ponder:

• A human can survive with about 2 liters of water per day. In the United States, the per capita use for all purposes is about 6000 liters per day.

• Industry uses about 38 percent and agriculture about 43 percent of the water withdrawn from our reservoirs.

• Per capita domestic water use in the United States is two to four times greater than in western Europe, where consumers pay as much as 350 percent more for their water.

• Although the states in the western United States receive one-fourth of the country's rainfall, their per capita water use (mostly for irrigation) is 10 times greater than that of the eastern states, and at much lower prices. In California, for example, which imports most of its water, 85 percent of the water is used for irrigation, 10 percent for municipalities, and 5 percent for industry. A 15 percent reduction in irrigation use would almost double the amounts of water available for use by cities and industries.

• The traditional ways of increasing the water supply, such as building dams and reservoirs and drilling, have become extremely costly because most of the good (and therefore cheaper) sites have already been used. The building of more dams to hold larger reservoirs carries environmental costs, such as the flooding of inhabited areas, detrimental changes in river flows above and below dams, and the disturbance of fish and other wildlife habitat. Factoring in these costs has led to delays or rejection of proposals for new dams.

• The fresh water used in the United States eventually returns to the hydrologic cycle, but it may return to a reservoir that is not well located for human use, and the quality may be degraded. Recycled irrigation water is often saltier than natural fresh water and is loaded with pesticides. Polluted urban waste water ends up in the oceans.

• Global climate change may lead to reduced rainfall in western states, exacerbating the problems there and making long-term solutions even more urgent.

HYDROLOGY AND CLIMATE

For most practical purposes, local hydrology—the amount of water there is in a region and how it flows from one reservoir to another—is more important than global hydrology. The strongest influence on local hydrology is the climate, which includes both temperature and precipitation. In warm areas where rain falls frequently throughout the year, water supplies—both at the surface and underground—are abundant. In warm arid or semiarid regions, it rarely rains and water is a precious resource. People who live in icy climates rely on meltwaters from snow and ice. In some parts of the world, seasons of heavy rain, called monsoons, alternate with long dry seasons during which water supplies shrink, the ground dries out, and vegetation shrivels.

Wherever we live, climate and the geology of the land strongly influence the amounts of water cycled from one reservoir to another. Geologists are especially interested in how changes in precipitation and evaporation affect water supplies by altering the amounts of infiltration and runoff, which determine groundwater levels. If sea level rises as a result of global warming, groundwaters in low-lying coastal regions may become salty as seawater invades formerly fresh groundwaters.

Humidity, Rainfall, and Landscape: Linking the Climate and Plate Tectonic Geosystems

Many differences in climate are related to the temperature of the air and the amount of water vapor it contains. The **relative humidity** is the amount of water vapor in the air, expressed as a percentage of the total amount of water the air could hold at that temperature if saturated. When the relative humidity is 50 percent and the temperature is 15°C, for example, the amount of moisture in the air is one-half the maximum amount the air could hold at 15°C.

Warm air can hold much more water vapor than cold air. When unsaturated warm air at a given relative humidity cools enough, it becomes supersaturated and some of the

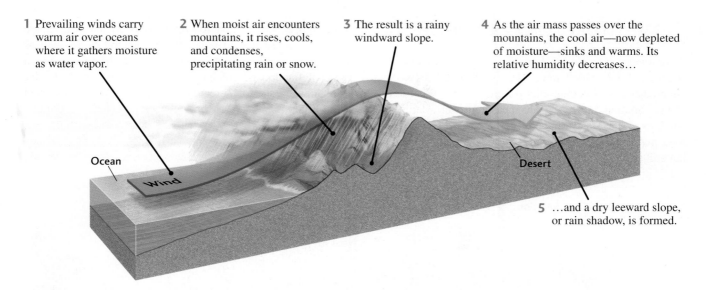

1 Prevailing winds carry warm air over oceans where it gathers moisture as water vapor.

2 When moist air encounters mountains, it rises, cools, and condenses, precipitating rain or snow.

3 The result is a rainy windward slope.

4 As the air mass passes over the mountains, the cool air—now depleted of moisture—sinks and warms. Its relative humidity decreases…

5 …and a dry leeward slope, or rain shadow, is formed.

Ocean

Wind

Desert

Figure 17.3 Rain shadows are areas of low rainfall on the leeward (downwind) slopes of a mountain range.

vapor condenses into water droplets. The condensed water droplets form clouds. We can see clouds because they are made up of visible water droplets rather than invisible water vapor. When enough moisture has condensed into clouds and the droplets have grown too heavy to stay suspended by air currents, they fall as rain.

Most of the world's rain falls in warm, humid regions near the equator, where both the air and the surface waters of the ocean are warm. Under these conditions, a great deal of the ocean water evaporates, resulting in high humidity. When water-laden winds from these oceanic regions rise over nearby continents, the air cools and becomes supersaturated. The result is heavy rainfall over the land, even at great distances from the coast.

Unlike tropical climates, polar climates tend to be very dry. The polar oceans and the air above them are cold, so evaporation from the sea surface is minimized and the air can hold little moisture. Between the tropical and polar extremes are the temperate climates, where rainfall and temperatures are moderate.

Tectonic processes help control climatic processes. Landscape can alter precipitation patterns, for example. Mountain ranges form **rain shadows,** areas of low rainfall on their leeward (downwind) slopes. Moisture-filled air rising over high mountains cools and precipitates rain on the windward slopes, losing much of its moisture by the time it reaches the leeward slopes (**Figure 17.3**). The air warms again as it drops to lower elevations on the other side of the mountain range. The relative humidity declines because warm air can hold more moisture before becoming saturated. This further decreases the moisture available for rain. There is a rain shadow on the eastern side of the Cascade Mountains of Oregon. Moist winds blowing over the Pacific Ocean hit the mountains' western slopes, causing heavy rainfall. The eastern slopes, on the other side of the range in the rain shadow, are

dry and barren. Ultimately, these variations in precipitation patterns control weathering and erosion rates, which influence rates of tectonic uplift. In Chapter 22, we will explore further how tectonics and climate control hydrologic patterns involved in landscape development.

Droughts

Droughts—periods of months or years when precipitation is much lower than normal—can occur in all climates. Arid regions are especially vulnerable to decreases in their water supplies during prolonged droughts. Lacking replenishment from precipitation, rivers may shrink and dry up, reservoirs may evaporate, and the soil may dry and crack while vegetation dies. As populations grow, demands on reservoirs increase; a drought can deplete already inadequate water supplies.

The severest droughts of the past few decades have affected lands along the southern border of the Sahara Desert, where tens of thousands of lives have been lost to famine. This long drought has expanded the desert and effectively destroyed farming and grazing in the area.

Another prolonged but less severe drought affected most of California from 1987 until February 1993, when torrential rains arrived. During the drought, groundwater and reservoirs dropped to their lowest levels in 15 years. Some control measures were instituted, but a move to reduce the extensive use of water supplies for irrigation encountered strong political resistance from farmers and the agricultural industry (see Feature 17.1).

The midwestern United States and parts of Canada experienced a severe but short-lived drought in 1988, when surface water supplies shrank and the Mississippi River, lowered by many feet, was closed to traffic. By 1989, precipitation over the region had returned to normal.

THE HYDROLOGY OF RUNOFF

A dramatic example of how precipitation affects local stream and river runoff can be seen when weather forecasters predict flash flooding after torrential rains. When levels of precipitation and runoff are measured over a large area (such as all the states drained by a major river) and over a long period (a year, say), the relationship is less extreme but still strong. The maps of precipitation and runoff shown in **Figure 17.4** illustrate this relationship. When we compare them, we see that in areas of low precipitation—such as southern California, Arizona, and New Mexico—only a small fraction of precipitation ends up as runoff. In dry regions, much of the precipitation is lost by evaporation and infiltration. In humid areas such as the southeastern United States, a much higher proportion of the precipitation runs off in rivers. A large river may carry great amounts of water from an area with high rainfall to an area with low rainfall.

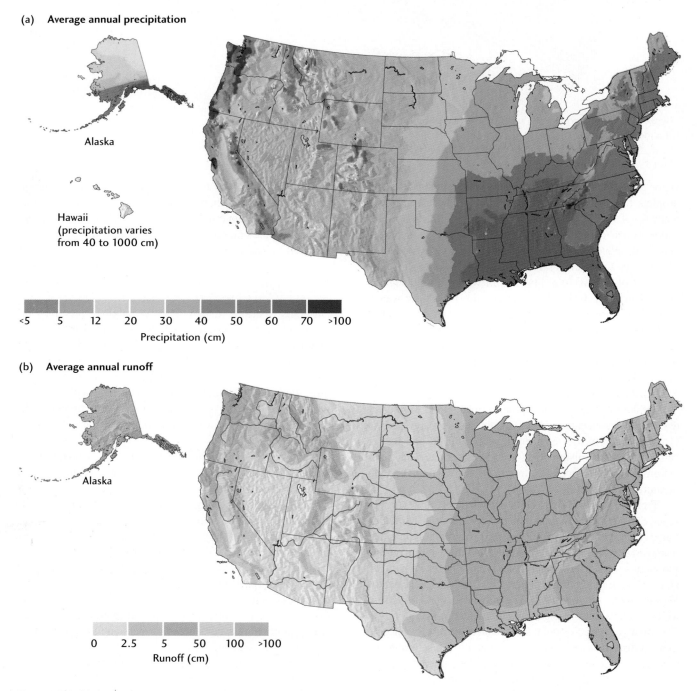

(a) Average annual precipitation

Alaska

Hawaii
(precipitation varies
from 40 to 1000 cm)

<5   5   12   20   30   40   50   60   70   >100

Precipitation (cm)

(b) Average annual runoff

Alaska

0   2.5   5   50   100   >100

Runoff (cm)

Figure 17.4 (a) Average annual precipitation in the United States. [Data from U.S. Department of Commerce, *Climatic Atlas of the United States,* 1968.] (b) Average annual runoff in the United States. [Data from USGS Professional Paper 1240-A, 1979.]

The Colorado River, for example, begins in an area of moderate rainfall in Colorado and then carries its water through arid western Arizona and southern California.

Major rivers carry most of the world's surface runoff. The millions of small and medium-sized rivers carry about half the world's entire runoff; about 70 major rivers carry the other half. And the Amazon River of South America carries almost half of that. The Amazon carries about 10 times more water than the Mississippi, the largest river of North America (Table 17.1).

Surface runoff collects and is stored in natural lakes and in artificial reservoirs created by the damming of rivers. Wetlands, such as swamps and marshlands, also act as storage depots for runoff (**Figure 17.5**). If these reservoirs are large enough, they can absorb short-term inflows from major rainfalls, holding some of the water that would otherwise spill over riverbanks. During dry seasons or droughts, the reservoirs release water to streams or to water systems built for human use. These reservoirs help to control flooding by smoothing out seasonal or yearly variations in runoff and releasing steady flows of water downstream. For this reason, some geologists work to stop the artificial draining of wetlands for real estate development. Destruction of wet-

Table 17.1	Water Flows of Some Great Rivers

River	Water Flow (m³/s)
Amazon, South America	175,000
La Plata, South America	79,300
Congo, Africa	39,600
Yangtze, Asia	21,800
Brahmaputra, Asia	19,800
Ganges, Asia	18,700
Mississippi, North America	17,500

lands also threatens biological diversity (see Chapter 11), because wetlands are breeding grounds for a great many types of plants and animals.

Wetlands are fast disappearing as land development continues. In the United States, more than half the original wetlands are gone. California and Ohio have kept only

DRY PERIOD: LOW RUNOFF

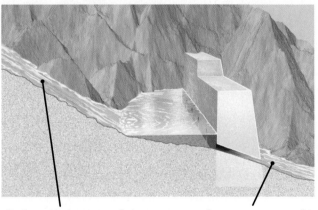

In dry periods, streams bring in small amounts of water...   ...and carry away small amounts.

WET PERIOD: HIGH RUNOFF

In wet periods, streams bring in large amounts of water,...   ...which is stored...   ...and slowly released during dry periods.

Figure 17.5 Like a natural lake or an artificial reservoir behind a dam, a wetland (such as a swamp or marsh) stores water during times of rapid runoff and slowly releases it during periods of little runoff.

10 percent of their original wetlands. The movement to protect wetlands has spawned heated controversy. The legal definition of wetland has been argued for years and has become a political football. A 1995 scientific study of the question by the National Academy of Sciences has been attacked as "political" by opponents of regulation. Some politicians who object to regulations designed to protect wetlands have asked for a 50 percent reduction in the extent of federally regulated wetlands.

GROUNDWATER

Groundwater forms as raindrops and melting snow infiltrate soil and other unconsolidated surface materials, sinking even into cracks and crevices of bedrock. We tap groundwater by drilling wells and pumping the water to the surface. Well drillers in temperate climates know they are most likely to find a good supply of water if they drill into porous sand or sandstone beds not far below the surface. Beds that store and transmit groundwater in sufficient quantity to supply wells are called **aquifers.**

The enormous reservoir of groundwater stored beneath Earth's surface equals about 22 percent of all the fresh water stored in lakes and rivers, glaciers and polar ice, and the atmosphere. For thousands of years, people have drawn on this resource, either by digging shallow wells or by storing water that flows out onto the surface at springs. Springs are direct evidence of water moving below the surface (**Figure 17.6**).

How Water Flows Through Soil and Rock

When water moves into and through the ground, what determines where and how fast it flows? With the exception of caves, there are no large open spaces for pools or rivers of water underground. The only spaces available for water are the pore space between grains of sand and other particles that make up the soil and bedrock and the space in fractures. Some pores, however small and few, are found in every kind of rock and soil, but large amounts of pore space are most often found in sandstones and limestones. Recall from Chapter 5 that the amount of pore space in rock, soil, or sediment is its *porosity*—the percentage of its total volume that is taken up by pores.

Porosity and Permeability

Pores occur as three types: the space between grains (*intergranular porosity*), the space in fractures (*fracture porosity*), and the space created by dissolution (*vuggy porosity*). Intergranular porosity depends on the size and shape of the grains that make up soils and siliciclastic sedimentary rocks and on how they are packed together. The more loosely packed the grains, the greater the pore space between them.

Figure 17.6 Groundwater exits from a cliff in Vasey's Paradise, Marble Canyon, Grand Canyon National Park, Arizona. This is a dramatic example of a spring formed where hilly topography allows water in the ground to flow out onto the surface. [Larry Ulrich.]

In many sandstones, porosity is as high as 30 percent (**Figure 17.7**). Minerals that cement grains reduce intergranular porosity. The smaller the particles and the more they vary in shape, the more tightly they fit together. Porosity is higher in soils, sediments, and sedimentary rocks (10 to 40 percent) than in igneous or metamorphic rocks, where porosity is created mostly by fractures, including joints and cleavage planes. Fractured rocks may contain appreciable pore space—as much as 10 percent of the volume—in their many cracks, though values commonly are as low as 1 to 2 percent. Pore space in limestones and other highly solu-

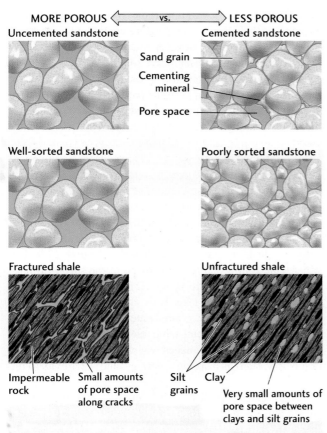

MORE POROUS ⟨⟵ vs. ⟶⟩ LESS POROUS

Uncemented sandstone   Cemented sandstone

Sand grain

Cementing mineral

Pore space

Well-sorted sandstone   Poorly sorted sandstone

Fractured shale   Unfractured shale

Impermeable rock   Small amounts of pore space along cracks   Silt grains   Clay   Very small amounts of pore space between clays and silt grains

Figure 17.7 Pores in rocks are normally filled partly or entirely with water. (Pores in oil- or gas-bearing sandstones and limestones are filled with oil or gas.)

ble rocks such as evaporites may be created when groundwater interacts with the rock and partially dissolves it, leaving irregular voids known as vugs. Vuggy porosity can be very high (over 50 percent); caves are examples of extremely large vugs.

Although porosity tells us how much water a rock can hold if all its pores are filled, it gives us no information about how rapidly water can flow through the pores. Water travels through a porous material by winding between grains and through cracks. The smaller the pore spaces and the more tortuous the path, the slower the water travels. The capacity of a solid to allow fluids to pass through it is its **permeability.** Generally, permeability increases as porosity increases. Permeability also depends on the sizes of the pores, how well they are connected, and how tortuous a path the water must travel to pass through the material. Vuggy pore networks in carbonate rocks may have extremely high permeabilities. Cave systems are so permeable that they allow people as well as water to move through them!

Both porosity and permeability are important factors when searching for a groundwater supply. In general, a good groundwater reservoir is a body of rock, sediment, or soil with both high porosity (so that it can hold large amounts of water) and high permeability (so that the water can be pumped from it easily). A rock with high porosity but low permeability may contain a great deal of water, but because the water flows so slowly, it is hard to pump it out of the rock. Table 17.2 summarizes the porosity and permeability of various rock types.

The Groundwater Table

As well drillers bore deeper into soil and rock, the samples they bring up become wetter. At shallow depths, the material is unsaturated—the pores contain some air and are not completely filled with water. This level is called the **unsaturated zone** (often termed the *vadose zone*). Below it is the **saturated zone** (often termed the *phreatic zone*), the level at which the pores of the soil or rock are completely filled with water. The saturated and unsaturated zones can be in unconsolidated material or bedrock. The boundary between the

Table 17.2	Porosity and Permeability of Aquifer Rock Types	
Rock Type	**Porosity (Pore Space That May Hold Fluid)**	**Permeability (Ability to Allow Fluids to Pass Through)**
Gravel	Very high	Very high
Coarse- to medium-grained sand	High	High
Fine-grained sand and silt	Moderate	Moderate to low
Sandstone, moderately cemented	Moderate to low	Low
Fractured shale or metamorphic rocks	Low	Very low
Unfractured shale	Very low	Very low

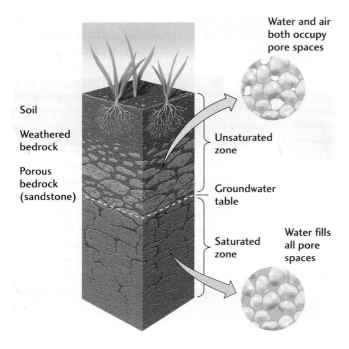

Soil

Weathered
bedrock

Porous
bedrock
(sandstone)

Water and air
both occupy
pore spaces

Unsaturated
zone

Groundwater
table

Saturated
zone

Water fills
all pore
spaces

Figure 17.8 The groundwater table is the boundary between the unsaturated zone and the saturated zone. The saturated and unsaturated zones can be in either unconsolidated material or bedrock.

two zones is the **groundwater table,** usually shortened to *water table* (**Figure 17.8**). When a hole is drilled below the water table, water from the saturated zone flows into the hole and fills it to the level of the water table.

Groundwater moves under the force of gravity, and so some of the water in the unsaturated zone may be on its way down to the water table. A fraction of the water, however, remains in the unsaturated zone, held in small pore spaces by surface tension—the attraction between the water molecules and the surfaces of the particles. Surface tension, you may recall from Chapter 16, keeps the sand on a beach moist, even though there are spaces below to which water could be driven by gravity. The evaporation of water in pore spaces in the unsaturated zone is slowed both by the effect of surface tension and by the relative humidity of the air in the pore spaces, which can be close to 100 percent.

If we were to drill wells at several sites and measure the elevations of the water levels in the wells, we could construct a map of the water table. A cross section of the landscape might look like the one shown in **Figure 17.9**. The water table follows the general shape of the surface topography, but the slopes are gentler. The water table is at the surface in river and lake beds and at springs. Under the influence of gravity, groundwater moves downhill from an area where the water-table elevation is high—under a hill, for example—to places where the water-table elevation is low, such as a spring where groundwater exits to the surface.

Water enters and leaves the saturated zone through recharge and discharge. **Recharge** is the infiltration of water into any subsurface formation, often from the surface by rain or melting snow. Recharge may also take place through the bottom of a stream where the stream channel lies at an elevation above that of the water table (see Figure 17.9). Streams

Figure 17.9 Dynamics of the groundwater table in permeable shallow formations in a temperate climate. The depth of the water table fluctuates in response to the balance between water added from precipitation (recharge) and water lost by evaporation and from wells, springs, and streams (discharge).

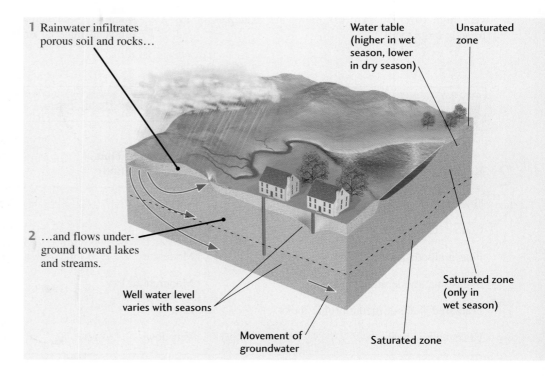

1 Rainwater infiltrates porous soil and rocks...

Water table (higher in wet season, lower in dry season)

Unsaturated zone

2 ...and flows underground toward lakes and streams.

Well water level varies with seasons

Movement of groundwater

Saturated zone

Saturated zone (only in wet season)

that recharge groundwater in this way are called *influent streams,* and they are most characteristic of arid regions, where the water table is deep. **Discharge** is the exit of groundwater to the surface, the opposite of recharge. When a stream channel intersects the water table, water discharges from the groundwater to the stream. Such an *effluent stream* is typical of humid areas. Effluent streams continue to flow long after runoff has stopped because they are fed by groundwater. Thus, the reservoir of groundwater may be increased by influent streams and depleted by effluent streams.

Aquifers

Groundwater may flow in unconfined or confined aquifers. In *unconfined aquifers,* the water travels through beds of more or less uniform permeability that extend to the surface in both discharge and recharge areas. The level of the reservoir in an unconfined aquifer is the same as the height of the water table.

Many permeable aquifers, typically sandstones, are bounded above and below by low-permeability beds, such as shales. These relatively impermeable beds are **aquicludes,** and groundwater either cannot flow through them or flows through them very slowly. When aquicludes lie both over and under an aquifer, they form a *confined aquifer.*

The impermeable beds above a confined aquifer prevent rainwater from infiltrating directly into the aquifer. Instead, a confined aquifer is recharged by precipitation over the recharge area, often characterized by outcropping rocks in a topographically higher upland. Here rainwater can enter the ground because there is no aquiclude preventing infiltration. The water then travels down the aquifer underground (**Figure 17.10**). Water moving in a confined aquifer—known as an **artesian flow**—is under pressure. At any point in the aquifer, the pressure is equivalent to the weight of all the water in the aquifer above that point.

If we drill a well into a confined aquifer at a point where the elevation of the ground surface is lower than that of the water table in the recharge area, the water will flow out of the well spontaneously. Such wells are called *artesian wells,* and they are extremely desirable because no energy is required to pump the water to the surface. The water is brought up by its own pressure.

In more complex geological environments, the water table may be more complicated. For example, if there is a relatively impermeable clay layer that creates an aquiclude in an otherwise permeable sand formation, the aquiclude may lie below the water table in a shallow aquifer and above the water table in a deeper aquifer (**Figure 17.11**). The water table in the shallow aquifer is called a *perched water table* because it is above the main water table in the lower aquifer. Many perched water tables are small, only a few meters thick and restricted in area, but some extend for hundreds of square kilometers.

Balancing Recharge and Discharge

When recharge and discharge are balanced, the reservoir of groundwater and the water table remain constant, even

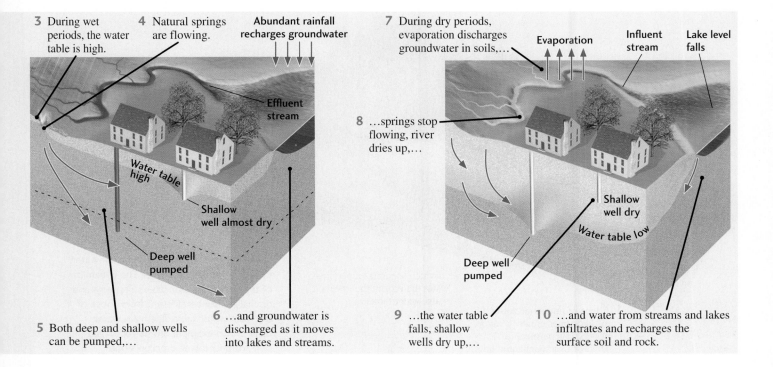

3 During wet periods, the water table is high.

4 Natural springs are flowing.

Abundant rainfall recharges groundwater

Effluent stream

Water table high

Shallow well almost dry

Deep well pumped

5 Both deep and shallow wells can be pumped,…

6 …and groundwater is discharged as it moves into lakes and streams.

7 During dry periods, evaporation discharges groundwater in soils,…

Evaporation

Influent stream

Lake level falls

8 …springs stop flowing, river dries up,…

Shallow well dry

Water table low

Deep well pumped

9 …the water table falls, shallow wells dry up,…

10 …and water from streams and lakes infiltrates and recharges the surface soil and rock.

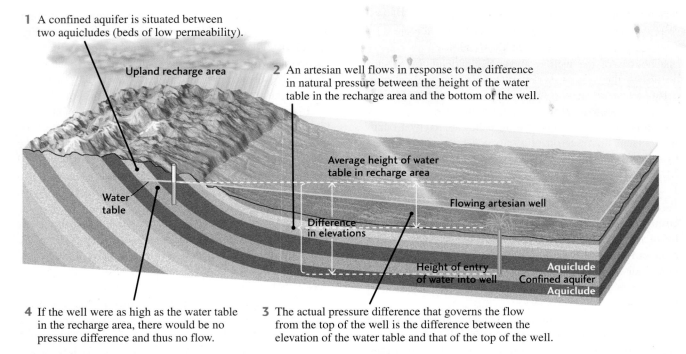

1 A confined aquifer is situated between two aquicludes (beds of low permeability).

Upland recharge area

2 An artesian well flows in response to the difference in natural pressure between the height of the water table in the recharge area and the bottom of the well.

Water table

Average height of water table in recharge area

Flowing artesian well

Difference in elevations

Height of entry of water into well

Aquiclude
Confined aquifer
Aquiclude

4 If the well were as high as the water table in the recharge area, there would be no pressure difference and thus no flow.

3 The actual pressure difference that governs the flow from the top of the well is the difference between the elevation of the water table and that of the top of the well.

Figure 17.10 A confined aquifer is created when an aquifer is situated between two aquicludes (beds of low permeability).

though water is continually flowing through the aquifer. For recharge to balance discharge, rainfall must be frequent enough to compensate for the runoff from rivers and the outflow from springs and wells.

But recharge and discharge will not always be equal, because rainfall varies from season to season. Typically, the water table drops in dry seasons and rises during wet periods. A decrease in recharge, such as during a prolonged drought, will be followed by a longer-term imbalance and a lowering of the water table.

An increase in discharge, usually from increased well pumping, can produce the same imbalance. Shallow wells may end up in the unsaturated zone and go dry. When water is pumped out of an aquifer faster than recharge can replenish it, the water level in the aquifer is lowered in a cone-shaped area around the well, called a *cone of depression*

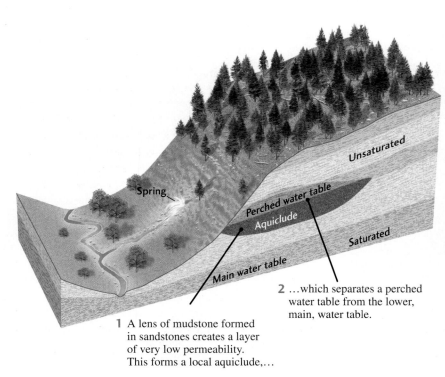

Spring

Unsaturated

Perched water table

Aquiclude

Saturated

Main water table

2 ...which separates a perched water table from the lower, main, water table.

1 A lens of mudstone formed in sandstones creates a layer of very low permeability. This forms a local aquiclude,...

Figure 17.11 A perched water table forms in geologically complex situations—in this case, where a shale aquiclude is located above the main water table in a sandstone aquifer. The dynamics of the perched water table's recharge and discharge may be different from those of the main water table. The main water table in this example can be recharged only from its lower outcrop slopes.

(**Figure 17.12**). The water level in the well is lowered to the depressed level of the water table. If the cone of depression extends below the bottom of the well, that well goes dry. If the bottom of the well is above the base of the aquifer, extending the well deeper into the aquifer may allow more water to be withdrawn, even at continued high pumping rates. If the rate of pumping is maintained and the well is deepened so much that the entire aquifer is tapped, however, the cone of depression can reach the bottom of the aquifer and deplete it. The aquifer will recover only if the pumping rate is reduced enough to give it time to recharge.

The extreme withdrawal of water not only can deplete the aquifer but also can cause another undesirable environmental effect. As the pressure of the water in pore space falls, the ground surface overlying the aquifer may subside, creating sinklike depressions (**Figure 17.13**). As water in some sediments is removed, the sediments compact, and the loss of volume lowers the surface. Subsidence caused by overpumping has occurred in Mexico City and in Venice, Italy, as well as in many other regions of heavy pumping, such as the San Joaquin Valley in California. In these places, the rate of subsidence of the surface has reached almost 1 m every 3 years. Although there have been a few attempts to reverse the subsidence by pumping water back into the groundwater system, they have not been very successful, because most compacted materials do not expand easily to their former state. The best measure to halt further subsidence is to restrict pumping.

People who live near the ocean's edge may face a different problem when pumping rates are high in relation to recharge: the flow of salt water into the well. Near shorelines or a little offshore, an underground boundary separates salt water under the sea from fresh water under the land. This

Figure 17.13 In Antelope Valley, California, overpumping of groundwater has led to fissures and sinklike depressions on Rogers Lakebed at Edwards Air Force Base. This fissure, formed in January 1991, is about 625 m long. [James W. Borchers/USGS.]

boundary slopes down and inland from the shoreline in such a way that salt water underlies the fresh water of the aquifer (**Figure 17.14**). Under many ocean islands, a lens of fresh groundwater (shaped like a simple double-convex lens) floats on a base of seawater. The fresh water floats because it is less dense than seawater (1.00 g/cm^3 versus 1.02 g/cm^3, a small but significant difference). Normally, the pressure of fresh water keeps the saltwater margin slightly offshore.

The balance between recharge and discharge in the freshwater aquifer maintains this freshwater-seawater boundary. As long as recharge by rainwater is at least equal to discharge by pumping, the well will provide fresh water. If water is withdrawn faster than it is recharged, however, a cone of depression develops at the top of the aquifer, mirrored by an inverted cone rising from the freshwater-seawater boundary below. The cone of depression at the upper part of the aquifer makes it more difficult to pump fresh water, and the inverted cone below leads to an intake of salt water at the bottom of the well (see Figure 17.14). People living closest to the shore are the first affected. Some towns on Cape Cod in Massachusetts, on Long Island in New York, and in many other nearshore areas have had to post notices that town drinking water contains more salt than is considered healthful by environmental agencies. There is no ready solution to this problem other than to slow the pumping or,

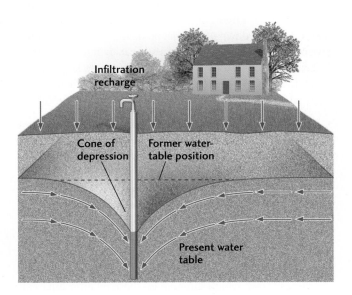

Figure 17.12 Excessive pumping in relation to recharge draws down the water table into a cone-shaped depression around a well. The water level in the well is lowered to the depressed level of the water table.

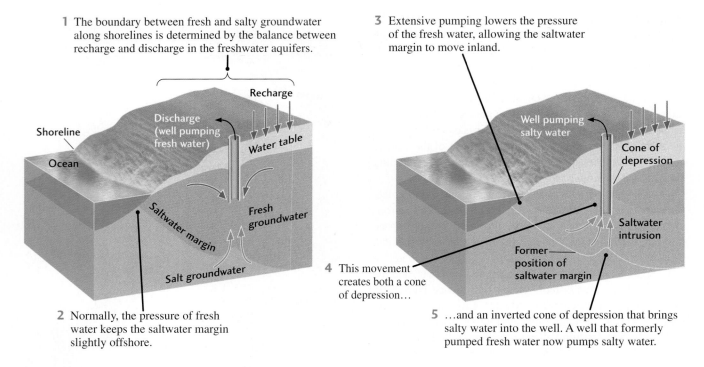

1 The boundary between fresh and salty groundwater along shorelines is determined by the balance between recharge and discharge in the freshwater aquifers.

3 Extensive pumping lowers the pressure of the fresh water, allowing the saltwater margin to move inland.

2 Normally, the pressure of fresh water keeps the saltwater margin slightly offshore.

4 This movement creates both a cone of depression…

5 …and an inverted cone of depression that brings salty water into the well. A well that formerly pumped fresh water now pumps salty water.

Figure 17.14 The balance between recharge and discharge maintains the freshwater-seawater boundary.

in some places, to recharge the aquifer artificially by funneling runoff into the ground.

We can see that a rise in sea level, which has been predicted as a result of global warming, would significantly alter the seawater margin. As sea level rises, the margin also rises. Seawater can then invade coastal aquifers and turn fresh groundwater into salt water.

The Speed of Groundwater Flows

The *speed* at which water moves in the ground strongly affects the balance between discharge and recharge. Most groundwaters move slowly, a fact of nature responsible for our groundwater supplies. If groundwater moved as rapidly as rivers, aquifers would run dry after a period of time without rain, just as many small streams do. The slow-moving groundwater flow also makes rapid recharge impossible if groundwater levels are lowered by excessive pumping.

Although all groundwaters flow through aquifers slowly, some flow slower than others. In the middle of the nineteenth century, Henri Darcy, town engineer of Dijon, France, proposed an explanation for the difference in flow rates. While studying the town's water supply, Darcy measured the elevations of water in various wells and mapped the varying heights of the water table in the district. He calculated the distances that the water traveled from well to well and measured the permeability of the aquifers. Here are his findings:

• For a given aquifer and distance of travel, the rate at which water flows from one point to another is directly propor-

tional to the drop in elevation of the water table between the two points. As the difference in elevation increases, the rate of flow increases.

• The rate of flow for a given aquifer and a given difference in elevation is inversely proportional to the flow distance that the water travels. As the distance increases, the rate decreases. The ratio between the elevation difference and the flow distance is known as the **hydraulic gradient.** Just as a ball rolls faster down a steep slope than down a gentle one, groundwater flows faster down a steeper hydraulic gradient. In general, groundwater does not run down the slope of the groundwater table but follows the hydraulic gradient of the flow, which may travel various paths below the water table.

• Darcy reasoned that the relationship between flow and hydraulic gradient should hold whether the water is moving through a porous sandstone aquifer or an open pipe. You might guess that the water would move faster through a pipe than through the tortuous turns of pore spaces in an aquifer. Darcy recognized this factor and included a measure of permeability in his final equation, and so, other things being equal, the greater the permeability and thus the greater the ease of flow, the faster the flow.

Darcy's law, which summarizes these relationships, can be expressed in a simple equation (**Figure 17.15**): the volume of water flowing in a certain time (Q) is proportional to the vertical drop in the water table (h) divided by the flow distance (l). The two remaining symbols are A, the cross-

Key Figure 17.15 | Darcy's law describes the rate of groundwater flow.

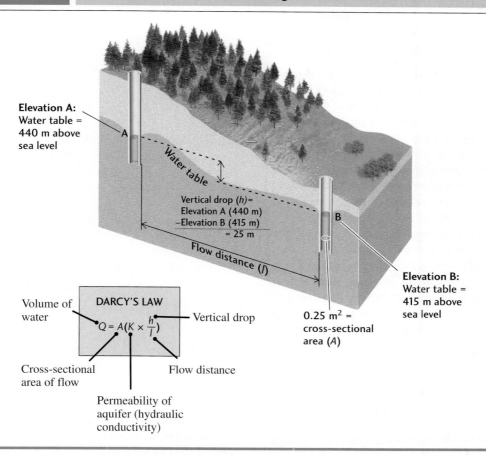

Elevation A:
Water table = 440 m above sea level

Water table

Vertical drop (*h*)=
Elevation A (440 m)
−Elevation B (415 m)
= 25 m

Flow distance (*l*)

DARCY'S LAW

Volume of water

$$Q = A\left(K \times \frac{h}{l}\right)$$

Vertical drop

Cross-sectional area of flow

Flow distance

Permeability of aquifer (hydraulic conductivity)

0.25 m² = cross-sectional area (*A*)

Elevation B:
Water table = 415 m above sea level

---

sectional area through which the volume of water flows, and *K,* which reflects permeability.

$$Q = A\left(K \times \frac{h}{l}\right)$$

Velocities calculated by Darcy's law have been confirmed experimentally by measuring how long it takes a harmless dye introduced into one well to reach another well. In most aquifers, groundwater moves at a rate of a few centimeters per day. In very permeable gravel beds near the surface, groundwater may travel as much as 15 cm/day. (This speed is still much slower than the speeds of 20 to 50 cm/s typical of river flows.)

WATER RESOURCES FROM MAJOR AQUIFERS

Large parts of North America rely on groundwater for all their water needs. The demand on groundwater resources has grown as populations have increased and uses such as irrigation have expanded (**Figure 17.16**). Many areas of the Great Plains and other parts of the Midwest rest on sandstone formations, most of which are confined aquifers like

the one shown in Figure 17.10. Thousands of wells have been drilled into these formations, most of which transport water over hundreds of kilometers and constitute a major resource. The aquifers are recharged from outcrops in the western high plains, some very close to the foothills of the Rocky

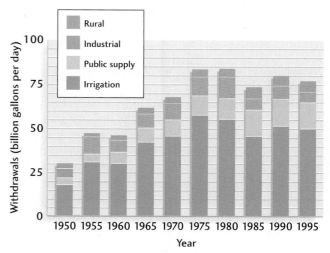

Figure 17.16 Groundwater withdrawals, United States, 1950–1995. [U.S. Geological Survey.]

EARTH ISSUES

17.2 When Do Groundwaters Become Nonrenewable Resources?

For more than 100 years, water from the Ogallala aquifer, a formation of sand and gravel, has supplied the cities, towns, ranches, and farms of much of the southwestern United States (see map). The population of the region has climbed from a few thousand late in the nineteenth century to about a million today. The Ogallala continues to provide the irrigation water needed to support agriculture, which is the area's economic base, but the water pressure in the wells has declined steadily and the water table has dropped by 30 m or more.

Natural recharge of the Ogallala aquifer of the southern plains is very slow because rainfall is sparse, the degree of evaporation is high, and the recharge area is small. Pumping, primarily for irrigation, has been so extensive—about 6 billion cubic meters of water per year from 170,000 wells—that recharge cannot keep up. At current rates of recharge, if all pumping were to stop, it would take several thousand years for the water table to recover its original position and well pressure to be restored. Some scientists have attempted to recharge the aquifer artificially by injecting water from shallow lakes that form in wet seasons on the high plains. These experiments have managed to increase recharge, but the aquifer is still in danger over the long term.

It is estimated that the remaining supplies of water in the Ogallala will last only into the early years of this century. As this valuable underground reservoir is drained, about 5.1 million acres of irrigated land in western Texas and eastern New Mexico will dry up—and so will 12 percent of the country's supply of cotton, corn, sorghum, and wheat and a significant fraction of the feedlots for the nation's cattle.

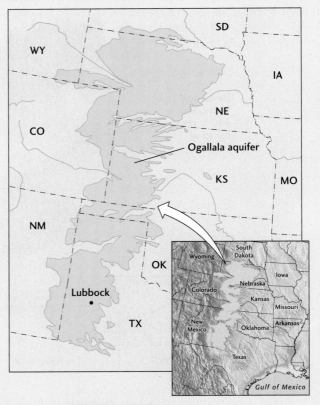

Much of the Southwest is underlain by the Ogallala aquifer. The blue region represents the aquifer. The general recharge area is located along the western margin of the aquifer. [U.S. Geological Survey.]

Other aquifers in the northern plains and elsewhere in North America are in a similar condition. In three major areas of the United States—Arizona, the high plains, and California—groundwater supplies are significantly depleted. As water use increases, we all must eventually adopt sensible conservation practices.

Mountains. From there, the water runs downhill in an easterly direction.

Darcy's law tells us that water flows at rates proportional to the slope of an aquifer between its recharge area and a given well. In the western plains, the slopes are gentle and water moves through the aquifers slowly, recharging them at low rates. At first, many of these wells were artesian and the water flowed freely. As more wells were drilled, the water levels dropped, and the water had to be pumped to the surface. Extensive pumping withdraws water from some aquifers faster than the slow recharge from far away can fill them, so the reservoirs are being depleted (see Feature 17.2).

Efforts to reduce excessive discharge have been supplemented by attempts to increase the recharge of aquifers arti-

ficially in some areas. On Long Island, New York, for example, the water authority drilled a large system of recharge wells to put water into the aquifer from the surface. These wells pumped used water, which had been treated to purify it, back into the ground.

The water authority also constructed large, shallow basins over natural recharge areas to augment infiltration from surface waters by catching and diverting runoff, including storm and industrial waste drainage. Officials in charge of the program knew that urban development can decrease recharge by interfering with infiltration. As urbanization progresses, the impermeable materials used to pave large areas for streets, sidewalks, and parking lots prevent water from infiltrating the ground. Rainwater runoff increases, and the decrease in natural infiltration into the ground may deprive the aquifers of much of their recharge. One remedy is to catch and use the storm runoff in a systematic program of artificial recharge, as the Long Island water authority did. The multiple efforts of the water authority helped rebuild the Long Island aquifer, though not to its original levels.

EROSION BY GROUNDWATER

Every year, thousands of people visit caves, either on tours of popular attractions such as Mammoth Cave, Kentucky, or in adventurous explorations of little-known caves. These underground open spaces are actually enormous vugs produced by the dissolution of limestone—or, rarely, of other soluble rocks such as evaporites—by groundwater. Huge amounts of limestone have dissolved to make some caves. Mammoth Cave, for example, has tens of kilometers of large and small interconnected chambers, and the large room at Carlsbad Caverns, New Mexico, is more than 1200 m long, 200 m wide, and 100 m high. Limestone formations are widespread in the upper parts of the crust, but caves form only where these relatively soluble rocks are at or near the surface and enough carbon dioxide– or sulfur dioxide–rich water infiltrates the surface to dissolve extensive areas of limestone.

As noted in Chapter 16, the atmospheric carbon dioxide contained in rainwater enhances the dissolution of limestone. Water that infiltrates soil may pick up even more carbon dioxide from plant roots, microbes, and other soil-dwelling organisms that give off this gas. As this carbon dioxide–rich water moves down to the water table, through the unsaturated zone to the saturated zone, it creates openings as it dissolves carbonate minerals. These openings enlarge as limestone dissolves along joints and fractures, forming a network of rooms and passages. Such networks form extensively in the saturated zone, where—because the caves are filled with water—dissolution takes place over all surfaces, including floors, walls, and ceilings.

We can explore caves that were once below the water table but are now in the unsaturated zone because the water table fell. In these caves, now air-filled, water saturated with calcium carbonate may seep through the ceiling. As each drop of water drips from the cave's ceiling, some of its dissolved carbon dioxide evaporates, escaping to the cave's atmosphere. Evaporation makes the calcium carbonate in the groundwater solution less soluble, and each water droplet precipitates a small amount of calcium carbonate on the ceiling. These deposits accumulate, just as an icicle grows, in a long, narrow spike of carbonate called a *stalactite* suspended from the ceiling. When some of the water falls to the cave floor, more carbon dioxide escapes and another small amount of calcium carbonate is precipitated on the cave floor below the stalactite. These deposits also accumulate, forming a *stalagmite*. Eventually, a stalactite and a stalagmite may grow together to form a column (**Figure 17.17**).

Microbial extremophiles (see Chapter 11) have been discovered in caves. Some geologists think that the Carlsbad Caverns were formed partly when these extremophiles used the sulfur in gypsum ($CaSO_4$) evaporites as an energy source and created sulfuric acid as a by-product. The sulfuric acid then helped to dissolve the limestone to form the caves.

In some places, dissolution may thin the roof of a limestone cave so much that it collapses suddenly, producing a **sinkhole**—a small, steep depression in the land surface above a cavernous limestone formation (**Figure 17.18**). Sinkholes are characteristic of a distinctive type of topography known as *karst*, named for a region in the northern part of the former Yugoslavia. **Karst topography** is an irregular hilly terrain

Figure 17.17 Chinese Theater, Carlsbad Caverns, New Mexico. Stalactites from the ceiling and stalagmites from the floor have joined to form a column. [David Muench.]

Figure 17.18 A large sinkhole formed by the collapse of a shallow underground cavern. Such collapses can occur so suddenly that moving cars are buried. Winter Park, Florida. [Leif Skoogfors/Woodfin Camp.]

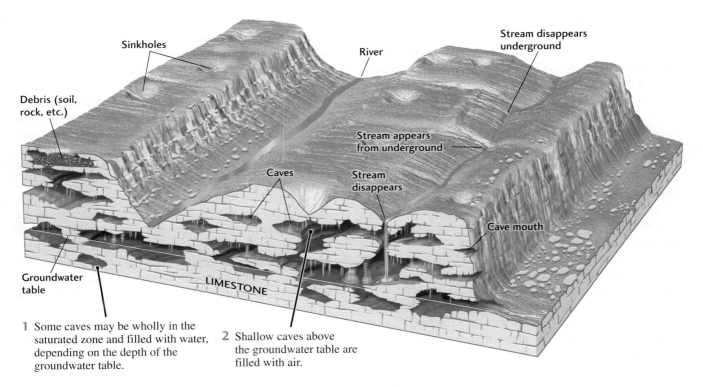

1 Some caves may be wholly in the saturated zone and filled with water, depending on the depth of the groundwater table.

2 Shallow caves above the groundwater table are filled with air.

Figure 17.19 Some major features of karst topography are caves, sinkholes, and disappearing streams.

characterized by sinkholes, caverns, and a lack of surface streams (**Figure 17.19**). Underground drainage channels replace the normal surface drainage system of small and large rivers. Short, scarce streams often end in sinkholes, detouring underground and sometimes reappearing miles away. Karst topography is found in regions with three characteristics:

1. A high-rainfall climate, with abundant vegetation (providing carbon dioxide–rich waters)

2. Extensively jointed limestone formations

3. Appreciable hydraulic gradients

Karst terrains often have environmental problems, including potentially catastrophic cave-ins and surface subsidence from the collapse of underground space. In North and Central America, karst topography is found in limestone terrains of Indiana, Kentucky, and Florida and on the Yucatán Peninsula of Mexico. Karst is well developed on uplifted coral limestone of late Cenozoic terrains of tropical volcanic island arcs.

WATER QUALITY

Most residents of the United States are beginning to see their supply of fresh, pure water as a limited resource. Many people now travel with their own supply of bottled water, supplied by either home-installed purification systems or commercially available spring water. Fortunately, almost all public water supplies in North America are free

of bacterial contamination, and the vast majority are free enough of chemicals to drink safely. Earlier, we discussed the problem of seawater incursion into the water supplies of some shoreline communities, which results in unacceptable levels of sodium in the water. A more common problem is the pollution of rivers and aquifers by toxic wastes from surface dumps.

Contamination of the Water Supply

The quality of groundwater is often threatened by a variety of contaminants. Generally, these are chemicals, though microorganisms can have bad effects under certain conditions.

Lead Pollution Lead is a well-known pollutant produced by industrial processes that inject contaminants into the atmosphere. When water vapor condenses in the atmosphere, lead is incorporated into raindrops, which then transport it to Earth's surface. Lead is routinely eliminated from public water supplies by chemical treatment before the water is distributed through the water mains. In older homes with lead pipes, however, lead can leach into the water. Even in newer construction, the lead solder used to connect copper pipes and the metals used in faucets are sources of contamination. Replacing old lead pipes with durable plastic pipes can reduce lead contamination. Even letting the water run for a few minutes to clear pipes can help.

Radioactive Wastes There is no easy solution to the problem of contamination from radioactive waste. When radioactive waste is buried underground, it may be leached by groundwaters and find its way into water-supply aquifers. Storage tanks and burial sites at the atomic weaponry plants in Oak Ridge, Tennessee, and Hanford, Washington, have already leaked radioactive wastes into shallow groundwaters.

Microorganisms in Groundwater As we learned earlier (contrary to all expectations), microbes can and do live in large numbers at great depths (as much as several thousand meters) in groundwaters, constituting a huge biomass. These microbes, well out of the reach of sunlight, draw their energy from reactions involving chemicals and minerals in rocks (see Chapter 11). These reactions, aside from serving as a source of energy for the bacteria, continue the weathering process underground. The chemicals released during these reactions can cause deterioration of water quality.

However, the most widespread causes of groundwater contamination by microbes are leaky residential septic tanks and cesspools. These containers, widely used in neighborhoods that lack full sewer networks, are settling tanks buried at shallow depths in which bacteria decompose the solid wastes from house sewage. To prevent contamination of potable water, cesspools should be replaced by septic tanks, and these must be installed at sufficient distance from water wells in shallow aquifers.

Other Chemical Contaminants Human activities also contaminate groundwaters (**Figure 17.20**). The disposal of chlorinated solvents—such as trichloroethylene (TCE), widely used as a cleaner in industrial processes—poses a

formidable problem. These solvents persist in the environment because they are difficult to remove from contaminated waters. Buried gasoline storage tanks can leak, and road salt inevitably drains into the soil and ultimately into aquifers. Rain can wash agricultural pesticides, herbicides, and fertilizers into the soil. From the soil, they percolate downward into aquifers. In some agricultural areas where nitrate fertilizers are heavily used, groundwaters may contain high quantities of nitrate. In one recent study, 21 percent of the shallow wells sampled for drinking water exceeded the maximum amounts of nitrate (10 ppm) allowed in the United States. Such high nitrate levels pose a danger of "blue baby" syndrome (the inability to maintain healthy oxygen levels) to infants 6 months old and younger.

Reversing Contamination Can we reverse the contamination of water supplies? Yes, but the process is costly and very slow. The faster an aquifer recharges, the easier it is to clean. If the recharge is rapid, fresh water moves into the aquifer once we close off the sources of contamination, and in a short time the water quality is restored. Even a fast recovery, however, can take a few years.

The contamination of slowly recharging reservoirs is more difficult to reverse. The rate of groundwater movement may be so slow that contamination from a distant source may take a long time to appear. By the time it does, it is too late for rapid recovery. Even with cleaned-up recharge, some contaminated deep reservoirs hundreds of kilometers from the recharge area may not respond for many decades.

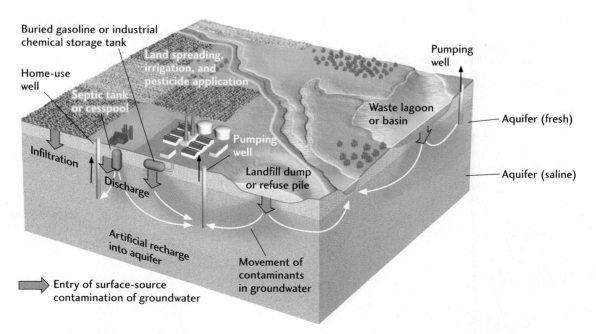

Figure 17.20 Human activities can contaminate groundwater. Contaminants from surface sources such as dumps and subsurface sources such as septic tanks enter aquifers through normal groundwater flow. Contaminants may be introduced into water supplies through pumping wells. Waste-disposal wells are designed to pump contaminants into deep saline aquifers, but they may accidentally leak into freshwater aquifers above. [Modified from U.S. Environmental Protection Agency.]

When public water supplies are polluted, we can pump the water and then treat it chemically to make it safe, but this is an expensive procedure. Alternatively, we can try to treat the water while it remains underground. In one moderately successful experimental procedure, contaminated water was funneled into a buried bunker full of iron filings that detoxified the water by reacting with contaminants. The reactions of the iron filings and contaminants produced new, nontoxic compounds that attached themselves to the iron filings.

Is the Water Drinkable?

Water that tastes agreeable and is not dangerous to health is called **potable water.** The amounts of dissolved substances in potable waters are very small, usually measured by weight in parts per million (ppm). Potable groundwaters of good quality typically contain about 150 ppm total dissolved materials, because even the purest natural waters contain some dissolved substances derived from weathering. Only distilled water contains less than 1 ppm dissolved substances.

The many cases of groundwater contamination have led to the establishment of water-quality standards based on medical studies. These studies have concentrated on the effects of ingesting average amounts of water containing various quantities of contaminant elements and compounds. For example, the Environmental Protection Agency has set the maximum allowable concentration of arsenic, a well-known poison, at 0.05 ppm.

Groundwater is almost always free of solid particles when it seeps into a well from a sand or sandstone aquifer. The tortuous passageways of the rock or sand act as a fine filter, removing small particles of clay and other solids and even straining out microbes and some large viruses. Limestone aquifers may have larger pores and so may filter less efficiently. Any microbial contamination found at the bottom of a well is usually introduced from the surface by the pump materials or from nearby underground sewage disposal, often when septic tanks leak or are located too close to the well.

Some groundwaters, although perfectly safe to drink, simply taste bad. Some have a disagreeable taste of "iron" or are slightly sour. Groundwaters passing through limestone dissolve carbonate minerals and carry away calcium, magnesium, and bicarbonate ions, making the water "hard." Hard water may taste fine but does not lather readily when used with soap. Water passing through waterlogged forests

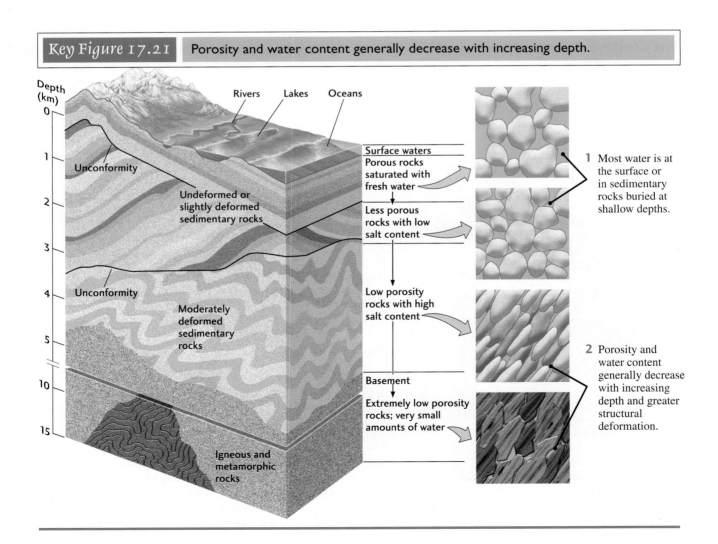

Key Figure 17.21 — Porosity and water content generally decrease with increasing depth.

Depth (km)

Rivers   Lakes   Oceans

Unconformity

Undeformed or slightly deformed sedimentary rocks

Unconformity

Moderately deformed sedimentary rocks

Igneous and metamorphic rocks

Surface waters
Porous rocks saturated with fresh water

Less porous rocks with low salt content

Low porosity rocks with high salt content

Basement

Extremely low porosity rocks; very small amounts of water

1 Most water is at the surface or in sedimentary rocks buried at shallow depths.

2 Porosity and water content generally decrease with increasing depth and greater structural deformation.

or swampy soils may contain dissolved organic compounds and hydrogen sulfide.

How do these differences in taste and quality arise in safe drinking waters? Some of the highest-quality, best-tasting public water supplies come from lakes and artificial surface reservoirs, many of which are simply collecting places for rainwater. Some groundwaters taste just as good, and these tend to be waters that pass through rocks that weather only slightly. Sandstones made up largely of quartz, for example, contribute little in dissolved substances, and thus waters passing through them taste fresh.

As we have seen, the contamination of groundwaters in relatively shallow aquifers is a problem, and recovery is difficult. But are there deeper groundwaters that we can use?

WATER DEEP IN THE CRUST

All rocks below the groundwater table are saturated with water. Even in the deepest wells drilled for oil, some 8 or 9 km deep, we always find water in permeable formations. At these depths, waters move so slowly—probably less than a centimeter per year—that they have plenty of time to dissolve even very insoluble minerals from the rocks through which they pass. Thus, dissolved materials become more concentrated in these waters than in near-surface waters, making them unpotable. For example, groundwaters that pass through salt beds, which dissolve quickly, tend to contain large concentrations of sodium chloride.

At depths greater than 12 to 15 km, deep into the basement igneous and metamorphic rocks that underlie the sedimentary formations of the upper crust, porosities and permeabilities are very low. The only pore spaces in these basement

rocks are distributed along small cracks and the boundaries between crystals. Although they are saturated, they contain very little water because their porosity is so low (**Figure 17.21**). In some deeper regions of the crust, such as along subduction zones, hot waters containing dissolved carbon dioxide play an important role in the chemical reactions of metamorphism. These waters help to dissolve some minerals and precipitate others (see Chapter 6). Even some mantle rocks are presumed to contain water, although in very minute quantities.

Hydrothermal Waters

Natural hot springs are found in Yellowstone National Park; in Hot Springs, Arkansas; in Banff Sulfur Springs, Alberta; in Reykjavík, Iceland; in New Zealand; and in many other places. Hot springs exist where **hydrothermal waters**—hot waters deep in the crust—migrate rapidly upward without losing much heat and emerge at the surface, sometimes at boiling temperatures.

Hydrothermal waters are loaded with chemical substances dissolved from rocks at high temperatures. As long as the water remains hot, the dissolved material stays in solution. As hydrothermal waters reach the surface and cool quickly, however, they may precipitate various minerals, such as opal (a form of silica) and calcite or aragonite (forms of calcium carbonate). Crusts of calcium carbonate produced at some hot springs build up to form the rock travertine, which can form impressive deposits such as those seen at Mammoth Hot Spring in Yellowstone National Park (**Figure 17.22**). Amazingly, microbial extremophiles that can withstand temperatures above the boiling point of water have been discovered in these environments, where they

Figure 17.22 Travertine deposits at Mammoth Hot Springs, Yellowstone National Park, form large lobelike masses made of aragonite and calcite. [John Grotzinger.]

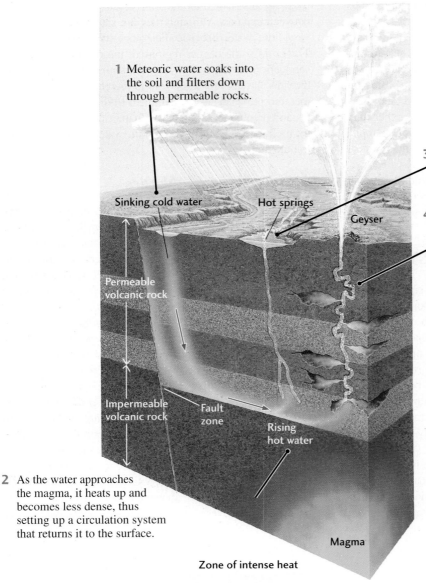

Figure 17.23 Circulation of water over a magma body produces geysers or hot springs.

1 Meteoric water soaks into the soil and filters down through permeable rocks.

3 Hot springs occur where heated groundwater is discharged at the surface.

4 The water in a geyser follows an irregular network of pores and cracks, which slows down and complicates the flow of water. Steam and boiling water are released to the surface under pressure, resulting in intermittent eruptions.

Sinking cold water

Hot springs

Geyser

Permeable volcanic rock

Impermeable volcanic rock

Fault zone

Rising hot water

2 As the water approaches the magma, it heats up and becomes less dense, thus setting up a circulation system that returns it to the surface.

Magma

Zone of intense heat

may contribute to the formation of calcium carbonate crusts. While still below the surface, hydrothermal waters also deposit some of the world's richest metallic ores as they cool, as we learned in Chapter 3.

Most hydrothermal waters of the continents come from surface waters that percolate downward to deeper regions of the crust (**Figure 17.23**). These surface waters originate primarily as **meteoric waters**—rain, snow, or other forms of water derived from the atmosphere (from the Greek *meteoron,* "phenomenon in the sky," which also gives us the word *meteorology*). Meteoric waters may be very old. It has been determined that the water at Hot Springs, Arkansas, derives from rain and snow that fell more than 4000 years ago and slowly infiltrated the ground. The waters in the Ogallala aquifer may be even older and were supplied about 10,000 years ago during the last episode of continen-

tal glaciation when the climate was wetter over the Great Plains.

Water that escapes from magma can also contribute to hydrothermal waters. In areas of igneous activity, sinking meteoric waters are heated as they encounter hot masses of rock. The hot meteoric waters then mix with water released from the nearby magma. Hydrothermal waters return to the surface as hot springs or geysers. Hot springs flow steadily; geysers erupt hot water and steam intermittently.

The theory explaining the intermittent eruptions of geysers is an example of geological deduction. We cannot observe the process directly because the dynamics of the underground hot-water system are hidden from sight hundreds of meters below the surface. Geysers are probably connected to the surface by a system of very irregular and crooked fractures, recesses, and openings—in contrast to the more regular and direct plumbing of hot springs. The irregular geyser plumbing sequesters some water in recesses, thus helping to prevent the bottom waters from mixing with shallower waters and cooling. The bottom waters are heated by contact with hot rock. When they reach the boiling point, steam starts to ascend and heats the shallower waters, increasing the pressure and triggering an eruption. After the pressure is released, the geyser then becomes quiet as the fractures slowly refill with water.

In 1997, geologists reported the results of a novel technique to learn about geysers. They lowered a miniature video camera to about 7 m below the surface of a geyser. They found that the geyser shaft was constricted at that point. Farther down, the shaft widened to a large chamber containing a wildly boiling mixture of steam, water, and what appeared to be carbon dioxide bubbles. These direct observations dramatically confirmed the previous theory of how geysers work.

Geologists have turned to hydrothermal waters in the search for new and clean sources of energy. Northern California, Iceland, Italy, and New Zealand have already harnessed the steam produced by hydrothermal activity in hot springs and geysers to drive electricity-generating turbines. Hydrothermal waters may soon be put to wider use for producing power, as we will see in Chapter 23.

Although hydrothermal waters are important for power generation and ore deposits, these waters do not contribute to surface water supplies, primarily because they contain so much dissolved material.

Ancient Microbes in Deep Aquifers

In recent years, exploration for potable groundwater in deep underground sources has unveiled another remarkable discovery of how life and Earth interact. In some cases, aquifers may not have been in contact with the surface of the Earth for hundreds of millions of years, yet they contain active colonies of microorganisms. Geobiologists think that the ancestors of these microbes were enclosed within the pores of sediments, which were then buried to great depth, where they became sealed off from the surface. There they existed for hundreds of millions of years, living solely on chemicals provided by the dissolution of minerals and evolving new generations of descendents without interference from any other organisms. These ecosystems, involving only microbes, are probably the most ancient on Earth and testify to the remarkable balance that can be achieved between life and environment.

SUMMARY

How does water move on and in the Earth in the hydrologic cycle? Water movements maintain a constant balance among the major reservoirs of water at or near Earth's surface. These reservoirs include oceans, lakes, and rivers; glaciers and polar ice; and groundwater. Evaporation from the oceans, evaporation and transpiration from the continents, and sublimation from glaciers transfer water to the atmosphere. Precipitation as rain and snow returns water from the atmosphere to the oceans and the land surface. Runoff in rivers returns part of the precipitation that falls on land to the ocean. The remainder infiltrates the ground to become groundwater. Differences in climate produce local variations in the balance among evaporation, precipitation, runoff, and infiltration.

How does water move below the ground? Groundwater forms as precipitation infiltrates the surface of the ground and travels through pore spaces in the soil, sediment, or rock that serves as an aquifer. Water moves through the upper unsaturated zone through the water table into the lower saturated zone. Groundwater moves downhill under the influence of gravity, eventually emerging at springs, where the water table intersects the ground surface. Over the long term, recharge and discharge of a groundwater aquifer are in dynamic balance. Groundwater may flow in unconfined aquifers, which are continuous to the surface, or in confined aquifers, which are bounded by aquicludes. Confined aquifers produce artesian flows and spontaneously flowing artesian wells. Darcy's law describes the groundwater flow rate in relation to the slope of the water table and the permeability of the aquifer.

What factors govern our use of groundwater resources? As the population grows, the demands on groundwaters increase greatly, particularly where irrigation is widespread. Many aquifers, such as those of the western plains of North America, have such slow recharge rates that continued pumping in the past century has reduced the pressure in artesian wells. As pumping discharge continues to exceed recharge, such aquifers are being depleted, and there is no prospect of renewal for many years. Artificial recharge may help to renew some aquifers, but conservation will be required to preserve others. The contamination of groundwater by sewage, industrial effluents, and radioactive wastes reduces the potability of some waters and limits our resources.

What geological processes are affected by groundwater? Erosion by groundwater in humid limestone terrains produces karst topography, with caves, sinkholes, and disappearing streams. The heating of downward-percolating meteoric waters by magma bodies leads to a circulation that brings hydrothermal waters to the surface as geysers and hot springs. At great depths in the crust, more than 12 to 15 km, rocks contain extremely small quantities of water because their porosities are significantly reduced. This extreme reduction in porosity results from the tremendous weight of the overlying rocks.

KEY TERMS AND CONCEPTS

aquiclude (p. 411)
aquifer (p. 408)
artesian flow (p. 411)
Darcy's law (pp. 414–415)
discharge (p. 411)
groundwater (p. 402)
groundwater table
   (p. 410)
hydraulic gradient
   (p. 414)
hydrologic cycle
   (p. 402)
hydrology (p. 401)
hydrothermal water
   (p. 421)

infiltration (p. 402)
karst topography (p. 417)
meteoric water (p. 422)
permeability (p. 409)
potable water (p. 420)
precipitation (p. 402)
rain shadow (p. 405)
recharge (p. 410)
relative humidity (p. 404)
reservoir (p. 402)
runoff (p. 403)
saturated zone (p. 409)
sinkhole (p. 417)
unsaturated zone (p. 409)

EXERCISES

1. What are the main reservoirs of water at or near Earth's surface?

2. How do mountains form rain shadows?

3. What is an aquifer?

4. What is the difference between the saturated and unsaturated zones of groundwater?

5. How do aquicludes make a confined aquifer?

6. How are recharge and discharge balanced to make a groundwater table stable?

7. How does Darcy's law relate groundwater movement to permeability?

8. How does dissolution of limestone relate to karst topography?

9. What are the sources of water in hot springs?

10. What are some common contaminants in groundwater?

11. How do microbes live in the subsurface?

THOUGHT QUESTIONS

1. If the Earth warmed, causing evaporation from the oceans to increase greatly, how would the hydrologic cycle of today be altered?

2. If you lived near the seashore and started to notice that your well water had a slightly salty taste, how would you explain the change in water quality?

3. Why would you recommend against extensive development and urbanization of the recharge area of an aquifer that serves your community?

4. If it were discovered that radioactive waste had seeped into groundwater from a nuclear processing plant, what kind of information would you need to predict how long it would take for the radioactivity to appear in well water 10 km from the plant?

5. What geological processes would you infer are taking place below the surface at Yellowstone National Park, which has many hot springs and geysers?

6. Why should communities ensure that septic tanks are maintained in good condition?

7. Why are more and more communities in cold climates restricting the use of salt to melt snow and ice on highways?

8. Your new house is built on soil-covered granite bedrock. Although you think that prospects for drilling a successful water well are poor because of the granite, the well driller familiar with the area says that he has drilled many good water wells in this granite. What arguments might each of you offer to convince the other?

9. How might the hydrologic cycle have differed during the maximum glaciation of 19,000 years ago, when North America, Europe, and Asia had significant ice cover?

10. You are exploring a cave and notice a small stream flowing on the cave floor. Where could the water be coming from?

SHORT-TERM PROJECT

Urban Groundwater Use

In the U.S. Senate and House of Representatives, recent college graduates work as legislative assistants. Their main job is to brief a senator or a representative about pressing issues on which the official must cast an informed vote.

You and a partner are legislative assistants to a U.S. senator from Texas. The senator has been contacted by constituents in San Antonio, which is located near the southern end of the Ogallala aquifer, one of the largest in the United States. San Antonio is the largest city in the United States to obtain 100% of its drinking water from a groundwater source.

But things are starting to change. Heavy withdrawal because of ever-increasing urbanization has resulted in drawdown of the aquifer, and salt water from the Gulf of Mexico is beginning to encroach on the pumping region. In addition, scientists from a nearby university have discovered a new species of fish that lives within caves populated by extremophile microbes. This may be the first ecosystem of its type ever discovered, where higher organisms (not just microbes) have been living out of contact with other organisms—including humans—for over 10 million years. The citizens are concerned about the future of their water supply, and the scientists are concerned about the significance of their new discovery. The San Antonio town planning board needs advice on how to plan for the future, when federal assistance may be required.

The senator wants to know:

1. How fast is the aquifer being drawn down, and how fast is the salty Gulf-fed groundwater migrating northward?

2. What might be done to curtail the saltwater intrusion?

3. What other sources of water, besides groundwater, are available? Does a dam need to be built to capture runoff water? If so, what new issues might arise?

4. How unique is this ecosystem? Does it need to be preserved? How certain can we be that there aren't others like it that just haven't been discovered yet?

The senator wants these questions answered in a concise two- to three-page memo. To prepare the senator to deal with the issue, your team should investigate the nature and quality of work undertaken by the San Antonio water board and the nearby university scientists and understand the geologic and ecologic nature of the problem.

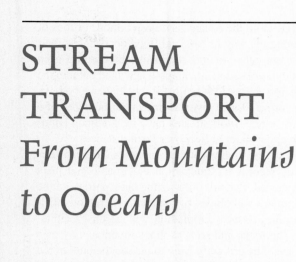

STREAM TRANSPORT
From Mountains to Oceans

Before cars and airplanes, people traveled on streams. In 1803, the United States purchased the Louisiana Territory from France. It was a huge tract of over 2 million square kilometers, taking in all of what today is Texas and Louisiana up to Montana and North Dakota. In 1804, U.S. President Thomas Jefferson asked Meriwether Lewis and William Clark to lead an expedition across this new territory and into western North America. One of their most important goals was to map the western rivers, for these provided the key to opening up this uncharted frontier. Lewis and Clark decided to explore the Missouri River and its headwaters to their source and westward to the Pacific Ocean. The total trip was 6000 km, with the section along the Missouri River alone extending over 3200 km—and upstream all the way.

The writings and maps produced by Lewis and Clark created a body of knowledge that could have been obtained only by following one of the great rivers that drain the interior of North America. On other continents and in other countries, other big rivers evoke a similar sense of excitement in exploration: in South America, the Amazon; in Asia, the Yangtze and Indus; and in Africa, the fabled Nile.

Streams and rivers not only are the access routes for legendary exploration adventures but also are major geological agents of change operating on Earth's land surface. As they erode bedrock and transport and deposit sand, gravel, and mud, streams of all sizes—from tiny rills to major rivers—are preeminent carvers of the landscape. They erode mountains, carry the products of

Aerial view of the meandering Adelaide River, in Australia. This appearance is typical for meandering streams in plains environments. [Peter Bowater/Photo Researchers.]

weathering down to the oceans, and drop billions of tons of sediment along the way in bars and flood deposits. At their mouths, on the edges of the continents, they dump even greater quantities of sediment, building new land out into the oceans. We use the word **stream** for any flowing body of water, large or small, and **river** for the major branches of a large stream system. Most streams have well-defined **channels,** which allow water to flow over great distances.

Streams are the lifelines of the continents. Their appearance directly records the interaction of climatic and plate tectonic processes. Tectonics lifts up the land, producing the steep topography and slopes of mountainous regions. Climate determines where precipitation will occur. Precipitation erodes the rocks and soils of those mountains by producing valleys. Streams carry back to the sea the bulk of the rainwater that falls on land and much of the sediment produced by erosion of the land surface. Streams are so important to understanding the role of climate and water on Earth that their discovery on Mars has fueled a generation of missions to search for evidence of water—and a different climate in its very ancient past.

A body of water flows through almost every town and city in most parts of the world. These streams have served as commercial waterways for barges and steamers and as water resources for resident populations and industries. The river Nile, for example, was vital to the agricultural economy of ancient Egypt and remains important to Egypt today. Living near a river also entails risk. When rivers flood, they destroy lives and property, sometimes on a huge scale.

Worldwide, streams carry about 16 billion tons of siliciclastic sediment and an additional 2 to 4 billion tons of dissolved matter each year. Humans are responsible for much of the present stream load. According to some estimates, prehuman sediment transport was about 9 billion tons per year, less than half the present value. In some places, we increase the sediment load of streams through agriculture and accelerated erosion. In other places, we decrease the sediment load by constructing dams, which trap sediment behind their retaining walls.

In this chapter, we focus on how streams form as a result of interactions between the plate tectonic and climate systems and how they accomplish their geological work: how, on a large scale, streams carve valleys and develop vast networks of channels; and how, at a smaller scale, streams break up and erode solid rock; how water flows in currents; and how currents carry sediment.

STREAM VALLEYS, CHANNELS, AND FLOODPLAINS

As streams move across Earth's surface—in some places bedrock, in others unconsolidated sediment—they erode these materials and create valleys. Identifying and mapping stream valleys were essential to Lewis and Clark during their mission 200 years ago. As they traveled upstream and the river branched, they had to choose which branch was the larger of the two. They used several observations about streams to help them make this choice. Two observations were the width of the stream valley and the depth of the stream channel. Was the valley wide enough, and the channel deep enough, for their boats? Narrow valleys and shallow channels would mean that the branch led into a much shorter and therefore less desirable route; wider valleys and deep channels, on the other hand, promised a longer passage up the main branch of the river.

A stream **valley** encompasses the entire area between the tops of the slopes on both sides of the river. The cross-sectional profile of many river valleys is V-shaped, but many other valleys have a broad, low profile like that shown in **Figure 18.1.** At the bottom of the valley is the *channel,* the trough through which the water runs. The channel carries all the water during normal, nonflood times. At low water level, the stream may run only along the bottom of the channel. At high water levels, the stream occupies most of the channel. In broad valleys, a **floodplain**—a flat area about level with the top of the channel—lies on either side of the channel. It is the part of the valley that is flooded when the river spills over its banks, carrying with it silt and sand from the main channel.

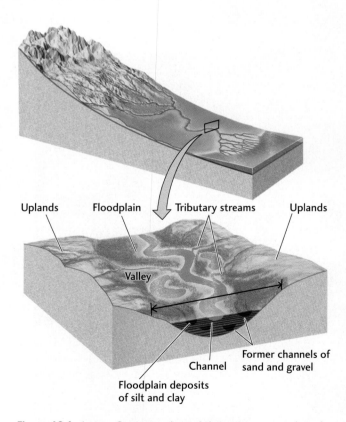

Uplands Floodplain Tributary streams Uplands

Valley

Former channels of
Channel sand and gravel

Floodplain deposits
of silt and clay

Figure 18.1 A river flows in a channel that moves over a broad, flat floodplain in a wide valley eroded from uplands. Floodplains may be narrow or absent in steep valleys.

Stream Valleys

In high mountains, stream valleys are narrow and have steep walls, and the channel may occupy most or all of the valley bottom. A small floodplain may be visible only at low water levels. In such valleys, the stream is actively cutting into the bedrock, a characteristic of tectonically active, newly uplifted highlands. In lowlands, where tectonic uplift has long since ceased, stream erosion of valley walls is helped by chemical weathering and mass wasting. With a long time to operate, these processes produce gentle slopes and floodplains many kilometers wide.

Channel Patterns

As a stream channel makes its way along the bottom of a valley, it may run straight in some stretches and take a snaking, irregular path in others, sometimes splitting into multiple channels. The channel may flow along the center of the floodplain or hug one edge of the valley. In addition to straight stretches, the two other types of channel patterns are meandering and braided.

Meanders On a great many floodplains, channels follow curves and bends called **meanders,** so named for the Maiandros (now Menderes) River in Turkey, known in ancient times for its winding, twisting course. Meanders are usual in streams flowing on low slopes in plains or lowlands, where channels typically cut through unconsolidated sediments— fine sand, silt, or mud—or easily eroded bedrock. Meanders are less pronounced but still common where the channel flows on higher slopes and harder bedrock. In such terrain, meandering stretches may alternate with long, relatively straight ones.

A stream that has cut deeply into the curves and bends of its channel may produce incised meanders (**Figure 18.2**). Other streams may meander on somewhat wider floodplains bounded by steep, rocky valley walls. We are not sure why these two different patterns appear. We do know that meandering is widespread not only in streams but also in a great many other kinds of flows. For example, the Gulf Stream, a powerful current in the western North Atlantic Ocean, meanders. Lava flows on Earth meander, and planetary geologists have found meanders in dry water channels (see Chapter 9, Figures 9.22 and 9.23) and lava flows on Mars and in lava flows on Venus.

Meanders on a floodplain migrate over periods of many years, eroding the outside banks of bends, where the current is strongest (**Figure 18.3**). Meanders shift position from side to side, as well as downstream, in a snaking motion something like that of a long rope being snapped (see Figure 18.3a). As the outside banks are eroded, curved sandbars called **point bars** are deposited along the inside banks, where the current is slower. Migration may be rapid: some meanders on the Mississippi shift as much as 20 m/year. As

Incised meanders   Point bar

Figure 18.2 This section of the San Juan River, Utah, is a good example of an incised meander belt, a deeply eroded, meandering, V-shaped valley with virtually no floodplain. [Tom Bean.]

meanders move, so do the point bars, building up an accumulation of sand and silt over the part of the floodplain across which the channel migrated.

As meanders migrate, sometimes unevenly, the bends may grow closer and closer together, until finally the river bypasses the next loop, often during a major flood. The river takes a new, shorter course like that shown in Figure 18.3a. In its abandoned path, it leaves behind an **oxbow lake**—a crescent-shaped, water-filled loop.

Engineers sometimes artificially straighten and confine a meandering river, channeling it along a straight path with the aid of concrete abutments. The Army Corps of Engineers has been channeling the Mississippi River since 1878. In a period of 13 years, it decreased the length of the lower Mississippi by 243 km. Part of the severity of the Mississippi River flood of 1993 has been ascribed to channelization and the height of the artificial riverbanks designed by flood-control engineers. Without channelization, floods are more frequent but less damaging. With it, damage may be catastrophic when a flood breaches the artificially high banks, as it did in 1993. Channelization has also been criticized for destroying wetlands and much of the natural vegetation and animal life of the floodplain. Such environmental concerns stimulated action to restore one channelized river, the Kissimmee in central Florida, to its original meandering course. Today, restoration projects are well under way. If left to its own natural processes, the Kissimmee

Key Figure 18.3 | Channel patterns depend on flow velocity and sediment load.

(a)

LOW-SEDIMENT LOAD, LOW VELOCITY

1 Low-velocity, low-sediment streams, flowing on nearly flat floodplains, form meanders.

2 Meanders shift from side to side in a snaking motion.

3 The current is faster at outside banks, which are eroded,…

4 …and sediments get deposited in inside banks where the current is slower, forming point bars.

Point bars

5 As the erosion and deposition process continues, the bends grow closer and the point bars bigger.

6 During a major flood, when velocity and water volume increase, the river takes a new, shorter course, cutting across the loop.

7 The abandoned loop remains as an oxbow lake.

Meanders in an Alaskan river

Point bar

High-velocity flow in channel

[Peter Kresan.]

Meanders in the Mississippi River delta

[Nathan Benn/Corbis.]

(b) HIGH-SEDIMENT LOAD, HIGH VELOCITY

Low-water period (e.g., summer)

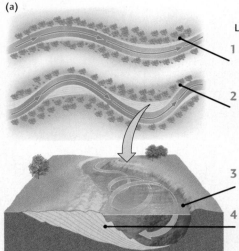

High-water period (e.g., spring snowmelt)

1 Where high-velocity, high-sediment streams flow over nearly flat, easily eroded terrain (e.g., at the mouths of canyons or the terminal ends of melting glaciers),…

2 …the fast-moving, sediment-laden water does not form oxbow bends…

3 …but cuts across the soft sediments at the edges of existing channels, creating shallow, crisscrossed braided channels.

A braided stream of the Chitina River, Alaska

Braided channels

[Tom Bean.]

might have taken many decades or hundreds of years to restore itself.

Braided Streams Some streams have many channels instead of a single one. A **braided stream** is one whose channel divides into an interlacing network of channels, which then rejoin in a pattern resembling braids of hair (see Figure 18.3b). Braided streams are found in many settings, from broad lowland valleys to wide, downfaulted, sediment-filled valleys adjacent to mountain ranges. Braids tend to form in rivers with large variations in volume of flow combined with a high sediment load and easily erodible banks. They are well developed, for example, in sediment-choked streams formed at the edges of melting glaciers.

The Stream Floodplain

A stream channel migrating over the floor of a valley creates a floodplain. Point bars formed during migration build up the surface of the floodplain, as does sediment deposited by floodwaters when the stream overflows its banks. Erosional floodplains, covered with a thin layer of sediment, can form when a stream erodes bedrock or unconsolidated sediment as it migrates.

As floodwaters spread out over the floodplain, the velocity of the water slows and the current loses its ability to carry sediment. The floodwater velocity drops most quickly along the immediate borders of the channel. As a result, the current deposits much coarse sediment, typically sand and gravel, along a narrow strip at the edge of the channel. Successive floods build up **natural levees,** ridges of coarse material that confine the stream within its banks between floods, even when water levels are high (**Figure 18.4**). Where natural levees have reached a height of several meters and the stream almost fills the channel, the floodplain level is below the stream level. You can walk the streets of an old river town built on a floodplain, such as Vicksburg, Mississippi, and look up at the levee, knowing that the river waters are rushing by above your head.

During floods, fine sediments—silts and muds—are carried well beyond the channel banks, often over the entire floodplain, and are deposited there as floodwaters continue to lose velocity. Receding floodwaters leave behind standing ponds and pools of water. The finest clays are deposited there as the standing water slowly disappears by evaporation and infiltration. Fine-grained floodplain deposits have been a major resource for agriculture since ancient times. The fertility of the floodplains of the Nile and other rivers of the Middle East, which contributed to the evolution of the cultures that flourished there thousands of years ago, depended on frequent flooding. Today the great, broad floodplain of the Ganga (Ganges) in northern India continues to play an important role in India's life and agriculture. Many ancient and modern cities are sited on floodplains (see Feature 18.1).

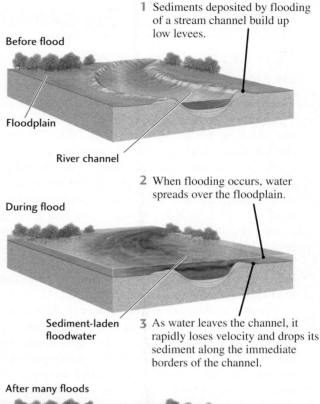

1 Sediments deposited by flooding of a stream channel build up low levees.

Before flood

Floodplain

River channel

2 When flooding occurs, water spreads over the floodplain.

During flood

Sediment-laden floodwater

3 As water leaves the channel, it rapidly loses velocity and drops its sediment along the immediate borders of the channel.

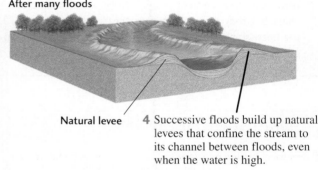

After many floods

Natural levee

4 Successive floods build up natural levees that confine the stream to its channel between floods, even when the water is high.

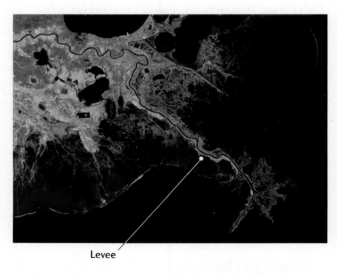

Levee

Figure 18.4 The formation of natural levees by river floods. Natural levees along the main channel of the Mississippi River near South Pass, Louisiana. [USGS National Wetlands Research Center.]

EARTH ISSUES

18.1 The Development of Cities on Floodplains

Floodplains are natural sites for urban settlements because they combine easy transportation along a river with access to fertile agricultural lands. Such sites, however, are subject to the floods that formed the plains.

About 4000 years ago, cities began to dot floodplains in Egypt along the Nile, in the ancient land of Mesopotamia along the Tigris and Euphrates rivers, and in Asia along the Indus River of India and the Yangtze and Huang Ho of China. Later, many of the capital cities of Europe were built on floodplains: London on the Thames, Paris on the Seine, Rome on the Tiber. Floodplain cities in North America include St. Louis on the Mississippi, Cincinnati on the Ohio, and Montreal on the St. Lawrence.

Floods periodically destroyed sections of these ancient and modern cities on the lower parts of the floodplains, but each time the inhabitants rebuilt them. Today, most large cities are protected by artificial levees that strengthen and heighten the river's natural levees. Extensive systems of dams can help control flooding that would affect these cities, but they cannot eliminate the risk entirely. In 1973, the Mississippi went on a rampage with a flood that continued for 77 consecutive days at St. Louis. It reached a record 4.03 m above flood stage (the height at which the river first overflows the channel banks). In 1993, the Mississippi and its tributaries broke loose again, shattering the old record in a flood that has been officially designated the second worst flood in U.S. history (behind the New Orleans flooding of 2005 by hurricane Katrina; see Chapter 20). The flood resulted in 487 deaths and $15 billion to $20 billion in property damage. At St. Louis, the Mississippi stayed above flood stage for 144 of the 183 days between April and September. An unexpected result of this flood was widespread pollution as floodwaters leached agricultural chemicals from farmlands and then deposited them in flooded areas. Some geologists believe that the construction of levees and dikes to confine the Mississippi contributes to the record high floods. The river can no longer erode its banks and widen its channel to accommodate some of the additional water flowing during times of high discharge.

What are cities and towns in this position to do? Some have urged a halt to all construction and development on the lowest parts of the floodplains. Some have called for the elimination of federally subsidized disaster funds for rebuilding in such areas. Harrisburg, Pennsylvania, hit hard by a flood in 1972, turned some of its devastated riverfront area into a park. In a dramatic step after the 1993 Mississippi flood, the town of Valmeyer, Illinois, voted to move the

DRAINAGE NETWORKS

Every topographic rise between two streams, whether it measures a few meters or a thousand, forms a **divide**—a ridge of high ground along which all rainfall runs off down one side of the rise or the other. A **drainage basin** is an area of land, bounded by divides, that funnels all its water into the network of streams draining the area (**Figure 18.5**). A drainage basin may be a small area, such as a ravine surrounding a small stream, or a great region drained by a major river and its tributaries (**Figure 18.6**).

A continent has several major drainage basins separated by major divides. In North America, the continental divide along the Rocky Mountains separates all waters flowing into the Pacific Ocean from all those entering the Atlantic. Lewis and Clark followed the Missouri River to its headwaters at the continental divide in western Montana. When they crossed over it, they descended from the headwaters of the Columbia River, which they followed to the Pacific Ocean.

Drainage Patterns

A map showing the courses of large and small streams reveals a pattern of connections called a **drainage network.** If you followed a stream from its mouth to its head, you would see that it steadily divides into smaller and smaller tributaries, forming drainage networks that show a characteristic branching pattern (**Figure 18.7**). Branching is a general property of many kinds of networks in which material is collected and distributed. The network of the human circulatory

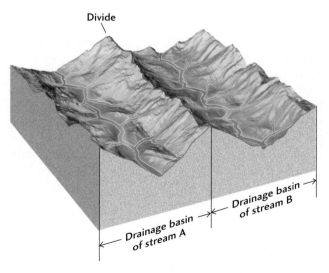

Figure 18.5 Stream valleys and drainage basins are separated by divides, which may be ridges, gentle uplands, or mountain ranges.

entire town to high ground several miles away. The new site was chosen with the help of a team of geologists from the Illinois Geological Survey. Yet some people who have lived on floodplains all their lives want to stay and are prepared to live with the risk. The costs of protecting some river-bottom areas are prohibitive, and these places will continue to pose public policy problems.

Like many cities built on river floodplains, Liuzhou, China, is subject to flooding. This flood, in July 1996, was the largest recorded in the city's 500-year history. [Xie Jiahua/China Features/Corbis Sygma.]

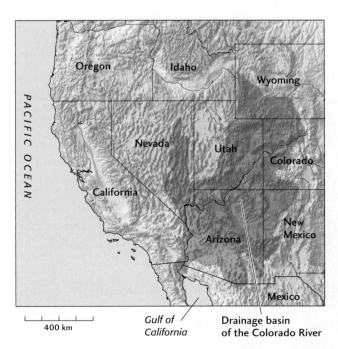

Figure 18.6 The natural drainage basin of the Colorado River covers about 630,000 km², a large part of the southwestern United States. The basin is surrounded by divides that separate it from the neighboring drainage basins. [After U.S. Geological Survey.]

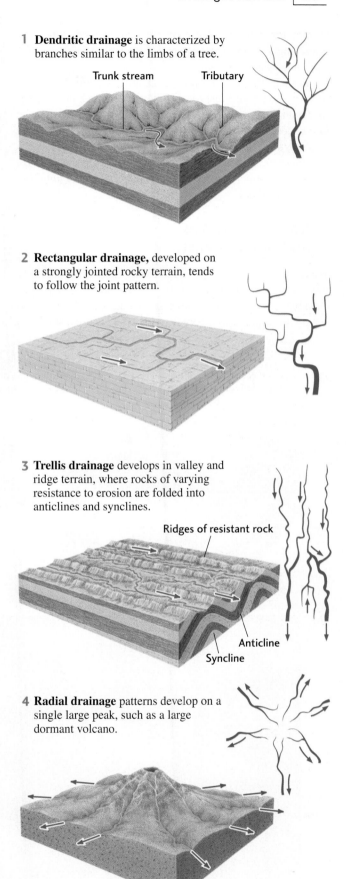

1 Dendritic drainage is characterized by branches similar to the limbs of a tree.

2 Rectangular drainage, developed on a strongly jointed rocky terrain, tends to follow the joint pattern.

3 Trellis drainage develops in valley and ridge terrain, where rocks of varying resistance to erosion are folded into anticlines and synclines.

4 Radial drainage patterns develop on a single large peak, such as a large dormant volcano.

Figure 18.7 Typical drainage networks.

system, for example, distributes blood to the body through a branching system of arteries and collects it through a corresponding system of veins.

Perhaps the most familiar branching networks are those of trees and roots. Most rivers follow the same kind of irregular branching pattern, called **dendritic drainage** (from the Greek *dendron,* for "tree"). This fairly random drainage pattern is typical of terrains where the bedrock is uniform, such as horizontally bedded sedimentary rocks or massive igneous or metamorphic rocks. Other patterns are *rectangular, trellis,* and *radial.*

Drainage Patterns and Geologic History

We can observe directly or judge from historical records how most stream drainage patterns evolved. Some streams, for example, cut through erosion-resistant bedrock ridges to form steep-walled notches or gorges. What could cause a stream to

cut a narrow valley directly through a ridge rather than running along the lowland on either side of it? The geologic history of the region provides the answers. If a ridge is formed by structural deformation while a preexisting stream is flowing over it, the stream may erode the rising ridge to form a steep-walled gorge, as in **Figure 18.8**. Such a stream is called an **antecedent stream** because it existed before the present topography was created, and it maintained its original course despite changes in the underlying rocks and in topography.

In another geological situation, a stream may flow in a dendritic drainage pattern over horizontal beds of sedimentary rocks that overlie folded and faulted rocks with varying resistance to erosion. Over time, the stream cuts into the underlying rocks and erodes a gorge in the resistant bed. Such a **superposed stream** flows through resistant formations because its course was established at a higher level, on uniform rocks, before downcutting began. A superposed stream tends to continue the pattern that it developed earlier rather than adjusting to its new conditions. In this case, a

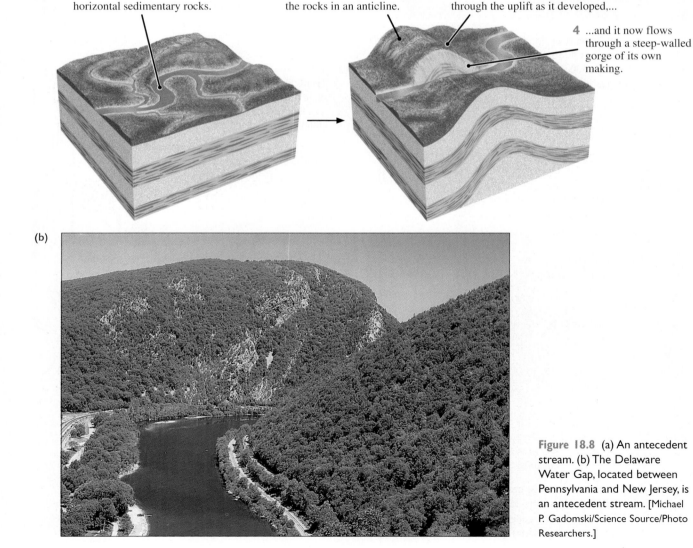

(a) **1** An antecedent stream flowed on horizontal sedimentary rocks.

2 Slow tectonic uplift folded the rocks in an anticline.

3 The stream maintained its course, cutting through the uplift as it developed,...

4 ...and it now flows through a steep-walled gorge of its own making.

(b)

Figure 18.8 (a) An antecedent stream. (b) The Delaware Water Gap, located between Pennsylvania and New Jersey, is an antecedent stream. [Michael P. Gadomski/Science Source/Photo Researchers.]

(a) **1** A superposed dendritic stream developed on horizontal beds.

(b) **2** Most horizontal beds were stripped away by erosion.　**3** A stream cut a gorge—or water gap—through resistant beds of a buried anticline.

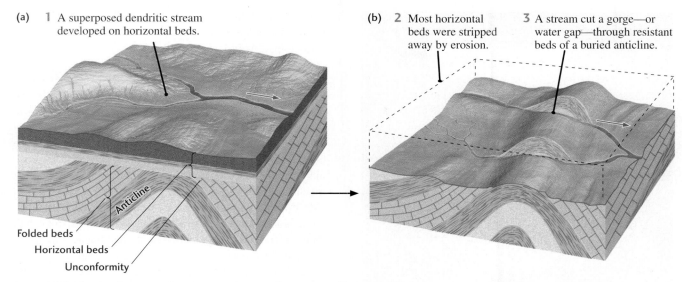

Folded beds
Horizontal beds
Unconformity
Anticline

Figure 18.9 The development of a superposed stream by erosion of horizontal beds overlying folded beds of varying resistance to erosion.

dendritic pattern is forced onto a surface that would otherwise have evolved a rectangular network (**Figure 18.9**).

WHERE DO CHANNELS BEGIN? HOW RUNNING WATER ERODES SOLID ROCK

We can easily see a rapid current picking up loose sand from its bed and carrying it away, thus eroding the bed. At high water levels and during floods, streams can even scour and cut into banks of unconsolidated sediment, which then slump into the flow and are carried away. *Gullies*—valleys made by small streams eroding soft soils or weak rocks—cut their way upstream into higher land (**Figure 18.10**). The process

by which rivers cut upstream, rather than downstream, is called *headward erosion*. Headward erosion commonly accompanies widening and deepening of valleys and may be extremely rapid—as much as several meters in a few years in easily erodible soils.

We cannot so easily see the erosion of solid rock. Running water erodes solid rock by abrasion, by chemical and physical weathering, and by the undercutting action of currents.

Abrasion

One of the major ways a river breaks apart and erodes rock is by abrasion. The sand and pebbles the river carries create a sandblasting action that wears away even the hardest rock. On some riverbeds, pebbles and cobbles rotating inside swirling eddies grind deep **potholes** into the bedrock (**Figure 18.11**). At low water, we can see pebbles and sand lying quietly at the bottom of exposed potholes.

Channels begin (see Figure 18.10) where rainwater, draining off the surface of the land, flows so fast that it abrades the soil and bedrock and carves into it. Once a small gully forms, it will capture more of the flow of water over the land surface, and thus the tendency to cut down will increase. As the gully progressively deepens, the rate of downcutting increases, because more water is captured. The ability of a river to erode its bed depends on the river's discharge and slope. Both of these attributes are discussed later

Figure 18.10 Streams create gullies when the action of water flowing across Earth's surface causes erosion of bedrock. The smallest gullies converge to form larger gullies, and farther downslope these become river channels. These gullies form in the desert of Oman when occasional rainstorms inundate the surface with rapidly flowing water that erodes the bedrock. [Petroleum Development Oman.]

Figure 18.11 Potholes in river rock along McDonald Creek, Glacier National Park, Montana. The pebbles rotate inside the potholes, grinding deep holes in the bedrock. [Carr Clifton/Minden Pictures.]

in this chapter, and in Chapter 22 we will see how they combine to shape landscapes.

Chemical and Physical Weathering

Chemical weathering, which alters a rock's minerals and weakens it along joints and cracks, helps destroy rocks in streambeds just as it does on the land surface. Physical weathering can be violent, as the crashes of boulders and the constant smaller impacts of pebbles and sand split the rock along cracks. Such impacts in a river channel break up rock much faster than slow weathering on a gently sloping hillside does. When impacts and weathering have loosened large blocks of bedrock, strong upward eddies may pull them up and out in a sudden violent plucking action.

Rock erosion is particularly strong at rapids and waterfalls. *Rapids* are places in a stream where the flow is extremely fast because the slope of the riverbed suddenly steepens, typically at rocky ledges. The speed and turbulence of the water quickly break blocks into smaller pieces that are carried away by the strong current.

Stream-channel erosion can be a major source of sediment in extensively urbanized areas. The erosion is accelerated by changing land uses, such as reservoir impoundments, dams, and recreational activities.

The Undercutting Action of Currents

The tremendous impact of huge volumes of plunging water and tumbling boulders quickly erodes rock beds below waterfalls. Waterfalls also erode the underlying rock of the cliff that forms the falls. As erosion undercuts these cliffs, the upper beds collapse and the falls recede upstream (**Figure 18.12**). Erosion by falls is fastest where the rock layers are horizontal, with erosion-resistant rocks at the top and

Figure 18.12 This waterfall on the Iguaçú River, Brazil, is retreating upstream as falling water and sediment pound the cliff's base and undercut it. From the center to the upper left, one can see steep walls, the remnants of the waterfall's retreat upstream. [Donald Nausbaum.]

softer rocks, such as shales, making up the lower layers. Historical records show that the main section of Niagara Falls, perhaps the best-known falls in North America, has been moving upstream at a rate of a meter per year.

HOW STREAM WATERS FLOW AND TRANSPORT SEDIMENT

All streams, large and small, move in accordance with some basic characteristics of flowing fluids. We can illustrate two kinds of fluid flow by using lines of motion called *streamlines* (**Figure 18.13**). In **laminar flow,** the simplest kind of movement, straight or gently curved streamlines run parallel to one another without mixing or crossing between layers. The slow movement of thick syrup over a pancake, with strands of unmixed melted butter flowing in parallel but separate paths, is a laminar flow. **Turbulent flow** has a more complex pattern of movement, in which streamlines mix, cross, and form swirls and eddies. Fast-moving river waters typically show this kind of motion. Turbulence—the degree to which there are irregularities and eddies in the flow—may be low or high. Whether a flow is laminar or turbulent depends on three factors:

1. Its velocity (rate of movement).

2. Its geometry (primarily its depth).

3. Its *viscosity,* which is a measure of a fluid's resistance to flow. The more viscous (the thicker) a fluid is, the more it resists flow. The higher the viscosity, the greater the tendency for laminar flow.

Viscosity arises from the attractive forces between the molecules of a fluid. These forces tend to impede the slipping and sliding of molecules past one another. The greater the attractive forces, the greater the resistance to mixing with neighboring molecules and the higher the viscosity. For example, when a cold syrup or a viscous cooking oil is poured, its flow is sluggish and laminar. The viscosity of most fluids, including water, decreases as the temperature increases. Given enough heat, a fluid's viscosity may decrease sufficiently to change a laminar flow into a turbulent one.

Water has low viscosity in the common range of temperatures at Earth's surface. For this reason alone, most watercourses in nature tend to turbulent flow. In addition, the rapid movement of water in most streams makes them turbulent. In nature, we are likely to see laminar flows of water only in

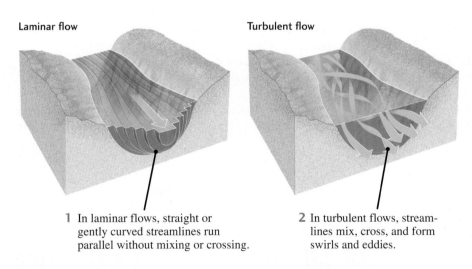

1 In laminar flows, straight or gently curved streamlines run parallel without mixing or crossing.

2 In turbulent flows, streamlines mix, cross, and form swirls and eddies.

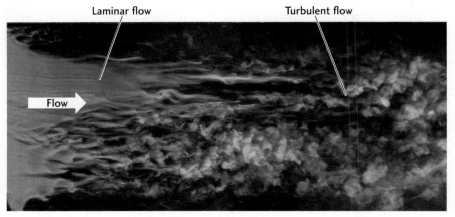

Figure 18.13 Laminar flows and turbulent flows. The photograph shows the transition from laminar to turbulent flow in water along a flat plate, revealed by the injection of a dye. Flow is from left to right. [ONERA.]

thin sheets of rain runoff flowing slowly down nearly level slopes. In cities, we may see small laminar flows in street gutters. Because most streams and rivers are broad and deep and flow quickly, their flows are almost always turbulent.

A stream may show turbulent flow over much of its width and laminar flow along its edge, where the water is shallow and is moving slowly. The flow velocity is highest near the center of a stream, and we commonly refer to a rapid flow as a strong current.

Erosion and Sediment Transport

Streams vary in their ability to erode and carry sand grains and other sediment. Laminar flows of water can lift and carry only the smallest, lightest clay-sized particles. Turbulent flows, depending on their speed, can move particles ranging from clay size to pebbles and cobbles. As turbulence lifts particles from the streambed into the flow, the flow carries them downstream. Turbulence also rolls and slides larger particles along the bottom. A stream's **suspended load** includes all the material temporarily or permanently suspended in the flow. Its **bed load** is the material the stream carries along the bed by sliding and rolling (**Figure 18.14**).

The faster the current, the larger the particles carried as suspended and bed load. A flow's ability to carry material of a given size is its **competence.** As a current increases in velocity and coarser particles are suspended, the suspended load grows. At the same time, more of the bed material is in motion, and the bed load also increases. As we would expect, the larger the volume of a flow, the more suspended load and bed load it can carry. The total sediment load carried by a flow is its **capacity.**

The velocity and the volume of a flow affect both the competence and the capacity of a stream. The Mississippi River, for example, flows at moderate speeds along most of its length and carries only fine to medium-sized particles (clay to sand), but it carries huge quantities of them. A small, steep, fast-flowing mountain stream, in contrast, may carry boulders, but only a few of them.

Settling of Sediment from Suspension

A stream's ability to carry sediment depends on a balance between turbulence, which lifts particles, and the competing downward pull of gravity, which makes them settle out of the current and become part of the bed. The speed with which suspended particles of various weights settle to the bottom is called the **settling velocity.** Small grains of silt and clay are easily lifted into the stream and settle slowly, so they tend to stay in suspension. The settling velocity of larger particles, such as medium- and coarse-grained sand, is much faster. Most larger grains therefore stay suspended in the current only a short time before they settle.

Sand grains in a flow typically move by **saltation**—an intermittent jumping motion along the streambed. The grains are sucked up into the flow by turbulent eddies, move with the current for a short distance, and then fall back to the bottom (see Figure 18.14). If you were to stand in a rapidly flowing sandy stream, you might see a cloud of saltating sand grains moving around your ankles. The bigger the grain, the longer it will tend to remain on the bed before it is picked up. Once a large grain is in the current, it will settle quickly. The smaller the grain, the more frequently it will be picked up, the higher it will "jump," and the longer it will take to settle.

Thus, turbulent flows transport sediment by suspension (clays), saltation (sands), and rolling and sliding along the

1 The hydrosphere and lithosphere interact to transport sediments in streams.

2 Current flowing over a bed of gravel, sand, silt, and clay carries a **suspended load** of finer particles…

4 As current velocity increases, the suspended load grows,…

6 Particles move by saltation, jumping along the bed.

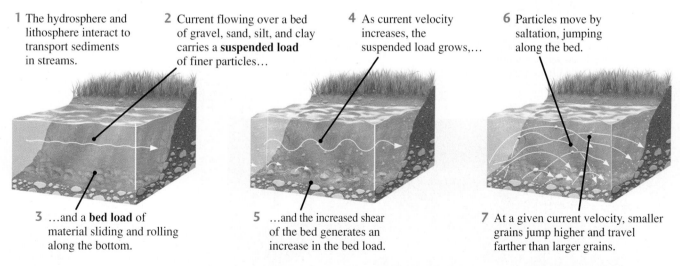

3 …and a **bed load** of material sliding and rolling along the bottom.

5 …and the increased shear of the bed generates an increase in the bed load.

7 At a given current velocity, smaller grains jump higher and travel farther than larger grains.

Figure 18.14 A current flowing over a bed of sand, silt, and clay transports particles in two ways: as bed load, the material sliding and rolling along the bottom; and as suspended load, the material temporarily or permanently suspended in the flow itself. Saltation is an intermittent jumping motion of grains. In general, the smaller the particle, the higher it jumps and the farther it travels.

Key Figure 18.15	Current velocity varies with particle size.

1 When stream velocity is high, large and small particles erode from the bed (blue area).

2 As the velocity decreases, particles may erode or settle as sediment (gray area).

3 At still lower stream velocity, particles of all sizes settle onto the streambed (brown area).

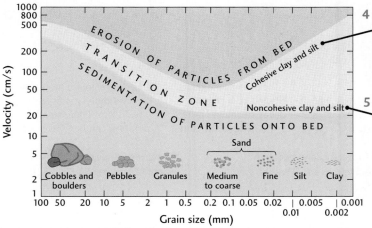

4 Cohesive particles such as clays require a high water velocity to erode. Smaller particles are more cohesive than larger particles.

5 Fine, clay-sized particles such as quartz and some other minerals are not charged and have a low surface-to-volume ratio. They erode from the streambed at low velocities.

[After F. Hjulstrom, as modified by A. Sundborg. "The River Klariven." *Geografisk Annaler*, 1956.]

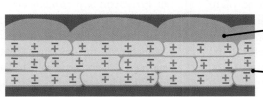

6 Clay particles are platy and have a high surface-to-volume ratio. Water tension acts on their large surface area to cause cohesion.

7 Charges on the clay attract opposite charges on other particles, holding them in place.

bed (sand and gravel). To study how a particular river carries sediment, geologists and hydraulic engineers measure the relationship between particle size and the force the flow exerts on the particles in the suspended and bed loads. Engineers use these data to calculate how much sediment a particular flow can move and how rapidly it can move it. This information allows them to design dams and bridges or to estimate how quickly artificial reservoirs behind dams will fill with sediment. Geologists can infer the velocities of ancient currents from the sizes of grains in sedimentary rocks.

Figure 18.15 graphs the relationship between grain size and flow velocity. This relationship, in which higher current velocities are required to transport finer grains, exists because it is easier for the flow to lift noncohesive particles (particles that do not stick together) from the bed than it is to lift cohesive particles (particles that stick together, as many clay minerals do). The finer the cohesive particles, the greater the velocity of the flow required to erode them. For these small grains, settling velocities are so slow that even a gentle current, about 20 cm/s, can keep the particles in suspension and transport sediment.

Sediment Bed Forms: Dunes and Ripples

When sand grains on a streambed are transported by saltation, they tend to form cross-bedded dunes and ripples (see

Chapter 5). **Dunes** are elongated ridges of sand that can be many meters high in large rivers. **Ripples** are very small dunes—with heights ranging from less than a centimeter to several centimeters—whose long dimension is formed at right angles to the current. Although underwater ripples and dunes are harder to observe than those produced on land by air currents, they form in the same way and are just as common. As a current moves sand grains by saltation, they are eroded from the upstream side of ripples and dunes and deposited on the downstream side. The steady downstream transfer of grains across the ridges causes the ripple and dune forms to migrate downstream. The speed of this migration is much slower than the movement of individual grains and very much slower than the current. (We will look at ripple and dune migration in more detail in Chapter 19.)

The shapes of ripples and dunes and their migration speeds change as the velocity of the current increases. At the lowest current velocities, few grains are saltating and the stream's sand bed is flat. At slightly higher velocities, the number of saltating grains increases. A rippled bed forms, and ripples migrate downstream (**Figure 18.16**). As the velocity increases further, the ripples grow larger and migrate faster until, at a certain point, dunes replace the ripples. Both ripples and dunes have a cross-bedded structure, and as the current flows over their tops it can actually reverse and flow backward along their lee (downstream) side. As the dunes grow larger, small ripples form. These ripples tend to climb

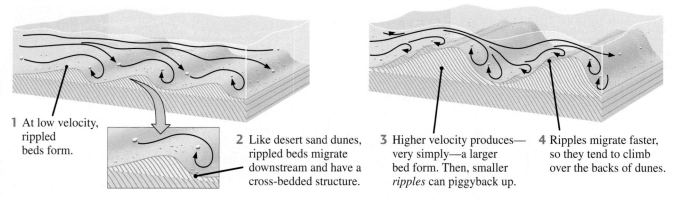

1 At low velocity, rippled beds form.

2 Like desert sand dunes, rippled beds migrate downstream and have a cross-bedded structure.

3 Higher velocity produces— very simply—a larger bed form. Then, smaller *ripples* can piggyback up.

4 Ripples migrate faster, so they tend to climb over the backs of dunes.

Figure 18.16 The change in the form of a sand bed with increasing flow velocity. [After D. A. Simmons and E. V. Richardson, "Forms of Bed Roughness in Alluvial Channels." *American Society of Civil Engineers Proceedings* 87: 87–105 (1961).]

over the backs of the dunes because they migrate more quickly than dunes. Very high current velocities will erase the dunes and form a flat bed below a dense cloud of rapidly saltating sand grains. Most of these grains hardly settle to the bottom before they are picked up again. Some are in permanent suspension.

DELTAS: THE MOUTHS OF RIVERS

Sooner or later, all rivers end as they flow into a lake or an ocean, mix with the surrounding water, and—no longer able to travel downhill—gradually lose their forward momentum. The largest rivers, such as the Amazon and the Mississippi, can maintain some current many kilometers out to sea. Where smaller rivers enter a turbulent, wave-swept coast, the current disappears almost immediately beyond the river's mouth.

Delta Sedimentation

As its current gradually dies out, a river progressively loses its power to transport sediment. The coarsest material, typically sand, is dropped first, right at the mouth of most rivers. Finer sands are dropped farther out, followed by silt and, still farther out, by clay. As the floor of the lake or sea slopes to deeper water away from the shore, all the dropped materials of various sizes build up a depositional platform called a **delta.** (We owe the name *delta* to the Greek historian Herodotus, who traveled through Egypt about 450 B.C. The roughly triangular shape of the sediments deposited at the mouth of the Nile prompted him to name it after the Greek letter Δ, delta.)

As a river approaches its delta, where the slope profile is almost level with the sea, it reverses its upstream-branching drainage pattern. Instead of collecting more water from tributaries, it discharges water into **distributaries**—smaller streams that receive water and sediment from the main chan-

nel, branch off *downstream,* and thus distribute the water and sediment into many channels. Materials deposited on top of the delta, typically sand, make up horizontal **topset beds.** Downcurrent, on the outer front of the delta, fine-grained sand and silt are deposited to form gently inclined **foreset beds,** which resemble large-scale cross-beds. Spread out on the seafloor seaward of the foreset beds are thin, horizontal **bottomset beds** of mud, which are eventually buried as the delta continues to grow. **Figure 18.17** shows a typical large marine delta.

The Growth of Deltas

As a delta builds out into the sea, the mouth of its river advances seaward, leaving new land in its wake. Much of this land is a delta plain just a few meters above sea level. Such plains include large areas of wetlands—which, as noted in Chapter 17, are valuable because they store water and form the habitat of many diverse species of plants and animals. In many areas, delta wetlands have suffered a two-pronged attack. First, the extensive flood-control dams built since the 1930s have decreased the volume of sediment brought to the edge of the delta, thereby reducing the sediment supply to the wetlands. Second, massive artificial levees have prevented the small but frequent floods that nourish the delta wetlands.

A growing delta shifts the flows from some distributaries to others with shorter routes to the sea. As a result of such shifts, the delta grows in one direction for some hundreds or thousands of years and then breaks out into a new distributary and begins to grow into the sea in another direction. A major river, such as the Mississippi or the Nile, forms a large delta thousands of square kilometers in area. The delta of the Mississippi, like many other major river deltas, has been growing for millions of years. About 150 million years ago, it started out around what is now the junction of the Ohio and the Mississippi rivers, at the southern tip of Illinois. It has advanced about 1600 km since then,

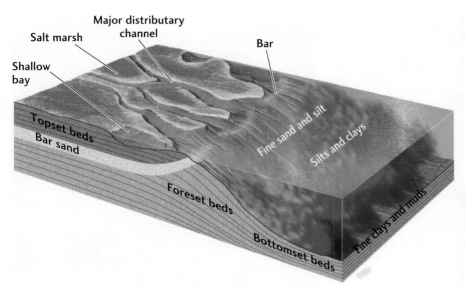

Major distributary channel

Salt marsh

Shallow bay

Bar

Topset beds

Bar sand

Fine sand and silt

Silts and clays

Foreset beds

Bottomset beds

Fine clays and muds

Some currents follow the slope and hug the bottom. As they decelerate, they drop their sediment load at the bottom.

Some currents flow out in shallow water, then decelerate. Suspended sediment particles settle out of shallow water.

Figure 18.17 A typical large marine delta, many kilometers in extent, in which the fine-grained foreset beds are deposited at a very low angle, typically only 4° to 5° or less. Sandbars form at the mouths of the distributaries, where the currents' velocity suddenly decreases. The delta builds forward by the advance of the bar and the topset, foreset, and bottomset beds. Between distributary channels, shallow bays fill with fine-grained sediment and become salt marshes. This general structure is found on the Mississippi delta.

creating almost the entire states of Louisiana and Mississippi as well as major parts of adjacent states. **Figure 18.18** shows the growth of the Mississippi delta over the past 6000 years.

Deltas grow by the addition of sediment, and they sink as the sediment becomes compacted and the crust subsides under the weight of the sediment load. Venice, built on part of the Po River delta in northern Italy, has been subsiding steadily. Both crustal subsidence and depression of the ground attributable to the pumping of water from aquifers beneath the city are responsible.

The Effects of Waves, Tides, and Tectonics

Strong waves, shoreline currents, and tides affect the growth and shape of marine deltas. Waves and shoreline currents may move sediment along the shore almost as rapidly as it is dropped there by a river. The delta front then becomes a long beach with only a slight seaward bulge at the mouth. Where tidal currents move in and out, they redistribute deltaic sediment into elongate bars parallel to the direction of the currents, which in most places is at approximately right angles to the shore.

In some places, waves and tides are strong enough to prevent a delta from forming. Instead, the sediment a river brings down to the sea is dispersed along the shoreline as beaches and bars and is transported into deeper waters off-

shore. The east coast of North America lacks deltas for this reason. The Mississippi has been able to build its delta because neither waves nor tides are very strong in the Gulf of Mexico.

Tectonics also exerts some control over where deltas form, because deltas require

• Uplift in the drainage basin, which provides abundant sediment

• Crustal subsidence in the delta region to accommodate the great weight and volume of sediment

Two of the world's large deltas—the Mississippi and the Rhône (in France)—derive their large sediment loads primarily from distant mountain ranges: the Rockies for the Mississippi and the Alps for the Rhône. Both are in the same type of plate tectonic setting—a passive margin originally formed from a rifted continental margin.

Few large deltas are associated with active subduction zones. The reason may be that it is unusual for a large river, such as the Columbia of Washington State and Oregon, to carry abundant sediment through a volcanic arc (the Cascade Range) to the sea. In addition, oceanic island arcs are too small in land area to provide much clastic sediment.

The continental convergence that elevated the Himalaya also formed the great deltas of the Indus and Ganga. Ultimately, plate tectonic settings influence delta location and formation.

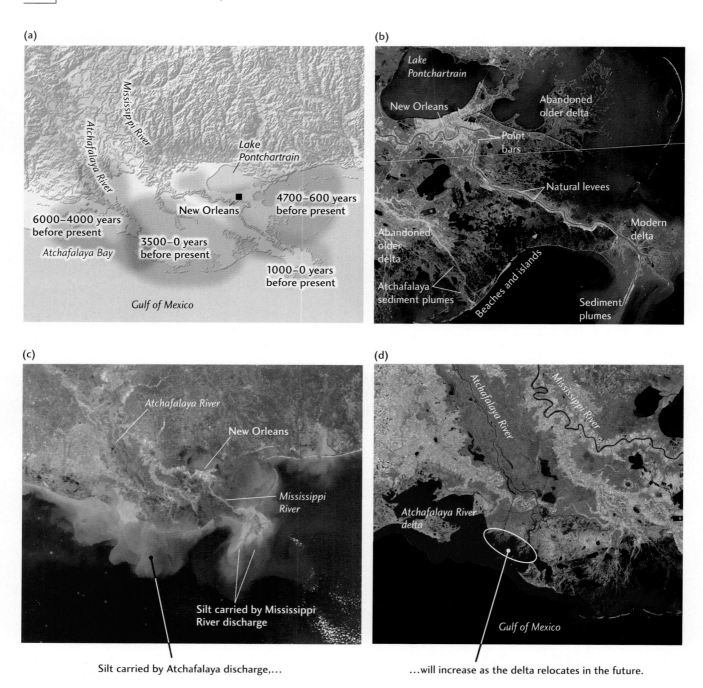

(a)

Mississippi River

Atchafalaya River

Lake Pontchartrain

New Orleans

4700–600 years before present

6000–4000 years before present

3500–0 years before present

Atchafalaya Bay

1000–0 years before present

Gulf of Mexico

(b)

Lake Pontchartrain

New Orleans

Abandoned older delta

Point bars

Natural levees

Abandoned older delta

Modern delta

Atchafalaya sediment plumes

Beaches and islands

Sediment plumes

(c)

Atchafalaya River

New Orleans

Mississippi River

Silt carried by Mississippi River discharge

Silt carried by Atchafalaya discharge,...

(d)

Atchafalaya River

Mississippi River

Atchafalaya River delta

Gulf of Mexico

...will increase as the delta relocates in the future.

Figure 18.18 (a) The Mississippi delta. Over the past 6000 years, the river has built its delta first in one direction and then in another as water flow shifted from one major distributary to another. The modern delta was preceded by deltas deposited to the east and west. (b) The infrared-sensitive film used to record this satellite image of the Mississippi delta causes the vegetation to appear red, relatively clear water to appear dark blue, and water with suspended sediment to appear light blue. At the upper left are New Orleans and Lake Pontchartrain. Well-defined natural levees and point bars are at the center. At the lower left are beaches and islands that formed as waves and currents transported river-deposited sand from the delta. [From G. T. Moore, "Mississippi River Delta from Landsat2." *Bulletin of the American Association of Petroleum Geologists* (1979).] (c) Satellite photograph of the Mississippi delta. [NASA.] (d) This image shows the discharge of sediment to the Gulf of Mexico from the Mississippi River delta and the Atchafalaya River. A major flood could divert the main flow of the Mississippi into the Atchafalaya, enabling a new delta to form. Construction of artificial levees by the Army Corps of Engineers has prevented this so far. [USGS National Wetlands Research Center.]

STREAMS AS GEOSYSTEMS

The flow of a stream appears steady when you view it from a bridge for a few minutes or canoe along it for a few hours, but its volume and velocity at a single place may change appreciably from month to month and season to season. Streams are dynamic geosystems, moving from low waters to floods over a few hours or days and reshaping their valleys over longer periods (**Figure 18.19**). The flow and channel dimensions of a stream also change as it moves downstream, from narrow valleys in the upland headwaters to broader floodplains in the middle and lower courses. Most of these longer-term changes are adjustments in the normal (nonflood) volume and velocity of flow as well as the depth and width of the channel.

From their headwaters near drainage divides to their termination points in *deltas* in lakes and oceans, all streams react to changes in climate (such as rainfall) and tectonics (uplift or subsidence of Earth's crust). Near the headwaters, streams gather to form a single large strand, as in the case of the Mississippi River. Precipitation far up in the headwaters may be felt far downriver, where river volume may exceed the volume of the channel and then spill over its levees to create a flood. In this way, processes and events in one part of the system are propagated through the system to affect the behavior of a different part of the system.

Sediment behaves in a similar way, though over a longer time scale. If precipitation in the headwaters occurred over a long period of time—say, because long-term climate became rainier or tectonic uplift rates increased—erosion rates and sediment yields would increase. The stream network would propagate a "wave" of sediment until it eventually reached the delta, where it might be preserved in the rock record as an interval of unusually high sediment accumulation. We will further explore this relationship between climate and tectonics and its effect on landscape development in Chapter 22.

Several processes are important in controlling how water and sediment move through stream geosystems. These include stream discharge, longitudinal profile, and changes in base level.

Discharge

We measure the size of a stream's flow by its **discharge**—the volume of water that passes a given point in a given time as it flows through a channel of a certain width and depth. (In Chapter 17, we defined discharge as the volume of water leaving an aquifer in a given time. These definitions are consistent, because they both describe volume of flow per unit of time.) Stream discharge is usually measured in cubic meters per second or cubic feet per second. Discharge in small streams may vary from about 0.25 to 300 m³/s. The discharge of a well-studied medium-sized river in Sweden, the Klarälven, varies from 500 m³/s at low water levels to 1320 m³/s at high water levels. The discharge of the Missis-

sippi River can vary from as low as 1400 m³/s at low water levels to more than 57,000 m³/s during floods.

To calculate discharge, we multiply the cross-sectional area (the width multiplied by the depth of the part of the channel occupied by water) by the velocity of the flow (distance traveled per second):

$$\text{discharge} = \text{cross section} \times \text{velocity}$$
$$(\text{width} \times \text{depth}) \quad (\text{distance traveled per second})$$

Figure 18.20 illustrates this relationship. For discharge to increase, either velocity or cross-sectional area, or both, must increase. Think of increasing the discharge of a garden hose by opening the valve more, which increases the velocity of water coming out the end of the hose. The cross-sectional area of the hose, measured by its diameter, cannot change, so the discharge must increase. In a stream, as discharge at a particular point increases, both the velocity and the cross-sectional area tend to increase. (The velocity is also affected by the slope of the channel and the roughness of the river bottom and sides, which we can neglect for the purposes of this explanation.) The cross-sectional area increases because the flow occupies more of the channel's width and depth.

Discharge in most rivers increases downstream as more and more water flows in **tributaries,** streams that release water into larger streams. Increased discharge means that width, depth, or velocity must increase, too. Velocity does not increase downstream as much as the increase in discharge might lead us to expect, because the slope along the lower courses of a stream decreases (decreasing slope reduces velocity). Where discharge does not increase significantly downstream and slope decreases greatly, a river will flow more slowly.

Floods

In 1996, continued heavy rains swelled the Yangtze, China's longest river, to flood stage. Large parts of the countryside were inundated. More than 700 people were killed, and hundreds of thousands who fled to the nearby mountains from the lowlands of the floodplain were trapped there. Such a flood is an extreme case of increased discharge that results from a short-term imbalance between input and output. As the discharge increases, the flow velocity in the channel increases and the water gradually fills the channel. As the discharge continues to increase, the water spills over the banks. Rivers flood regularly, some at infrequent intervals, others almost every year. Some floods are large, with very high water levels lasting for days. At the other extreme are minor floods that barely break out from the channel before they recede. Small floods are more frequent, occurring every 2 or 3 years on average. Large floods are generally less frequent, usually occurring every 10, 20, or 30 years.

No one can know exactly how high—either in water height or in discharge—a flood will be in any given year, so

Earth System Figure 18.19   Stream networks transport water and sediments from the mountains to the sea.

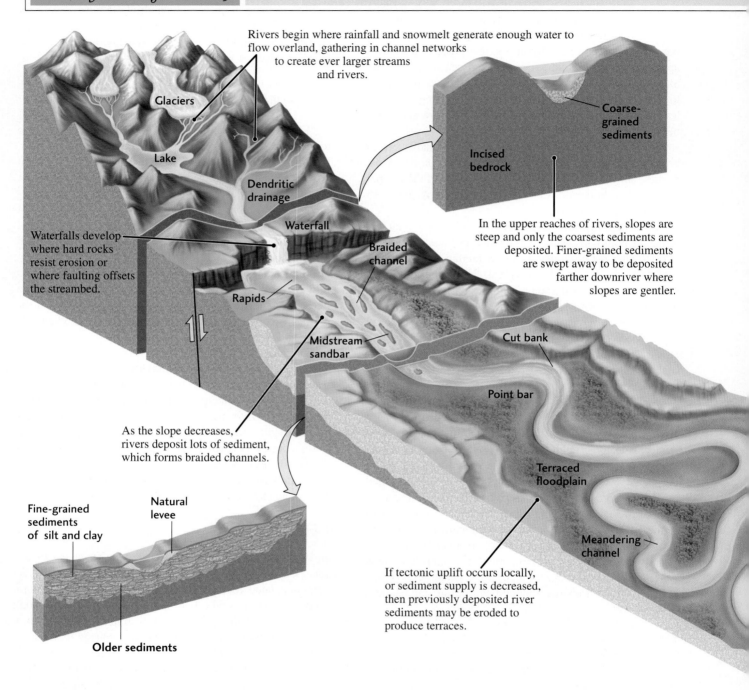

Rivers begin where rainfall and snowmelt generate enough water to flow overland, gathering in channel networks to create ever larger streams and rivers.

Glaciers

Lake

Dendritic drainage

Waterfall

Incised bedrock

Coarse-grained sediments

In the upper reaches of rivers, slopes are steep and only the coarsest sediments are deposited. Finer-grained sediments are swept away to be deposited farther downriver where slopes are gentler.

Waterfalls develop where hard rocks resist erosion or where faulting offsets the streambed.

Braided channel

Rapids

Midstream sandbar

Cut bank

Point bar

As the slope decreases, rivers deposit lots of sediment, which forms braided channels.

Terraced floodplain

Fine-grained sediments of silt and clay

Natural levee

Meandering channel

Older sediments

If tectonic uplift occurs locally, or sediment supply is decreased, then previously deposited river sediments may be eroded to produce terraces.

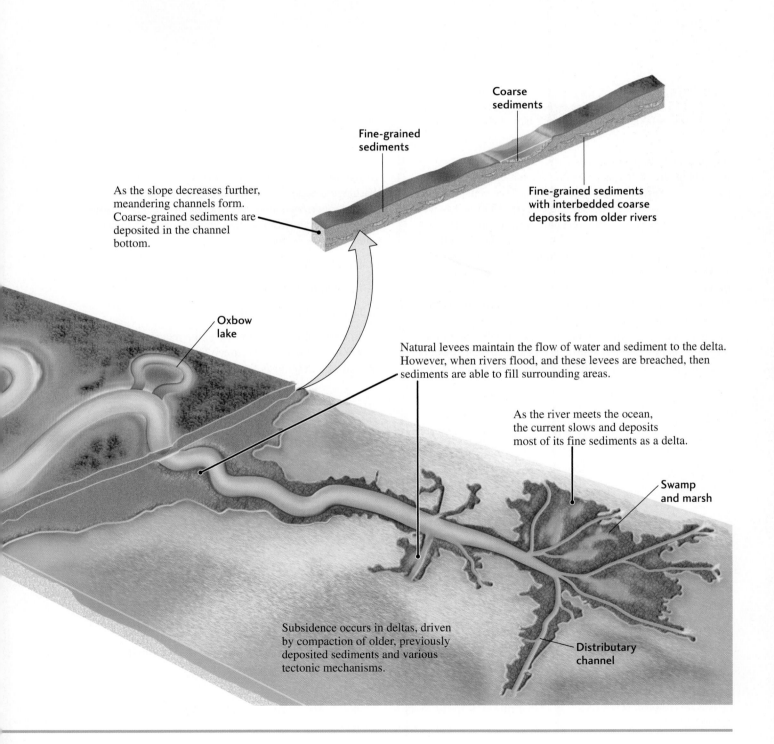

Coarse
sediments

Fine-grained
sediments

Fine-grained sediments
with interbedded coarse
deposits from older rivers

As the slope decreases further,
meandering channels form.
Coarse-grained sediments are
deposited in the channel
bottom.

Oxbow
lake

Natural levees maintain the flow of water and sediment to the delta.
However, when rivers flood, and these levees are breached, then
sediments are able to fill surrounding areas.

As the river meets the ocean,
the current slows and deposits
most of its fine sediments as a delta.

Swamp
and marsh

Subsidence occurs in deltas, driven
by compaction of older, previously
deposited sediments and various
tectonic mechanisms.

Distributary
channel

(a)

1 A river with a lower cross-sectional area and a lower velocity has a lower discharge ($3 \text{ m} \times 10 \text{ m} = 30 \text{ m}^2 \times 1 \text{ m/s} = 30 \text{ m}^3/\text{s}$ discharge)…

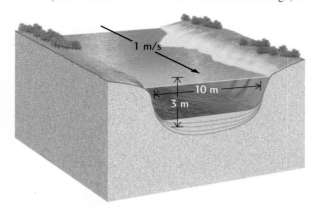

Figure 18.20 Discharge depends on velocity and cross-sectional area. A river at (a) low discharge and (b) high discharge. [Data

(b)

2 …than a river with a higher cross-sectional area and a higher velocity which has a higher discharge ($9 \text{ m} \times 10 \text{ m} = 90 \text{ m}^2 \times 2 \text{ m/s} = 180 \text{ m}^3/\text{s}$).

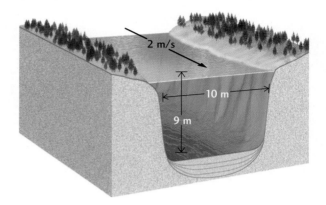

from T. Dunne and L. B. Leopold, *Water in Environmental Planning.* San Francisco: W. H. Freeman, 1978.]

predictions are stated as *probabilities,* not certainties. For a particular stream, we might say there is a 20 percent probability that a flood of a given discharge—say, 1500 m^3/s—will occur in any one year. This probability corresponds to an average time interval—in this case, 5 years (1 in 5 = 20 percent) —that we expect between two floods with a discharge of 1500 m^3/s. We speak of a flood of this discharge as a 5-year flood. The average time interval between the occurrence of two geological events of a given magnitude is called a **recurrence interval.** A flood of greater magnitude—say, 2600 m^3/s —on the same stream is likely to happen only once every 50 years, and it would therefore be called a 50-year flood. A graph of the annual probabilities and recurrence intervals for

a range of flood discharges in one river—the Skykomish, in Washington State—is shown in **Figure 18.21.**

The recurrence interval of floods in streams of different discharge varies from stream to stream. The interval depends on three factors:

1. The climate of the region

2. The width of the floodplain

3. The size of the channel

In a dry climate, for example, the recurrence interval of a 2600 m^3/s flood may be much longer than the recurrence interval of a 2600 m^3/s flood on a similar stream in an area

Figure 18.21 The flood-frequency curve for annual floods on the Skykomish River at Gold Bar, Washington. This curve predicts the probability that a flood of a certain discharge will occur in any given year. [After T. Dunne and L. B. Leopold, *Water in Environmental Planning.* San Francisco: W. H. Freeman, 1978.]

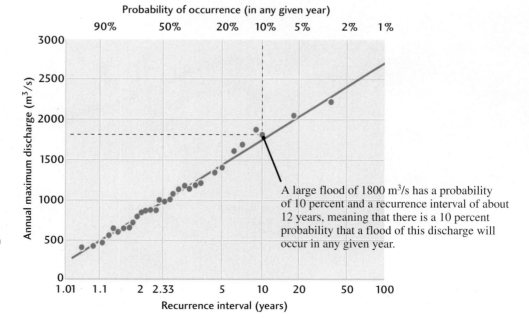

A large flood of 1800 m^3/s has a probability of 10 percent and a recurrence interval of about 12 years, meaning that there is a 10 percent probability that a flood of this discharge will occur in any given year.

that gets intermittent rain. For this reason, individual graphs of recurrence intervals of major rivers are necessary if towns along these rivers are to be prepared to cope with floods of various heights.

The prediction of river floods and their heights has become much more reliable as automated rainfall and river measurements coupled with new computer models have come into use. These methods of forecasting could allow geologists to predict rises and falls of rivers as much as several months in advance. Today, however, predictions are chiefly used for short-term forecasts, such as flood warnings of only a few days.

Longitudinal Profile and the Concept of Grade

We have seen that streamflow at any locality balances inputs and outputs, which become temporarily out of balance during floods. Studies of changes in discharge, velocity, channel dimensions, and topography (especially slope) along the entire length of a stream, from its headwaters (where it begins) to its mouth (where it ends), reveal a larger-scale and longer-term balance. A stream is in dynamic equilibrium between erosion of the streambed and sedimentation in the channel and floodplain over its entire length. This equilibrium is controlled by five factors:

1. Topography (including slope)

2. Climate

3. Streamflow (including both discharge and velocity)

4. The resistance of rock to weathering and erosion

5. Sediment load

A particular combination of factors—such as high topography, humid climate, high discharge and velocity, hard rocks, and low sediment load—would make a stream erode a steep valley into bedrock and carry downstream all sediment derived from that erosion. Downstream, where the topography is lower and the stream might flow over easily erodible sediments, it would deposit bars and floodplain sediments, building up the elevation of the streambed by sedimentation.

We describe the slope of a river from headwaters to mouth by plotting the elevation of its streambed against distances from its headwaters. **Figure 18.22** plots the slope of the Platte and South Platte rivers from the headwaters of the South Platte in central Colorado to the mouth of the Platte in Nebraska. This smooth, concave-upward curve, which represents a cross-sectional view of the river, is its **longitudinal profile.** All streams, from small rills to large rivers, show this same general concave-upward profile, from notably steep near a stream's head to low, almost level, near its mouth.

Why do all streams follow this profile? The answer lies in the combination of factors that control erosion and sedi-

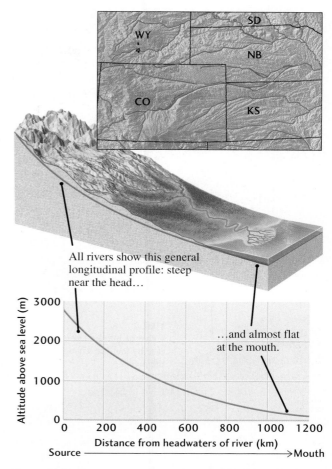

Figure 18.22 The longitudinal profile of the Platte and South Platte rivers from the headwaters of the South Platte in central Colorado to the mouth of the Platte at the Missouri River in Nebraska. [Data from H. Gannett, in *Profiles of Rivers in the United States*. USGS Water Supply Paper 44, 1901.]

mentation. All streams run downhill from their headwaters to their mouths. Erosion is greater in the higher parts of a stream's course than in the lower parts because slopes are steeper and flow velocities can be very high, which has an important influence on the erosion of bedrock (see Chapter 22). In a stream's lower courses, where it carries sediments derived from erosion of the upper courses, sedimentation becomes more significant. Differences in topography and the other factors listed above may make the longitudinal profile steeper or shallower in the upper and lower courses of a stream, but the general shape remains concave upward.

The longitudinal profile is controlled at its lower end by a stream's **base level,** the elevation at which it ends by entering a large standing body of water, such as a lake or ocean. Streams cannot cut below base level, because base level is the "bottom of the hill"—the lower limit of the longitudinal profile.

Profiles Change When Base Levels Change Changes in natural base level affect a longitudinal profile in predictable ways. **Figure 18.23** illustrates the longitudinal profile for the natural local and regional base levels of a river flowing into a lake and from the lake into the ocean. If the regional base level rises—perhaps because of faulting—the profile will show the effects of sedimentation as the river builds up channel and floodplain deposits to reach the new, higher

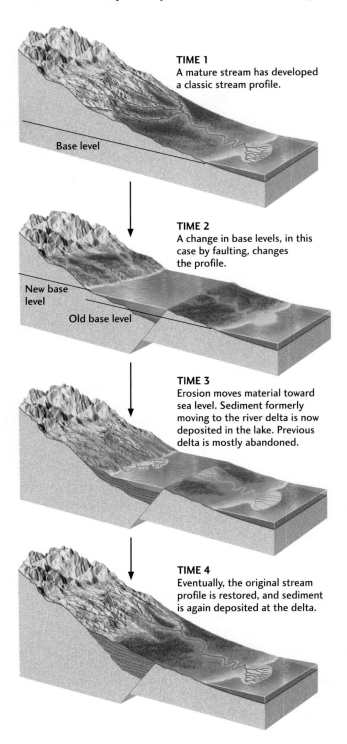

TIME 1
A mature stream has developed a classic stream profile.

Base level

TIME 2
A change in base levels, in this case by faulting, changes the profile.

New base level

Old base level

TIME 3
Erosion moves material toward sea level. Sediment formerly moving to the river delta is now deposited in the lake. Previous delta is mostly abandoned.

TIME 4
Eventually, the original stream profile is restored, and sediment is again deposited at the delta.

Figure 18.23 The base level of a stream controls the lower end of its longitudinal profile. The profiles illustrated here are for natural regional and local base levels of a river flowing into a lake and from the lake into an ocean. In each river segment, the profile adjusts to the lowest level that the river can reach.

regional base-level elevation. Damming a river artificially can create a new local base level, with similar effects on the longitudinal profile (**Figure 18.24**). The slope of the river upstream from the dam decreases, because the new local base level artificially flattens the river's profile at the location of the reservoir formed behind the dam. The decrease in slope lowers the river's velocity, decreasing its ability to transport sediment. The stream deposits some of the sediment on the bed, which makes the concavity somewhat shallower than it was before the dam was built. Below the dam, the river, now carrying much less sediment, adjusts its profile to the new conditions and typically erodes its channel in the section just below the dam.

This kind of erosion has severely affected sandbars and beaches in Grand Canyon National Park downstream from the Glen Canyon Dam. The erosion threatens animal habitats and archaeological sites as well as beaches used for recreation. River specialists have calculated that if river discharge during floods were increased by a certain amount, enough sand would be deposited to prevent depletion by erosion. This calculation was confirmed by an experiment in which a controlled flood was staged at the Glen Canyon Dam in 1996. As the gates of the dam were opened, about 38 billion liters of water spilled into the canyon at a rate fast enough to fill Chicago's 100-story Sears Tower in 17 minutes. This experiment showed that eroded areas could be brought back by sedimentation during floods.

Falling sea level also alters regional base levels and longitudinal profiles. The regional base levels of all streams flowing into the ocean are lowered, and their valleys are cut into former stream deposits. When the drop in sea level is large, as it was during the last glacial period, rivers erode steep valleys into coastal plains and continental shelves.

Graded Streams Over a period of years, a stream's profile becomes stable as the stream gradually fills in low spots and erodes high spots, thereby producing the smooth curve that represents the balance between erosion and sedimentation. That balance is governed not only by the stream's base level but also by the elevation of its headwaters and by all the other factors controlling the equilibrium of the stream profile, as discussed earlier in the chapter. At equilibrium, the stream is a **graded stream**—one in which the slope, velocity, and discharge combine to transport its sediment load, with neither sedimentation nor erosion. If the conditions that produce a particular graded-stream profile change, the stream's profile will change to reach a new equilibrium. Such changes may include depositional and erosional patterns and alterations in the shape of the channel.

In places where regional base level is constant over geologic time, the longitudinal profile represents the balance between tectonic uplift and erosion on the one hand and transport and deposition on the other. If uplift is dominant, typically in the upper courses of a stream, the profile is steep and expresses the dominance of erosion and transport. As uplift slows and the headwater region is eroded, the profile is shallower.

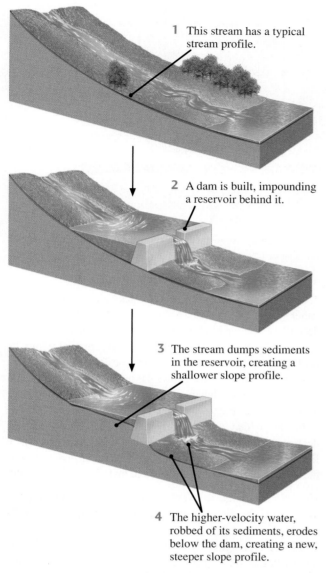

1 This stream has a typical stream profile.

2 A dam is built, impounding a reservoir behind it.

3 The stream dumps sediments in the reservoir, creating a shallower slope profile.

4 The higher-velocity water, robbed of its sediments, erodes below the dam, creating a new, steeper slope profile.

Figure 18.24 A change in the base level of a river caused by human intervention, such as the construction of a dam, alters a river's profile.

Alluvial Fans One place where a river must adjust suddenly to changed conditions is at a mountain front. Here, streams leave narrow mountain valleys and enter broad, relatively flat valleys at lower elevations. Along such fronts, typically at steep fault scarps, streams drop large amounts of sediment in cone- or fan-shaped accumulations called **alluvial fans (Figure 18.25)**. This deposition results from the sudden decrease in velocity that occurs as the channel widens abruptly. To a minor extent, a lowering of the slope below the front also slows the stream velocity. The surface of the alluvial fan typically has a concave-upward shape connecting the steeper mountain profile with the gentler valley profile. Coarse materials, from boulders to sand, dominate on the steep upper slopes of the fan. Lower down, finer sands, silts, and muds are deposited. Fans from many adjacent streams along a mountain front may merge to form a long wedge of sediment whose appearance may mask the outlines of the individual fans that make it up.

Terraces Tectonic uplift can change the equilibrium of a stream valley, resulting in flat, steplike surfaces that line the stream above the floodplain. These **terraces** mark former floodplains that existed at a higher level before regional uplift or an increase in discharge caused the stream to erode into the former floodplain. Terraces are made of floodplain deposits and are often paired, one on each side of the stream, at the same level **(Figure 18.26)**. Terrace formation starts when a stream creates a floodplain. Rapid uplift then changes the stream's equilibrium, causing it to cut down into the floodplain. In time, the stream reestablishes a new equilibrium at a lower level. It may then build another floodplain, which will also undergo uplift and be sculpted into another, lower pair of terraces.

The Effect of Climate Climate also strongly affects the longitudinal profile, primarily through the influence of temperature and precipitation on weathering and erosion (see Chapter 16). Warm temperatures and high rainfall promote weathering and erosion of soils and hillslopes and thus

Playa lake

Alluvial fan

Road

Figure 18.25 An alluvial fan (Tucki Wash) in Death Valley, California. Alluvial fans are large cone- or fan-shaped accumulations of sediment deposited when a stream must suddenly adjust to changed conditions, as at a mountain front. [Martin Miller.]

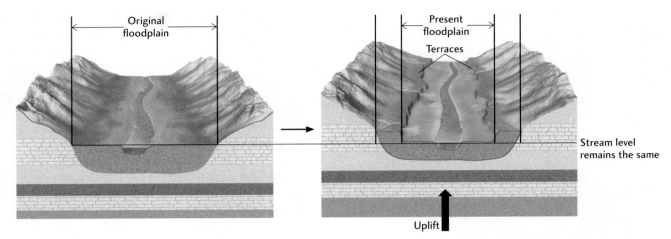

Figure 18.26 Terraces form when the land surface is uplifted and a river erodes into its floodplain and establishes a new floodplain at a lower level. The terraces are remnants of the former floodplain.

enhance sediment transport by streams. High rainfall also leads to greater river discharge, which results in more erosion of riverbeds. An analysis of sediment transport over the entire United States provides evidence that global climate change over the past 50 years is responsible for a general increase in streamflow. Although tectonics plays a dominant role in the formation of alluvial fans at fault scarps, short-term buildup of sediment or trenching is the result of climatic change, primarily variations in temperature.

Lakes

Lakes are accidents of the longitudinal profile, as we can see easily where a lake has formed behind a dam (see Figure 18.24). Lakes range in size from ponds only 100 m across to the world's largest and deepest lake, Lake Baikal in southwestern Siberia. This lake holds approximately 20 percent of the total amount of fresh water in the world's lakes and rivers. It is located in a continental rift zone, a typical plate tectonic setting for lakes. The damming that takes place in a rift valley results from faulting that blocks a normal exit of water. Streams can flow into a rift valley easily but cannot flow out until water builds up to a high enough level to allow them to exit. There are a great many lakes in the northern United States and Canada because glacial ice and glacial sedimentary debris have disrupted normal drainage. Sooner or later, if tectonics and climate remain stable, such lakes will drain away as new outlets form and the longitudinal profile becomes smooth. Because lakes are so much smaller than the oceans, they are more likely than the oceans to be affected by water pollution. Chemical and other industries have polluted Lake Baikal, and Lake Erie has been polluted for many years, although there has been some improvement recently.

SUMMARY

How do stream valleys and their channels and floodplains evolve? As a stream flows, it carves a valley with steep to gently sloping walls and a more or less broad floodplain on either side of its channel. The channel may be straight, meandering, or braided. During normal, nonflood periods, a stream's channel carries all the water and sediment within its banks. When stream discharge increases to flood stage, the sediment-laden water overflows the banks of the channel and inundates the floodplain. The velocity of the floodwater decreases as it spreads over the floodplain. The water drops sediment, which builds up natural levees and floodplain deposits. Recurrence intervals relate the probability that a flood of a given discharge or height will occur in any year to the interval of time between floods of that discharge or height.

How do drainage networks work as collection systems and deltas as distribution systems for water and sediment? Rivers and their tributaries constitute an upstream-branching drainage network that collects the water and sediment running off a specific drainage basin. Each drainage basin is separated from its neighbors by a divide. Drainage networks show various branching patterns—dendritic, rectangular, trellis, or radial—depending on topography, rock type, and geologic structure. Near the mouth of a river, where it forms its delta, the river tends to branch downstream into distributaries. Rivers drop their sediment load at deltas, in topset, foreset, and bottomset beds. Deltas are modified or even absent where waves, tides, and shoreline currents are strong. Tectonics controls delta formation by uplift in the drainage basin and subsidence in the delta region.

How does flowing water in streams erode solid rock and transport and deposit sediment? Any fluid can move in either laminar or turbulent flow, depending on its velocity, viscosity, and flow geometry. The turbulence that characterizes most streams is responsible for transporting sediment by suspension (clays), saltation (sands), and rolling and sliding along the bed (sand and gravel). The tendency for particles to be carried in suspension is countered by the gravitational force that pulls them to settle to the bottom. The settling velocity measures the speed with which suspended particles settle to the bottom. When a streamflow slows, it loses its competence to carry sediment and deposits it, in many places as beds of rippled, cross-bedded sand. Running water erodes solid rock by abrasion; by chemical weathering that enlarges and opens cracks; by physical weathering as sand, pebbles, and boulders crash against rock; and by the plucking and undercutting actions of currents.

How does a stream's longitudinal profile represent the equilibrium between erosion and sedimentation? A stream is in dynamic equilibrium between erosion and sedimentation over its entire length. Topography, discharge, velocity, slope, and sediment load affect this equilibrium. A stream's longitudinal profile, always concave upward, is a cross section of the stream's elevation from its headwaters to the base level at its mouth in a lake or the ocean. Uplift at the upper end of a stream and the rise and fall of sea level at its lower end will change the profile. Alluvial fans form at mountain fronts, primarily as a result of an abrupt widening of the valley and secondarily as a result of a change in slope.

KEY TERMS AND CONCEPTS

alluvial fan (p. 449)

antecedent stream (p. 434)

base level (p. 447)

bed load (p. 438)

bottomset bed (p. 440)

braided stream (p. 431)

capacity (p. 438)

channel (p. 428)

competence (p. 438)

delta (p. 440)

dendritic drainage (p. 434)

discharge (p. 443)

distributary (p. 440)

divide (p. 432)

drainage basin (p. 432)

drainage network (p. 432)

dune (p. 439)

floodplain (p. 428)

foreset bed (p. 440)

graded stream (p. 448)

laminar flow (p. 437)

longitudinal profile (p. 447)

meander (p. 429)

natural levee (p. 431)

oxbow lake (p. 429)

point bar (p. 429)

pothole (p. 435)

recurrence interval (p. 446)

ripple (p. 439)

river (p. 428)

saltation (p. 438)

settling velocity (p. 438)

stream (p. 428)

superposed stream (p. 434)

suspended load (p. 438)

terrace (p. 449)

topset bed (p. 440)

tributary (p. 443)

turbulent flow (p. 437)

valley (p. 428)

EXERCISES

1. How does velocity determine whether a given flow is laminar or turbulent?

2. How does the size of a sediment grain affect the speed with which it settles to the bottom of a flow?

3. What kind of bedding characterizes a ripple or a dune?

4. How do braided and meandering river channels differ?

5. Why is a floodplain so named?

6. What is a natural levee, and how is it formed?

7. What is the discharge of a stream, and how does it vary with velocity?

8. How is a river's longitudinal profile defined?

9. What is the most common kind of drainage network developed over horizontally bedded sedimentary rocks?

10. What is a delta distributary?

THOUGHT QUESTIONS

1. Why might the flow of a very small, shallow stream be laminar in winter and turbulent in summer?

2. Describe and compare the river floodplain and valley above and below the waterfall shown in Figure 18.12.

3. You live in a town on a meander bend of a major river. An engineer proposes that your town invest in new and higher artificial levees to prevent the meander from being cut off. Give the arguments, pro and con, for and against this investment.

4. In some places, engineers have artificially straightened a meandering stream. If such a straightened stream were then left free to adjust its course naturally, what changes would you expect?

5. If global warming produces a significant rise in sea level as polar ice melts, how will the longitudinal profiles of the world's rivers be affected?

6. In the first few years after a dam was built on it, a stream severely eroded its channel downstream of the dam. Could this erosion have been predicted?

7. Your hometown, built on a river floodplain, experienced a 50-year flood last year. What are the chances that another flood of that magnitude will occur next year?

8. Define the drainage basin where you live in terms of the divides and drainage networks.

9. What kind of drainage network do you think is being established on Mount St. Helens since its violent eruption in 1980?

10. The Delaware Water Gap is a steep, narrow valley cut through a structurally deformed high ridge in the Appalachian Mountains. How could it have formed?

11. A major river, which carries a heavy sediment load, has no delta where it enters the ocean. What conditions might be responsible for the lack of a delta?

CHAPTER NINETEEN

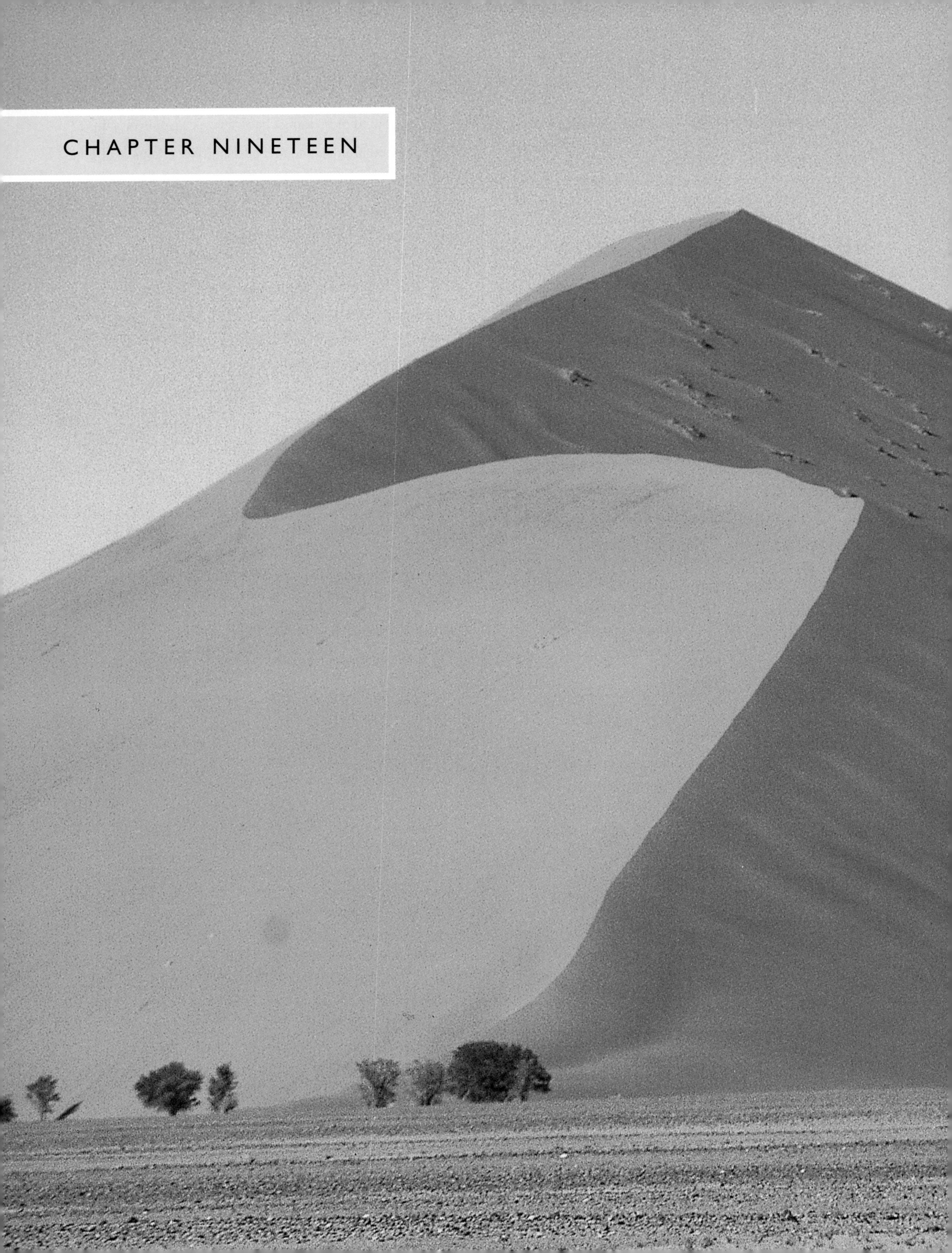

WINDS AND DESERTS

At one time or another, we've all been caught in a wind strong enough to have blown us over, had we not leaned into it or held onto something solid. London, England, which rarely gets strong winds, experienced a major windstorm on January 25, 1990. Winds blowing at more than 175 km/hour ripped roofs off buildings, blew trucks over, and made it virtually impossible to walk on the streets. Some areas of the world experience hurricanes, which bring torrential rains and high winds that erode the land and drive ocean waves that erode and transport huge tonnages of sediment on beaches and in shallow waters. Many winds are strong enough to blow sand grains into the air, creating sandstorms.

Recently, concern over the expansion of Earth's deserts has increased. The process of **desertification,** in which land is degraded by decreased rainfall resulting from various factors such as climatic variations and human activities, has become a major focus of scientists trying to understand Earth's climate system. The population of southern Spain increasingly wonders if the Sahara Desert has jumped the Mediterranean Ocean and is now encroaching on southern Europe.

The wind is a major agent of erosion and deposition, moving enormous quantities of sand, silt, and dust over large regions of the continents and oceans. Wind is much like water in its ability to erode, transport, and deposit sediment. This is not surprising, because the general laws of fluid motion that govern liquids also govern gases. As we will see, however, there are differences that make the wind less powerful than water currents. Unlike a stream, whose discharge is increased by rainfall, wind works most effectively in the absence of rain. We will consider Earth's deserts in particular detail because so many of the geological processes there are related to the work of the wind. The ancient Greeks called the god of winds Aeolus, and geologists today use the term **eolian** for the geological processes powered by the wind.

In this chapter, we focus on these eolian processes, which shape the surface of the land, particularly in deserts, where strong winds can howl for days on end. We will also look at desert environments—where they are located, what elements they are made up of, and how they may be spreading across the globe.

Sand dunes in the Namib Desert, southwest Africa, are among the tallest in the world. [John Grotzinger.]

WIND AS A FLOW OF AIR

Wind is a natural flow of air that is parallel to the surface of the rotating planet. As with water flows, streamlines can be used to describe air flows. Although winds obey all the laws of fluid flow that apply to water in streams (see Chapter 14), there are some differences. Unlike the flow of water in river channels, winds are generally unconfined by solid boundaries, except for the ground surface and narrow valleys. Air flows are free to spread out in all directions, including upward into the atmosphere.

Turbulence

Like the water flowing in rivers, air flows are nearly always turbulent. As we saw in Chapter 14, turbulence depends on three characteristics of a fluid: density, viscosity, and velocity. The extremely low density and viscosity of air make it turbulent even at the velocity of a light breeze.

Wind Belts

Winds vary in speed and direction from day to day, but over the long term they tend to come mainly from one direction. In the temperate latitudes, which include most of North America, the prevailing winds come from the west and so are referred to as the *westerlies* (**Figure 19.1**). Temperate climates are located at latitudes between 30° and 60° north and 30° and 60° south. In the tropics, which are between 30° south and 30° north of the equator, the *trade winds* (named for an archaic use of the word *trade* meaning "track" or "course") blow from the east, which is where most of the world's deserts are.

These wind belts arise because the Sun warms a given amount of land surface most intensely at the equator, where

Key Figure 19.1 Earth's atmosphere circulates in wind belts created by solar input and Earth's rotation.

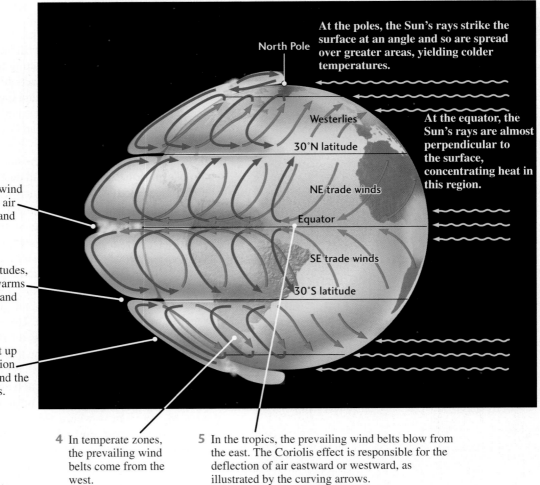

1 There is little surface wind at the equator, and the air rises, forming clouds and rain as it cools.

2 At 30°N and 30°S latitudes, the cooled air sinks, warms up, absorbs moisture, and yields clear skies.

3 These two motions set up the horizontal circulation between the equator and the North and South Poles.

At the poles, the Sun's rays strike the surface at an angle and so are spread over greater areas, yielding colder temperatures.

At the equator, the Sun's rays are almost perpendicular to the surface, concentrating heat in this region.

North Pole

Westerlies

30°N latitude

NE trade winds

Equator

SE trade winds

30°S latitude

4 In temperate zones, the prevailing wind belts come from the west.

5 In the tropics, the prevailing wind belts blow from the east. The Coriolis effect is responsible for the deflection of air eastward or westward, as illustrated by the curving arrows.

the Sun's rays are almost perpendicular to the surface. The Sun heats the Earth less quickly at high latitudes and at the poles because there the rays strike the surface at an angle (see Figure 19.1). Hot air, which is less dense than cold air, rises at the equator and flows toward the poles, gradually sinking as it cools. The cold, dense air at the poles then flows back along the surface of the globe toward the equator.

The simple circulatory pattern of air flow between the equator and the poles is complicated by Earth's rotation, which deflects any current of air or water to the right in the Northern Hemisphere and to the left in the Southern Hemisphere. This effect on Earth's air flow is called the *Coriolis effect,* named after its discoverer.

The Coriolis effect on atmospheric circulation deflects warm and cold air flows moving both northward and southward in both hemispheres (see Figure 19.1). For example, as surface winds in the Northern Hemisphere blow southward into the hot equatorial belt, they are deflected to the right (westward) and hence blow from the northeast rather than from the north. These are the northeast trade winds. The Northern Hemisphere westerlies are flows that initially moved northward but were deflected to the right (eastward) and thus blow from the southwest. Near the equator, air movement is mostly upward, so there is little wind at the surface. As the air rises, it cools, causing the cloudiness and abundant rain of the tropics. At about 30°N and 30°S latitude, some of the high-altitude poleward flow starts to sink to Earth's surface. The cold, dry air warms and absorbs moisture as it sinks, producing clear skies and arid climates.

This dry, sinking air overlies many of the world's deserts, such as the Sahara. As global climate changes, these belts of sinking, dry air may also change, expanding and shifting their margins in some places and perhaps contracting them in others. In this way, the long-term climate of a given region—perhaps already suffering from a shortage of rain—may begin to emerge as a persistent desertlike environment. Eventually, the region may become part of the desert.

WIND AS A TRANSPORT AGENT

Most of us are familiar with rainstorms or snowstorms—high winds accompanied by heavy precipitation. We may have less experience of dry storms, during which high winds blowing for days on end carry enormous tonnages of sand, silt, and dust. The amount of material the wind can carry depends on the strength of the wind, the sizes of the particles, and the surface materials of the area over which the wind blows.

Wind Strength

Figure 19.2 shows how much sand that winds of various speeds can erode from a 1-m-wide strip across a sand dune's surface. A strong wind of 48 km/hour can move half a ton of sand (about equivalent in volume to two large suitcases)

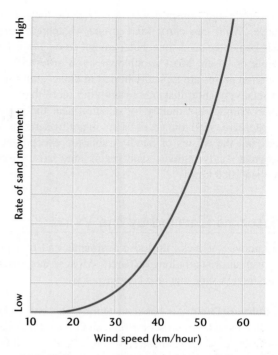

Figure 19.2 The amount of sand moved daily across each meter of width of a dune's surface in relation to wind speed. High-speed winds blowing for several days can move enormous quantities of sand. [After R. A. Bagnold, *The Physics of Blown Sand and Desert Dunes.* London: Methuen, 1941.]

from this small surface area in a single day. At higher wind speeds, the amounts of sand that can be moved increase rapidly. No wonder entire houses can be buried by a sandstorm lasting several days!

Particle Size

The wind exerts the same kind of force on particles on the land surface as a river current exerts on particles on its bed. Turbulence and forward motion combine to lift particles into the wind and carry them along, at least temporarily. Even the lightest breezes carry dust, the finest-grained material. Dust usually consists of particles less than 0.01 mm in diameter but often includes somewhat larger silt particles. Moderate winds can carry dust to heights of many kilometers, but only strong winds can carry particles larger than 0.06 mm in diameter, such as sand grains. Moderate breezes can roll and slide these grains along a sandy bed, but it takes a stronger wind to lift sand grains into the air flow. A wind usually cannot transport the largest particles, however, because air has low viscosity and density. Even though winds can be very strong, only rarely can they move pebbles in the way that rapidly flowing rivers do.

Surface Material

The wind can transport sand and dust only from dry surface materials such as dry soil, sediment, or bedrock. The wind cannot erode and transport wet soils because they are too cohesive. Wind can carry sand grains weathered from a loosely cemented sandstone, but it cannot erode grains from a granite or basalt. Most windblown sands are locally derived. Sand grains are typically buried in dunes after traveling a relatively short distance (usually no more than a few hundred kilometers), mainly by saltation near the ground. The extensive sand dunes of such major deserts as the Sahara and the wastes of Saudi Arabia are exceptions. In those great sandy regions, sand grains may have traveled more than 1000 km.

Materials Carried by the Wind

As air moves, it picks up loose sediments and transports them over surprisingly long distances. Most of this material is dust, though sand can also be transported.

Windblown Dust Air has a staggering capacity to hold dust. Dust includes microscopic rock and mineral fragments of all kinds, especially silicates, as might be expected from their abundance as rock-forming minerals. Two of the most important sources of silicate minerals in dust are clays from soils on dry plains and volcanic dust from eruptions. Organic materials, such as pollen and bacteria, are also common components of dust. Charcoal dust is abundant downwind of forest fires. When found in buried sediments, it is evidence of forest fires in earlier geologic times. Since the beginning of the industrial revolution, we have been pumping new kinds of synthetic dust into the air—from ash produced by burning coal to the many solid chemical compounds produced by manufacturing processes, incineration of wastes, and motor vehicle exhausts.

In large dust storms, 1 km³ of air may carry as much as 1000 tons of dust, equivalent to the volume of a small house. When such storms cover hundreds of square kilometers, they may carry more than 100 million tons of dust and deposit it in layers several meters thick. (See Feature 19.1 for a discussion of similar dust storms on Mars.) Fine-grained particles from the Sahara have been found as far away as England and have been traced across the Atlantic Ocean to Barbados in the Caribbean. Wind annually transports about 260 million tons of material, mostly dust, from the Sahara to the Atlantic Ocean. Scientists on oceanographic research vessels have measured airborne dust far out to sea, and today it can be observed directly from space (**Figure 19.3**). Comparison of the composition of this dust with that of deep-sea sediments in the same region indicates that windblown dust is an important contributor to oceanic sediment, supplying up to a billion tons each year. A large part of this dust comes from volcanoes, and there are individual ash beds on the seafloor marking very large eruptions.

Volcanic dust is abundant because much of it is very fine grained and is erupted high into the atmosphere, where it can travel farther than nonvolcanic dust blown by winds

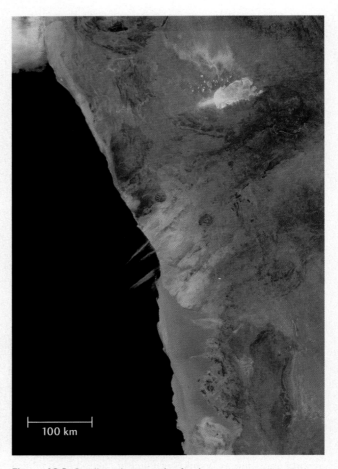

Figure 19.3 Satellite photograph of a dust storm originating in the Namib Desert in September 2002. Dust and sand are being transported from right (east) to left (west) in response to strong winds blowing out to sea. The sediment can be transported for hundreds to thousands of kilometers across the ocean. [NASA.]

closer to the surface. Volcanic explosions inject huge quantities of dust into the atmosphere. The volcanic dust from the 1991 eruption of Mount Pinatubo in the Philippine Islands circled the Earth, and most of the finest-grained particles did not settle until 1994 or 1995. By partly reflecting the Sun's rays, dust from the Pinatubo eruption cooled Earth's surface by a small amount for a few years.

Mineral dust in the atmosphere increases when agriculture, deforestation, erosion, or changing land use disrupts soils. A large amount of mineral dust may come from the Sahel region on the southern border of the Sahara, where drought and overgrazing are responsible for a heavy load of dust.

Windblown dust has complex effects on climate. Mineral dust scatters the incoming visible radiation from the Sun and absorbs the outgoing infrared (heat) radiation from the Earth. Thus, mineral dust has a net cooling effect in the visible part of the spectrum and a net warming effect in the infrared.

Scientists use the term *aerosol* for atmospheric particles of natural or synthetic origin that are small enough to be sus-

EARTH ISSUES

19.1 Martian Dust Storms and Dust Devils

Of all the planets in the solar system, Mars is the most like Earth. Although Mars has a thinner atmosphere, it has weather that changes seasonally and an Earth-like day of 24 hours and 37 minutes. Mars also has a complex surface environment including ice, soil, and sediment. A variety of features provide evidence that water once flowed on the Martian surface.

Today, Mars is cold and dry, and its surface environment is dominated by various eolian processes. These processes have created a variety of wind-sculpted landforms and deposited a wide variety of sediments. Sediments range from very widespread dust layers to more localized dune fields made up of sand and silt. Coarser deposits, composed of basaltic and hematitic granules, have been observed by the Mars Exploration Rovers. Planetary geologists estimate that the winds that produced these eolian deposits blew at velocities of up to 30 m/s (108 km/hour) during great dust storms that covered the entire planet (see Figure 9.21). Although these winds are not as fast as Earth's strongest winds and Mars's atmosphere is less dense than Earth's, Martian winds are still strong enough to form an array of eolian depositional and erosional features identical to those observed on Earth. Even the pink color of the Martian atmosphere owes its origin to suspended quantities of windblown dust entrained during dust storms.

As the seasons change on Mars, so does the weather. When the Mars Exploration Rover *Spirit* landed in January 2004, it explored for months on end without seeing any significant evidence of recent eolian activity. However, in March 2005—more than a year after landing—*Spirit*'s cameras began to observe dust devils in action, a sign that the seasons were changing on the surface of Mars. During the windy and dusty season, global dust storms are also accompanied by local dust devils. Dust devils occur when the Sun heats the surface of Mars. Warmed soil and rocks heat the layer of the atmosphere closest to the surface, and the warm air rises in a whirling motion, stirring up the dust from the surface like a miniature tornado.

Martian dust storms and dust devils directly affect our ability to study the Martian surface. The rovers we've sent to Mars depend on solar power to move and conduct their explorations. Ultimately, the life span of these rovers is limited by the time it takes for enough wind-blown dust to settle out on the rovers' solar panels and terminate power generation. Global dust storms contribute to the fallout of dust on the solar panels and the demise of power. However, when they move over the rovers, dust devils are thought to help clean the dust off the solar panels.

View of two dust devils on the floor of Gusev Crater taken by the *Spirit* rover from the summit of Husband Hill on August 21, 2005. These dust devils move across the landscape at about 10 to 15 km/hour. [NASA/JPL.]

---

pended in the lower levels of the atmosphere for days or weeks. Aerosols affect climate in the same ways as mineral dust does.

Windblown Sand The sand the wind transports may consist of almost any kind of mineral grain produced by weathering. Quartz grains are by far the most common because quartz is such an abundant constituent of many sur-

face rocks, especially sandstones. Many windblown quartz grains have a frosted or matte (roughened and dull) surface (**Figure 19.4**) like the inside of a frosted light bulb. Some of the grain frosting is produced by wind-driven impacts, but most of it results from slow, long-continued dissolution by dew. Even the tiny amounts of dew found in arid climates are enough to etch microscopic pits and hollows into sand grains, creating the frosted appearance. Frosting is found only in

Figure 19.4 Photomicrograph of frosted and rounded grains of quartz from sand dunes in Saudi Arabia. [Walter N. Mack.]

Figure 19.5 A ventifact. This wind-faceted pebble from Antarctica has been shaped by windblown sand in a frigid environment. [E. R. Degginger.]

eolian environments, so it is good evidence that a sand grain has been blown by the wind.

Windblown calcium carbonate grains accumulate where there are abundant fragments of shells and coral, such as in Bermuda and on many coral islands in the Pacific Ocean. The White Sands National Monument in New Mexico is a prominent example of sand dunes made of gypsum sand grains eroded from evaporite deposits formed in nearby *playa lakes* (see the "Evaporite Sediments" section near the end of this chapter).

WIND AS AN AGENT OF EROSION

By itself, the wind can do little to erode large masses of solid rock exposed at Earth's surface. Only when rock is fragmented by chemical and physical weathering can the wind pick up particles. In addition, the particles must be dry, because wet soils and damp fragmented rock are held together by moisture. Thus, wind erodes most effectively in arid climates, where winds are strong and dry and any moisture quickly evaporates.

Sandblasting

Windblown sand is an effective natural **sandblasting** agent. The common method of cleaning buildings and monuments with compressed air and sand works on exactly the same principle: the impact of high-speed particles wears away the solid surfaces. Natural sandblasting mainly works close to the ground, where most sand grains are carried. Sandblasting rounds and erodes rock outcrops, boulders, and pebbles and frosts the occasional glass bottle.

Ventifacts are wind-faceted pebbles having several curved or almost flat surfaces that meet at sharp ridges

(**Figure 19.5**). Each surface or facet is made by sandblasting of the pebble's windward side. Occasional storms roll or rotate the pebbles, exposing a new windward side to be sandblasted. Many ventifacts are found in deserts and in glacial gravel deposits, where the necessary combination of gravel, sand, and strong winds is present.

Deflation

As particles of dust, silt, and sand become loose and dry, blowing winds can lift and carry them away, gradually eroding the ground surface in a process called **deflation** (**Figure 19.6**). Deflation, which can scoop out shallow depressions

Figure 19.6 A shallow deflation hollow in the San Luis Valley, Colorado. Wind has scoured the surface and eroded it to a slightly lower elevation. Deflation occurs in dry areas where the vegetation cover is absent or broken. [Breck P. Kent.]

Key Figure 19.7 Desert pavement is created by climate and microbial interactions with windblown sediment and soil.

1 According to a new theory, the evolution of desert pavement begins when the wind blows fine-grained materials into heterogeneous soil or sediment.

2 During rainstorms, the fine, windblown sediments infiltrate beneath the coarse layer of pebbles.

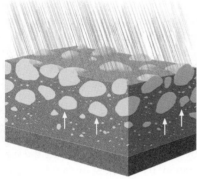

3 Microbes living beneath the pebbles produce bubbles that help raise the pebbles and maintain their position at the surface.

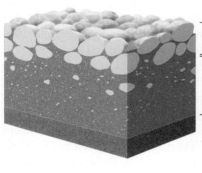

Stays consistent thickness

Gets thicker as more windblown dust is supplied over the millennia

4 Over time, these processes lead to thickening of the dust accumulating beneath the pebble layer.

5 A continued supply of windblown dust makes the deposit thicker.

[William E. Ferguson.]

or hollows, occurs on dry plains and deserts and on temporarily dried-up river floodplains and lake beds. Firmly established vegetation, even the sparse vegetation of arid and semiarid regions, retards it. Deflation occurs slowly in areas with plants because their roots bind soil and their stems and leaves disrupt the air currents and shelter the ground surface. Deflation works fast where the vegetation cover is broken, either naturally by killing drought or artificially by cultivation, construction, or motor vehicle tracks.

When deflation removes the finer-grained particles from a mixture of gravel, sand, and silt in sediments and soils, it produces a remnant surface of gravel too large for the wind to transport. Over thousands of years, as deflation removes the finer-grained particles from successive stream deposits, the gravel accumulates as a layer of **desert pavement**—a coarse, gravelly ground surface that protects the soil or sediments below from further erosion (**Figure 19.7**).

This theory of pavement formation is not completely accepted, because a number of pavements seem not to have formed in this way. A new theory is that some of them are formed by the deposition of windblown sediments. The coarse rock pavement stays at the surface, while windblown dust infiltrates below the surface layer of pavement, is modified by soil-forming processes, and accumulates there.

WIND AS A DEPOSITIONAL AGENT

When the wind dies down, it can no longer transport the sand, silt, and dust it carries. The coarser material is deposited in sand dunes of various shapes, ranging in size from low knolls to huge hills more than 100 m high (see the chapter opener photo on page 452). The finer silt and dust fall as a more or less uniform blanket of silt and clay. By observing these depositional processes working today, geologists have been able to link them with such characteristics as bedding and texture to deduce past climates and wind patterns from ancient sandstones and dust falls.

Where Sand Dunes Form

Sand dunes occur in relatively few environmental settings. Many of us have seen the dunes that form behind ocean beaches or along large lakes. Some dunes are found on the sandy floodplains of large rivers in semiarid and arid regions. Most spectacular are the fields of dunes that cover large expanses of some deserts (**Figure 19.8**). Such dunes

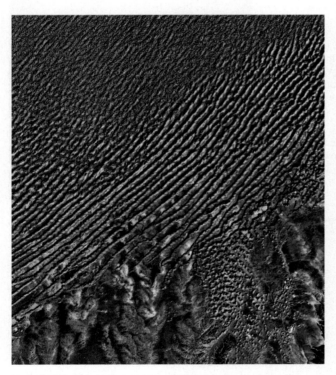

Figure 19.8 A vast expanse of sand dunes on the southern Arabian Peninsula. These linear dunes trend northeast-southwest, parallel with the prevailing northeasterly winds. At the lower right is an eroding plateau of horizontally bedded sedimentary rock. [Satellite image data processing by ZERIM, Ann Arbor, Michigan.]

may reach heights of several hundred meters, truly mountains of sand.

Dunes form only in settings that have a ready supply of loose sand: beach sands along coasts, sandy river bars or floodplain deposits in river valleys, and sandy bedrock formations in deserts. Another common factor in dune formation is wind power. On oceans and lakes, strong winds blow onshore off the water. Strong winds, sometimes of long duration, are common in deserts.

The wind cannot pick up wet materials easily, so most dunes are found in dry climates. The exception is dune belts along a coast, where sand is so abundant and dries so quickly in the wind that dunes can form even in humid climates. In such climates, soil and vegetation begin to cover the dunes only a little way inland from the beaches, and the winds no longer pick up the sand.

Dunes may stabilize and become vegetated when the climate grows more humid and then start moving again when an arid climate returns. There is geological evidence that during droughts two to three centuries ago and earlier, sand dunes in the western high plains of the United States and Canada were reactivated and migrated over the plains.

How Sand Dunes Form and Move

The wind moves sand by sliding and rolling it along the surface and by causing saltation, the jumping motion that temporarily suspends grains in a current of water or air. Saltation in air flows works the same way as it does in rivers (see Figure 18.14), except that the jumps in air flows are higher and longer. Sand grains suspended in an air current often rise to heights of 50 cm over a sand bed and 2 m over a pebbly surface—much higher than grains of the same size can jump in water. The difference arises partly from the fact that air is less viscous than water and therefore does not inhibit the bouncing of the grains as much as water does. In addition, the impact of grains falling in air induces higher jumps as they hit the surface. These collisions, which the air hardly cushions, kick surface grains into the air in a sort of splashing effect. As saltating grains impact a sand bed, they can push forward grains too large to be thrown up into the air, causing the bed to creep in the direction of the wind. A sand grain striking the surface at high speed can propel another grain as far as six times its own diameter.

When the wind moves sand along a bed, it almost inevitably produces ripples and dunes much like those formed by water (**Figure 19.9**). Ripples in sand, like those under water, are transverse; that is, at right angles to the current. At low to moderate wind speeds, small ripples form. As the wind speed increases, the ripples become larger. Ripples migrate in the direction of the wind over the backs of larger dunes. Some wind is almost always blowing, so a sand bed is almost always rippled to some extent.

Given enough sand and wind, any obstacle—such as a large rock or a clump of vegetation—can start a dune.

Figure 19.9 Wind ripples in sand at Stovepipe Wells, Death Valley, California. Although complex in form, these ripples are always transverse (at 90°) to the wind direction. [Tom Bean.]

Streamlines of wind, like those of water, separate around obstacles and rejoin downwind, creating a wind shadow downstream of the obstacle. Wind velocity is much lower in the wind shadow than in the main flow around the obstacle. In fact, it is low enough to allow sand grains blown into the shadow to settle there. The wind is moving so slowly that it can no longer pick up these grains, and they accumulate as a *sand drift,* a small pile of sand in the lee of the obstacle (**Figure 19.10**). As the process continues, the sand drift itself becomes an obstacle. If there is enough sand and the wind continues to blow in the same direction long enough, the drift grows into a dune. Dunes may also grow by the enlargement of ripples, just as underwater dunes do.

As a dune grows, the whole mound starts to migrate downwind by the combined movements of a host of individual grains. Sand grains constantly saltate to the top of the low-angle windward slope and then fall over into the wind shadow on the lee slope, as shown in **Figure 19.11**. These grains gradually build up a steep, unstable accumulation on the upper part of the lee slope. Periodically, the steepened buildup gives way and spontaneously slips or cascades down this **slip face,** as it is called, to a new slope at a lower angle. If we overlook the short-term, unstable steepenings of the slope, the slip face maintains a stable, constant slope angle—its angle of repose. As we saw in

(a) Early stage: small sand drifts form in wind shadow

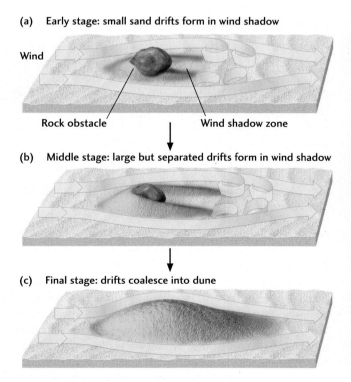

(b) Middle stage: large but separated drifts form in wind shadow

(c) Final stage: drifts coalesce into dune

Figure 19.10 Sand dunes may form in the lee of a rock or other obstacle. By separating the wind streamlines, the rock creates a wind shadow in which the eddies are weaker than the main flow. The windborne sand grains are thus able to settle and pile up in drifts that eventually coalesce into a dune. [After R.A. Bagnold, *The Physics of Blown Sand and Desert Dunes.* London: Methuen, 1941.] Sand dunes, Owens Lake, California. [Martin Miller.]

Key Figure 19.11	Sand-dune formation depends on wind velocity and amount of sand.

1 A ripple or dune advances by the movements of individual grains of sand. The whole form moves forward slowly as sand erodes from the windward slope and is deposited on the leeward slope.

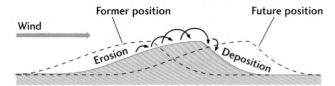

2 Particles of sand arriving on the windward slope of the dune move by saltation over the crest,...

3 ...where the wind velocity decreases and the sand deposited slips down the leeward slope.

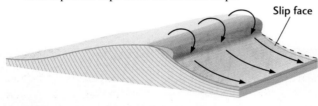

4 This process acts like a conveyor belt that moves the dune forward.

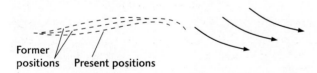

5 As the dune grows higher, the wind streamlines become compressed and their velocity increases. The dune stops growing vertically when it reaches a height at which the wind is so fast that it blows the sand grains off the dune as quickly as they are brought up.

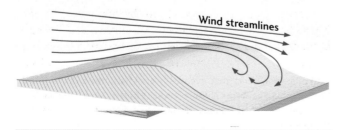

Chapter 16, this angle increases with the size and angularity of the particles.

Successive slip faces deposited at the angle of repose create the cross-bedding that is the hallmark of windblown dunes. Cross-bedding preserved on Mars (see Figure 9.26) is also used as evidence for ancient windblown dunes. As dunes accumulate, interfere with one another, and become buried in a sedimentary sequence, the cross-bedding is preserved even though the original shapes of the dunes are lost. Sets of sandstone cross-bedding many meters thick are evidence of high, windblown dunes. From the directions of these eolian cross-beds, geologists can reconstruct wind directions of the past (see Figure 5.7).

As more sand accumulates on the windward slope of a dune than blows off onto the slip face, the dune grows in height. Heights of 30 m are common, and huge dunes in Saudi Arabia reach 250 m, which seems to be the limit. The explanation for the limited dune height lies in the relationship among wind streamline behavior, velocity, and topography. Wind streamlines advancing over the back of a dune become more compressed as the dune grows higher (see Figure 19.11). As more air rushes through a smaller space, the wind velocity increases. Ultimately, the air speed at the top of the dune becomes so great that sand grains blow off the top of the dune as quickly as they are brought up the windward slope. When this balance is reached, the height of the dune remains constant.

Dune Types

A person standing in the middle of a large expanse of dunes might be bewildered by the seemingly orderless array of undulating slopes. It takes a practiced eye to see the dominant pattern, and it may even require observation from the air. The general shapes and arrangements of sand dunes depend on the amount of sand available and the direction, duration, and strength of the wind. We cannot yet be certain of the specific mechanisms by which a particular wind regime results in one dune form or another. Geologists recognize four main types of dunes: barchans, blowout dunes, transverse dunes, and linear dunes (**Figure 19.12**).

Dust Falls and Loess

As the velocity of dust-laden wind decreases, the dust settles to form **loess,** a blanket of sediment composed of fine-grained particles. Beds of loess lack internal stratification. In compacted deposits more than a meter thick, loess tends to form vertical cracks and to break off along sheer walls during erosion (**Figure 19.13**). Geologists theorize that the vertical cracking may be caused by a combination of root penetration and uniform downward percolation of groundwater, but the exact mechanisms are still unknown.

Loess covers as much as 10 percent of Earth's land surface. The major loess deposits are found in China and North America. China has more than a million square kilometers

1 Barchans are crescent-shaped dunes, usually but not always found in groups. The horns of the crescent point downwind. Barchans are the products of limited sand supply and unidirectional winds.

Wind

2 Blowout dunes are almost the reverse of barchans. The slip face of a blowout dune is convex downwind, whereas the barchan's is concave downwind.

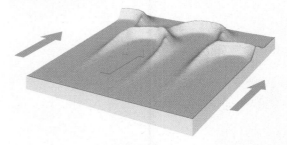

3 Transverse dunes are long ridges oriented at right angles to the wind direction. These dunes form in arid regions where there is abundant sand and vegetation is absent. Typically, sand-dune belts behind beaches are transverse dunes formed by strong onshore winds.

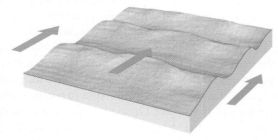

4 Linear dunes are long ridges of sand whose orientation is parallel to the wind direction. These dunes may reach heights of 100 m and extend many kilometers. Most areas covered by linear dunes have a moderate sand supply, a rough pavement, and winds that are always in the same general direction.

Figure 19.12 Dune types in relation to prevailing winds.

of loess deposits (**Figure 19.14**). The great loess deposits of China extend over wide areas in the northwest; most are 30 to 100 m thick, although some exceed 300 m. The winds blowing over the Gobi Desert and the arid regions of central Asia provided the dust, which still blows over eastern Asia and the Chinese interior. Some of the loess deposits in China are 2 million years old. They formed after an increase in the elevation of the Himalaya and related mountain belts of western China introduced rain shadows and dry climates to the continental interior. The tectonic uplift of these mountain belts was responsible for the cold, dry climates of the Pleistocene epoch in much of Asia. These climates inhibited vegetation and dried out soils, causing extensive wind erosion and transportation.

The best-known loess deposit in North America is in the Upper Mississippi Valley. It originated as silt and clay deposited on extensive floodplains of rivers draining the edges of melting glaciers of the Pleistocene epoch. Strong winds dried the floodplains, whose frigid climate and rapid rates of sedimentation inhibited vegetation, and blew up tremendous amounts of dust, which then settled to the east. Geologists recognize that this loess deposit is distributed as a blanket of more or less uniform thickness on both hills and valleys, all in or near formerly glaciated areas. Changes in the regional thickness of the loess in relation to the prevailing westerly winds confirm its eolian origin. The loess on the eastern sides of major river floodplains is 8 to 30 m thick, greater than on the western sides, and the thickness decreases rapidly downwind to 1 to 2 m farther east of the floodplains.

Figure 19.13 Pleistocene loess, Catalina Mountains, Arizona, showing vertical cracking. [E. R. Degginger.]

Figure 19.14 Comfortable dwelling caves hand-carved into steep cliffs of loess in central China. These deposits of windblown dust accumulated in the past 2.5 million years, reaching a thickness of as much as 400 m. [Stephen C. Porter.]

Soils formed on loess are fertile and highly productive. They also pose environmental problems, because they are easily eroded into gullies by small streams and are deflated by the wind when they are poorly cultivated.

THE DESERT ENVIRONMENT

The hot, dry deserts of the world are among the most hostile environments for humans. Yet many of us are fascinated by the strange forms of animal and plant life and the bare rocks and sand dunes found there. Of all Earth's environments, the wind is best able to do its work of erosion and sedimentation in the desert.

All told, arid regions amount to one-fifth of Earth's land area, about 27.5 million square kilometers. Semiarid plains account for an additional one-seventh. Given the reasons for the existence of large areas of deserts in the modern world —mountain building by plate tectonics, the transport of

continental regions to low latitudes by continental drift, and Earth's climatic belts—we can be confident that, by the principle of uniformitarianism, extensive deserts have existed throughout geologic time.

Where Deserts Are Found

Rainfall is the major factor determining the location of the world's great deserts. The Sahara and Kalahari deserts of Africa and the Great Australian Desert get extremely low amounts of rainfall, normally less than 25 mm/year and in some places less than 5 mm/year. These deserts are found in Earth's warmest regions, between 30°N and 30°S latitudes (**Figure 19.15**). The deserts lie under virtually stationary areas of high atmospheric pressure. The Sun beats down through a cloudless sky week after week, and the relative humidity is extremely low.

Deserts also exist in the midlatitudes—between 30°N and 50°N and between 30°S and 50°S—in regions where rainfall is low because moisture-laden winds either are blocked by mountain ranges or must travel great distances from the ocean, their source of moisture. The Great Basin and Mojave deserts of the western United States, for example, lie in rain shadows created by the western coastal mountains. As we saw in Chapter 17, wind descending from the mountains warms and dries, leading to low precipitation (see Figure 17.2). The Gobi and other deserts of central Asia are so far inland that the winds reaching them have precipitated all their ocean-derived moisture long before they arrive at the interior of the continent.

Another kind of desert is found in polar regions. There is little precipitation in these cold, dry areas because the frigid air can hold only extremely small amounts of moisture. The dry valley region of southern Victoria Land in Antarctica is so dry and cold that its environment resembles that of Mars.

The Role of Plate Tectonics In a sense, deserts are a result of plate tectonics. The mountains that create rain shadows are made by collisions between converging continental and oceanic plates. The great distance separating central Asia from the oceans is a consequence of the size of the continent, a huge landmass assembled from smaller plates by continental drift. Large deserts are found at low latitudes because continental drift powered by plate tectonics moved them there from higher latitudes. If, in some future plate tectonic scenario, the North American continent were to move south by 2000 km or so, the northern Great Plains of the United States and Canada would become a hot, dry desert. Something like that happened to Australia. About 20 million years ago, Australia was far to the south of its present position, and its interior had a warm, humid climate. Since then, Australia has moved northward into an arid subtropical zone, where its interior has become a desert.

The Role of Climate Change Changes in a region's climate may transform semiarid lands into deserts, a process

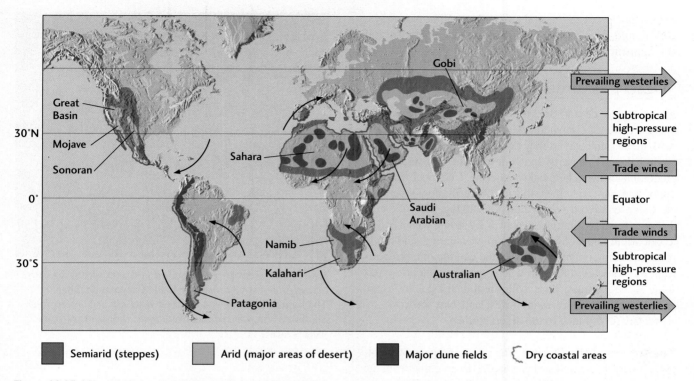

Figure 19.15 Major desert areas of the world (exclusive of polar deserts) in relation to prevailing wind directions and major mountain and plateau areas. Sand dunes make up only a small proportion of the total desert area. [After K. W. Glennie, *Desert Sedimentary Environments.* New York: Elsevier, 1970.]

called *desertification*. Climatic changes that we do not fully understand may decrease precipitation for decades or even centuries. After a dry period, a region may return to a milder, wetter climate. Over the past 10,000 years, climates of the Sahara appear to have oscillated between drier and wetter conditions. We have evidence from radar imaging of the Sahara from the space shuttle *Endeavor* that an extensive system of river channels existed a few thousand years ago. Now dry and buried by more recent sand deposits, these ancient drainage systems carried abundant running water across the northern Sahara during wetter periods.

The Sahara Desert may be expanding northward. The Desert Watch project, led by the European Space Agency, reports that over 300,000 km² of Europe's Mediterranean coast—an area almost as large as the state of New York—with a population of 16 million, is threatened by desertification. During 2005, fires raged along the southern Spanish coast as the current drought reached historic duration. Temperatures set new record highs for weeks on end. Is this merely a long, hot summer, or are these the initial symptoms of enduring climate change, made worse by overpopulation and overdevelopment within the fragile ecologies of dry landscape? Evidence to support the latter scenario is building.

Soils are weakening due to prolonged dryness, making them more susceptible to wind transport and deflation. Groundwater levels have reached new lows. And there is little question that Europe is getting warmer; during the last century, average temperatures have increased about 0.7°C .

The 1990s marked the hottest decade since record-keeping began in the mid-1800s, registering two of the five hottest years ever recorded.

The Role of Humans Oscillating climates in the Sahara occurred naturally, but human activities are responsible for some desertification today. The unrestrained growth of human populations and their agriculture and increased animal grazing may result in the expansion of deserts. When population growth and periods of drought coincide, the results in semiarid regions can be disastrous. In Spain, the greatest urban and agricultural expansion is taking place in the driest region. Former farmlands are now stripped of vegetation due to overfarming (up to four crops per year), which depletes water and strips soils. Development and the tourism boom are literally paving over the dry lands and desiccating the countryside that's left. In 2004, more than 350,000 new homes were built on Spain's Mediterranean coast, many with backyard swimming pools and nearby golf courses. In isolation, any one of these human activities may not have a negative effect. However, in concert they add up to desertification.

"Making the desert bloom," the opposite of desertification, has been a slogan of some countries with desert lands. They irrigate on a massive scale to convert marginal or desert areas into productive farmlands. The Great Valley of California, where much of North America's fruits and vegetables are grown, is one example. The growth of cities such as Phoenix, Arizona, which depend on imported water, has produced islands of urbanization—complete with humid haze and air pollution—in an otherwise arid environment.

Desert Weathering

As unique as deserts are, the same geological processes operate there as elsewhere. Weathering and transportation work in the same way as they do everywhere, but with a different balance in deserts. In the desert, physical weathering predominates over chemical weathering. Chemical weathering of feldspars and other silicates proceeds slowly because the water required for the reaction is lacking. The little clay that does form is usually blown away by strong winds before it can accumulate. Slow chemical weathering and rapid wind transport combine to prevent the buildup of any significant thickness of soil, even where sparse vegetation binds some of the particles. Thus, soils are thin and patchy. Sand, gravel, rock rubble of many sizes, and bare bedrock are characteristic of much of the desert surface.

The Colors of the Desert The rusty, orange-brown colors of many weathered surfaces in the desert come from the ferric iron oxide minerals hematite and limonite. These minerals are produced by the slow weathering of iron silicate minerals such as pyroxene. The iron oxides, even when present only in small amounts, stain the surfaces of sands, gravels, and clays.

Desert varnish is a distinctive dark brown, sometimes shiny coating found on many rock surfaces in the desert. It is a mixture of clay minerals with smaller amounts of manganese and iron oxides. Geologists hypothesize that desert varnish forms very slowly from a combination of dew, the chemical weathering that produces clay minerals and iron and manganese oxides, and the adhesion of windblown dust to exposed rock surfaces. The process is so slow that Native American inscriptions scratched in desert varnish hundreds of years ago still appear fresh, with a stark contrast between the dark varnish and the light unweathered rock beneath (**Figure 19.16**). Varnish requires thousands of years to form, and some particularly ancient varnishes in North America are of Miocene age. However, recognizing varnish as such in ancient sandstones is difficult.

Streams: The Primary Agents of Erosion Wind plays a larger role in erosion in the desert than it does elsewhere, but it cannot compete with the erosive power of streams. Even though it rains so seldom that most desert streams flow only intermittently, streams do most of the erosional work in the desert when they do flow.

Even the driest desert gets occasional rain. In sandy, gravelly areas of deserts, infiltration of the infrequent rainfall into soil and permeable bedrock temporarily replenishes groundwater in the unsaturated zone. There, some of it evaporates very slowly into pore spaces between the grains. A smaller amount eventually reaches the groundwater table far

Figure 19.16 Petroglyphs scratched in desert varnish by early Native Americans. Newspaper Rock, Canyonlands, Utah. The scratches are several hundred years old and appear fresh on the varnish, which accumulated over thousands of years. [Peter Kresan.]

below—in some places, as much as hundreds of meters below the surface. Desert oases form where the groundwater table comes close enough to the surface that roots of palms and other plants can reach it.

When rain occurs in heavy cloudbursts, so much water falls in such a short time that infiltration cannot keep pace, and the bulk of the water runs off into streams. Unhindered by vegetation, the runoff is rapid and may cause flash floods along valley floors that have been dry for years. Thus, a large proportion of streamflows in the desert consist of floods. When floods occur, they have great erosive power because most loose, weathering debris is not held in place by vegetation. Streams may become so choked with sediment that they look more like fast-moving mudflows than like rivers. The abrasiveness of this sediment load moving rapidly at flood velocities makes such streams efficient eroders of bedrock valleys.

Desert Sediment and Sedimentation

Deserts are composed of a diverse set of depositional environments. These environments may change dramatically when rain suddenly forms raging rivers and widespread lakes. Prolonged dry periods intervene, and sediments are blown into sand dunes.

Alluvial Sediments As sediment-laden flash floods dry up, they leave distinctive deposits on the floors of desert valleys. A flat fill of coarse debris often covers the entire valley floor, and the ordinary differentiation of a stream into channel, levees, and floodplains is absent (**Figure 19.17**). The sediments of many other desert valleys clearly show the intermixing of stream-deposited channel and floodplain sed-

(a)

(b)

Figure 19.17 (a) Dry-wash flooding during a summer thunderstorm in Saguaro National Monument, Arizona. (b) The same dry wash a day after flooding. The coarse debris deposited by sudden desert floods may cover the entire valley floor. [Peter Kresan.]

iments with eolian sediments. The combination of stream and eolian processes occurring in the past formed extensive sheets of eolian sandstone separated by stream flood surfaces and floodplain sandstones deposited between the sheets of eolian sandstone. Large alluvial fans (see Chapter 18) are prominent features at mountain fronts in deserts because desert streams deposit much of their high sediment load on the fans. The rapid infiltration of stream water into the permeable fan material deprives the streams of the water required to carry the sediment load any farther downstream. Debris flows and mudflows make up large parts of the alluvial fans of arid, mountainous regions.

Eolian Sediments By far the most dramatic sedimentary accumulations in the desert are sand dunes. Dune fields range in size from a few square kilometers to "seas of sand" found in major deserts such as those of Namibia. These sand seas—or *ergs*—may cover as much as 500,000 km², twice the size of the state of Nevada. Although film and television portrayals lead one to think that deserts are mostly sand, actually only one-fifth of the world's desert area is covered by sand. The other four-fifths are rocky or covered with desert pavement. Sand covers only a little more than one-tenth of the Sahara Desert, and sand dunes are far less common in the deserts of the southwestern United States.

Evaporite Sediments **Playa lakes** are permanent or temporary lakes that occur in arid mountain valleys or basins (**Figure 19.18**). As the lake water evaporates, dissolved weathering products are concentrated and gradually precipitated.

Figure 19.18 A desert playa lake in Death Valley, California. Playa lakes are sources of evaporite minerals such as borax. Their waters may be deadly to drink because of high quantities of dissolved minerals. [David Muench.]

Playa lakes are sources of evaporite minerals: sodium carbonate, borax (sodium borate), and other unusual salts. The water in playa lakes may be deadly to drink. Desert streams contribute large amounts of dissolved salts, and these salts accumulate in playa lakes when the streams redissolve evaporite minerals deposited by evaporation from earlier runoff. If evaporation is complete, the lakes become **playas,** flat beds of clay that are sometimes encrusted with precipitated salts.

Desert Landscape

Desert landscapes are some of the most varied on Earth. Large low, flat areas are covered by playas, desert pavements, and dune fields. Uplands are rocky, cut in many places by steep river valleys and gorges. The lack of vegetation and soil makes everything seem sharper and harsher than it would seem in landscapes of more humid climates. (We will discuss landscapes in more detail in Chapter 22.) In contrast with the rounded, soil-covered, vegetated slopes found in most humid regions, the coarse fragments of varying size produced by desert weathering form steep cliffs with masses of angular talus slopes at their bases (**Figure 19.19**).

Figure 19.19 This desert landscape at Kofa Butte, Kofa National Wildlife Refuge, Arizona, shows the steep cliffs and masses of talus slopes produced by desert weathering. [Peter Kresan.]

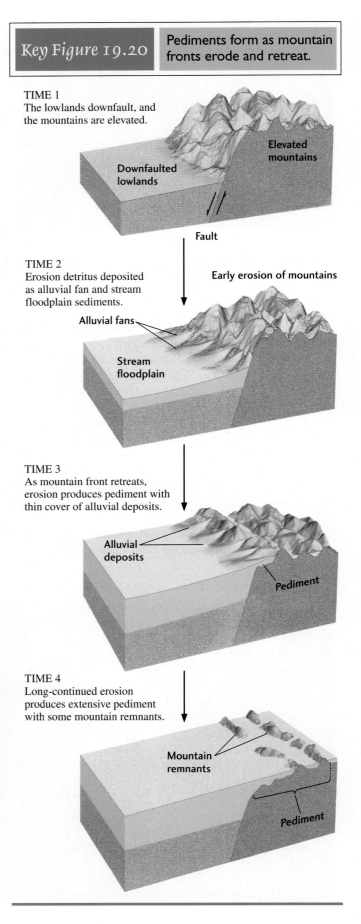

Key Figure 19.20 Pediments form as mountain fronts erode and retreat.

TIME 1
The lowlands downfault, and the mountains are elevated.

Elevated mountains

Downfaulted lowlands

Fault

TIME 2
Erosion detritus deposited as alluvial fan and stream floodplain sediments.

Early erosion of mountains

Alluvial fans

Stream floodplain

TIME 3
As mountain front retreats, erosion produces pediment with thin cover of alluvial deposits.

Alluvial deposits

Pediment

TIME 4
Long-continued erosion produces extensive pediment with some mountain remnants.

Mountain remnants

Pediment

Figure 19.21 Cima Dome is a pediment in the Mojave Desert. The surface of the dome is covered by a thin veneer of alluvial sediments. The two knobs, on the left and right sides of the dome, are regarded as the final remnants of a former mountain. [Martin Miller.]

Valleys in deserts have the same range of profiles as valleys elsewhere, but far more of them have steep walls produced by the rapid erosion caused by mass movements and streams. Much of the landscape of deserts is shaped by rivers, but the valleys—called **dry washes** in the western United States and **wadis** in the Middle East—are usually dry.

Desert streams are widely spaced because of the relatively infrequent rainfall, but drainage patterns are generally similar to those of other terrains—with one difference. Many desert streams die out before they can reach across the desert to join larger rivers flowing to the oceans. This may happen because of infrequent rainfall and the loss of water by evaporation and infiltration or because dunes dam the streams. Damming by dunes may lead to the development of playa lakes.

A special type of eroded bedrock surface, called a **pediment,** is a characteristic landform of the desert. Pediments are broad, gently sloping platforms of bedrock left behind as a mountain front erodes and retreats from its valley (**Figure 19.20**). The pediment spreads like an apron around the base of the mountains as thin deposits of alluvial sands and gravels accumulate. Long-continued erosion eventually forms an extensive pediment below a few mountain remnants (**Figure 19.21**). A cross section of a typical pediment and its mountains would reveal a fairly steep mountain slope abruptly leveling into the gentle pediment slope. Alluvial fans deposited at the lower edge of the pediment merge with the sedimentary fill of the valley below the pediment.

There is much evidence that a pediment is formed by running water that both cuts the erosional platform above and deposits an alluvial fan apron below. At the same time, the mountain slopes at the head of the pediment maintain their steepness as they retreat, instead of becoming the rounded, gentler slopes found in humid regions. We do not know how the specific rock types and erosional processes interact in an arid environment to keep slopes steep while the pediment enlarges.

SUMMARY

Where do winds form and how do they flow? Earth is encircled by belts of wind that develop in response to warming of the atmosphere at the equator, which causes the air to rise and flow toward the poles. As the air moves toward the poles, it gradually cools and begins to sink. The cold, dense air at the poles then flows back along Earth's surface to the equator. Winds vary in speed and direction from day to day, but over the long term, they tend to come mainly from one direction. The Coriolis effect, produced by Earth's rotation, deflects these moving winds to the right in the Northern Hemisphere and to the left in the Southern Hemisphere.

How do winds erode and transport sand and finer-grained sediments? Wind can pick up and transport dry sediment particles in the same way that running water transports sediment. Air flows are limited both in the size of particles they can carry (rarely larger than coarse-grained sand) and in their restricted ability to keep particles in suspension for long times. These limitations result from air's low viscosity and density. Almost all air flows are turbulent, and high winds may reach velocities of 100 km/hour or more, enhancing the wind's capacity to carry sediment. Windblown materials include volcanic ash, quartz and other mineral fragments such as clay minerals, and organic materials such as pollen and bacteria. The wind can carry great amounts of sand and dust. It moves sand grains primarily by saltation and carries the finer-grained silt- and clay-sized particles (dust) in suspension. Winds blow sand into cross-bedded ripples and dunes. Deflation and sandblasting are the primary ways in which winds erode Earth's surface, thereby producing desert pavement and ventifacts.

How do winds deposit sand dunes and dust? When winds die down, they deposit sand in dunes of various shapes and sizes. Dunes form in sandy desert regions, behind beaches, and along sandy floodplains, all of which are places with a ready supply of loose sand and moderate to strong winds. Dunes start as sand drifts in the lee of obstacles and grow to heights of many meters or even hundreds of meters. Dunes migrate downwind as sand grains saltate up the gentler windward slopes and fall over onto the steeper downwind slip faces. The various kinds of dunes—transverse, linear, barchan, and blowout—are determined by the speed of the wind, its constancy or variability of direction, and the abundance of sand. As the velocity of dust-laden winds decreases, the dust settles to form loess, a blanket of dust. Loess layers in many recently glaciated areas were deposited by winds blowing over the floodplains of muddy streams formed by meltwater. Loess can accumulate to great thicknesses downwind of dusty desert regions.

How do wind and water combine to shape the desert environment and its landscape? Deserts occur in the rain shadows of mountain ranges, in subtropical regions of constant high atmospheric pressure, and in the interiors of some continents. In all these places, originally moisture-laden winds become dry, and rainfall is rare. Weathering mechanisms are the same in deserts as in more humid regions, but in deserts physical breakdown of rocks is predominant, with chemical weathering at a minimum because of the lack of water. Most desert soils are thin, and bare rock surfaces are common. Streams may run only intermittently, but they are responsible for much of the erosion and sedimentation in the desert, carrying away heavy loads of coarse sediment and depositing them on alluvial fans and floodplains. Naturally dammed rivers in mountainous deserts can form playa lakes, which deposit evaporite minerals as they dry up.

Desert landscapes comprise dunes formed by eolian sedimentation; desert pavements; and pediments, which are broad, gently sloping platforms eroded from bedrock as mountains retreat while maintaining the steepness of their slopes.

KEY TERMS AND CONCEPTS

deflation (p. 458)
desertification (p. 453)
desert pavement (p. 459)
desert varnish (p. 466)
dry wash (p. 469)
eolian (p. 453)
loess (p. 463)

pediment (p. 469)
playa (p. 468)
playa lake (p. 467)
sandblasting (p. 458)
slip face (p. 461)
ventifact (p. 458)
wadi (p. 469)

EXERCISES

1. What types of materials and sizes of particles can the wind move?

2. What is the difference between the way wind transports dust and the way it transports sand?

3. How is the wind's ability to transport sedimentary particles linked to climate?

4. What are the main features of wind erosion?

5. Where do sand dunes form?

6. Name three types of sand dunes and show the relationship of each to wind direction.

7. What typical desert landforms are composed of sediment?

8. What are the geologic processes that form playa lakes?

9. What is desertification?

10. Where are loess deposits found?

THOUGHT QUESTIONS

1. You have just driven a truck through a sandstorm and discover that the paint has been stripped from the lower parts of the truck but the upper parts are barely scratched. What process is responsible, and why is it restricted to the lower parts of the truck?

2. What evidence might you find in an ancient sandstone that would point to its eolian origin?

3. Compare the heights to which sand and dust are carried in the atmosphere and explain the differences or similarities.

4. Trucks continually have to haul away sand covering a coastal highway. What do you think might be the source of the sand? Could its encroachment be stopped?

5. What features of a desert landscape would lead you to believe it was formed mainly by streams, with secondary contributions from eolian processes?

6. Which of the following options would be a more reliable indication of the direction of the wind that formed a barchan: cross-bedding or the orientation of the dune's shape on a map? Why?

7. What factors determine whether sand dunes will form on a stream floodplain?

8. There are large areas of sand dunes on Mars. From this fact alone, what can you infer about conditions on the Martian surface?

9. What aspects of an ancient sandstone would you study to show that it was originally a desert sand dune?

10. What kinds of landscape features would you ascribe to the work of the wind, to the work of streams, or to both?

11. How does desert weathering differ from or resemble weathering in more humid climates?

12. What evidence would cause you to infer that dust storms and strong winds were common in glacial times?

SHORT-TERM PROJECT

See Chapter 12.

COASTLINES AND OCEAN BASINS

For most of human history, the 71 percent of Earth's surface covered by the oceans was a mystery. The large populations who lived at the edge of the sea knew well the impact of the waves, the rise and fall of the tides, and the devastating effects of powerful storms. But they could only guess at the forces that caused these important processes. Coastal processes that affect shorelines are now known to result from interactions within the climate system and the solar system. Tides are caused by gravitational interactions between Earth and the Sun and Moon, and coastal surf and storms—such as hurricanes—result from interactions between the atmosphere and hydrosphere.

How can we monitor these processes and gauge their effects? Ordinary waves and astronomical tides are responsible for the slow, steady change of the shape of coastlines and the movement of valuable materials, such as sand, up and down the coastline. What is present today may be gone tomorrow, and vice versa. These patterns may be modified or accelerated when storms arrive at the coast.

And what about the deep sea, which is invisible to humans without the aid of remote observation tools? The nature of the seafloor deeper than the shallowest coastal waters remained a mystery until the middle of the nineteenth century. In 1872, the *Challenger*, a small wooden British warship converted and fitted out for scientific study of the seas, became the first research vessel to explore the oceans scientifically. The *Challenger* expedition discovered great areas of submerged hills and flat plains, extraordinarily deep trenches, and submarine volcanoes.

Today, Earth scientists still search for answers to the questions first raised by the early discoveries. What tectonic forces raised the submarine mountain ranges and depressed the trenches? Why are some areas of the seafloor flat and others hilly? Although oceanographers made many important discoveries in the first half of the twentieth century, the answers to most of these questions had to await the plate tectonics revolution of the late 1960s. As we saw in

Waves pounding a rocky shoreline at Cape Arago State Park, Coos County, Oregon. [Steve Terrill/Corbis.]

Chapter 2, it was geological and geophysical observations of the ocean floors, not of the continents, that led to the theory of plate tectonics.

In this chapter, we examine the processes that affect shorelines and coastal areas and consider the impact of waves, tides, and damaging storms. In the second half of the chapter, we explore the geology of Earth's ocean basins, with their submerged mountains and valleys, underwater volcanoes, and many kinds of rocks and sediments.

The word *ocean* applies both to the five major oceans (Atlantic, Pacific, Indian, Arctic, and Antarctic) and to the single connected body of water called the *world ocean.* The term *sea* includes both the oceans and the smaller bodies of water set off somewhat from the world ocean. Thus, the Mediterranean Sea is narrowly connected with the Atlantic Ocean by the Strait of Gibraltar and with the Indian Ocean by the Suez Canal. Other seas, such as the North Sea and the Atlantic Ocean, are broadly connected. In the world ocean, seawater—the salty water of the oceans and seas—is remarkably constant in its general chemical composition from year to year and from place to place. The equilibrium maintained by the oceans is determined by the general composition of river waters entering the sea, the composition of the sediment brought into the oceans, and the formation of new sediment in the oceans.

BASIC DIFFERENCES BETWEEN OCEAN BASINS AND CONTINENTS

Plate tectonics has provided us with a basic understanding of the differences between the geology of continents and the geology of ocean basins. Away from continental margins, the deep seafloor has no folded and faulted mountains like those on the continents. Instead, plate tectonic deformation is largely restricted to the faulting and volcanism found at mid-ocean ridges and subduction zones. Moreover, the weathering and erosion processes discussed in previous chapters are much less important in the oceans than on the land because the oceans lack efficient fragmentation processes, such as freezing and thawing, and major erosive agents, such as streams and glaciers. Deep-sea currents can erode and transport sediment but cannot effectively attack the plateaus and hills of basaltic rock that form the oceanic crust.

Because tectonic deformation, weathering, and erosion are minimal over much of the seafloor, volcanism and sedimentation dominate the geology of the oceans. Volcanism creates mid-ocean ridges, island groups (such as the Hawaiian Islands) in the middle of the oceans, and arcs of volcanic islands near deep oceanic trenches. Sedimentation shapes much of the rest of the ocean floor. Soft sediments of mud and calcium carbonate blanket the low hills and plains of the sea bottom and accumulate on oceanic plates as they spread

from mid-ocean ridges. As the plates move farther and farther from a ridge, they accumulate sediments. Deep-sea sedimentation is more continuous than the sedimentation in most continental environments, and it therefore preserves a better record of geologic events—for example, a more detailed history of Earth's climate changes.

This record is limited in time, however, because subduction is continually swallowing the oceanic plates back into the mantle, thereby destroying oceanic sediments by metamorphism and melting. On average, it takes only a few tens of millions of years for the crust created at a mid-ocean ridge to spread across an ocean and come to a subduction zone. As we saw in Chapter 2, the oldest parts of today's ocean floor were formed in the Jurassic period, about 180 million years ago; they are currently found near the western edge of the Pacific Plate (see Figure 2.12). In the next 10 million years or so, the sedimentary record that lies atop this crust will disappear down a subduction zone.

We will begin our exploration of the oceans with a close look at shorelines and coastal processes. Following that, we will move farther offshore to examine the submerged margins of the continents that bound the ocean basins and finish with a discussion of the deep ocean floor.

COASTAL PROCESSES

Coasts, the broad regions where land meets sea, present striking contrasts of landscape (**Figure 20.1**). On the coast of North Carolina, for example, long, straight, sandy beaches stretch for miles along low coastal plains. In most of New England, by contrast, rocky cliffs bound elevated shores, and the few beaches that exist are made of gravel. Many of the seaward edges of islands in the tropics, such as those in the Caribbean Sea, are coral reefs, the delight of divers. As we will see, tectonics, erosion, and sedimentation work together to create this great variety of shapes and materials.

The major geological forces operating at the *shoreline—* the line where the water surface intersects the shore—are waves and tides. Together, they erode even the most resistant rocky shores. Waves and tides create currents, which transport sediment produced by erosion of the land and deposit it on beaches and in shallow waters along the shore.

Wave Motion: The Key to Shoreline Dynamics

Centuries of observation have taught us that waves are changeable. In quiet weather, waves with calm troughs between them roll regularly into shore. In the high winds of a storm, however, waves move in a confusion of shapes and sizes. They may be low and gentle far from the shore yet become high and steep as they approach land. High waves can break on the shore with fearful violence, shattering concrete seawalls and tearing apart houses built along the beach. To understand the dynamics of shorelines and to make sen-

(a)

(b)

(c)

(d)

Figure 20.1 Coastlines exhibit a variety of geologic forms.
(a) Long, straight, sandy beach coastline, Pea Island, North
Carolina. [Courtesy of Bill Birkemeier/U.S. Army Corp of Engineers.]
(b) Rocky coastline, Mount Desert Island, Maine. This glaciated
coastline is being uplifted following the end of the last ice age,
about 10,000 years ago. [Neil Rabinowitz/Corbis.] (c) The Twelve

Apostles, a group of stacks at Port Campbell, Australia, developed
in horizontal beds of sedimentary rock. These remnants of shore
erosion are left as the shoreline retreats under the action of
waves. [Kevin Schafer.] (d) Coral reef along the Florida coastline.
[Hays Cummins, Miami University.]

sible decisions about shore development, we need to understand how waves work.

The wind blowing over the surface of the water creates
waves by transferring the energy of motion from air to
water. As a gentle breeze of 5 to 20 km/hour starts to blow
over a calm sea surface, ripples—little waves less than a
centimeter high—take shape. As the speed of the wind
increases to about 30 km/hour, the ripples grow to full-
sized waves. Stronger winds create larger waves and blow
off their tops to make whitecaps. The height of the waves
increases as

• The wind speed increases.

• The wind blows for longer times.

• The distance over which the wind blows the water
increases.

Storms blow up large, irregular waves that radiate outward from the storm center, like the ripples moving outward
from a pebble dropped into a pond. As the waves travel out
from the storm center in ever-widening circles, they become
more regular, changing into low, broad, rounded waves
called *swell,* which can travel hundreds of kilometers. Several storms at different distances from a shoreline, each producing its own pattern of swell, account for the often irregular intervals between waves approaching the shore.

If you have seen waves in an ocean or a large lake, you
have probably noticed how a piece of wood or other light
material floating on the water moves a little forward as
the crest of a wave passes and then a little backward as the
trough between waves passes. While moving back and forth,
the wood stays in roughly the same place, and so does the
water around it.

We describe a wave form in terms of the following three characteristics (**Figure 20.2**):

1. *Wavelength,* the distance between crests

2. *Wave height,* the vertical distance between the crest and the trough

3. *Period,* the time that it takes for successive waves to pass

We measure the velocity at which a wave moves forward by using a basic equation:

$$V = L/T$$

where V is the velocity, L is the wavelength, and T is the period. Thus, a typical wave with a length of 24 m and a period of 8 s would have a velocity of 3 m/s. The periods of most waves range from just a few seconds to as long as 15 or 20 s, with wavelengths varying from about 6 m to as much as 600 m. Consequently, wave velocities vary from 3 to 30 m/s. Wave motion becomes very small below a depth of about one-half the wavelength. That is why deep divers and submarines are unaffected by the waves at the surface.

The Surf Zone

Swell becomes higher as it approaches the shore. There it assumes the familiar sharp-crested wave shape. These waves are called *breakers* because, as the waves come closer to shore, they break and form surf—a foamy, bubbly surface. The *surf zone* is the offshore belt along which breaking waves collapse as they approach the shore. Breaking waves pound the shore, eroding and carrying away sand, weathering and breaking up solid rock, and destroying structures built close to the shoreline.

The transformation from swell to breakers starts where the bottom shallows to less than one-half the wavelength of the swell. At that point, the wave motion just above the bottom becomes restricted because the water can only move back and forth horizontally. Above that, the water can move vertically just a little (see Figure 20.2). The restricted motion of the water particles slows the whole wave. Although the wave slows, its period remains the same because the swell keeps coming in from deeper water at the same rate. From the wave equation, we know that if the period remains constant but the wavelength decreases, then the velocity must also decrease. The typical wave that we used as our example earlier might keep the same period of 8 s while changing to a length of 16 m, in which case it would have a velocity of 2 m/s. Thus, as the waves approach the shore, they become more closely spaced, higher, and steeper, and their wave crests become sharper.

As a wave rolls toward the shore, it becomes so steep that the water can no longer support itself, and the wave breaks with a crash in the surf zone (see Figure 20.2). Gently sloping bottoms cause waves to break farther from shore, and steeply sloping bottoms make waves break closer to shore. Where rocky shores are bordered by deep water, the waves break directly on the rocks with a force amounting to tons per square meter, throwing water high into the air. It is not surprising that concrete seawalls built to protect buildings along the shore quickly start to crack and must be repaired constantly.

After breaking at the surf zone, the waves, now reduced in height, continue to move in, breaking again right at the shoreline. They run up onto the sloping front of the beach, forming an uprush of water called *swash.* The water then runs back down again as *backwash.* Swash can carry sand and even large pebbles and cobbles if the waves are high enough. The backwash carries the particles seaward again.

The back and forth motion of the water near the shore is strong enough to carry sand grains and even gravel. Wave action in water as deep as about 20 m can move fine sand. Large waves caused by intense storms can scour the bottom at much greater depths, down to 50 m or more.

Wave Refraction

Far from shore, the lines of swell are parallel to one another but are usually at some angle to the shoreline. As the waves approach the beach over a shallowing bottom, the rows of waves gradually bend to a direction more parallel to the shore. This bending of lines of wave crests as they approach the shore from an angle is called *wave refraction* (see Figure 20.2). It is similar to the bending of light rays in optical refraction, which makes a pencil half in and half out of water appear to bend at the water surface. Wave refraction begins as a wave approaches the shore at an angle. The part of the wave closest to the shore encounters the shallowing bottom first, and the front of the wave slows. Then the next part of the wave meets the bottom and it, too, slows. Meanwhile, the parts closest to shore have moved into even shallower water and slowed even more. Thus, in a continuous transition along the wave crest, the line of waves bends toward the shore as it slows.

Wave refraction results in more intense wave action on projecting headlands and less intense action in indented bays, as Figure 20.2 illustrates. The water becomes shallow more quickly around headlands than in the surrounding deeper water on either side. Waves are refracted around headlands—that is, they are bent toward the projecting part of the shore from both sides. The waves converge around the point of land and expend proportionately more of their energy breaking there than at other places along the shore. Thus, erosion by waves is concentrated at headlands and tends to wear them away more quickly than it does straight sections of shoreline.

The opposite happens as a result of wave refraction in a bay. The waters in the center of the bay are deeper, so the waves are refracted on either side into shallower water. The energy of wave motion is diminished at the center of the bay, which makes bays good harbors for ships.

Key Figure 20.2   Wave motion is influenced by water depth and shape of the shoreline.

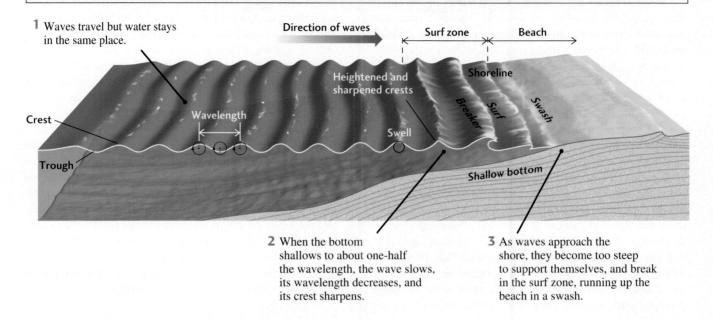

1 Waves travel but water stays in the same place.

Direction of waves

Surf zone   Beach

Heightened and sharpened crests

Shoreline

Breaker

Surf

Swash

Crest

Wavelength

Swell

Trough

Shallow bottom

2 When the bottom shallows to about one-half the wavelength, the wave slows, its wavelength decreases, and its crest sharpens.

3 As waves approach the shore, they become too steep to support themselves, and break in the surf zone, running up the beach in a swash.

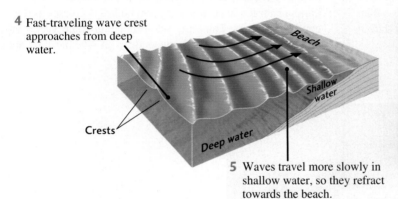

4 Fast-traveling wave crest approaches from deep water.

Beach

Crests

Shallow water

Deep water

5 Waves travel more slowly in shallow water, so they refract towards the beach.

[John S. Shelton.]

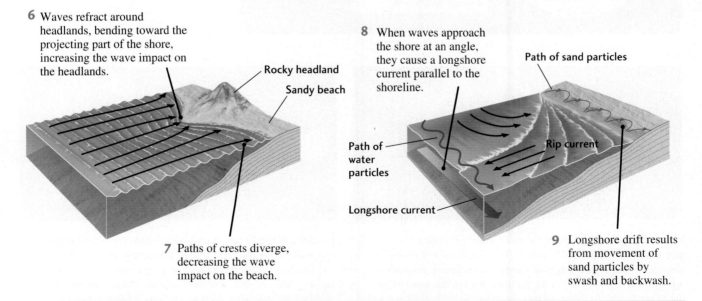

6 Waves refract around headlands, bending toward the projecting part of the shore, increasing the wave impact on the headlands.

Rocky headland

Sandy beach

7 Paths of crests diverge, decreasing the wave impact on the beach.

8 When waves approach the shore at an angle, they cause a longshore current parallel to the shoreline.

Path of sand particles

Path of water particles

Rip current

Longshore current

9 Longshore drift results from movement of sand particles by swash and backwash.

Although refraction makes waves more parallel to the shore, many waves still approach at some small angle. As the waves break on the shore, the swash moves up the beach slope perpendicular to this small angle. The backwash runs down the slope in the opposite direction at a similar small angle. The combination of the two motions moves the water a short way down the beach (see Figure 20.2). Sand grains carried by swash and backwash are thus moved along the beach in a zigzag motion known as *longshore drift.*

Waves approaching the shoreline at an angle can also cause a **longshore current,** a shallow-water current that is parallel to the shore. The water movement of swash and backwash in and out from the shore at an angle creates a zigzag path of water particles that adds up to a net transport along the shore in the same direction as the longshore drift. Much of the net transport of sand along many beaches comes from this kind of current. Longshore currents are prime determiners of the shape and extent of sandbars and other depositional shoreline features. At the same time, because of their ability to erode loose sand, longshore currents may remove much sand from a beach. Longshore drift and longshore currents working together are potent processes in the transport of large amounts of sand on beaches and in very shallow waters. In deeper but still shallow waters (less than 50 m), longshore currents—especially those running during large storms—strongly affect the bottom.

Some types of flows related to longshore currents can pose a threat to unwary swimmers. A *rip current,* for example, is a strong flow of water moving outward from the shore at right angles to the shore. It occurs when a longshore current builds up along the shore and the water piles up imperceptibly until a critical point is reached. At that point, the water breaks out to sea, flowing through the oncoming waves in a fast current. Swimmers can avoid being carried out to sea by swimming parallel to the shore to get out of the rip.

The Tides

For thousands of years, mariners and shoreline dwellers have known the twice-daily rise and fall of the sea that we call **tides.** Many observers noticed a relationship among the position and phases of the Moon, the heights of the tides, and the times of day at which the water reaches high tide. Not until the seventeenth century, however, when Isaac Newton formulated the laws of gravitation, did we begin to understand that the tides result from the gravitational pull of the Moon and the Sun on the water of the oceans.

The Moon, the Sun, Gravity, and the Tides The gravitational attraction between any two bodies decreases as they get farther apart. Thus, the tide-producing force varies on different parts of the Earth, depending on whether they are

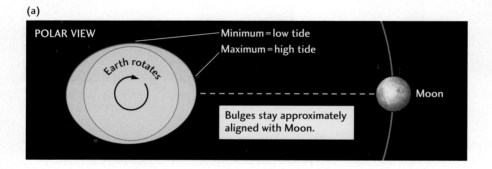

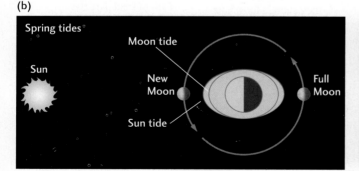 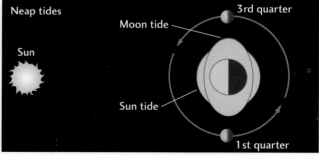

Figure 20.3 Origin of the ocean tides from gravitational attraction of the Sun and Moon. (a) The Moon's gravitational attraction causes two bulges of water on the Earth's oceans, one on the side nearest the Moon and the other on the side farthest from the Moon. As Earth rotates, these bulges remain aligned and pass over Earth's surface, creating the high tides. The relative positions of the Earth, Moon, and Sun determine the heights of high tide during the lunar month. (b) At new and full Moon, Sun and Moon tides reinforce each other and make the highest (spring) tides. At first- and third-quarter Moon, Sun and Moon tides are in opposition, causing the lowest (neap) tides.

closer to or farther from the Moon. On the side of the Earth closest to the Moon, the water, being closer to the Moon, experiences a larger gravitational attraction than the average attraction for the whole of the solid Earth. This produces a bulge in the water, seen as a tide. On the side of the Earth farthest from the Moon, the solid Earth, being closer to the Moon than the water, is pulled toward the Moon more than the water is, and the water therefore appears to be pulled away from the Earth as another bulge.

Thus, two bulges of water occur on Earth's oceans: one from the side nearest the Moon, where the net attraction is toward the Moon, and the other on the side farthest from the Moon, where the net attraction is away from the Moon (**Figure 20.3**). The net gravitational attraction between the oceans and the Moon is at a maximum on the side of Earth facing the Moon and at a minimum on the side facing away from the Moon. As Earth rotates, the bulges of water stay approximately aligned. One always faces the Moon, the other is always directly opposite. These bulges passing over the rotating Earth are the high tides.

The Sun, although much farther away, has so much mass (and thus so much gravity) that it, too, causes tides. Sun tides are a little less than half the height of Moon tides. Sun tides are not synchronous with Moon tides. Sun tides occur as the Earth rotates once every 24 hours, the length of a solar day. The rotation of the Earth with respect to the Moon is a little longer because the Moon is moving around the Earth, resulting in a lunar day of 24 hours and 50 minutes. In that lunar day, there are two high tides, with two low tides between them.

When the Moon, Earth, and Sun line up, the gravitational forces of the Sun and the Moon reinforce each other. This produces the *spring tides,* which are the highest tides; their name is not related to the season, but to the German verb *springen,* meaning "to leap up," They appear every two weeks at full and new Moon. The lowest tides, the *neap tides,* come in between, at the first- and third-quarter Moon, when the Sun and Moon are at right angles to each other with respect to the Earth (see Figure 20.3).

Although the tides occur regularly everywhere, the difference between high and low tides varies in different parts of the ocean. As the tidal bulges of water rise and fall, they also move along the surface of the ocean, encountering obstacles, such as continents and islands, that hinder the flow of water. In the middle of the Pacific Ocean—in Hawaii, for example, where there is little constriction or obstruction of the flow of the tides—the difference between low and high tides is only 0.5 m. Near Seattle, where the shore along Puget Sound is very irregular and the tides are constricted through narrow passageways, the difference between the two tides is about 3 m. Extraordinary tides occur in a few places, such as the Bay of Fundy in eastern Canada, where the tidal range can be more than 12 m. Many people living along the shore need to know when tides will occur, so governments publish tide tables showing predicted tide heights and times. These tables combine local experience with

Figure 20.4 Tidal flats, such as this one at Mont-Saint-Michel, France, may be extensive areas covering many square kilometers but most often are narrow strips seaward of the beach. When a very high tide advances on a broad tidal flat, it may move so rapidly that areas are flooded faster than a person can run. The beachcomber is well advised to learn the local tides before wandering. [Thierry Prat/Corbis Sygma.]

knowledge of the astronomical motions of Earth and the Moon with respect to the Sun.

Tides may combine with waves to cause extensive erosion of the shore and destruction of shoreline property. Intense storms passing near the shore during a spring tide may produce *tidal surges,* waves at high tide that can overrun the entire beach and batter sea cliffs. Do not confuse tidal surges with tsunamis, commonly but incorrectly termed "tidal waves." *Tsunamis* are large ocean waves caused by undersea events such as earthquakes, landslides, and the explosion of oceanic volcanoes (see Chapter 13).

Tidal Currents Tides moving near shorelines cause currents that can reach speeds of a few kilometers per hour. As the tide rises, the water flows in toward the shore as a *flood tide,* moving into shallow coastal marshes and up small streams. As the tide passes the high stage and starts to fall, the *ebb tide* moves out, and low-lying coastal areas are exposed again. Such tidal currents meander across the cut channels into **tidal flats,** the muddy or sandy areas that lie above low tide but are flooded at high tide (**Figure 20.4**).

Hurricanes and Coastal Storm Surge

On August 29, 2005, Hurricane Katrina made landfall in Louisiana and struck New Orleans as a nearly Category 4

EARTH ISSUES

20.1 The Great New Orleans Flood

On August 25, 2005, Hurricane Katrina struck south Florida as a Category 1 storm, killing 11 people. Three days later, the hurricane grew to a monster Category 5 storm, with maximum sustained winds of up to 280 km/hour (175 miles/hour) and gusts up to 360 km/hour (225 miles/hour). On August 28, the National Weather Service issued a bulletin predicting "devastating" damage, and the mayor of New Orleans ordered an unprecedented mandatory evacuation of the city. When Katrina made landfall just south of New Orleans on August 29, it was almost a Category 4 storm with sustained winds of 204 km/hour (127 miles/hour). It had a minimum pressure of 918 mbar (27.108 inches), making it the third strongest hurricane on record to make landfall in the United States. More than 100 people lost their lives during the early morning hours of August 29 as a result of the direct impact of the storm.

A 5- to 9-m storm surge came ashore on virtually the entire coastline of Louisiana, Mississippi, Alabama, and Florida. The 9-m storm surge at Biloxi, Mississippi, is the highest ever recorded in America. The effects of this storm surge were devastating and unprecedented. Lake Pontchartrain, which is really a coastal embayment that is easily influenced by ocean conditions, was inundated by the storm surge. By midday on August 29, several sections of the levee system that holds back Lake Pontchartrain

Water spills over a levee along the Inner Harbor Navigational Canal and floods the inner city of New Orleans. [Vincent Laforet-Pool/Getty Images.]

storm, with sustained winds of 204 km/hour (127 miles/hour) (see Feature 20.1). Later that day, the rise in sea level associated with the hurricane storm surge caused several sections of the levee system in New Orleans to collapse. Subsequent flooding of parts of the city to depths of up to 7 or 8 m claimed hundreds of lives and left the city submerged and abandoned for almost a month. The damaging effects of extremely high, sustained winds and torrential rains are intuitively easy to understand. However, the associated *storm surge,* in which major regions of the coastline become flooded, is potentially the most destructive influence of hurricanes. The disaster in New Orleans was not so much the direct impact of the hurricane itself as of the storm surge that ultimately caused the collapse of part of the city's levee system.

Hurricanes Few things in nature can compare with the destructive force of a hurricane. Regarded as the greatest storms on Earth, **hurricanes** can annihilate shorelines and coastal regions with sustained winds, lasting many hours, of 260 km/hour (160 miles/hour) or higher combined with intense rainfall and a storm surge. Earth scientists have estimated that during its life cycle, a hurricane can expend as much energy as 10,000 nuclear bombs!

from New Orleans had collapsed. Subsequent flooding of the city to depths of up to 7 or 8 m left 80 percent of New Orleans under water. The effects of flooding claimed at least another 300 lives, and by September 21, the death toll exceeded 1000 as disease and malnourishment indirectly caused by the flooding took effect.

The damage caused by Hurricane Katrina is expected to surpass that caused by Hurricane Andrew as the most costly natural disaster in U.S. history, with estimates ranging between $100 billion and $200 billion. In addition to the hundreds of lives lost, over 150,000 homes were destroyed and over 1 million people were displaced —a humanitarian crisis unrivaled in the United States since the Great Depression.

What happened, and what could have been done to limit the damage? As with most natural disasters, the outcome was a result of rare but powerful geologic forces coupled with a lack of human preparation. No one anticipated and planned for the worst-case scenario. Earth scientists predicted for decades that eventually a Category 4 or 5 hurricane would strike New Orleans, which was only prepared to resist the damaging effects of a Category 3 or smaller hurricane. Katrina was nearly a Category 4 storm when it struck New Orleans. The historical record of hurricanes told us that this event would happen with a high degree of certainty; as Figure 20.9 shows, New Orleans is about in the middle of the "catcher's mitt" for hurricane landfalls in the United States. The hard part is to predict exactly when such a hurricane would occur. Federal budget cuts provided only token support to maintain and reinforce major work on the New Orleans area's east bank of hurricane levees that hold back the waters of Lake Pontchartrain. This complex network of concrete walls, metal gates, and giant earthen berms was never completed, leaving the city vulnerable.

It is not easy to protect a city from hurricane storm surge when its sidewalks and houses are, on average, 4 m below sea level. New Orleans is equally susceptible to unusually large and rare floods of the Mississippi River, which also is held back by an artificial levee system. When events are rare, it is natural to question if they are worth worrying about, and human memory may fail in providing the necessary guidance. Over the short term, we may escape these threats by random good luck. But over the long term, recorded history and the geologic record show that these rare and devastating forces will eventually take their toll if we are not adequately prepared.

Residents wade through a flooded street in New Orleans in the aftermath of Hurricane Katrina. [James Nielsen/AFP/ Getty Images.]

The term *hurricane* originates from Huracan, a god of evil to the aboriginal people of Central America. In the western Pacific and China Sea, hurricanes are known as *typhoons,* from the Cantonese word *tai-fung,* meaning "great wind." In Australia, Bangladesh, Pakistan, and India, they are known as *cyclones;* in the Philippines, they are called *baguios.*

Hurricanes form over tropical parts of Earth's oceans, between 8° and 20° latitude, in areas of high humidity, light winds, and warm sea-surface temperatures (typically 26°C or greater). These conditions usually occur in the summer and early fall months of the tropical North Atlantic and North Pacific oceans. For this reason, hurricane "season" in the Northern Hemisphere runs from June through November (**Figure 20.5**).

The first sign of hurricane development is the appearance of a cluster of thunderstorms over the tropical oceans in a region where tropical winds converge on one another. Occasionally, a cluster of thunderstorms breaks out from this convergence zone and becomes better organized. Most hurricanes that affect the Atlantic Ocean and Gulf of Mexico originate in a convergence zone just off the coast of west Africa and intensify as they break out and move westward across the tropical Atlantic.

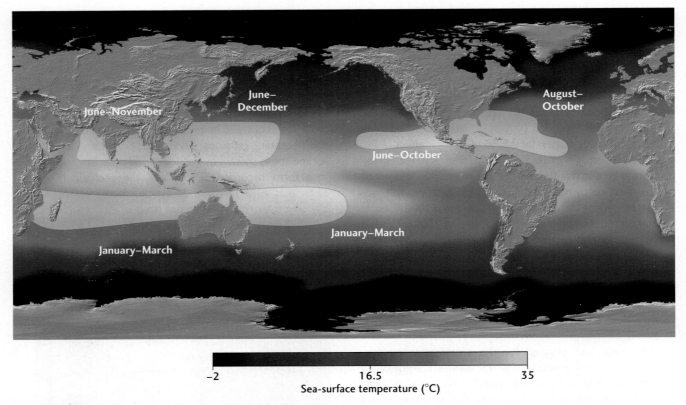

Sea-surface temperature (°C)
−2   16.5   35

Figure 20.5 Hurricanes arise in summer and early fall when ocean temperatures are warmest. [NASA/GSFC.]

As the storm develops, water vapor condenses to form rain, which releases heat energy. In response to atmospheric heating, the surrounding air becomes less dense and begins to rise. The atmospheric pressure at sea level drops in the region of heating. As the warm air rises, it triggers more condensation and rainfall, which in turn releases more heat. This causes more air to rise, which causes the surface pressure at sea level to fall even further. At this point a chain reaction occurs, as the rising temperatures in the center of the storm cause surface temperatures to fall to progressively lower levels. In the Northern Hemisphere, because of the *Coriolis effect* (see Chapter 19), the increasing winds will begin to circulate in a counterclockwise pattern around the storm's area of lowest pressure, which ultimately becomes the "eye" of the hurricane (**Figure 20.6**).

Once sustained wind speeds reach 37 km/hour (23 miles/ hour), the storm system is called a *tropical depression.* As winds increase to 63 km/hour (39 miles/hour), the system is called a *tropical storm* and receives a name. This naming tradition started with the use of World War II code names, such as Andrew, Bonnie, Charlie, and so forth. Finally, when wind speeds reach 119 km/hour (74 miles/ hour), the storm is classified as a hurricane. Once it becomes a hurricane, the storm is assigned a 1–5 rating based on its current intensity (Table 20.1). This **hurricane intensity scale** (*Saffir-Simpson scale*) is used to estimate the potential property damage and flooding expected along

Figure 20.6 Hurricane Katrina on August 28, 2005, a few hours before it struck New Orleans. In the Northern Hemisphere, winds circulate in a counterclockwise direction around the "eye" of the hurricane, which is the location of lowest surface pressure. Katrina was one of the most powerful hurricanes on record, with maximum sustained winds of up to 280 km/hour (175 miles/hour) and gusts up to 360 km/hour (225 miles/hour). [NASA/Jeff Schmaltz, MODIS Land Rapid Response Team.]

Storm Classification	Description
Category 1	Winds 119–153 km/hour (74–95 miles/hour). Storm surge generally 1–1.5 m (4–5 ft) above normal. No real damage to building structures. Damage primarily to unanchored mobile homes, shrubbery, and trees. Some damage to poorly constructed signs. Also, some coastal road flooding and minor pier damage.
Category 2	Winds 154–177 km/hour (96–110 miles/hour). Storm surge generally 2–2.5 m (6–8 ft) above normal. Some roofing material, door, and window damage to buildings. Considerable damage to shrubbery and trees, with some trees blown down. Considerable damage to mobile homes, poorly constructed signs, and piers. Coastal and low-lying escape routes flood 2–4 hours before arrival of the hurricane center. Small craft in unprotected anchorages break moorings. Hurricane Frances of 2004 made landfall over the southern end of Hutchinson Island, Florida, as a Category 2 hurricane.
Category 3	Winds 178–209 km/hour (111–130 miles/hour). Storm surge generally 2.5–3.5 m (9–12 ft) above normal. Some structural damage to small residences and utility buildings with a minor amount of curtain-wall failures. Damage to shrubbery and trees with foliage blown off trees and large trees blown down. Mobile homes and poorly constructed signs are destroyed. Low-lying escape routes are cut off by rising water 3–5 hours before arrival of the center of the hurricane. Flooding near the coast destroys smaller structures, with larger structures damaged by battering from floating debris. Terrain continuously lower than 1.5 m (5 ft) above mean sea level may be flooded inland 3 m (10 ft) or more. Evacuation of low-lying residences within several blocks of the shoreline may be required. Hurricanes Jeanne and Ivan of 2004 were Category 3 hurricanes when they made landfall in Florida and Alabama, respectively. Hurricane Katrina of 2005 made landfall near Buras-Triumph, Louisiana, with winds of 204 km/hour (127 miles/hour). Katrina will prove to be the costliest hurricane on record, with estimates of more than $100 billion in losses.
Category 4	Winds 210–250 km/hour (131–155 miles/hour). Storm surge generally 3.5–5 m (13–18 ft) above normal. More extensive curtain-wall failures with some complete roof structure failures on small residences. Shrubs, trees, and all signs are blown down. Complete destruction of mobile homes. Extensive damage to doors and windows. Low-lying escape routes may be cut off by rising water 3–5 hours before arrival of the center of the hurricane. Major damage to lower floors of structures near the shore. Terrain lower than 3 m (10 ft) above sea level may be flooded, requiring massive evacuation of residential areas as far inland as 15 km (9 miles).
Category 5	Winds greater than 250 km/hour (155 miles/hour). Storm surge generally greater than 5 m (18 ft) above normal. Complete roof failure on many residences and industrial buildings. Some complete building failures with small utility buildings blown over or away. All shrubs, trees, and signs blown down. Complete destruction of mobile homes. Severe and extensive window and door damage. Low-lying escape routes are cut off by rising water 3–5 hours before arrival of the center of the hurricane. Major damage to lower floors of all structures located less than 4.5 m (15 ft) above sea level and within 500 m (1650 ft) of the shoreline. Massive evacuation of residential areas on low ground within 15–20 km (9–12 miles) of the shoreline may be required. Only three Category 5 hurricanes have made landfall in the United States since records began. Hurricane Andrew of 1992 made landfall over southern Miami-Dade County, Florida, causing $26.5 billion in losses—the second costliest hurricane on record.

the coast from hurricane landfall. It is analogous to the Mercalli scale for earthquakes (see Table 13.1).

Storm Surge As a hurricane intensifies, a dome of seawater—known as **storm surge**—forms at levels higher than the surrounding ocean surface. The height of the storm surge is directly related to the extremely low atmospheric pressure and strong winds that encircle the eye of the hurricane. Large swells, high surf, and wind-driven waves ride atop this dome as it impacts coastal land areas, causing

Figure 20.7 Hurricane storm surge along coastlines may result in the complete destruction of residential buildings, which pile up as lines of debris well inland of the normal coastline. The damage seen here was caused by Hurricane Katrina. [U.S. Navy/Getty Images.]

extensive damage to facilities and the shoreline environment (**Figure 20.7**). This buildup of water can produce severe flooding in coastal areas, particularly when storm surge coincides with normal high tides. When the effects of a normal astronomical high tide and a storm surge coincide, it is known as a *storm tide* (**Figure 20.8**). Any landmass in the path of a storm surge will be affected to a greater or lesser extent, depending upon a number of factors. The stronger the storm and the shallower the offshore waters, the higher the storm surge.

Storm surge is the most deadly of a hurricane's associated hazards, as recently underscored by Hurricane Katrina. The magnitude of a hurricane is usually described in terms of its wind speed (see Table 20.1), but coastal flooding causes many more deaths than high wind. Boats ripped from their moorings, utility poles, and other debris floating atop hurricane surge often demolish buildings not destroyed by extreme winds. Even without the weight of floating debris, storm surge can severely erode beaches and highways and undermine bridges. Because much of the United States' densely populated Atlantic and Gulf Coast shorelines lie less than 3 m above mean sea level, the danger from storm tides is tremendous.

Hurricane Landfall Because hurricanes form above tropical waters and move across tropical waters to maintain their

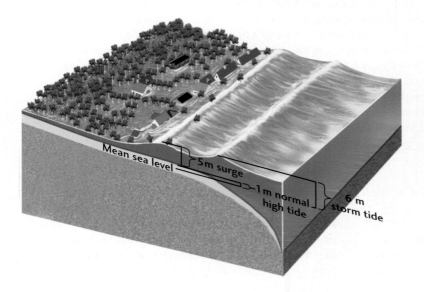

Figure 20.8 A storm tide is the combination of storm surge and the normal astronomical tide. If storm surge arrives at the same time as a high tide, the water height will be even greater. For example, if a normal astronomical tide is 1 m and a storm surge is 5 m, the resulting storm tide will be 6 m in height. As a hurricane nears land, a dome of water, topped by battering waves, moves ashore along an area of the coastline as wide as 160 km. The combination of storm surge, battering waves, and high winds can be life-threatening. The storm surge associated with the 2005 Katrina hurricane caused partial collapse of the levee system that protects New Orleans, resulting in extensive urban flooding and hundreds of deaths.

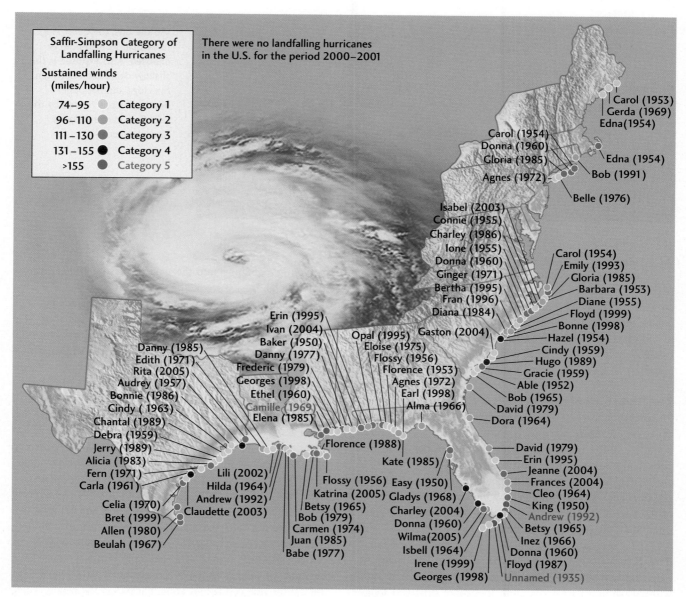

Figure 20.9 Hurricanes originating in the North Atlantic Ocean will usually make landfall in the coastal areas of the southeastern United States, including the Gulf Coast states. Hurricanes lose energy if they move across cold water, so the number of hurricanes that make landfall drops dramatically for the central and northeastern states. [NOAA.]

energy, most make *landfall* at low latitudes. In the Atlantic Ocean of the Northern Hemisphere, most hurricanes tend to wind up in Florida and the northern Gulf of Mexico (**Figure 20.9**). However, because there is a tendency for the winds to deflect northward (the Coriolis effect), hurricanes sometimes make landfall farther up the coast. In rare cases, they may reach landfall in New England but are always of lower intensity. The most powerful, Category 4 and 5 hurricanes are all restricted to lower latitudes.

SHORELINES

At shorelines, we can observe the constant motion of ocean waters and their effects on the shore. Current environmental problems such as coastal erosion and polluted shallow waters have made the geology of shorelines and shallow seas a critical area of research. Waves, longshore currents, and tidal currents interact with the rocks and tectonics of the coast to shape shorelines into a multitude of forms (see Figure 20.1). We can see these factors at work in the most popular of shorelines—beaches.

Beaches

A beach is a shoreline made up of sand and pebbles. Beaches may change shape from day to day, week to week, season to season, and year to year. Waves and tides sometimes broaden and extend a beach by depositing sand and sometimes narrow it by carrying sand away.

Figure 20.10 Tide terrace. At low tide, the outer ridge (a sandbar at high tide) is exposed. Also exposed is the shallow depression between the ridge and the upper beach, which is rippled by the tidal flow in many places. [James Valentine.]

Many beaches are straight stretches of sand that range from 1 km to more than 100 km long; others are smaller crescents of sand between rocky headlands. Belts of dunes border the landward edge of many beaches; bluffs or cliffs of sediment or rock border others. A beach may have a tide terrace—a flat, shallow area between the upper beach and an outer bar of sand—on its seaward side (**Figure 20.10**).

The Structure of Beaches **Figure 20.11** shows the major parts of a beach. These parts may not all be present at all times on any particular beach. Farthest out is the *offshore,* bounded by the surf zone, where the bottom begins to become shallow enough for waves to break. The *foreshore* includes the surf zone; the tidal flat; and, right at the shore, the swash zone, a slope dominated by the swash and backwash of the waves. The *backshore* extends from the swash zone up to the highest level of the beach.

The Sand Budget of a Beach A beach is a scene of incessant movement. Each wave moves sand back and forth with swash and backwash. Both longshore drift and longshore currents move sand down the beach. At the end of a beach and to some extent along it, sand is removed and deposited in deep water. In the backshore or along sea cliffs, sand and pebbles are freed by erosion and replenish the beach. The wind that blows over the beach transports sand,

sometimes offshore into the water and sometimes onshore onto the land.

All these processes together maintain a balance between the addition and removal of sand, resulting in a beach that may appear to be stable but is actually exchanging its material on all sides. **Figure 20.12** illustrates the **sand budget** of a beach—the inputs and outputs caused by erosion and sedimentation. At any point along a beach, it gains sand by the inputs: material along the backshore is eroded; longshore drift and longshore currents bring sand to the beach; and rivers that enter the sea along the shore bring in sediment. The beach loses sand from the outputs: winds carry sand to backshore dunes, longshore drift and currents carry it downcurrent, and deep-water currents and waves transport it during storms.

If the total input balances the total output, the beach is in equilibrium and keeps the same general form. If input and output are not balanced, the beach grows or shrinks. Temporary imbalances are natural over weeks, months, or years. A series of large storms, for example, might move large amounts of sand from the beach to somewhat deeper waters on the far side of the surf zone, narrowing the beach. Then, weeks of mild weather and low waves might move the sand onto shore and rebuild a wide beach. Without this constant shifting of the sands, beaches might be unable to recover from trash, litter, and some kinds of pollution. Within a year or two, even oil from spills will be transported or buried out of sight,

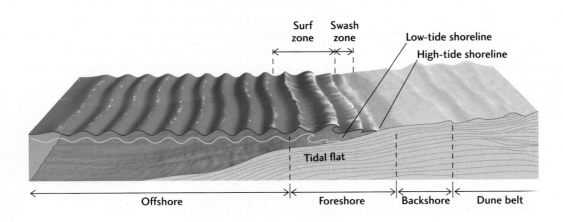

Figure 20.11 A profile of a beach, showing its major parts.

SAND BUDGET	
INPUTS	**OUTPUTS**
Sediments eroded from backshore cliffs by waves	Sediments transported to backshore dunes by offshore winds
Sediments eroded from upcurrent beach by longshore drift and current	Sediments transported downcurrent by longshore drift and current
Sediments brought in by rivers	Sediments transported to deep water by tidal currents and waves

Figure 20.12 The sand budget is a balance between inputs and outputs of sand by erosion, transport, and sedimentation.

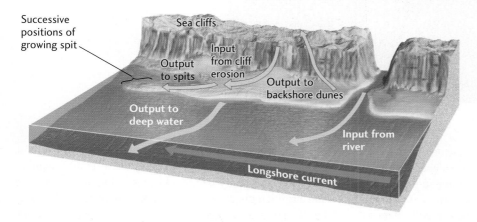

although the tarry residue may later be uncovered in spots. Beaches would clean up rapidly if the littering were to stop.

Some Common Forms of Beaches Long, wide, sandy beaches grow where sand inputs are abundant, often where soft sediments make up the coast. Where the backshore is low and the winds blow onshore, wide dune belts border the beach. If the shoreline is tectonically elevated and the rocks are hard, cliffs line the shore, and any small beaches that evolve are composed of material eroded from the cliffs. Where the shore is low-lying, sand is abundant, and tidal currents are strong, extensive tidal flats are laid down and are exposed at low tide.

What happens if one of the inputs is cut off—for example, by a concrete wall built at the top of the beach to prevent erosion? Because erosion supplies sand to the beach, preventing erosion cuts the sand supply and so shrinks the beach. Attempts to save the beach may actually destroy it (see Feature 20.2 on pages 490–491).

Erosion and Deposition at Shorelines

The topography of the shoreline, like that of the land interior, is a product of tectonic forces elevating or depressing Earth's crust, erosion wearing it down, and sedimentation filling in the low spots. Thus, the factors directly at work are

• Uplift of the coastal region, which leads to erosional coastal forms

• Subsidence of the coastal region, which produces depositional coastal forms

• The nature of the rocks or sediments at the shoreline

• Changes in sea level, which affect the drowning or emergence of a shoreline

• The average and storm wave heights

• The heights of the tides, which affect both erosion and sedimentation

Erosional Coastal Forms Erosion is active at tectonically uplifted rocky coasts. Along these coasts, prominent cliffs and headlands jut into the sea, alternating with narrow inlets and irregular bays with small beaches. Waves crash against rocky shorelines, undercutting cliffs and causing huge blocks to fall into the water, where they are gradually worn away. As the sea cliffs retreat by erosion, isolated remnants called *stacks* are left standing in the sea, far from the shore (see Figure 20.1c). Erosion by waves also planes the rocky surface beneath the surf zone and creates a **wave-cut terrace,** sometimes visible at low tide (**Figure 20.13**). Wave erosion continuing for long periods may straighten shorelines, as headlands retreat faster than recesses and bays.

Where relatively soft sediments or sedimentary rocks make up the coastal region, the slopes are gentler and the heights of shoreline bluffs are lower. Waves efficiently erode these softer materials, and erosion of bluffs on such shores may be extraordinarily rapid. The high sea cliffs of soft glacial materials along the Cape Cod National Seashore in

Figure 20.13 Multiple wave-cut terraces on the California coastline. [Photo by Dan Muhs/USGS. Daniel R. Muhs, Kathleen R. Simmons, George L. Kennedy, and Thomas K. Rockwell. "The Last Interglacial Period on the Pacific Coast of North America: Timing and Paleoclimate." *Geological Society of America Bulletin* (May 2002): 569–592.]

Massachusetts, for instance, are retreating about a meter each year. Since Henry David Thoreau walked the entire length of the beach below those cliffs in the mid-nineteenth century and wrote of his travels in *Cape Cod,* about 6 km² of coastal land have been eaten by the ocean, equivalent to about 150 m of beach retreat.

In recent decades, more than 70 percent of the total length of the world's sand beaches has retreated at a rate of at least 10 cm/year, and 20 percent of the total length has retreated at a rate of more than 1 m/year. Much of this loss can be traced to the damming of rivers, which decreases the sediment supply to the shoreline.

Depositional Coastal Forms Sediment builds up in areas where tectonic subsidence depresses the crust along a coast. Such coasts are characterized by long, wide beaches and wide, low-lying coastal plains of sedimentary strata. Shoreline forms include sandbars, low-lying sandy islands, and extensive tidal flats. Long beaches grow longer as longshore currents carry sand to the downcurrent end of the beach. There it builds up, first as a submerged bar, then rising above the surface and extending the beach by a narrow addition called a *spit* (**Figure 20.14**).

Offshore, long sandbars may build up into **barrier islands** that form a barricade between open ocean waves and the main shoreline. Barrier islands are common, especially along low-lying coasts composed of easily erodible and transportable sediments or of poorly cemented sedimentary rocks where longshore currents are strong. Some of the most prominent barrier islands are found along the coast of New Jersey; at Cape Hatteras, North Carolina; and along the Texas coast of the Gulf of Mexico, where one—Padre Island—is 130 km long. As the bars build up above the waves, vegetation takes hold, stabilizing the islands and helping them resist wave erosion during storms. Barrier islands are separated from the coast by tidal flats or shallow lagoons. Like beaches on the main shore, barrier islands are in dynamic equilibrium with the forces shaping them. If their equilibrium is disturbed by natural changes in climate or in wave and current regimes or by real estate development, they may be disrupted or devegetated, leading to increased erosion. Under such conditions, barrier islands may even disappear beneath the sea surface. Barrier islands may also grow larger and more stable if sedimentation increases.

Over hundreds of years, shorelines may undergo significant changes. Hurricanes and other intense storms, such as the "storm of the century" that hit the East Coast of the United States in March 1993, may form new inlets, elongate spits, or breach existing inlets and spits. Such changes have been documented by remapping from aerial photographs taken at various time intervals. The shoreline of Chatham, Massachusetts, at the elbow of Cape Cod, has changed enough in the past 160 years or so that a lighthouse has had to be moved. Figure 20.14 illustrates the many changes that have taken place in the configuration of the bars to the north and to the long spit of Monomoy Island, as well as several breaches of the bars. Many homes in Chatham are now at risk, but there is little that the residents or the state can do to prevent these beach processes from taking their natural course.

SEA-LEVEL CHANGE AS A MEASURE OF GLOBAL WARMING

Shorelines are sensitive to changes in sea level, which can alter tidal heights, change the approach of waves, and affect the path of longshore currents. The rise and fall of sea level can be local—the result of tectonic subsidence or uplift—or global—the result of glacial melting or growth, for example. One of the primary concerns about human-induced global warming is its potential for causing sea level to rise and flood coastlines.

(a) Beach near Chatham Light

The 1987 breach in the barrier spit, shown at the right below, closed again before this photo was taken.

(b)

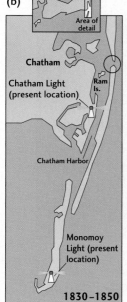

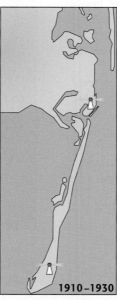
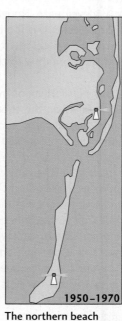
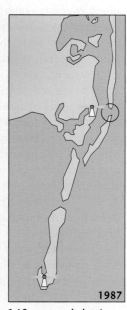

Circle shows approximate location of 1846 breach in barrier spit. Ram Island later disappears.

Beach south of the inlet breaks up and migrates southwest toward the mainland and Monomoy.

The southern beach has disappeared, and its remnants soon will connect Monomoy to the mainland.

The northern beach steadily grows with cliff sediment; Monomoy breaks from the mainland.

140-year cycle begins again with Jan. 2 breach in the barrier spit across from Chatham Light (circle).

Figure 20.14 Migrating barrier islands at the southern tip of Cape Cod, Massachusetts. (a) Aerial view of Monomoy Point, Massachusetts. This spit has advanced into deep water to the south (*foreground*) from barrier islands along the main body of the Cape to the north (*background*). [Steve Dunwell/The Image Bank.] (b) Changes in the shoreline at Chatham, Massachusetts, at the elbow of Cape Cod, over the past 160 years. [After Cindy Daniels, *Boston Globe* (February 23, 1987).]

In periods of lowered sea level, areas that were offshore are exposed to agents of erosion. Rivers extend their channels over formerly submerged regions and cut valleys into newly exposed coastal plains. When sea level rises, flooding the lands of the backshore, river valleys are drowned, marine sediments build up along former land areas, and erosion is replaced by sedimentation. Today, long fingers of the sea indent many of the shorelines of the northern and central Atlantic coast. These long indentations are former river valleys that were flooded as the last glacial age ended about 10,000 years ago and the sea level rose.

Sea-level variations on geologic time scales can be measured by studies of wave-cut terraces (see Figure 20.13), but detecting global changes on short (human) time scales can be difficult. Changes can be measured locally by using a tide gauge that records sea level relative to a land-based benchmark. The major problem is that the land itself moves vertically as a result of tectonic deformation, sedimentation, and other geologic changes, and this motion is incorporated into the tide-gauge observations. Through careful analysis, however, oceanographers have found that global sea level has risen by 10 to 25 cm over the last century.

EARTH ISSUES

20.2 Preserving Our Beaches

Orrin Pilkey of Duke University, a geologist and oceanographer, is in the forefront of scientists concerned about saving our beaches and halting massive development on fragile shorelines. Many houses built on the shore, battered by waves, could be saved by constructing concrete buttresses, seawalls, and other structures designed to protect shoreline property, but these structures would destroy the beach. Pilkey, a well-known researcher on coastal processes, is an advocate for the beaches of the Carolinas, which have come under heavy pressure from commercial developers. Knowing how the beach system works, he believes it is foolish to try to interfere with the natural processes by which beaches remain in dynamic equilibrium with the waves and currents.

Humans are altering this equilibrium on more and more beaches by building cottages on the shore; paving beach parking lots; erecting seawalls; and constructing groins, piers, and breakwaters. The consequence of poorly thought out construction is the shrinkage of the beach in one place and its growth in another. The classic example is a narrow groin built out from shore at right angles to it. In the subsequent months and years, the sand disappears from the beach on one side of the groin and greatly enlarges the beach on the other side. As landowners and developers bring suit against one another and against state governments, trial lawyers take the issue of "sand rights"—the beach's right to the sand that it naturally contains—into the courts.

The shrinkage and enlargement of beaches are the predictable results of a longshore current. The waves,

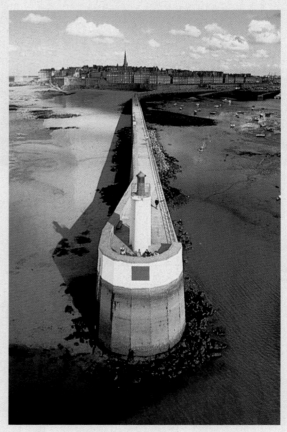

Construction of groins along a shore to control erosion of a beach may result in erosion downcurrent of the groin and loss of parts of the beach (*right of groin*) while sand piles up on the other side (*left of groin*). Longshore current flows from left to right. [Philip Plisson/Explorer.]

This increase correlates with a worldwide increase in temperatures, which most scientists now believe has been caused, at least in part, by human pollution of the atmosphere with carbon dioxide and other greenhouse gases (see Chapter 23). Global warming leads to a worldwide rise in sea level in two different ways. First, it causes continental glaciers and polar ice caps to melt, which increases the amount of water in the ocean basins. Second, higher temperatures cause the water to expand by a small fraction, increasing its volume (just as thermal expansion causes liquid in a thermometer to rise). These effects appear to be about equal in magnitude; that is, each can explain about half of the observed increase in sea level.

Satellite altimeters provide a sensitive new technique for determining the altitude of the sea surface relative to the or-

bit of the satellite (see the discussion in the "Charting the Seafloor by Satellite" section on page 495). The data indicate that sea level is rising at a rate of about 4 mm/year. Some of the rise may result from short-term variations, but the magnitude of the rise is consistent with climate models that take into account greenhouse warming. These models predict that without significant worldwide efforts to reduce greenhouse gas emissions, sea level will probably rise by another 30 to 60 cm during this century.

As global warming causes the sea level to rise, we will see the effects on our beaches. Indeed, the shorelines of the world serve as barometers for impending changes caused by many types of human activities. The pollution of our inland waterways sooner or later arrives at our beaches, as sewage from city dumping and oil from ocean tankers wash up on

current, and drift bring sand toward the groin from the upcurrent direction (usually the dominant wind direction). Stopped at the groin, they dump the sand there. On the downcurrent side of the groin, the current and drift pick up again and erode the beach. On this side, however, replenishment of sand is sparse because the groin blocks the current. As a result, the beach budget is out of balance, and the beach shrinks. If the groin is removed, the beach returns to its former state.

The only way to save a beach is to leave it alone. Even if concrete walls and piers can be kept in repair with large expenditures of money, many times at public expense, the beach itself will suffer. Along some beaches,

resort hotels truck in sand, but that expensive solution is temporary, too.

At Seabright, in Monmouth County, New Jersey, a beach has been built up by the U.S. Army Corps of Engineers, using half a million dump-truck loads of sand. This is the costliest beach "nourishment" project to date. As of 2002, the beach has been replenished twice, with a further supplement planned in another year. Sands are piped from several miles off shore and dumped in front of the beach, which helps to decrease the erosive effects of storm waves.

Sooner or later, we must learn to let the beaches remain in their natural state.

Beach-fill placement at the southern end of Monmouth Beach, Monmouth County, New Jersey. This erosion control project by the U.S. Army Corps of Engineers includes periodic nourishment of the restored beaches on a 6-year cycle for a period of 50 years. [U.S. Army Corps of Engineers, New York District.]

the shore. As real estate development and construction along shorelines expands, we will see the continuing contraction and even disappearance of some of our finest beaches.

CONTINENTAL MARGINS

The shorelines, shelves, and slopes of the continents are together called **continental margins.** There are two basic types of continental margins, passive and active (**Figure 20.15**). A **passive margin** is far from a plate boundary; those off the eastern coasts of North America and Australia and the western coast of Europe are examples. The name implies quiescence: volcanoes are absent and earthquakes are few and far between. In contrast, **active margins,** such as the one off the

western coast of South America, result from subduction. Occasionally, active margins are associated with transform faulting. The volcanic activity and frequent earthquakes give these continental margins their name. Active margins at subduction zones include an offshore trench and an active volcanic belt.

The continental shelves of passive margins consist of essentially flat-lying, shallow-water sediments, both terrigenous and carbonate, several kilometers thick (see Figure 20.15a). Although the same kinds of sediment can be found on active margin shelves, they are more likely to be structurally deformed and to include ash and other volcanic materials as well as deep-ocean sediments. Active margins on the eastern side of the Pacific Ocean usually feature a narrow continental shelf that falls off sharply into a deep-sea trench

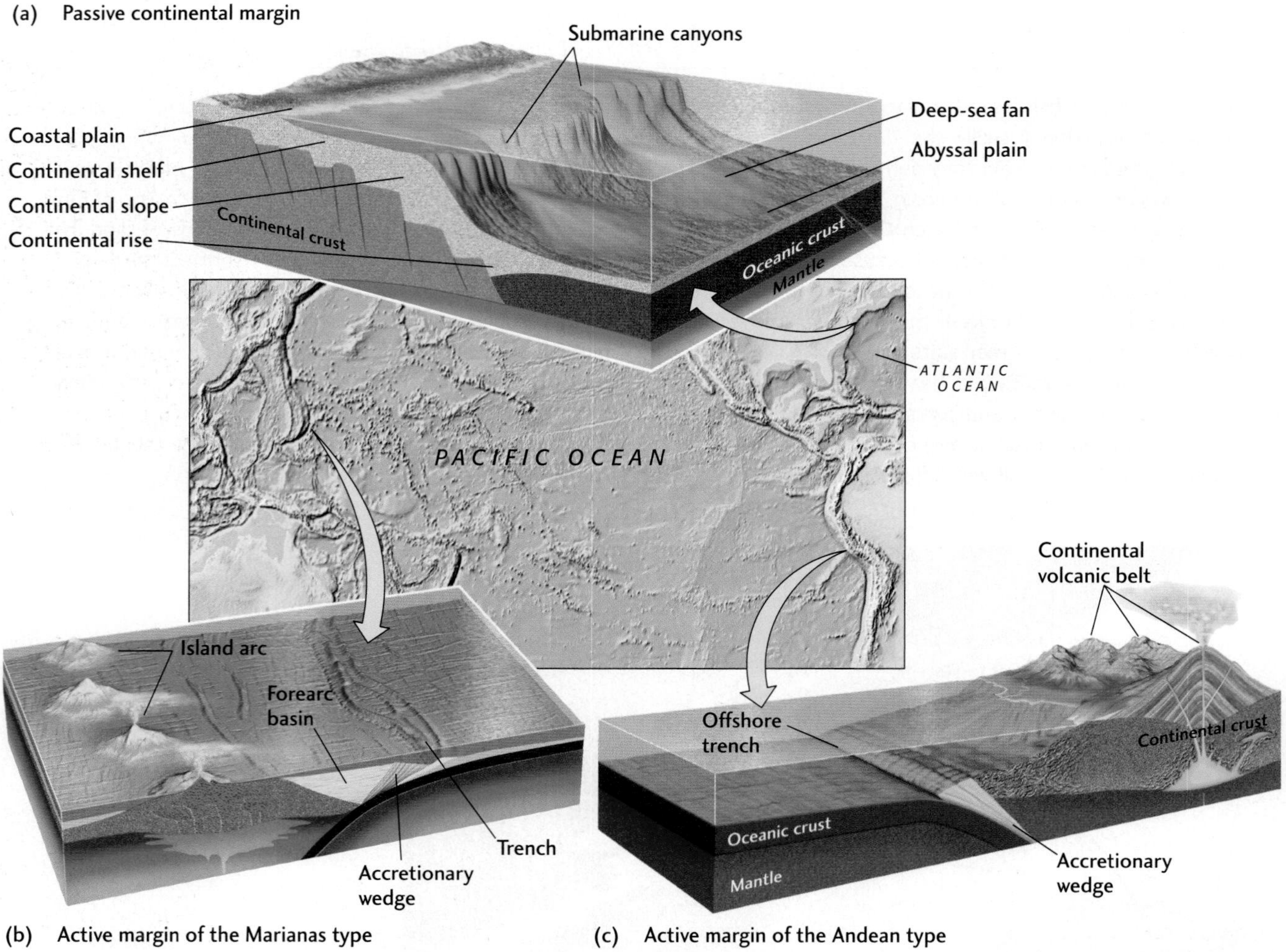

(a) Passive continental margin

Submarine canyons

Deep-sea fan

Abyssal plain

Coastal plain

Continental shelf

Continental slope

Continental rise

Continental crust

Oceanic crust

Mantle

ATLANTIC OCEAN

PACIFIC OCEAN

Island arc

Forearc basin

Trench

Accretionary wedge

Continental volcanic belt

Offshore trench

Oceanic crust

Mantle

Continental crust

Accretionary wedge

(b) Active margin of the Marianas type

(c) Active margin of the Andean type

Figure 20.15 Schematic profiles of three types of continental margins. (a) Passive margin. (b) Active margin of the Marianas type. (c) Active margin of the Andean type.

without much accumulation of sediments (see Figure 20.15c). Those on the western side of the Pacific (for example, off the Marianas Islands) have wider margins with substantial *fore-arc basins,* where thick sequences of sediment are deposited (see Figure 20.15b). These sediments come partly from erosion of the uplifted volcanic arc, but they also accumulate by being "scraped off" the subducting oceanic crust.

The Continental Shelf

The continental shelf is one of the most economically valuable parts of the ocean. Georges Bank off New England and the Grand Banks of Newfoundland, for example, are among the world's most productive fishing grounds. Recently, oil-drilling platforms have been used to extract huge quantities of oil and gas from the continental shelf, especially off the Gulf Coast of Louisiana and Texas. For economic and other reasons, most of the world's nations (though not the United States) signed the international Law of the Sea treaty in 1982.

The treaty governs the territorial and economic rights of nations in the world's oceans.

Because continental shelves lie at shallow depths, they are subject to exposure and submergence as a result of changes in sea level. During the Pleistocene glaciation, all of the shelves now at depths of less than 100 m were above sea level, and many of their features formed then. Shelves at high latitudes were glaciated, producing an irregular topography of shallow valleys, basins, and ridges. Those at lower latitudes are more regular, cut by occasional stream valleys.

The Continental Shelf and Rise: Turbidity Currents

The waters of the continental slope and rise are too deep for the seafloor to be affected by waves and tidal currents. As a consequence, sediments that have been carried across the shallow continental shelf by waves and tides come to rest as they are draped over the slope. Continental slope shows

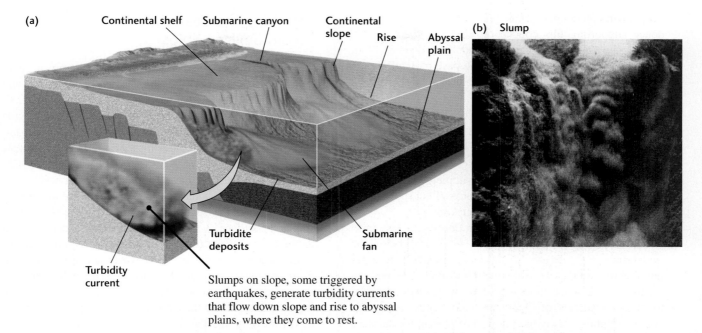

(a) Continental shelf   Submarine canyon   Continental slope   Rise   Abyssal plain   (b) Slump

Turbidity current   Turbidite deposits   Submarine fan

Slumps on slope, some triggered by earthquakes, generate turbidity currents that flow down slope and rise to abyssal plains, where they come to rest.

Figure 20.16 (a) How a turbidity current forms in the ocean. These currents can erode and transport large quantities of sand down the continental slope. Submarine canyons are deep valleys eroded into the continental shelf and slope by turbidity currents.

(b) Sandfall at the head of a submarine canyon at the edge of the continental shelf. These falls generate sandy flows, such as turbidity currents, that lay down fans of sandy sediment at the foot of the continental slope. [U.S. Navy.]

signs of sediment slumping and the erosional scars of gullies and submarine canyons. Deposits of sand, silt, and mud on both slope and rise indicate active sediment transport into deeper waters. For some time, geologists puzzled over what kind of current might cause both erosion and sedimentation on the slope and rise at such great depths.

The answer proved to be **turbidity currents**—flows of turbid, muddy water down a slope (**Figure 20.16**). Because of its suspended load of mud, the turbid water is denser than the overlying clear water and flows beneath it. Turbidity currents can both erode and transport sediment. Turbidity currents start when the sediment draped over the edge of the continental shelf slumps onto the continental slope (see Figure 20.16). The sudden submarine landslide, which can occur spontaneously or be triggered by an earthquake, throws mud into suspension, creating a dense, turbid layer of water near the bottom. This turbid layer starts to flow, accelerating down the slope. As the turbidity current reaches the foot of the slope and the gentler incline of the continental rise, it slows. Some of the coarser sandy sediment starts to settle, often forming a *submarine fan*—a deposit something like an alluvial fan on land. Some of the stronger currents continue across the rise, cutting channels into the submarine fans (see Figure 20.16a). Eventually, the currents reach the *abyssal plain,* the level bottom of the ocean basin, where they spread out and come to rest in graded beds of sand, silt, and mud called **turbidites.**

According to current research, submarine landslides that start turbidity currents are common. Some of them may be huge. One slide generated 8- to 10-m-thick turbidites over a large area of the western Mediterranean, accumulating a volume of 500 km³. Submarine slides may be caused by thawing of methane gas hydrates—crystalline solids composed of methane (found as a constituent of natural gas) and water. Gas hydrates are stable at the high pressures and low temperatures of many large areas of the oceans. In deeply buried sediment, the hydrate turns into methane gas (see Chapter 11). If sea level is lowered, as it was during the ice ages, the pressure at the bottom is reduced and the hydrate may gasify, triggering a landslide. The quantities of gas produced are enormous, and geologists have speculated about the possibility of exploiting these subsea gas hydrates.

Submarine Canyons

Submarine canyons are deep valleys eroded into the continental shelf and slope. They were discovered near the beginning of the twentieth century and were first mapped in detail in 1937. At first, some geologists thought they might have been formed by rivers. There is no question that the shallower parts of some canyons were river channels during periods of low sea level. But this hypothesis could not provide a complete explanation. Most of the canyon floors are thousands of meters deep. Even during the maximum lowering of sea level in the ice ages, rivers could have eroded only to a depth of about 100 m.

Although other types of currents have been proposed, turbidity currents are now the favored explanation for the deeper parts of submarine canyons (see Figure 20.16). Evidence supporting this conclusion comes in part from a comparison of modern canyons and their deposits with well-preserved similar deposits of the past, particularly the pattern of turbidites deposited on submarine fans.

THE DEEP OCEANS

A topographic map of Earth's surface (**Figure 20.17**) reveals the most important geologic features submerged beneath the oceans: mid-ocean ridges, volcanic tracks of hot spots, deep-sea trenches, island arcs, and continental margins.

Making a map of the deep ocean floor is no easy task. Because sunlight can penetrate only 100 m or so below the sea surface, the deep ocean is a very dark place. It is not possible to map the ocean floor using visible light, nor can we use radar beamed from spacecraft, which we have used to map the surface of cloud-shrouded Venus. Ironically, spacecraft photography has allowed us to map the surfaces of our planetary neighbors with much higher resolution than we have been able to map Earth's deep ocean floor, even to this day.

Probing the Seafloor from Surface Ships

It is possible to view the seafloor directly from a deep-diving submersible. These small ships can observe and photograph at great depths (**Figure 20.18**). With their mechanical arms, they can break off pieces of rock, sample soft sediments, and catch specimens of exotic deep-sea animals. Newer robotic submersibles are guided by scientists on the mother ship above. But submersibles are expensive to build and operate, and they cover small areas at best.

For most work, today's oceanographers use instrumentation to sense the seafloor topography indirectly from a ship at the surface. A shipboard echo sounder, developed in the early part of the twentieth century, sends out pulses of sound waves. When the sound waves are reflected back from the ocean bottom, they are picked up by sensitive microphones in the water. Oceanographers can measure the depth by determining the interval between the time the pulse leaves the ship and the time it returns as a reflection. The result is an automatically traced profile of the bottom topography. Powerful echo sounders are also used to probe the stratigraphy of sedimentary layers beneath the ocean floor.

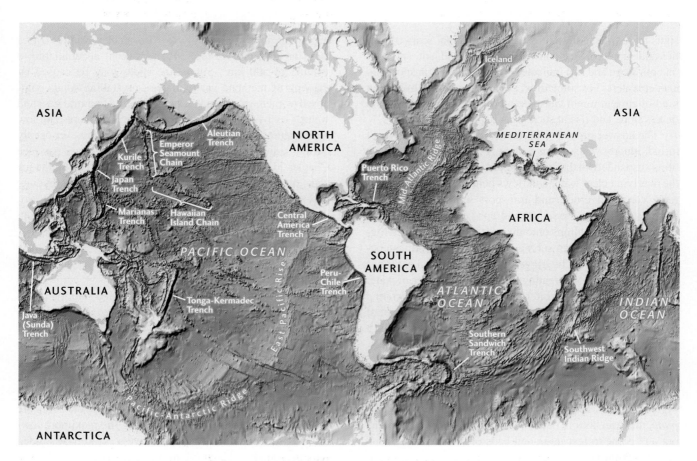

Figure 20.17 Earth's topography beneath the oceans showing the major features of the deep seafloor.

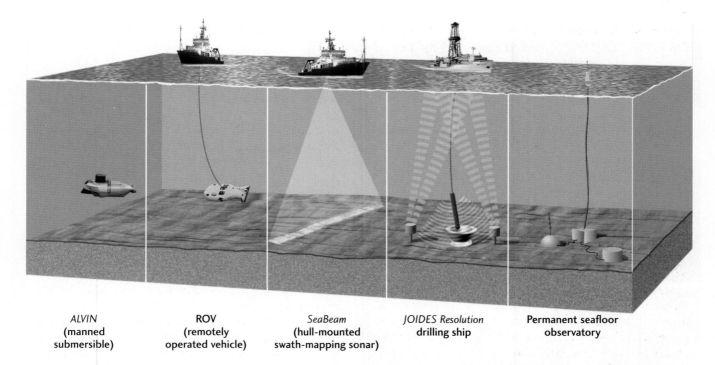

ALVIN
(manned
submersible)

ROV
(remotely
operated vehicle)

SeaBeam
(hull-mounted
swath-mapping sonar)

JOIDES Resolution
drilling ship

Permanent seafloor
observatory

Figure 20.18 High-technology methods for exploring the deep seafloor. The manned deep submersible *Alvin* and a remotely operated vehicle (ROV) are directed from a surface ship. *SeaBeam*, a hull-mounted multibeam echo sounder, continuously maps seafloor topography in a wide swath as a ship steams across the ocean surface. The drilling ship *JOIDES Resolution* uses bottom transponders to maneuver a drill pipe into a reentry hole on the seafloor. Permanent unmanned seafloor observatories monitor processes in the subsurface and the overlying water column for extended periods of time.

Many of today's oceanographic vessels are outfitted with hull-mounted arrays of echo sounders that can reconstruct a detailed image of seafloor topography in a swath extending as much as 10 km on either side of the ship as it steams along (see Figure 20.18). These systems can map the seafloor with unprecedented resolution of small-scale geologic features, such as undersea volcanoes, canyons, and faults. **Figure 20.19** shows several impressive images of the seafloor derived by this type of mapping.

Other types of instruments can be towed behind a ship or lowered to the bottom to evaluate the magnetism of the seafloor, the shapes of undersea cliffs and mountains, and the heat coming from the crust. Underwater cameras on sleds towed near the ocean bottom can photograph the details of the seafloor and the organisms that inhabit the deep. Since 1968, the U.S. Deep Sea Drilling Project and its successor, the international Ocean Drilling Program, have sunk hundreds of drill holes to depths of many hundreds of meters below the seafloor. Cores obtained from these drill holes have provided geologists with sediment and rock samples for detailed physical and chemical studies.

Charting the Seafloor by Satellite

Despite all this fancy gear, there are still many regions of the oceans that have not been surveyed, and our knowledge of the seafloor remains fragmentary. Recently, however, a new method has been used to chart the global topography of the seafloor indirectly.

The new method makes use of an altimeter mounted on a satellite. The altimeter sends pulses of radar beams that are reflected off the ocean surface, giving measurements of the distance between the satellite and the sea surface with a precision of a few centimeters. The height of the sea surface depends not only on waves and ocean currents but also on changes in gravity caused by the topography and composition of the underlying seafloor. The gravitational attraction of a seamount, for example, can cause water to "pile up" above it, producing a bulge in the sea surface of as much as 2 m above average sea level. Similarly, the diminished gravity over a deep-sea trench is evident as a depression in the sea surface of as much as 60 m below average sea level.

This method has allowed us to infer features of the ocean floor from satellite data and display them as if the seas were drained away. Marine geologists have used this technique to map new features of the seafloor not revealed by ship surveys, especially in the poorly surveyed southern oceans. The satellite data have also revealed deeper structures below the oceanic crust, including gravity anomalies associated with convection currents in the mantle, which we discussed in Chapter 14.

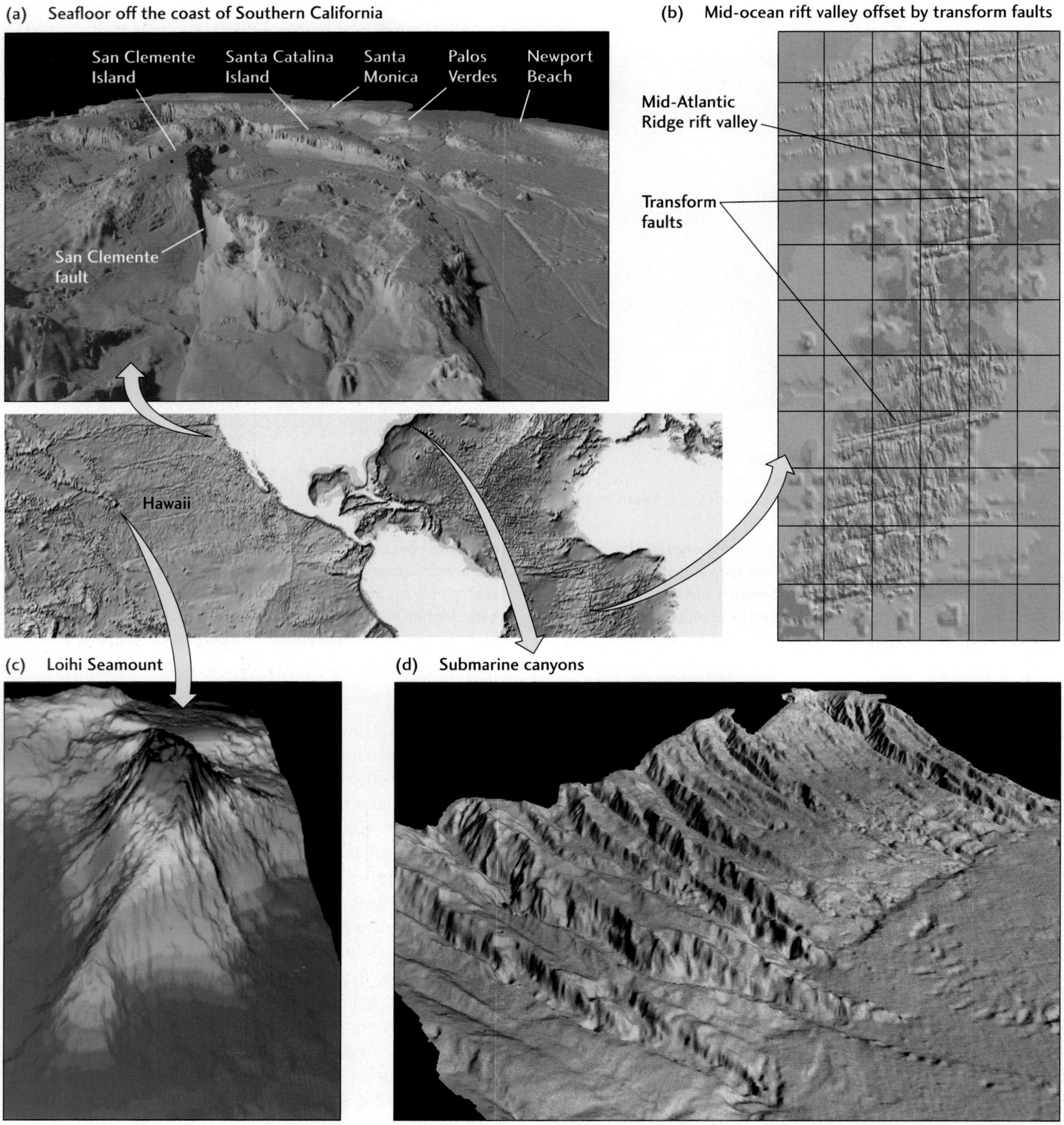

(a) Seafloor off the coast of Southern California

(b) Mid-ocean rift valley offset by transform faults

Figure 20.19 Three examples of topography of the seafloor obtained from high-resolution swath mapping by surface ships and rendered by computer processing as three-dimensional images. (a) The seafloor off the coast of southern California, showing fault-controlled structures of a geologic province known as the California Borderland. [Chris Goldfinger and Jason Chaytor, Oregon State University.] (b) Ridge between 25°S and 36°S latitudes showing the southeast-trending rift valley offset by northeast-trending transform faults. [Ridge Multibeam Database, Lamont-Doherty Earth Observatory, Columbia University.] (c) Loihi Seamount, just south and east of the Big Island of Hawaii, the newest in the string of hot-spot volcanoes that form the Hawaiian Islands chain. [Ocean Mapping Development Center, University of Rhode Island.] (d) The continental shelf (top), slope (central and upper area), and rise (lower left) off the coast of New England. Note the deep submarine canyons that incise the continental margin. [From L. Pratson and W. Haxby, Geology 24(1) (1996): 4. Courtesy of Lincoln Pratson and William Haxby, Lamont-Doherty Earth Observatory, Columbia University.]

Profiles Across Two Oceans

To gain an appreciation of the geologic features that lie beneath the oceans, we will take a brief tour across two of Earth's major ocean basins, the Atlantic and the Pacific, as if we were driving a deep-diving submarine along the ocean floor.

An Atlantic Profile The Atlantic profile shown in **Figure 20.20** extends from North America to Gibraltar. Starting from the coast of New England, we descend from the shoreline to depths of 50 to 200 m and travel eastward along the **continental shelf.** This broad, flat, sand- and mud-covered platform is a slightly submerged part of the continent. After traveling about 50 to 100 km across the shelf, down a very gently inclined surface, we find ourselves at the edge of the shelf. There, we start down a steeper incline, the **continental slope.** This mud-covered slope descends at an angle of about 4°, a drop of 70 m over a horizontal distance of 1 km, which would be a noticeable grade if we were driving on land.

The continental slope is irregular and is marked by gullies and **submarine canyons,** deep valleys eroded into the slope and the shelf behind it (see Figure 20.19). On the lower parts of the slope, at depths of about 2000 to 3000 m, the downward incline becomes gentler. Here it merges into a more gradual downward incline called the **continental rise,** an apron of muddy and sandy sediment extending into the main ocean basin.

The continental rise is tens to hundreds of kilometers wide, and it grades imperceptibly into a wide, flat **abyssal plain** that covers large areas of the ocean floor at depths of about 4000 to 6000 m. The plain is broken by occasional submerged volcanoes, mostly extinct, called **seamounts.** As we travel along the abyssal plain, we gradually climb into a province of low **abyssal hills** whose slopes are covered with fine sediment. Continuing up the hills, the sediment layer becomes thinner, and outcrops of basalt appear beneath it. As we rise along this steep, irregular topography to depths of about 3000 m, we are climbing the flanks and then the mountains of the Mid-Atlantic Ridge.

Abruptly, we come to the edge of a deep, narrow valley a few kilometers wide at the top of the ridge (**Figure 20.21**). This narrow cleft, marked by active volcanism, is the rift valley where two plates separate. As we cross the valley and climb its eastern side, we move from the North American Plate to the Eurasian Plate. Continuing eastward, we encounter topography similar to that on the western side of the ridge, only in reverse order, because the ocean floor is more or less symmetrical on either side of the ridge. On the path we have taken, this symmetry is disturbed by some large seamounts and the volcanic Azores Islands, which mark an active hot spot, perhaps caused by the heat from an upwelling mantle plume. Passing over the rough topography of the abyssal hills on the flank of the Mid-Atlantic Ridge, we descend to the abyssal plain and then ascend up the continental rise, slope, and shelf off the coast of Europe.

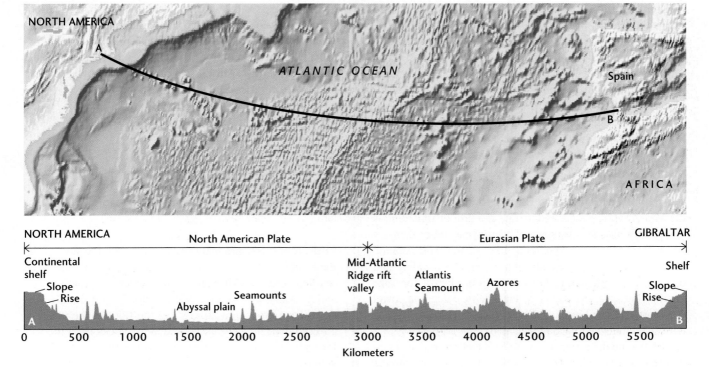

Figure 20.20 A topographic profile of the floor of the Atlantic Ocean traveling from New England (*left*) to Gibraltar (*right*).

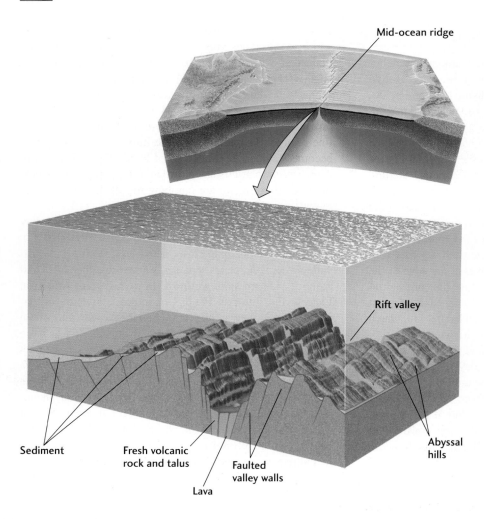

Figure 20.21 A profile of the central rift valley of the Mid-Atlantic Ridge in the FAMOUS (French-American Mid-Ocean Undersea Study) area southwest of the Azores Islands. The deep valley, where most of the basalt is extruded, is faulted. [After ARCYANA, "Transform Fault and Rift Valley from Bathyscaph and Diving Saucer." *Science* 190 (1975): 108.]

A Pacific Profile Our second virtual tour moves westward across the Pacific, from South America to Australia (**Figure 20.22**). Beginning on the western coast of Chile, we cross a narrow continental shelf only a few tens of kilometers wide. The edge of the shelf plunges off a continental slope that is much steeper than the one we traversed in the Atlantic, extending down to 8000 m as we enter the Peru-Chile Trench. This long, deep, narrow depression of the seafloor is the surface expression of the subduction of the Nazca Plate under the South American Plate.

Continuing across the trench and up onto the abyssal hills of the Nazca Plate, we come to the East Pacific Rise, an active mid-ocean ridge. The East Pacific Rise is lower than the Mid-Atlantic Ridge, and its rate of seafloor spreading is the world's fastest—at about 150 mm/year, more than six times faster than the spreading rate of the Mid-Atlantic Ridge—but it has the characteristic central rift valley and outcrops of fresh basalt. On the western side of the East Pacific Rise, we cross over to the Pacific Plate and travel westward over its broad central regions, which are studded with seamounts and volcanic islands.

We eventually arrive at another subduction zone, marked by the Tonga Trench, where the Pacific Plate returns into the mantle beneath the Australian Plate. This is one of the deepest places in all the oceans, almost 11,000 m below the ocean surface. On the western side of the trench, the seafloor rises to the volcanic Tonga and Fiji islands. Beyond this island arc, we return to the deep seafloor, now on the Australian Plate, and come to the continental rise, slope, and shelf of eastern Australia, which is similar to the eastern coast of North America.

Main Features of the Deep Ocean Floor

Away from the edges of continents and subduction zones, the deep seafloor is constructed primarily by volcanism related to plate tectonic motions and secondarily by sedimentation in the open sea.

Mid-Ocean Ridges Mid-ocean ridges are the sites of the most intense volcanic and tectonic activity on the deep seafloor. The main rift valley is the center of the action. The valley walls are faulted and intruded with basalt sills and dikes (see Figure 20.21), and the floor of the valley is covered with flows of basalt and talus blocks from the valley walls, mixed with a little sediment settling from surface waters.

Hydrothermal springs form on the rift valley floor as seawater percolates into fractures in the basalt on the flanks

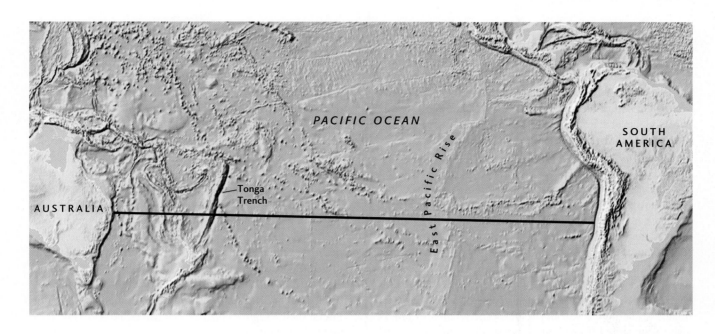

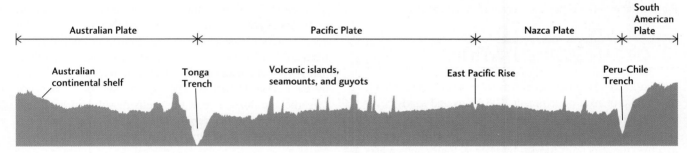

Figure 20.22 A topographic profile of the floor of the Pacific Ocean traveling from South America (*right*) to Australia (*left*).

of the ridges, is heated as it moves down to hotter basalt, and finally exits at the valley floor. There, it boils up at temperatures as high as 380°C . Some springs are "black smokers," full of dissolved hydrogen sulfide and metals the hot water has leached from the basalt (see Figure 11.16). Others are cooler "white smokers" that release different compounds made of barium, calcium, and silicon. Both black smokers and white smokers produce mounds of iron-rich clay minerals, iron and manganese oxides, and large deposits of iron-zinc-copper sulfides.

Mid-ocean ridges are offset at many places by transform faults that displace the rift valleys laterally (see Figure 20.19b). Large earthquakes occur on these faults as one plate slips past the other. Rocks collected from the walls of the transform faults often have the olivine-rich compositions typical of the mantle, rather than the basaltic composition of the oceanic crust. This suggests that the magmatic process that creates oceanic crust may operate less efficiently where the spreading center abuts a fault.

Abyssal Hills and Plains The floor of the deep ocean away from the mid-ocean ridges is a landscape of hills,

plateaus, sediment-floored basins, and seamounts. Abyssal hills are ubiquitous on the slopes of mid-ocean ridges. They are typically 100 m or so high and lineated parallel to the ridge crest (see Figure 20.21). They are formed primarily by normal faulting of the basaltic oceanic crust as it moves out of the rift valley. Almost all of this faulting occurs during the first million years of the plate's existence, after which the faults bounding the hills become inactive. As the new oceanic lithosphere drifts away from the spreading center, it cools and contracts, lowering the seafloor. Its hilly, subsiding surface receives a steady rain of sediment from surface waters, gradually becoming mantled with deep-sea muds and other deposits.

Near the continental margins, terrigenous sediments (sediments that originate on land) moving down the continental slopes add to this sediment cover and create the flat, unbroken expanses of the abyssal plains. These plains are the flattest solid surfaces on the planet.

Seamounts, Hot-Spot Island Chains, and Plateaus
The seafloor is littered with tens of thousands of seamounts. Most are submerged, but some rise to the sea surface as

volcanic islands. Seamounts and volcanic islands may be isolated or found in clusters or chains. Most seamounts, though not all, are created by eruptions near active spreading centers or where a plate overrides a mantle hot spot.

Some of the larger seamounts, called **guyots,** have flat tops, the result of erosion of an island volcano when it was above sea level. These islands have since submerged as the plate on which they ride cooled, contracted, and subsided as it passed away from the spreading center or hot spot that formed it.

Among the most surprising features of the deep ocean basins are the large basalt plateaus. Some appear to have formed near triple junctions where three spreading centers meet. Others are associated with massive eruptions at hot spots far from spreading centers. One of the largest examples of the latter type, and probably the oldest, is the Shatsky Rise in the northwestern Pacific Ocean, about 1600 km southeast of Japan. The origin of oceanic plateaus may be related to the plume volcanism that some scientists think is also responsible for other large igneous provinces, such as the continental flood basalts discussed in Chapter 12.

OCEAN SEDIMENTATION

Almost everywhere that oceanographers search the seafloor, they find a blanket of sediment. The muds and sands cover the topography of basalt originally formed at mid-ocean ridges. The ceaseless sedimentation in the world's oceans modifies the structures formed by plate tectonics and creates its own topography at sites of rapid deposition. The sediment is mainly of two kinds: terrigenous muds and sands eroded from the continents and biochemically precipitated shells of organisms that live in the sea. In parts of the ocean near subduction zones, sediments derived from volcanic ash and lava flows are abundant. In tropical arms of the sea where evaporation is intense, evaporite sediments are deposited.

Sedimentation on the Continental Shelf

Terrigenous sedimentation on the continental shelf is produced by waves and tides. The waves of large storms and hurricanes move sediment over the shallow and moderate depths of the shelf, and tidal currents flow over the shelf. The waves and currents distribute the sediment brought in by rivers into long ribbons of sand and layers of silt and mud.

Biochemical sedimentation on the shelf results from the buildup of layers of the calcium carbonate shells of clams, oysters, and many other organisms living in shallow waters. Most of these organisms cannot tolerate muddy waters and are found only where terrigenous materials are minor or absent, such as along the extreme southern coast of Florida or off the coast of Yucatán in Mexico. Here, coral reefs thrive and organisms build up large thicknesses of carbonate sediment (see Chapter 5).

Deep-Sea Sedimentation

Far from the margins of continents, fine-grained terrigenous and biochemically precipitated particles suspended in seawater slowly settle from the surface to the bottom. These open-ocean sediments, called **pelagic sediments,** are characterized by their great distance from continental margins, their fine particle size, and their slow-settling mode of deposition. The terrigenous materials are brownish and grayish clays, which accumulate on the seafloor at a very slow rate, a few millimeters every 1000 years. A small fraction, about 10 percent, may be blown by the wind to the open ocean.

Within the pelagic sediments, the most abundant biochemically precipitated particles are the shells of *foraminifers,* tiny single-celled animals that float in the surface waters of the sea. These calcium carbonate shells fall to the bottom after their occupants die. There they accumulate as **foraminiferal oozes,** sandy and silty sediments composed of foraminiferal shells (**Figure 20.23**). Other carbonate oozes are made up of shells of different microorganisms, called *coccoliths.*

Foraminiferal and other carbonate oozes are abundant at depths of less than about 4 km, but they are rare on the deeper parts of the ocean floor. This rarity cannot result from a lack of shells, because the surface waters are full of them everywhere and the living foraminifers are unaffected by the bottom far below. The explanation for the absence of carbonate oozes is that the shells dissolve in deep seawater below a certain depth, called the *carbonate compensation depth (CCD)* (**Figure 20.24**). Because of how oceans

~1 mm

Figure 20.23 Scanning electron micrograph of oceanic ooze. Shown here are shells of both carbonate- and silica-secreting unicellular organisms. [Scripps Institution of Oceanography, University of California, San Diego.]

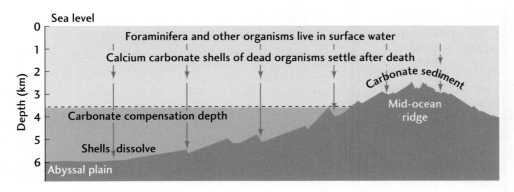

Figure 20.24 The carbonate compensation depth is the level in an ocean below which calcium carbonate dissolves. As the shells of dead foraminifers and other carbonate-shelled organisms settle into the deep waters, they enter an environment undersaturated in calcium carbonate and therefore dissolve.

circulate, deeper waters differ from shallower waters in three ways:

1. Deeper waters are colder. Colder, denser polar waters sink beneath warmer tropical waters and travel toward the equator along the ocean bottom.

2. Deeper waters contain more carbon dioxide. Not only do colder waters absorb more carbon dioxide than warmer waters, but any organic matter they carry tends to be oxidized to carbon dioxide in the course of their long circulation.

3. Deeper waters are under higher pressure. This increase in pressure results from the greater weight of overlying water.

These three factors make calcium carbonate more soluble in deep waters than in shallow ones. As the shells of dead foraminifers fall below the CCD, they enter an environment undersaturated in calcium carbonate, and they dissolve.

Another kind of biochemically precipitated sediment, **silica ooze,** is produced by sedimentation of the silica shells of diatoms and radiolarians. *Diatoms* are green algae abundant in the surface waters of the oceans. *Radiolarians* are unicellular organisms that secrete shells of silica. After burial on the seafloor, silica oozes are cemented into the siliceous rock *chert.*

Some components of pelagic sediments form by chemical reactions of seawater with sediment on the seafloor. The most prominent examples are manganese nodules—black, lumpy accumulations ranging from a few millimeters to many centimeters across. These nodules cover large areas of the deep ocean floor, as much as 20 to 50 percent of the Pacific. Rich in nickel and other metals, they are a potential commercial resource if some economical way can be found to mine them from the seafloor.

SUMMARY

How does the geology of the ocean basins differ from that of the continents? Continents have grown and been modified over the past 4 billion years as a result of multiple plate collisions, the creation of folded and faulted mountain belts, magmatism, erosion, and sedimentation. The geology of the continents is complex, and only a portion of their history is preserved in the rock record. In comparison, ocean basins are simple. They are created at mid-ocean ridges where plates separate and are destroyed by subduction over the brief period of less than 200 million years. Deep-sea sediments provide an almost complete record of the relatively brief geologic history of ocean basins.

What processes shape shorelines? At the edge of the sea, waves and tides, interacting with tectonics, control the formation and dynamics of shorelines, from beaches and tidal flats to uplifted rocky coasts. Winds blowing over the sea generate waves; as the waves approach the shore, they are transformed into breakers in the surf zone. Wave refraction results in longshore currents and longshore drift, which transport sand along beaches. Tides, generated by the gravitational pull of the Moon and Sun on the water of the oceans, are agents of sedimentation on tidal flats.

How do hurricanes impact coastal areas? Hurricanes are intense storms with extremely high winds (can be greater than 300 km/hour) and very low atmospheric pressure. The low atmospheric pressure results in formation of a dome of seawater, known as storm surge, beneath the storm center. As the storm surge moves onshore, it floods low-lying areas up to depths of 10 m. The direct effect of the high winds can be very damaging, but the impact of storm surge can be devastating over a very broad area of the coastline.

What are the major components of a continental margin? A continental margin is made up of a shallow continental shelf; a continental slope that descends more or less steeply into the depths of the ocean; and a continental rise, a gently sloping apron of sediment deposited at the lower edges of the continental slope and extending to the abyssal plain farther out in the ocean. Waves and tides affect the continental shelf, but the continental slope is shaped primarily by turbidity currents. These deep-water currents form as slumps and slides on the continental slope create turbid suspensions of muddy sediment in bottom waters. Turbidity currents also produce submarine fans and submarine canyons. Active continental margins form where oceanic lithosphere is subducted beneath a continent. Passive continental margins form where rifting and seafloor spreading carry continental margins away from plate boundaries.

How is the deep seafloor formed? The deep seafloor is constructed by volcanism at mid-ocean ridges and at oceanic hot spots such as Hawaii and by deposition of fine-grained clastic and biochemically precipitated sediments. Seafloor spreading and the extrusion of basalt, which produces new oceanic lithosphere, take place at mid-ocean ridges. Volcanic islands, submerged seamounts and guyots, plateaus, and abyssal hills are all accumulations of volcanic rock, most of which are mantled with sediment. Deep-sea trenches form as oceanic lithosphere is pulled downward into a subduction zone.

What are the characteristics of a mid-ocean ridge? A rift valley runs down the crest, where plates separate and new oceanic lithosphere is produced. Earthquakes and active basaltic volcanoes are concentrated near the ridge crest. Seawater percolates through cracks on the flanks of such ridges and emerges in the valley floors as hydrothermal springs called smokers. These heated waters, laden with dissolved metal oxides and sulfides, precipitate mounds of mineral-rich deposits when they mix with the cold ocean-bottom water.

What kinds of sedimentation occur in and near the oceans? The two main types are terrigenous sediments and biochemically precipitated sediments. Terrigenous sediments are primarily muds and sands eroded from the continents and deposited by wave action and tidal currents along the continental shelf and by turbidity currents that flow down the continental slope. Biochemical sedimentation on the shelf results from the buildup of calcium carbonate from shells and coral reefs. In shallow tropical seas where evaporation is intense, evaporite sediments can be deposited. Open-ocean (pelagic) sediments consist of reddish brown clays and foraminiferal and silica oozes composed of the biochemically precipitated calcium carbonate and silica shells of microscopic organisms living in surface ocean waters. In parts of the ocean near subduction zones, sediments derived from volcanic ash and lava flows are abundant.

KEY TERMS AND CONCEPTS

abyssal hill (p. 497)
abyssal plain (p. 497)
active margin (p. 491)
barrier island (p. 488)
continental margin (p. 491)
continental rise (p. 497)
continental shelf (p. 497)
continental slope (p. 497)
foraminiferal ooze (p. 500)
guyot (p. 500)
hurricane (p. 480)
hurricane intensity scale (p. 482)
longshore current (p. 478)
passive margin (p. 491)
pelagic sediment (p. 500)
sand budget (p. 486)
seamount (p. 497)
silica ooze (p. 501)
storm surge (p. 483)
submarine canyon (p. 497)
tidal flat (p. 479)
tide (p. 478)
turbidite (p. 493)
turbidity current (p. 493)
wave-cut terrace (p. 487)

EXERCISES

1. How are ocean waves formed?

2. How does wave refraction concentrate erosion at headlands?

3. What is hurricane storm surge?

4. How has human interference affected some beaches?

5. Along what kinds of continental margins do we find broad continental shelves?

6. Where and how is the deep seafloor created by volcanism?

7. What plate tectonic process is responsible for deep-sea trenches?

8. Where do turbidity currents form?

9. What are pelagic sediments?

THOUGHT QUESTIONS

1. Why would you want to know the timing of high tide if you wanted to observe a wave-cut terrace?

2. In a 100-year period, the southern tip of a long, narrow, north-south beach has become extended about 200 m to the south by natural processes. What shoreline processes could have caused this extension?

3. After a period of calm along a section of the Gulf Coast of North America, a severe storm with high winds passes over the shore and up the Mississippi Valley. Describe the state of the surf zone before the storm, during it, and after the storm has passed. What would happen to inland rivers?

4. What are the chief differences between the Atlantic and Pacific oceans with regard to topography, tectonics, volcanism, and other seafloor processes?

5. How might plate tectonics account for the contrast between the broad continental shelf off the eastern coast of North America and the narrow, almost nonexistent shelf off the western coast?

6. A major corporation hires you to determine whether New York City's garbage can be dumped at sea within 100 km offshore. What kinds of places would you explore, and what are your concerns?

7. There is very little sediment on the floor of the central rift valley of the Mid-Atlantic Ridge. Why is this so?

8. A plateau rising from the deep ocean floor to within 2000 m of the surface is mantled with foraminiferal ooze, whereas the deep seafloor below the plateau, about 5000 m deep, is covered with reddish brown clay. How can you account for this difference?

9. You are studying a sequence of sedimentary rocks and discover that the beds near the base are shallow marine sandstones and mudstones. Above these beds is an unconformity, overlying which are nonmarine sandstones. Above this group of beds is another unconformity, and overlying it are marine beds similar to those at the base. What could account for this sequence?

LONG-TERM PROJECT

Coastal Protection

In 1996, the Institute of Marine and Coastal Sciences at Rutgers University published a report that calls for a major shift in policy with regard to coastal protection. Rather than recommending that coastal beaches and shore properties in high-hazard areas be defended and rebuilt, the report urges the New Jersey Department of Environmental Protection to advocate the removal of storm-damaged buildings and a policy of letting nature reclaim the beach. The report also suggests that shore towns be encouraged not to zone vulnerable areas for development. Some geologists consider this policy very wise.

Working in teams of three or four, pick a nearby or favorite vacation beach. If feasible, each group should visit its beach at the beginning of the semester and after each storm event during the semester to measure and record the width of the beach and make other observations of how the beach changes. Other sources of information might include the state's department of environmental protection, the town's chamber of commerce, and U.S. Geological Survey Web sites. Using records available from the U.S. Army Corps of Engineers under the Freedom of Information Act, investigate the approach that has been used to maintain and preserve the local shore that you have chosen to study. Has the Corps constructed seawalls or groins? Alternatively, has the Corps recommended pumping sand from offshore back onto the beach? What measures has the state taken to protect its shoreline, and what effects have they had on the physical appearance of the beach? Has the town chosen to let shore processes proceed naturally? Using your knowledge of coastal processes; your observations, if possible; and information from other sources about changes in your beach, assess the wisdom of the approach taken to preserve the beach of your study.

GLACIERS
The Work of Ice

The view of Earth from space is painted with the colors of water: vast blue oceans, swirling white clouds, and the frozen whites of solid ice and snow. The Earth system is constantly moving water across the planetary surface in ever-changing patterns. Among the main reservoirs of water, it is the system's icy component—its *cryosphere*—that waxes and wanes most visibly during climate cycles. Nowhere are changes in the cryosphere more evident than on the continents carved by glacial ice.

Only about 10 percent of Earth's land surface is now covered by glaciers, almost all of it blanketed by the huge ice sheets of Greenland and Antarctica (**Figure 21.1**). However, as recently as 20,000 years ago, ice sheets covered almost three times more land than they do now. Within only a few decades from now, global warming could melt parts of these ice sheets with worldwide effects on human society. Sea level would rise and submerge low-lying cities. Climate zones would migrate, changing wet zones into deserts and vice versa. Given these threats, there is no doubt that understanding Earth's cryosphere—always an interesting scientific subject—has become an extremely practical goal.

In this chapter, we take a close look at Earth's glaciers, their role in the climate system, how they change over time, and how way they erode and deposit rock and leave their marks on Earth's surface as they advance and retreat.

The landscapes of many continents have been sculpted by glaciers now melted away. In mountainous regions, glaciers have eroded steep-walled valleys, scraped bedrock surfaces, and plucked huge blocks from their rocky floors. During the ice ages of the Pleistocene epoch, glaciers pushed across entire continents, carving far more topography than rivers and wind. Glacial erosion creates enormous amounts of debris, and ice transports huge tonnages of sediment, depositing them at the edges of glaciers, where the sediment may be carried away by meltwater streams. Glacial processes affect the water discharge and sediment loads of major river systems, the erosion and sedimentation of coastal areas, and the quantity of sediment delivered to the oceans.

Several glaciers flow together in the St. Elias Mountains, Kluane National Park, Yukon, Canada. [Stephen J. Krasemann/DRK PHOTO.]

505

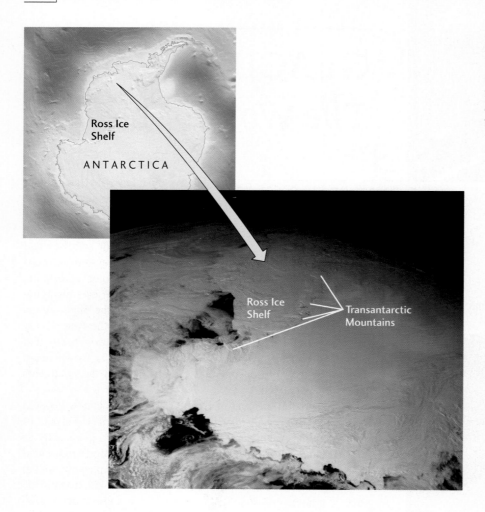

Figure 21.1 Satellite image of part of Antarctica taken on December 8, 1990. The view shows the Ross Ice Shelf. An occasional mountain can be seen poking through the ice. It is late spring in Antarctica, so the Sun never sets on the frigid, icy continent. [NASA.]

ICE AS A ROCK

To a geologist, a block of ice is a rock, a mass of crystalline grains of the mineral ice. Like igneous rocks, it is formed by the freezing of a fluid. Like sediments, it is deposited in layers at Earth's surface and can accumulate to great thicknesses. Like metamorphic rocks, it is transformed by recrystallization under pressure: glacial ice forms by the burial and metamorphism of the "sediment" snow. Loosely packed snowflakes—each a single crystal of the mineral ice—age and recrystallize into a solid rock of ice (**Figure 21.2**).

Ice has some unusual properties. Its melting temperature is very low (0°C), hundreds of degrees below the melting temperatures of silicate rocks. Most rocks are denser than their melts, which is why magma rises buoyantly through the lithosphere. But ice is less dense than its melt, which is why icebergs float on the ocean. And although ice may seem hard, it is much weaker than most rocks.

Because ice is weak, it flows readily downhill like a viscous fluid. **Glaciers** are large masses of ice on land that show evidence of being in motion or of once having moved under the force of gravity. We divide glaciers, on the basis of size and shape, into two basic types: *valley glaciers* and *continental glaciers*.

Valley Glaciers

Many skiers and mountain climbers are familiar with **valley glaciers,** sometimes called *alpine glaciers* (**Figure 21.3**). These rivers of ice form in the cold heights of mountain ranges where snow accumulates, and they flow down the

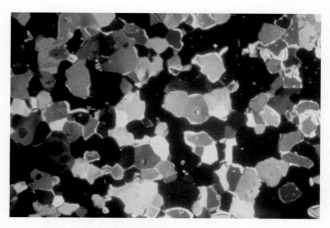

Figure 21.2 A typical mosaic of crystals of glacial ice. The tiny circular and tubular spots are bubbles of air. [Science VU/NOAA/USGS/Visuals Unlimited.]

Figure 21.3 Herbert Glacier, a valley glacier near Juneau, Alaska. [Greg Dimijian/Photo Researchers.]

bedrock valleys. A valley glacier usually occupies the complete width of the valley and may bury its bedrock base under hundreds of meters of ice. In warmer, low-latitude climates, valley glaciers are found only at the heads of valleys on the highest mountain peaks. An example is the glacial ice that covers the Mountains of the Moon at elevations over 5000 m along the Uganda-Zaire border in east-central Africa. In colder, high-latitude climates, valley glaciers may descend many kilometers, down the entire length of a valley. Broad lobes of ice may descend into lower lands bordering mountain fronts. Valley glaciers that flow down coastal mountain ranges may terminate at the ocean's edge, where masses of ice break off and form icebergs—a process called **iceberg calving** (see Figure 21.8).

Continental Glaciers

A **continental glacier** is an extremely slow moving, thick sheet of ice (hence, sometimes called an *ice sheet*) that covers a large part of a continent. Today, the world's largest ice sheets overlay much of Greenland and Antarctica (**Figure 21.4**). The

glacial ice of Greenland and Antarctica is not confined to mountain valleys but covers virtually the entire land surface. In Greenland, 2.8 million cubic kilometers of ice occupy 80 percent of the island's total area of 4.5 million square kilometers. The upper surface of the ice sheet resembles an extremely wide convex lens. At its highest point, in the middle of the island, the ice is more than 3200 m thick (**Figure 21.5**). From this central area, the ice surface slopes to the sea on all sides. At the mountain-rimmed coast, the ice

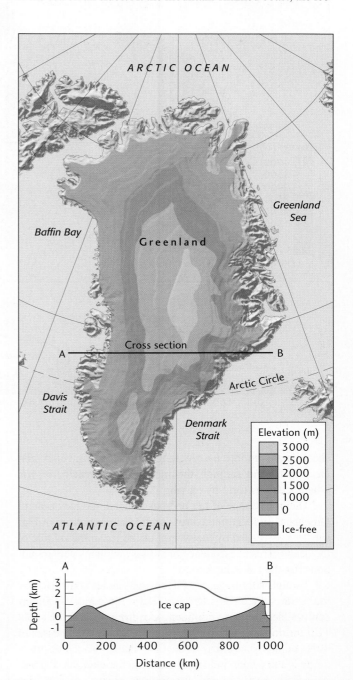

Figure 21.5 The extent of the glacial ice cap and the elevation of the ice surface on Greenland. The generalized cross section of south-central Greenland, A–B, shows the lenslike shape of the ice cap. The ice moves down and out from the thickest section. [After R. F. Flint, *Glacial and Quaternary Geology*. New York: Wiley, 1971.]

Figure 21.4 Sentinel Range, Antarctica. These mountains stick up through the thick ice of the Antarctic continental glacier. [Betty Crowell.]

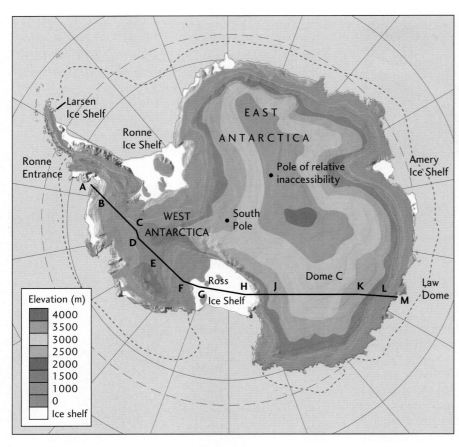

Figure 21.6 This contour map and the cross section of Antarctica show the topography of the ice sheet that covers the entire continent and the land beneath it. Ice shelves are shown in white. [After U. Radok, "The Antarctic Ice." *Scientific American* (August 1985): 100; based on data from the International Antarctic Glaciological Project.]

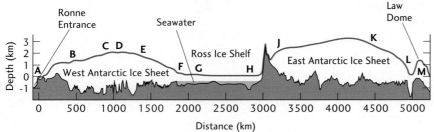

sheet breaks up into narrow tongues resembling valley glaciers that wind through the mountains to reach the sea, where icebergs form by calving.

The bowl shape of the bedrock beneath the ice dome of Greenland, evident in the cross section at the bottom of Figure 21.5, is caused by the weight of the ice in the middle of the island. This consequence of isostasy explains why mountains rim the Greenland coast.

Though very large, the Greenland glacier is dwarfed by the Antarctic ice sheet (**Figure 21.6**). Ice blankets 90 percent of the Antarctic continent, covering an area of about 12.5 million square kilometers and reaching thicknesses of 4000 m. In Antarctica, as in Greenland, the ice forms a dome in the center and slopes down to the margins. Parts of Antarctica are rimmed by thinner sheets of ice—**ice shelves**—floating on the ocean and attached to the main glacier on land. The best known of these is the Ross Ice Shelf, a thick layer of ice about the size of Texas that floats on the Ross Sea (see Figure 21.6).

Ice caps are the masses of ice that blanket Earth's North and South Poles. Most of the Arctic ice cap, located at the highest north latitudes, lies over water and is generally not referred to as a glacier. Almost all of the Antarctic ice cap lies on the continent of Antarctica and is therefore considered a continental glacier.

HOW GLACIERS FORM

A glacier starts with abundant winter snowfall that does not melt away in the summer. The snow is slowly converted into ice, and when the ice is thick enough, it begins to flow.

Basic Ingredients: Freezing Cold and Lots of Snow

For a glacier to form, temperatures must be low enough to keep snow on the ground year-round. These conditions occur near the poles, because the Sun's rays strike Earth at low angles at high latitudes, and at high altitudes, because the atmosphere becomes steadily cooler up to about 10 km. Therefore, the height of the snow line—the altitude above which snow does not completely melt in summer—generally decreases toward the poles, where snow and ice stay year-round even at sea level. Near the equator, glaciers form only on mountains that are higher than about 5000 m.

The precipitation of snow and the formation of glaciers require moisture as well as cold. Moisture-laden winds tend to drop most of their snow on the windward side of a high mountain range, so the leeward side is likely to be dry and unglaciated. Parts of the high Andes of South America, for instance, lie in a belt of prevailing easterly winds. Glaciers form on the moist eastern slopes, but the dry western side has little snow and ice.

Cold climates are not necessarily the snowiest. Nome, Alaska, has a polar arctic climate with an annual average maximum temperature of 9°C, but it gets only about 4.4 cm of precipitation a year, virtually all of it as snow. In comparison, Caribou, Maine, has a cool climate with an annual average maximum temperature of 25°C, and its average annual snowfall is a whopping 310 cm. Nevertheless, the conditions around Nome, where little of the snow melts, are better for the formation of glaciers than the conditions in Caribou, where the snow melts in the spring. In arid climates, glaciers are unlikely to form, unless the temperature is frigid enough throughout the year, so that very little snow melts, as in Antarctica.

Glacial Growth: Accumulation

A fresh snowfall is a fluffy mass of loosely packed snowflakes. As the small, delicate crystals age on the ground, they shrink and become grains (**Figure 21.7**). During this transformation, the mass of snowflakes compacts to form a dense, granular snow. As new snow falls and buries the older snow, the granular snow compacts further to an even denser form, called *firn*. Further burial and aging produce solid glacial ice as the smallest grains recrystallize, cementing all the grains together. The whole process may take only a few years, although 10 to 20 years is more likely. A typical glacier grows slightly during the winter, as snow falls on its surface. The amount of snow added to the glacier annually is its **accumulation.**

As glacial snow and ice accumulate, they entrap and preserve valuable relics of Earth's past. In 1992, Italian and Austrian scientists announced that they had discovered the body of a prehistoric human preserved in alpine ice on the border between their two countries. In northern Siberia, extinct animals such as the woolly mammoth, a great elephant-like creature that once roamed icy terrains, have been found frozen and preserved by ancient ice. Ancient dust particles and bubbles of atmospheric gases are also preserved in glacial ice (see Figure 21.2). Chemical analysis of air bubbles found in very old, deeply buried Antarctic and Greenland ice

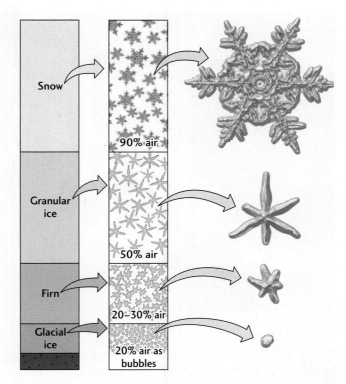

Figure 21.7 Stages in the transformation of snow crystals, first into granular ice, then into firn, and finally into glacial ice. A corresponding increase in density accompanies this transformation as air is eliminated from the crystals. [After H. Bader et al., "Der Schnee und seine Metamorphose." *Beiträge zur Geologie der Schweiz* (1939).]

tells us that levels of atmospheric carbon dioxide were lower during the last glaciation than they have been since the glaciers retreated (see Figure 15.10).

Glacial Shrinkage: Ablation

When ice accumulates to a thickness sufficient for movement to begin, a glacier is born. Ice, like water, flows downhill under the pull of gravity. The ice moves down a mountain valley or down from the dome of ice at the center of a continental ice sheet. In either case, it extends the glacier into lower altitudes where temperatures are warmer.

The total amount of ice that a glacier loses each year is called **ablation.** Four mechanisms are responsible for the loss of ice:

1. *Melting.* As the ice melts, the glacier loses material.

2. *Iceberg calving.* Pieces of ice break off and form icebergs when a glacier reaches a shoreline (**Figure 21.8**).

3. *Sublimation.* In cold climates, ice can be transformed directly from the solid into the gaseous state.

4. *Wind erosion.* Strong winds can erode the ice, primarily by melting and sublimation.

Figure 21.8 Iceberg calving at Wrangell-St. Elias National Park, Alaska. Calving occurs when huge blocks of ice break off at the edge of a glacier that has moved to a shoreline. [Tom Bean.]

Most of the glacial shrinkage that results from warming and melting takes place at the glacier's leading edge. Thus, even though a glacier is advancing downward or outward from its center, the ice edge may be retreating. The two mechanisms by which glaciers lose the most ice are melting and calving.

Glacial Budgets: Accumulation Minus Ablation

The difference between accumulation and ablation, called the glacial budget, results in the growth or shrinkage of a glacier (**Figure 21.9**). When accumulation minus ablation is zero over a long period, the glacier remains of constant size, even as it continues to flow downslope from the area where it formed. Such a glacier accumulates snow and ice in its upper reaches and ablates an equal amount in its lower parts. If accumulation exceeds ablation, the glacier grows; if ablation exceeds accumulation, the glacier shrinks.

Glacial budgets vary from year to year. In the past several thousand years, many glaciers have maintained a constant average size, though some show evidence of growth or shrinkage in response to short-term climate variations. It has recently become clear, however, that glaciers at lower lati-

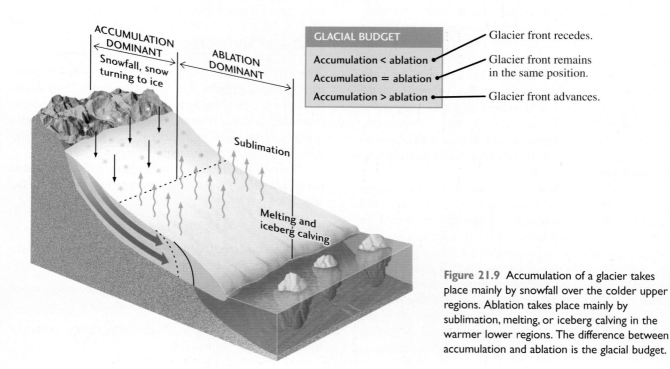

Figure 21.9 Accumulation of a glacier takes place mainly by snowfall over the colder upper regions. Ablation takes place mainly by sublimation, melting, or iceberg calving in the warmer lower regions. The difference between accumulation and ablation is the glacial budget.

LANDSCAPES
Tectonic and Climate Interaction

Have you ever gazed out over the horizon and wondered why Earth's surface has the shape it does and what forces created that shape? From high, snowcapped peaks to broad, rolling plains, Earth's landscape comprises a diverse array of landforms—large and small, rough and smooth. Landscapes evolve through slow changes as uplift, weathering, erosion, transportation, and deposition combine to sculpt the land surface.

In the past, these changes were imperceptible on a human time scale, but new technologies now permit us to measure the rates of many of these processes directly. **Geomorphology**—the study of landscapes and their evolution—is a revitalized branch of the Earth sciences that has benefited tremendously from the ability to measure these slow processes. Knowing how landscapes evolve helps us to manage our land resources and appreciate the linkages between tectonics and climate. Understanding the development of landscapes represents a great challenge to geologists because it requires them to integrate many branches of Earth science.

In the most basic sense, landscapes can be viewed as the results of a competition between processes that raise Earth's surface and those that lower it. Driven by the plate tectonic geosystem, Earth's crust is lifted up into mountain ranges. Uplifted rocks are exposed to weathering and erosion, which are driven by the climate geosystem. The geomorphology of uplifted rocks is an expression of the interaction between these two geosystems.

The uplifted areas of Earth's crust can be narrow or broad, and uplift rates can be fast or slow. Similarly, physical and chemical weathering may operate over narrow or broad areas, and the intensity of weathering can be high or low. Thus, landscapes themselves depend on the *proportion* of these tectonic and climatic influences. Moreover,

Grand Teton Mountains, Wyoming. The rugged mountains are made of granite that has been uplifted and weathered to form impressive spires. In the foreground, uplift of the plains has resulted in incision of the Snake River, which left behind a series of steplike benches in the topography. The topography of Earth records the ceaseless competition between uplift and erosion. [Lester Lefkowitz/Corbis.]

these influences interact. For example, uplift may drive regional (or even global) climate change and weathering rates, which in turn help to control further uplift of mountain ranges.

In this chapter, we look closely at how tectonics and climate—and their component processes, including uplift, weathering, erosion, and sediment transport and deposition—interact in the dynamic process that sculpts the landscape.

TOPOGRAPHY, ELEVATION, AND RELIEF

We begin our study of landscape evolution with the elementary facts of any terrain that are obvious as one examines

Earth's surface: the height and ruggedness, or roughness, of mountains and lowlands. **Topography** is the general configuration of varying heights that gives shape to Earth's surface. We compare the heights of landscape features with sea level —the average height of the oceans throughout the world. We then express the altitude, or vertical distance above or below sea level, as **elevation.** A topographic map shows the distribution of elevations in an area and most often represents this distribution by **contours,** lines that connect points of equal elevation (**Figure 22.1**). The more closely spaced the contour lines, the steeper the slope.

Centuries ago, geologists learned to survey topography and construct maps to plot and record geological information. Although land surveying based on age-old methods is still used for some purposes, modern mapmakers rely on satellite photographs, radar imaging, airborne laser range

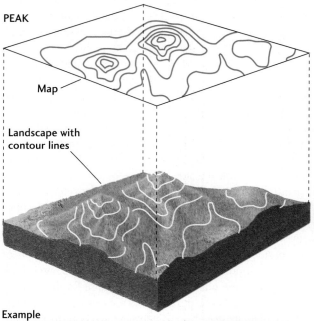

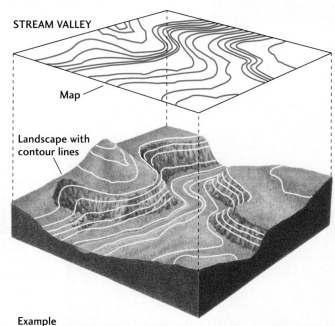

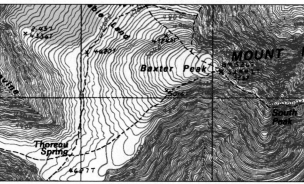

Mt. Katahdin, Maine

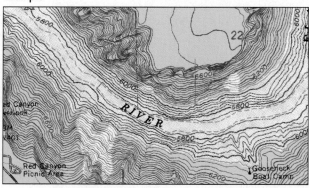

Flaming Gorge, Wyoming

Figure 22.1 The topography of a mountain peak (*left*) and a stream valley (*right*) can be accurately depicted on a flat topographic map by contours, lines that connect points of equal elevation. The more closely spaced the contour lines, the steeper the slope. [After A. Maltman, *Geological Maps: An Introduction.* New York: Van Nostrand Reinhold, 1990, p. 17. Topographical maps from USGS/DRG.]

Digital elevation map (DEM)

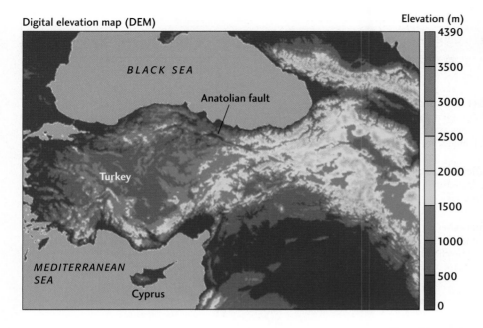

Slope map

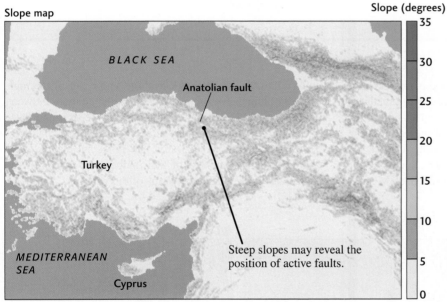

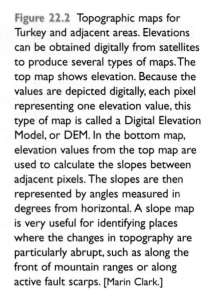

Figure 22.2 Topographic maps for Turkey and adjacent areas. Elevations can be obtained digitally from satellites to produce several types of maps. The top map shows elevation. Because the values are depicted digitally, each pixel representing one elevation value, this type of map is called a Digital Elevation Model, or DEM. In the bottom map, elevation values from the top map are used to calculate the slopes between adjacent pixels. The slopes are then represented by angles measured in degrees from horizontal. A slope map is very useful for identifying places where the changes in topography are particularly abrupt, such as along the front of mountain ranges or along active fault scarps. [Marin Clark.]

finders, and other technologies that enable them to discern elevation and other topographic properties (**Figure 22.2**).

One of the properties of topography is **relief**—the difference between the highest and lowest elevations in a particular area (**Figure 22.3**). As this definition implies, relief varies with the scale of the area over it which it is measured. In studies of geomorphology, it is useful to define three fundamental components of relief: hillslope relief (the decrease in elevation between mountain summits/ridgelines and the point at which channels begin), tributary channel relief (the decrease in elevation along tributaries), and trunk channel relief (the decrease in elevation along a trunk channel).

To estimate relief in an area of interest from contours on a topographic map, we subtract the elevation of the lowest

contour, usually at the bottom of a river valley, from that of the highest, at the top of the highest hill or mountain. Relief is a measure of the roughness of a terrain. The greater the relief, the more rugged the topography. Mount Everest, the highest mountain in the world with an elevation of 8861 m, is located in an area of extremely high relief (**Figure 22.4a**). In general, most regions of high elevation also have high relief, and most areas of low elevation have low relief. There are exceptions, however. For example, the Dead Sea in Israel and Jordan has the lowest land elevation in the world at 392 m below sea level, but it is flanked by impressive mountains that provide significant relief in this small area of the Earth (see Figure 22.4b). Other regions, such as the Tibetan Plateau of the Himalaya, may lie at high elevations but have relatively low relief.

Key Figure 22.3   Relief is the difference between the highest and lowest elevations in a region.

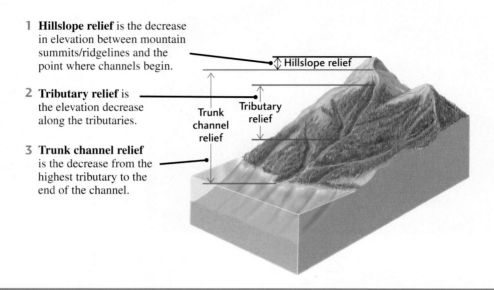

1 Hillslope relief is the decrease in elevation between mountain summits/ridgelines and the point where channels begin.

2 Tributary relief is the elevation decrease along the tributaries.

3 Trunk channel relief is the decrease from the highest tributary to the end of the channel.

If we fly over North America, we can see many sorts of topography. **Figure 22.5** is a computer-processed digital map that shows the details of large- and small-scale landforms. This digital map provides an overall view of the entire contiguous United States and Canada while showing features as small as 2.5 km across. The moderate elevations and relief of the elongate ridges and valleys of the Appalachian Mountains contrast with the low elevations and relief of the midwestern plains. Even more striking is the contrast between the plains and the Rocky Mountains. As we examine these different types of topography more closely, we can characterize them not only by elevation and relief but also by their landforms: the steepness of their slopes, the shapes of the mountains or hills, and the forms of the valleys.

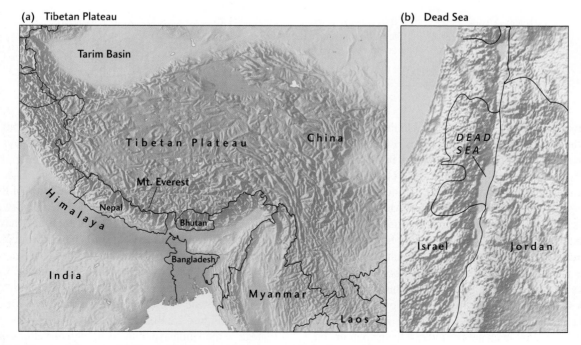

Figure 22.4 Topographic maps. (a) Mount Everest, the highest mountain in the world, and (b) the Dead Sea, the lowest land elevation in the world. [Marin Clark and Nathan Niemi.]

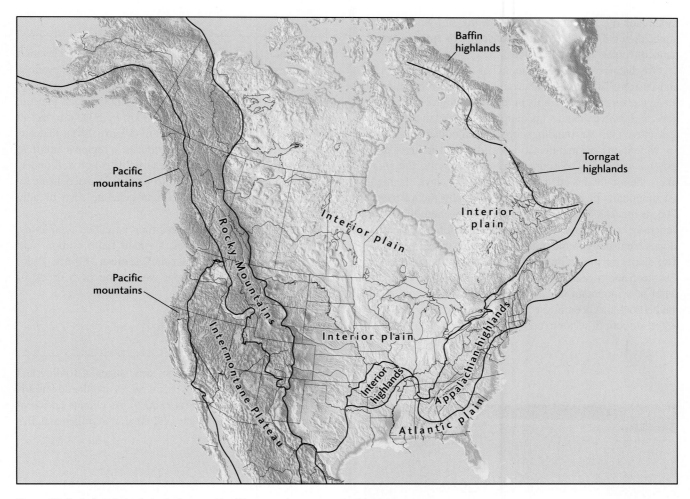

Figure 22.5 A digital shaded relief map of landforms in the contiguous United States and Canada. [Gail P. Thelin and Richard J. Pike/USGS, 1991.]

LANDFORMS: FEATURES SCULPTED BY EROSION AND SEDIMENTATION

Rivers, glaciers, and the wind leave their marks on Earth's surface in a variety of **landforms:** rugged mountain slopes, broad valleys, floodplains, dunes, and many others. The scale of landforms ranges from regional to very local. At the largest scale (tens of thousands of kilometers), mountain belts form topographic walls along the boundaries of lithospheric plates. At the smallest scale (meters), the topography of an individual outcrop may be caused by differential weathering of the rocks of differing hardnesses that compose it. This chapter focuses mostly on the regional-scale features that define the overall topography of Earth's surface.

Mountains and Hills

We have used the word *mountain* many times in this book, yet we can define it no more precisely than to say that a mountain is a large mass of rock that projects well above its surroundings. Most mountains are found with others in ranges, where peaks of various heights are easier to distinguish than distinct separate mountains (**Figure 22.6**). Where

Figure 22.6 Most mountains are found in ranges, not as individual peaks. In this glacially sculpted terrain in southern Argentina, all the peaks are sharp arêtes. [Galen Rowell/Peter Arnold.]

mountains rise as single peaks above the surrounding low-lands, they are usually isolated volcanoes or erosional remnants of former mountain ranges.

We distinguish between mountains and hills only by size and custom. Elevations that would be called mountains in lower terrains are called hills in regions of high elevation. In general, however, landforms more than several hundred meters above their surroundings are called mountains.

Mountains are manifestations of direct or indirect plate tectonic activity. The more recent the activity, the more likely the mountains are to be high. The Himalaya, the highest mountains in the world, are also among the youngest. The steepness of the slopes in mountainous and hilly areas generally correlates with elevation and relief. The steepest slopes are usually found in high mountains with great relief. The slopes of mountains with less elevation and relief are less steep and rugged. As we will see later in this chapter, the relief of a mountain range strongly depends on how much the bedrock has been incised by glaciers and rivers, relative to the amount of tectonic uplift.

Plateaus

A **plateau** is a large, broad, flat area of appreciable elevation above the neighboring terrain. Most plateaus have elevations of less than 3000 m, but the Altiplano of Bolivia lies at an elevation of 3600 m. The extraordinarily high Tibetan Plateau, which extends over an area of 1000 km by 5000 km, has an average elevation of almost 5000 m (**Figure 22.7**). Plateaus form where tectonic activity produces a regional uplift in response to vertical forces. Plateau uplift is not well understood, and geologists struggle to identify a mechanism by which plate tectonic processes can generate such broadly uplifted areas.

Smaller plateaulike features may be called *tablelands*. In the western United States, a small, flat elevation with steep slopes on all sides is called a **mesa** (**Figure 22.8**). Mesas result from differential weathering of bedrock of varying hardness.

Structurally Controlled Ridges and Valleys

In young mountains, during the early stages of folding and uplift, the upfolds (anticlines) form ridges and the downfolds (synclines) form valleys (**Figure 22.9**). As climate and weathering begin to predominate and erosional gullies and val-

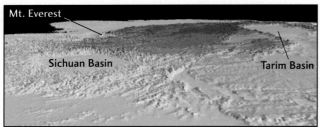

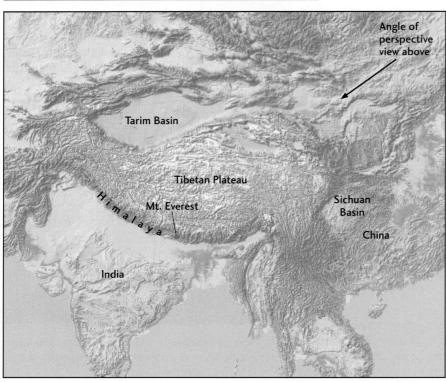

Figure 22.7 Map view of the Tibetan Plateau, the highest and most extensive on Earth. The tectonic mechanisms that produced this massive uplift are still unclear. Some geologists think the plateau formed when the Indian Plate was subducted beneath the Eurasian Plate, doubling the lithospheric thickness in the region where the plateau is today. Isostatic rebound of this doubly thick lithosphere would produce uplift of the plateau region. Other geologists argue that during the subduction of India beneath Asia, the Earth's lower crust may have flowed—just like the mantle—then pooled as a fluid cushion and pushed up the plateau region.

Figure 22.8 A mesa in Monument Valley, Arizona. The flat tops are held up by erosion-resistant beds. [Raymond Siever.]

leys bite deeper into the structures, the topography may become inverted, so that anticlines form valleys and synclines form ridges. This happens where the rocks—typically sedimentary rocks such as limestones, sandstones, and shales —exert strong control on the topography by their variable resistance to erosion. If the rocks beneath an anticline are mechanically weak, as shales are, the core of the anticline may be eroded to form an anticlinal valley (**Figure 22.10**). In a region that has been eroded for many millions of years, a pattern of linear anticlines and synclines produces a series of ridges and valleys such as those of the Appalachian Mountains in Pennsylvania and adjacent states (**Figure 22.11**).

River Valleys and Bedrock Erosion

Observations of river valleys in various regions led to one of the important early theories of geology: the idea that river valleys were created through erosion by the rivers that flowed in them. Geologists could see that sedimentary rock formations on one side of a valley matched the same kinds of formations on the opposite side. This led them to conclude that the formations had once been deposited as a continuous sheet of sediment. The river had removed enormous quantities of the original formation by breaking up the rock and carrying it away. Today, geologists are trying to figure

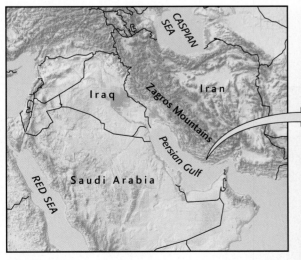

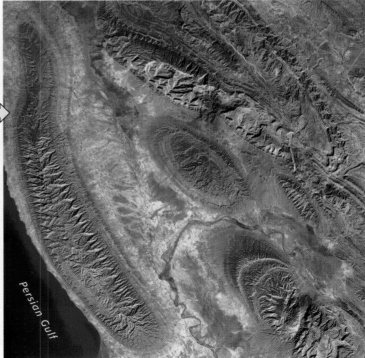

Figure 22.9 Valley and ridge topography formed on a folded terrain of sedimentary rock. The deformation is so recent (Pliocene) that erosion has not yet significantly modified the original structural forms of anticlines (ridges) and synclines (valleys). Zagros Mountains, Iran. [NASA.]

TIME 1
Harder, erosion-resistant rocks lie over softer, more erodable layers. Ridges are over anticlines and streams flow in valleys formed by synclines. Tributary streams on anticline slopes flow faster and with more power than valley streams. They erode the slopes faster than main streams erode the valleys.

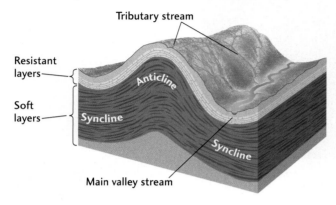

TIME 2
Tributaries over the synclines cut through resistant rock layers and start to quickly carve the softer underlying rock into steep valleys over the anticlines.

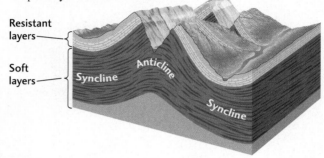

TIME 3
As the process continues, valleys form over the anticlines and ridges capped by resistant strata are left over the synclines.

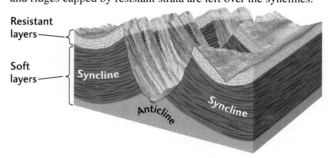

Figure 22.10 Stages in the development of ridges and valleys in folded mountains. In early stages, ridges are formed by anticlines. In later stages, the anticlines may be breached and ridges may be held up by caps of resistant rocks as erosion forms valleys in less resistant rocks.

out which of several physical processes contribute to this erosion of bedrock.

How a river erodes bedrock depends on the river's **stream power**—which is the product of river slope and river discharge—balanced by the bedrock's ability to resist erosion—which is the product of the volume and grain size of sed-

iment in the river channel (**Figure 22.12**). If stream power is high enough to wash away the sediment load, resistance to erosion is a function mostly of bedrock hardness. As it turns out, rates of bedrock erosion increase dramatically as stream power increases. On most days, a flowing river accomplishes little erosion because discharge, and thus stream power, is

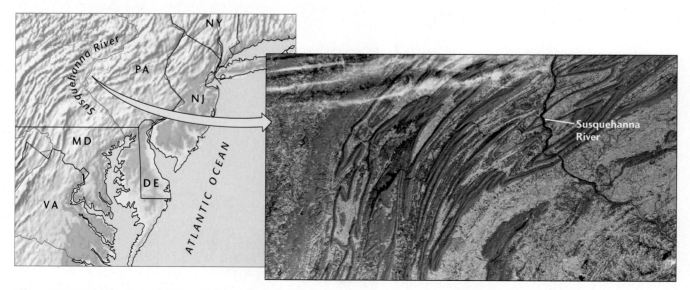

Figure 22.11 The Appalachian Valley and Ridge province has the tectonically controlled topography of linear anticlines and synclines produced by millions of years of erosion. The prominent ridges, shown in reddish orange, are held up by erosion-resistant sedimentary rocks that have been folded into an intricate series of anticlines and synclines. [Earth Satellite Corp.]

1 Increasing sediment size, sediment volume, and bedrock hardness increase the resistance to erosion.

2 Increasing river slope and river discharge increase stream power and thus erosion.

3 Sediment particles protect the bedrock from erosion. When they are removed, bedrock erosion results from saltation abrasion ("sandblasting"),…

(a)

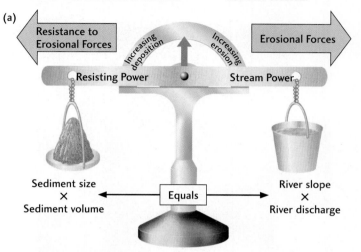

Resistance to Erosional Forces
Increasing deposition
Increasing erosion
Erosional Forces

Resisting Power
Stream Power

Sediment size × Sediment volume
Equals
River slope × River discharge

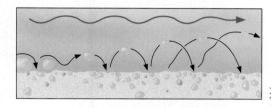

4 …or plucking of the bedrock along zones of weakness, such as fractures.

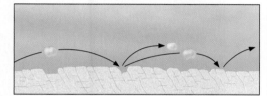

5 In steep, wet terrain, stream power overcomes resistance to erosion. Sediment particles are transported away, and bedrock hardness becomes the principal factor in resistance to erosion.

(b)

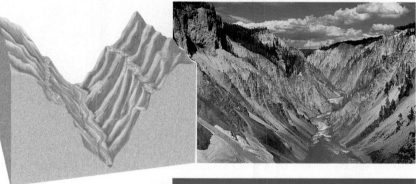

6 Where slopes are gentler, stream discharge is lower, and, therefore, stream power is lower. Thus, sediment begins to be deposited, armoring the streambed and stopping erosion. At this point, stream power and resistance to erosion are in balance.

(c)

7 Where slopes are much flatter, stream power is so decreased that much sediment is deposited and the streambed builds up and fills the valley with sediment.

(d)

[After D. W. Burbank and R. S. Anderson, *Tectonic Geomorphology*. Oxford: Blackwell, 2001. (*b*) Yellowstone River, Yellowstone National Park (Jeff Henry/RocheJaune Pictures). (*c*) Snake River, Suicide Point, Idaho (David G. Houser/Corbis). (*d*) Mulchatna River, Alaska (Glenn Oliver/Visuals Unlimited.)]

Figure 22.13 Gully erosion in the Badlands of South Dakota, formed in easily eroded sedimentary rocks. Sage Creek Wilderness, Badlands National Park. [Willard Clay.]

low. However, on the rare days when discharge (and stream power) is very large, erosion rates can be dramatically high. This relationship illustrates a fundamental characteristic shared by many of Earth's geosystems: rare, large events often create much more change than frequent but small ones.

Three principal processes erode bedrock. The first is **abrasion** of the bedrock by suspended and saltating sediment particles that move along the bottom and sides of the river channel. (In Chapter 18, we learned that saltation is the movement of sediment particles by bounces and impacts along the bottom and sides of a channel.) Second, the drag force of the current itself abrades the bedrock as it plucks rock fragments from the channel. Third, at higher elevations, glacial erosion forms valleys that can then be occupied by rivers. Determining the relative importance of these three processes in mountainous terrains is one way geologists can distinguish between the influences of climate and tectonics on landscape evolution.

River valleys have many names—canyons, gulches, arroyos, gullies—but all have the same general geometry. A vertical cut through a young mountain river valley with little or no floodplain has a simple V-shaped profile (see Figure 22.12b). A broad, low river valley with a wide flood-

plain has a cross section that is more open but is still distinct from the U-shaped profile of a glacial valley. Regions of different general topography and type of bedrock produce river valleys of varying shape and width (see Figure 22.12c,d). Valleys range from the narrow gorges of mountain belts and erosion-resistant rock types to the wide, shallow valleys of plains and easily eroded rock types. Between these extremes, the width of a valley generally corresponds to the erosion state of the region. Valleys are somewhat broader in mountains that have begun to be lowered and rounded from erosion and are much broader in low-lying hilly topography.

A **badland** is a deeply gullied topography resulting from the fast erosion of easily erodible shales and clays, such as those of the South Dakota Badlands (**Figure 22.13**) Virtually the entire area is a proliferation of gullies and valleys, with little flat land between them.

Structurally Controlled Cliffs

The folds and faults produced by rock deformation during mountain building leave their marks on Earth's surface. **Cuestas** are asymmetrical ridges in a tilted and eroded series

(a)

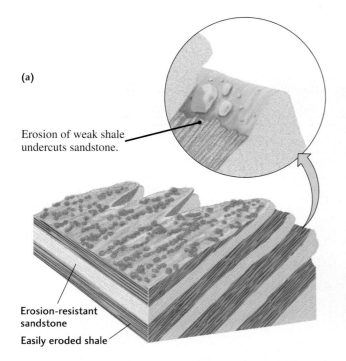

Erosion of weak shale undercuts sandstone.

Erosion-resistant sandstone

Easily eroded shale

(b)

Figure 22.14 (a) Cuestas form where gently dipping beds of erosion-resistant rock, such as sandstone, are undercut by erosion of an easily eroded underlying rock, such as shale. (b) Cuestas formed on structurally tilted sedimentary rocks in Dinosaur National Monument, Colorado. [Martin Miller.]

of beds with alternating weak and strong resistance to erosion. One side of a cuesta has a long, gentle slope determined by the dip of the erosion-resistant bed. The other side is a steep cliff formed at the edge of the resistant bed where it is undercut by erosion of a weaker bed beneath (**Figure 22.14**).

Figure 22.15 Hogbacks are narrow ridges formed by layers of erosion-resistant sedimentary rocks that are tectonically turned up so that the beds are vertical or nearly so. These hogback ridges are in the Rocky Mountains near Denver, Colorado. [Tom Till/DRK/PHOTO.]

Much more steeply dipping or vertical beds of hard strata erode more slowly to form **hogbacks**—steep, narrow, more or less symmetrical ridges (**Figure 22.15**). Steep cliffs are also produced by nearly vertical faults in which one side rises higher than the other.

INTERACTING GEOSYSTEMS CONTROL LANDSCAPE

Broadly speaking, the interaction of Earth's internal and external heat engines controls landscape. The internal heat engine drives tectonics, which elevates mountains and volcanoes. The external heat engine, powered by the Sun, controls climate and weathering, which wear away the mountains and fill basins with sediment. Sunlight causes the motions of the atmosphere that produce climate, Earth's different temperature regimes, and the rainwater that runs off the continents as rivers. Thus, landscape is controlled by Earth's interacting geosystems (**Figure 22.16**).

Feedback Between Uplift and Erosion

The ceaseless competition between tectonic processes, which tend to create mountains and build topography, and surface processes, which tend to tear them down, is the focus of intensive study by geomorphologists. The interaction between the two is a **negative feedback.** In this kind of process, one action produces an effect (the feedback) that tends to slow the original action and stabilize the process at a lower rate. For example, if you are thirsty, you

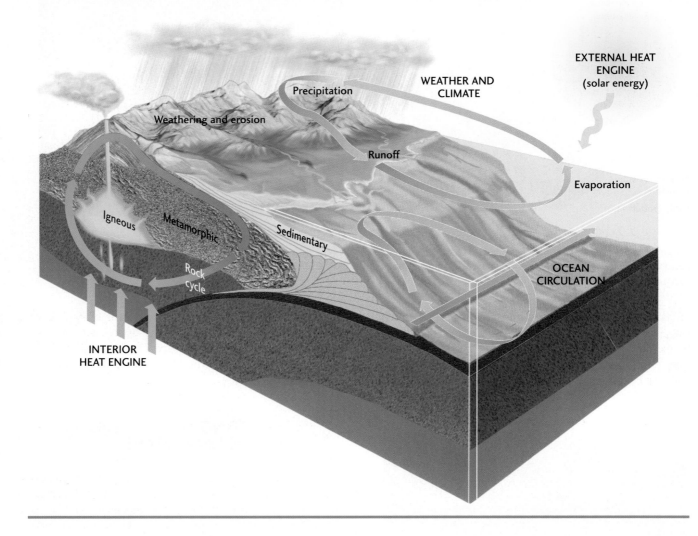

Key Figure 22.16 Landscape is controlled by interactions between the plate tectonics and climate geosystems.

EXTERNAL HEAT ENGINE (solar energy)

WEATHER AND CLIMATE

Precipitation

Weathering and erosion

Runoff

Evaporation

Igneous

Metamorphic

Sedimentary

Rock cycle

OCEAN CIRCULATION

INTERIOR HEAT ENGINE

will drink a glass of water quickly at first. As the drinking reduces your thirst (the feedback), you will drink more slowly, until finally your thirst is completely satisfied and you stop drinking. The process of drinking has stabilized at a rate of zero.

A similar negative-feedback process begins when strong tectonic action elevates mountains. Tectonic uplift provokes an increase in erosion rate (**Figure 22.17**). The higher the mountains grow, the faster erosion wears them down. As long as mountain building continues, elevations stay high or increase. When mountain building slows, perhaps because of a change in the rate of plate tectonic movements, the mountains rise more slowly or stop rising entirely. As growth slows or stops, erosion starts to dominate, and the mountains are worn down to lower elevations. This process explains why old mountains, such as the Appalachians, are relatively low

compared to much younger mountains, such as the Rockies. As the mountains continue to be worn down, erosion also slows, and the whole process eventually tapers off. Elevation is thus a balance between the rate of tectonic uplift and the rate of erosion.

Curiously, over the shorter time scales of thousands to millions of years, tectonics and climate can interact in a **positive feedback** (see Figure 22.17b and Feature 22.1) in which mountain summits may become *higher* as a result of erosion. This occurs because continents and mountains are buoyed up and float in the Earth's mantle, much as the icebergs discussed in Chapter 21 float in seawater.

Continents and mountains float because the large volume of less dense continental crust that projects into the denser mantle provides buoyancy. The crust is thicker under a mountain because a deeper root is needed to float the addi-

Key Figure 22.17 | Elevation is a balance between the rate of tectonic uplift and the rate of erosion.

(a) Uplift stimulates erosion (negative feedback)

1 Tectonic action elevates mountains. Uplift rate is greater than the rate of erosion. Uplift generates weathering (negative feedback).

2 Uplift slows, and erosion is in balance with uplift. Elevations stay high.

3 Uplift slows more, and erosion starts to dominate and elevation begins to decrease.

4 Uplift is almost stopped. Lower elevations—low hills—generate less weather, and erosion slows.

5 Uplift stops and erosion slows further. The elevation decreases further as the landscape evolves into lowlands and plains.

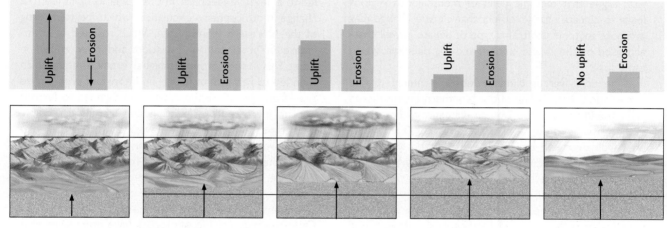

(b) Isostatic mantle rebound raises mountain elevation (positive feedback)

1 Continents collide, uplifting a high plateau.

2 Uplift creates a weather pattern that increases erosion rate.

3 Crust rebounds isostatically (positive feedback) and causes mountain summits to be uplifted above pre-erosion elevation.

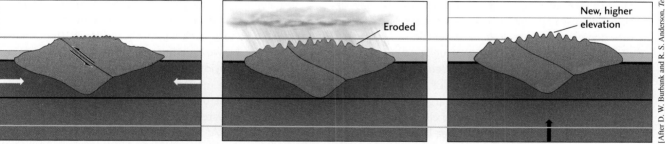

[After D. W. Burbank and R. S. Anderson, *Tectonic Geomorphology.* Oxford: Blackwell, 2001, p. 9.]

tional weight of the mountain. In Chapter 14, we saw that although the mantle just beneath the crust is solid rock, the mantle flows very slowly when forces are applied over thousands to millions of years. The *principle of isostasy* implies that, over this period of time, the mantle has little strength and behaves like a viscous fluid when it is forced to support the weight of continents and mountains.

Isostasy also implies that as a large mountain range forms, it slowly sinks under gravity and the crust bends downward. When enough of a root bulges into the mantle, the mountain floats. When the valleys in a mountain range are deepened by erosion (see Figure 22.17b), the weight on the crust is lightened and less root is needed for flotation. As

the valleys erode down, the root floats upward. This process, called *isostatic rebound,* forms the basis of positive feedback between climate and tectonics, which results in mountain summits being elevated to new heights because of erosion (see Figure 22.17b). Over longer time scales, however, the mountain summits will be worn down as described above and in Figure 22.17a.

Feedback Between Climate and Topography

Glaciers, rivers, and landslides—powerful agents of erosion—operate at different rates at different altitudes. Thus

---

EARTH ISSUES

22.1 Uplift and Climate Change: A Chicken-and-Egg Dilemma

One of the clearest illustrations of how Earth's climate and plate tectonic systems are coupled is provided by the feedbacks between climate change and the mean elevation of mountain belts. Currently, there is controversy over the directionality of these feedbacks. Some geologists argue that tectonic uplift of mountainous regions leads to climate change, others that climate change may promote tectonic uplift. This type of debate is well characterized by the classic chicken-and-egg dilemma: which came first?

The uplift versus climate debate is fueled by the observation that cooling of the Northern Hemisphere's climate and uplift of the Tibetan Plateau may have been synchronous. With a mean elevation of 5000 m and an area well over half the size of the United States, the Tibetan Plateau is the most imposing topographic feature on the surface of the Earth (see Figure 22.7). The Tibetan Plateau is so high and so extensive in area that it not only drives the Asian monsoon but may also influence the broader patterns of atmospheric circulation in the Northern Hemisphere. The absence of such a topographic feature would no doubt result in significant climatological differences in Earth's Northern Hemisphere.

Unfortunately, although the timing of Northern Hemisphere cooling is well calibrated by the age of glacial deposits and by the deep-sea record of temperature decrease that signals the construction of major ice sheets and glaciers, the timing of uplift of the Tibetan Plateau is not. This is where the debate begins. If the Tibetan uplift preceded the onset of Northern Hemisphere glaciation, then it might be argued that tectonically induced uplift indirectly caused climate change. On the other hand, if the uplift of Tibet lagged behind the onset of the Northern Hemisphere glaciation, then it might be argued that climate change promoted uplift as an isostatic response to enhanced erosion rates.

Point: Negative Feedback

The possibility that mountain building may have promoted glaciation of the Northern Hemisphere has been recognized for more than 100 years. Geologists who currently advocate this point of view believe that several important processes occurred in the uplift of Tibet that led to a negative feedback. In this scenario, uplift led to a change in atmospheric circulation; which led to cooling of the Northern Hemisphere; which resulted in an increase in precipitation, glaciation, and river runoff in Tibet. This in turn produced higher erosion rates; which then led to removal of CO_2—an important greenhouse gas—from the atmosphere; which in turn led to further cooling, increased precipitation, and increased erosion. Over time, the mountains will wear down and their elevations will decrease. Effectively, an elevation increase—which then modulates climate—results in an elevation decrease. This response is a negative feedback.

Counterpoint: Positive Feedback

Over the past decade, geologists have discovered that climate change might lead to uplift of mountainous regions such as Tibet. In this unexpected and counterintuitive scenario, an initial cooling of climate stimulated an increase in precipitation rates, which in turn led to enhanced erosion by glaciers and rivers. In the absence of isostatic responses, an increase in erosion would be predicted to feed back negatively and lower the mountain ranges. However, when the influence of isostasy is considered, erosion is predicted to result in a mass decrease in the mountain range as a whole, with the result being that the mountains would be uplifted and their peaks elevated to new and higher positions (see Figure 22.17b). At these higher elevations, the mountains would feed back positively and help to modify climate further, thereby increasing precipitation and erosion rates and further enhancing uplift.

---

climate, which changes with altitude, modulates weathering and therefore also modulates erosion and uplift of mountain ranges.

In Chapter 16, we saw that the effects of climate on weathering include freezing and thawing and expansion and contraction due to heating and cooling. Climate also affects the rate at which water dissolves minerals. Rainfall and temperature, the components of climate, affect weathering and erosion through the rain that falls on bedrock and soil, the

infiltration of water into the soil, mass wasting, rivers, and glaciers—all of which help to break up the rock and mineral particles eroded from slopes and carry them downhill. Thus, climate plays a role in topography.

High elevation and relief enhance the fragmentation and mechanical breakup of rocks, partly by promoting freezing and thawing. At high elevations, where the climate is cool, mountain glaciers scour bedrock and erode deep valleys. Fragmented debris on mountain slopes moves downhill

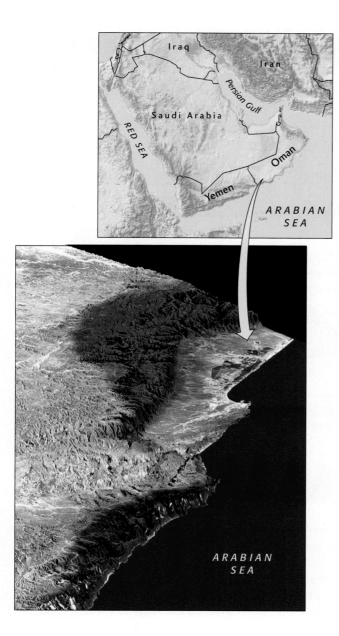

Figure 22.18 Easterly view of the escarpment along the Arabian Sea at the Yemen-Oman border. This three-dimensional satellite image illustrates how topography determines local climate and, in turn, controls erosion and landscape development. Although the Arabian Peninsula is very arid, the steep escarpment of the Qara Mountains wrings moisture from the monsoons (seasonal rains). The moisture allows natural vegetation to grow (green along the mountain fronts and in the canyons) and soil to develop (dark brown areas). In contrast, the light-colored areas are mostly dry desert. This climate focuses erosion on the ocean side of the mountain range. The intense erosion, in turn, has caused the escarpment to retreat landward, from right to left. [NASA.]

Most rivers run over broad floodplains and do little mechanical cutting of bedrock. Glaciers are absent, except in cold polar regions. Even in lowland deserts, strong winds merely facet and round rock fragments and outcrops rather than breaking them up. A lowland thus tends to have a gentle topography with rounded slopes, rolling hills, and flat plains.

Just as climate affects topography, topography can affect climate. For example, mountains cause rain shadows, dry areas on the leeward slopes of mountain ranges (see Figure 17.3). Rain shadows result in preferential erosion of one side of a mountain belt (**Figure 22.18**). In places such as New Guinea, where the difference between rainfall on the windward and leeward sides of the mountain belt is extreme, geologists believe that the uplift history of metamorphic rocks buried deep in the crust is influenced by the history of rainfall at Earth's surface.

MODELS OF LANDSCAPE EVOLUTION

The strong contrasts in the morphology of landscapes stimulated early geologists to speculate about their causes. Three prominent and influential geologists were William Morris Davis, Walther Penck, and John Hack. Davis believed that an initial burst of tectonic uplift is followed by a long period of erosion, with landscape morphology depending mostly on geologic age. Davis's view was so dominant during the early 1900s that it overshadowed the contemporaneous hypothesis of Walther Penck, who argued that tectonic uplift competes with erosion to control landscape morphology. Penck's ideas were not given the attention they deserved until the 1950s, more than two decades after Davis's death. In the 1960s, another conceptual breakthrough occurred when John Hack recognized that uplift could not raise elevation above some critical limit, even if sustained for long periods of time. Mountains, in the absence of erosion, will collapse under their own weight because of the limited strength of rocks.

Modern views of landscape evolution incorporate parts of all these early ideas and acknowledge that there is a natural,

quickly in slides and other mass movements, exposing fresh rock to attack by weathering. Rivers run faster in mountains than in lowlands and therefore erode and transport sediment more rapidly. Chemical weathering plays an important role in the erosion of high mountains, but the mechanical breakup of rocks is so rapid that most of the debris appears to be almost unweathered. The products of chemical decay—dissolved material and clay minerals—are carried down from steep mountain slopes as soon as they form. The intense erosion that occurs at high elevations produces a topography of steep slopes; deep, narrow river valleys; and narrow floodplains and drainage divides.

In lowlands, by contrast, weathering and erosion are slow, and the clay mineral products of chemical weathering accumulate as thick soils. Mechanical breakup occurs, but its effects are small compared with those of chemical weathering.

> ### Key Figure 22.19   Classic models of landscape evolution result from tectonic uplift and erosion.

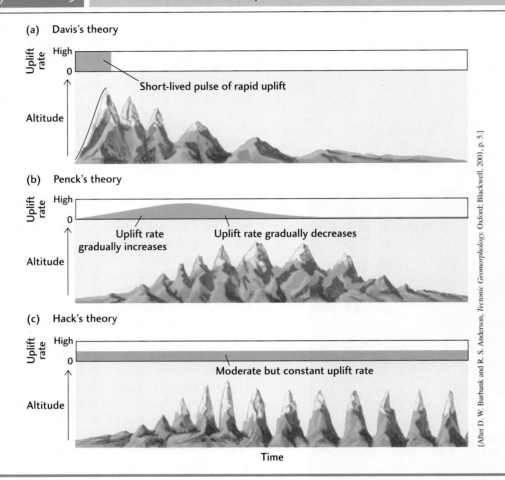

(a) Davis's theory

Uplift rate — High / 0

Short-lived pulse of rapid uplift

Altitude

(b) Penck's theory

Uplift rate — High / 0

Uplift rate gradually increases   Uplift rate gradually decreases

Altitude

(c) Hack's theory

Uplift rate — High / 0

Moderate but constant uplift rate

Altitude

Time

[After D. W. Burbank and R. S. Anderson, *Tectonic Geomorphology.* Oxford: Blackwell, 2001, p. 5.]

time-dependent progression of landscape form. Geologists now understand that the evolutionary path of landscapes depends strongly on the *time scale* over which geomorphic change occurs. The importance of the different landscape-modifying processes varies as a function of the interval of time over which the landscape is observed to change. For example, variations in climate are a very important factor in landscape evolution over the past 100,000 years but only a minor factor over time scales of 100 million years. Over these longer intervals of geologic time, the history of tectonic uplift is probably most important.

Davis's Cycle of Uplift and Erosion

William Morris Davis, a Harvard geologist in the early 1900s, studied mountains and plains throughout the world. Davis proposed a *cycle of erosion* that progresses from the high, rugged, tectonically formed mountains of youth to the rounded forms of maturity and the worn-down plains of old age and tectonic stability (**Figure 22.19a**). Davis believed

that a strong, rapid pulse of tectonic uplift begins the cycle. All the topography is built during this first stage. Erosion eventually wears down the landscape to a relatively flat surface, leveling all structures and differences in bedrock. Davis saw the flat surfaces of extensive unconformities as evidence of such plains in past geologic times. Here and there, an isolated hill might stand as an uneroded remnant of former heights. Most geologists at that time accepted Davis's assumption that mountains are elevated suddenly over short geologic times and then stay tectonically fixed as erosion slowly wears them down. Davis's cycle was accepted partly because geologists could find many examples of what seemed to be the different stages of youth, maturity, and old age.

Erosion Competes with Uplift

Davis's view was challenged by his contemporary, Walther Penck, who argued that the magnitude of tectonic deformation and uplift gradually increases to a climax and slowly wanes (see Figure 22.19b). Unfortunately, Davis, with his greater professional stature and prolific publication style, was able to promote his own ideas more effectively. Penck's ideas did not really become broadly noticed until about 50 years after their publication.

Penck proposed that geomorphic surface processes attack the uplifting mountains throughout the interval of uplift. Eventually, as the rate of deformation wanes, rates of erosion predominate over rates of uplift, resulting in a gradual decrease in both relief and mean elevation. This model was a conceptual breakthrough because it recognized that landscape evolution may result from a competition between uplift and erosion. Davis's model, in contrast, emphasized the temporal distinction between these two processes. In his model, landscape age was the primary determinant of form.

Measuring Rates of Uplift and Erosion

Over the past decade, geologists have shown that Penck's view of the competition between uplift and erosion is correct, especially for tectonically active landscapes. Choosing between alternative theories of landscape evolution requires that we determine rates of uplift and erosion in regions of mountain building. New technologies such as the Global Positioning System (GPS) and radar interferometry have produced spectacular maps of crustal deformation and uplift rates. New dating methods (Table 22.1) have helped us determine the age of geomorphically important surfaces, such as river terraces that date back 1 million years.

A promising new dating scheme is based on the fact that cosmic rays penetrating the uppermost meter of exposed rock or soil lead to the production of very small quantities of certain radioactive isotopes. One of them is beryllium-10, which accumulates to a greater degree the longer the rock or soil is exposed and to a lesser degree the deeper it is buried. Geologists used beryllium-10 to compare river-terrace ages along the course of the Indus River in the Himalaya. They plotted height changes against time to find average erosion and uplift rates. Stream erosion rates in the Himalaya were found to vary between 2 and 12 mm/year. Elsewhere, high mountain tectonic uplift rates have been measured in the same general range, from 0.8 to 12 mm/year.

Landscapes Achieve Dynamic Equilibrium

John Hack elaborated on the idea that erosion competes with uplift. He believed that when uplift and erosion rates are sustained for a long period of time, landscape evolution will achieve a balance, or dynamic equilibrium (see Figure 22.19c). During the period of equilibrium, landforms may undergo minor adjustments, but the overall landscape will look more or less the same.

Table 22.1	Methods for Absolute Dating of Landscape	
Method	**Useful Range (Years Ago)**	**Materials Needed**
RADIOISOTOPIC		
Carbon-14	35,000	Wood, shell
Uranium/thorium	10,000–350,000	Carbonate (corals)
Thermoluminescence (TL)	30,000–300,000	Quartz silt
Optically stimulated luminescence	0–300,000	Quartz silt
COSMOGENIC		
In situ beryllium-10, aluminum-26	3–4 million	Quartz
Helium, neon	Unlimited	Olivine, quartz
Chlorine-36	0–4 million	Quartz
CHEMICAL		
Tephrochronology	0–several million	Volcanic ash
PALEOMAGNETIC		
Identification of reversals	>700,000	Fine sediments, volcanic lava flows
Secular variations	0–700,000	Fine sediments
BIOLOGICAL		
Dendrochronology	10,000	Wood

Source: Modified from D. W. Burbank and R. S. Anderson, *Tectonic Geomorphology.* Oxford: Blackwell, 2001, p. 39.

Hack recognized that the height of mountains could not increase forever, even if uplift rates were extremely high. Rocks break if large enough stresses are imposed, so it stands to reason that if mountains become too high and steep, they will collapse under their own weight simply because of gravitational stresses. Thus, with continued uplift beyond some critical limit, slope failures and mass wasting alone will prevent further increases in elevation. Consequently, rates of uplift and rates of erosion come into a long-term balance. Unlike the models of Davis and Penck, Hack's model does not require uplift rates to decrease.

A fascinating implication of Hack's model is that a landscape does not have to evolve at all as long as uplift and erosion rates are balanced. Nevertheless, Earth's history teaches us that whatever goes up eventually does come down. Over very long time scales, the models of Davis and Penck are more accurate descriptions of how landscapes will ultimately change form (see Figure 22.17a). When erosion exceeds uplift, slopes become lower and more rounded. Because few areas of the world remain tectonically quiescent for as long as 100 million years, the perfectly flat erosion plain that Davis proposed could form only rarely in Earth's history. The model of dynamic equilibrium is perhaps most appropriate for landscapes in tectonically active areas where a given uplift rate can be sustained for more than a million years or so.

SUMMARY

What are the principal components of landscape? Landscape is described in terms of topography, which includes elevation, the altitude of the Earth's surface above or below sea level, and relief, the difference between the highest and the lowest spots in a region. Landscape comprises the varied landforms produced through erosion and sedimentation by rivers, glaciers, mass wasting, and the wind. The most common landforms are mountains and hills, plateaus, and structurally controlled cliffs and ridges—all of which are produced by tectonic activity modified by erosion.

How do the climate and plate tectonic systems interact to control landscape? Landscape is determined by tectonics, erosion, climate, and the hardness of bedrock. Tectonics, driven by plate motions, elevates mountains and lowers tectonic valleys and basins. Erosion carves bedrock into valleys and slopes. Climate affects weathering and erosion and makes glacial and desert landscapes possible. The varying erosion resistance of bedrock types partly accounts for differences in slope and valley profiles, steep slopes being found in rocks with greater resistance.

How do landscapes evolve? The evolution of landscapes depends strongly on the competition between the forces of uplift and the forces of erosion. Landscapes begin their evolution with tectonic uplift, which in turn stimulates erosion. While tectonic uplift rates are high, erosion rates may also be high, but mountains will be high and steep. As uplift rates wane, erosion rates will still be high and thus become relatively more important; the land surface is lowered and slopes are rounded. As uplift rates decrease to nil, erosion becomes dominant and wears down the former mountains to gentle hills and broad plains. In the unusual event of long-continued tectonic inactivity, the land surface may be worn to a level plain. Climate and bedrock type strongly modify the evolutionary path in various surface environments, making desert and glacial landscapes very different.

Why don't mountains sink? Mountains, like icebergs, float —but they float on the mantle instead of the ocean. Over long time scales (thousands to millions of years), the mantle behaves as a fluid and exerts a buoyancy force on the base of mountains that counteracts the gravitational force. Thus, many mountain ranges are in isostatic equilibrium with the mantle. During rapid erosion of mountain ranges, summits may be uplifted to new heights because the mass of mountains is reduced by erosion, resulting in isostatic uplift.

KEY TERMS AND CONCEPTS

abrasion (p. 540)

badland (p. 540)

contour (p. 532)

cuesta (p. 540)

elevation (p. 532)

geomorphology (p. 531)

hogback (p. 541)

landform (p. 535)

mesa (p. 536)

negative feedback (p. 541)

plateau (p. 536)

positive feedback (p. 542)

relief (p. 533)

stream power (p. 538)

topography (p. 532)

EXERCISES

1. Give three examples of landforms.

2. What is topographic relief, and how is it related to altitude?

3. Why does relief vary with the scale of the area over it which it is measured?

4. How do faulting and uplift control topography?

5. Compare the different erosional processes in topographically high and low areas.

6. How do river slope and discharge affect stream power?

7. How does climate affect topography, and vice versa?

8. How does the balance between tectonics and erosion affect mountain heights?

9. In what regions of North America do active plate tectonic movements currently affect landscape?

THOUGHT QUESTIONS

1. The summits of two mountain ranges lie at different elevations: range A at about 8 km and range B at about 2 km. Without knowing anything else about these ranges, could you make an intelligent guess about the relative ages of the mountain-building processes that formed them?

2. If you were to climb 1 km from a river valley to a mountaintop 2 km high in a tectonically active area versus a tectonically inactive area, which would probably be the more rugged climb?

3. A young mountain range of uniform age, rock type, and structure extends from a far northern frigid climate through a temperate zone to a southern tropical rainy climate. How would the topography of the mountain range differ in each of the three climates?

4. Describe the main landforms in a low-lying humid region where the bedrock is limestone.

5. In what landscapes would you expect to find lakes?

6. What changes could you predict for the landscape of the Himalaya and the Tibetan Plateau in the next 10 million years? The next 100 million years?

7. What changes in the landscape of the Rocky Mountains of Colorado might result from a change in the present temperate but somewhat dry climate to a warmer climate with a large increase in rainfall?

8. Over short time scales (thousands of years), isostatic uplift may temporarily raise mountain summits to higher elevations. However, over the longer term (millions of years), continued erosion will reduce these summits to progressively lower elevations. As this happens, what does the concept of isostatic equilibrium predict for the depth of the base of continental lithosphere beneath mountains? Remember that mountains float on the mantle at the base of the continental lithosphere. Should this depth increase or decrease as mountains are worn down?

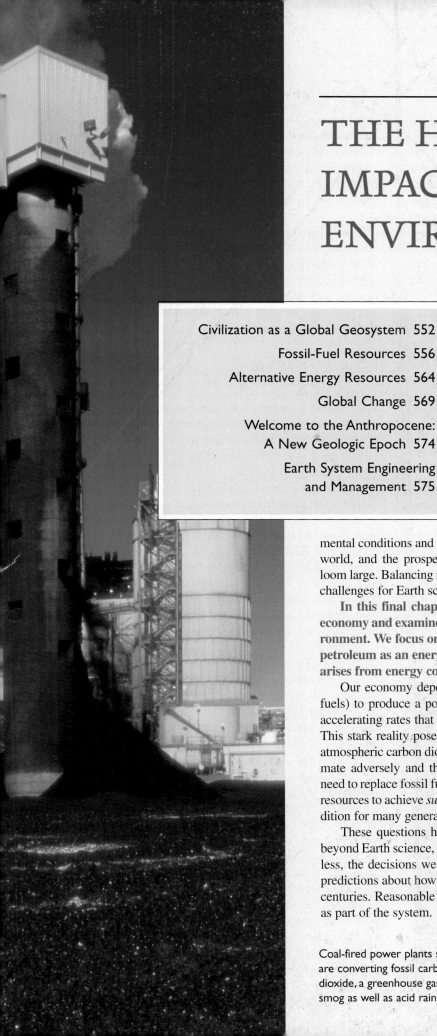

THE HUMAN IMPACT ON EARTH'S ENVIRONMENT

The human habitat is a thin interface between Earth and sky, where the global geosystems of climate, plate tectonics, and the geodynamo interact to provide a life-sustaining environment. We have increased our standard of living by discovering clever ways to exploit our natural environment: grow food, extract minerals, build structures, transport materials, and manufacture goods of all kinds.

In previous chapters, you have learned how understanding Earth can improve the human condition. Progress cannot be taken for granted, however. Human population has exploded into the billions, and our appetites for energy and other resources have become voracious. Environmental conditions and overall prosperity are not improving in some parts of the world, and the prospects for detrimental changes to the global environment loom large. Balancing resource usage against environmental damage raises new challenges for Earth science and society.

In this final chapter, we survey the energy resources that power our economy and examine how the activities of our civilization impact our environment. We focus on two of the most pressing problems: the depletion of petroleum as an energy resource and the potential for climate change that arises from energy consumption.

Our economy depends on the burning of a nonrenewable resource (fossil fuels) to produce a potentially dangerous greenhouse gas (carbon dioxide) at accelerating rates that will be difficult to sustain for more than a few decades. This stark reality poses difficult questions. To what extent will the increase in atmospheric carbon dioxide caused by fossil-fuel burning change the global climate adversely and thereby reduce global prosperity? How quickly will we need to replace fossil fuels with alternative energy sources? How can we use our resources to achieve *sustainable development*—improvement of the human condition for many generations into the future?

These questions have political and economic dimensions that extend far beyond Earth science, so they do not have strictly scientific answers. Nevertheless, the decisions we will make as a society must be informed by scientific predictions about how the Earth system will change over the next decades and centuries. Reasonable predictions can be made only if we include civilization as part of the system.

Coal-fired power plants such as this one in San Juan County, New Mexico, are converting fossil carbon previously buried in Earth's crust into carbon dioxide, a greenhouse gas. Fossil-fuel burning also produces gases that cause smog as well as acid rain. [Larry Lee Photography/Corbis.]

CIVILIZATION AS A GLOBAL GEOSYSTEM

Human activities are multiplying at phenomenal rates. From 1930 to 2000, the global population rose by 300 percent, from 2 billion to 6 billion. The population is currently increasing by 100 million per year, and the total is projected to exceed 8 billion within the next 30 years. Our energy usage has increased by 1000 percent over the last 70 years and is now rising twice as fast as the population. A glowing lattice of highly energized urbanization is spreading rapidly across the planet's surface (**Figure 23.1**).

Humans have altered our environment by deforestation, agriculture, and other types of land usage since the first civilizations arose in the warmer climate of the Holocene epoch about 11,000 years ago. But the effects in earlier times were usually restricted to local or regional habitats. Our heavily industrialized society is now changing its environment on a truly global scale, as illustrated by some startling observations:

• Reservoirs built by humans now trap about 30 percent of the sediment transported down the world's rivers.

• In most developed countries, civil engineers move more tons of soil and rock each year than do all natural processes of erosion combined.

• Within 50 years after the invention of the coolant *freon*, enough man-made coolants had leaked out of refrigerators and air conditioners and floated up into the stratosphere to damage Earth's protective ozone layer.

• Humans have converted about one-third of the world's forested area to other land use, primarily agriculture, in the last half-century.

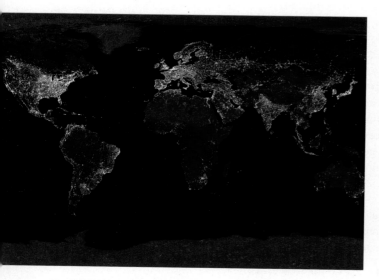

Figure 23.1 Earth at night, showing the lights of a globalized, energy-intensive civilization. From data collected by the Defense Meteorological Satellite Program. [NASA Visible Earth.]

• Since the industrial revolution began in the early nineteenth century, deforestation and the burning of fossil fuels have increased the amount of carbon dioxide in the atmosphere by almost 40 percent. Atmospheric carbon dioxide content is rising at the unprecedented rate of 4 percent per decade and is likely to cause significant global climate change during the lifetimes of most people now in college.

Energy production on an industrial scale makes it possible for humans to compete with plate tectonics and the climate system in modifying Earth's surface environment. We are not just part of the Earth system; we are transforming how the Earth system works, perhaps in fundamental ways. In a geological instant, civilization has developed into a full-fledged global geosystem!

Energy Sources

Geosystems are open systems that exchange energy and matter with the environment. Solar radiation energizes the climate system, whereas the heat from Earth's deep interior energizes the plate tectonics and geodynamo systems. The human geosystem draws its energy in various forms from both of these basic sources.

A century and a half ago, most of the energy used in the United States came from the burning of wood (**Figure 23.2**). A wood fire, in chemical terms, is the combustion of *biomass*—organic matter consisting of carbon and hydrogen compounds. Biomass is produced by plants and animals from a food web that is based on photosynthesis. The ultimate source of energy is the sunlight green plants use to convert carbon dioxide and water to carbohydrates. Burning wood or other biomass produces heat energy and returns the carbon dioxide and water back into the environment.

In this capacity, the biomass acts as a short-term reservoir for storing solar energy. It is a **renewable energy source** because the biosphere is constantly producing new biomass; for example, a forest chopped down for wood can be regrown and harvested again. Humans have used a variety of renewable energy sources to power mills and other machinery for thousands of years, including moving wind; falling water; and the work of oxen, horses, and elephants.

Some of the biomass that was buried in sedimentary rock formations of the Carboniferous period, more than 300 Ma, has been transformed into the combustible rock called *coal*. When we burn coal, we are using energy stored by photosynthesis from Paleozoic sunlight. Thus, the primary source of this "fossilized" energy was the same solar power that drives the climate system. As we will see, however, energy from Earth's deep interior was also needed to power the plate tectonic deformation and heating of organic-rich rocks to produce the coal.

Our other major fuels, crude oil (petroleum) and natural gas, were also created by the metamorphism of dead organic matter. These **fossil fuels** are considered to be **nonrenewable energy sources** because geological processes produce them

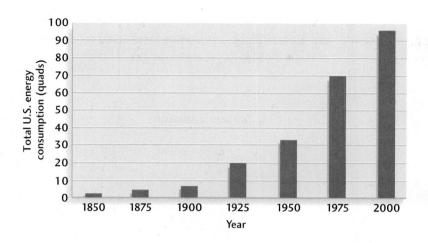

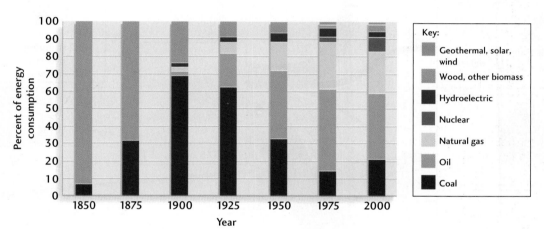

Figure 23.2 (a) Total energy consumption in the United States from 1850 to 2000. (b) Percentages of various types of energy consumed in the United States from 1850 to 2000. [U.S. Energy Information Agency.]

much more slowly than we are using them up. Our reserves of oil, natural gas, and coal will be exhausted long before the Earth system can replenish them.

Rise of the Fossil-Fuel Economy

Industrialization increased the demand for energy beyond what the traditional renewable sources could supply. By the late eighteenth century, James Watt and others had developed steam engines fired by coal that could do the work of hundreds of horses. Steam technology lowered the price of energy dramatically, in part because it made coal mining possible on a very large scale—in Britain, from the coalfields of England and Wales; in continental Europe, from the coal basins of western Germany and bordering countries; and in North America, from the Appalachian coalfields of Pennsylvania and West Virginia. The availability of cheap energy sparked the industrial revolution. By the end of the nineteenth century, coal accounted for more than 60 percent of the U.S. energy supply (see Figure 23.2).

The first oil well was drilled in Pennsylvania by Colonel Edwin Drake in 1859. The idea that petroleum could be profitably mined like coal provoked skeptics to call the project "Drake's Folly." They were wrong, of course; by the

early twentieth century, oil and natural gas were beginning to displace coal as the fuels of choice. Not only did they burn more cleanly, producing no ash, but they could be transported by pipeline as well as by rail and ship. Moreover, gasoline and diesel fuels derived from oil were suitable for burning in the newly invented internal combustion engine. By 1950, the energy from oil and gas exceeded that from coal, and the fossil fuels taken together supplied over 90 percent of U.S. energy needs.

Nuclear reactors that could produce electricity from the fission of uranium-235 were developed in the last half of the twentieth century. The early expectations that nuclear fuels would provide a large, low-cost, environmentally safe source of energy have not been realized, however. Safety concerns, the inability to dispose of nuclear wastes, and the escalating costs of stringent safety and security measures have slowed the construction of nuclear plants. Nuclear power supplies a substantial fraction of the electrical energy used by some countries, such as France (76 percent) and Sweden (52 percent), but this proportion is much smaller in the United States (21 percent). Overall, nuclear power accounts for less than a tenth of the total U.S. energy budget.

Today the engine of civilization runs primarily on fossil fuels, as illustrated by the energy system of the United States (**Figure 23.3**). Taken together, oil, natural gas, and coal

Earth System Figure 23.3 The energy system of the United States consumes fossil fuels to produce useful energy, emitting carbon dioxide, a greenhouse gas.

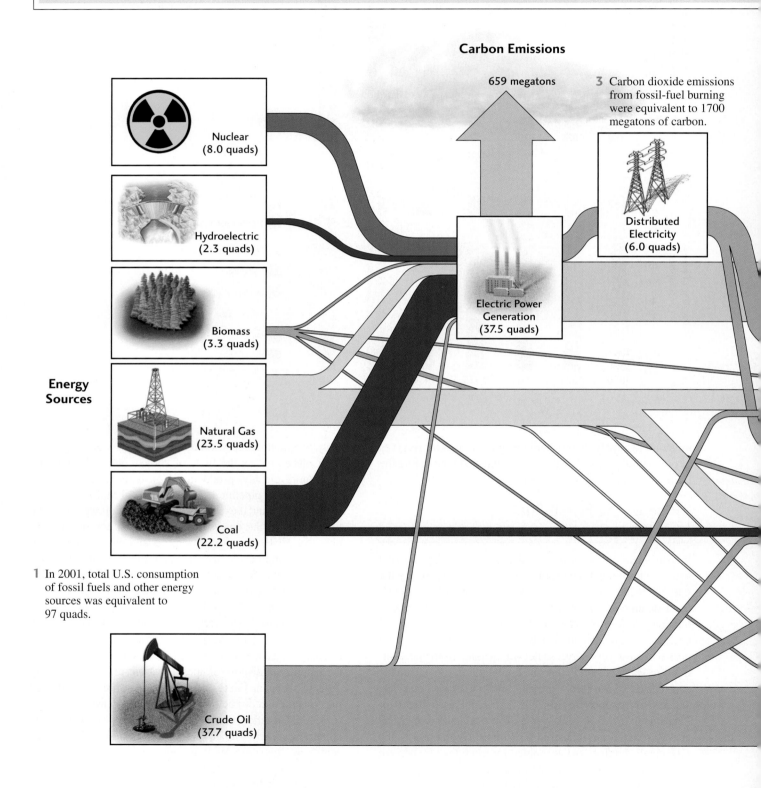

Carbon Emissions

659 megatons

3 Carbon dioxide emissions from fossil-fuel burning were equivalent to 1700 megatons of carbon.

Nuclear
(8.0 quads)

Hydroelectric
(2.3 quads)

Biomass
(3.3 quads)

**Energy
Sources**

Natural Gas
(23.5 quads)

Coal
(22.2 quads)

Electric Power
Generation
(37.5 quads)

Distributed
Electricity
(6.0 quads)

1 In 2001, total U.S. consumption of fossil fuels and other energy sources was equivalent to 97 quads.

Crude Oil
(37.7 quads)

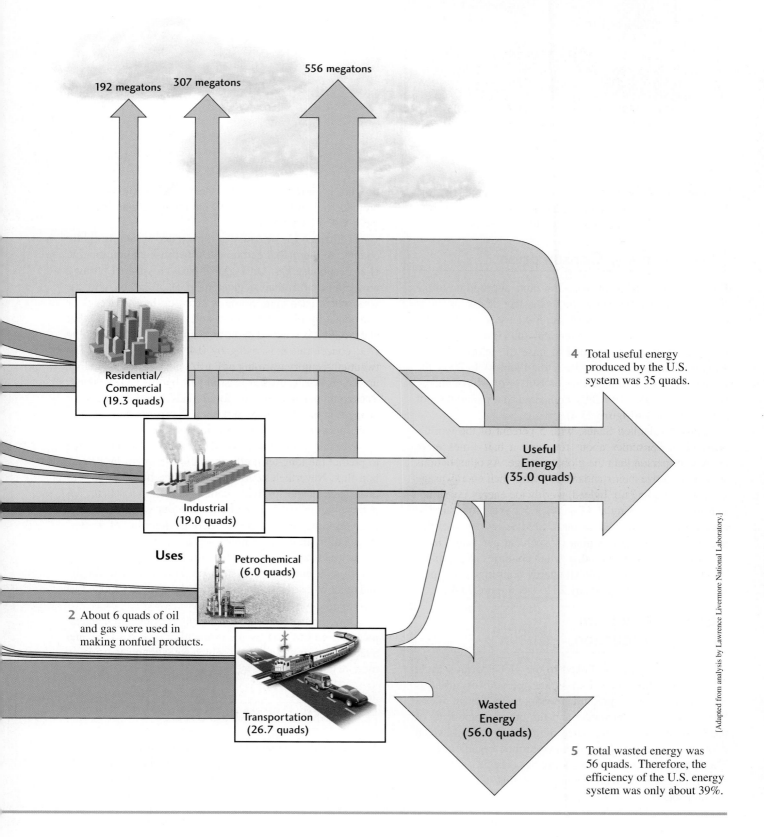

192 megatons

307 megatons

556 megatons

Residential/
Commercial
(19.3 quads)

Industrial
(19.0 quads)

Uses

Petrochemical
(6.0 quads)

2 About 6 quads of oil
and gas were used in
making nonfuel products.

Transportation
(26.7 quads)

Useful
Energy
(35.0 quads)

4 Total useful energy
produced by the U.S.
system was 35 quads.

Wasted
Energy
(56.0 quads)

5 Total wasted energy was
56 quads. Therefore, the
efficiency of the U.S. energy
system was only about 39%.

[Adapted from analysis by Lawrence Livermore National Laboratory.]

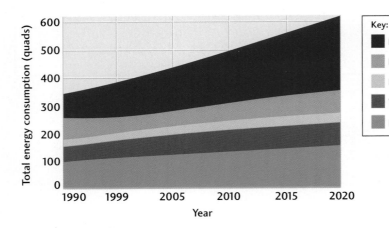

Figure 23.4 Actual and projected energy consumption, 1990–2020, by regional groupings. [U.S. Energy Information Agency.]

Key:
- Developing countries
- Eastern Europe/former USSR
- Industrialized Asia
- Western Europe
- North America

account for 86 percent of the energy consumed by the United States and the world at large. Not surprisingly, the fossil-fuel industry is by far the world's largest business.

Global Energy Consumption

Energy usage is often measured in units appropriate to the fuel; for example, barrels of oil, cubic feet of gas, and tons of coal. But comparisons are easier if we use a standard unit of energy such as the British thermal unit (Btu). One Btu is the amount of energy needed to raise the temperature of 1 pound of water by 1°F (1 Btu = 1054 joules). In 2004, the United States used 100 quadrillion Btu, or **quads,** of energy per year (1 quad = 10^{15} Btu), compared to a global total of about 440 quads (**Figure 23.4**).

Thus, the United States, with 5 percent of the world's population, consumes about four and a half times more energy per person than the global average. As other nations strive to improve their standards of living, their energy usage per capita is going up. Indeed, worldwide energy consumption is currently rising at 4.3 percent per year, compared to only 1.4 percent per year in the United States. Most of this growth is occurring in China (15 percent per year), India (7 percent per year), and other Asian countries with rapidly developing economies. Global energy consumption is projected to exceed 600 quads by 2020 (see Figure 23.4).

Carbon Flux from Energy Production

Our energy economy is linked to the global carbon cycle because fossil fuels, like biomass, are reservoirs of carbon as well as energy. In the prehuman world, the exchange of carbon between the lithosphere and the other components of the Earth system was regulated by the slow rates at which geological processes buried and unearthed organic matter. The rise of the fossil-fuel economy has disrupted this natural carbon cycle by creating a huge new flux of carbon from the lithosphere to the atmosphere. The effect of this sudden

flux has been to increase atmospheric carbon dioxide from preindustrial levels of about 280 ppm to more than 380 ppm today (see Figure 15.16).

In Chapter 15, we saw that the climate system is tightly coupled to the global carbon cycle because carbon dioxide is a greenhouse gas. On a global basis, fossil-fuel burning now emits more than 6 gigatons (Gt) of carbon into the atmosphere each year. Some is taken up by the oceans or consumed by plant growth on land (see Figure 15.15), but about 3.3 Gt is retained in the atmosphere, which is why the CO_2 concentration is rising by 0.4 percent per year. The twentieth-century warming of Earth's climate, shown in Figure 15.16, was caused at least in part by this CO_2 increase. The continued burning of fossil fuels could force major changes in the climate system, such as perturbations of thermohaline circulation in the oceans.

We really don't understand the Earth system well enough to predict the long-term effects of these changes in the carbon flux. Nevertheless, it is clear that the future of the climate system and its living component, the biosphere, depends to an increasing degree on how our society manages its energy resources and engineers its dumping of wastes. In this difficult task, we must consider not only our energy needs and social concerns but also the effects of energy usage on the entire Earth system, prompting Braden Allenby and other industrial ecologists to label the subject *Earth systems engineering and management.*

Our options for Earth systems engineering and management are constrained by the availability of fossil fuels and alternative sources of energy, which we will now consider in more detail.

FOSSIL-FUEL RESOURCES

The supply of any material taken from Earth's crust is finite, and its availability depends on its distribution in accessible deposits, as well as on how much we are willing to pay to get it out of the ground. Geologists use two measures of supply. **Reserves** are deposits that have already been discovered and can be mined economically and legally at the present time. In contrast, **resources** constitute the entire amount of a given material that may become available for use in the future. Resources include reserves, plus discovered but currently unprofitable deposits, plus undiscovered deposits that geologists think they may be able to find eventually. Often, resources that are too poor in quality or quantity to be worth mining now or that are too difficult to retrieve become profitable when new technology is developed or prices rise. A recent example is the discovery and production of oil and gas from large reservoirs in the continental margin of the Gulf of Mexico, in water depths reaching 3000 m.

Reserves are considered a dependable measure of supply as long as economic and technological conditions remain constant. As conditions change, some resources become reserves, and vice versa. The conversion of oil resources in the North Sea into productive oil fields, for example, was accelerated by new technology and by price increases during the oil crisis of the mid-1970s. The assessment of resources is much less certain than the assessment of reserves. Any figure cited as the resources of a particular material represents only an educated guess of how much will be available in the future.

Figure 23.5 shows one estimate of the world's remaining nonrenewable energy resources of all types—about 360,000 quads. Calculations based on these numbers can be deceptive, however. For example, simply dividing total resources by annual consumption might lead us to conclude

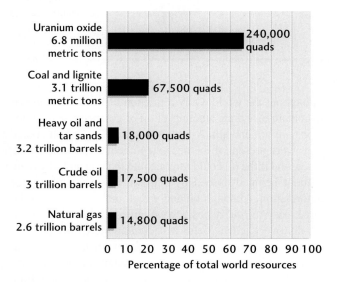

Figure 23.5 A rough estimate of total remaining nonrenewable world energy resources amounts to about 360,000 quads. Amounts are given in conventional units of weight (metric tons), volume (barrels), and energy content (quads). A barrel of oil is 42 gallons. Coal and lignite resources, for example, amount to 3.1 trillion metric tons, equivalent to 67,500 quads, or 19 percent of total energy resources. [World Energy Council.]

(mistakenly) that many hundreds of years of resources remain before we have to worry about depletion of energy supplies. The economics are more complicated, however, because some energy sources will give out before others, the various sources of energy are not readily interchangeable, and the environmental costs of resource conversion may be too great.

How Oil and Gas Form

Economic deposits of our most important energy resources, crude oil (petroleum) and natural gas, develop under special environmental and geological conditions. Both come from the organic debris of former life—plants and microorganisms (such as bacteria and algae) that have been buried, transformed, and preserved in sediments.

Oil and gas begin to form when more organic matter is produced than is destroyed by scavengers and decay. This condition exists in environments where the production of organic matter is high and the supply of oxygen in bottom sediments is inadequate to decompose all the organic matter. Many offshore sedimentary basins on continental shelves satisfy both these conditions. In such environments, and to a lesser degree in some river deltas and inland seas, the rate of sedimentation is high and organic matter is buried and protected from decomposition.

During millions of years of burial, chemical reactions triggered by the elevated temperatures at depth slowly transform some of the organic material in these *source beds* into the combustible compounds of hydrogen and carbon (hydrocarbons). The simplest hydrocarbon is methane gas, CH_4, the compound we call **natural gas.** Raw petroleum or **crude oil** includes a diverse class of liquids composed of more complex hydrocarbons, including long molecular chains comprising dozens of carbon and hydrogen atoms.

Crude oil forms in a limited range of pressures and temperatures, known as the **oil window,** usually located at depths between about 2 and 5 km. The temperatures above the oil window are too low to transform organic material into crude oil, whereas the temperatures below the oil window are so high (greater than 150°C) that the hydrocarbons in crude oil are broken down into methane, producing only natural gas.

Compaction of the source beds forces the hydrocarbon-containing fluids and gases into adjacent beds of permeable rock (such as sandstones or porous limestones), which act as *oil reservoirs.* The low density of oil and gas causes them to rise to the highest place that they can reach, where they float atop the water that almost always occupies the pores of permeable formations.

Geological conditions that favor the large-scale accumulation of oil and natural gas are combinations of structures and rock types that create an impermeable barrier to upward migration—an **oil trap.** Some oil traps are caused by structural deformation and are called *structural traps.*

Key Figure 23.6 — Oil is trapped in reservoirs formed by geologic structures.

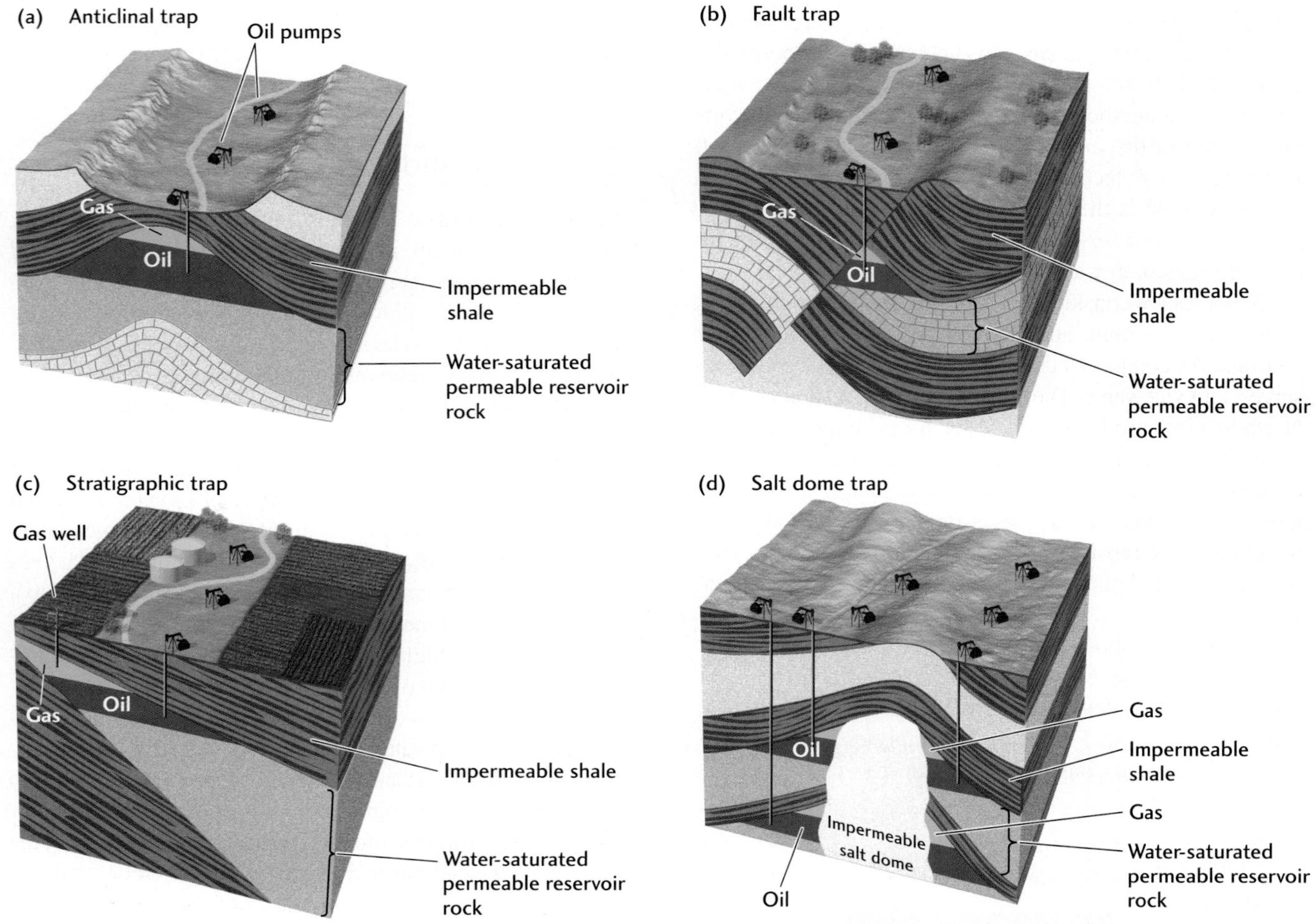

(a)   Anticlinal trap

Oil pumps

Gas

Oil

Impermeable shale

Water-saturated permeable reservoir rock

(b)   Fault trap

Gas

Oil

Impermeable shale

Water-saturated permeable reservoir rock

(c)   Stratigraphic trap

Gas well

Gas   Oil

Impermeable shale

Water-saturated permeable reservoir rock

(d)   Salt dome trap

Oil

Impermeable salt dome

Oil

Gas

Impermeable shale

Gas

Water-saturated permeable reservoir rock

One type of structural trap is formed by an anticline in which an impermeable shale overlies a permeable bed of sandstone (**Figure 23.6a**). The oil and gas accumulate at the crest of the anticline—the gas highest, the oil next—both floating on the groundwater that saturates the sandstone. Similarly, an angular unconformity or displacement at a fault may place a dipping permeable limestone bed opposite an impermeable shale, creating a structural trap for oil (see Figure 23.6b). Other oil traps are created by the original pattern of sedimentation, such as when a dipping permeable sandstone bed thins out against an impermeable shale (see Figure 23.6c). These are called *stratigraphic traps*. Oil can also be trapped against an impermeable mass of salt in a *salt dome trap* (see Figure 23.6d).

Distribution of Oil Reserves

In their search for petroleum resources, geologists have mapped thousands of structural and stratigraphic traps throughout the world. Only a fraction of them have proved to contain economic amounts of oil or gas, because the traps alone are not enough. A trap will contain oil only if source beds were present, if the necessary chemical reactions took place, and if the oil could migrate into the trap and stay there without being disturbed by subsequent severe heating or deformation. Although oil and gas are not rare, most of the large, easy-to-find deposits have already been located, and the discovery of new fields is becoming more difficult.

The worldwide reserves of oil are estimated to be about 1.2 trillion barrels; these are summarized by region in **Figure 23.7**. The oil fields of the Middle East—including Iran, Kuwait, Saudi Arabia, Iraq, and the Baku region of Azerbaijan—contain nearly two-thirds of the world's total. Here organic-rich sediments have been folded and faulted by closure of the ancient Tethys Ocean, forming a nearly ideal environment for oil accumulation. The extensive reservoirs discovered in this vast convergent zone include the enormous Ghawar field in Saudi Arabia, the world's largest. Ghawar

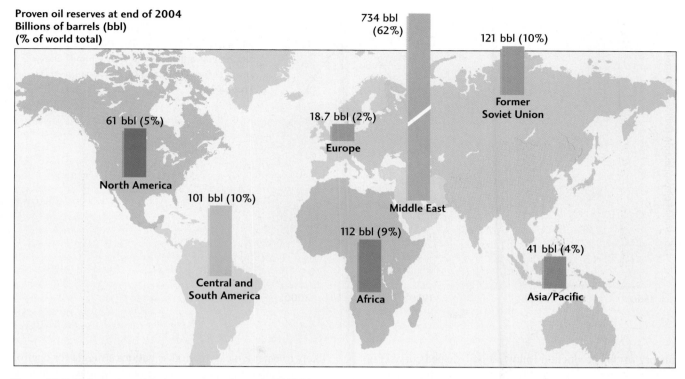

Figure 23.7 Estimated world oil reserves at the end of 2004 by region. [*British Petroleum Statistical Review of World Energy 2005*, June 2005.]

began production in 1948 and is expected to produce at least 88 billion barrels of oil over its lifetime (**Figure 23.8**).

Most of the oil reserves in the Western Hemisphere are located in the highly productive Gulf Coast–Caribbean area, which includes the Louisiana-Texas region, Mexico, Colombia, and Venezuela. Saudi Arabia ranks first among oil-producing nations with 22 percent of the world's reserves; the United States ranks tenth with 2.5 percent.

Oil Production and Consumption

In 2004, the amount of oil pumped out of the ground worldwide was about 30 billion barrels, an increase of 3.4 percent from the previous year. The United States produced 2.6 billion barrels, more than any other nation except Saudi Arabia and Russia, but it consumed almost three times that much—7.5 billion barrels, one-quarter of the world total. The gap between production and consumption is growing at 5 percent per year, and it must be filled by importing oil, at an annual cost of hundreds of billions of dollars. This imbalance, $174 billion in 2004, contributes more than any other factor to the massive U.S. foreign trade deficit.

Thirty-one U.S. states have commercial oil resources, and small, noncommercial occurrences are known in most of the others. However, the United States is a "mature" oil producer in the sense that most of the petroleum resources within its borders have already been exploited; production reached a maximum in 1970 and is now in decline. The

Figure 23.8 The Ghawar oil field, Saudi Arabia, is the world's largest with a total resource of at least 88 billion barrels. [Robert Azzi/Woodfin Camp Associates.]

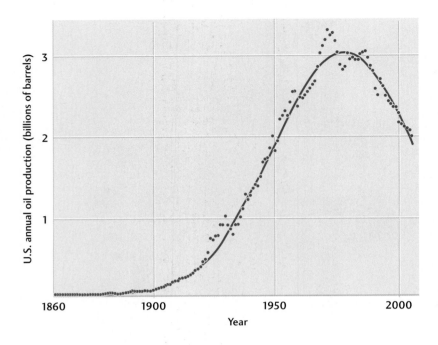

Figure 23.9 Data on U.S. oil production from 1860 to 2000, showing a peak production in 1970 and subsequent decline. The solid line is a bell-shaped curve that fits the data. [From K. Deffeyes, *Hubbert's Peak*. Princeton, N.J.: Princeton University Press, 2001.]

history of this production follows a bell-shaped curve (**Figure 23.9**), now known as **Hubbert's peak** after a famous petroleum geologist, M. King Hubbert. In 1956, Hubbert used a simple mathematical model to predict that U.S. oil production, which was growing rapidly at the time, would actually begin to decline sometime in the early 1970s. His arguments were roundly dismissed as overly pessimistic, but history has proved him right.

When Will We Run Out of Oil?

At the current production rate, the world will consume all of the known oil reserves in just 40 years. Does that mean we will run out of oil before mid-century? No, because oil resources are much greater than oil reserves.

In fact, we will never really "run out" of oil. As the resources diminish, prices will eventually rise to such high levels that buyers cannot afford to waste oil by burning it as a fuel. Its main use will then become the manufacturing of indispensable organic chemicals. This petrochemical industry, which uses oil as a feedstock to produce plastics, fertilizers, and a host of other products, is already a very big business, consuming 7 percent of the global oil production. As the oil geologist Ken Deffeyes has noted, future generations will probably look back on the Petroleum Age with a certain amount of disbelief: "They burned it? All those lovely organic molecules, and they just burned it?"

The key issue is not when oil runs out, but when oil production stops rising and begins to decline, as clearly it must someday. This milestone—Hubbert's peak for world oil production—is the real tipping point; once it is reached, the gap between supply and demand will grow rapidly, driving oil prices sky-high. If our society is not already well along in the transition to a postpetroleum energy system, we are then

likely to enter a messy period as nations struggle for control of oil supplies, increasing international tensions, while they try to adapt their economies to alternative energy sources. Energy conversion may require some countries to accept lower standards of living.

So how close are we to Hubbert's peak? The answer to this critical question is the subject of considerable debate. Oil optimists believe there is enough undiscovered oil to satisfy world demand for several decades into the future. The U.S. Geological Survey, the federal agency responsible for estimating energy resources, sits squarely in the optimist camp. It issued a worldwide assessment in 2000 that pegged total oil resources at more than twice the proved reserves, boosting by 20 percent their estimate made six years earlier. This is the resource value used in Figure 23.5.

At the other end of the spectrum are the oil pessimists, who believe that we are fast approaching Hubbert's peak, if not already there. Their views are supported by the same type of analysis Hubbert used to predict the 1970 peak in U.S. oil production. In particular, the pessimists note that the rate of oil discovery, which determines the growth of reserves, is declining too rapidly to be consistent with the U.S. Geological Survey's optimistic scenario. One projection, by the petroleum geologist Colin Campbell, shows global oil production declining after 2006 (**Figure 23.10**). If this projection is correct, the estimate in Figure 23.5 may be high by a factor of two.

A respected economist and spokesman for the optimists, Morris Adelman of the Massachusetts Institute of Technology (MIT), challenges pessimistic estimates of remaining supply: "Nobody knows how much hydrocarbon exists or what percentage of that will be recoverable. The tendency to deplete a resource is counteracted by increases in knowledge." He believes that with improved world oil

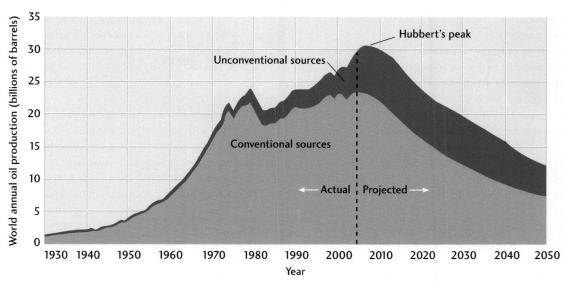

Figure 23.10 A pessimistic projection of world oil production, showing Hubbert's peak occurring in 2007. [Colin J. Campbell, "The General Depletion Picture," *Association for the Study of Peak Oil and Gas Newsletter*, 2004.]

exploration and production technology, oil supplies will continue to increase.

Whether the oil optimists or pessimists are correct is a central issue for Earth systems engineering and management. In particular, the answer will substantially affect how much CO_2 will be produced by the human geosystem in the future.

Oil and the Environment

In addition to the global effects of carbon dioxide emissions, the environmental effects of oil production involve a number of regional and local issues. Pollution is the major problem of offshore drilling. The environment around Santa Barbara, California, was damaged in 1969 when oil was accidentally released from an offshore drilling platform. In 1979, a well that was being drilled in the Gulf of Mexico off the Yucatán coast "blew out," spilling as much as 100,000 barrels of oil a day for many weeks before it could be capped. In 1988, an explosion destroyed a drilling platform in the North Sea, killing many oil workers and marine animals. The grounding of the tanker *Exxon Valdez* off the coast of Alaska in 1989, with the release of 240,000 barrels of crude oil in pristine coastal waters, was covered widely by television and newspapers and heightened public awareness of the severe ecological damage that can result from an oil spill (**Figure 23.11**). Despite such incidents and the difficulty of guaranteeing the safety of a well or a tanker, proponents of oil development believe that careful design of equipment and safety procedures can greatly reduce the chances of a serious accident.

There are large resources of oil and gas under the coastal plain of northern Alaska and under the submerged continental shelves of North America. The deep water of the conti-

nental margin in the Gulf of Mexico is one of the best new prospects for oil and gas. Many people argue that eventually these areas will have to be drilled to satisfy the world's growing energy needs. A skeptical public, however, is not

Figure 23.11 The effect on wildlife of an oil spill from an oil tanker, the *Exxon Valdez*, in Prince William Sound, Alaska. [UPI/Corbis-Bettmann.]

convinced that drilling can be done without serious threat to pristine environments.

The nation is currently embroiled in a strident political debate about whether to allow drilling for oil and natural gas in the Arctic National Wildlife Refuge (ANWR). The total ANWR resource has not been fully evaluated, but it could be as much as 20 to 30 billion barrels of oil. The U.S. Geological Survey estimates that, if oil prices were high enough, 4 to 12 billion barrels of this total could be produced using current technologies. There is no doubt that these resources would contribute to the national economy. But oil and gas production requires the building of roads, pipelines, and housing in a very delicate ecological environment that is a particularly important breeding area for caribou, musk-oxen, snow geese, and other wildlife. The decision to drill ANWR depends on whether the short-term economic benefits are valued more highly than the long-term environmental losses.

Natural Gas

The resources of natural gas are comparable to those of crude oil (see Figure 23.5) and may exceed them in the decades ahead. Estimates of natural gas resources have been rising in recent years. Exploration for this relatively clean

fuel has increased, and geologic traps have been identified in new settings, such as very deep formations, overthrust belts, coal beds, tight (somewhat impermeable) sandstones, and shales. The world's resources of natural gas are less depleted than oil resources because natural gas is a relative newcomer on the energy scene.

Natural gas is mostly methane (CH_4). In combustion, it combines with atmospheric oxygen, releasing energy in the form of heat and producing carbon dioxide and water. It therefore burns much more cleanly than coal or oil, which also produce ash, sulfur dioxide (a precursor of acid rain), and other pollutants. As a fuel, natural gas emits 30 percent less carbon dioxide per unit of energy than oil and more than 40 percent less than coal, and it is easily transported across continents through pipelines. Getting it from source to market across oceans has been more difficult, but the construction of tanker ships and ports that can handle liquefied natural gas (LNG) is beginning to solve this problem.

For these reasons, natural gas is a premium fuel. It accounts for about 24 percent of all fossil fuels consumed in the United States each year, most of it for industry and commerce (55 percent), followed by residential use (24 percent) and the generation of electric power (21 percent). More than half of American homes and a great majority of commercial

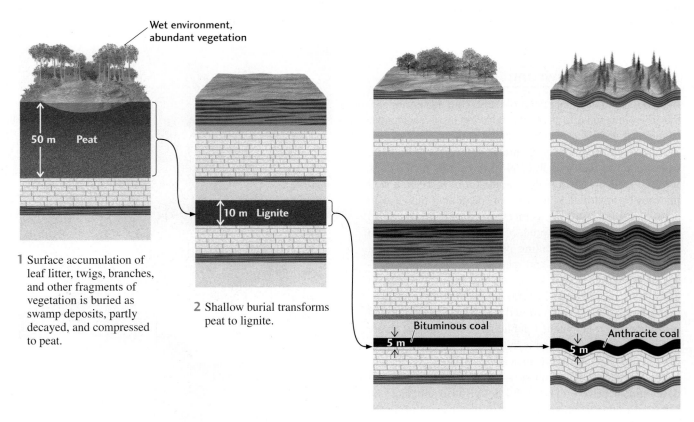

Wet environment, abundant vegetation

50 m Peat

10 m Lignite

Bituminous coal
5 m

Anthracite coal
5 m

1 Surface accumulation of leaf litter, twigs, branches, and other fragments of vegetation is buried as swamp deposits, partly decayed, and compressed to peat.

2 Shallow burial transforms peat to lignite.

3 Further burial under hundreds to thousands of meters of sediment transforms lignite to soft (bituminous) coal.

4 Continued burial and structural deformation, plus heat, metamorphose soft coal to hard (anthracite) coal.

Figure 23.12 The process by which coal beds form begins with the deposition of vegetation.

and industrial buildings are connected to a network of underground pipelines that draw gas from fields in the United States, Canada, and Mexico. At current rates of use, natural gas reserves in the United States will last about 10 years, and the resources are likely to provide a substantial fraction of our energy budget for several decades longer.

Coal

The abundant plant fossils found in **coal beds** are evidence that coal forms from large accumulations of plant materials of the sort that exist in wetlands. As the luxuriant plant growth of a wetland dies, it falls to the waterlogged soil. Rapid burial by falling leaves and immersion in water protect the dead twigs, branches, and leaves from complete decay because the bacteria that decompose vegetative matter are cut off from the oxygen they need. The vegetation accumulates and gradually turns into peat, a porous brown mass of organic matter in which twigs, roots, and other plant parts can still be recognized (**Figure 23.12**). The accumulation of peat in an oxygen-poor environment can be seen in modern swamps and peat bogs. When dried, peat burns readily because it is 50 percent carbon.

Over time, with continued burial, the peat is compressed and heated. Chemical transformations increase the peat's already high carbon content, and it becomes *lignite,* a very soft, brownish black, coal-like material containing about 70 percent carbon. The higher temperatures and structural deformation that accompany greater depths of burial may metamorphose the lignite into *subbituminous* and *bituminous coal,* or soft coal, and ultimately into *anthracite,* or hard coal. The greater the metamorphism, the harder and brighter the coal

and the higher its carbon content, which increases its heat value. Anthracite is more than 90 percent carbon.

Coal Resources According to some estimates, the amount of coal remaining in the world is about 3.1 trillion metric tons, capable of producing 67,500 quads of energy (see Figure 23.5). About 85 percent of the world's coal resource is concentrated in the former Soviet Union, China, and the United States, which has extensive deposits in many states (**Figure 23.13**), and these areas are the largest producers. Domestic coal resources in the United States would last for a few hundred years at current rates of use—about a billion tons a year. Coal has supplied a larger proportion of U.S. energy needs since 1975, when the price of oil began to rise; it currently accounts for about 22 percent of the energy consumed in this country.

Environmental Costs of Coal There are serious problems with the recovery and use of coal that make it less desirable than oil or gas, whether the coal is burned or converted into synthetic liquid fuel. On average, coal combustion releases 25 percent more carbon dioxide per unit of energy than oil and 70 percent more than natural gas. Much coal contains appreciable amounts of sulfur, which vaporizes during combustion and releases noxious sulfur oxides into the atmosphere. Acid rain, which forms when these gases combine with rainwater, has become a severe problem in Canada, Scandinavia, the northeastern United States, and eastern Europe.

Coal ash is the inorganic residue that remains after coal is burned. It contains metal impurities from the coal, some of which are toxic. Ash can amount to several tons for every 100

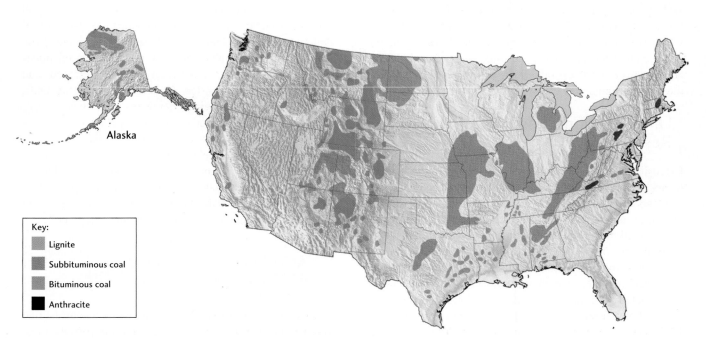

Figure 23.13 Coalfields of the United States. [U.S. Bureau of Mines.]

Figure 23.14 *Left:* Coal strip mine, Buskin, Indiana. *Right:* Coal strip mine after reclamation, Buskin, Indiana. [Photos courtesy of Vigo Coal Company.]

tons of coal burned and poses a significant disposal problem. It can escape from smokestacks, creating a health risk to people downwind. Underground mining accidents take the lives of miners each year, and many more suffer from black lung, a debilitating inflammation of the lungs caused by the inhalation of coal particles. Strip mining, the removal of soil and surface sediments to expose coal beds, can ravage the countryside if the land is not restored (**Figure 23.14**).

Government regulations now require that technologies for the "clean" combustion of coal be phased in. The law mandates the restoration of land disrupted by strip mining and the reduction of danger to miners. But these measures are expensive and add to the cost of coal. This drawback, however, is unlikely to prevent the increased use of coal, which is so much more abundant than oil. Many countries have no other fuel resources, and some countries will not be able to afford to import oil, which will increase in price as the supply diminishes.

Unconventional Fossil-Fuel Resources

Extensive deposits of hydrocarbons occur in two other forms: organic-rich source beds that never reached the pressure and temperature conditions in the oil window, such as *oil shales,* and formations that once contained oil but have since "dried out," losing many of their volatile components to form *extra-heavy oils* or a tarlike substance called *natural bitumen* (not to be confused with bituminous coal). One deposit of the latter type, the **tar sands** of Alberta, Canada, is estimated to have a hydrocarbon reserve equivalent to 180 billion barrels of oil and a total resource perhaps 10 times that amount. Almost 400 million barrels of oil are now recovered from the Alberta tar sands each year, and

Canadian production is projected to increase fivefold by 2030, providing as much as 5 percent of world demand.

Developing the tar sands raises important environmental concerns, however. It takes two tons of mined sand to produce one barrel of synthetic crude oil, leaving lots of waste sand. Because of the energy-intensive production, oil sands are large sources of greenhouse gases, emitting six times more carbon dioxide than conventional oil production. Extracting oil from tar sands sucks up about two-thirds of the energy they ultimately render—a very low energy efficiency.

A second unconventional source of fossil fuel is **oil shale,** a fine-grained, clay-rich sedimentary rock containing relatively large amounts of organic matter. In the 1970s, oil companies began trying to commercialize the extensive oil shales of western Colorado and eastern Utah, but those efforts were largely abandoned by the 1980s as oil prices fell, concerns over environmental damage increased, and technical problems persisted. As is the case with tar sands, the efficiency of producing energy from oil shale is very low, and the environmental costs per unit of energy are very high. For example, shale oil and combustible gas can be extracted from these rocks, but the process requires huge amounts of water, a scarce resource in the western United States. Nevertheless, renewed interest in energy production from domestic oil shale has been sparked by higher oil prices and is being encouraged by national energy policies.

ALTERNATIVE ENERGY RESOURCES

If crude oil and gas continue to be the major resources used to satisfy the world's voracious appetite for energy, the great bulk of the world's supply will be exhausted within a century.

Coal will eventually become the predominant fossil fuel in many countries. It may be reassuring to know that at modest annual energy growth rates—say, 3 percent per year—coal and other fossil fuels can meet the world's energy needs for about 100 years or longer. The environmental costs of coal may be unacceptable, however. In particular, the large amounts of carbon dioxide released during the combustion of coal may trigger climate changes that could force a shift away from these traditional fuels before they are depleted.

These estimates do not consider the possibility that we may begin to meet our energy needs in nontraditional ways: through increased efficiency in the use of fossil fuels and by the development and use of alternative energy sources such as nuclear energy, solar energy, geothermal energy, and energy derived from biomass. To the extent that alternative sources can be used, we can reduce the pressure on our fossil-fuel resources and their negative environmental effects.

Nuclear Energy

Although the first large-scale use of the radioactive isotope uranium-235 was in an atomic bomb in 1944, the nuclear physicists who first observed the vast energy released when its nucleus splits spontaneously (called *fission*) foresaw the possibility of peaceful applications of this new energy source. After World War II, their predictions were fulfilled as countries around the world built nuclear reactors to produce **nuclear energy:** the fission of uranium-235 releases heat to make steam, which then drives turbines to create electricity. In the United States, some 110 nuclear reactors now provide about 22 percent of the electric energy used. More than 400 nuclear reactors in 25 countries account for about 6 percent of global energy production. If its full potential were to be realized, nuclear energy could meet the world's needs for electric energy for hundreds of years.

The geologic problems of nuclear energy include the assessment of uranium reserves and the underground disposal of nuclear wastes.

Uranium Reserves Uranium constitutes only 0.00016 percent of the average continental crustal rock. Only 1 in 139 atoms of this element is uranium-235; its other, much more abundant isotopes (for example, uranium-238) are not radioactive enough to be used as fuel in nuclear reactors. However, uranium still turns out to be by far our largest minable energy resource, with a potential generating capacity of at least 240,000 quads (see Figure 23.5). It is typically found as small quantities of uraninite, a uranium oxide mineral (also called pitchblende), in veins in granites and other felsic igneous rocks. Under near-surface groundwater conditions, uranium in igneous rocks may oxidize and dissolve, be transported in groundwater, and later be reprecipitated as uraninite in sedimentary rocks.

Nuclear Energy Hazards The biggest drawbacks of nuclear energy are reactor safety, environmental contamination, and the potential use of radioactive fuels for making nuclear weapons.

Two accidents have raised questions about the safety of nuclear energy. The first was at the Three Mile Island reactor in Pennsylvania in 1979. A reactor was destroyed, and radioactive debris was released, although it was confined within the containment building. No one was harmed, but most experts agree that it was a close call. Much more serious was the destruction of a nuclear reactor in the town of Chernobyl in Ukraine in 1986. The reactor went out of control because of poor design and human error. Radioactive debris spilled into the atmosphere and was carried by winds over Scandinavia and western Europe. Contamination of buildings and soil has made hundreds of square miles of land surrounding Chernobyl uninhabitable. Food supplies in many countries were contaminated by the fallout and had to be destroyed. Deaths from cancer caused by exposure to the fallout may be in the thousands.

The uranium consumed in nuclear reactors leaves behind dangerous radioactive wastes (see Feature 23.1). A system of safe long-term waste disposal is not yet available, and reactor wastes are being held in temporary storage at reactor sites. In a few years, the limits of space available for temporary storage in the United States will be reached. Although many scientists believe that geological containment—the burial of nuclear wastes in deep, stable, impermeable rock formations—is a workable solution, there is not yet a generally approved plan for the safe storage of the most dangerous wastes for the hundreds of thousands of years required before they cease to be radioactive. France and Sweden have built underground nuclear waste depositories. A major facility is also being developed in the United States—the Yucca Mountain Nuclear Waste Repository at the Nevada Test Site (**Figure 23.15**)—but it is embroiled in litigation as the state of Nevada battles the federal government to keep it from

Figure 23.15 Aerial view of the Yucca Mountain Nuclear Waste Repository being developed at the Nevada Test Site, north of Las Vegas. [U.S. Department of Energy.]

EARTH POLICY

23.1 Subsurface Toxic and Nuclear Waste Contamination

Some decades ago, when we knew much less about the health and environmental effects of toxic wastes, industrial and military wastes now known to be hazardous were dumped on the ground; disposed of in ponds, lakes, and rivers; or discharged underground. Among the most hazardous and difficult to clean up are federal government facilities that were engaged in the production of nuclear weapons.

The Hanford site in Washington State is an example. The toxic and radioactive wastes come from leaking storage tanks, buried boxes and drums filled with radioactive solids, and contaminated liquids that were discharged into the soil and groundwater in the past. Hanford is located on an alluvial plain composed of sands and gravels, underlain by folded and fractured basalt. In general, waste-bearing fluids move downward into the unsaturated zone and enter the aquifer, where they flow toward the Columbia River. The flow rates vary according to local geological conditions, such as the permeability of a formation and its mineral and organic content (see the accompanying diagram). Some contaminants become chemically bound to minerals or adhere to the sediments and hardly move; others move readily with the groundwater. Seeps and springs in the area are contaminated. In a 20-year period, some chemically toxic and radioactive wastes have migrated to the Columbia River, prompting the worry that much larger amounts could follow if effective cleanup is not pursued.

Hanford and the other toxic waste sites raise a number of challenging policy questions. For example, what should be the cleanup goal for these sites? Op-

tions include (1) the pristine original "green field" condition; (2) partial cleanup to achieve a state that poses acceptable health risk and does not endanger an important ecological resource such as the salmon spawning grounds; (3) incomplete "brown field" restoration to a state that permits commercial or industrial use but not residential use; and (4) transition to a "stewardship" state in which the land is set aside and the contaminants are contained and monitored to prevent their spread outside the site.

Estimating the effect of a particular clean-up path on health or the environment is called *risk assessment*. The uncertainty of exposure that will result and disagreement about what levels of health and environmental effects are acceptable make risk assessment controversial, but it is an important tool of decision makers who, by law, must take cleanup action.

Cleanup decisions are made by collective agreement of the U.S. Department of Energy (DOE) and federal and state environmental regulators. In earlier years, nuclear weapons manufacturing activities were kept secret, and the government was insensitive to health risks and environmental effects. DOE now considers socioeconomic, cultural, and ethical concerns in its decisions. It also keeps local governments, affected communities, and other interest groups informed and considers their views.

This 35-km cross section of the Hanford site illustrates how liquid waste enters the ground and moves through the unsaturated zone into and through the underlying aquifer. Height in this drawing is exaggerated to show vertical detail. [After R. E. Gephart, *An Overview of Hanford's Waste Generation History and the Challenges Facing Waste Cleanup*. Richland, Wash.: Pacific Northwest National Laboratory, 1998.]

being built. No state wants to be the dumping ground for a long-term environmental hazard.

In the United States, these unresolved problems, plus the danger that nuclear fuels could be used to build nuclear weapons, have essentially halted the installation of new nuclear power plants, and new installations have been slowed in other countries as well. Concerns about the safety of nuclear reactors, their high cost, and the safe disposal of radioactive waste must be allayed before we can light up the world with nuclear energy.

Solar Energy

An enthusiast for solar energy recently reminded us that "every 20 days Earth receives from sunlight the energy

equivalent of the entire planetary reserves of coal, oil, and natural gas!" Because all our fossil energy sources ultimately come from the Sun anyway, why not convert its rays into energy? In principle, the Sun can provide us with all the energy we need, in all the forms we use. Light from the Sun can be converted into heat and electricity. It can even be used to obtain hydrogen, which can be used as a gaseous fuel, from water. **Solar energy** is risk-free and non-depletable—the Sun will continue to shine for at least the next several billion years. The bad news is that the technology currently available for the large-scale conversion of solar energy into useful forms is inefficient and expensive; the good news is that it is improving.

In the near term, the only form of solar energy likely to be available at costs nearly competitive with those of other

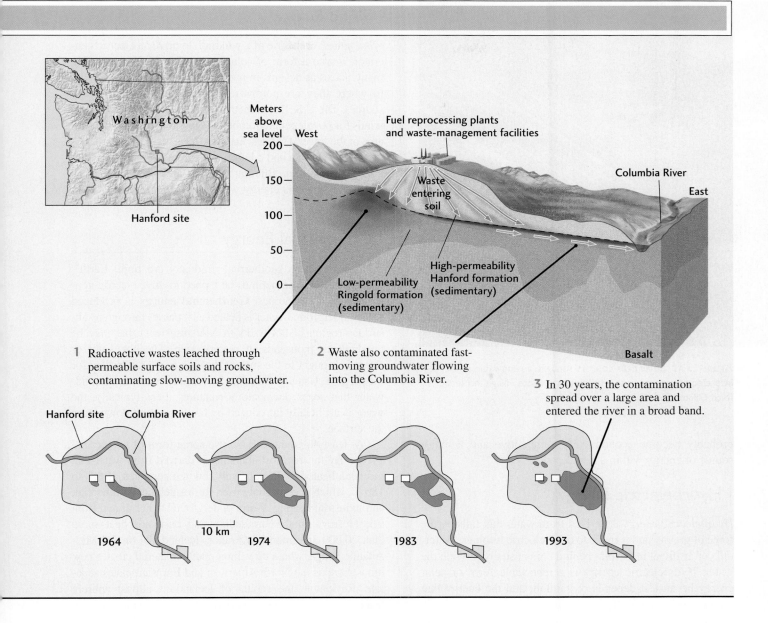

1 Radioactive wastes leached through permeable surface soils and rocks, contaminating slow-moving groundwater.

2 Waste also contaminated fast-moving groundwater flowing into the Columbia River.

3 In 30 years, the contamination spread over a large area and entered the river in a broad band.

sources is heat for homes, water, and industrial and agricultural processes. Some homes and factories use solar energy for these purposes, motivated in part by government tax credits and other incentives.

Solar energy can be used to generate electricity in several ways. Solar generating systems are being used for installations where costs are no impediment, such as demonstration projects, or in remote areas where alternatives are not available (**Figure 23.16**). But large-scale solar-generated electricity is not yet an alternative to conventional sources in general use. The efficiency of converting sunlight into electricity is improving but is still too low, and the costs of installing and maintaining the systems are still too high. Moreover, commercial-scale solar electric power plants can present significant environmental problems. A plant with a

100-megawatt electric capacity (about 10 percent the capacity of a nuclear power plant) located in a desert region would require at least a square mile of land and might alter the local climate by significantly changing the balance of solar radiation in the area.

In 2004, solar energy supplied only 0.06 percent of U.S. energy consumption. Solar energy enthusiasts believe that this amount could increase to some 20 quads per year in a decade or so, equivalent to about half the oil we now use. A more realistic figure is probably less than 10 quads per year. Even this more conservative scenario would yield important social benefits: conservation of other energy resources, diversification of energy supply so that we are not overly dependent on a single source, and reduction of fuel imports. With adequate research and development, solar energy can

Figure 23.16 Solar cells convert sunlight, a renewable resource, into electric energy at this utility in a remote village in Nepal. [Ned Gillette/Corbis.]

probably become economically competitive and a major source of energy within a few decades.

Hydroelectric Energy

Hydroelectric energy is derived from water that falls by the force of gravity and is made to drive electric turbines. Waterfalls or artificial reservoirs behind dams usually provide the water. Hydroelectric energy is a renewable form of solar energy because it depends on rainfall, and the energy that drives weather comes from the Sun.

Hydroelectric energy is clean, relatively risk free, and cheap. It delivers about 3 quads annually, or about 3 percent of the annual energy consumption in the United States. The U.S. Department of Energy recently identified more than 5000 sites in the United States where new hydroelectric plants could be built and economically operated, boosting the national hydroelectric capacity by as much as 40 percent. Significant expansion of the present capacity would be resisted in the United States, however, because it would drown farmlands and wilderness areas under reservoirs behind dams but add only a few percentage points to U.S. energy production. This is one energy issue on which most people agree that the environmental costs would greatly exceed the economic gain. For this reason, energy experts expect almost no new hydroelectric capacity to be added through 2020, and its share of the nation's electrical generation capacity will decline to about 6 percent from about 10 percent today.

Wind Power

Wind power, or the use of a windmill to drive an electric generator, is also a form of solar energy. Its use is growing in many places as designs improve and costs are brought down to where they are competitive with conventional energy sources. The U.S. Department of Energy estimates that good winds for power generation blow across 6 percent of the land area of the lower 48 states, and they have the potential to supply more than one and a half times the nation's current electricity consumption. But that's a lot of area—550,000 km²— and harvesting this energy would require millions of 30-m windmills.

Geothermal Energy

In regions where geothermal gradients are high, Earth's internal heat can sometimes be tapped to drive electric generators and heat homes. **Geothermal energy** is produced when underground water is heated as it passes through a subsurface region of hot rocks (a *heat reservoir*) that may be hundreds or thousands of meters deep. The hot water or steam is brought to the surface through boreholes drilled for the purpose. Usually, the water is naturally occurring groundwater that seeps down along fractures. Less typically, the water is artificially introduced by being pumped down from the surface.

At least 46 countries now use some form of geothermal energy. By far the most abundant is derived from water that has been heated to the relatively low temperatures of 80° to 180°C, which is commonly used for residential, commercial, and industrial heating. Warm underground water drawn from a heat reservoir in the Paris sedimentary basin now heats more than 20,000 apartments in France. Iceland sits on the Mid-Atlantic Ridge, where upwelling mantle material creates new lithosphere as the North American and Eurasian plates separate. Reykjavík, the capital of Iceland, is almost entirely heated by geothermal energy derived from this volcanic heat.

Geothermal reservoirs with temperatures above 180°C are useful for generating electricity. They are present primarily in regions of recent volcanism as hot, dry rock; natural hot water; or natural steam. The latter two sources are limited to those few areas where surface water seeps down through underground faults and fractures to reach deep rocks heated by recent magmatic activity. Naturally occurring water heated above the boiling point and naturally occurring steam are highly prized resources. The world's largest facility for producing electricity from natural steam is at The Geysers, 120 km north of San Francisco (see Figure 12.27).

According to one Icelandic estimate, as much as 40 quads of electricity could be generated each year from accessible geothermal energy sources, but so far only a tiny fraction of that amount, about 0.15 quad per year, is actually being used. Another 0.12 quad is used for direct heating.

Like most of the other energy sources we have looked at, geothermal energy presents some environmental problems.

Regional subsidence can occur if hot groundwater is withdrawn without being replaced. In addition, geothermally heated waters can contain salts and toxic materials dissolved from the hot rock.

GLOBAL CHANGE

The expression **global change** entered the world's vocabulary when it became clear that emissions from fossil-fuel burning and other human activities could alter the chemistry of the atmosphere. The worldwide consequences might include

• Mass die-offs due to acid rain

• Increased exposure to ultraviolet rays due to stratospheric ozone depletion

• Global warming due to an enhanced greenhouse effect

These changes are all **anthropogenic,** or human-generated (from *anthropos,* the Greek word for "man"), but they differ considerably in scope and magnitude. Acid rain is primarily a regional problem, although if left unabated, it could have global consequences. Stratospheric ozone depletion is clearly a global problem, but its cause is mainly chlorofluorocarbons (CFCs) of industrial origin, which are now strictly regulated through an international treaty, the Montreal Protocol—a major environmental success. In contrast, there appears to be little agreement over what we should do about global warming and related aspects of human-induced climate change (**Figure 23.17**).

The potentially dire consequences of anthropogenic global change are motivating politicians to work together in ways they never have before, as we all try to avoid the "tragedy of the commons"—the spoiling of our commonly held environmental resources. Neighboring nations are enacting mutually beneficial regulations to address regional environmental problems, and new multinational treaties are being formulated in attempts to manage human impacts on global geosystems.

We will use the problems of acid rain, stratospheric ozone depletion, and climate change to illustrate the problems of anthropogenic global change and some of their possible solutions.

Acid Rain

In many industrialized areas of the world, the air is greatly polluted with sulfur-containing gases such as sulfur dioxide (SO_2). These gases are emitted from the smokestacks of power plants that burn coal containing large amounts of the mineral pyrite (iron sulfide, FeS_2), from smelters of sulfide ores, and from some factories. Coals mined in the eastern and midwestern regions of the United States contain more of these pollutants than coals mined in the western states. Although

---

Key Figure 23.17   Human activities can change many components of Earth's climate system.

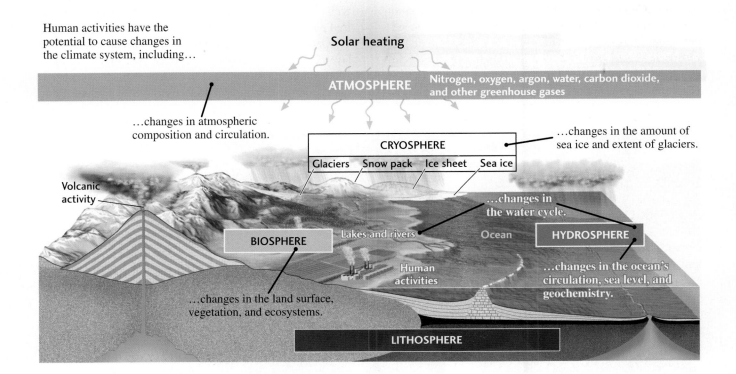

Human activities have the potential to cause changes in the climate system, including…

Solar heating

ATMOSPHERE   Nitrogen, oxygen, argon, water, carbon dioxide, and other greenhouse gases

…changes in atmospheric composition and circulation.

CRYOSPHERE

Glaciers   Snow pack   Ice sheet   Sea ice

…changes in the amount of sea ice and extent of glaciers.

Volcanic activity

…changes in the water cycle.

Lakes and rivers   Ocean   HYDROSPHERE

BIOSPHERE

Human activities

…changes in the ocean's circulation, sea level, and geochemistry.

…changes in the land surface, vegetation, and ecosystems.

LITHOSPHERE

volcanoes and coastal marshes also add sulfur gases to the atmosphere, more than 90 percent of the sulfur emissions in eastern North America are of anthropogenic origin.

Chemistry and Effects Sulfur gases in the atmosphere react with oxygen and rainwater to form sulfuric acid, which is far stronger than the carbonic acid formed by carbon dioxide and rainwater. Some nitric acid is formed in the same way from nitrogen oxide gases (NO_x) emitted from smokestacks and automobile exhausts. Small amounts of sulfuric and nitric acids turn harmless rainwater into **acid rain.** Although it is much too weak to sting human skin, acid rain causes widespread damage to delicate organisms and solid rock.

By acidifying sensitive lakes and streams, particularly those underlain by soils with limited ability to neutralize acidic compounds, acid rain has caused massive kills of fish in many lakes in Canada, the northeastern United States, and Scandinavia. A survey of more than 1000 lakes and thousands of miles of streams in the northeastern United States showed that acid rain has affected 75 percent of the lakes and 50 percent of the streams. In some acidified lakes and streams, fish species such as the brook trout have been completely eradicated. The salmon habitat of eastern Canada has been greatly affected. Acid rain also damages mountain forests, particularly those at higher elevations. Acid moisture in the air reduces visibility.

Acid rain causes noticeable damage to fabrics, paints, metals, and rocks, and it rapidly weathers stone monuments and outdoor sculptures (**Figure 23.18**). In Canada alone, acid rain causes about $1 billion in damage every year to buildings and monuments.

Coal Burning The relationship between acid rain and the burning of coal has been firmly established. Agencies of the U.S. and Canadian governments, as well as independent scientific panels, have tracked sulfur gases emitted by smokestacks to downwind locations where they are precipitated as acid rain. Careful tracing of pollutants from their sources to the sites of acid rain is necessary to demonstrate

that coal-burning power plants are responsible for much of the problem.

Most of the sulfurous emissions in North America come from coal-fired electric utility plants in the midwestern and eastern United States, and most of the fallout occurs in these regions (**Figure 23.19**) and in eastern Canada. About 50 percent of the sulfate deposits in Canada originate in the United States—a source of friction between the two countries. The problem is equally critical in Europe and Asia, where scientists have noted damage to lakes and forests caused by acid rain. The damage is particularly heavy in China, in eastern European countries, and in Russia.

The Clean Air Act Concerned scientists have recommended the restriction of sulfur emissions from power plants and smelters. For a long time, however, political resistance prevented effective control. Finally, in 1990, after many years of wrangling, the U.S. Congress passed major amendments to the 1970 Clean Air Act. The legislation required coal-burning power plants to reduce their annual SO_2 emissions by 10 million tons and their annual NO_x emissions by 2 million tons by the year 2000. Deregulation of U.S. railroads has lowered the costs of transporting cleaner western coal to midwestern and eastern power plants, thus making it easier to meet the new emissions standards. The current federal administration's proposal to relax the 1990 standards, however, has reopened this contentious debate.

Stratospheric Ozone Depletion

Near Earth's surface, ozone gas (O_3^+) is a major constituent of smog. It undermines health, damages crops, and corrodes materials. Low-lying ozone forms when sunlight interacts with nitrogen oxides and other chemical wastes from industrial processes and automobile exhausts.

Ozone in Earth's stratosphere, which is concentrated in a layer 25 to 30 km above the surface (see Figure 15.2), is another matter. There, solar radiation ionizes oxygen gas (O_2) into ozone, forming a protective layer that shields Earth

Figure 23.18 A monument before and after deterioration caused by acid rain. [Westfälisches Amt für Denkmalpflege.]

1 Burning of high-sulfur coal creates atmospheric sulfuric acid when smokestack emissions react with water vapor.

2 The acid was carried northeastward by prevailing winds, where it fell as acid rain.

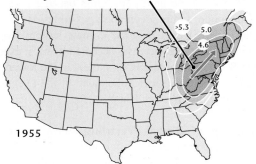

1955

3 Rain became more acidic and affected a broad area before regulations to reduce sulfur emissions were introduced.

1985

4 While acid rain has been reduced in the Northeast, coal-burning power plants in the Southwest have increased acid rainfall there.

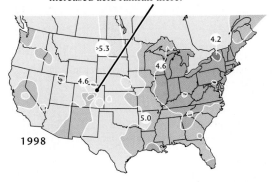

1998

Figure 23.19 Acidity of precipitation across the contiguous United States and Canada. Units are in pH; the lower the value, the higher the acidity. The most acidic precipitation occurs in the east, primarily as a result of burning high-sulfur coal. [National Atmospheric Deposition Program/National Trends Network, 2003.]

reactions involving human-made compounds. One class of compounds, CFCs—which are used as refrigerants, spray-can propellants, and cleaning solvents—are stable and harmless except when they migrate to the stratosphere. High above Earth, the intense sunlight breaks down CFCs, releasing their chlorine. Molina and Rowland proposed that chlorine reacts with the ozone molecules in the stratosphere and thins the protective ozone layer. Molina and Rowland's hypothesis was confirmed when a large hole in the ozone layer was discovered over Antarctica in 1985 (**Figure 23.20**). Subsequently, **stratospheric ozone depletion** was found to be a global phenomenon. Elevated surface levels of UV radiation have been observed to correspond to low stratospheric ozone levels since the early 1990s.

In the 1980s, when scientists were trying to convince government and industry officials that the ozone layer was possibly being depleted due to CFCs, a senior government official remarked that the solution was for people to wear hats, sunscreen, and dark glasses. Fortunately, political wisdom prevailed. In 1989, a group of nations entered into a global treaty to protect the ozone layer. This treaty, called the Montreal Protocol, now includes 140 countries. The Montreal Protocol phased out CFC production in 1996 and set

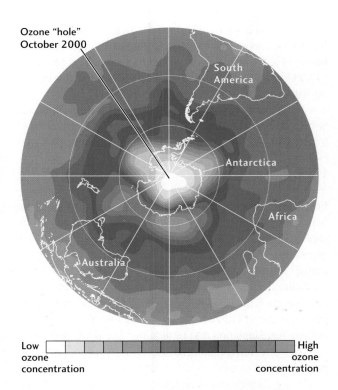

Figure 23.20 The ozone hole is a well-defined, large-scale destruction of the ozone layer over Antarctica that occurs each Antarctic spring. The word *hole* is a misnomer; the hole is really a significant thinning or reduction of ozone concentrations, which results in the destruction of up to 70 percent of the ozone normally found over Antarctica. The whiter areas represent thinning in the ozone layer that appeared over Antarctica in October 2000. [NOAA TOVS Satellite.]

by absorbing cell-damaging UV radiation. Skin cancer, cataracts, impaired immune systems, and reduced crop yields are attributable to excessive UV exposure.

In 1995, the Nobel Prize in chemistry was awarded to Paul Crutzen, Mario Molina, and Sherwood Rowland for the hypotheses they had advanced more than 20 years earlier about how the protective ozone layer can be depleted by

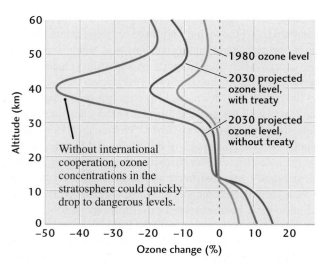

Figure 23.21 Comparison of ozone levels in the atmosphere in 1980 with projected ozone levels for 2030 with and without the Montreal Protocol. The 140 nations that signed this international treaty have been successful in eliminating emissions of CFCs, which attack the ozone layer. Without such international cooperation, ozone concentrations in the stratosphere would have quickly dropped to dangerous levels. [After T. E. Graedel and P. J. Crutzen, *Atmospheric Change.* New York: W. H. Freeman, 1993.]

up a fund, paid for by developed nations, to help developing nations switch to ozone-safe chemicals. The manufacturing of safer alternatives has led to a successful phase-out of CFCs, and the long-term projections indicate diminished destruction of the ozone layer (**Figure 23.21**).

The Montreal Protocol has become a model for how scientists, industrial leaders, and government officials can work together to head off an environmental disaster.

Global Warming

Fossil-fuel burning and the changes in land use since the beginning of the industrial revolution have caused significant CO_2 pollution of the atmosphere and the rise of other greenhouse gases (see Figure 15.16). Recognizing the potential problems that these trends pose for global climate, the United Nations established an Intergovernmental Panel on Climate Change (IPCC) in 1988 to assess the risk of human-induced climate change, its potential impacts, and options for adaptation and mitigation. The IPCC provides a continuing forum for hundreds of scientists, economists, and policy experts to work together to understand these issues.

In a major assessment report, published in 2001, the IPCC drew the following conclusions:

• Human-induced increases in greenhouse gases are probably responsible for much of the twentieth-century warming.

• Levels of greenhouse gases will continue to rise throughout the twenty-first century primarily because of human activities, although the increase depends on a number of socioeconomic factors that will govern the rate of greenhouse gas emissions,

including active steps by society to limit these emissions. Examples of some scenarios are illustrated in **Figure 23.22**.

• The increase in greenhouse gases will cause significant global warming during this century. Projections of the amount of warming are highly uncertain because of doubt about the amount of future greenhouse gas emissions and an incomplete knowledge of how the climate system works. The range of temperature increases accepted by most experts is 1.4° to 5.8°C (see Figure 23.22). Note that the lower end of this range is more than twice the value of the twentieth-century warming discussed in Chapter 15.

Consequences of Global Warming According to the IPCC, global warming during this century is likely be accompanied by regional climate changes and other effects. The warming will probably be greater over land than over the ocean, and regional climates will probably become more variable. Other possible effects include increasing rainfall in

(a) 1 Rate of atmospheric CO_2 increase depends on growth in fossil fuel usage, especially in the developing world. Here are three plausible scenarios.

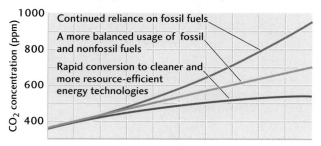

(b) 2 Climate models can be used to predict the temperature increase expected in each of these three scenarios (curves).

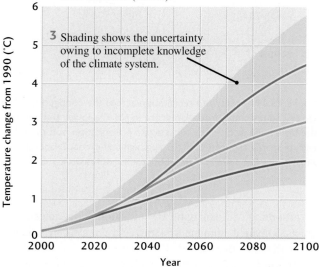

Figure 23.22 Projections of (a) atmospheric carbon dioxide concentrations and (b) average surface temperatures, for the next 100 years. [IPCC, *Climate Change 2001: The Scientific Basis.* Cambridge University Press, 2001.]

the tropics, decreasing rainfall at midlatitudes, stronger hurricanes and typhoons, more frequent and intense El Niño events, continued thinning of polar sea ice, melting of continental glaciers, and sea level rise. Table 23.1 summarizes the effects of climate change on various systems.

Some of the effects of global warming can already be seen and felt. The 10 warmest years ever recorded have all occurred since 1989. In 1998 and 2005, the two warmest years since accurate records began in 1880, the global mean temperature was 0.7°C above the 1961–1990 annual average.

Hurricane intensities may have already increased because of warming in the tropical oceans. The number of very strong hurricanes (category 4 and 5 storms) has almost doubled over the past three decades, and increases in storm intensity have been observed in all tropical storm basins around the world. The devastation to the Gulf Coast caused by Hurricane Katrina in 2005 (see Figure 20.7) is a sobering reminder of what this intensification might imply as ocean temperatures continue to rise.

The amount of sea ice in the Arctic Ocean is decreasing, and the downward trend seems to be accelerating. The sea ice cover in September 2005 was the lowest for that month since satellite records began in 1978: 5.5 million square kilometers, down by 25 percent from the 1978–1988 average of 7.4 million square kilometers. According to climate models, the north polar ice cap will continue to shrink rapidly, and much of the Arctic Ocean will become ice-free within a few decades (**Figure 23.23**). Continental glaciers are also receding, some at rapid rates (see Figure 21.10), and portions of the Antarctic ice shelves have collapsed (see

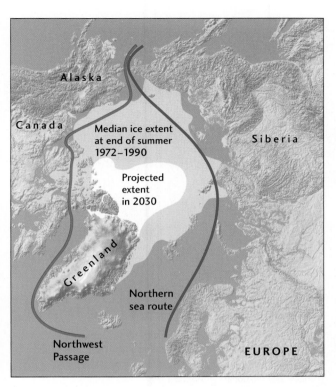

Figure 23.23 Global warming is melting the north polar ice cap. This map of the Arctic compares the extent of the polar ice at the end of the summer for the period 1972–1990 with the projected extent in 2030. The change is expected to disrupt arctic ecosystems. One benefit for humans may be the opening of the Northwest Passage and other shorter sea routes between the Atlantic and Pacific oceans within the twenty-first century (red and blue lines). [U.S. Navy.]

Table 23.1	Potential Climate Change Effects on Various Systems

Systems	Potential Effects
Forests and terrestrial vegetation	Migration of vegetation; reduction in inhabited range; altered ecosystem composition
Species diversity	Loss of diversity; migration of species; invasion of new species
Coastal wetlands	Inundation of wetlands; migration of wetlands
Aquatic ecosystems	Loss of habitat; migration to new habitats; invasion of new species
Coastal resources	Inundation of coastal development; increased risk of flooding
Water resources	Changes in supplies; changes in droughts and floods; changes in water quality and hydropower production
Agriculture	Changes in crop yields; shifts in relative productivity and production
Human health	Shifts in range of infectious diseases; changes in heat-stress and cold-weather afflictions
Energy	Increase in cooling demand; decrease in heating demand; changes in hydropower output
Transportation	Fewer disruptions of winter transportation; increased risk for summer inland navigation; risks to coastal roads

Source: Office of Technology Assessment, U.S. Congress.

Figure 23.24 The Bay of Bengal, Bangladesh. Low-lying Bangladesh is subject to disastrous flooding in the event of a rise in sea level, which could result from global warming. [James P. Blair/National Geographic.]

Figure 21.14). The surface melt of the Greenland ice sheet in 2005 was the largest on record.

As we discussed in Chapter 21, the melting of sea ice does not affect sea level, but the melting of continental glaciers causes sea level to rise. Sea level also rises because global warming heats up the ocean, increasing its overall volume of water by a tiny fraction. The IPCC estimates that sea level rose about 120 mm because of global warming during the twentieth century. Sea level is currently rising at about 2 mm/year, and this rate is expected to accelerate. Climate models indicate that sea level could rise by as much as 1 m during this century—creating serious problems for low-lying countries such as Bangladesh (**Figure 23.24**), as well as the eastern seaboard and Gulf Coast of the United States, where flooding during storms and tidal surges could become much worse.

Because human-induced warming will be rapid, many plant and animal species will have difficulty adjusting or migrating. Those that cannot cope with rapid warming might become extinct. Global warming is already being blamed for a variety of adverse ecological effects, such as the disruption of arctic ecosystems as the ice lines move northward and permafrost begins to melt, and the spread of malaria as more of the world experiences a tropical climate.

WELCOME TO THE ANTHROPOCENE: A NEW GEOLOGIC EPOCH

We know from the study of geology that the Earth system does not evolve uniformly. Its behavior is instead punctuated by extreme events—short periods of rapid global change. Geologists have ranked these changes according to their impact on the fossil record, and they have used major transitions in life-forms to delineate the eras, periods, and epochs of the geologic time scale (see Figure 8.14). The time scale's most recent division occurred about 11,000 years ago, at the end of the Pleistocene epoch. A cold climate with sporadic ice ages gave way to a warm, stable period—the Holocene epoch. Average global temperatures jumped by 7°C over an interval as short as a few decades, sea level rose by 20 meters per thousand years, ecosystems migrated with the receding ice, and many large mammals went extinct. Also at about this time, our own species began to alter the environment in the novel landscaping process we call civilization.

No doubt humans had a hand in Pleistocene-Holocene global change. The extinction of the ice-age megafauna, which included saber-toothed tigers, woolly mammoths, giant sloths, and a host of other large mammals, may have resulted from overhunting as human tribes pushed across Asia and into Australia and the Americas. The deforestation by agricultural communities during this expansion released carbon dioxide that may have helped to stabilize the Holocene climate.

Now, 11 millennia later, our civilization has suddenly harnessed tremendous sources of energy, increasing its power usage by a hundredfold in less than a century and transforming itself into a highly interactive global geosystem. If our current reliance on fossil fuels continues, the carbon flux from industrialized society will double the atmosphere's CO_2 content before the end of the twenty-first century (see Figure 23.22).

In 2003, the atmospheric chemist and Nobel Laureate Paul Crutzen proposed to recognize a new geologic epoch —the *Anthropocene,* or Age of Man—beginning around 1780, when James Watt's steam engine launched the industrial revolution. The global changes that mark the Holocene-Anthropocene boundary are just now getting under way, so a future scientist with a full record of the next few thousand years may place the geologic boundary at a somewhat later date. Crutzen's main point is that global changes are proceeding so rapidly that such quibbles are likely to be minor. As with many previous geologic boundaries, the main marker will be a mass extinction.

Between 1850 and 1880, as much as 15 percent of the land surface was deforested, and rates of deforestation have continued to rise. According to the United Nations, over 150,000 km^2 of tropical rain forests, about 1 percent of the total resource, are being converted each year to other land uses, mostly agricultural. In 1950, forests covered approximately 25 percent of Haiti (a Caribbean island country the size of Maryland); by 1987, the forested area had shrunk to 10 percent, and it now stands at less than 2 percent (**Figure 23.25**). Other poor nations face a similar problem.

Not surprisingly, the number of extant species, the most important measure of *biodiversity,* is declining. Biologists estimate there are over 10 million different species alive on

Figure 23.25 The Caribbean island country of Haiti is now 98 percent deforested.
[Daniel Morel/AP World Wide.]

the planet today, although only 1.5 million have been officially classified. Extinction rates are difficult to quantify, but most knowledgeable scientists believe that up to one-fifth of all species will disappear during the next 30 years, and as many as one-half may go extinct during this century. One respected biologist, Peter Raven, has put the problem bluntly:

We are confronting an episode of species extinction greater than anything the world has experienced for the past 65 million years. Of all the global problems that confront us, this is the one that is moving the most rapidly and the one that will have the most serious consequences. And, unlike other global ecological problems, it is completely irreversible.

Some observers, such as the sociobiologist E. O. Wilson, have gone so far as to call the rapid worldwide decline in biodiversity the "Sixth Extinction," placing it in the same rank as the Big Five mass extinctions of the Phanerozoic eon (see Figure 1.13 and Chapter 11). Others consider this extrapolation to be premature, however, because even the rapid losses in biodiversity we see today will not necessarily affect the fossil record as profoundly as, say, the impact-related mass extinction at the end of the Cretaceous period or even the less severe mass extinction associated with global warming at the Paleocene-Eocene boundary.

EARTH SYSTEM ENGINEERING AND MANAGEMENT

By any measure, the problems we face in confronting anthropogenic global change are daunting. A major challenge is to provide civilization with enough energy in the face of oil depletion and the probable need to reduce carbon emissions. No one can say at this time how much global warming will ultimately occur or to what effect, but a climate crisis could force us, sooner rather than later, to shift away from fossil fuels to an energy budget drawn primarily from natural gas, nuclear energy, and renewable resources such as solar and biomass energy. Yet even a rapid move in this direction would not necessarily avert such a crisis: if all human activities that generate carbon dioxide were to stop today, it would still take about 200 years for atmospheric carbon dioxide to return to its preindustrial level.

Energy Policy

The uncertainty of our situation presents a problem for policy makers. How much money should we spend to curb anthropogenic emissions, and will the benefits justify the costs? On the one hand, too much spending could depress the economy and cause job losses. On the other hand, prevention might be less costly than coping with a disaster after it happens.

A partial solution, perhaps the cheapest one, is to improve energy efficiency. Right now, the U.S. energy system wastes about 60 quads of energy for every 100 quads it generates; only 40 percent of our energy production actually ends up doing useful work (see Figure 23.3). In a real sense, using energy more efficiently is like discovering a new source of fuel. Some experts believe the United States could reduce emissions of greenhouse gases by as much as 50 percent from 1990 levels by implementing efficiency measures that cost little—for example, insulating buildings; replacing incandescent lights with fluorescent lights; increasing the fuel efficiency of motor vehicles; and making greater use of natural gas. These modest steps constitute a low-cost insurance policy against global warming that also confers substantial benefits, including reductions in expensive oil imports, lowered manufacturing costs, and increasingly healthful air quality. The savings in the United States could amount to hundreds of billions of dollars a year, much of it now being spent on imported oil.

Many observers would say that fossil fuels are simply too cheap in the United States. Carbon emissions are certainly not taxed as much as they are in many other advanced nations. Consequently, there is little incentive for conservation and conversion to other energy sources. If the full economic costs of fossil fuels were included in the prices—the cost of cleaning up acid rain, oil spills, and other environmental damage; the cost of trade deficits; the cost of global warming; the military cost of defending oil supplies—alternative energy resources might compete much better against fossil fuels. Such full-cost accounting has not been politically popular in this country.

We also face the issue of fairness in international politics. The United States, Canada, the European Union, and Japan—with only 25 percent of the world's population—are responsible for about 75 percent of the global increase in CO_2 and other greenhouse gases. These rich, industrially advanced nations will be able to adjust to standards for reducing emissions more easily than the developing countries. China, for example, depends on its huge coal deposits for its economic growth. Developing nations argue for financial and technological support to help them cope with the demand to reduce emissions.

People have come to agree that the problems of Earth system engineering and management cannot be solved on a national scale and will have to be addressed through international cooperation. New types of treaties, such as the Kyoto Protocol, have been formulated to manage human carbon emissions, but some nations—most notably the United States —have refused to participate, emphasizing the potential costs of moving too quickly (see Feature 23.2).

Engineering the Carbon Cycle

We are inadvertently changing the climate system by emitting greenhouse gases, thereby perturbing the carbon cycle and enhancing the greenhouse effect. What about the possi-

bility of engineering the carbon cycle to reduce the accumulation of greenhouse gases in the atmosphere? Several promising technologies aim to reduce greenhouse gas emissions by pumping the carbon dioxide generated by fossil-fuel burning into reservoirs other than Earth's atmosphere—a procedure known as *carbon sequestration.*

One obvious reservoir is the biosphere. In Chapter 15, we saw that deforestation feeds CO_2 into the atmosphere, whereas reforestation withdraws it, in surprisingly large amounts. Land-use policies that encourage reforestation and other biomass production, discourage deforestation, suppress fires, and reduce the loss of carbon from soils might help to mitigate anthropogenic climate change.

Some far-out biotechnologies could possibly improve the ability of terrestrial ecosystems to absorb more carbon, such as genetically modifying bacteria to make them more efficient in fixing carbon in soils. Extremophile organisms (see Chapter 11) provide models for engineering biological entities that may be able ingest a feedstock such as methane, sequester the carbon dioxide, and give off hydrogen. Hydrogen is the ultimate clean fuel; burning it produces only water.

Another controversial possibility is fertilization of the marine biosphere. We know that small marine plants (plankton) pump CO_2 out of the atmosphere by photosynthesis. In most regions of the ocean, the limits to plankton productivity come from a lack of nutrients, such as iron. Preliminary experiments in the 1990s suggested that the growth of plankton can be stimulated by dumping modest amounts of iron into the ocean. Unfortunately, it appears that fertilizing the ocean in this manner also stimulates the growth of animals that eat the plants and quickly return the CO_2 back into the atmosphere.

The last example suggests that using biotechnology to remove CO_2 from the atmosphere on a large scale will require a much better understanding of terrestrial and marine ecosystems to avoid unforeseen effects. However, this type of Earth system engineering is not as far-fetched as it once might have seemed.

In fact, one straightforward technology for carbon sequestration—underground storage of CO_2—may offer the best hope. Carbon dioxide captured from oil and gas wells is already being pumped back into the ground as a means of moving oil toward the wells, thereby enhancing oil recovery. If long-term underground storage of the carbon dioxide from coal-fired power plants were economically feasible, the world's abundant coal resources (see Figure 23.5) would become much more attractive as a replacement for petroleum. For example, "clean coal" could be used to generate hydrogen fuel for transportation.

Sustainable Development

The term **sustainable development** appears with increasing frequency in newspapers, public debates, classroom discussions, and scholarly journals. The concept was popularized in *Our Common Future,* a 1987 report by the World Commis-

EARTH POLICY

23.2 The Kyoto Accords and the Politics of Climate Change

After a long debate, most world political leaders are now convinced that using the atmosphere as a dump for carbon dioxide and other greenhouse gases will lead to global climate change in the next 50 to 100 years, with potentially adverse consequences. The readiness of governments to take joint action first became clear in a historic conference that took place on December 10, 1997, in Kyoto, Japan, where 160 nations agreed to limit emissions of greenhouse gases. Despite the fact that curtailing the use of fossil fuels—the major source of greenhouse gas emissions—could slow economic growth, all agreed to the Kyoto Protocol. The treaty finally came into force on February 16, 2005, following formal ratification by Russia in November 2004. As of September 2005, a total of 156 countries have ratified the treaty (representing over 61 percent of global emissions). Notable exceptions include the United States and Australia.

The Kyoto Protocol requires 39 industrial countries currently responsible for three-fourths of fossil-fuel burning to limit their annual CO_2 emissions between 2008 and 2012 to a percentage of their 1990 levels. For the European Union, the reduction is 8 percent below 1990 releases; for Japan, 6 percent; for the United States,

7 percent; and for developing countries, no reduction. The United States is the largest emitter of CO_2 from fossil-fuel usage, accounting for about 25 percent of the world total.

The Clinton administration signed the protocol in 1997, but President Bush rejected it in 2002 and refused to submit it to the Senate for ratification. The Bush administration and large segments of business and labor oppose the treaty, claiming that deep reductions in emissions would reduce energy usage by 25 to 35 percent below that expected for the year 2010. They argue that the dire predictions of adverse climate change are unproved and that the economic costs—as much as 2.4 percent of the gross domestic product (GDP), or about $260 billion each year—would be too high for uncertain benefits to the environment. The critics complain that the treaty does not include emission limits for the developing countries, which, with their growing populations and increasing use of energy, will become the major polluters of the future (see Figure 23.4).

Even if fully implemented, the Kyoto Protocol will only slow the increase of CO_2 and, at most, delay warming for about 10 years. However, it does provide a first step toward international cooperation in managing energy use and other practices that increase CO_2 in the atmosphere.

---

sion on Environment and Development (also known as the Brundtland Commission), where it was defined as "development that meets the needs of the present without compromising the ability of future generations to meet their own needs."

Sustainable development is difficult to define more precisely, but it offers an appealing if somewhat utopian vision: a civilization that maintains a viable habitat for many generations by carefully managing its interactions with the Earth system as a whole.

Sustainability involves many economic and political issues about which nations at differing stages of development do not agree, so forging a global strategy that moves civilization toward this goal may prove difficult. As a prerequisite, Earth science will have to provide better knowledge of how geosystems operate, interact, and can be perturbed by human activities.

SUMMARY

In what sense is civilization a global geosystem? Human society has harnessed the means of energy production on a planetary scale and can now compete with plate tectonics and the climate system in modifying Earth's surface envi-

ronment. Like other geosystems, civilization exchanges mass and energy with its environment. In particular, the rise of the fossil-fuel economy has disrupted the natural carbon cycle by creating a huge new flux of carbon from the lithosphere to the atmosphere, increasing atmospheric carbon dioxide from preindustrial levels of about 280 ppm to more than 380 ppm today.

What is the origin of oil and natural gas? Oil and gas form from organic matter deposited in marine sediments, typically in the coastal waters of the sea. The organic materials are buried as the sedimentary layers grow in thickness. Under heat and high pressure, organic carbon is transformed into liquid and gaseous hydrocarbons, compounds of carbon and hydrogen. Oil and gas accumulate in geologic traps that confine the fluids within impermeable barriers.

Why is there concern about the world's oil supply? Oil is a nonrenewable resource: it will be depleted faster than nature can replenish it. Therefore, as the supply is withdrawn from the oil reservoirs of the world, its availability diminishes, its price rises, and other types of energy sources will have to be found. The key issue is not when oil runs out but when global oil production reaches Hubbert's peak—when it stops rising and begins to decline. Oil optimists argue that oil resources will meet demand for decades to come; oil pessimists think we are within a few years of Hubbert's peak.

What is the origin of coal and how big a resource is it? Coal is formed by the compaction and chemical alteration of wetland vegetation. There are huge resources of coal in sedimentary rocks. We have used only about 2.5 percent of the world's minable coal. Coal mining and pollution caused by coal burning are risky to human life and to the environment. Coal combustion is a major source of carbon dioxide and the acid emissions that are the precursors of acid rain. Because of its abundance and low cost, however, the use of coal is likely to increase in the next decades to generate electric power and to be converted into liquid and gaseous fuels.

What are the prospects for alternative energy sources? Alternative sources include nuclear, hydroelectric, wind, solar, biomass, and geothermal. Taken together, these sources currently provide only about 14 percent of world energy needs. Nuclear power from the fission of uranium-235 can be a major energy source, but only if its costs do not keep escalating and the public can be assured of its safety. It has the advantage that it does not release carbon dioxide and the disadvantage that safe repositories must be found to store radioactive wastes for hundreds of thousands of years. With advances in technology and reductions in cost, renewable sources of energy could become major sources in this century.

What is acid rain, and what is being done about it? Acid rain is rain that contains sulfuric acid or nitric acid or both, formed when sulfur and nitrogen gases react with rainwater. The sulfur gases are waste gases from power plants that burn coal containing pyrite and from smelters of sulfide ores. Nitrogen gases are products of coal combustion and automobile exhausts. Acid rain can destroy fish stocks in lakes and rivers, damage mountain forests, and damage buildings and monuments. In the United States, the Clean Air Act was amended in 1990 to reduce sulfur and nitrogen gas emissions.

What is ozone depletion, and why should it concern us? CFCs are chemicals used as refrigerants and in industrial processes. When they reach the upper atmosphere, their breakdown products react with ozone (O_3^+) and convert it into oxygen (O_2). Ozone absorbs harmful ultraviolet radiation streaming in from the Sun. Without the ozone shield in the upper atmosphere, many life-forms would die off. An international agreement has been reached to stop the production of chemicals that destroy the ozone shield.

How much global warming will there be in the twenty-first century, and what will be its consequences? Levels of greenhouse gases will continue to rise throughout the twenty-first century primarily because of human activities, although the increase depends on a number of socioeconomic factors that will govern the rate of greenhouse gas emissions, including active steps by society to limit these emissions. Projections of the amount of warming during the next century are highly uncertain because of the uncertainty in greenhouse gas emissions as well as an incomplete knowledge of how the climate system works, but the range accepted by most experts is from 1.4° to 5.8°C. This warming will disrupt plant and animal habitats, changing ecosystems and increasing the rate of species extinctions. The oceans will warm and expand, raising sea level as much as 1 m. The north polar ice cap will continue to shrink rapidly, and much of the Arctic Ocean is expected to become ice-free.

What should be the goal of energy and environmental policies? Policies should guide the nations of the world through the transition from hydrocarbon fuels to less polluting, sustainable energy sources. In particular, carbon dioxide emissions should be reduced to decrease the impact of global climate change. The more efficient use of energy, the greater use of natural gas, and the introduction of safer nuclear energy and clean coal technology would facilitate this changeover.

KEY TERMS AND CONCEPTS

acid rain (p. 570)
anthropogenic (p. 569)
coal bed (p. 563)
crude oil (p. 557)
fossil fuel (p. 552)
geothermal energy (p. 568)
global change (p. 569)
Hubbert's peak (p. 560)
natural gas (p. 557)
nonrenewable energy
  source (p. 552)
nuclear energy (p. 565)
oil shale (p. 564)

oil trap (p. 557)
oil window (p. 557)
quad (p. 556)
renewable energy source
  (p. 552)
reserve (p. 557)
resource (p. 557)
solar energy (p. 566)
stratospheric ozone
  depletion (p. 571)
sustainable development
  (p. 576)
tar sand (p. 564)

EXERCISES

1. Describe how the human geosystem is fundamentally different from the natural geosystems studied in this book.

2. Which fossil fuel produces the least amount of carbon dioxide per unit of energy: oil, natural gas, or coal?

3. What are the prerequisites for oil traps to contain oil?

4. Which of the following factors are most important in estimating the future supply of oil and gas? (a) Rate of oil accumulation, (b) rate of natural seepage of oil, (c) rate of pumping of oil from known reserves, (d) rate of discovery of new reserves, (e) total amount of oil now present in the Earth.

5. An aggressive drilling program in the Arctic National Wildlife Reserve could produce as much as 12 billion barrels of oil. At current consumption rates, how many years would this resource supply U.S. oil needs?

6. Which three countries have the largest coal reserves?

7. Why is ozone depletion a problem in the stratosphere but not in the troposphere?

8. If we keep pumping carbon dioxide into the atmosphere and Earth's climate warms significantly in the next 100 years, how might the global carbon cycle be affected?

9. An economist once wrote: "The predicted change in global temperature due to human activity is less than the difference in winter temperature between New York and Florida, so why worry?" Explain why he should worry.

10. Do you think a geologist several thousand years in the future will consider the industrial revolution the beginning of a new geologic epoch?

THOUGHT QUESTIONS

1. In what ways does Earth's internal heat contribute to the formation of fossil-fuel resources?

2. Are you an oil optimist or an oil pessimist? Explain why.

3. What issues related to the use of nuclear power can be addressed by a geologist?

4. Contrast the risks and benefits of nuclear fission and coal combustion as energy sources.

5. What do you think will be the major sources of the world's energy in the year 2030? In the year 2100?

6. Do you think we should act now to reduce carbon emissions or delay until the science of climate change is better understood?

7. Is the United States justified in insisting that developing countries that now use much less fossil fuel than developed countries agree to limit their future carbon emissions?

8. Do you think that future scientists and engineers will be able to intervene and prevent catastrophic changes in the Earth system?

SHORT-TERM PROJECT

Energy Assessment for Your Home State

Investigate the energy resources of your state by compiling data on fossil-fuel resources, nuclear energy production, and the usage of renewable energy sources. A good place to get information is the U.S. Department of Energy's Annual Energy Review, posted at http://www.eia.doe.gov/emeu/aer/. Based on your assessment, make policy recommendations to the state governor that might address the current issues of Earth system engineering and management.

APPENDIX I

Conversion Factors

LENGTH

1 centimeter	0.3937 inch
1 inch	2.5400 centimeters
1 meter	3.2808 feet; 1.0936 yards
1 foot	0.3048 meter
1 yard	0.9144 meter
1 kilometer	0.6214 mile (statute); 3281 feet

LENGTH

1 mile (statute)	1.6093 kilometers
1 mile (nautical)	1.8531 kilometers
1 fathom	6 feet; 1.8288 meters
1 angstrom	10^{-8} centimeter
1 micrometer	0.0001 centimeter

VELOCITY

1 kilometer/hour	27.78 centimeters/second
1 mile/hour	17.60 inches/second

AREA

1 square centimeter	0.1550 square inch
1 square inch	6.452 square centimeters
1 square meter	10.764 square feet; 1.1960 square yards
1 square foot	0.0929 square meter
1 square kilometer	0.3861 square mile
1 square mile	2.590 square kilometers
1 acre (U.S.)	4840 square yards

VOLUME

1 cubic centimeter	0.0610 cubic inch
1 cubic inch	16.3872 cubic centimeters
1 cubic meter	35.314 cubic feet
1 cubic foot	0.02832 cubic meter
1 cubic meter	1.3079 cubic yards
1 cubic yard	0.7646 cubic meter
1 liter	1000 cubic centimeters; 1.0567 quarts (U.S. liquid)
1 gallon (U.S. liquid)	3.7853 liters

MASS

1 gram	0.03527 ounce
1 ounce	28.3495 grams
1 kilogram	2.20462 pounds
1 pound	0.45359 kilogram

PRESSURE

1 kilogram/square centimeter	0.96784 atmosphere; 0.98067 bar; 14.2233 pounds/square inch
1 bar	0.98692 atmosphere; 10^5 pascals

ENERGY

1 joule	0.239 calorie; 9.479×10^{-4} Btu
1 quad	10^{15} Btu

POWER

1 watt	0.001341 horsepower (U.S.); 3.413 Btu/hour

Degrees

F	C
210	100
200	90
190	
180	80
170	
160	70
150	
140	60
130	
120	50
110	
100	40
90	30
80	
70	20
60	
50	10
40	
30	0
20	
10	10
0	20

APPENDIX 2

Numerical Data Pertaining to Earth

Equatorial radius	6378 kilometers
Polar radius	6357 kilometers
Radius of sphere with Earth's volume	6371 kilometers
Volume	1.083×10^{27} cubic centimeters
Surface area	5.1×10^{18} square centimeters
Percent surface area of oceans	71
Percent surface area of land	29
Average elevation of land	623 meters
Average depth of oceans	3.8 kilometers
Mass	5.976×10^{27} grams
Density	5.517 grams/cubic centimeters
Gravity at equator	978.032 centimeters/second/second
Mass of atmosphere	5.1×10^{21} grams
Mass of ice	$25–30 \times 10^{21}$ grams
Mass of oceans	1.4×10^{24} grams
Mass of crust	2.5×10^{25} grams
Mass of mantle	4.05×10^{27} grams
Mass of core	1.90×10^{27} grams
Mean distance to Sun	1.496×10^{8} kilometers
Mean distance to Moon	3.844×10^{5} kilometers
Ratio: Mass of Sun/mass of Earth	3.329×10^{5}
Ratio: Mass of Earth/mass of Moon	81.303
Total geothermal energy reaching Earth's surface each year	10^{21} joule; 2.39×10^{20} calories; 949 quads
Earth's daily receipt of solar energy	1.49×10^{22} joules
U.S. energy consumption, 2004	99.7 quads

Chemical Reactions

Electron Shells and Ion Stability

Electrons surround the nucleus of an atom in a unique set of concentric spheres called electron shells. Each shell can hold a certain maximum number of electrons. In the chemical reactions of most elements, only the electrons in the outermost shells interact. In the reaction between sodium (Na) and chlorine (Cl) that forms sodium chloride (NaCl), the sodium atom loses an electron from its outer shell of electrons, and the chlorine atom gains an electron in its outer shell (see Figure 3.4, page 48).

Before reacting with chlorine, the sodium atom has one electron in its outer shell. When it loses that electron, its outer shell is eliminated and the next shell inward, which has eight electrons (the maximum that this shell can hold), becomes the outer shell. The original chlorine atom had seven electrons in its outer shell, with room for a total of eight. By gaining an electron, its outer shell is filled. Many elements have a strong tendency to acquire a full outer electron shell, some by gaining electrons and some by losing them in the course of a chemical reaction. The stability of ions with fully occupied outer shells is related to the interactions of electrons in various orbitals around the nucleus.

Many chemical reactions entail gains and losses of several electrons as two or more elements combine. The element calcium (Ca), for example, becomes a doubly charged cation, Ca^{2+}, as it reacts with two chlorine atoms to form calcium chloride. (In the chemical formula for calcium chloride, $CaCl_2$, the presence of two chloride ions is symbolized by the subscript 2.) Chemical formulas thus show the relative proportion of atoms or ions in a compound.

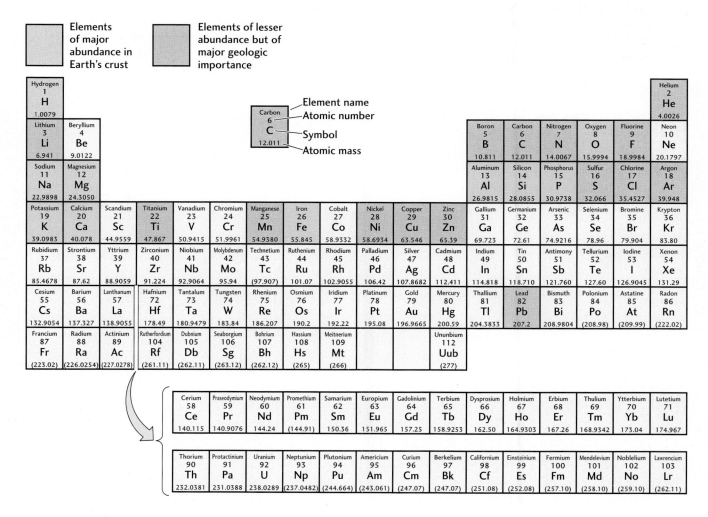

The periodic table.

Common practice is to omit the subscript 1 next to single ions in a formula.

The periodic table organizes the elements (from left to right along a row) in order of atomic number (the number of protons), which also means increasing the numbers of electrons in the outer shell. The third row from the top, for example, starts at the left with sodium (atomic number 11), which has one electron in its outer shell. The next is magnesium (atomic number 12), which has two electrons in its outer shell, followed by aluminum (atomic number 13), with three, and silicon (atomic number 14), with four. Then come phosphorus (atomic number 15), with five; sulfur (atomic number 16), with six; and chlorine (atomic number 17), with seven. The last element in this row is argon (atomic number 18), with eight electrons, the maximum possible, in its outer shell. Each column in the table forms a vertical grouping of elements with similar electron-shell patterns.

Elements That Tend to Lose Electrons

The elements in the leftmost column of the table all have a single electron in their outer shells and have a strong tendency to lose that electron in chemical reactions. Of this group, hydrogen (H), sodium (Na), and potassium (K) are found in major abundance at Earth's surface and in its crust.

The second column from the left includes two more elements of major abundance, magnesium (Mg) and calcium (Ca). Elements in this column have two electrons in their outer shells and a strong tendency to lose both of them in chemical reactions.

Elements That Tend to Gain Electrons

Toward the right side of the table, the two columns headed by oxygen (O), the most abundant element in the Earth, and fluorine (F), a highly reactive toxic gas, group the elements that tend to gain electrons in their outer shells. The elements in the column headed by oxygen have six of the possible eight electrons in their outer shells and tend to gain two electrons. Those in the column headed by fluorine have seven electrons in their outer shells and tend to gain one.

Other Elements

The columns between those farthest on the left and those farthest on the right have varying tendencies to gain, lose, or share electrons. The column toward the right side of the table headed by carbon (C) includes silicon (Si), of major abundance in the Earth. Both silicon and carbon tend to share electrons.

The elements in the last column on the right, headed by helium (He), have full outer shells and thus no tendency either to gain or to lose electrons. As a result, these elements, in contrast with those in other columns, do not react chemically with other elements, except under very special conditions.

APPENDIX 4

Properties of the Most Common Minerals of Earth's Crust

	Mineral or Group Name	Structure or Composition	Varieties and Chemical Composition	Form, Diagnostic Characteristics	Cleavage, Fracture	Color	Hardness
LIGHT-COLORED MINERALS, VERY ABUNDANT IN EARTH'S CRUST IN ALL MAJOR ROCK TYPES	FELDSPAR	FRAMEWORK SILICATES	*POTASSIUM FELDSPARS* $KAlSi_3O_8$ Sanidine Orthoclase Microcline	Cleavable coarsely crystalline or finely granular masses; isolated crystals or grains in rocks, most commonly not showing crystal faces	Two at right angles, one perfect and one good; pearly luster on perfect cleavage	White to gray, frequently pink or yellowish; some green	6
			PLAGIOCLASE FELDSPARS $NaAlSi_3O_8$ Albite $CaAl_2Si_2O_8$ Anorthite		Two at nearly right angles, one perfect and one good; fine parallel striations on perfect cleavage	White to gray, less commonly greenish or yellowish	
	QUARTZ		SiO_2	Single crystals or masses of 6-sided prismatic crystals; also formless crystals and grains or finely granular or massive	Very poor or nondetectable; conchoidal fracture	Colorless, usually transparent; also slightly colored smoky gray, pink, yellow	7
	MICA	SHEET SILICATES	*MUSCOVITE* $KAl_3Si_3O_{10}(OH)_2$	Thin, disc-shaped crystals, some with hexagonal outlines; dispersed or aggregates	One perfect; splittable into very thin, flexible, transparent sheets	Colorless; slight gray or green to brown in thick pieces	$2-2\frac{1}{2}$
DARK-COLORED MINERALS, ABUNDANT IN MANY KINDS OF IGNEOUS AND METAMORPHIC ROCKS			*BIOTITE* $K(Mg, Fe)_3AlSi_3O_{10}(OH)_2$	Irregular, foliated masses; scaly aggregates	One perfect; splittable into thin, flexible sheets	Black to dark brown; translucent to opaque	$2\frac{1}{2}-3$
			CHLORITE $(Mg,Fe)_5(Al,Fe)_2Si_3O_{10}(OH)_8$	Foliated masses or aggregates of small scales	One perfect; thin sheets flexible but not elastic	Various shades of green	$2-2\frac{1}{2}$
	AMPHIBOLE	DOUBLE CHAINS	*TREMOLITE-ACTINOLITE* $Ca_2(Mg,Fe)_5Si_8O_{22}(OH)_2$	Long, prismatic crystals, usually 6-sided; commonly in fibrous masses or irregular aggregates	Two perfect cleavage directions at 56° and 124° angles	Pale to deep green Pure tremolite white	5–6
			HORNBLENDE Complex Ca, Na, Mg, Fe, Al silicate				
	PYROXENE	SINGLE CHAINS	*ENSTATITE-HYPERSTHENE* $(Mg,Fe)_2Si_2O_6$	Prismatic crystals, either 4- or 8-sided; granular masses and scattered grains	Two good cleavage directions at about 90°	Green and brown to grayish or greenish white	5–6
			DIOPSIDE $(Ca,Mg)_2Si_2O_6$			Light to dark green	
			AUGITE Complex Ca, Na, Mg, Fe, Al silicate			Very dark green to black	

(Continued)

	Mineral or Group Name	Structure or Composition	Varieties and Chemical Composition	Form, Diagnostic Characteristics	Cleavage, Fracture	Color	Hardness
LIGHT-COLORED MINERALS, TYPICALLY AS ABUNDANT CONSTITUENTS OF SEDIMENTS AND SEDIMENTARY ROCKS	OLIVINE	ISOLATED TETRAHEDRA	$(Mg,Fe)_2SiO_4$	Granular masses and disseminated small grains	Conchoidal fracture	Olive to grayish green and brown	$6\frac{1}{2}$–7
	GARNET		Ca, Mg, Fe, Al silicate	Isometric crystals, well formed or rounded; high specific gravity, 3.5–4.3	Conchoidal and irregular fracture	Red and brown, less commonly pale colors	$6\frac{1}{2}$–7
	CALCITE	CARBONATES	$CaCO_3$	Coarsely to finely crystalline in beds, veins, and other aggregates; cleavage faces may show in coarser masses; calcite effervesces rapidly in acid, but dolomite effervesces slowly and only if crushed into powder	Three perfect cleavages, at oblique angles; splits to rhombohedral cleavage pieces	Colorless, transparent to translucent; variously colored by impurities	3
	DOLOMITE		$CaMg(CO_3)_2$				$3\frac{1}{2}$–4
	CLAY MINERALS	HYDROUS ALUMINO-SILICATES	*KAOLINITE* $Al_2Si_2O_5(OH)_4$ *ILLITE* Similar to muscovite +Mg,Fe *SMECTITE* Complex Ca, Na, Mg, Fe, Al silicate + H_2O	Earthy masses in soils; bedded; in association with other clays, iron oxides, or carbonates; plastic when wet; montmorillonite swells when wet	Earthy, irregular	White to light gray and buff; also gray to dark gray, greenish gray, and brownish depending on impurities and associated minerals	$1\frac{1}{2}$–$2\frac{1}{2}$
	GYPSUM	SULFATES	$CaSO_4 \cdot 2H_2O$	Granular, earthy, or finely crystalline masses; tabular crystals	One perfect, splitting to fairly thin slabs or sheets; two other good cleavages	Colorless to white; transparent to translucent	2
	ANHYDRITE		$CaSO_4$	Massive or crystalline aggregates in beds and veins	One perfect, one nearly perfect, one good; at right angles	Colorless, some tinged with blue	3–$3\frac{1}{2}$
	HALITE	HALIDES	$NaCl$	Granular masses in beds; some cubic crystals; salty taste	Three perfect cleavages at right angles	Colorless, transparent to translucent	$2\frac{1}{2}$
	OPAL-CHALCEDONY	SILICA	SiO_2 [Opal is an amorphous variety; chalcedony is a formless microcrystalline quartz.]	Beds in siliceous sediments and chert; in veins or banded aggregates	Conchoidal fracture	Colorless or white when pure, but tinged with various colors by impurities in bands, especially in agates	5–$6\frac{1}{2}$
DARK-COLORED MINERALS, COMMON IN MANY ROCK TYPES	MAGNETITE	IRON OXIDES	Fe_3O_4	Magnetic; disseminated grains, granular masses; occasional octahedral isometric crystals; high specific gravity, 5.2	Conchoidal or irregular fracture	Black, metallic luster	6

Mineral or Group Name	Structure or Composition	Varieties and Chemical Composition	Form, Diagnostic Characteristics	Cleavage, Fracture	Color	Hardness
HEMATITE	IRON OXIDES	Fe_2O_3	Earthy to dense masses, some with rounded forms, some granular or foliated; high specific gravity, 4.9–5.3	None; uneven, sometimes splintery fracture	Reddish brown to black	5–6
"LIMONITE"		GOETHITE [the major mineral of the mixture called "limonite," a field term] $FeO(OH)$	Earthy masses, massive bodies or encrustations, irregular layers; high specific gravity, 3.3–4.3	One excellent in the rare crystals; usually an early fracture	Yellowish brown to dark brown and black	5–5½
LIGHT-COLORED MINERALS, MAINLY IN IGNEOUS AND METAMORPHIC ROCKS AS COMMON OR MINOR CONSTITUENTS						
KYANITE	ALUMINO-SILICATES	Al_2SiO_5	Long, bladed or tabular crystals or aggregates	One perfect and one poor, parallel to length of crystals	White to light-colored or pale blue	5 parallel to crystal length 7 across crystals
SILLIMANITE		Al_2SiO_5	Long, slender crystals or fibrous, felted masses	One perfect parallel to length, not usually seen	Colorless, gray to white	6–7
ANDALUSITE		Al_2SiO_5	Coarse, nearly square prismatic crystals, some with symmetrically arranged impurities	One distinct; irregular fracture	Red, reddish brown, olive-green	7½
FELDSPATHOIDS		NEPHELINE $(Na,K)AlSiO_4$	Compact masses or as embedded grains, rarely as small prismatic crystals	One distinct; irregular fracture	Colorless, white, light gray; gray-greenish in masses, with greasy luster	5½–6
		LEUCITE $KAlSi_2O_6$	Trapezohedral crystals embedded in volcanic rocks	One very imperfect	White to gray	5½–6
SERPENTINE		$Mg_6Si_4O_{10}(OH)_8$	Fibrous (asbestos) or platy masses	Splintery fracture	Green; some yellowish brownish or gray; waxy or greasy luster in massive habit; silky luster in fibrous habit	4–6
TALC		$Mg_3Si_4O_{10}(OH)_2$ masses or aggregates	Foliated or compact masses or aggregates	One perfect, making thin flakes or scales; soapy feel	White to pale green; pearly or greasy luster	1
CORUNDUM		Al_2O_3	Some rounded, barrel-shaped crystals; most often as disseminated grains or granular masses (emery)	Irregular fracture	Usually brown, pink, or blue; emery black. Gemstone varieties: ruby, sapphire	9

(Continued)

	Mineral or Group Name	Structure or Composition	Varieties and Chemical Composition	Form, Diagnostic Characteristics	Cleavage, Fracture	Color	Hardness
DARK-COLORED MINERALS, COMMON IN METAMORPHIC ROCKS	EPIDOTE	SILICATES	$Ca_2(Al,Fe)Al_2Si_3O_{12}(OH)$	Aggregates of long prismatic crystals, granular or compact masses, embedded grains	One good, one poor at greater than right angles; conchoidal and irregular fracture	Green, yellow-green, gray, some varieties dark brown to black	6–7
	STAUROLITE		$Fe_2Al_9Si_4O_{22}(O,OH)_2$	Short prismatic crystals, some cross-shaped, usually coarser than matrix of rock	One poor	Brown, reddish, or dark brown to black	$7–7\frac{1}{2}$
METALLIC LUSTER, COMMON IN MANY ROCK TYPES, ABUNDANT IN VEINS	PYRITE	SULFIDES	FeS_2	Granular masses or well-formed cubic crystals in veins and beds or disseminated; high specific gravity, 4.9–5.2	Uneven fracture	Pale brass-yellow	$6–6\frac{1}{2}$
	GALENA		PbS	Granular masses in veins and disseminated; some cubic crystals; very high specific gravity, 7.3–7.6	Three perfect cleavages at mutual right angles, giving cubic cleavage fragments	Silver-gray	$2\frac{1}{2}$
	SPHALERITE		ZnS	Granular masses or compact crystalline aggregates; high specific gravity, 3.9–4.1	Six perfect cleavages at 60° to one another	White to green, brown, and black; resinous to submetallic luster	$3\frac{1}{2}–4$
	CHALCOPYRITE		$CuFeS_2$	Granular or compact masses; disseminated crystals; high specific gravity, 4.1–4.3	Uneven fracture	Brassy to golden-yellow	$3\frac{1}{2}–4$
	CHALCOCITE		Cu_2S	Fine-grained masses; high specific gravity, 5.5–5.8	Conchoidal fracture	Lead-gray to black; may tarnish green or blue	$2\frac{1}{2}–3$
MINERALS FOUND IN MINOR AMOUNTS IN A VARIETY OF ROCK TYPES AND IN VEINS OR PLACERS	RUTILE	TITANIUM OXIDES	TiO_2	Slender to prismatic crystals; granular masses; high specific gravity, 4.25	One distinct, one less distinct; conchoidal fracture	Reddish brown, some yellowish, violet, or black	$6–6\frac{1}{2}$
	ILMENITE		$FeTiO_3$	Compact masses, embedded grains, detrital grains in sand; high specific gravity, 4.79	Conchoidal fracture	Iron-black; metallic to submetallic luster	5–6
	ZEOLITES	SILICATES	Complex hydrous silicates; many varieties of minerals, including analcime, natrolite, phillipsite, heulandite, and chabazite	Well-formed radiating crystals in cavities in volcanics, veins, and hot springs; also as fine-grained and earthy bedded deposits	One perfect for most	Colorless, white, some pinkish	4–5

Glossary

Words in *italic* have separate entries in the Glossary. Specific minerals are defined in Appendix 4.

ablation   The total amount of ice that a *glacier* loses each year. (Compare *accumulation.*)

abrasion   The erosive action that occurs when suspended and saltating *sediment* particles move along the bottom and sides of a stream *channel.*

absolute age   The actual number of years elapsed from a geologic event until now. (Compare *relative age.*)

abyssal hill   A hill on the slope of a mid-ocean ridge, typically 100 m or so high and lineated parallel to the ridge crest, formed primarily by normal *faulting* of the basaltic oceanic *crust* as it moves out of the *rift valley.*

abyssal plain   A wide, flat plain that covers large areas of the ocean floor at depths of about 4000 to 6000 m.

accreted terrane   An individual, geologically coherent fragment of an amalgamation of odd pieces of *crust—island arcs, seamounts;* remnants of thickened oceanic *plateaus;* old mountain ranges; and other slivers of continental crust—that were plastered onto the leading edge of a continent as it moved across Earth's surface.

accretion   A process of continental growth in which buoyant fragments of *crust* are attached (accreted) to continents during plate motions.

accumulation   The amount of snow added to a *glacier* annually. (Compare *ablation.*)

acid rain   Acidic *precipitation* caused by the pollution of air with sulfur gases emitted by the smokestacks of power plants that burn *coal* containing large amounts of the *mineral* pyrite, by smelters of sulfide *ores,* and by some factories; and with nitrogen oxide gases emitted by smokestacks and automobile exhausts. These gases react with water to form sulfuric acid and nitric acid.

active margin   A *continental margin* characterized by volcanic activity and frequent *earthquakes* and associated with *subduction* or *transform faulting.*

aerobic zone   The oxygen-rich upper zone of most *sediment* layers. (Compare *anaerobic zone.*)

aftershock   An *earthquake* that occurs as a consequence of a previous earthquake of larger magnitude.

albedo   The fraction of *solar energy* reflected by the surface of a planet or moon.

alluvial fan   A cone- or fan-shaped accumulation of *sediment* deposited where a *stream* widens abruptly as it leaves a mountain front for an open *valley.*

amphibolite   A usually nonfoliated *metamorphic rock* made up mainly of amphibole and plagioclase feldspar. Foliated amphibolites can be produced by deformation.

anaerobic zone   The oxygen-depleted zone at the bottom of water bodies created by *microbes* at the air-water interface that use up all the oxygen during respiration. (Compare *aerobic zone.*)

andesite   A volcanic *rock* type intermediate in composition between *rhyolite* and *basalt.* The extrusive equivalent of *diorite.*

andesitic lava   A lava of intermediate composition that has a higher silica content than *basalt,* erupts at lower temperatures, and is more viscous. Extrusive equivalent of *diorite.*

angle of repose   The maximum angle at which a slope of loose material will lie without sliding.

anion   A negatively charged ion.

antecedent stream   A *stream* that existed before the present *topography* was created and so maintained its original course despite changes in the structure of the underlying *rocks* and in topography. (Compare *superposed stream.*)

anthropogenic   Arising from human activity.

anticline   A large upfold or arch of layered *rocks.* (Compare *syncline.*)

aquiclude   A relatively impermeable bed that bounds an *aquifer* above or below and acts as a barrier to the flow of *groundwater.*

aquifer   A bed that stores and transmits *groundwater* in sufficient quantity to supply wells.

arkose   A sandstone containing more than 25 percent feldspar.

artesian flow   A water flow produced by the greater pressure of *groundwater* in a confined aquifer than in an unconfined aquifer.

ash-flow deposit   An extensive sheet of hard volcanic *tuff.*

asteroid   One of the small celestial bodies orbiting the Sun, most of them between Mars and Jupiter.

asthenosphere   The weak layer of soft but solid *rock* comprising the lower part of the upper *mantle* (below the *lithosphere*) and over which the plates slide. Movement in the asthenosphere occurs by plastic deformation. (From the Greek *asthenes,* meaning "weak.")

astrobiologist A scientist who searches for evidence of life on other worlds.

atom The smallest unit of an element that retains the element's physical and chemical properties.

atomic mass The sum of an element's *protons* and *neutrons.*

atomic number The number of *protons* in the *nucleus* of an *atom.*

autotroph An organism that makes its own food by manufacturing organic compounds, such as carbohydrates, that it uses as sources of energy and nutrients. (Compare *heterotroph.*)

axial plane An imaginary surface that divides a fold as symmetrically as possible, with one limb on either side of the plane.

badland A deeply gullied *topography* resulting from the fast *erosion* of easily erodible *shales* and *clays.*

banded iron formation A *sedimentary rock* composed of alternating thin layers formed of iron oxide minerals and silica-rich minerals, precipitated from seawater when oxygen, produced by *cyanobacteria,* reacted with iron dissolved in seawater. Banded iron formations mark the rise of oxygen in the atmosphere between 2.7 and 2.1 Ga.

barrier island A long offshore sandbar that builds up to form a barricade between open ocean waves and the main shoreline.

basal slip The sliding of a *glacier* along its base. (Compare *plastic flow.*)

basalt A fine-grained, dark, *mafic igneous rock* composed largely of plagioclase feldspar and pyroxene. The extrusive equivalent of *gabbro.*

basaltic lava A lava of mafic composition that has a low silica content, erupts at high temperatures, and flows readily. Extrusive equivalent of *gabbro.*

base level The *elevation* at which a *stream* ends by entering a large standing body of water, such as a lake or ocean.

basin A bowl-shaped depression of *rock* layers in which the beds *dip* radially toward a central point.

batholith A great irregular mass of coarse-grained *igneous rock* that covers at least 100 km^2; the largest *pluton.*

bed load The material a *stream* carries along its bed by sliding and rolling. (Compare *suspended load.*)

bedding The formation of parallel layers of *sediment* as particles settle to the bottom of the sea, a *river,* or a land surface.

bedding sequence A pattern of interbedded and vertically stacked layers of *sandstone, shale,* and other *sedimentary rock* types.

bioclastic sediment A shallow-water *sediment* consisting primarily of two calcium carbonate *minerals*—calcite and aragonite—in variable proportions.

biogeochemical cycle A pathway through which a chemical element or molecule moves between the biologic ("bio") and environmental ("geo") components of an *ecosystem.*

biological sediment A *sediment* that forms near its place of deposition as a result of *mineral precipitation* within organisms as they grow.

biosphere The sum total of all living organisms and the organic matter they produce.

bioturbation The process by which organisms rework existing *sediments* by burrowing through *muds* and *sands.*

blueschist A *metamorphic rock* formed under conditions of high pressure and moderate temperature, often containing glaucophane, a blue amphibole.

bottomset bed A thin, horizontal bed of *mud* deposited seaward of a *delta* and then buried by continued delta growth.

braided stream A *stream* whose *channel* divides into an interlacing network of channels, which then rejoin in a pattern resembling braids of hair.

brittle material A material that undergoes little deformation under increasing force, until it breaks suddenly. (Compare *ductile material.*)

building code Standards for the design and construction of new buildings that specify the level of shaking a structure must be able to withstand in an *earthquake,* based on the maximum intensity expected from the *seismic hazard.*

calcium cycle A geochemical cycle that includes all fluxes of calcium into and out of its *reservoirs.*

caldera A large, steep-walled, basin-shaped depression formed after a violent eruption in which large volumes of magma are discharged, when the overlying volcanic structure collapses catastrophically through the roof of the emptied *magma chamber.*

capacity The total *sediment* load carried by a flow. (Compare *competence.*)

carbon cycle The *geosystem* that describes the continual movement of carbon among the atmosphere and its other principal *reservoirs*—the *lithosphere, hydrosphere,* and *biosphere.*

carbonate environment A marine setting where calcium carbonate, principally of biochemical origin, is the main *sediment.*

carbonate platform An extensive flat, shallow area where both biological and nonbiological carbonates are deposited.

carbonate rock A *sedimentary rock* formed from the accumulation of carbonate *minerals precipitated* organically or inorganically.

carbonate sediment A *sediment* formed from the accumulation of carbonate *minerals precipitated* organically or inorganically.

cation A positively charged ion.

cementation A major chemical diagenetic change in which *minerals* are *precipitated* in the pores of *sediments,* forming cements that bind *clastic sediments* and *rocks.*

channel The trough through which the water in a *stream* flows.

chemical and biochemical sediments New chemical substances that form by *precipitation* when some of a *rock's* components dissolve during *weathering* and are carried in *river* waters to the sea.

chemical reaction The interaction of the *atoms* of two or more chemical elements in certain fixed proportions that produces a new chemical substance.

chemical sediment The dissolved product of *weathering precipitated* from water (usually seawater) by *chemical reactions* and formed at or near its place of deposition.

chemical stability A measure of a substance's tendency to remain in a given chemical form rather than reacting spontaneously to become a different chemical substance.

chemical weathering The *weathering* that occurs when the *minerals* in a *rock* are chemically altered or dissolved.

chemofossil The chemical remains of organic compounds made by an organism while it was alive.

chert A *sedimentary rock* made up of chemically or biochemically *precipitated* silica.

cirque An amphitheater-like hollow, usually shaped like half of an inverted cone, formed at the head of a glacial valley by the plucking and tearing action of ice.

clastic particle A physically transported *rock* fragment produced by the *weathering* of preexisting rocks.

clastic sediment An accumulation of *clastic particles* laid down by running water, wind, or ice and forming layers of *sand, silt,* or *gravel.*

clay The most abundant component of fine-grained *sediments* and *sedimentary rocks,* consisting largely of clay *minerals.* Clay-sized particles are less than 0.0039 mm in diameter.

claystone A *rock* made up exclusively of *clay*-sized particles.

cleavage (1) The tendency of a *crystal* to break along flat planar surfaces. (2) The geometric pattern produced by such breakage.

climate The average surface conditions and their variation during cycles of solar forcing. Solar forcing is the cyclical variations in the input of solar energy that force changes in the conditions of the surface *environment.*

climate model Any representation of the *climate system* that can reproduce one or more aspects of *climate* behavior.

climate system A *geosystem* that includes all parts of the *Earth system* and all the interactions among these components needed to describe how *climate* behaves in space and time.

coal A biochemically produced *sedimentary rock* composed almost entirely of organic carbon formed by the *diagenesis* of swamp vegetation.

coal bed A layer or stratum of *coal.*

color A property of a *mineral* imparted by light—either transmitted through or reflected by *crystals,* irregular masses, or a *streak.*

compaction A decrease in the volume and *porosity* of a *sediment* that occurs when the grains are squeezed closer together by the weight of overlying sediment.

competence The ability of a flow to carry material of a given size. (Compare *capacity.*)

compressional wave A *seismic wave* that propagates by a push-pull motion of a material. (Compare *shear wave.*)

compressive force A force that squeezes together or shortens a body. (Compare *tensional force.*)

concordant intrusion An *intrusive igneous rock* whose boundaries lie parallel to layers of preexisting bedded rock. (Compare *discordant intrusion.*)

conduction The mechanical transfer of heat energy (the vibrational energy of thermally agitated *atoms* and molecules) by the mechanism of atomic or molecular impact. (Compare *convection.*)

conglomerate A *sedimentary rock* composed of pebbles, cobbles, and boulders. The lithified equivalent of *gravel.*

consolidated material *Sediment* that is compacted and bound together by *mineral* cements. (Compare *unconsolidated material.*)

contact metamorphism Changes in the *mineralogy* and *texture* of *rock* resulting from the heat and pressure in a small area, such as the rocks near and in contact with an igneous intrusion.

continental drift The large-scale motion of continents across Earth's surface driven by plate tectonics.

continental glacier A thick, extremely slow moving sheet of ice that covers a large part of a continent. (Compare *valley glacier.*)

continental margin The shoreline, shelf, and slope of a continent.

continental rise An apron of muddy and sandy *sediment* extending from the *continental slope* into the main ocean basin.

continental shelf A broad, flat, *sand*- and *mud*-covered platform that is a slightly submerged part of a continent and extends to the edge of the *continental slope.*

continental slope A steep, *mud*-covered slope between the *continental shelf* and *continental rise.*

contour A line that connects points of equal *elevation* on a topographic map.

convection A mechanism of heat transfer in which a heated fluid expands and rises because it has become less dense than the surrounding material. Colder material flows in to take the place of the hot rising fluid, is itself heated, and then rises to continue the cycle.

core The central part of the Earth below a depth of 2900 km, comprising a liquid *outer core* and a solid *inner core.* The core is composed of iron and nickel, with minor amounts of some lighter element, such as oxygen or sulfur.

core-mantle boundary The boundary between the *core* and lower *mantle,* about 2890 km below Earth's surface, where S-wave speed drops from about 7.5 km/s to zero, and P-wave speed drops from more than 13 km/s to about 8 km/s.

country rock The rock surrounding an *intrusive igneous rock.*

covalent bond A bond between *atoms* in which the outer *electrons* are shared. (Compare *ionic bond.*)

crater (1) A bowl-shaped pit found at the summit of most *volcanoes,* centered on the vent. (2) A depression caused by the impact of a *meteorite.*

craton A stable nucleus composed of the eroded remnants of ancient deformed *rocks* that comprises the continental *shields* and platforms.

cratonic keel A part of the *lithosphere* that extends into the convecting *asthenosphere* at 100 to 200 km beneath the cratons like the hull of a boat into water.

creep The slow downhill *mass movement* of *soil* or other debris at a rate ranging from about 1 to 10 mm/year.

crevasse A large vertical crack in the surface of a *glacier* caused by the movement of brittle surface ice as it is dragged along by the *plastic flow* of the ice below.

cross-bedding A *sedimentary structure* consisting of bedded material deposited by currents of wind or water and inclined at angles as large as 35° from the horizontal.

crude oil A diverse class of liquids composed of complex hydrocarbons, including long molecular chains comprising dozens of carbon and hydrogen atoms.

crust The thin outer layer of the Earth, averaging about 8 km thick under the oceans to about 40 km thick under the continents, consisting of relatively light materials that melt at low temperatures. Continental crust consists largely of *granite* and *granodiorite.* Oceanic crust is mostly *basalt.*

crystal An ordered three-dimensional array of *atoms* in which the basic arrangement is repeated in all directions.

crystal habit The shape in which a *mineral's* individual *crystals* or aggregates of crystals grow.

crystallization The growth of a solid from a gas or liquid whose constituent *atoms* come together in the proper chemical proportions and ordered three-dimensional arrangement.

cuesta An asymmetrical ridge in a tilted and eroded series of beds with alternating weak and strong resistance to *erosion.* One side has a long, gentle slope; the other is a steep cliff formed at the edge of the resistant bed where it is undercut by erosion of a weaker bed beneath.

cyanobacteria A group of microorganisms that produce oxygen and carbohydrates by *photosynthesis.* It is believed that photosynthesis originated in cyanobacteria.

dacite A light-colored, fine-grained *extrusive igneous rock* with the same general composition as *andesite.* The extrusive equivalent of *granodiorite.*

Darcy's law A summary of the relationships among the volume of water flowing through an *aquifer* in a certain time, the vertical drop of the flow, the flow distance, and the *permeability* of the aquifer.

decompression melting Melting that occurs when *mantle* material rises to an area of lower pressure at a mid-ocean ridge. As the mantle material rises and the pressure decreases below a critical point, solid *rocks* melt spontaneously, without the introduction of any additional heat. (Compare *fluid-induced melting.*)

deflation The removal of dust, *silt,* and *sand* from dry *soil* by strong winds that gradually scoop out shallow depressions in the ground.

delta A depositional platform built of *sediments* deposited in an ocean or lake at the mouth of a *stream.*

dendritic drainage An irregular *stream drainage network* that resembles the limbs of a branching tree.

density The mass per unit volume of a substance, commonly expressed in grams per cubic centimeter (g/cm^3). (Compare *specific gravity.*)

depositional remanent magnetization A weak magnetization created in *sedimentary rocks* by the parallel alignment of magnetic particles in the direction of Earth's *magnetic field* as they settle.

desert pavement A remanent ground surface of *gravel* too large for the wind to transport, left when continued *deflation* removes the finer-grained particles from a mixture of *gravel, sand,* and *silt* in *sediments* and *soils.*

desert varnish A distinctive dark brown, sometimes shiny coating found on many *rock* surfaces in the desert that is a mixture of *clay minerals* with smaller amounts of manganese and iron oxides. Desert varnish is hypothesized to form very slowly from a combination of dew, *chemical weathering,* and the adhesion of windblown dust to exposed rock surfaces.

desertification Changes in a region's *climate* that transform semiarid lands into deserts.

diagenesis The physical and chemical changes—including pressure, heat, and *chemical reactions*—by which buried *sediments* are lithified into *sedimentary rocks.*

diatreme A volcanic vent formed by the explosive escape of gases and often filled with breccia.

differentiation The transformation of random chunks of primordial matter into a body whose interior is divided into concentric layers that differ both physically and chemically.

dike A tabular igneous intrusion that cuts across layers of *bedding* in *country rock.*

diorite A coarse-grained *intrusive igneous rock* with composition intermediate between *granite* and *gabbro.* The intrusive equivalent of *andesite.*

dip The amount of tilting of a *rock* layer; the angle at which the bed inclines from the horizontal.

dipole Two oppositely polarized magnetic poles.

dip-slip fault A fault on which there has been relative movement of the blocks up or down the dip of the fault plane.

discharge (1) The volume of *groundwater* leaving an *aquifer* in a given time. (Compare *recharge.*) (2) The volume of water that passes a given point in a given time as it flows through a *channel* of a certain width and depth.

discordant intrusion An *intrusive igneous rock* that cuts across the layers of the *country rock* that it intrudes. (Compare *concordant intrusion.*)

distributary A smaller *stream* that receives water and *sediment* from the main *channel,* branches off downstream, and thus distributes the water and sediment into many channels. (Compare *tributary.*)

divide A ridge of high ground along which all rainfall runs off down one side of the rise or the other.

dolostone An abundant *carbonate rock* composed primarily of dolomite and formed by the *diagenesis* of *carbonate sediments* and *limestones.*

dome A broad circular or oval upward bulge of *rock* layers.

drainage basin An area of land, bounded by *divides,* that funnels all its water into the network of *streams* draining the area.

drainage network The pattern of connections of tributaries, large and small, of a *stream* system.

drift All material of glacial origin found anywhere on land or at sea.

drumlin A large streamlined hill of *till* and bedrock that parallels the direction of ice movement in a *continental glacier* terrain.

dry wash A desert *valley* that carries water only briefly after a rain. Called a *wadi* in the Middle East.

ductile material A material that undergoes smooth and continuous plastic deformation under increasing force and does not spring back to its original shape when the deforming force is released. (Compare *brittle material.*)

dune An elongated ridge of *sand* formed by wind or water.

Earth system All the parts of our planet and all their interactions, taken together.

earthquake The violent motion of the ground that occurs when *rocks* being stressed suddenly break along a new or pre-existing fault.

eclogite A *metamorphic rock* formed under very high pressure and moderate to high temperature, typically containing *minerals* such as coesite (a very dense, high-pressure form of quartz).

ecosystem An organizational unit composed of organisms and geologic components that function in a balanced, interrelated fashion.

El Niño An anomalous warming of the eastern tropical Pacific Ocean that occurs every 3 to 7 years and lasts for a year or so. El Niño events can be calamitous for local *ecosystems,* which depend on upwellings of cold water for their nutrient supply.

elastic rebound theory A theory of fault movement and *earthquake* generation holding that faults remain locked while strain energy accumulates in the *rock formations* on both sides, temporarily deforming them until a sudden slip along the fault releases the energy.

electron A negatively charged particle that moves around the *nucleus* of an *atom.*

electron sharing The mechanism by which a *covalent bond* is formed between the elements in a *chemical reaction.*

electron transfer The mechanism by which an *ionic bond* is formed between the elements in a *chemical reaction.*

elevation The altitude, or vertical distance above or below sea level.

enhanced greenhouse effect Global warming caused by the human-induced increase of atmospheric carbon dioxide (CO_2).

environment The complex of physical, chemical, and biological factors (such as climate, soil, and living things) that act upon an organism or an ecological community and ultimately determine its form and survival.

eolian Pertaining to geological processes powered by the wind.

eon The largest division of geologic time, embracing several *eras;* for example, the Phanerozoic eon, from 543 Ma to the present.

epeirogeny The gradual downward and upward movements of broad regions of *crust* without significant *folding* or *faulting.*

epicenter The geographic point on Earth's surface directly above the *focus* of an *earthquake.*

epoch A subdivision of a geologic *period,* often chosen to correspond to a stratigraphic sequence. Also used for a division of time corresponding to a paleomagnetic interval.

era A division of geologic time including several *periods,* but smaller than an *eon.* Commonly recognized eras are the Paleozoic, Mesozoic, and Cenozoic.

erosion The set of processes that loosen *soil* and *rock* and move them downhill or downstream, where they are deposited as layers of *sediment.*

esker A long, narrow, winding ridge of *sand* and *gravel* found in the middle of a ground *moraine.*

Eukarya One of three domains of organisms that includes all eukaryotes. All organisms within the Eukarya share a common origin and are distinct from the Bacteria and Archaea. The Eukarya include animals, plants, and fungi, most of which are multicellular.

eukaryote An organism with complex cells, in which the genetic material is organized within distinct nuclei. Eukaryotes include animals, plants, and fungi, most of which are multicellular.

evaporite rock A *sedimentary rock* formed from *evaporite sediments.*

evaporite sediment An accumulation of materials *precipitated* inorganically from evaporating seawater and from water in arid-region lakes that have no *river* outlets.

evolution The process by which organisms change through time, driven by the process of *natural selection.*

exfoliation A *physical weathering* process in which large flat or curved sheets of *rock* fracture and are detached from an *outcrop.*

exhumation A process in which subducted *metamorphic rocks* rise to the surface because of buoyancy and circulation in the *subduction* zone.

exoplanet A planet that orbits a star other than the Sun.

extinct No longer in existence.

extremophile A *microbe* that lives in *environments* that would kill any other organisms.

extrusive igneous rock A fine-grained or glassy *igneous rock* formed from a rapidly cooled magma that erupts at the surface through a *volcano.*

fault mechanism The type of fault rupture (normal, thrust, or strike-slip) that produced an *earthquake;* it is determined by the orientation of the fault rupture and the direction of slip.

fault slip The distance of the displacement between two blocks that occurs during an *earthquake* when elastic rebound causes the blocks on either side of a fault to spring back to their undeformed state.

faulting The processes by which tectonic forces cause the lithosphere to break and slip along a fault.

faunal succession, principle of The principle that the layers of *sedimentary rocks* in an *outcrop* contain *fossils* in a definite sequence.

felsic rock A light-colored *igneous rock* that is poor in iron and magnesium and rich in high-silica *minerals* such as quartz, orthoclase feldspar, and plagioclase feldspar. (Compare *mafic rock, ultramafic rock.*)

fissure eruption A volcanic eruption emanating from an elongate fissure rather than a central vent.

fjord A former glacial valley with steep walls and a U-shaped profile, now occupied by the sea.

flake tectonics The unique geology of Venus, which seems to be dominated by vigorous *convection* currents in the mantle beneath a *crust* formed of a very thin layer of frozen lava. As these convection currents push and stretch the surface, the crust breaks up into flakes or crumples like a rug. As the *mantle* of Venus moves around, blobs of hot lava bubble up to form large landmasses, mountains, and volcanic deposits. (Compare *plate tectonics.*)

flexural basin A *basin* that develops within zones of tectonic convergence, where one lithospheric plate pushes up over the other and the weight of the overriding plate causes the overridden plate to bend or flex downward.

flood basalt An immense *basaltic lava plateau* extending many kilometers in flat, layered flows originating from *fissure eruptions.*

floodplain A flat area about level with the top of a *channel* that lies on either side of the channel. It is the part of the *valley* that is flooded when a *stream* spills over its banks, carrying with it *silt* and *sand* from the main channel.

fluid-induced melting   Melting that takes place when water-laden *sediments* on a subducting oceanic plate are carried downward into the *subduction* zone. The increase in pressure squeezes water out of the *minerals* in the outer layers of the descending slab. The water rises buoyantly into the *mantle* wedge above the slab. Because water lowers the melting temperature of *rock,* it induces melting in the mantle wedge. (Compare *decompression melting.*)

focus   The point along a fault at which slip initiates an *earthquake.*

folding   The processes by which crustal forces deform an area of *crust* so that layers of *rock* are pushed into folds.

foliated rock   A *metamorphic rock* that displays *foliation.* Foliated rocks include *slate, phyllite, schist,* and *gneiss.* (Compare *granoblastic rock.*)

foliation   A set of flat or wavy parallel planes produced by deformation.

foraminifer   One of a group of tiny single-celled organisms that live in surface waters and whose secretions and calcite shells account for most of the ocean's *carbonate sediments.*

foraminiferal ooze   A sandy and silty *sediment* composed of the shells of dead *foraminifers.*

foreset bed   A gently inclined deposit of fine-grained *sand* and *silt,* resembling large-scale cross-beds, on the outer front of a *delta.*

foreshock   A small *earthquake* that occurs in the vicinity of, but before, a main shock.

formation   A distinctive set of *rock* layers in a region that can be recognized and mapped as a unit, having the same physical properties and possibly containing the same assemblage of *fossils.*

fossil   Trace of an organism of past geologic ages that has been preserved in the *crust.*

fossil fuel   An energy *resource* derived from natural organic materials, from *coal* to *oil* and *natural gas.*

fractional crystallization   The process by which the *crystals* formed in a cooling magma are segregated from the remaining liquid at progressively lower temperatures.

fracture   The tendency of a *crystal* to break along irregular surfaces other than *cleavage* planes.

frost wedging   A *physical weathering* process in which the expansion of freezing water in a crack breaks a *rock.*

gabbro   A dark gray, coarse-grained *intrusive igneous rock* with an abundance of mafic *minerals,* particularly pyroxene. The intrusive equivalent of *basalt.*

gas   A fluid *organic sediment* formed by the *diagenesis* of organic material in the pores of *sedimentary rocks,* mainly sandstones and *limestones.*

gene   A large molecule within the cells of every organism encoding all the information that determines what the organism will look like, how it will live and reproduce, and how it differs from all other organisms.

geobiology   The study of how organisms have influenced and been influenced by Earth's *environment.*

geochemical reservoir   One of the components of the *Earth system*—atmosphere, hydrosphere, *lithosphere,* and *biosphere*—that is a *reservoir* for holding terrestrial chemicals.

geodesy   The ancient science of measuring the shape of the Earth and locating points on its surface.

geodynamo   The *geosystem* that sustains Earth's *magnetic field,* driven by *convection* in the *outer core.*

geologic cross section   A diagram showing the geologic features that would be visible if vertical slices were made through part of the *crust.*

geologic map   A map representing the *rock formations* exposed at Earth's surface.

geologic time scale   (1) A timeline based on a *stratigraphic succession* that provides a chronological record of the history of a region. (2) The entire span of time since the Earth formed.

geomorphology   The study of landscapes and their evolution.

geosystem   A specialized subsystem of the *Earth system* that encompasses specific types of terrestrial behavior. Geosystems include the *plate tectonic system,* the *climate system,* the *geodynamo,* and other smaller subsystems. The Earth system can be thought of as the collection of all these open, interacting (and often overlapping) geosystems.

geotherm   The curve that describes how temperature increases with depth in the Earth.

geothermal energy   Energy produced when underground water is heated as it passes through a subsurface region of hot *rocks.*

glacier   A large mass of ice on land that shows evidence of being in motion or of once having moved. (See also *continental glacier; valley glacier.*)

global change   A change in *climate* that has worldwide effects on the *biosphere,* atmosphere, and other components of the *Earth system.*

gneiss   A light-colored, poorly *foliated rock* with coarse bands of segregated light and dark *minerals* throughout, produced by high-pressure, high-temperature *metamorphism.*

graded bedding   A bed that formed horizontal or nearly horizontal layers at the time of deposition, in which the coarsest particles are concentrated at the bottom and grade gradually upward into fine *silt*.

graded stream   A *stream* in which the slope, velocity, and *discharge* combine to transport its *sediment* load, with neither sedimentation nor *erosion*.

granite   A felsic, coarse-grained, *intrusive igneous rock* composed of quartz, orthoclase feldspar, sodium-rich plagioclase feldspar, and micas. The intrusive equivalent of *rhyolite*.

granite-greenstone terrain   Areas of massive granite intrusions in Archean cratons that surround smaller pockets of deformed, metamorphosed volcanic rocks (greenstones), primarily of mafic composition.

granoblastic rock   A nonfoliated *metamorphic rock* composed mainly of *crystals* that grow in equant shapes, such as cubes and spheres, rather than in platy or elongate shapes. Granoblastic rocks include *hornfels, quartzite, marble, greenstone, amphibolite,* and *granulite*.

granodiorite   A light-colored, coarse-grained *intrusive igneous rock* similar to *granite* in having abundant quartz, but whose predominant feldspar is plagioclase, not orthoclase. The intrusive equivalent of *dacite*.

granulite   A medium- to coarse-grained *granoblastic rock* formed under conditions of relatively high pressure and temperature.

gravel   The coarsest *clastic sediment,* consisting of particles larger than 2 mm in diameter and including pebbles, cobbles, and boulders.

graywacke   A sandstone composed of a heterogeneous mixture of *rock* fragments and angular grains of quartz and feldspar, the sand grains being surrounded by a fine-grained *clay* matrix.

greenhouse effect   A global warming effect in which a planet with an atmosphere containing *greenhouse gases* radiates *solar energy* back into space less efficiently than it would without an atmosphere. The net effect is to trap heat within the atmosphere by increasing the temperature of the surface relative to the temperature at higher levels of the atmosphere.

greenhouse gas   A gas that absorbs heat emitted by the Earth and traps it in the atmosphere. Greenhouse gases include water vapor, carbon dioxide, and methane.

greenschist   A *schist* containing chlorite and epidote (which are green) and formed by low-pressure, low-temperature *metamorphism* of *mafic rocks*.

greenstone   A *granoblastic rock* produced by the *low-grade metamorphism* of *mafic rock*. Chlorite accounts for the greenish cast.

groundwater   The mass of water stored beneath Earth's surface.

groundwater table   The boundary between the *unsaturated zone* and the *saturated zone*.

guyot   A large, flat-topped *seamount* resulting from *erosion* of an island *volcano* when it was above sea level.

habitable zone   In the context of astronomy, a habitable zone is the space surrounding a star where water is stable as a liquid. If a planet is within the habitable zone, there is a chance that life could exist there.

half-life   The time required for one-half of the original number of radioactive *atoms* in an element to decay.

hanging valley   A *valley* left by a melted *tributary glacier* that enters a larger glacial valley above its base, high up on the valley wall.

hardness   A measure of the ease with which the surface of a *mineral* can be scratched.

Heavy Bombardment   A time early in the history of the solar system when crater-forming impacts were very frequent.

hematite   The principal iron *ore;* the most abundant iron oxide at Earth's surface.

heterotroph   An organism that gets its food by feeding directly or indirectly on *autotrophs*.

high-grade metamorphic terrain   Areas of high-grade (granulite facies) metamorphic rocks in Archean cratons, derived primarily from the compression, burial, and subsequent erosion of ancient granitic crust.

high-pressure (ultra-high-pressure) metamorphism   *Metamorphism* occurring at high pressure (8 to 12 kbar) and ultrahigh pressure (greater than 28 kbar).

hogback   A *formation* similar to a *cuesta* with steep, narrow, more or less symmetrical ridges.

hornfels   A *granoblastic rock* of uniform grain size that has undergone little or no deformation. Usually formed by *contact metamorphism* at high temperatures.

hot spot   A volcanic center found at the beginning of progressively older aseismic ridges or within a continent far from a plate boundary. Hypothesized to be the surface expression of a *mantle plume*.

Hubbert's peak   A bell-shaped curve developed by M. King Hubbert predicting that U.S. *oil* production would begin to decline sometime in the early 1970s.

humus   The remains and waste products of the many plants, animals, and bacteria living in a *soil*.

hurricane   A great storm that forms over tropical parts of Earth's oceans, between 8° and 20° latitude, in areas of high humidity, light winds, and warm sea-surface temperatures. Hurricanes produce winds of at least 119 km/hour (74 miles/hour) and large amounts of rainfall.

hurricane intensity scale   A scale used to estimate the potential property damage and flooding expected along the coast from *hurricane* landfall. It is analogous to the Mercalli scale for *earthquakes*.

hydration   The absorption of water by a *mineral,* usually in *weathering.*

hydraulic gradient   The ratio between the difference in *elevation* of the *groundwater table* at two points and the flow distance that the water travels between those two points.

hydrologic cycle   The cyclical movement of water from the ocean to the atmosphere by evaporation, to the surface through rain, to *streams* through *runoff* and *groundwater,* and back to the ocean.

hydrology   The science that studies the movements and characteristics of water on and under Earth's surface.

hydrothermal activity   The circulation of water through hot volcanic *rocks* and magmas, producing hot springs and geysers on the surface.

hydrothermal solution   A hot solution formed around bodies of molten rock when circulating *groundwater* comes into contact with a hot intrusion, reacts with it, and carries off significant quantities of elements and ions released by the reaction. These elements and their ions then interact with one another to deposit ore minerals, usually as the fluid cools.

hydrothermal vein   A *vein* filled with *minerals* that contain large amounts of chemically bound water and are known to crystallize from hot-water solutions.

hydrothermal water   Hot water in the *crust.*

ice age   A time interval during which Earth cools and water is transferred from the hydrosphere to the cryosphere. The amount of sea ice increases, and more snow falls on the continents in winter than melts in summer, increasing the volume of the continental ice sheets and decreasing sea level. As the polar ice caps expand, they reflect more *solar energy* back into space, and the surface temperatures fall further.

ice shelf   Sheets of ice floating on the ocean that are attached to a continental glacier on land; common on the periphery of Antarctica.

ice stream   A current of ice in a *continental glacier* that flows faster than the surrounding ice.

iceberg calving   The process by which pieces of ice break off and form icebergs when a glacier reaches a shoreline.

igneous rock   A *rock* formed by the solidification of a magma, before or after it reaches the surface.

infiltration   The movement of *groundwater* or *hydrothermal water* into *rock* or *soil* through pores and *joints.*

inner core   The central part of the Earth from a depth of 5150 to 6370 km, composed of iron and nickel. A solid metallic sphere with a radius of 1220 km—about two-thirds the size of the Moon—suspended within the liquid *outer core.*

intensity scale   A scale for estimating the intensity of *earthquake* shaking directly from an event's destructive effects.

intermediate igneous rock   An *igneous rock* midway in composition between mafic and felsic, neither as rich in silica as the *felsic rocks* nor as poor in it as the *mafic rocks.*

intrusive igneous rock   A coarse-grained *igneous rock* that crystallized slowly when magma intruded into *country rock* deep in Earth's *crust.*

ionic bond   A bond between *atoms* formed by electrical attraction between ions of the opposite charge. (Compare *covalent bond.*)

iron formation   A *sedimentary rock* that usually contains more than 15 percent iron in the form of iron oxides and some iron silicates and iron carbonates.

island arc   A linear or arc-shaped chain of volcanic islands formed on the seafloor at a convergent plate boundary. The islands are formed in the overriding plate from rising melt derived from *fluid-induced melting* of the mantle wedge above the downgoing lithospheric slab.

isochron   A *contour* that connects points of equal age on the world's ocean floors as determined by magnetic reversal data and *fossils* from deep-sea drilling.

isostasy, principle of   The principle that the buoyancy force that pushes a lower-density body (such as a continent or an iceberg) floating in a higher-density medium (such as the *asthenosphere* or seawater) upward must be balanced by the gravitational force that pulls it downward.

isotope   One of two or more species of *atoms* of the same *atomic number* that have differing *atomic masses.*

isotopic dating   The use of naturally occurring radioactive elements to determine the ages of *rocks.*

joint   A crack in a *rock* along which there has been no appreciable movement.

kaolinite   The white to cream-colored *clay* produced by the *weathering* of feldspar.

karst topography   An irregular hilly terrain characterized by *sinkholes,* caverns, and a lack of surface *streams;* formed in regions of high rainfall with extensively jointed *limestone formations* and an appreciable *hydraulic gradient.*

kettle A hollow or undrained depression that often has steep sides and may be occupied by ponds or lakes; formed in glacial deposits when *outwash* was deposited around a residual block of ice that later melted.

La Niña An anomalous cooling of the eastern tropical Pacific Ocean, characterized by stronger trade winds, that is complementary to and sometimes follows *El Niño* episodes.

lahar A torrential mudflow of wet volcanic debris produced when pyroclastic or lava deposits mix with rain or the water of a lake, *river,* or melting *glacier.*

laminar flow A flow in which straight or gently curved streamlines run parallel to one another without mixing or crossing between layers. (Compare *turbulent flow.*)

landform A characteristic landscape feature on Earth's surface that attained its shape through the processes of *erosion* and sedimentation; for example, a hill or a *valley.*

large igneous province (LIP) A voluminous emplacement of predominantly mafic *extrusive* and *intrusive igneous rock* whose origins lie in processes other than "normal" *seafloor spreading.* LIPs include continental *flood basalts,* ocean basin flood basalts, and aseismic ridges.

limestone A biochemical *sedimentary rock* lithified from *carbonate sediments* and composed mainly of calcium carbonate in the form of the *mineral* calcite.

liquefaction The temporary transformation of solid material to a fluid state caused by the pressure of water in the pores of the material.

lithic sandstone A sandstone that contains many fragments derived from fine-grained rocks, mostly *shales,* volcanic rocks, and fine-grained *metamorphic rocks.*

lithification The process that converts *sediments* into solid *rock* by *compaction* or *cementation.*

lithosphere The strong, rigid outer shell of the Earth that encases the *asthenosphere* and contains the *crust* and the uppermost part of the *mantle* down to an average depth of about 100 km and forms the rigid plates.

loess A blanket of unstratified, wind-deposited, fine-grained *sediment* rich in *clay minerals.*

longitudinal profile The smooth, concave-upward curve that represents a cross-sectional view of a *stream,* from notably steep near its head to almost level near its mouth.

longshore current A shallow-water current that runs parallel to the shore.

lower mantle A relatively homogeneous region, beginning nearly 660 km below the surface and more than 2000 km thick,

where *S-wave* speed abruptly increases, indicating a major *phase change* to a more closely packed *crystal* structure.

low-grade (burial) metamorphism *Metamorphism* in which buried *sedimentary rocks* are altered by the progressive increase in pressure exerted by overlying *sediments* and sedimentary rocks and by the increase in heat associated with increased depth of burial in the Earth.

low-velocity zone A layer near the base of the *lithosphere* where the *S-wave* speed abruptly decreases.

lunar highlands The light-colored mountainous landscape that covers about 80 percent of the surface of the Moon. Because numerous *craters* are found in these highlands, they are known to be older than the *lunar maria.*

lunar maria The dark-colored, low-lying plains of the Moon, named after the Latin word for "seas" (*maria*). Because few craters are found on these plains, they are known to be younger than the *lunar highlands.*

luster The way in which the surface of a *mineral* reflects light to produce the shine of its surface, described by such subjective terms as dull, glassy, or metallic. (See Table 3.3.)

mafic rock A dark-colored *igneous rock* containing *minerals* (such as pyroxenes and olivines) rich in iron and magnesium and relatively poor in silica. (Compare *felsic rock, ultramafic rock.*)

magma chamber A magma-filled cavity in the *lithosphere* that forms as ascending drops of melted *rock* push aside surrounding solid rock.

magma ocean An outer molten layer hundreds of kilometers thick that surrounded Earth after it collided with a Mars-sized body about 4.5 Ga.

magmatic differentiation A process by which *rocks* of varying composition can arise from a uniform parent magma. Various *minerals* crystallize at different temperatures, and the composition of the magma changes as it is depleted of the chemical elements withdrawn to make the crystallized minerals.

magnetic anomaly A pattern of long, narrow bands of high and low *magnetic fields* on the seafloor that are parallel to and almost perfectly symmetrical with respect to the crest of a mid-ocean ridge. The detection of such patterns was one of the great discoveries that confirmed *seafloor spreading* and led to *plate tectonics* theory.

magnetic field The region of influence of a magnetized body or an electric current.

magnetic stratigraphy The time sequence of reversals of Earth's *magnetic field* as indicated in the *fossil* magnetic record of layered lava flows.

magnetic time scale The detailed history of Earth's *magnetic-field* reversals going back into geologic time, as determined by measuring the *thermoremanent magnetization* of *rock* samples.

magnitude scale   A scale for estimating the size of an *earth-quake* using the logarithm of the largest ground motion registered by a *seismograph* (Richter magnitude) or the logarithm of the area of the fault rupture (moment magnitude).

mantle   The region that forms the main bulk of the solid Earth, between the *crust* and the *core,* ranging from depths of about 40 km to 2900 km. It is composed of *rocks* of intermediate *density,* mostly compounds of oxygen with magnesium, iron, and silicon.

mantle plume   A narrow, cylindrical jet of hot, solid material rising from deep within the *mantle* and thought to be responsible for intraplate *volcanism.*

marble   A *granoblastic rock* produced by the *metamorphism* of *limestone* or *dolostone.*

mass extinction   A short interval during which many species simply disappear from the geologic record.

mass movement   A downhill movement of masses of *soil* or *rock* under the force of gravity.

mass wasting   All the processes by which masses of *rock* and *soil* move downhill under the influence of gravity.

meander   A curve or bend in a *stream* that develops as the stream erodes the outer bank of a bend and deposits *sediment* against the inner bank. Meanders are usual in streams flowing on low slopes in plains or lowlands, where *channels* typically cut through *unconsolidated* sediments or easily eroded bedrock.

mélange   A heterogeneous mixture of *rock* materials produced by high-pressure, low-temperature *metamorphism,* found where a plate carrying a continent on its leading edge converges with a subducting oceanic plate.

mesa   A small, flat elevation with steep slopes on all sides created by differential *weathering* of bedrock of varying hardness.

metabolism   The process that all organisms use to convert inputs (such as sunlight, water, and carbon dioxide) to outputs (such as oxygen and carbohydrates).

metallic bond   A *covalent bond* in which freely mobile *electrons* are shared and dispersed among ions of metallic *elements,* which have the tendency to lose electrons and pack together as *cations.*

metamorphic facies   Groupings of *metamorphic rocks* of various *mineral* compositions formed under different grades of *metamorphism* from different parent rocks.

metamorphic P-T path   The history of the changing conditions of pressure (P) and temperature (T) that occurred during *metamorphism.* The P-T path can be a sensitive recorder of many important factors that influence metamorphism and thus may provide insight into the plate tectonic settings responsible for metamorphism.

metamorphic rock   A *rock* formed by the transformation of preexisting solid rocks under the influence of high pressure and temperature.

metamorphism   The transformation of preexisting solid *rocks* under the influence of high pressure and temperature.

metasomatism   A type of metamorphism in which a rock's bulk composition changes only because of fluid transport of chemical substances into or out of the rock.

meteoric water   Rain, snow, or other forms of water derived from the atmosphere.

meteorite   A solid particle from interplanetary space that survives the flight through Earth's atmosphere and falls to the ground in one or more pieces. Most meteorites that strike Earth are tiny pieces of *asteroids* ejected during collisions with one another.

microbe   A single-celled organism. Microbes include bacteria, some fungi and algae, and protozoa.

microbial mat   A layered microbial community commonly occurring in tidal flats, hypersaline lagoons, and thermal springs.

microfossil   A *fossil* that can only be studied microscopically. The oldest microfossils that preserve possible morphologic evidence for life are tiny threads similar in size and appearance to modern *microbes,* encased in *chert.*

migmatite   A mixture of *igneous* and *metamorphic rock* produced by incomplete melting. Migmatites are badly deformed and contorted, and they are penetrated by many *veins,* small pods, and lenses of melted rock.

Milankovitch cycle   An astronomical cycle that causes periodic variations in the amount of heat Earth receives from the Sun. Such cycles include the eccentricity of Earth's orbit; the tilt of Earth's axis of rotation; and precession, Earth's wobble about its axis of rotation.

mineral   A naturally occurring, solid crystalline substance, generally inorganic, with a specific chemical composition.

mineralogy   (1) The branch of geology that studies the composition, structure, appearance, stability, occurrence, and associations of *minerals.* (2) The relative proportions of a *rock's* constituent minerals.

Mohorovičić discontinuity (Moho)   The boundary between the *crust* and the *mantle,* at a depth of 5 to 45 km, marked by an abrupt increase in *P-wave* velocity to more than 8 km/s.

Mohs scale of hardness   An empirical, ascending scale of mineral *hardness* based on the ability of one *mineral* to scratch another. (See Table 3.2.)

moraine   An accumulation of rocky, sandy, and clayey material carried by glacial ice and deposited as *till.*

mud   A *clastic sediment,* mixed with water, in which most of the particles are less than 0.062 mm in diameter.

mudstone   A blocky, poorly bedded, fine-grained *sedimentary rock* produced by the *lithification* of *mud.*

natural gas Methane gas, CH_4, the simplest hydrocarbon.

natural levee A ridge of coarse material that confines a *stream* within its banks between floods, even when water levels are high.

natural selection The adaptation of populations of organisms to new *environments*. Natural selection is responsible for the development of new species.

nebular hypothesis The idea that the solar system originated from a diffuse, slowly rotating cloud of gas and fine dust (a "nebula") that contracted under the force of gravity and eventually evolved into the Sun and planets.

negative feedback A process in which one action produces an effect (the feedback) that tends to slow the original action and stabilize the process at a lower rate. (Compare *positive feedback*.)

neutron An electrically neutral elementary particle in the *nucleus* of an *atom,* having an *atomic mass* of 1.

nonrenewable energy source An energy source produced by geological processes at a rate much slower than the rate at which we are using it up; for example, fossil fuels. (Compare *renewable energy source*.)

normal fault A *dip-slip fault* in which the rocks above the fault plane move down relative to the rocks below the fault plane, extending the structure horizontally.

nuclear energy Energy produced by the fission of uranium-235, which can be used to heat steam and drive turbines to create electricity.

nucleus The center of an *atom,* comprising *protons* and *neutrons* and containing virtually all the mass of the atom.

obsidian A dense, dark, glassy volcanic *rock,* usually of felsic composition.

oil An organic fluid formed by the *diagenesis* of buried organic material that migrates into reservoirs in porous crustal rocks.

oil shale A fine-grained, clay-rich *sedimentary rock* containing relatively large amounts of organic matter.

oil trap An impermeable barrier that blocks the upward migration of *oil* or *gas,* allowing them to collect beneath the barrier.

oil window A limited range of pressures and temperatures, usually located at depths between about 2 and 5 km, at which *crude oil* forms.

ophiolite suite An unusual assemblage of *rocks,* characteristic of the seafloor but found on land, consisting of deep-sea *sediments,* submarine *basaltic lavas,* and mafic igneous intrusions. The assemblage comprises fragments of oceanic *crust* that were transported by *seafloor spreading* and then raised above sea level and thrust onto a continent in a later episode of plate collision.

ore A *mineral* deposit from which valuable metals can be recovered profitably.

organic sedimentary rock A *sedimentary rock* that consists entirely or partly of organic carbon-rich deposits formed by the decay of once-living material that has been buried.

original horizontality, principle of The principle that *sediments* are deposited as essentially horizontal beds.

orogen An elongated mountain belt arrayed around a continental *craton* and formed by a later episode of compressive deformation.

orogeny Mountain building, particularly by the *folding* and thrusting of *rock* layers, often with accompanying magmatic activity.

outer core A liquid zone—composed of iron, nickel, and minor amounts of some lighter element, such as oxygen or sulfur—below the *mantle,* from a depth of 2890 to 5190 km, that surrounds a solid iron-nickel *inner core.*

outwash *Drift* that has been caught up and modified, sorted, and distributed by meltwater *streams.*

oxbow lake A crescent-shaped, water-filled loop created in the former path of a *stream* when it abandons a *meander* and takes a new course.

P wave The first *seismic wave* to arrive from the *focus* of an *earthquake;* a type of *compressional* wave.

paleomagnetism The remanent magnetization recorded in ancient *rocks,* which allows the reconstruction of Earth's ancient *magnetic field* and the positions of continents and supercontinents.

Pangaea A supercontinent that coalesced in the latest Paleozoic *era* and comprised all present continents. The breakup of Pangaea began in Mesozoic time, as inferred from paleomagnetic and other data.

partial melting The incomplete melting of a *rock* that occurs because the *minerals* that compose it melt at different temperatures.

passive margin A *continental margin* far from a plate margin, with no active *volcanoes* and few *earthquakes.* (Compare with *active* margin.)

peat A rich organic material that contains more than 50 percent carbon.

pediment A broad, gently sloping platform of bedrock left behind as a mountain front erodes and retreats from its *valley.*

pegmatite A *vein* of extremely coarse grained *granite,* crystallized from a water-rich magma in the late stages of solidification, that cuts across a much finer grained *country rock.* Pegmatites provide *ores* of many rare elements, such as lithium and beryllium.

pelagic sediment　An open-ocean *sediment* composed of fine-grained terrigenous and biochemically *precipitated* particles that slowly settle from the surface to the bottom.

peridotite　A coarse-grained *ultramafic intrusive igneous rock* composed of olivine with small amounts of pyroxene and amphibole. The dominant rock in Earth's *mantle* and the source rock of basaltic melts.

period　A division of geologic time representing one subdivision of an *era.*

permafrost　A permanently frozen aggregate of ice and *soil* occurring in very cold regions. Any *rock* or soil remaining at or below 0°C for 2 or more years.

permeability　The ability of a solid to allow fluids to pass through it.

phase change　A transition in a *rock*'s *mineralogy* (but probably not its composition) to a denser, more closely packed *crystal* structure, signaled by a change in *seismic-wave* velocity.

phosphorite　A chemical or biochemical *sedimentary rock* composed of calcium phosphate *precipitated* from phosphate-rich seawater and formed diagenetically by the interaction between muddy or *carbonate sediments* and the phosphate-rich water.

photosynthesis　The process by which organisms such as green plants and algae use energy from sunlight to convert water and carbon dioxide to carbohydrates and oxygen.

phyllite　A *foliated rock* that is intermediate in grade between *slate* and *schist.* Small *crystals* of mica and chlorite give it a more or less glossy sheen.

physical weathering　*Weathering* in which solid *rock* is fragmented by mechanical processes that do not change its chemical composition.

planetesimal　Any of numerous kilometer-sized solid celestial bodies that aggregated through the clumping by gravitational attraction of dust and smaller chunks of material at an early stage of the development of the solar system.

plastic flow　The total of all the small movements of the ice *crystals* that make up a *glacier,* resulting in a large movement of the whole mass of ice. (Compare *basal slip.*)

plate tectonics　The theory proposing that the *lithosphere* is broken into about a dozen large plates that move over Earth's surface. Each plate acts as a distinct rigid unit that rides on the *asthenosphere,* which also is in motion. The theory attempts to explain seismicity, *volcanism,* mountain building, and evidence of *paleomagnetism* in terms of these plate motions. (Compare *flake tectonics.*)

plate tectonic system　A *geosystem* that includes all parts of the *Earth system* and all the interactions among these components needed to describe how *plate tectonics* works in space and time.

plateau　A large, broad, flat area of appreciable *elevation* above the neighboring terrain.

playa　A flat bed of *clay,* sometimes encrusted with *precipitated* salts, that forms by the complete evaporation of a *playa lake.*

playa lake　A permanent or temporary lake that occurs in arid mountain *valleys* or *basins.* As the lake water evaporates, dissolved *weathering* products are concentrated and gradually *precipitated.*

pluton　A large igneous intrusion ranging in size from a cubic kilometer to hundreds of cubic kilometers, formed at depth in the *crust.*

point bar　A curved sandbar deposited along the inside bank of a *stream,* where the current is slower.

polymorph　One of two or more alternative possible *crystal* structures for a single chemical compound; for example, the *minerals* quartz and cristobalite are polymorphs of silica (SiO_2).

porosity　The percentage of a *rock*'s volume consisting of open pores between grains.

porphyroblast　A large *crystal* surrounded by a much finer grained matrix of other *minerals* in a *metamorphic rock.*

porphyry　A lava of mixed *texture* in which large *crystals* (phenocrysts) "float" in a predominantly fine crystalline matrix.

positive feedback　A process in which one action produces an effect (the feedback) that tends to speed up the original action and stabilize the process at a faster rate. (Compare *negative feedback.*)

potable water　Water that tastes agreeable and is not dangerous to health.

pothole　A hemispherical hole in the bedrock of a streambed, formed by *abrasion* of small pebbles and cobbles in a strong current.

precipitate　(1) (verb) To drop out of a saturated solution as *crystals.* (2) (noun) The crystals that drop out of a saturated solution.

precipitation　(1) A deposit on the Earth of hail, mist, rain, sleet, or snow. Precipitation is generated in clouds, which reach a point of saturation; at this point larger and larger droplets form, which then fall to Earth under gravity. (2) The condensation of a solid from a solution during a chemical reaction.

pressure-temperature-time path　The changing regimes of pressure and temperature recorded by changes in the assemblages and chemical compositions of the *minerals* in regionally metamorphosed *rocks.*

principle of faunal succession See *faunal succession, principle of.*

principle of isostasy See *isostasy, principle of.*

principle of original horizontality See *original horizontality, principle of.*

principle of superposition See *superposition, principle of.*

principle of uniformitarianism See *uniformitarianism, principle of.*

proton An elementary particle in the *nucleus* of an *atom,* having an *atomic mass* of 1 and a positive electrical charge of +1.

pumice A frothy mass of volcanic glass with a great number of holes (vesicles) that remain after trapped gas has escaped from the solidifying melt.

pyroclast A volcanic *rock* fragment ejected into the air during an eruption.

quad A measure of energy consisting of 1 quadrillion British thermal units (Btu).

quartz arenite A sandstone made up almost entirely of quartz grains, usually well sorted and rounded.

quartzite A very hard, white *granoblastic rock* derived from quartz-rich *sandstones.*

radiation The relatively rapid development of new types of organisms that derive from a common ancestor.

rain shadow An area of low rainfall on the leeward slope of a mountain range.

recharge The *infiltration* of water into any subsurface *formation,* often by rain or snow meltwater from the surface.

recurrence interval The average time interval between the occurrence of two ruptures on the same fault that produce earthquakes of approximately equal magnitude.

red bed A bed of sandstones and shales that is red because of the presence of iron oxide cement that binds the grains together. The presence of red beds around 2.7 to 2.1 Ga indicates that oxygen must have been present in the atmosphere to precipitate the iron oxide cements.

reef A moundlike or ridgelike organic structure constructed of the carbonate skeletons of millions of organisms.

regional metamorphism *Metamorphism* caused by high pressures and temperatures that extend over large regions, as happens where plates collide.

rejuvenation Renewed uplift in a mountain chain on the site of earlier uplifts, returning the area to a more youthful stage.

relative age The age of one geologic event in relation to another. (Compare *absolute age.*)

relative humidity The amount of water vapor in the air, expressed as a percentage of the total amount of water the air could hold at that temperature if saturated. When the relative humidity is 50 percent and the temperature is 15°C, for example, the amount of moisture in the air is one-half the maximum amount the air could hold at 15°C.

relative plate velocity The velocity at which one plate moves relative to another plate.

relief The difference between the highest and lowest *elevations* in a particular area.

renewable energy source An energy source that can be renewed quickly enough to compensate for our use of it; for example, a forest chopped down for wood can be regrown and harvested again. (Compare *nonrenewable energy source.*)

reserve A deposit that has already been discovered and can be mined economically and legally at the present time. (Compare *resource.*)

reservoir (1) A place in which water is stored, including the oceans; *glaciers* and polar ice; *groundwater;* lakes and *rivers;* the atmosphere; and the *biosphere.* (2) A *geochemical reservoir:* a place of residence for elements in a geochemical cycle.

residence time The average time that an *atom* of a particular element spends in a *reservoir* before leaving.

resource The entire amount of a given resource that may become available for use in the future, including *reserves,* plus discovered but currently unprofitable deposits, plus undiscovered deposits that geologists think they may be able to find eventually. (Compare *reserve.*)

respiration The process that allows an organism to release the energy stored within carbohydrates. All organisms use oxygen to burn, or respire, carbohydrates to create energy.

rhyolite A light-brown to gray, fine-grained *extrusive igneous rock* with a felsic composition. The extrusive equivalent of *granite.*

rhyolitic lava The lava that is richest in silica, making it the stickiest and least fluid kind of lava. It erupts at temperatures of only 600° to 800°C. (Compare *andesitic lava; basaltic lava.*)

rift valley A deep, narrow, elongate *basin* with thick successions of *sedimentary rocks* and also *extrusive* and *intrusive igneous rocks.* Current examples include the rift valleys of East Africa, the rifted valley of the Rio Grande, and the Jordan Valley in the Middle East.

ripple A *sedimentary structure* consisting of a very small *dune* of *sand* or *silt* whose long dimension is at right angles to the current.

river A major branch of a *stream* system.

rock A naturally occurring solid aggregate of minerals or, in some cases, nonmineral solid matter.

rock cycle The set of geologic processes that convert each type of *rock* into the other two types.

Rodinia A supercontinent that formed about 1.1 Ga and began to break up about 750 Ma.

runoff The sum of all rainwater that flows over the surface, including the fraction that may temporarily infiltrate near-surface *formations* and then flow back to the surface.

S wave The second *seismic wave* to arrive from the *focus* of an *earthquake;* a type of *shear* wave. S waves do not exist in liquids or gases.

salinity The total amount of dissolved substances in a given volume of water.

saltation The transportation of *sand* or fine *sediment* by currents of water or air in such a manner that they move along in a series of short intermittent jumps.

sand A *clastic sediment* consisting of medium-sized particles, ranging from 0.062 to 2 mm in diameter.

sand budget The inputs and outputs caused by *erosion* and sedimentation of a beach.

sandblasting The *erosion* of *rock outcrops,* boulders, and pebbles by *abrasion* caused by the impact of high-speed *sand* grains carried by the wind.

sandstone A clastic *rock* composed of grains of quartz, feldspar, and rock fragments, bound together by a cement of quartz, carbonate, or other *minerals,* or by a matrix of *clay* minerals. The lithified equivalent of *sand.*

saturated zone The level in the *groundwater table* in which the pores of the *soil* or *rock* are completely filled with water. (Compare *unsaturated zone.*)

schist A *metamorphic rock* characterized by the pervasive coarse, wavy *foliation* known as schistosity.

scientific method A general research plan, based on methodical observations and experiments, by which scientists propose and test hypotheses that explain some aspect of how the universe works.

seafloor metamorphism *Metamorphism* associated with mid-ocean ridges, in which changes in a *rock*'s bulk chemical composition are produced by fluid transport of chemical components into or out of the rock.

seafloor spreading The mechanism by which new seafloor is created along the rift at the crest of a mid-ocean ridge as adjacent plates move apart. The *crust* separates along the rift, and new seafloor forms as hot new crust upwells into these cracks. The new seafloor spreads laterally away from the rift and is replaced by even newer crust in a continuing process of plate creation.

seamount A submerged *volcano,* usually extinct, found on the seafloor.

sediment A material deposited at Earth's surface by physical agents (wind, water, and ice), chemical agents (*precipitation* from oceans, lakes, and *rivers*), or biological agents (organisms, living and dead).

sedimentary basin A region of considerable extent (at least 10,000 km^2) where the combination of deposition and *subsidence* has formed thick accumulations of *sediment* and *sedimentary rock.*

sedimentary environment A geographic location characterized by a particular combination of climate conditions and physical, chemical, and biological processes.

sedimentary rock A *rock* formed as the burial product of layers of *sediments* (such as *sand, mud,* and calcium carbonate shells), whether they were laid down on the land or under the sea.

sedimentary structure Any kind of *bedding* or other surface (such as *cross-bedding, graded bedding,* or *ripples*) formed at the time of deposition.

seismic hazard The intensity of shaking and ground disruption in an *earthquake* that can be expected over the long term at some specified location, expressed in the form of a seismic hazard map.

seismic risk The *earthquake* damage that can be expected over the long term for a specified region, such as a county or state, usually measured in terms of average dollar loss per year.

seismic tomography A method that uses the *seismic waves* from *earthquakes* recorded on thousands of *seismographs* all over the world to sweep Earth's interior in many different directions and construct a three-dimensional image of what's inside.

seismic wave A ground vibration produced by *earthquakes* or explosions. (See also *P wave; S wave; surface wave.*)

seismicity map A map that shows the *epicenters* of *earthquakes* recorded around the world or in a particular region over a given period of time.

seismograph An instrument that magnifies and records the motions of Earth's surface caused by *seismic waves.*

settling velocity The speed at which particles of various weights suspended in a *stream* settle to the bottom.

shadow zone (1) A zone beyond 105° from the *focus* of an *earthquake* where S waves are absent because the waves are not transmitted through the liquid *outer core.* (2) A zone at an angular distance of 105° to 142° from the focus of an earthquake where *P waves* are absent because they are refracted downward into the core and emerge at greater distances after the delay caused by their detour through the core.

shale A fine-grained clastic *rock* composed of *silt* plus a significant component of *clay,* which causes it to break readily along *bedding* planes.

shear wave A wave that propagates by a side-to-side motion. Shear waves cannot propagate through any fluid—air, water, or the liquid iron in Earth's *outer core.* (Compare *compressional wave.*)

shearing force A force that pushes two sides of a body in opposite directions. (Compare *compressive force; tensional force.*)

shield A large region of stable, ancient crystalline basement *rocks* within a continent.

shield volcano A broad, shield-shaped *volcano* many tens of kilometers in circumference and more than 2 km high built by successive flows of fluid *basaltic lava* from a central vent.

shock metamorphism *Metamorphism* that occurs when *minerals* are subjected to the high pressures and temperatures of shock waves generated when a *meteorite* collides with Earth.

silica ooze A biochemically *precipitated pelagic sediment* consisting of the remains of the silica shells of diatoms and radiolarians.

siliciclastic sediment *Clastic sediment* produced by the *weathering* of rocks composed largely of silicate *minerals.*

siliciclastic sedimentary environment Those environments dominated by silclastic sediments. They include the continental alluvial (stream), desert, lake, and glacial environments, as well as the shoreline environments transitional between continental and marine: deltas, beaches, and tidal flats. They also include oceanic environments of the continental shelf, continental margin, and deep-ocean floor where siliciclastic sands and muds are deposited.

sill A sheetlike *concordant intrusion* formed by the injection of magma between parallel layers of preexisting bedded *rock.*

silt A *clastic sediment* in which most of the grains are between 0.0039 and 0.062 mm in diameter.

siltstone A clastic *rock* that contains mostly *silt* and looks similar to *mudstone* or very fine grained *sandstone.* The lithified equivalent of silt.

sinkhole A small, steep depression in the land surface above a cavernous *limestone formation.*

slate The lowest grade of *foliated rock,* easily split into thin sheets; produced primarily by the *metamorphism* of *shale.*

slip face The steep lee slope of a *dune* on which *sand* is deposited in cross-beds at the *angle of repose.*

slump A slow *mass movement* of *unconsolidated material* that travels as a unit.

soil A *weathering* product composed of fragments of bedrock, *clay minerals,* and organic matter.

soil profile The final composition and appearance of a soil.

solar energy Energy derived from the Sun.

solar nebula A disk of gas and dust that surrounded the Sun during the formation of the early solar system. Once formed, the disk began to cool, and many of the gases condensed. Gravitational attraction caused the dust and condensing material to clump together into *planetesimals,* then Moon-sized bodies, and finally planets.

sorting The tendency for variations in current velocity to segregate *sediments* according to size.

specific gravity The weight of a *mineral* in air divided by the weight of an equal volume of pure water at 4°C. (Compare *density.*)

spectrum A continuum of color formed when a beam of white light is dispersed (as by passage through a prism) so that its component wavelengths are arranged in order. The colors of these spectra reveal the chemical composition of light-producing or light-reflecting materials. Geologists can look at the spectra of the planets and know which gases are in their atmospheres and which chemicals and minerals are in their rocks and soils.

spreading center The region at the crest of a mid-ocean ridge, where new *crust* is being formed by *seafloor spreading.*

steady state A state in which the inflow of an element into a *reservoir* equals the outflow.

stock An irregular mass of coarse-grained *igneous rock* less than 100 km^2 in area.

storm surge A dome of seawater formed by a *hurricane* at levels higher than the surrounding ocean surface.

stratigraphic succession A vertical set of rock strata that provides a chronological record of the geologic history of a region.

stratigraphy The description, correlation, and classification of strata in *sedimentary rocks.*

stratosphere A cold, dry layer of the atmosphere above the *troposphere* that extends from about 11 to 50 km in altitude.

stratospheric ozone depletion The thinning of the layer of ozone in the *stratosphere* caused by reactions involving human-made compounds, such as chlorofluorocarbons (CFCs). The stratospheric ozone forms a protective layer that shields Earth by absorbing cell-damaging ultraviolet radiation.

stratovolcano A concave-shaped *volcano* containing alternating layers of lava flows and beds of *pyroclasts.*

streak The *color* of the fine deposit of mineral dust left on an abrasive surface, such as a tile of unglazed porcelain, when a *mineral* is scraped across it.

stream Any flowing body of water, large or small.

stream power   The rate at which a *stream* can erode bedrock and transport its load; the product of stream slope and stream *discharge*.

stress   The force per unit area acting on any surface within a solid body. Stress consists of confining pressure and directed pressure.

striation   A scratch or groove left on bedrock and boulders by overriding ice, showing the direction of glacial motion.

strike   The compass direction of a rock layer, fault, or other planar structure where it intersects with a horizontal surface.

strike-slip fault   A fault on which the movement has been horizontal, parallel to the *strike* of the fault plane.

stromatolite   A rock with distinctive lamination, believed to have been formed by ancient *microbial mats*. Stromatolites range in shape from flat sheets to domal structures with complex branching patterns. They are one of the most ancient types of *fossils* on Earth and give us a glimpse of a world once ruled by microorganisms.

subduction   The sinking of an oceanic plate beneath an overriding oceanic or continental plate at a convergent plate boundary.

submarine canyon   A deep *valley* eroded into the *continental shelf.*

subsidence   A depression or sinking of the *crust* induced partly by the additional weight of *sediments* on the crust but driven mostly by tectonic mechanisms, such as regional downfaulting or cooling of the lithosphere

superposed stream   A *stream* that erodes a gorge in resistant *formations* because its course was established at a higher level on uniform *rocks* before downcutting began. A superposed stream tends to continue the pattern it developed earlier rather than adjusting to its new conditions. (Compare *antecedent stream.*)

superposition, principle of   The principle that each layer of *sedimentary rock* in a tectonically undisturbed sequence is younger than the one beneath it and older than the one above it.

surface wave   *Seismic waves* that travel around Earth's surface that arrive later than the *S wave.*

surge   A sudden period of fast movement of a *valley glacier,* sometimes occurring after a long period of little movement and lasting several years.

suspended load   All the material temporarily or permanently suspended in the flow of a *stream.* (Compare *bed load.*)

sustainable development   Development that meets the needs of the present without compromising the ability of future generations to meet their own needs.

syncline   A large downfold whose limbs are higher than its center. (Compare *anticline.*)

talus   An accumulation of blocks of fallen *rock* at the foot of a steep bedrock cliff.

tar sand   A deposit of sand or sandstone containing a tarlike substance called natural bitumen. Natural bitumen is petroleum from which the lighter portions have escaped.

tectonic age   The age of a *rock* that corresponds to the last major episode of crustal deformation; that is, to the last time the radiometric "clock" within the rock was reset by tectonic activity.

tensional force   A force that stretches a body and tends to pull it apart. (Compare *compressive force; shearing force.*)

terrace   A flat, steplike surface that lines a *stream* above the *floodplain,* often paired one on each side of the stream, marking a former floodplain that existed at a higher level before regional uplift or an increase in *discharge* caused the stream to erode into the former floodplain.

terrigenous sediment   *Sediment* eroded from the land surface.

texture   The sizes and shapes of a *rock's mineral crystals* and the way they are put together.

thermal subsidence basin   A *basin* produced in the later stages of rifting, when newly formed continental plates are drifting away from each other. The *lithosphere* that was thinned and heated during the earlier rifting stage cools, leading to an increase in *density,* which in turn leads to *subsidence* below sea level, where *sediments* can accumulate.

thermohaline circulation   A three-dimensional pattern of ocean circulation driven by differences in temperature and salinity that is an important component of the ocean-atmosphere climate system. In the Atlantic, winds transport warm tropical surface water northward where it cools, becomes denser, and sinks into the deep ocean. It then reverses direction and migrates back to the tropics, where it eventually warms and returns to the surface.

thermoremanent magnetization   A permanent magnetization acquired by *minerals* in *igneous rocks* when groups of *atoms* of the mineral align themselves in the direction of the *magnetic field* that existed when the material was hot. When the material has cooled, these atoms are locked in place and therefore are always magnetized in the same direction.

thrust fault   A low-angle reverse fault—one with a dip of less than 45°—so that the overlying block moves mainly horizontally.

tidal flat   A muddy or sandy area that lies above low *tide* but is flooded at high tide.

tide   The twice-daily rise and fall of the sea caused by the gravitational attraction of the Moon and, to a lesser degree, the Sun.

till   An unstratified and poorly sorted *sediment* containing all sizes of fragments from *clay* to boulders and deposited by glacial action.

tillite Ancient lithified *till.*

topography The general configuration of varying heights that gives shape to Earth's surface.

topset bed A horizontal bed of *sediments* formed at the top of a *delta* and overlying the *foreset beds.*

transform fault A plate margin at which the plates slide past each other and *lithosphere* is neither created nor destroyed. Relative displacement occurs along the fault as horizontal slip between the adjacent plates.

tributary A *stream* that *discharges* water into a larger stream.

troposphere The lowest layer of the atmosphere, which has an average thickness of about 11 km and convects vigorously owing to surface heating by the Sun, causing storms and other short-term disturbances in the weather. (Compare *stratosphere.*)

tsunami A fast-moving, towering sea wave generated by an *earthquake* that occurs beneath the ocean.

tuff Any volcanic *rock* lithified from *pyroclasts.*

turbidite A *graded bed* of *sand, silt,* and *mud* deposited by a *turbidity current* on the *abyssal plain.*

turbidity current A flow of turbid, muddy water down a slope. The *suspended load* of *mud* makes the turbid water denser than the overlying clear water, so that the turbidity current flows beneath the clear water.

turbulent flow A flow in which streamlines mix, cross, and form swirls and eddies. (Compare *laminar flow.*)

twentieth-century warming A rise in Earth's mean surface temperature of about 0.6°C between the end of the nineteenth century and the beginning of the twenty-first.

ultramafic rock An *igneous rock* consisting primarily of mafic *minerals* and containing less than 10 percent feldspar. (Compare *felsic rock, mafic rock.*)

unconformity A surface between two rock layers that were not laid down in an unbroken sequence. An unconformity represents the passage of time.

unconsolidated material *Sediment* that is loose and uncemented. (Compare *consolidated material.*)

uniformitarianism, principle of The principle that the geologic processes we see in action today have worked in much the same way throughout geologic time; that is, "the present is the key to the past."

universal ancestor The single root of the *universal tree of life.* This universal ancestor led to three major domains of descendants: the Bacteria, the Archaea, and the Eukarya.

universal tree of life A family tree of the hierarchy of ancestors and descendants of all organisms, derived by comparing the genes of different organisms within a particular group.

unsaturated zone The level in the *groundwater table* in which the pores of the *soil* or *rock* are not completely filled with water. (Compare *saturated zone.*)

U-shaped valley A deep *valley* with steep upper walls that grade into a flat floor; the typical shape of a valley eroded by a *glacier.*

valley The entire area between the tops of the slopes on both sides of a *stream.*

valley glacier A river of ice that forms in the cold heights of mountain ranges where snow accumulates, usually in preexisting *valleys,* and flows down the bedrock valleys. (Compare *continental glacier.*)

varve A pair of alternating coarse and fine layers formed in one year by the seasonal freezing of a lake surface.

vein deposit (vein) A sheetlike deposit of *minerals precipitated* in fractures or *joints* that are foreign to the host *rock.*

ventifact A wind-faceted pebble having several curved or almost flat surfaces that meet at sharp ridges. Each surface or facet is made by *sandblasting* of the pebble's windward side.

viscosity A measure of a fluid's resistance to flow.

volcanic ash Extremely small fragments, usually of glass, that form when escaping gases force a fine spray of magma from a *volcano.*

volcanic geosystem The total system of *rocks,* magmas, and interactions needed to describe the entire sequence of events from melting to eruption.

volcano A hill or mountain constructed from the accumulation of lava and other erupted materials.

wadi A desert *valley* that carries water only briefly after a rain. Called a *dry wash* in the western United States.

wave-cut terrace A level surface formed by wave *erosion* of coastal bedrock beneath the surf zone. May be visible at low *tide.*

weathering The general process that breaks up *rocks* into fragments of various sizes by a combination of physical fracturing and chemical decomposition.

Wilson cycle A general plate tectonic cycle that comprises (1) rifting during the breakup of a supercontinent, (2) *passive-margin* cooling and *sediment* accumulation during *seafloor spreading* and ocean opening, (3) *active-margin* volcanism and terrane *accretion* during *subduction* and ocean closure, and (4) *orogeny* during the continent-continent collision that forms the next supercontinent.

zeolite A class of silicate *minerals* containing water in cavities within the *crystal* structure and formed by *metamorphism* at very low temperatures and pressures.

Suggested Readings

Chapter 1

Allegre, C. J., and S. H. Schneider. 2005. Evolution of Earth. In *Scientific American,* Special Edition: *Our Ever Changing Earth,* pp. 4–13.

Alley, R. B. 2001. The key to the past. *Nature* 409: 289.

Becker, L. 2002. Repeated blows. *Scientific American* (March): 77–83.

Dalrymple, G. B. 2005. *Ancient Earth, Ancient Skies: The Age of Earth and Its Cosmic Surroundings.* Stanford University Press.

Hallam, A. 1991. *Great Geologic Controversies.* Oxford University Press.

Hallam, T. 2005. *Catastrophes and Lesser Calamities: The Causes of Mass Extinctions.* Oxford University Press.

McSween, H. Y. 1999. *Meteorites and Their Parent Planets.* Cambridge University Press.

Menking, K., D. Merritts, and A. de Wet. 1998. *Environmental Geology: An Earth System Science Approach.* W. H. Freeman.

National Aeronautics and Space Administration. 2002. *Living on a Restless Planet.* Jet Propulsion Laboratory. See also http:\\solidearth.jpl.nasa.gov.

National Research Council. 1993. *Solid Earth Sciences and Society.* National Academy Press.

Roberts, F. 2001. The origin of water on Earth. *Science* 293: 1056–1058.

Stanley, S. M. 2005. *Earth System History,* 2nd ed., W. H. Freeman.

Westbroek, P. 1991. *Life as a Geologic Force.* W. W. Norton.

Chapter 2

Carlson, R. W. 1997. Do continents part passively or do they need a shove? *Science* 278: 240–241.

Conrad, C. P., and C. Lithgow-Bertelloni. 2002. How mantle slabs drive plate tectonics. *Science* 298: 207–209.

Cox, A. 1989. *Plate Tectonics: How It Works.* Blackwell Scientific.

Dalziel, I. W. D. 1995. Earth before Pangaea. *Scientific American* (January): 58–63.

Hallam, A. 1973. *A Revolution in the Earth Sciences: From Continental Drift to Plate Tectonics.* Oxford University Press, Clarendon Press.

Kearey, P., and F. J. Vine. 1990. *Global Tectonics.* Blackwell Scientific.

Macdonald, K. C. 1998. Exploring the mid-ocean ridge. *Oceanus* 41: 2–8.

Menard, H. W. 1986. *The Ocean of Truth: A Personal History of Global Tectonics.* Princeton University Press.

Muller, R. D., et al. 1997. Digital isochrons of the world's ocean floor. *Journal of Geophysical Research* 102: 3211–3214.

National Academy of Sciences. 1999. When the Earth moves: Seafloor spreading and plate tectonics. In *Beyond Discovery: The Path from Research to Human Benefit.* www.beyonddiscovery.org.

Norabuena, E., et al. 1998. Space geodetic observations of Nazca-South America convergence across the central Andes. *Science* 279: 358–362.

Oreskes, N. 1999. *The Rejection of Continental Drift.* Oxford University Press.

Oreskes, N., ed., with Homer Le Grand. 2001. *Plate Tectonics: An Insider's History of the Modern Theory of Earth.* Westview Press.

Richards, M. A. 1999. Prospecting for Jurassic slabs. *Nature* 397: 203–204.

Scotese, C. R. 2001. *Atlas of Earth History,* vol. 1: Paleogeography. PALEOMAP Project.

Smith, D. K., and J. R. Cann. 1993. Building the crust at the Mid-Atlantic Ridge. *Nature* 365: 707–714.

Van der Hilst, R. D., S. Widiyantoro, and E. R. Engdahl. 1997. Evidence for deep mantle circulation from global tomography. *Nature* 386: 578–584.

Wessel, G. R. 1986. *The Geology of Plate Margins.* Geological Society of America, Map and Chart Series MC-59.

Chapter 3

Barnes, H. L., and A. W. Rose. 1998. Origins of hydrothermal ores. *Science* 279: 2064–2065.

Blatt, H., R. J. Tracy, and B. E. Owens. 2005. *Petrology: Igneous, Sedimentary, and Metamorphic,* 3rd ed. W. H. Freeman.

Briskey, J. A., et al. 2001. It's time to know the planet's mineral resources. *Geotimes* (March): 14–22.

Deer, W. A., Howie, R. A., and Zussman, J., 1996. *An Introduction to Rock-Forming Minerals.* Prentice Hall.

Dietrich, R. V., and B. J. Skinner. 1990. *Gems, Granites, and Gravels.* Cambridge University Press.

Hall, C. 2002. *Gemstones.* Smithsonian Handbooks.

Keller, P. C. 1990. *Gemstones and Their Origins.* Chapman & Hall.

Klein, C. 2001. *Manual of Mineral Science,* 22nd ed. Wiley.

Patch, S. S. 1999. *Blue Mystery: The Story of the Hope Diamond,* 2nd ed. Abrams.

Robb, L. J. 2004. *Introduction to Ore-Forming Processes.* Blackwell Scientific.

Varon, L. 2001. Internet gems: Web sites for rockhounds and lapidaries. *Rock and Gem* 31(6): 34–36, 90.

Chapter 4

Barker, D. S. 1983. *Igneous Rocks.* Prentice Hall.

Blatt, H., R. J. Tracy, and B. E. Owens. 2005. *Petrology: Igneous, Sedimentary, and Metamorphic,* 3rd ed. W. H. Freeman.

Coffin, M. F., and O. Eldholm. 1993. Large igneous provinces. *Scientific American* (June): 42–49.

Philpotts, A. R. 2003. *Petrography of Igneous and Metamorphic Rocks.* Waveland Press.

Raymond, L. A. 1995. *Petrology.* Wm. C. Brown.

Chapter 5

Boggs, S. 2001. *Principles of Sedimentology and Stratigraphy,* 3rd ed. Prentice Hall.

Brookfield, M. E. 2004. *Principles of Stratigraphy.* Blackwell Scientific.

Goreau, T. F., N. I. Goreau, and T. J. Goreau. 1979. Corals and coral reefs. *Scientific American* (August): 124–136.

Julien, P. Y. 1998. *Erosion and Sedimentation.* Cambridge University Press.

Komar, P. D. 1997. *Beach Processes and Sedimentation,* 2nd ed. Prentice Hall.

Leeder, M. R. 1999. *Sedimentology and Sedimentary Basins.* Blackwell Scientific.

Mack, W. N., and E. A. Leistikow. 1996. Sands of the world. *Scientific American* (August): 62–67.

Prothero, D. R., and F. Schwab. 1996. *Sedimentary Geology.* W. H. Freeman.

Siever, R. 1988. *Sand.* Scientific American Library.

Stanley, G. D., Jr. 2001. *The History and Sedimentology of Ancient Reef Systems.* Springer.

Chapter 6

Barker, A. J. 1998. *Introduction to Metamorphic Textures and Microstructures,* 2nd ed. Stanley Thornes.

Blatt, H., R. J. Tracy, and B. E. Owens. 2005. *Petrology: Igneous, Sedimentary, and Metamorphic,* 3rd ed. W. H. Freeman.

Coleman, R. G., X. Wang, P. C. Hess, and A. B. Thompson. 1995. *Ultrahigh Pressure Metamorphism.* Cambridge University Press.

Frey, M., and D. Robinson. 1998. *Low-Grade Metamorphism.* Blackwell Scientific.

Wynn, J. C., and E. Shoemaker. l998. The day the sands caught fire. *Scientific American* (November): 64–21.

Chapter 7

McPhee, J. 2000. *Annals of the Former World.* Farrar, Straus & Giroux.

Pinter, N., and M. T. Brandon. 2005. How erosion builds mountains. In *Scientific American,* Special Edition: *Our Ever Changing Earth,* pp. 74–81.

Ramsay, J. F. 1987. *Techniques of Modern Structural Geology: Folds and Fractures.* Academic Press.

Suppe, J. 1985. *Principles of Structural Geology.* Prentice Hall.

Twiss, R. J., and E. M. Moores. 1992. *Structural Geology.* W. H. Freeman.

Chapter 8

Berry, W. B. N. 1987. *Growth of a Prehistoric Time Scale.* Blackwell Scientific.

Faure, G., and T. M. Mensing. 2005. *Isotopes: Principles and Applications,* 3rd ed. Wiley.

Palmer, A. R. 1984. *Decade of North American Geologic Time Scale.* Geological Society of America. Map and Chart Series MC-50.

Simpson, G. G. 1983. *Fossils. Scientific American Books.*

Spencer, E. W. 1999. *Geologic Maps: A Practical Guide to the Preparation and Interpretation of Geologic Maps,* 2nd ed. Prentice Hall.

Stanley, S. M. 2005. *Earth System History,* 2nd ed. W. H. Freeman.

Winchester, S. 2002. *The Map That Changed the World: William Smith and the Birth of Modern Geology.* HarperCollins, Perennial.

Chapter 9

Ahrens, T. J. 1994. The origin of the Earth. *Physics Today* 47: 38–45.

Albee, A. 2003. The unearthly landscapes of Mars. *Scientific American* (June): 46–53.

Allegre, C. 1992. *From Stone to Star.* Harvard University Press.

Ardilla, D. R., 2004. Planetary systems. *Scientific American* (April): 62–69.

Becker, L. 2002. Repeated blows. *Scientific American* (March): 77–83.

Bullock, M. A., and D. H. Grinspoon. 2003. Global climate change on Venus. In *Scientific American,* Special Edition: *New Light on the Solar System,* pp. 20–27.

Christensen, P. R. 2005. The many faces of Mars. *Scientific American* (July): 32–39.

Cole, G. H. A. 2001. Exoplanets. *Astronomy and Geophysics* (February): 1.13–1.17.

Doyle, L. R., H.-J. Deeg, and T. M. Brown. 2000. Searching for shadows of other Earths. *Scientific American* (July): 60–65.

Golombek, M. P. 1998. The Mars *Pathfinder* mission. *Scientific American* (July): 40–49.

Halliday, A. N., and M. J. Drake. 1999. Colliding theories (origin of Earth and Moon). *Science* 283: 1861–1864.

Kargel, J. S., and R. G. Strom. 1996. Global climatic change on Mars. *Scientific American* (November): 80–88.

Kasting, J. F. 1998. Origins of water on Earth. *Scientific American* (October): 16–22.

Kasting, J. F. 2004. When methane made climate. *Scientific American* (July): 80–85.

Lunine, J. I. 2004. Saturn at last! *Scientific American* (June): 56–63.

Nelson, R. M. 1997. Mercury: The forgotten planet. *Scientific American* (November): 56–67.

Pappalardo, R. T., Head, J. W., and Greeley, R., 1999. The Hidden Ocean of Europa. *Scientific American* (October): 54–63.

Scientific American. 1990. Special Issue: Exploring Space. Scientific American, Inc. 2000. *The Frontiers of Space.*

Spudis, P. D. 2003. New moon. *Scientific American* (December): 86–93.

Squyres, S. 2005. *Roving Mars: Spirit, Opportunity, and the Exploration of the Red Planet.* Hyperion.

Wetherill, G. W. 1990. Formation of the Earth. *Annual Review of Earth and Planetary Sciences* 8: 205–256.

Woolfson, M. 2000. The origin and evolution of the solar system. *Astronomy and Geophysics* 41: 1.2–1.18.

York, D. 1993. The earliest history of the Earth. *Scientific American* (January): 90–96.

Zorpette, G. 2000. Why go to Mars? *Scientific American* (March): 40–43.

Chapter 10

Bally, A. W., and A. R. Palmer (eds.). 1989. *The Geology of North America: An Overview.* Geological Society of America.

Burchfiel, B. C. 1983. The continental crust. *Scientific American* (September): 130.

Dalziel, I. W. D. 2005. Earth before Pangaea. In *Scientific American,* Special Edition: *Our Ever Changing Earth,* pp. 14–21.

Geissman, J. W., and A. F. Glazner (eds.). 2000. Focus on the Himalayas. *Bulletin of the Geological Society of America* 112: 323–511.

Gurnis, M. 2005. Sculpting Earth inside out. In *Scientific American,* Special Edition: *Our Ever Changing Earth,* pp. 56–63.

Jones, D. L., A. Cox, P. Coney, and M. Beck. 1982. The growth of western North America. *Scientific American* (November): 70.

Jordan, T. H. 1979. The deep structure of continents. *Scientific American* (January): 92–107.

Kearey, P., and F. J. Vine. 1990. *Global Tectonics.* Blackwell Scientific.

Molnar, P. 1997. The rise of the Tibetan Plateau: From mantle dynamics to the Indian monsoon. *Astronomy and Geophysics* 36(3): 10–15.

Murphy, J. B., and R. D. Nance. 1992. Mountain belts and the supercontinent cycle. *Scientific American* (April): 84.

Murphy, J. B., G. L. Oppliger, G. Brimhall, Jr., and A. J. Hynes. 1999. Mantle plumes and mountains. *American Scientist* 87 (March-April): 146–153.

Taylor, S. R., and S. L. McLennan. 2005. The evolution of continental crust. In *Scientific American,* Special Edition: *Our Ever Changing Earth,* pp. 44–49.

Twiss, R. J., and E. M. Moores. 1992. *Structural Geology.* W. H. Freeman.

Chapter 11

American Academy of Microbiology, 2000. *Geobiology: Exploring the Interface Between the Biosphere and the Geosphere.* American Academy of Microbiology.

Becker, L. 2002. Repeated blows. *Scientific American* (March): 78–83.

Fredrickson, J. K., and T. C. Onstott. 1996. Microbes deep within the Earth. *Scientific American* (October): 68–73.

Gould, S. J. 1994. The evolution of life on Earth. *Scientific American* (October): 85–91.

Hallam, T. 2005. *Catastrophes and Lesser Calamities: The Causes of Mass Extinctions.* Oxford University Press.

Hazen, R. M. 2001. Life's rocky start. *Scientific American* (April): 76–85.

Kasting, J. F. 2004. When methane made climate. *Scientific American* (July): 80–85.

Kring, D. A., and D. D. Durda. 2003. The day the world burned. *Scientific American* (December): 100–105.

Madigan, M. T., and B. L. Marrs. 1997. Extremophiles. *Scientific American* (April): 82–87.

Simpson, S. 2003. Questioning the oldest signs of life. *Scientific American* (April): 70–77.

Stanley, S. M. 2005. *Earth System History,* 2nd ed. W. H. Freeman.

Chapter 12

American Geophysical Union. 1992. *Volcanism and Climatic Change,* Special Report.

Bruce, V. 2001. *No Apparent Danger: The True Story of Volcanic Disaster at Galeras and Nevado del Ruiz.* HarperCollins.

Decker, R. W., and B. Decker. 2006. *Volcanoes,* 3rd ed. W. H. Freeman.

Dvorak, J. J., and D. Dzurisin. 1997. Volcano geodesy: The search for magma reservoirs and the formation of eruptive vents. *Reviews of Geophysics* 35: 343–384.

Dvorak, J. J., C. Johnson, and R. I. Tilling. 1982. Dynamics of Kilauea volcano. *Scientific American* (August): 46–53.

Edmond, J. M., and K. L. Von Damm. 1992. Hydrothermal activity in the deep sea. *Oceanus* (Spring): 74–81.

Fisher, R. V., G. Heiken, and J. B. Hulen. 1997. *Volcanoes: Crucibles of Change.* Princeton University Press.

Francis, P. 1983. Giant volcanic calderas. *Scientific American* (June): 60–70.

Heiken, G. 1979. Pyroclastic flow deposits. *American Scientist* 67: 564–571.

Krakaner, J. 1996. Geologists worry about dangers of living "under the volcano." *Smithsonian* (July): 33–125.

Larson, R. L. 2005 The Mid-Cretaceous superplume episode. In *Scientific American,* Special Edition: *Our Ever Changing Earth,* pp. 21–27.

Lauber, P. 1993. *Volcano: The Eruption and Healing of Mount St. Helens.* Alladin.

McPhee, J. 1990. Cooling the lava. In *The Control of Nature.* Farrar, Straus & Giroux.

MELT Seismic Team. 1998. Imaging the deep seismic structure beneath a mid-ocean ridge: The MELT experiment. *Science* 280: 1215–1218.

National Research Council. 1994. *Mount Rainier: Active Cascade Volcano.* National Academy Press.

Sigurdsson, H., ed. 2000. *The Encyclopedia of Volcanoes.* Academic Press.

Simkin, T., L. Siebert, and R. Blong. 2001. Volcano fatalities: Lessons from the historical record. *Science* 291: 255.

Smith, R. B., and L. J. Siegel. 2000. *Windows into the Earth.* Oxford University Press.

Tilling, R. I. 1989. Volcanic hazards and their mitigation: Progress and problems. *Reviews of Geophysics* 27: 237–269.

Vink, G. E., and W. J. Morgan. 1985. The Earth's hot spots. *Scientific American* (April): 50–57.

White, R. S., and D. P. McKenzie. 1989. Volcanism at rifts. *Scientific American* (July): 62–72.

Williams, S., and F. Montaigne. 2001. Surviving *Galeras.* Houghton Mifflin.

Winchester, S. 2003. *Krakatoa: The Day the World Exploded, August 27, 1883.* HarperCollins.

Wright, T. L., and T. C. Pierson. 1992. Living with Volcanoes. U.S. Geological Survey Circular 1073. U.S. Government Printing Office.

Chapter 13

Bolt, B. A. 2006. *Earthquakes,* 4th ed. W. H. Freeman.

Clague, J. J., et al. 2000. Great Cascadia Earthquake Tricentennial. *GSA Today* (November): 14–15.

Earthquakes and Volcanoes. A bimonthly periodical. U.S. Government Printing Office.

Federal Emergency Management Agency. 2001. *HAZUS99 Estimated Annualized Earthquake Losses for the United States.* FEMA Report 366.

Hough, S. E. 2002. *Earthquake Science: What We Know (and Don't Know) About Earthquakes.* Princeton University Press.

Kanamori, H., and E. E. Brodsky. 2001. The physics of earthquakes. *Physics Today* (June): 34–40.

Kanamori, H., E. Hauksson, and T. Heaton. 1997. Real-time seismology and earthquake hazard mitigation. *Nature* 390: 461–464.

National Research Council, Committee on the Science of Earthquakes (T. H. Jordan, chair). 2003. *Living on an Active Earth: Perspectives on Earthquake Science.* U.S. Government Printing Office.

Normile, D. 1995. Quake builds case for strong codes. *Science* 267: 444–446.

Prager, E. J. 1999. *Furious Earth: The Science and Nature of Earthquakes, Volcanoes, and Tsunamis.* McGraw-Hill.

Shedlock, K. M., D. Giardini, G. Grünthal, and P. Zhang. 2000. The GSHAP Global Seismic Hazard Map. *Seismological Research Letters* 71: 679–686.

Southern California Earthquake Center. 2004. *Putting Down Roots in Earthquake Country.* http://www.scec.org /resources/ catalog/roots.html.

Stein, R. S. 2005. Earthquake conversations. In *Scientific American,* Special Edition: *Our Ever Changing Earth,* pp. 82–89.

Tibballs, G. 2005. *Tsunami: The Most Terrifying Disaster.* Carlton.

Yeats, R. S. 2001. *Living with Earthquakes in California.* Oregon State University Press.

Yeats, R. S., K. Sieh, and C. R. Allen. 1996. *The Geology of Earthquakes.* Oxford University Press.

Chapter 14

Bolt, B. A. 1993. *Earthquakes and Geological Discovery.* Scientific American Library.

Glatzmaier, G. A., and P. Olsen. 2005. Probing the geodynamo. In *Scientific American,* Special Edition: *Our Ever Changing Earth,* pp. 28–35.

Gurnis, M. 2005. Sculpting the Earth from inside out. In *Scientific American,* Special Edition: *Our Ever Changing Earth,* pp. 56–63.

Hager, B. H., and M. A. Richards. 1989. Long-wavelength variations in Earth's geoid: Physical models and dynamical implications. *Philosophical Transactions of the Royal Society of London, Series A* 328: 309–327.

Helffrich, G. R., and B. J. Wood. 2001. The Earth's mantle. *Nature* 412: 501–507.

Jeanloz, R., and T. Lay. 2005. The core-mantle boundary. In *Scientific American,* Special Edition: *Our Ever Changing Earth,* pp. 36–43.

Kellog, L. H. 1997. Mapping the core-mantle boundary. *Nature* 277: 646–647.

Kerr, R. A. 2001. A lively or stagnant lowermost mantle? *Science* 292: 841.

Lay, T., and T. C. Wallace. 1995. *Modern Global Seismology.* Academic Press.

Masters, T. G., and P. M. Shearer. 1995. Seismic models of the Earth. In T. J. Ahrens (ed.), *A Handbook of Physical Constants: Global Earth Physics* (vol. 1, pp. 88–103). American Geophysical Union.

McKenzie, D. P. 1983. The Earth's mantle. *Scientific American* (September): 66.

Olson, P., P. G. Silver, and R. W. Carlson. 1990. The large-scale structure of convection of the Earth's mantle. *Nature* 344: 209–215.

Stein, S., and M. Wysession. 2003. *An Introduction to Seismology, Earthquakes, and Earth Structure.* Blackwell.

Wysession, M. 1995. The inner workings of the Earth. *American Scientist* 83: 134–146.

Chapter 15

Alley, R. B. 2004. *The Two-Mile Time Machine: Ice Cores, Abrupt Climate Change, and Our Future.* Princeton University Press.

American Geophysical Union. 1999. Climate change and greenhouse gases. *EOS* 80: 453–458.

Berner, R. A., and A. C. Lasaga. 1989. Modeling the geochemical carbon cycle. *Scientific American* (March): 74–81.

Bonnet, S., and A. Crave. 2003. Landscape response to climate change: Insights from experimental modeling and implications for tectonic versus climactic uplift of topography. *Geology* 31: 123–126.

Cook, E. 1996. *Marking a Milestone in Ozone Protection: Learning from the CFC Phase-Out.* World Resources Institute.

Cox, J. D. 2005. *Climate Crash: Abrupt Climate Change and What It Means for Our Future.* National Academies Press, Joseph Henry Press.

Graedel, T. E., and P. J. Crutzen. 1993. *Atmospheric Change: An Earth System Perspective.* W. H. Freeman.

Intergovernmental Panel on Climate Change. 2001. *Climate Change 2001: The Scientific Basis.* Cambridge University Press.

National Academy of Sciences. 1992. *Policy Implications of Greenhouse Warming.* Washington, D.C.: National Academy Press.

National Research Council. 2001. *Climate Change Science: An Analysis of Some Key Questions.* National Academy Press.

Ruddiman, W. F. 2001. *Earth's Climate: Past and Future.* W. H. Freeman.

Webster, P. J., and J. A. Curry. 1998. The oceans and weather. In Scientific American Presents: *The Oceans: The Ultimate Voyage Through Our Watery Home* (Fall): 38–43.

Chapter 16

Birkeland, P. 1999. *Soils and Geomorphology.* Oxford University Press.

Bloom, A. L. 2004. *Geomorphology: A Systematic Analysis of Late Cenozoic Landforms,* 3rd ed. Waveland Press.

Brevik, C. 2002. Problems and suggestions related to soil classification as presented in introduction to physical geology textbooks. *Journal of Geoscience Education* 50: 541.

Colman, S. M., and D. P. Dethier. 1986. *Rates of Chemical Weathering of Rocks and Minerals.* Academic Press.

Chapman, D. 1995. *Natural Hazards.* Oxford University Press.

Cooke, R. U., R. J. Inkpen, and G. F. S. Wiggs. 1995. Using gravestones to assess changing rates of weathering in the United Kingdom. *Earth Surface Processes and Landforms* 20: 531–546.

Lee, E. M. 2005. *Landslide Risk Assessment.* Thomas Telford.

Miller, R. W., and D. T. Gardiner. 2001. *Soils in Our Environment,* 9th ed. Prentice Hall.

Nahon, D. B. 1991. *Introduction to the Petrology of Soils and Chemical Weathering.* Wiley.

Retallack, G. J. 1990. *Soils of the Past: An Introduction to Paleopedology.* Blackwell Scientific.

Singer, M. J., and D. N. Munns. 2002. *Soils: An Introduction,* 5th ed. Prentice Hall.

Soil Survey Staff. 1975. *Soil Taxonomy: A Basic System of Soil Classification for Making and Interpreting Soil Surveys.* U.S. Government Printing Office.

Tyler, M. B. 1995. Look Before You Build: Geologic Studies for Safer Land Development in the San Francisco Bay Area. *U.S. Geological Survey Circular 1130.* U.S. Geological Survey.

Chapter 17

de Villiers, M. 2001. *Water: The Fate of Our Most Precious Resource.* Mariner Books.

Dolan, R., and H. G. Goodell. 1986. Sinking cities. *American Scientist* 74: 38–47.

Gleick, P. H. 2005. *World's Water, 2004–2005: The Biennial Report on Freshwater Resources (World's Water).* Island Press.

Gunn, J. 2004. *Encyclopedia of Caves and Karst Science.* Fitzroy Dearborn.

National Research Council. 1993. *Solid-Earth Sciences and Society.* National Academy Press.

Schwartz, F. W., and H. Zhang. 2002. *Fundamentals of Ground Water.* Wiley.

Todd, D. W., and L. W. Mays. 2004. *Groundwater Hydrology,* 3rd ed. Wiley.

U.S. Geological Survey. 1990. Hydrologic Events and Water Supply and Use. National Water Summary 1987. *U.S. Geological Survey Water-Supply Paper 2350.*

Chapter 18

Leopold, L. B. 2003. *A View of the River.* Harvard University Press.

Leopold, L. B., M. G. Wolman, and J. P. Miller. 1995. *Fluvial Processes in Geomorphology.* Dover.

McPhee, J. 1989. *The Control of Nature.* Farrar, Straus & Giroux.

National Research Council, 1995. *Flood Risk Management and the American River Basin: An Evaluation.* National Academy Press.

Schumm, S. A. 2003. *The Fluvial System.* Blackburn Press.

Shelby, A. 2004. *Red River Rising: The Anatomy of a Flood and the Survival of an American City.* Borealis Books.

Chapter 19

Bagnold, R. A. 2005. *The Physics of Blown Sand and Desert Dunes.* Dover.

Cooke, R. U., A. Warren, and A. Goudie. 1993. *Desert Geomorphology.* UCL Press.

Geist, H. 2005. *The Causes and Progression of Desertification.* Ashgate.

Goudie, A. S., I. Livingstone, and S. Stokes. 2000. *Aeolian Environments, Sediments and Landforms.* Wiley.

Pye, K. 1989. *Aeolian Dust and Dust Deposits.* Academic Press.

Reisner, M. 1993. *Cadillac Desert: The American West and Its Disappearing Water.* Penguin.

Stallings, F. L. 2001. *Black Sunday: The Great Dust Storm of April 14, 1935.* Eakin Press.

Winger, C., and D. Winger. 2003. *The Essential Guide to Great Sand Dunes National Park and Preserve.* Colorado Mountain Club Press.

Chapter 20

Cone, J. 1991. *Fire Under the Sea.* William Morrow.

Davis, R. A. 1994. *The Evolving Coast.* W. H. Freeman.

Dolan, R., and H. Lins. 1987. Beaches and barrier islands. *Scientific American* (July): 146.

Erikson, J. 2003. *Marine Geology: Exploring the New Frontiers of the Ocean.* Checkmark Books.

Editors of Time Magazine, 2005. *Time: Hurricane Katrina: The Storm That Changed America.* Time Inc.

Fischetti, M. 2001. Drowning New Orleans. *Scientific American* (October): 78–85.

Hardisty, J. 1990. *Beaches: Form and Process.* HarperCollins Academic.

Komar, P. 1998. *The Pacific Northwest Coast.* Duke University Press.

Menard, H. W. 1986. *The Ocean of Truth: A Personal History of Global Tectonics.* Princeton University Press.

Moore, R. 2004. *Faces from the Flood: Hurricane Floyd Remembered.* University of North Carolina Press.

National Research Council. 2000. *Clean Ocean Coastal Waters.* National Academy Press.

National Science Foundation. 2001. *Ocean Sciences at the New Millennium.*http://www.joss.ucar.edu/joss_psg/publications /decadal/Decadal.low.pdf.

Psuty, N. P., and D. D. Ofiara. 2002. *Coastal Hazard Management: Lessons and Future Directions from New Jersey.* Rutgers University Press.

Schlee, J. S., H. A. Karl, and M. E. Torresan. 1995. Imaging the Sea Floor. *U.S. Geological Survey Bulletin 2079.*

U.S. Army Corps of Engineers. 2004. *Coastal Geology.* University Press of the Pacific.

Chapter 21

Alley, R. B. 2004. *The Two-Mile Time Machine: Ice Cores, Abrupt Climate Change, and Our Future.* Princeton University Press.

Bindshadler, R. A., and C. R. Bentley. 2002. On Thin Ice. *Scientific American* (December): 98–105.

Broecker, W. S., and G. H. Denton. 1990. What drives glacial cycles? *Scientific American* (January): 48–56.

Covey, C. 1984. The Earth's orbit and the ice ages. *Scientific American* (February): 58.

Denton, G. H., and T. J. Hughes. 1981. *The Last Great Ice Sheets.* Wiley.

Hambrey, M. J., and J. Alean. 1992. *Glaciers.* Cambridge University Press.

Hoffman, P., and D. Schrag. 2000. Snowball Earth. *Scientific American* (January): 68.

Imbrie, J., and K. P. Imbrie. 1979. *Ice Ages: Solving the Mystery.* Enslow Press.

Menzies, J. (ed.). 1995. *Modern Glacial Environments: Processes, Dynamics and Sediments.* Butterworth-Heinemann.

Sharp, R. P. 1988. *Living Ice: Understanding Glaciers and Glaciation.* Cambridge University Press.

Sturm, M., D. K. Perovich, and M. C. Serreze. 2004. Meltdown in the north. *Scientific American* (March): 60–67.

Chapter 22

Allen, P. 2005. Striking a chord. *Nature* 434: 961.

Bloom, A. L. 2004. *Geomorphology: A Systematic Analysis of Late Cenozoic Landforms,* 3rd ed., Waveland Press.

Burbank, D. W., and R. S. Anderson. 2001. *Tectonic Geomorphology.* Blackwell.

Goudie, A. 1995. *The Changing Earth: Rates of Geomorphological Processes.* Blackwell.

Merritts, D., and M. Ellis. 1994. Introduction to special section on tectonics and topography. *Journal of Geophysical Research* 99: 12135–12141.

Pinter, N., and M. T. Brandon. 1997. How erosion builds mountains. *Scientific American* (April): 74–79.

Pratson, L. F., and W. F. Haxby. 1997. Panoramas of the seafloor. *Scientific American* (June): 83–87.

Strain, P., and F. Engle. 1992. *Looking at Earth.* Turner Publishing.

Sullivan, W. 1984. *Landprints: On the Magnificent American Landscape.* Times Books.

Summerfield, M. A. 1991. *Global Geomorphology.* Longman.

Chapter 23

Allenby, B. 2005. *Reconstructing Earth: Technology and Environment in the Age of Humans.* Island Press.

Alley, R. B. 2004. Abrupt climate change. *Scientific American* (November): 62–69.

Baldwin, S. F. 2002. Renewable energy: Progress and prospects. *Physics Today* (April): 62–68.

Boyle, G., B. Everett, and J. Ramage. 2003. *Energy Systems and Sustainability.* Oxford University Press.

Brundtland Commission. 1987. *Our Common Future.* Oxford University Press.

Campbell, C. J., and J. H. Laherrère. 1998. The end of cheap oil. *Scientific American* (March): 78–84.

Costanza, R., et al. 1997. The value of the world's ecosystem services and natural capital. *Nature* 387: 253–260.

Cox, J. D. 2005. *Climate Crash: Abrupt Climate Change and What It Means for Our Future.* National Academies Press, Joseph Henry Press.

Deffeyes, K. S. 2003. *Hubbert's Peak: The Impending World Oil Shortage.* Princeton University Press.

Diamond, J. 2005. *Collapse: How Societies Choose to Fail or Succeed.* Viking.

Ehlers, E., and T. Krafft. 2005. *Earth System Science in the Anthropocene: Emerging Issues and Problems.* Springer.

Energy Information Administration. 2004. *Annual Energy Review.* U.S. Department of Energy.

Hansen, J. 2003. Defusing the global warming time bomb. *Scientific American* (March): 69–77.

Heaton, G., R. Repetto, and R. Sobin. 1991. *Transforming Technology: An Agenda for Environmentally Sustainable Growth in the 21st Century.* World Resources Institute.

Houghton, J. T. 2004. *Global Warming: The Complete Briefing,* 3rd ed. Cambridge University Press.

Intergovernmental Panel on Climate Change. 2001. *Climate Change 2001: The Scientific Basis.* Cambridge University Press.

Lovins, A. B. 2005. More profit with less carbon. *Scientific American* (September): 74–83.

Moniz, E. J., and M. A. Kenderdine. 2002. Meeting energy challenges: Technology and policy. *Physics Today* (April): 40–46.

Parsons, E. A., and D. W. Keith. 1998. Fossil fuels without CO_2 emissions. *Science* 282: 1053–1054.

Physics Today. 2002. Special Issue: The Energy Challenge (April).

Ruddiman, W. F. 2005. How Did Humans First Alter Global Climate? *Scientific American* (March): 46–53.

Sawkins, F. J. 1984. *Metal Deposits in Relation to Plate Tectonics.* Springer.

Socolow, R. H. 2005. Can we bury global warming? *Scientifiic America* (July): 49–55.

Sorenson, B. 2004. *Renewable Energy,* 3rd ed. Academic Press.

Victor, D. G. 2004. *Climate Change: Debating America's Policy Options.* Council on Foreign Relations Press.

Wilson, E. O. 2002. *The Future of Life.* Knopf.

in granite-greenstone terrains, 236, *236*
Grenville orogeny, 233
GRIP. *See* Greenland Ice Core Project (GRIP)
Groins, *490*
Gros Ventre slide, 395–397, *396*
Ground moraines, *518, 519t*
Groundwater, **402,** 408–415
  deep crustal, *420, 421–423, 422*
  in deserts, 466
  erosion by, *417,* 417–418, *418*
  and hydrothermal veins, 91
  and ore deposits, 71–72
  quality of, 418–420, *419*
  resources of, *415,* 415–417, *416*
  and sandstones, 118
  and sea-level changes, 414
Groundwater tables, 409–411, **410,** *410–411*
Gulches. *See* Valleys
Gulf of California
  as divergent boundary, 23, *29*
Gulf of Mexico
  oil and production in, 557
  as sedimentary basin, 102, 106, 160
  sequence stratigraphy in, 185, *185*
Gulf Stream, 349, 351
  meanders in, 429
Gullies, *435, 435,* 540, *540*
Gunflint formation, 256, *256*
Gutenberg, Beno, 7, 327
Guyots, *499,* **500**
Gypsum
  and calcium cycle, 364
  as chemical sediment, 104, 119t, *120*
  dunes of, 458
  and extremophiles, 417
  as marine evaporite, 119t, 123, 124
  and Mohs scale of hardness, 56t
  in sedimentary rocks, 64t, 66
  stability of, 377, 377t
  as sulfate, 55, *55*

Habit. *See* Crystal habit
Habitable zones, 265–**266,** *266*
Hack, John, 545–547
Hadean eon, *183,* 183–184
  and Late Heavy Bombardment, 199
Hager, Brad, 338
Haicheng earthquake, 321
Half-life, *181,* **181**
Halite, *48*
  as chemical sediment, 104, 119t, *120*
  cleavage in, 58
  deposits of, 125
  as halide, 52, 52t
  as marine evaporite, 119t, 123, 125
  precipitation of, 51, *51,* 106, 111
  in sedimentary rocks, 64t, 65, 66
  stability of, 377, 377t

Halley's comet, 197
Hallucigenia [eukaryote], *259*
Halophiles, 249, *249,* 249t
Hanford nuclear weapons site, 566, *567*
Hanging valleys, **516,** *517*
Hard water, 420
Hardness, **56**–57, 56t, 61t
Hawaii
  lava types in, *273,* 273–274, *274*
  mantle-plume hot spot under, 41, 97, 286, 287, *287,* 332
  sedimentary environments of, 108
Hazards. *See also* Emergency management
  from hurricanes, 484
  from mass movements, 388
  from nuclear energy, 565–566
  seismic, *316,* 316–317, *318*
  volcanic, 289–291, *290, 291*
Headward erosion, 435, *436*
Health, and mineral dust diseases, 61, 564
Heat engines, 10, *11*
  and metamorphism, 132–133
  and plate tectonics, 13, 37, 40–42
Heat reservoirs, 568
Heat transport. *See also* Core convection; Internal Heat; Mantle convection
  in the climate system, 348, 349–350, *350*
  interior, 333–334
  via conduction, 333, *333*
Heavy Bombardment, 184, **196,** *197,* 198, 203, 208, 235
Hector Mine earthquake, 313
Heimaey volcano, 293
Hematite, **378**
  in acidic water, 57
  in banded iron formations, 257, *258*
  in deserts, 466
  extraterrestrial, 206, *208*
  formation of, 378, *378*
  as ore mineral, 69
  as oxide, 52, 52t, 54, *55*
  and respiration, 246
  in sedimentary rocks, 119t, *379*
  in soils, 378
  stability of, 377t
  streak of, 60, *60*
Herbert Glacier, *507*
Hercynian orogeny, 230, *231*
Hermit Shale, 176, *177*
Herodotus, 440
Hess, Harry, 22
Heterotrophs, **243**–244, *244,* 244t
High-grade metamorphic terrains, **236,** *236*
High-grade metamorphism, 133
High-pressure, low-temperature metamorphism, *67, 135,* **136**

Hills
  as landforms, 535–536
Hillslope relief, 533, *534*
Himalaya Mountains
  as convergent boundary, *27,* 28
  and Earth's topography, 4, *4*
  fold and thrust belts in, 164
  formation of, 146, 147, 228–230, *229, 230*
  and glaciations, 365
  metamorphism in, 135
  topography of, 536, *536*
Hogbacks, **541,** *541*
Holmes, Arthur, 21, 37
Holocene epoch, 178, *180, 183*
  and climate system, 347
  as interglacial period, 357, 360–361
Homo sapiens. See Humans
Horizons, soil, 383
Horizontality, original. *See* Principle of original horizontality
Hornblende, 54
Hornfels, **139,** 140t
  and contact metamorphism, *67*
Hot spots, 41–42, *93,* 97, **286**–288, *287,* 497, 499–500
  and basalt plateaus, 500
  and internal heat, 332
Hot springs, *421,* 421–423, *422*
  microbes in, *240–241,* 249
  and vein deposits, 70
Huang Ho [river]
  floodplains of, 432
  loess deposits near, 462–463, *464*
Hubbert, M. King, 560
Hubbert's peak, **560,** *560*
Hubble, Edwin Powell, 189
Hubble Space Telescope, 190, 210
  images from, *190*
Human affairs. *See also* Environmental geology
  and biodiversity, 574–575
  and the carbon cycle, 365, *366*
  and the climate system, 352, *569*
  and desertification, 465–466
  and the environment, 550–579
  and erosion, 436
  and global change, *569,* 569–574
  and the hydrologic cycle, 404, 416–417
  and longitudinal profiles, 448, *449*
  and mass extinctions, 574–575
  and mass movements, 384, 385, 397, *397*
  and volcanism, 288–293
  and water pollution, 418–420, *419*
Humans
  agricultural development of, 361, 432
  cultural development of, 431, 432, 523
  evolution of, 15, 184, 265
  preserved in glaciers, 509

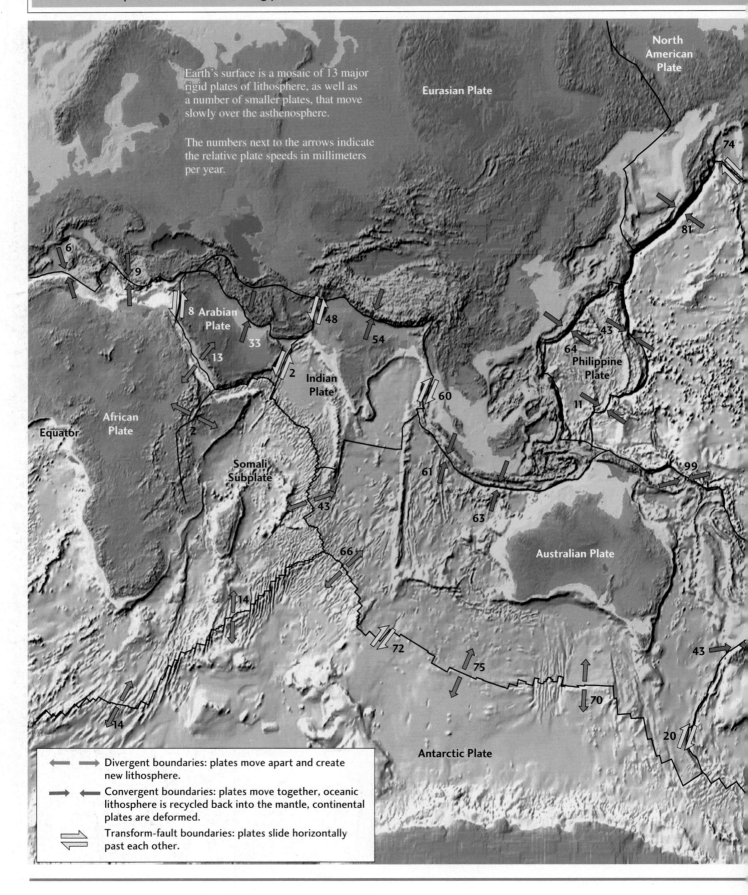

Earth's lithosphere is made of moving plates.

Earth's surface is a mosaic of 13 major rigid plates of lithosphere, as well as a number of smaller plates, that move slowly over the asthenosphere.

The numbers next to the arrows indicate the relative plate speeds in millimeters per year.

North American Plate

Eurasian Plate

74

81

6

9

8 Arabian Plate

48

54

43

64 Philippine Plate

33

13

2

Indian Plate

60

11

African Plate

Equator

2

99

Somali Subplate

61

43

63

66

Australian Plate

14

72

75

70

43

14

20

Antarctic Plate

←— —→ Divergent boundaries: plates move apart and create new lithosphere.

—→ ←— Convergent boundaries: plates move together, oceanic lithosphere is recycled back into the mantle, continental plates are deformed.

⇒ Transform-fault boundaries: plates slide horizontally past each other.